Medical Geology

Medical Geology

En route to One Health

Edited by

Majeti Narasimha Vara Prasad
University of Hyderabad
Hyderabad
India

Meththika Vithanage
University of Sri Jayewardenepura
Nugegoda
Sri Lanka

WILEY

Published by John Wiley & Sons, Inc., Hoboken, New Jersey.
Published simultaneously in Canada.

For general information on our other products and services or for technical support, please contact our Customer Care Department within the United States at (800) 762-2974, outside the United States at (317) 572–3993 or fax (317) 572–4002.

Wiley also publishes its books in a variety of electronic formats. Some content that appears in print may not be available in electronic formats. For more information about Wiley products, visit our web site at www.wiley.com.

Library of Congress Cataloging-in-Publication Data Applied for:
Hardback ISBN: 9781119867340

Cover Design: Wiley
Cover Image: © Miha Creative/Shutterstock

Set in 9.5/12.5pt STIXTwoText by Straive, Pondicherry, India
Printed and bound by CPI Group (UK) Ltd, Croydon, CR0 4YY
C9781119867340_230623

Contents

List of Contributors

Nadeesh M. Adassooriya
Department of Chemical and Process Engineering
Faculty of Engineering
University of Peradeniya
Peradeniya, Sri Lanka

Seda Alp
Ministry of Health
Oral and Dental Health Department
Ankara, Turkey

Ebenezer Ebo Yahans Amuah
Department of Environmental Science
Kwame Nkrumah University of Science and Technology
Kumasi, Ghana

José Ângelo Sebastião Araujo dos Anjos
Department of Geology
Institute of Geosciences
Federal University of Bahia
Salvador, BA, Brazil

Ayse Dilek Atasoy
Environmental Engineering Department
Faculty of Engineering
Harran University
Sanliurfa, Turkey

Pasan Chinthana Bandara
Department of Biosystems Technology
Faculty of Technology
University of Sri Jayewardenepura
Homagama, Sri Lanka

Johannes A. C. Barth
GeoZentrum Nordbayern
Department Geographie und Geowissenschaften
Friedrich-Alexander- Universität
Erlangen-Nürnberg (FAU)
Erlangen, Germany

Hamna Bashir
Institute of Soil and Environmental Sciences
University of Agriculture Faisalabad
Faisalabad, Pakistan

İbrahim Bayhan
Public Health Services Department
Environmental Health Unit
Sanliurfa, Turkey

Prosun Bhattacharya
KTH-International Groundwater Arsenic
Research Group
Department of Sustainable Development
Environmental Science and Engineering
KTH Royal Institute of Technology
Stockholm, Sweden

Irshad Bibi
Institute of Soil and Environmental Sciences
University of Agriculture Faisalabad
Faisalabad, Pakistan

Rohana Chandrajith
Department of Geology
Faculty of Science
University of Peradeniya
Peradeniya, Sri Lanka

GeoZentrum Nordbayern
Department Geographie und Geowissenschaften
Friedrich-Alexander- Universität
Erlangen-Nürnberg (FAU)
Erlangen, Germany

Fernanda Gonçalves da Cunha
Geological Survey of Brazil (CPRM)
Rio de Janeiro
Bahia, Brazil

Jeevani Maheshika Dahanayake
Institute of Indigenous Medicine
University of Colombo
Colombo, Sri Lanka

Paul Dankwa
Department of Environmental Science
C. K. Tedam University of Technology and
Applied Sciences
Navrongo, Ghana

Rohan S. Dassanayake
Department of Biosystems Technology
Faculty of Technology
University of Sri Jayewardenepura
Homagama, Sri Lanka

Ayşegul Demir-Yetis
Medical Services and Techniques Department
Bitlis Eren University
Bitlis, Turkey

Perihan Derin
Environmental Engineering Department
Faculty of Engineering
Harran University
Sanliurfa, Turkey

M. W. C. Dharma-wardana
National Research Council of Canada
Ottawa, Ontario, Canada

Departement de Physique
Université de Montréal
Montréal, Quebec, Canada

Jalathge Isurika Dilanthi Diddeniya
Institute of Indigenous Medicine
University of Colombo
Colombo, Sri Lanka

C. B. Dissanayake
Department of Geology
Faculty of Science
University of Peradeniya
Peradeniya, Sri Lanka

Niwanthi Dissanayake
Department of Chemistry and Physics
McNeese State University
Lake Charles, LA, USA

Rohan D'Souza
Department of Botany
St. John's College
Dr. B.R. Ambedkar University
Agra, Uttar Pradesh, India

Elikplim Abla Dzikunoo
Earth Science Department
University of Ghana
Accra, Ghana

Mohamed N. M. Farhath
Department of Chemical Sciences
Faculty of Applied Sciences
South Eastern University of Sri Lanka
Sammanthurai, Sri Lanka

Paulo J. C. Favas
School of Life Sciences and the Environment
University of Trás-os-Montes e Alto Douro (UTAD)
Vila Real, Portugal

MARE – Marine and Environmental Sciences Centre/
ARNET – Aquatic Research Network,
Department of Life Sciences, University of Coimbra,
Coimbra, Portugal

R. B. Finkelman
Department of Geosciences
The University of Texas at Dallas
Richardson, TX, USA

Zhipeng Gao
MOE Key Laboratory of Groundwater Circulation
and Environmental Evolution & School of
Water Resources and Environment
China University of Geosciences (Beijing)
Beijing, China

State Key Laboratory of Biogeology
and Environmental Geology
China University of Geosciences (Beijing)
Beijing, China

Huaming Guo
MOE Key Laboratory of Groundwater Circulation and
Environmental Evolution & School of Water Resources
and Environment
China University of Geosciences (Beijing)
Beijing, China

State Key Laboratory of Biogeology
and Environmental Geology
China University of Geosciences (Beijing)
Beijing, China

Rangika S. Hikkaduwa Koralege
Department of Chemistry and Physics
College of Arts and Sciences
Western Carolina University
Cullowhee, NC, USA

Muhammad Mahroz Hussain
Institute of Soil and Environmental Sciences
University of Agriculture Faisalabad
Faisalabad, Pakistan

Lander de Jesus Alves
Postgraduate Program in Biology and Biotechnology
of Microorganisms
State University of Santa Cruz (UESC)
Ilhéus, Bahia, Brazil

C. C. Kadigamuwa
Department of Chemistry
Faculty of Science
University of Kelaniya
Kelaniya, Sri Lanka

Abdullah İzzeddin Karabulut
Environmental Engineering Department
Faculty of Engineering
Harran University
Sanliurfa, Turkey

Raymond Webrah Kazapoe
Department of Geological Engineering
University for Development Studies
Tamale, Ghana

S. Keerthanan
Ecosphere Resilience Research Center
Faculty of Applied Sciences
University of Sri Jayewardenepura
Nugegoda, Sri Lanka

Pabasari Arundathi Koliyabandara
Department of Civil and Environmental Technology
Faculty of Technology
University of Sri Jayewardenepura
Homagama, Sri Lanka

Udayagee Kumarasinghe
Department of Biosystems Technology
Faculty of Technology
University of Sri Jayewardenepura
Homagama, Sri Lanka

Nadun H. Madanayake
Department of Botany
Faculty of Science
University of Peradeniya
Peradeniya, Sri Lanka

Danushika C. Manatunga
Department of Biosystems Technology
Faculty of Technology
University of Sri Jayewardenepura
Nugegoda, Sri Lanka

Mapa S. T. Mapa
Department of Biosystems Technology
Faculty of Technology
University of Sri Jayewardenepura
Homagama, Sri Lanka

Hector Hugo Silva Medrado
Security
Radioprotection and Environment Management
Indústrias Nucleares do Brasil (INB) – Uranium
Concentration Unit
Caetité, Bahia, Brazil

Natasha
Department of Environmental Sciences
COMSATS University Islamabad
Vehari Campus
Vehari, Pakistan

Nabeel Khan Niazi
Institute of Soil and Environmental Sciences
University of Agriculture Faisalabad
Faisalabad, Pakistan

N. T. Nichols
Department of Geosciences
The University of Texas at Dallas
Richardson, TX, USA

Fábio Carvalho Nunes
Academic Department
Federal Institute Baiano (IFBAIANO)
Santa Inês
Bahia, Brazil

Hazal Ozer
Pediatric Dentistry Department
Faculty of Dentistry
Necmettin Erbakan University
Konya, Turkey

Manoj S. Paul
Department of Botany
St. John's College
Dr. B.R. Ambedkar University
Agra, Uttar Pradesh, India

Dinusha Peramune
Department of Biosystems Technology
Faculty of Technology
University of Sri Jayewardenepura
Homagama, Sri Lanka

B. S. S. Perera
Department of Chemistry
Faculty of Science
University of Kelaniya
Kelaniya, Sri Lanka

Nuwan T. Perera
Department of Chemistry and Physics
College of Arts and Sciences
Western Carolina University
Cullowhee, NC, USA

Pathirage Kamal Perera
Institute of Indigenous Medicine
University of Colombo
Colombo, Sri Lanka

W. P. R. T. Perera
Department of Indigenous Medical Resources
Faculty of Indigenous Health Sciences and Technology
Gampaha Wickramarachchi University of Indigenous Medicine
Yakkala, Sri Lanka

Joel Augusto Moura Porto
Environment Coordination
Indústrias Nucleares do Brasil (INB) – Uranium Concentration Unit
Caetité, Bahia, Brazil

Majeti Narasimha Vara Prasad
School of Life Sciences
University of Hyderabad (an Institution of Eminence)
Hyderabad, Telangana, India

Dilan Ranaweera
Department of Civil and Environmental Technology
Faculty of Technology
University of Sri Jayewardenepura
Homagama, Sri Lanka

Sandali Ranaweera
Department of Biosystems Technology
Faculty of Technology, University of Sri Jayewardenepura
Nugegoda, Sri Lanka

Edvaldo Cruz dos Santos
Environment Coordination
Indústrias Nucleares do Brasil (INB) – Uranium Concentration Unit
Caetité, Bahia, Brazil

Chathuri Senanayake
Department of Biosystems Technology
Faculty of Technology
University of Sri Jayewardenepura
Homagama, Sri Lanka

Muhammad Shahid
Department of Environmental Sciences
COMSATS University Islamabad
Vehari Campus
Vehari, Pakistan

Sachira Sadhana Hewawardhana Silva
Sri Lanka Institute of Nanotechnology
Nanotechnology and Science Park Mahenwatta
Homagama, Sri Lanka

Rodrigo Gibaut de Souza Gois
Environment Coordination
Indústrias Nucleares do Brasil (INB) – Uranium Concentration Unit
Caetité, Bahia, Brazil

Vidura D. Thalangamaarachchige
Department of Chemistry and Physics
McNeese State University
Lake Charles, LA, USA

Engin Tutkun
Department of Public Health
Faculty of Medicine
Bozok University
Yozgat, Turkey

Mayank Varun
Department of Botany
Hislop College
Nagpur, Maharashtra, India

Eduardo Paim Viglio
Geological Survey of Brazil (CPRM)
Rio de Janeiro
Bahia, Brazil

Meththika Vithanage
Ecosphere Resilience Research Center
Faculty of Applied Sciences
University of Sri Jayewardenepura
Nugegoda, Sri Lanka

Shiping Xing
MOE Key Laboratory of Groundwater Circulation and Environmental Evolution & School of Water Resources and Environment
China University of Geosciences (Beijing)
Beijing, China

State Key Laboratory of Biogeology and Environmental Geology
China University of Geosciences (Beijing)
Beijing, China

Pelin Yapıcıoğlu
Environmental Engineering Department
Faculty of Engineering
Harran University
Sanliurfa, Turkey

Benan Yazici-Karabulut
Environmental Engineering Department
Faculty of Engineering
Harran University
Sanliurfa, Turkey

Mehmet İrfan Yeşilnacar
Environmental Engineering Department
Faculty of Engineering
Harran University
Sanliurfa, Turkey

Preface

Medical geology studies exposure to or deficiency of macro- and micronutrients that may lead to health problems in humans and animals. The condition of our environment influences our health as mostly the water and food come from environmental sources. Medical geology helps us to understand the relationship between the aerosphere, hydrosphere, and geosphere on human and animal health. The elements and minerals in soil, water, and atmospheric dust affect people and animals as there is a close relationship between the two. Naturally occurring geogenic/pedogenic elements such as fluoride, selenium, iodide, and water hardness are extremely detrimental to the health of those people who come in close contact with these elements. Specially, people from developing and least developed nations depend heavily on land due to the human–animal-based farming and less technological advancements. As an example, people in Maputaland, South Africa, have a number of diseases due to the exposure to heavily impoverished soils caused by mineral imbalances, whereas drinking arsenic-rich water in parts of India and Bangladesh often causes toxicity and diseases. Various regions in the world and in countries such as India, Sri Lanka, and the Balkan region are seriously affected by chronic kidney disease of unknown etiology due to some mineral imbalance in water. Fluoride, selenium, and iodide are some of the geogenic and pedogenic examples for such imbalanced elements that can cause health issues in the public. Therefore, the field of medical geology called "bioavailability" receives greatest attention, as the total concentration is not significant for health.

Medical geology is an emerging field in ecosystem–human health interaction. It reveals the impact of geological materials and geological processes on public health and offers scientific support for the prevention and treatment of metabolic disorders and endemic diseases. It also explores the broader relationships between geoenvironmental elements and health or occurrence of diseases in humans, animals, and plants living in the environment.

Medical geology emphasizes internal interactions of particular diseases within physical and social settings, in addition to the geographical distribution characteristics of geochemical elements and diseases. It is an interdisciplinary field that integrates concepts from biomedicine, geography, hydrogeology, geochemistry, and public health. Its primary goal is to use multidisciplinary knowledge and resources to identify, address, and improve health and illness issues associated with the geological environment. A subset of medical geology called medical geochemistry focuses on how the geochemical reactions of elements affect human and animal health.

Geogenic/pedogenic contaminants and human health bring scientists together from various fields, including geochemistry, geography, soil science, biology, and animal science to understand interactions among them. There is a dramatic increase in the development of novel, transdisciplinary approaches including One Health. Specially in the tropical environment, due to the high rate of weathering, the unique geochemistry has a noticeable impact on the health of humans and animals.

Groundwater and public health, exposure pathways and public health, and other relevant subjects can be loosely grouped together in the 24 chapters' topics. The titles and keywords of the contributory chapters were used to create a word cloud map (Figure 1). According to Figure 1, *water*, *geology*, *trace elements*, *release*, *health*, and *deficiency* are most frequently mentioned in various chapters. These words implied that the primary medium or exposure channel endangering the public's health is water, particularly groundwater via waterborne diseases. Additionally, among all the constituents in water, metals are probably the primary indicator.

This book focuses on the present level of knowledge on the connections between geological/geochemical processes and human health to report the most recent results of medical geology and medical geochemistry. This work is a compilation of compelling insights into medical geology in the

Figure 1 The titles and keywords of the contributory chapters in a word cloud map.

recent Anthropocene and its characteristics regarding the chemistry of the aerosphere, geosphere, and hydrosphere. Most of the published books are almost a decade old, and no new knowledge is being incorporated. This book further highlights the macro- and molecular-level interactions of the three domains of the One Health triad: ecosystem, human, and animal. Also complete with illustrations and case studies, this book highlights a comprehensive overview of medical geology.

The following are the different sections and various contributory chapters in this work:

Section 1 – Geochemistry and Health

Medical Geology: Geosphere, human and animal interface

Biogeochemistry: Essential link between geosphere and biosphere

Geochemical release and environmental interfaces

Section 2 – Dust Storms and Health

Minerogenic dust: Trace elements

Silicosis and asbestosis

Radon and health

Section 3 – Medical Geology of the Hydrosphere

Water–rock interactions: Mineral dissolution

Water hardness and health

Medical geology of fluoride

Iodine deficiency disorders (IDDs) and strategies for alleviation

Medical geology of arsenic hydrogeochemical cycling

Potentially toxic metals and health

Section 4 – Medical Pedology: Health Effects from Soils and Sediments

Bioavailability of trace elements in soils

Geochemical provenance of metalloids and their release: Implications on medical geology

Cobalt and copper deficiency and molybdenosis

Biomedical applications of clay materials: Structure and functions

Section 5 – Case Studies

Chronic kidney disease of unknown etiology in humid tropics

Uraniferous province of Lagoa Real: Routes, dispersion, and impacts on health

Defluoridation

Pharmacology, toxicology, and therapeutic effect of metals and minerals used in traditional medicine

Understanding the aetiology of trace element related non-communicable diseases – reviewing the Ghanaian situation

Dental fluorosis in Turkey

Environmental and medical geology of the lead mining and metallurgical complex of Bahia: The case of lead metallurgy, Santo Amaro, Bahia, Brazil

Uncontrolled coal fires: How medical geology can save lives

Majeti Narasimha Vara Prasad
University of Hyderabad
Hyderabad
India

Meththika Vithanage
University of Sri Jayewardenepura
Nugegoda
Sri Lanka

Acknowledgments

We would like to thank all the authors of this volume for their cogent and comprehensive contributions. We are thankful to Dr. Rituparna Bose, Dr. Frank Weinreich, Ms. Sarah Higginbotham, Ms. Stefanie Volk, Mr. Sakthivel Kandaswamy, Mr. Kavin Shanmughasundaram, and the team at John Wiley & Sons for excellent technical help in many ways that resulted in the publication of this edition. Last but not least, we wish to acknowledge numerous colleagues from overseas, our collaborators for sharing knowledge, ideas, and assistance that helped in developing and shaping this book.

Section 1

Geochemistry and Health

1

Medical Geology

Biosphere, Geosphere, and Noosphere Interface

Majeti Narasimha Vara Prasad

School of Life Sciences, University of Hyderabad (an Institution of Eminence), Hyderabad, Telangana, India

"All substances are poisons; there is none which is not a poison. Only the dose differentiates a poison and a remedy."
Paracelsus (1493–1541).

1.1 Introduction

Medical geology is now developed into a truly fascinating new science, with potentially enormous consequences for the well-being of people around the world. On our planet, the chemical elements flow through the different planetary compartments, including the atmosphere, hydrosphere, lithosphere, and biosphere. Humans and animals are part of these cycles. The chemical elements pass into and out of them in complex biogeochemical cycles (sedimentary, gaseous, and hydrologic).

Dissanayake (2005) envisioned rocks, minerals, and health as medical geology. The basic building blocks of the Earth – the rocks and minerals – must therefore have a bearing on the health of the human and animal populations that live on these earthy materials. International Union of Geological Sciences in early 1990 defined medical geology as "the science dealing with the influence of ordinary environmental factors on the geographical distribution of health problems in man and animals."

1.2 Medical Geology in Russia and Newly Independent States (NIS)

Russia and the Newly Independent States (NIS) have a rich knowledge of medical geology (Selinus et al. 2021). The mineral rich Ural region of Russia has rich knowledge on the valuable properties of these minerals for human health (Figure 1.1). These minerals are used in cosmetics, preventive medicine and treatment of different diseases, such as skin lesions, and many diseases of internal organs, traumas, and fractures and to treat damages from ecological catastrophes. The medical geological traditions in Russia and NIS are still popular.

Physicians and academicians from Central Asia and Kazakhstan are currently making significant progress with the use of natural minerals and their mixtures. It is believed that at present about 500 known minerals are used in medicine, and more than 350 are well defined in the territory of NIS and particularly in Kazakhstan. Among these are ore minerals and economic minerals such as clays, shungites, flints, zeolites, and salts (Volfson et al 2010).

Selected examples of domestic minerals

Emerald: "Who often look at the emerald, then human eye is strengthened, protected and healthy. And besides emerald brings a cheer to whom who carries it."

Turquoise: "After it is ground, take it, and it helps from bite of snake."

Quartz: "Crystal-stone to grind and stirred with fresh honey, that mother's milk multiplies"

Obviously, the chemistry of any local geological environment must have a direct influence on the chemical makeup of those living there. This is most readily seen in places where humans live in particularly intimate contact with the local physical environment, as is the case with rural people living in tropical countries. Those living on lands with heavily impoverished soils, such as in

Medical Geology: En route to One Health, First Edition. Edited by Majeti Narasimha Vara Prasad and Meththika Vithanage.

Figure 1.1 Different colored stones with diverse minerals and chemicals are the natural wealth from geological environment.

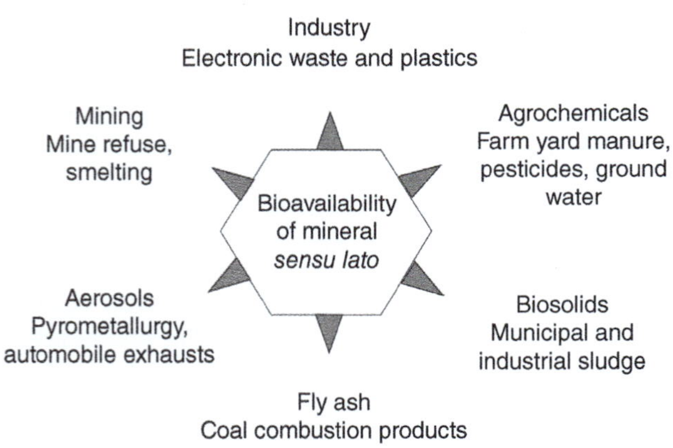

Figure 1.2 Bioavailability of minerals (*sensu lato, sensu stricto*).

Maputaland, South Africa (Gomes and Rautureau 2021; Gwenzi 2021; Hassina 2021), have such a low intake of essential elements that a very large percentage of the population suffers from a variety of diseases caused by severe mineral imbalances (Figure 1.2).

Likewise, in other areas, there is an excess intake of elements due mainly to the abundance of certain elements in the environment. This leads to high incidences of mineral toxicity, such as the widespread and tragic arsenic poisoning in India and Bangladesh (Table 1.8) (Kumaresan and Riyazuddin 2001; Li and Chen 2005; Silva et al. 2005; Sadeghfam et al. 2021). Even though many other factors among them life-style, sex, age, migrations, and food habits affect health, optimal amounts of trace elements are needed to maintain good health and an imbalance in these elements can lead to disease.

Imbalances in the supply of inorganic elements exert marked influences on both human and animal health. Anomalies in the local abundances of trace elements, for example, have a large impact on food chains. As it was more than 500 years ago, it remains relevant to bear in mind the basic law of toxicity as defined by Paracelsus (1493–1541), the father of pharmacology: "All substances are poisons; there is none which is not a poison. The right dosage differentiates a poison and a remedy." Even water, when consumed too quickly and in inordinate amounts, can be lethal.

One of the primary objectives of medical geologists, therefore, is to determine the optimal exposures for people to the essential elements in order to maintain or improve health (Figures 1.3 and 1.4). Diseases such as hypealernia (due to excess potassium), hypercalcemia (due to excess calcium), and hyperphosphatemia (due to excess phosphorus) exist in various parts of the world even though potassium, calcium, and phosphorus are essential dietary elements.

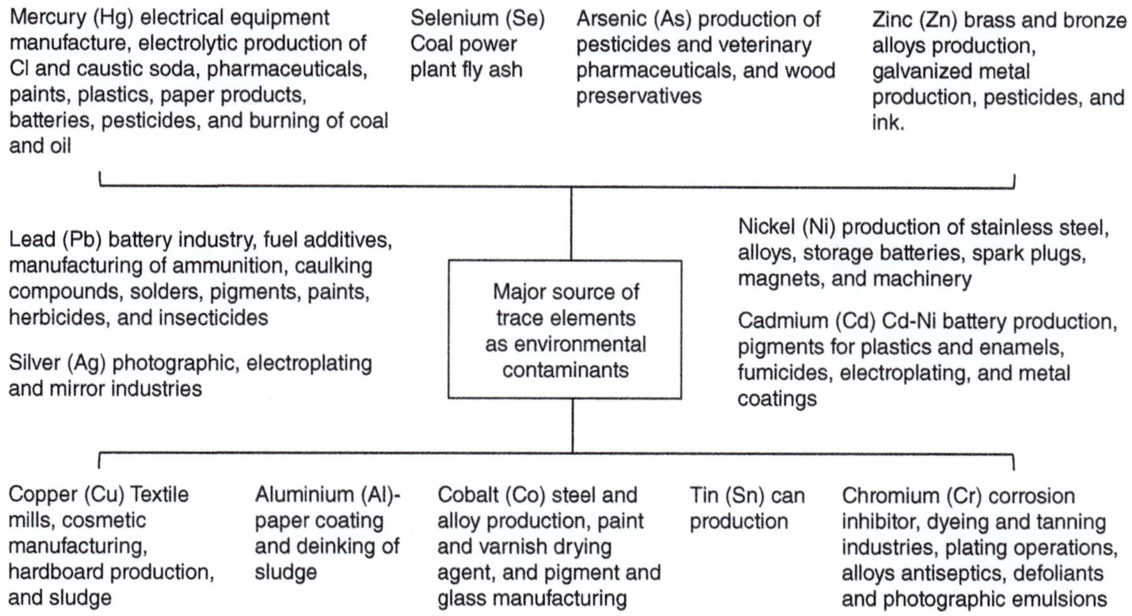

Figure 1.3 Trace elements as nutrients and contaminants.

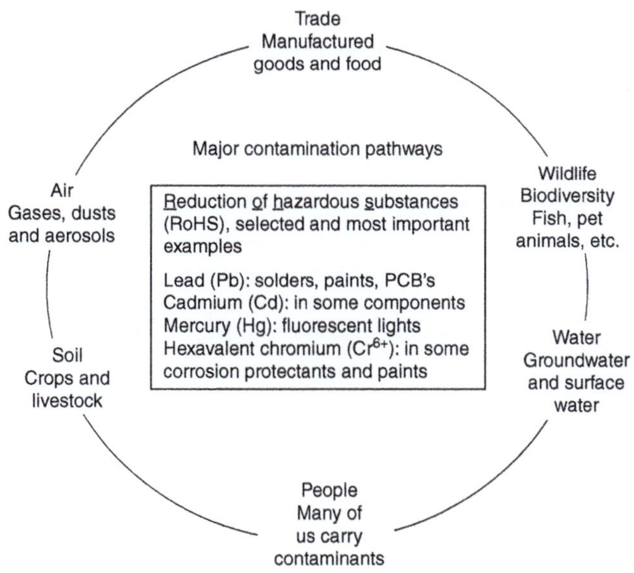

Figure 1.4 Major contamination pathways of minerals (*sensu lato, sensu stricto*).

Minerals *sensu lato* are classified in the following three categories and are detailed in Tables 1.2–1.4.

1) Medicinally important (Table 1.2)
2) Nutritionally important (Table 1.3)
3) Toxic elements (Table 1.4)

1.2.1 Linking Geology to Soils – Minerals and HealthCare (Medicine)

Nutrinogenomics and pharmacogenomics, medical geology (geomedical science), and metallomics are emerging as new areas of research with challenging tasks ahead. The relevance of these cutting-edge fields and their relevance to nutritional security and quality of agricultural produce including healing plants play a significant role in human health.

Several of the trace elements serve as essential nutrients and are constituents of bimolecules as cofactors for various enzymes and in a variety of metabolic functions (Figure 1.5). Certain trace elements accumulated in medicinal plants have the healing function to some ailments and disorders, for example, zinc – neurochemical transmission; Cr and Mn can be correlated with therapeutic properties against diabetic and cardiovascular diseases. Certain transition group elements regulate hepatic synthesis of cholesterol (Table 1.1). Bioorganic farming takes total micronutrient care and is an important soil management practice for fortification of crop produce with essential nutrients for the health of humans as well as livestock.

Humanity depended on natural resources (air, water, and soil/sediment-containing minerals) for survival (Table 1.5).

Industrial minerals are crucial in water management, whether considering drinking water preparation or wastewater treatment. Thus, silica sands are used as filters, perlite, zeolites, or talc as flocculants or adsorbents, bentonite as industrial mineral-based liners and geosynthetic liners, either basal or superficial, are increasingly used to avoid escape of leachate from landfill sites. Air treatment of industrial effluents also largely relies on minerals. Activated carbon is the best-known technique, but other minerals are used as well. For instance, flue gas desulfurization of power station fumes is achieved with calcium carbonate. Industrial minerals such as clays, sand, feldspar,

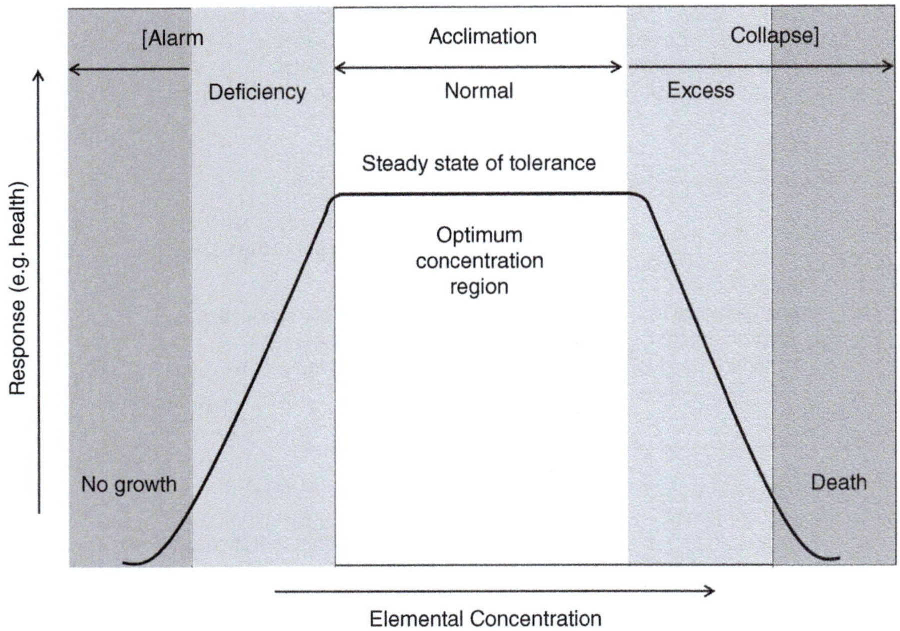

Figure 1.5 Triphasic kinetics of general adaptation syndrome (GAS) similar to dose/functional response in biological system to the dose of the elements as the nutrients. Phase 1 – alarm reaction: this is the somatic response to the first exposure to a localized or general stressor and initially lowers basal state of resistance. If the "stressor" is sufficiently strong (e.g. extremes of temperature, salinity, etc.), death may result; phase 2 – stage of resistance or acclimation: if continued exposure to the stressor is compatible with adaptation, resistance mechanisms of either a "syntoxic" (coexistence with stress) or a "catatoxic" (stressor-removal) nature ensues. The cellular manifestation characteristics of the alarm reaction have virtually disappeared or are strictly held in abeyance and resistance rises above normal; phase 3 – state of collapse or exhaustion: following long-term exposure to the stressor, adaptation capability is eventually exhausted. Signs of the alarm reaction reappear and the organ or the whole organism senesces and dies.

Table 1.1 Diseases/symptoms associated with deficiency or efficiency of essential nutrients.

Element	Symptoms or disease due to deficiency	Element	Symptom or disease due to excess availability
Selenium	Keshena (China), bone, arthritis, cardiovascular, and cancer	Mercury	Minamata disease, neurological disorders
Zinc	Dwarfness (Iran and Egypt) infertility, impaired taste, and smell	Cadmium	Itai-Itai, bone crippling, cancer → prostrate, lung, renal, and heart problems
Copper	Anemia, skeletal defects	Fluorine	Yellowing of teeth, skeletal deformities
Chromium	Glucose metabolism, kidney function, cholesterol disorders	Lead	Lung cancer, neurological disorders
Iodine	Goiter	Manganese	Paraplegia = skeletal deformations → paralysis
Fluorine	Dental caries	Arsenic	Skin, lung cancer
Magnesium	Depression, nervous system. Every organ in the body, especially the heart, muscles, and kidneys, needs the nutritional element Mg		
Molybdenum	Mouth/esophageal cancer		
Cobalt	Anemia		
Iron	Anemia		
Sodium	Coma		

kaolin, and other minerals are basic to all construction materials, from bricks to tiles and from cement to limes and plastics, and cleaning of contaminated water is one of the most difficult and expensive goals in environmental engineering. Clay minerals are the most important inorganic components in soil for sorption of contaminants, but without modification, they are ineffective sorbents for contaminant compounds. The treatment with surfactants

Table 1.2 Medicinally important elements and functions.

Element	Functions
Bi Bismuth Atomic number: 83 Atomic mass: 208	Use – antacids, diarrhea Source – mining, consumer products Absorption – intestine Toxicity – kidney, chronic use results in range of effects Facts – discovered in 1753, used to treat syphilis and malaria
F Fluorine Atomic number: 9 Atomic mass: 19	Use – tooth protection Source – drinking water, food supply Absorption – intestine Toxicity – excess causes mottled teeth enamel (fluorosis) Facts – common water level 0.5–1.5 ppm, 3 ppm effects teeth
Ga Gallium Atomic number: 31 Atomic mass: 69.72	Use – visualization tool for soft tissues in X-rays Source – mining, medical injection Absorption – very poor Toxicity – kidney Facts – liquid at room temperature, half-life four–five days
Au Gold Atomic number: 79 Atomic mass: 196.97	Use – treat rheumatoid arthritis, range of industrial uses Source – mining, medical injection Absorption – poor Toxicity – kidney, skin, and mouth lesions Facts – long half-life
Li Lithium Atomic number: 3 Atomic mass: 6.939	Use – treat psychiatric disorders Source – food supply, plants and meat Absorption – intestine Toxicity – wide range, e.g. tremor, seizures, slurred speech, cardiovascular, nausea, vomiting Facts – daily intake about 2 mg
Pt Platinum Atomic number: 78 Atomic mass: 195.02	Use – anticancer agent (cisplatin), catalytic converters, metal alloy Source – mining, road dust Absorption – poor, as a drug intravenous administration Toxicity – neuromuscular, kidney Facts – inhibits cell division, treats ovarian and testicular cancer

Table 1.3 Nutritionally important elements and functions.

Element	Functions
Cr Chromum Atomic number: 24 Atomic mass: 52	Use – essential element, associated with insulin, stainless steel, tanning leather Source – food supply, inhalation Recommended daily – 50–200 μg Absorption – intestine Toxicity – acute exposure causes kidney damage, lung cancer Facts – comes in different oxidized forms – Cr^{3+}, Cr^{6+}
Cu Copper Atomic number: 29 Atomic mass: 63.55	Use – essential element, widely used Source – readily available in food Recommended daily – 1.5–3.0 mg Absorption – intestine Toxicity – deficiency – anemia-excess rare, Wilson's disease Facts – excess treated with penicillamine – can be toxic to grazing animals

(Continued)

Table 1.3 (Continued)

Element	Functions
Fe Iron Atomic number: 26 Atomic mass: 55.85	Use – oxygen carrying hemoglobin Source – food Recommended daily – 10–15 mg Absorption – intestine Toxicity – excess causes bloody feces bloody vomit, liver damage Facts – 3–5 g in the body, 67% associated with hemoglobin
Mg magnesium Atomic number: 12 Atomic mass: 24.3	Use – essential nutrient, associated with many enzymes, antacids Recommended daily – 280–350 mg Source – food supply, nuts, cereals, seafood, meats, drinking water Absorption – small intestine Toxicity – deficiency – convulsions – excess – nervous system Facts – 20 g in body
Mn Manganese Atomic number: 25 Atomic mass: 54.94	Use – trace element, associated with many enzymes Source – food supply, grains, nuts Recommended daily – 2–5 mg Absorption – intestine poor (5%) Toxicity – inhalation – respiratory disease, nervous system, Parkinson's-like syndrome, psychiatric disorders Facts – half-life 37 days
Se Selenium Atomic number: 34 Atomic mass: 78.96	Use – essential element, present in most tissue, anticancer, reduces toxicity of metal mercury and cadmium Source – food supply, shrimp, meat Recommended daily – 55–70 μg/day, not to exceed 200 μg/day Absorption – intestine Toxicity – deficiency – heart disorders - excess – "blind staggers," neurological effects
S Sulfur Atomic number: 16 Atomic mass: 32.07	Use – sulfuric acid, fertilizer, fine chemicals, pesticides, pharmaceuticals, winemaking, and food preservation Source – sulfur is the fifth most common on Earth
Zn Zinc Atomic number: 30 Atomic mass: 65.39	Use – essential element, cofactor with several enzymes, and proteins Source – food supply, drinking water Recommended daily – 12–25 mg Absorption – intestine Toxicity – deficiency – impaired growth, neurological disorders – inhalation can cause metal fume fever
Si Silica Atomic number: 53 Atomic mass: 126.9	Use – clay minerals and other aluminosilicates Have several medicinal uses Source – Earth crust Silica rich plant leaves are resistant to bacterial and fungal pathogens as they cannot penetrate
I Iodine Atomic number: 53 Atomic mass: 126.9	Use – chemicals and pharmaceuticals. In chemistry, I compounds are used as catalyst Source: crustal average abundance is estimated as 0.5 mg/kg
Mo Molybdenum Atomic number: 42 Atomic mass: 95.95	Use – the essential component of several enzymes, such as nitrate reductase, xanthine dehydrogenase, aldehyde oxidase, and sulfide oxidase Source: Earth's crust as 1.5 mg/kg

Table 1.4 Toxic elements that do not have metabolic significance in humans.

Element	Functions
Al Aluminum Atomic number: 13 Atomic mass: 26.98	Use – wide range of consumer products, airplanes to cans Source – food, drinking water Absorption – poor Toxicity – Dialysis dementia, possibly neurotoxic Facts – non-essential, intake 1–10 mg/day
As Arsenic Atomic number: 33 Atomic mass: 74.92	Use – pesticide and herbicide Source – food, drinking water Absorption – intestine Toxicity – cancer, heart, liver, neurological Facts – exists in different states – trivalent (most common), pentavalent, arsenic trioxide, organic, and inorganic, etc.
Be Beryllium Atomic number: 4 Atomic mass: 9.01	Use – metal alloy, nuclear power plants Source – workplace, coal combustion Absorption – lung, skin Toxicity – lung, can be delayed and is progressive, contact dermatitis probable carcinogen Facts – discovered in 1828, more than 1250 tons from oil and coal combustion
Cd Cadmium Atomic number: 48 Atomic mass: 112.41	Use – alloy in metal, paint Source – shellfish, cigarette smoke, workplace – welding, paints Absorption – intestine, lungs Toxicity – lung, emphysema, kidney, calcium metabolism, possible lung carcinogen Facts – "Itai-Itai" is Japanese for "ouch-ouch" – refers to bone pain related to calcium loss
Co Cobalt Atomic number: 27 Atomic mass: 58.93	Use – component of vitamin B_{12} Source – alloy in metals, magnets Recommended daily – none Absorption – intestine Toxicity – excessive heart failure, inhalation – "hard metal" lung disease Facts – once used as a foaming agent in beer
Pb Lead Atomic number: 82 Atomic mass: 207.20	Use – not essential, batteries, old paint and previously gasoline, hobbies Source – home, paint, dust, kids – hands to mouth, workplace Absorption – intestine (50% kids, 10% adults) Toxicity – developmental and nervous system Facts – developing nervous system very sensitive to low levels of exposure
Hg Mercury Atomic number: 80 Atomic mass: 200	Use – consumer products, industry, dental amalgams, switches, thermometers Source – mining, environment Absorption – inhalation, intestine poor Toxicity – nervous system toxicant, "Mad Hatters" disease Facts – liquid silver evaporates at room temperature, bacteria convert to organic methyl mercury (see next slide)
Organic Mercury ($Hg\text{-}CH_3$)	Use – limited laboratory use – most common is methyl mercury ($Hg\text{-}CH_3$) Source – contaminates some fish (e.g. tuna, shark, pike) Absorption – intestine very good (90%) Toxicity – nervous system toxicant and developmental toxicant Facts – bacteria convert inorganic mercury to methyl mercury then into food supply (bioaccumulation)
Ni Nickel Atomic number: 28 Atomic mass: 58.96	Use – not essential, metal alloy, stainless steel Source – food supply, jewelry, workplace Absorption – intestine, skin Toxicity – carcinogen (lung), contact dermatitis Facts – discovered in 1751, 200 000 metric tons used yearly
Sn Tin Atomic number: 50 Atomic mass: 118.71	Use – inorganic – consumer products – organic – fungicide, bactericides Source – food packaging Absorption – intestine (low inorganic, high organic) Toxicity – inorganic – little – organic – central nervous system Facts – triethyltin and trimethyltin most toxic

Table 1.5 The elements contained in humans and their biological functions.

	Element	Total amount (70 kg weight)	Functions and diseases
Major elements	O	45.5 kg	Compositions of water and organics
	C	12.6	Compositions of organics
	H	7.0	Compositions of water and organics
	N	2.1	Compositional elements of amino acids, proteins, nucleic acids, etc.
	Ca	1.05	Main constituents of bones, compositions of cell membrane, blood coagulation factors
	P	0.70 (98.5%)	Main constituents of bones, compositional elements of nucleic acids and ATP minor elements
Minor elements	S	175 g	Compositions of amino acids (cysteine, methionine), vitamin B_1, biotin, heparin, etc.
	K	140	Compositions of inner-cell electrolytes Na 105 Compositions of outer-cell electrolytes
	Cl	105	Main anions existing inside and outside cells
	Mg	105 (99.4%)	Cofactors of enzymes, composition of chlorophylls, trace elements
Trace elements	Fe[a]	6	Transportation and storage of oxygen, transportation and storage of iron, antioxidation
	F[b]	3	Concomitant elements in bone
	Si[a]	2	Unknown
	Zn[a]	2	Cofactors of enzymes, cell division, nucleic acids polymerases
	Sr	32 mg	Unknown
	Rb[b]	320	Unknown
	Pb[b]	120	Metabolism of iron blood making
	Mn[a]	100	Cofactors (superoxide dismutase, pyruvate kinase), metabolism of fats
	Cu[a]	80	Transportation of oxygen, redox reactions, electron transfer, oxygen addition, ultratrace elements
Ultratrace elements	Al	60	Alzheimer's disease
	Cd	50	Auch-auch disease
	Sn[b]	20	Redox reactions
	Ba	17	Unknown
	Hg	13	Minamata disease (methylmercury)
	Se[a]	12	Antioxidation (glutathione peroxyalase), mercury toxicity reduction
	I[a]	11	Activation of thyroid gland
	Mo[a]	10	Cofactors of enzymes (xanthine oxidase, nitrate reductase), metabolism of uric acid
	Ni[b]	10	Cofactor of enzyme (urease), stabilization of RNAs
	Cr[a]	2	Glucose tolerance factor, metabolisms of fats and proteins
	As[b]	2	Metabolic activation of zinc Co[a] 1.5 Blood making (vitamin B_{12})
	V[b]	1.5	Metabolism of cholesterols, inhibitors of Na1- and K1-ATPase a

Source: Haraguchi (2004).
[a] Essential elements in man.
[b] Essential elements in experimental mammal.

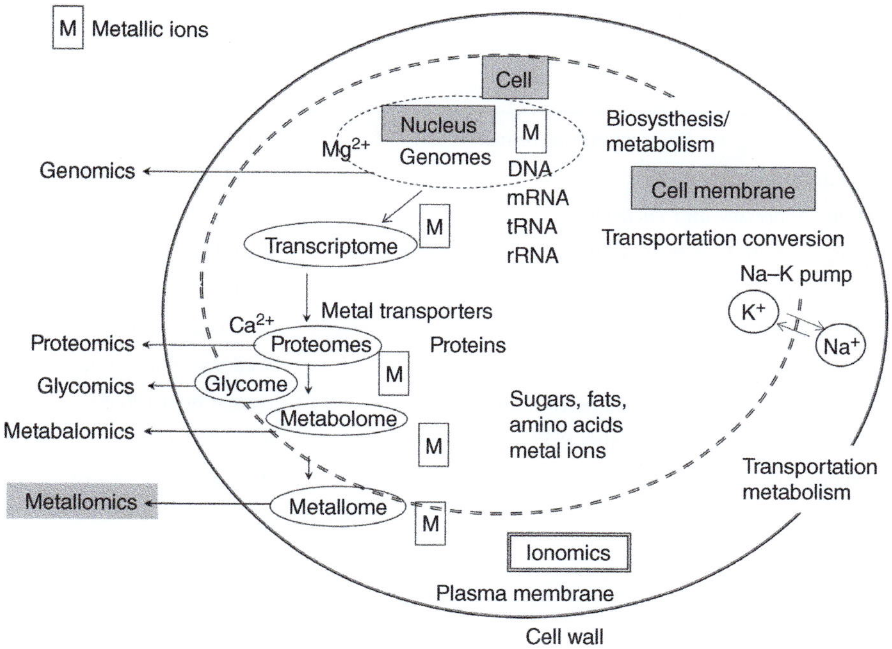

Figure 1.6 A schematic model of the biological system elucidating *metallomics* as a multidisciplinary metal-assisted functional biogeochemistry, showing the relationships among genomics, proteomics, metabolomics, and metallomics, where metallic ions are shown as M. *Source:* Redrawn from Haraguchi (2004).

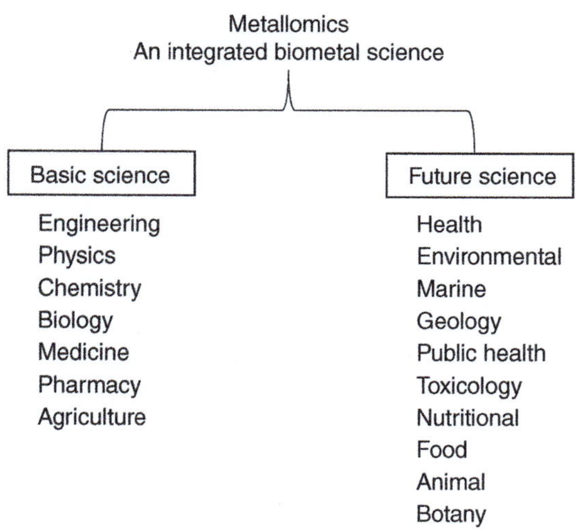

Figure 1.7 Significance of metallomics for basic and future sciences.

changes the surface properties from hydrophilic to hydrophobic and greatly increases the basal spacing of the layers. This new material, named organoclay, is used in some environmental applications, such as the removal of contaminants from polluted waters, offering a promising device in decontamination planning and environmental remediation.

The Ayurvedic system of medicine has been in practice for millennia. The ancient seers found that drugs of different origin (herbal, metal, or animal) in addition to codes of conduct and dietary regulations are suitable tools to address several diseases. Use of metallic preparations [(Bhasmas = ash and bhasmikarana (=process of preparation of metallic ash which is zero valent)] in healthcare is unique. Bhasmikarana process with mercury, gold, silver, lead, zinc, copper, etc. was frequently used by seers of the Indian tradition to treat different diseases with great certainty of cure. It is generally claimed that these metals are detoxified during the highly complex bhasmikarana process described in Ayurveda especially Rasashastra manuscripts. Charaka Samhita, one of the scheduled books of Ayurveda, also holds ample references regarding the use of metals for different purposes; selected examples are enumerated in this chapter (Table 1.6).

Nutrinogenomics and pharmacogenomics, medical geology (geomedical science), and metallomics are emerging as new areas of research with challenging tasks ahead. The relevance (Figures 1.10 and 1.11) of these cutting-edge fields and their relevance to nutritional security and quality of agricultural produce including healing plants will be discussed with selected examples.

"Metallomics" is the metal-assisted functional biogeochemistry; a new scientific field was first proposed by Haraguchi (2004) to integrate the research fields related to metal biomolecules and their metals (Figures 1.6 to 1.9).

Metallomics is the scientific field of symbiosis with genomics and proteomics, because syntheses and metabolic functions of genes (DNA and RNA) and proteins cannot function without the coordination of various metal

Table 1.6 Bibliographic resume of significant findings in medical geology.

Significant findings / observations	Author(s)/year
Chemical and pharmacological evaluation of different ayurvedic preparations of iron	Pandit et al. (1999)
Geochemical assessment of the heavy metal pollution of urban soils	Grzebisz et al. (2002)
Handbook on Ayurvedic Medicines with Formulae, Processes, and Their Uses	Panda (2004)
Concentrations of additive arsenic in Beijing pig feeds and the residues in pig manure resources	Li and Chen (2005)
Determination of metal content in some herbal drugs – empirical and chemometric approach	Ražić et al. (2005)
Dietary intake and health effects of selected toxic elements	Silva et al. (2005)
Preparation and characterization of a copper-based Indian traditional drug: Tamra bhasma	Wadekar et al. (2005)
Bhasmas: unique ayurvedic metallic-herbal preparations, chemical characterization	Kumar et al. (2006)
Estimation of trace elements in some anti-diabetic medicinal plants using PIXE technique	Raju et al. (2006)
Effect of calcination cycles on the preparation of tin oxide-based traditional drug: studies on its formation and characterization	Wadekar et al. (2006)
Minerals and clay minerals in medical geology	Gomes and Silva (2007)
Medicinal clay and spiritual healing	Ferrell (2008)
Medical geology curriculum for African geoscience institutions	Davies (2009)
Medical geology of fluoride in introduction to medical geology	Dissanayake (2005)
A medical geology curriculum for African geoscience institutions	Davies (2009)
Some observations on the metal-based preparations in the Indian system of medicine	Kapoor (2010)
Physicochemical characterization of Ayurvedic Bhasma (Swarna makshika bhasma): an approach to standardization	Mohaptra and Jha (2010)
Geochemical speciation and risk assessment of heavy metals in the river estuarine sediments – a case study: Mahanadi basin, India	Sundaray et al. (2011)
Assessment of heavy metal contamination in Hindon River sediments: a chemometric and geochemical approach	Chabukdhara and Nema (2012)
Elucidation of a core-shell model for lauha bhasma through physicochemical characterization	Krishnamachary et al. (2012)
Structural analysis of metallic medicines	Parivallal (2012)
Geochemical variables as plausible aetiological cofactors in the incidence of some common environmental diseases in Africa	Davies (2013)
Geology and health: A brief history from the Pleistocene to today	Hasan et al. (2013)
Geochemical variables as plausible aetiological cofactors in the incidence	Davies (2013)
Phytoextraction as a tool for green chemistry	Hunt et al. (2014)
Blood minerals	Marks 2014
Climate change, food security, and trade: an overview of global assessments and policy insights	Elbehri et al. (2015)
Agromining: farming for metals in the future	van der Ent et al. (2015)
Geochemical normalization and assessment of heavy metals (Cu, Pb, Zn, and Ni) in sediments from the Huaihe River, Anhui, China	Wang et al. (2015)
Geomedicine and history of science	Rodrigues et al. (2016)
Recent aspects of uranium toxicology in medical geology	Bjørklund et al. (2017)
Progress in Medical Geology	Ibaraki and Mori (2017)
Geochemical assessment of heavy metals pollution in surface sediments of Vellar and Coleroon estuaries, Southeast coast of India	Nethaji et al. (2017)
The effect of portable HEPA filter air cleaners on indoor PM 2.5 concentrations and second-hand tobacco smoke exposure among pregnant women in Ulaanbaatar, Mongolia: the UGAAR randomized controlled trial	Barn et al. (2018)
Medical geology in India	Lohar (2019)
Natural antibacterial clays: historical uses and modern advances	Williams (2019)

Table 1.6 (Continued)

Significant findings / observations	Author(s)/year
Clays in Pelotherapy: a review, part I – mineralogy, chemistry, physical and physicochemical properties	Carretero (2020a)
Clays in Pelotherapy: a review, part II – organic compounds, microbiology, and medical application	Carretero (2020b)
A comprehensive analysis to understand the mechanism of action of balneotherapy: why, how, and where they can be used? Evidence from *in vitro* studies performed on human and animal samples	Cheleschi et al. (2020)
Determination of Metals and Anions in Soils, Sediments, and Sludges	Crompton (2020)
On the why's and how's of clay minerals importance in life's emergence	Duval et al. (2020)
Effect of dead sea climatotherapy on psoriasis: a prospective cohort study	Emmanuel et al. (2020)
Historical aspects of medical geology	Farkhutdinov et al. (2020)
Wound healing activity of nanoclay/spring water hydrogels. Pharmaceuticals	Garcia-Villén et al. (2020)
Biotechnology applied to cosmetics and aesthetic medicines. Cosmetics	Gomes et al. (2020a)
Bacteriostatic and bactericidal clays: an overview. *Environ Geochem Health*	Gomes et al. (2020b)
Bacteriostatic and bactericidal clays: an overview	Gomes et al. (2020b)
Evidence that Vitamin D supplementation could reduce risk of influenza and COVID-19 infections and deaths	Grant et al. (2020)
Endemic diseases of geochemical origin and methodological approaches toward their prevention and elimination	Korobova et al. (2020)
Selenium deficiency risk in central Kenya highlands: an assessment from the soil to the body	Ngigi et al. (2020)
Adsorption of Salmonella in clay minerals and clay-based materials	Pardo et al. (2020)
Halloysite-and montmorillonite-loaded scaffolds as enhancers of chronic wound healing	Sandri et al. (2020)
Human health risk assessment of cadmium exposure through rice consumption in cadmium contaminated areas of the Mae Tao sub-district, Tak, Thailand	Suwatvitayakorn et al. (2020)
Bridging the gaps: bole and terra sigillata as artefacts, as simples and as antibacterial clays, *Minerals*	Venieri et al. (2020)
Environmental engineers and scientists have important roles to play in stemming outbreaks and pandemics caused by enveloped viruses	Wigginton and Boehm (2020)
Metal accumulation potential, human health risks, and yield attributes of hundred bread wheat genotypes on irrigation with municipal and remediated wastewater	Akbar et al. (2021)
Techniques for assessing metal mobility in the environment: a geochemical perspective, *Practical Applications of Medical Geology*	Avasarala (2021)
Carcinogen soil radon enrichment in a geothermal area: Case of Güzelçamlı-Davutlar district of Aydın city, Western Turkey	Aydar and Diker (2021)
Naturally occurring arsenic and boron in geothermal systems and their health effects: a case study from Turkey, *Practical Applications of Medical Geology*	Baba et al. 2021
Fluoride hazard and risk enumeration of Chhota Nagpur Gneissic complex.	Bera et al. (2021)
Hydrochemical and radiometric evaluation of fresh and thermal waters from Araxá city (Minas Gerais, Brazil)	Bonotto (2021)
Biological Remediation of Selenium in Soil and Water	Borah et al. (2021)
Toxic metal species and 'Endogenous' metalloproteins at the blood–organ Interface: analytical and bioinorganic aspects	Bridle et al. (2021)
Application of bioavailability measurements in medical geology, *Practical Applications of Medical Geology*	Cave and Wragg (2021)
With bare feet in the soil: podoconiosis, a neglected cause of tropical lymphedema	Chandler et al. (2021)
Hydrogeochemical Characteristics and Geothermometry of Hot Springs in the Mongolian Altai Region, Mongolia	Chimeddorj et al. (2021)
Knowledge, attitude, practice, and associated factors of health professionals toward podoconiosis in Gamo zone, Ethiopia	Churko et al. (2021)
Analysis of some concepts related to the environment and health with the N-gram method	Çiftçi et al. (2021)

(Continued)

Table 1.6 (Continued)

Significant findings / observations	Author(s)/year
Biominerals: formation, function, properties	Cölfen et al. (2021)
Editorial for the special issue "development of w-sn and rare-metal metallogenic systems in an orogenic belt"	Cuney (2021)
Geophagic clays from Cameroon: provenance, metal contamination, and health risk assessment	Ekosse et al. (2021)
Ecological study of fractures in pediatric Melanesian communities with varying endemic environmental fluoride exposure	Elizabeth et al. (2021)
Chemical pollution of drinking water in Haiti: an important threat to public health	Emmanuel and Emmanuel (2021)
Exploring the intersections of environmental health and urban medical geology, *Practical Applications of Medical Geology*	Filippelli and Finkelman (2021)
Fluoride polluted groundwaters in Calabria region (Southern Italy): natural source and remediation	Fuoco et al. (2021)
A review on the occurrence of some potentially harmful elements in the natural environment and their health implications in the South-Eastern Kenya	Gevera et al. (2021)
A review on the occurrence of some potentially harmful elements in the natural environment and their health implications: examples of fluoride, iron, and salinity in the South-Eastern Kenya region	Gevera et al. (2021)
Residential radon in Manizales, Colombia: results of a pilot study	Giraldo-Osorio et al. (2021)
Minerals, natural environment, and medical geology, *Minerals Latu Sensu and Human Health*	Gomes et al. (2021)
Interactions of clay and clay minerals with the human health, *Minerals Latu Sensu and Human Health*	Gomes et al. (2021)
Historical evolution of the use of minerals in human health	Gomes and Rautureau (2021)
Interactions of clay and clay minerals with the human health, *Minerals Latu Sensu and Human Health*	Gomes et al. (2021)
Toxic geogenic contaminants in Serpentinitic Geological Systems: Occurrence, Behavior, Exposure Pathways, and Human Health Risks	Gwenzi (2021)
Occurrence, behavior, and human exposure and health risks of potentially toxic elements in edible mushrooms with focus on Africa	Gwenzi et al. (2021)
Toxic geogenic contaminants in serpentinitic geological systems: occurrence, behavior, exposure pathways, and human health risks. Geochemistry: concepts and applications	Gwenzi (2021)
Occurrence, behavior, and human exposure and health risks of potentially toxic elements in edible mushrooms with focus on Africa	Gwenzi et al. (2021)
Medical geology in *Encyclopedia of Geology*	Hasan (2021)
Medical geology in Africa: an example of a successful educational and research initiative at the University of Johannesburg, South Africa, *Practical Applications of Medical Geology*	Hassina (2021)
The improvement of indoor air quality in residential buildings in Dubai, UAE	Jung and Awad (2021)
Known and anticipated medical geology issues in Ghana. *Ecofeminism and Climate Change*	Kazapoe et al. (2021)
Coal combustion residuals and health, *Practical Applications of Medical Geology*	Kravchenko and Ruhl (2021)
Designing an indoor radon risk exposure indicator (IRREI): an evaluation tool for risk management and communication in the IoT age	Lopes et al. (2021)
Gold-Mining Pollution Exposure, Health Effects and Private Healthcare Expenditure in Tanzania	Magambo et al. (2021)
Concurrent heavy metal exposures and idiopathic dilated cardiomyopathy: a case-control study from the Katanga mining area of the Democratic Republic of Congo	Malamba-Lez et al. (2021)
Perspectives on complexity, chaos, and thermodynamics in environmental pathology	Manera (2021)
Veterinary geology. In *Practical Applications of Medical Geology*	Myburgh et al. (2021)
Hydrogeochemical and ecological risk assessments of trace elements in the coastal surface water of the Southern Caspian Sea.	Nematollahi et al. (2021)
Balneotherapy: an overview of healing with natural waters	Nghargbu (2021)
Health effects of exposure to specific geologic materials: summary of clinical findings, treatment, and prevention, *Practical Applications of Medical Geology*	Nichols and Nelson (2021)

Table 1.6 (Continued)

Significant findings / observations	Author(s)/year
Hydrochemistry and Fluoride Contamination in Ndali-Kasenda Crater Lakes, Albertine Graben: Assessment Based on Multivariate Statistical Approach and Human Health Risk	Ojok et al. (2021)
Geochemical and health risk assessment of heavy metals concentration in soils around Oke-Ere mining area in Kogi State, Nigeria.	Olumayowa et al. (2021)
Anthropocene: new geological era or new biophysical	Ortega (2021)
Balneotherapy with the use of radon–sulfide water: the mechanisms of therapeutic effect	Pawlik-Sobecka et al. (2021)
Medical geology of soil ecology, *Practical Applications of Medical Geology*	Pereg et al. 2021
Geology and environment: a problem-based learning study in higher education	Pinto et al. (2021)
Human health risk assessment by exposure to arsenic through groundwater in a rural area of Canelones, Uruguay	Pizzorno et al. (2021)
Selenium biofortification of three wild species, *Rumex acetosa* L., *Plantago coronopus* L., and *Portulaca oleracea* L., grown as microgreens	Puccinelli et al. (2021)
Selenium biofortification of wheat as a strategy to improve human nutrition	Radawiec et al. (2021)
Selenium as a bioactive micronutrient in the human diet and its cancer chemopreventive activity	Radomska et al. (2021)
The importance of minerals in medical geology: impacts of the environment on health	Ríos-Reyes et al. (2021)
Selenium status, its interaction with selected essential and toxic elements, and a possible sex-dependent response	Röllin et al. (2021)
Indoor radon concentration and a diffusion model in dwellings situated in a subalkaline granitoid area, Southern Brazil	Romero-Mujalli et al. (2021)
An investigation into uncertainties within human health risk assessment to gain an insight into plans to mitigate impacts of arsenic contamination	Sadeghfam et al. (2021)
Role of soil in the regulation of human and plant pathogens: soils' contributions to people	Samaddar et al. (2021)
Sources, pathways, and health effects of iodine in the environment, *Practical Applications of Medical Geology*	Sanyaolu et al. (2021)
Acid precipitation and the prevalence of Parkinson's disease: an ecologic study in US states. *Brain Sciences*, 11(6): 779	Schwartz and Williamson (2021)
A guide to the medical geology literature, *Practical Applications of Medical Geology*	Selinus et al. (2021)
A framework for applied medical geology:	Siegel (2021)
practical applications of medical geology	Siegel et al. (2021)
A framework for applied medical geology: Part I. The environmental pathways analysis, *Practical Applications of Medical Geology*	Siegel (2021a)
A framework for applied medical geology: Part II. The biological impact analysis, *Practical Applications of Medical Geology*	Siegel (2021b)
Mapping of podoconiosis cases and risk factors in Kenya: a nationwide cross-sectional study	Sultani et al. (2021)
Artisanal, small-scale, and large-scale mining in Lao PDR	Tappe (2021)
Assessing the contamination level, sources, and risk of potentially toxic elements in urban soil and dust of Iranian cities using secondary data of published literature	Tashakor et al. (2022)
Assessing the contamination level, sources, and risk of potentially toxic elements in urban soil and dust of Iranian cities using secondary data of published literature	Tashakor et al. (2022)
Co-occurrence of metal contaminants in United States public water systems in 2013–2015	Thompson et al. (2021)
Environmental contamination from uranium mining and milling in the Western US, *Practical Applications of Medical Geology*	Thomson (2021)
Environmental contamination from uranium mining and milling in the Western US, *Practical Applications of Medical Geology*	Thomson (2021)
Method for determining regional reference values of metal content in biological substrates and their intake into the body via drinking water	Tunakova et al. (2021)
Occupational health and medical geology. In practical applications of medical geology, *Practical Applications of Medical Geology*	Wålinder (2021)

(*Continued*)

Table 1.6 (Continued)

Significant findings / observations	Author(s)/year
Iodine excretion and intake in women of reproductive age in South Australia eating plant-based and omnivore diets: a pilot study.	Whitbread et al. (2021)
Distribution of groundwater arsenic in Uruguay using hybrid machine learning and expert system approaches	Wu et al. 2021
Fluoride contamination in groundwater, impacts, and their potential remediation techniques, *Groundwater Geochemistry: Pollution and Remediation Methods*	Yadav et al. (2021)
Quantitative-qualitative method for quick assessment of geodiversity.	Zakharovskyi and Németh (2021)
Lung cancer disease and arsenic: the medical geology as key	Ziarati et al. (2021)

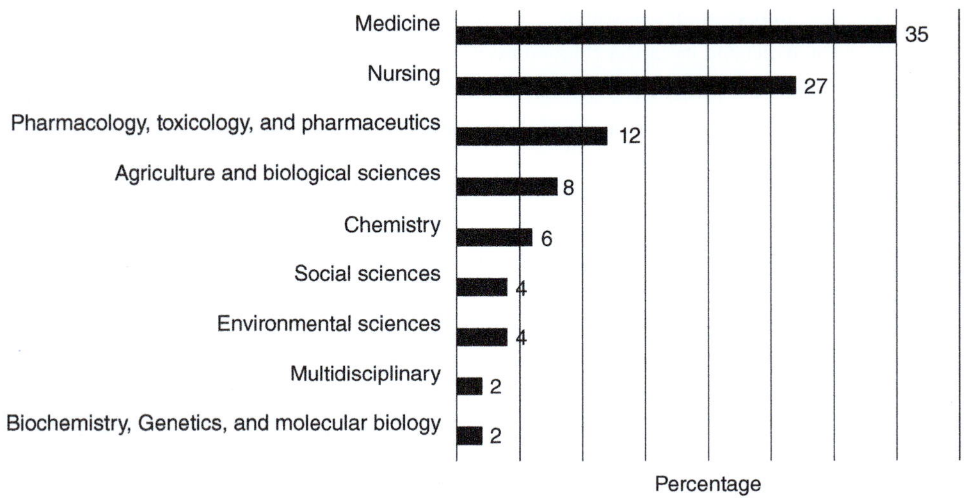

Figure 1.8 Different scientific fields (percentage) studied in relation with metallomics.

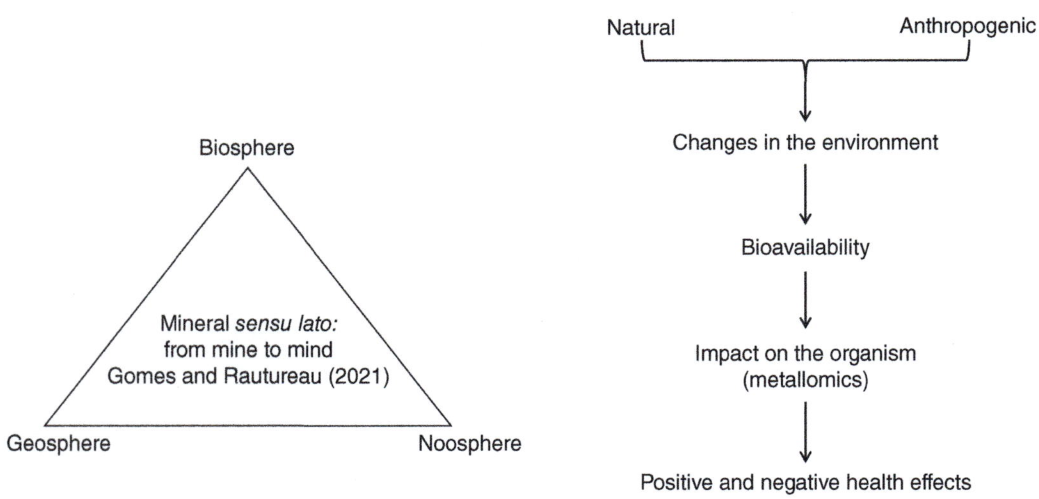

Figure 1.9 Medical geology: biosphere, geosphere, and noosphere interface. Minerals *sensu latu* and *sensu strict* for human health. Natural and anthropogenic sources of elements, their bioavailability, and fate.

ions and metalloenzymes. In metallomics, metalloproteins, metalloenzymes, and other metal-containing biomolecules are defined as "metallomes," in a similar manner to genomes in genomics as well as proteomes in proteomics. Since the identification of metallomes and the elucidation of their biological or physiological functions in biological systems are the main research target of metallomics, chemical speciation for specific identification of bioactive metallomes is the crux to establish metallomics as an integrated bio-metal science.

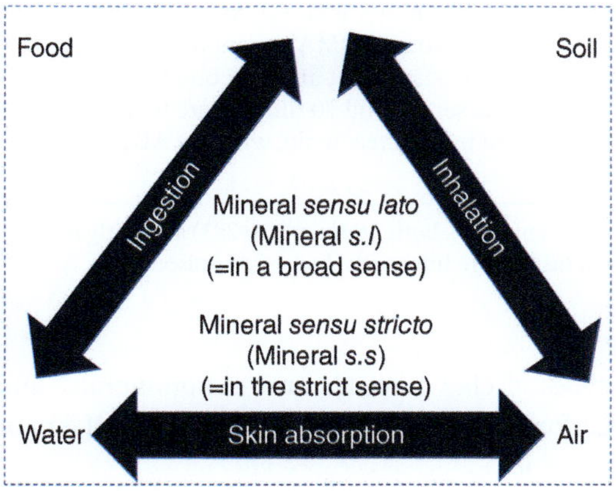

Figure 1.10 There are four major sources of minerals for human intake, viz. food, soil, water, and air. There are three pathways for the human intake of minerals: ingestion, inhalation, and skin absorption.

Table 1.7 Heavy metal contamination in facial cosmetics.

Country	Traditional cosmetic	Composition
Saudi Arabia	Henna	Amorphous carbon
Saudi Arabia, India	Kohl	Cuprite (Cu_2O) and
Middle East	Eyebrow liners	Galena (PbO)
Morocco, USA, Mauritania, Pakistan, India, Great Britain, Saudi Arabia	Kohl	Goethite (FeO(OH)
Bulgaria	Eye shadow, lipstick	Magnetite (Fe_3O_4)
Oman and UAE	Bint al dhahab	Minium (Pb_3O_4)
Bahrain	Surma and kohl	Sassolite (H_3BO_3)
Israel	Kwali (galena based)	Talc ($Mg_3Si_4O.10(OH)_2$)
Nigeria	Kwali eye shadow (graphite based)	Zincite (ZnO)

Source: Adapted from Nnorom et al. (2005).

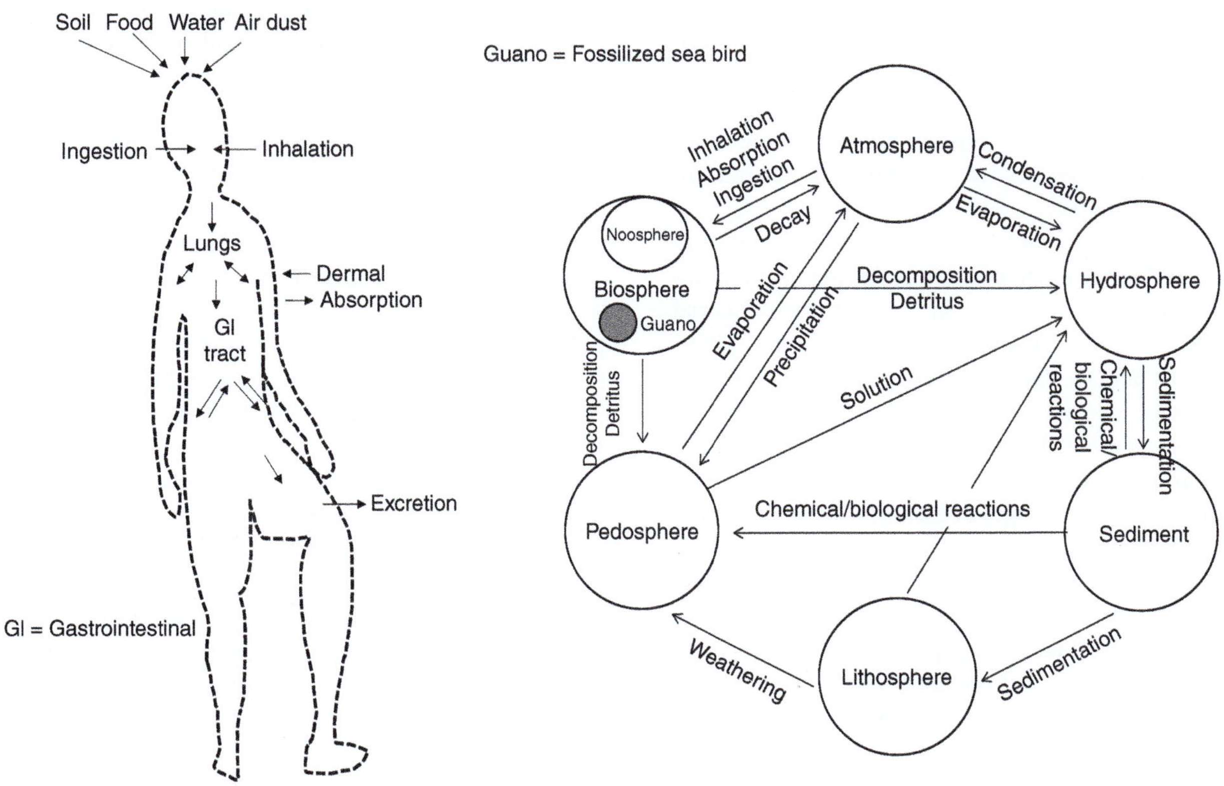

Figure 1.11 Sources and availability pathways of minerals *latu sensu* and human health benefits and toxicity.

Table 1.8 Sources of arsenic and toxicity (Silva et al. 2005, Kumaresan and Riyazuddin 2001, Li and Chen 2005).

Natural sources	Windblown dusts, sea salt spray, volcanoes, forest fires, continental particulates and volatiles, marine
Anthropogenic sources	Fossil fuel combustion, nonferrous metal production, iron and steel production, cement production, poultry and pig manure, waste disposal
Food	Bovine, chicken meat, sheep, seafood, Indian mustard, tomato, grape juice, wine, spinach, carrots, schrimps, green papaya, rice, milk, soyasauce, etc.
Industrial and veterinary products	Pigments, insecticides and herbicides, hemostasis additives, wood preservatives, manufacturers of glass, alloys, and semiconductor materials, electronics, catalysts, and veterinary chemicals
Toxicity	The major arsenic species are arsenite As(III), arsenate As(V), arsenious acids (H_3AsO_3, H_2AsO^{3-}, $HAsO_3^{2-)}$, arsenic acids (H_3AsO_4, H_2AsO_4–, $HAsO_4^{2-)}$ dimethylarsinate (DMA), monomethylarsonate (MMA), arsenobetaine (AB), and arsenocholine (AC). Oxidation states that arsenic commonly exhibits (– III, 0, III, V). Arsenite is 10 times more toxic than arsenate and 70 times more toxic than the methylated species, DMA and MMA. DMA and MMA are moderately toxic, whereas AB and AC are virtually nontoxic

The most poisonous minerals known in the Greek and Roman times were the minerals of arsenic (As), both orpiment (As2S3) and realgar (AsS), which could be lethal when inhaled. Solutions of these minerals, without smell and taste, were intentionally used as poisons.

Gomes and Rautureau (2021) treated minerals *latu sensu* and human health benefits, toxicity, and pathologies in 12 categories and they are defined as follows:

1) The natural, inorganic, solid, and crystalline chemical elements or compounds that participate in the composition of rocks and soils and can be present too dissolved or suspended in both mineral water and air; some of these minerals, particularly those suspended in respirable air and being either naturally or anthropogenically produced, can cause serious diseases such as silicosis and asbestosis depending upon concentration and exposure duration in the respiratory system, whereas some other minerals can play positive functions in pharmaceutics and cosmetics.
2) The natural, inorganic, solid, and crystalline compounds that are amenable to physic-chemical modifications. These are important for certain industrial applications, viz. delamination, intercalation, pillaring, and acid or alkaline activation of certain clay minerals.
3) The bio-essential chemical elements of natural origin named as macronutrients or mineral salts and micronutrients or oligominerals. These are essential for human nutrition and wellness.
4) The chemical elements that in ionic form are constituents of potable mineral water play decisive role in conditioning of human health and living quality is well established. Mineral water used for therapeutical purposes, e.g. in crenotherapy, hydrotherapy, and thalassotherapy.
5) The chemical elements (=minerals, oligoelements) are important in the formulations of the so-called mineral supplements.
6) Some types of clay minerals are "nanominerals" and are used for the controlled and sustainable delivery of pharmaceutics.
7) The biominerals are bio-essential for humans for some physiological processes and parts such as bones and teeth, since they are involved in metabolic functions such as lithiasis – urinary, salivary, renal, and biliary – due to the formation of the so-called calculi or stone.
8) The special clays that are the natural associations of minerals dominated by the so-called clay minerals – hydrous phyllosilicates – are having healing functions and are used since ancient times as cosmetics and therapeutic materials for internal and external applications. For example, geophagy, mud therapy, and pelotherapy.
9) The mineral–organic hybrid complexes are of paramount importance in the field of biomedical applications.
10) The special sands, such as biogenic carbonate minerals, those of radioactive minerals yielding radon gas, and those of natural volcaniclasts, are used in sand baths, either outdoors in the natural environment or indoors inside spa facilities in a therapeutic application known as "psammotherapy."
11) The chemical elements of natural and inorganic origin usually are called metals; some could participate in the composition of metalloproteins and of metalloenzymes acting as catalysts that are essential for metabolic functions. Technitium (Tc)-labeled antibodies are used for disease diagnosis and treatment in cancer therapy.
12) Synthetic minerals that can be used in pharmaceutics and cosmetics.

These 12 categories belong to medicinally important, nutritionally important, and toxic elements (see Tables 1.3–1.5).

1.3 Medicinal Value of Metals in Ancient Indian System of Medicine (After Charaka Samhita)

The ancient seers found that drugs of different origin (herbal, metal, or animal) in addition to codes of conduct and dietary regulations are suitable tools to address several diseases. Use of metallic preparations [(Bhasmas = ash and bhasmikarana (=process of preparation of metallic ash which is zero valent)] in healthcare is unique. Bhasmikarana process with mercury, gold, silver, lead, zinc, copper, etc. was frequently used by seers of the Indian tradition to treat different diseases with great certainty of cure. It is generally claimed that these metals are detoxified during the highly complex bhasmikarana process described in Ayurveda especially Rasashastra manuscripts. The Ayurvedic system of medicine has been in practice in India for millennia. Charaka Samhita, one of the scheduled books of Ayurveda, also holds ample references regarding the use of metals for different purposes; selected examples are enumerated in this chapter.

The Ayurvedic system of medicine dates back to 5000 years BCE. Its Materia Medica contains resources in the form of drugs derived from plant, animal, metal, and mineral sources; the use of which has been advocated in various different pathological manifestations. These drugs have also been converted into poly-herbal, herbo-mineral, and metallic compound formulations by the seers, who have documented their clinical experiences and passed on the knowledge to further generations.

Use of certain heavy metals and minerals in Ayurvedic therapeutics increased. Rasashastra, an integral part of Ayurveda, deals with the drugs of mineral origin and details their varieties, characteristics, processing techniques, properties, therapeutic uses, possibilities of developing adverse effects and their management, etc. in a comprehensive way. Although the roots of this science (Rasa Shastra) exist in the ancient texts of Indian civilization, its development as an independent system of therapy started around the eighth century CE. Ayurvedic classics written before that time, like Charaka Samhita and Sushruta Samhita, contain descriptions of metals and minerals, their processing techniques, and their utilization in therapeutics.

In due course of time, herbo-mineral and metallic preparations came to occupy a significant seat in Ayurvedic pharmacopoeia and have routinely been used in practice in different parts of India for many centuries. Such preparations are held to be safe, efficacious even in minute doses, and, when manufactured and used following specified classical guidelines, not to lead to any significant untoward effects. The past decade, however, has witnessed concerns by the western scientific community, regarding the safety of Ayurvedic herbal, herbo-mineral, and metallic preparations, which is a major concern for the age-old Ayurvedic heritage.

1.3.1 Parada (Mercury)

In Ayurveda, mercury is called Parada or Rasa. Formulations containing mercury are recommended for external application – Kushtha (skin disorders) and to enhances complexion of the skin.

1.3.2 Swarna (Gold) Bhasma

Swarna (gold) is an important, precious metal known to Indians since antiquity. The "Bhasma" form of gold is in metallic state. Quantitatively, it is a combination of metallic gold (96.76%), silica (1.14%), ferric oxide (0.14%), phosphates (0.78%), potash (0.16%), salt (0.078%), and traces of copper and magnesium.

In its elemental form, gold has been employed for centuries as an antipruritic agent to relieve itching palms. In 1980, Robert Koch observed that gold inhibits Mycobacterium tuberculosis in vitro. This led to trials on arthritis and lupus erythematosus. Also, previous studies carried out in 1973 on gold and its compounds observed beneficial activities at different levels. Gold compounds have the ability to decrease concentrations of rheumatic factors and influence the immunological responses [53]. It has also been established [54] that gold suppresses the anaphylactic release of histamine more effectively than glucocorticoids. Sodium aurothiomalate (water soluble preparation) was introduced around 25 years ago to treat arthritis and is administered through intramuscular injections. Its effects are probably due to its antimicrobial effects and stimulation of the reticuloendothelial system [8].

1.3.3 Rajata (Silver) Bhasma

Silver is another precious metal like gold, which attracted the attention of the ancient Acharyas. The use of silver in therapeutics dates back to the period of Charaka and his contemporaries. Though its therapeutic applications are not as extensive as other metals like Tamra or Loha, the ancient classics reveal that silver also enjoyed an important place in Ayurveda therapeutics.

Quantitatively, "Rajata Bhasma" is a combination of metallic silver (52–59%), free sulfur (0.675%), ferric oxide (14.33%), calcium (10.769%), silver chloride (0.479%), and traces of sodium, potassium, and aluminum.

Therapeutic uses of Rajata Bhasma are the following: Rasayana (rejuvenators), Pandu Roga (hematinics), Hicca (hiccup), Swasa (respiratory distress), tongue scrappers, nozzle of enema pot, silver containers idol of male gender, and sharp instruments to be placed in labor room, e.g. scissors for cutting umbilical cord.

1.3.4 Tamra (Copper) Bhasma

Copper is another ancient metal known to human civilization. During pre-Vedic times, the metal was part of day-to-day livelihood functions. Further, it is the earlier known metal for the preparation of the stronger alloy metals brass and bronze of which it is a component. Charaka uses the term Arka in a few places which Chakrapani clarifies as synonymous with Tamra. According to the descriptions of Rasa in Vagbhata Ref. there are two forms of Tamra, viz. Nepaliya and Mleccha, only the former being acceptable. Samples with characteristic metallic sheen (Snigdham), soft (Mridulam), bright reddish in color (Shonam), having high tensile strength (Ghanaghata Ksamam), heavy (Guru), and devoid of impurities (Nirvikaram) are identified as best used for medicinal purposes.

Formulations of "Tamra" are useful in a wide range of diseases like Krimi, Sthaulya, Arsha, Ksaya, Pandu, Kusta, Swasa, Kasa, Amlapitta, Sotha, Sula, Yakrit Roga, and Grahani dosha. In addition, Charaka advocates the use of Tamra Patra (copper).

Therapeutic uses of Tamra Bhasma are the following: Rasayana (rejuvenators), Hicca (hiccup), Swasa (respiratory distress), Visha hara (antipoisonous), Kusta (skin disorders), Granthi (abscess), Collyrium for Akshi Roga (eye disorders) – probable translation, tongue scrappers, copper containers, and nozzle of enema pot.

1.3.5 Aayasa or Loha (Iron)

Aayasa or Loha is another metal known to ancient civilizations. Iron compounds were particularly employed in diseases such as anemia and other debilitating conditions, where functions of hematopoietic systems are disturbed and the blood has consequently become deficient in iron. Rasa Shastra classics explain that Loha is par excellence a rejuvenator as it stimulates functional activity of all the organs, promotes life, strength, destroys a number of diseases, and acts as a restorative. The utility of this metal in therapeutics was only identified modern medicine in the first half of the seventeenth century, when its salts were recognized as the best hematinics.

Malabsorption syndrome, abdominal lump, abscess of kapha, alopecia, ascities, disorders of female genital tract, eye, hemorrhoids, anemia, hiccup, respiratory distress, jaundice, leukoderma, obesity, rejuvenators, skin diseases, inflammation of hemorrhoids, anemia, and chronic inflammation provide complexion to psychosis and skin diseases.

1.3.6 Mandura Bhasma

Mandura, the second form of iron, has been used for a wide range of therapeutic procedures in classical Ayurveda since antiquity. Purified Mandura Bhasma, when administered with proper justification, is beneficial in inflammations, edematous conditions, jaundice, etc. It is the drug of choice in cases of anemia (Pandu). Chemically, Mandura is a combination of ferric oxide (59.14%), ferrous oxide (26.7%), chlorides (4.4%), magnesium (3.9%), sodium (1.7%), and a few other elements in trace quantities.

1.3.7 Naga/Sisaka (lead) Bhasma

Quantitatively, Naga Bhasma is a combination of lead oxide (75.6%) and ferric oxide (7.5%), together with traces of calcium and magnesium chlorides and carbonates. Different formulations of "Naga" are beneficial in diseases, Hicca (hiccup), respiratory distress, and Kusta (skin diseases).

1.3.8 Vanga/Trapu (tin) Bhasma

Vanga, one of the Puti Lohas, was known to ancient Indian physicians by the name of Trapu. In Charaka Samhita, the metal is categorized under Parthiva Dravyas. According to descriptions in Rasa Vagbhata, there are two varieties of Vanga, viz. Khuraka and Mishraka, the former being acceptable for therapeutic applications. Samples with the characteristics, bright white in color (Dhavalam), soft (Mridulam), shiny, smooth (Snigdham), easily melts (Drutadravam), and heavy (Guru), are identified as Khura Vanga and should be preferred for therapeutic purposes. Quantitatively, Vanga Bhasma is a combination of stannic oxide (i.e. of tin) (91.4%), ferric oxide (2.9%), potassium (2.9%), calcium oxide (2%), aluminum (2%), and magnesium (0.6%) oxide. Formulations of "Vanga" are beneficial in diseases such as skin disease, tongue scrappers, and nozzle of enema pot. Singly or in combination with other metals/minerals, it is beneficial in disorders of the genitourinary tract.

Kamsya is another important Mishra Loha (mixture of metals), an alloy of copper and tin known since the period of Samhita Kala. Charaka used this metal to prepare Vasti netra. It is known as bell metal or bronze. According to the descriptions given in Ayurveda Prakasha, there are two varieties of Kamsya, viz. Pushpa and Tailika, only the former being acceptable for therapeutic applications. Samples giving a sharp sound (Teekshna Shabdam), soft (Mridu), smooth to touch (Snigdha), slightly grayish (Eshat Shyamalam), clear form.

1.3.9 Pittala (Brass)

Pittala is an important Mishra Loha, an alloy of copper and zinc, known since the period of Samhita Kala. Charaka used this metal to prepare Vasti netra. It is known as brass. As per the descriptions available in Rasa Ratna Samuchaya, there are two varieties of Pittala, viz. Ritika and Kakatundi.

Formulations of "Pittala" are beneficial in diseases like Krimi, Kusta, Pandu, etc. The therapeutic doses given for "Pittala Bhasma" range from 60 to 120 mg. Very few references mentioned in Charaka Samhita that hold Pittala as one of the components are depicted in Table 1.9. Bhasma ranges from 60 to 120 mg. Very few references mentioned in Charaka Samhita that hold Kamsya as one of the active components are depicted in Table 1.9.

Humanity depended on natural resources (air/water/soil/sediment-containing minerals) for survival. Construction materials, viz. aluminum, iron, steel, and cement, are produced from mining activity. Several of the excipients are minerals. Minerals in pharmaceuticals fall into one of two main categories: excipient or active substances. The excipients have no intrinsic health benefit of their own; they are used solely as carriers, allowing the intake of minute amount of active substances. pH regulation or adsorbents are the kind of applications for which some minerals are used as active ingredients.

Thus, antacid pills are composed mainly of calcium carbonate, lithium used in antidepressants is derived from industrial minerals, and many excipients are minerals such as talc, magnesium carbonate, or silica. Many cosmetics incorporate industrial minerals such as talc, although others like mica, silica, or borates are utilized for their abrasive, visual, or stabilizing properties (Table 1.7). It is necessary to remember that earliest civilizations (e.g. the Romans) already made use of earth pigments for body painting.

Industrial minerals are crucial in water management, whether considering drinking water preparation or wastewater treatment. Thus, silica sands are used as filters, perlite, zeolites, or talc as flocculants or adsorbents, bentonite as industrial mineral-based liners and geosynthetic liners, either basal or superficial, are increasingly used to avoid escape of leachate from landfill sites. Air treatment of industrial effluents also largely relies on minerals. Activated carbon is the best-known technique, but other minerals are used as well. For instance, flue gas desulfurization of power station fumes is achieved with calcium carbonate.

Industrial minerals such as clays, sand, feldspar, kaolin, and other minerals are basic to all construction materials, from bricks to tiles and from cement to limes and plastics.

1.4 Linking Geology to Medicine?

- Increases in sediment erosion and transport, which can uncover toxic wastes deposited earlier ("chemical time bombs").
- Permafrost and gas-hydrate melting, which can lead to methane degassing and influx of insects and diseases.
- Changes in groundwater and surface water quality resulting from accelerated erosion, deforestation, pollution, river diversion, and groundwater mining.
- Drop in lake levels and increase in salinity causing loss of water supplies and increase in airborne particulates.
- Catastrophic natural events – landslides and avalanches, earthquakes, tsunamis, and volcanic eruptions, with many effects on settlements, water, soils, and air quality (Figures 1.12 and 1.13).

1.5 Mineral-Enriched Yeast: Vehicles of Nutrition

Yeasts are microorganisms, which represent valuable supplements to the human diet. This is not simply due to provision of cellular proteins, carbohydrates, vitamins, etc. by yeast but also to the propensity of yeast cells to accumulate a variety of nutritionally useful metal cations. Easily grown and readily available species such as the baker's yeast, *Saccharomyces cerevisiae*, are excellent natural sources of essential metals (e.g. K, Mg, Ca, Mn, Fe, Zn) and this yeast can be further "enriched" with other inorganic micronutrients, including selenium and chromium. Such mineral-enriched yeasts may serve as effective carriers of key trace elements in alleviation of occasional inorganic dietary deficiencies in humans and animals.

In order to maintain their viability and vitality, yeast cells adopt several strategies for transporting and localizing growth-requiring metal ions and additional strategies exist for neutralizing potentially cytotoxic metals, depending on the prevailing environmental conditions. The molecular biology of several yeast transmembrane ion transporters has been studied, and it has recently been established that some metal ion transport mechanisms (e.g. for iron and

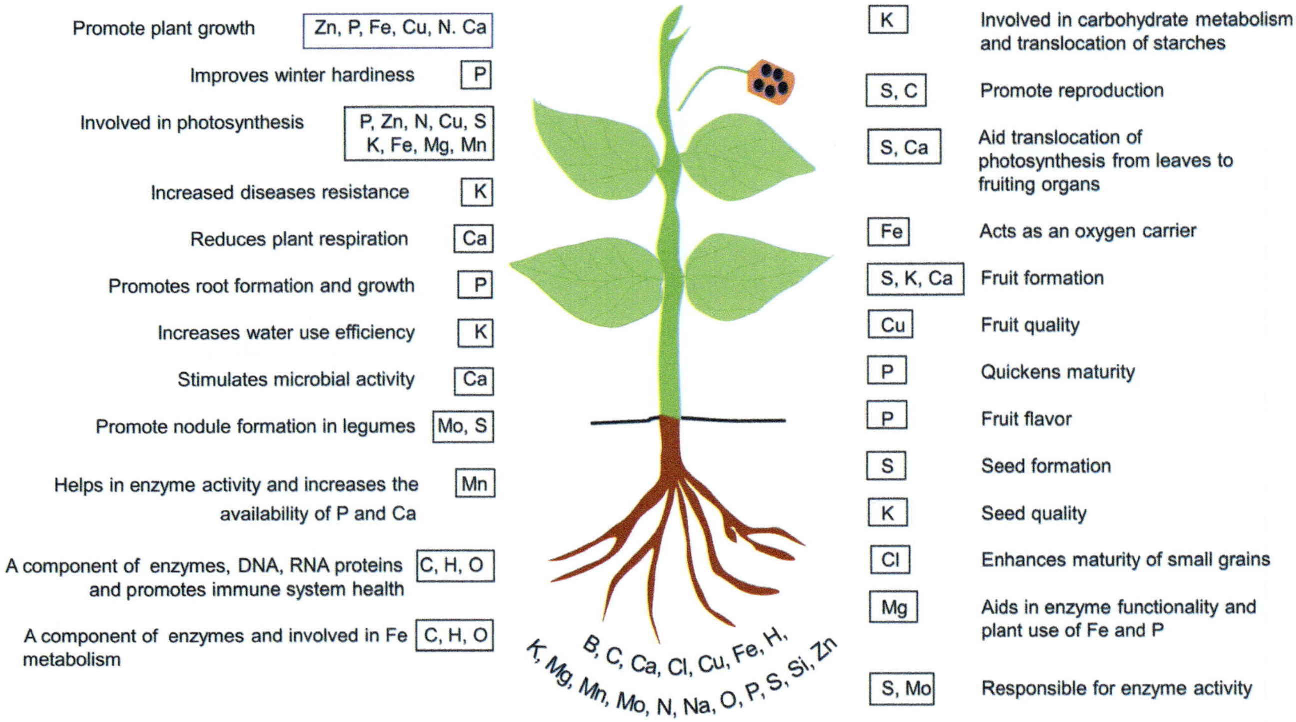

Figure 1.12 Soil degradation leads to the loss of micro- and macronutrients. Thus, nutrient poor soils are unable to produce healthy food necessary for well-being of human and livestock.

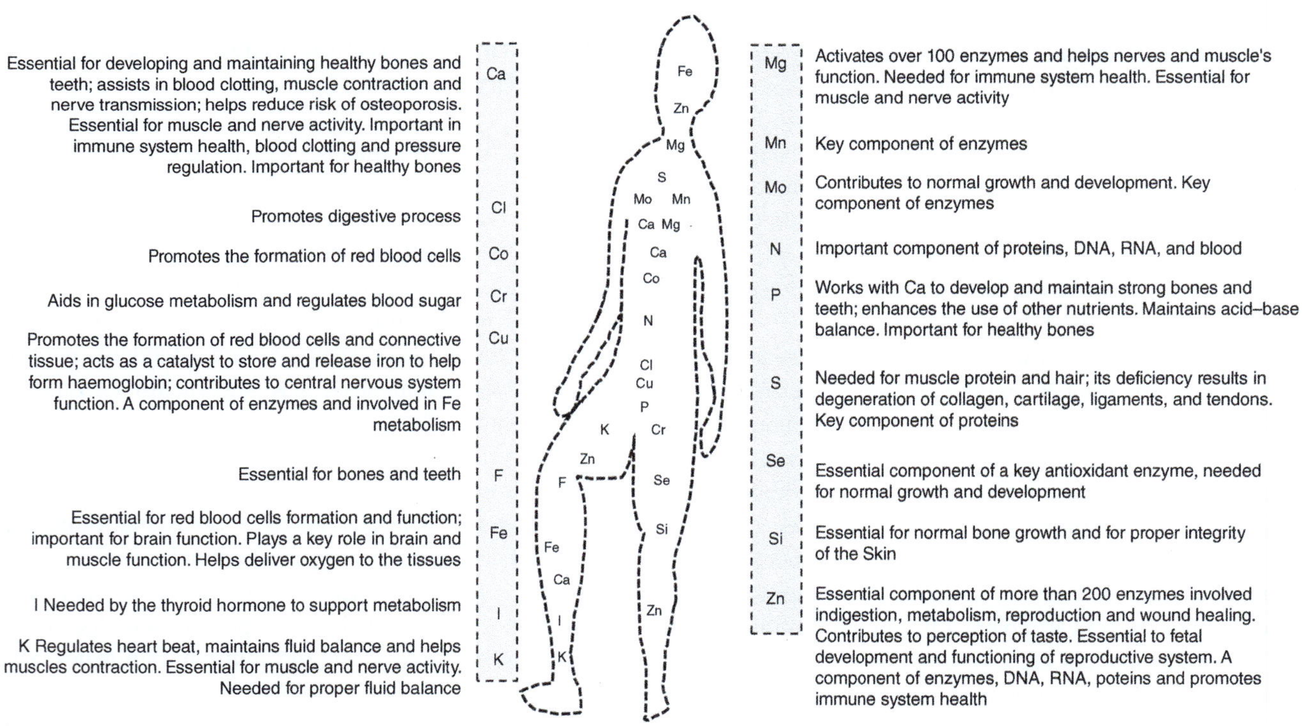

Figure 1.13 Soil is the foundation of human health via balanced nutrition. Over two billion people are estimated to suffer from micronutrient deficiencies. The essential nutrients that are necessary for human health and metabolism are depicted in this figure.

copper) are remarkably similar in both yeast and human cells. In fact, *S. cerevisiae* is now regarded as an invaluable model eukaryotic organism for studies of human disorders linked to dysfunction of metal ion homeostasis (e.g. Menkes disease).

1.6 Trace Elements/Functional Foods

1.6.1 Antidiabetic Plants – The Chromium Connection

Insulin insensitivity, leading to glucose intolerance and ultimately diabetes, is widespread in the United States and other westernized countries.

Diabetes is one of the leading causes of death and increases in blood glucose are associated with increases in the incidence of cardiovascular diseases, the leading cause of death in the western world.

Diet and lifestyle seem to be major contributing factors in the rapid increases in glucose intolerance and diabetes. Chromium is low in many modern diets that are high in refined sugars and fats. High sugar diets not only are low in chromium but also enhance chromium losses. Chromium has been shown to reverse the signs and symptoms of varying stages of glucose intolerance ranging from hypoglycemia to type 2, gestational, and steroid-induced diabetes. Chromium functions by improving insulin function, which leads to a normalization of blood glucose. Improved chromium nutrition leads to increases in insulin binding, increases in insulin receptor number, and activation of insulin receptor kinase leading to increased insulin sensitivity.

1.6.2 Lithium

Lithium is normally present in trace amounts in both food and water, but daily intakes vary significantly depending on location and the type of food consumed. In rats maintained on lithium deficient diets, deleterious effects on reproduction, lactation, activity, and social interactions have been observed.

In goats, lithium deficiency resulted in reproductive abnormalities, increased abortion rates, diminished weight gains, and shorter life expectancy.

In humans, defined lithium deficiency syndromes so far have not been described. However, inverse associations have been observed between lithium concentrations in drinking water and the rate of mental hospital admissions, homicide, suicide, rape, arrests for drug use, and in juveniles, with the rate of runaways from home. Results of supplementation studies indicate that lithium, in the nutritional dosage range, has mood-improving and stabilizing effects.

Boron, nickel, silicon, vanadium, and arsenic as elements of nutritional and pharmacological relevance.

If the lack of an element cannot be shown to cause death or interrupt the life cycle, many scientists do not consider that element essential unless it has a defined biochemical function. Furthermore, the nutritional community is reluctant to provide dietary guidance for any element without a defined function, which can help indicate status. Not being accepted as essential and not having dietary guidance such as a dietary reference intake often are equated to not being nutritionally important.

Several mineral elements have recently been shown to have beneficial effects at nutritionally relevant intakes in animals and sometimes in humans; these elements include boron, nickel, silicon, vanadium, and arsenic.

Nutritionally relevant intakes of boron have been found to affect biochemical indicators in humans related to bone turnover, physiological indicators of psychomotor and cognitive function, and blood cellular composition. Recent research with animals has shown that boron is needed in the early stage of life; this includes the demonstration that the lack of boron adversely affects reproduction and embryo development in both the African clawed frog and zebrafish.

Findings from animal experiments with nickel suggest it is beneficial to cardiovascular health. Nickel deprivation affects the response of the rat to excessive intakes of methionine and deficient intakes of vitamin B_{12}. Recently, it was found that nickel deprivation of rats increased blood pressure and exacerbated the response to a high salt intake and an acute salt load. Some of the findings suggest that nickel is physiologically important in functions involving the cyclic guanosine monophosphate (cGMP) signal transduction system.

The distribution of silicon in the body and the biochemical changes in bone caused by silicon deprivation indicate that silicon influences bone formation by affecting cartilage composition, and ultimately the initiation of cartilage calcification and the regulation of bone crystal growth. Evidence is emerging suggesting that there are nutritionally relevant dietary silicon intakes that positively influence bone mineral density in postmenopausal women. Also, because silicon affects collagen formation, it may influence wound healing.

On the basis of its functional roles in lower forms of life, vanadium possibly is important for some enzyme reactions in higher animals and humans. Vanadium is receiving much attention at present as an element that can be used therapeutically for diabetes.

The responses of experimental animals to arsenic deprivation suggest that this element affects the utilization of labile methyl groups arising from methionine. Low arsenic intakes have been suggested to result in hypomethylation of DNA, which has been associated with an increased risk for some types of cancer. These elements might be of more practical nutritional importance that the nutritional community acknowledges at present. Dietary Supplement Health and Education Act and structure/function statements must meet the truthful and not misleading standard.

1.7 Public Health Informatics (PHI)

Whenever we teach a session on the introduction to GIS, we refer to Dr. John Snow. Dr. John Snow is known as the "father of modern epidemiology" and the "father of GIS" because of the famous case of the 1854 cholera outbreak in London's Broad Street region.

In the 1850s, cholera was very poorly understood and massive outbreaks were a common in major industrial cities. An outbreak in London in 1854 was typical. Several deaths it caused are shown in the map on the right. The map was made by Dr. John Snow, who has conceived the hypothesis that cholera was transmitted through the drinking of polluted water, rather than through the air, as was commonly believed. He noticed that the outbreak appeared to be centered on a public drinking water pump in Broad Street – and if this hypothesis was correct, the pattern shown on the map would reflect the locations of people who drank the pump's water. There were anomalies, in the sense that deaths had occurred in households that were located closer to other sources of water, but he was able to confirm that these households also drew their water from the Broad Street pump. Snow had the handle of the pump removed, and the outbreak subsided, providing direct causal evidence in favor of his hypothesis (Longley et al. 2005).

Public health informatics (PHI) is an emerging discipline that studies the application of information science and technology to public health practice and research. As part of that effort, a GIS – or more generally a Spatial Decision Support System (SDSS) – which offers improved geographic visual is at ion techniques, leading to faster, better, and more robust understanding and decision-making capabilities in the public health arena.

Geoenvironmental variables were implicated as plausible etiological cofactors in the incidence of some common environmental diseases in Africa (Davies 2013). Trace elements found in soil, water, and air are found to be carcinogenic.

1.7.1 Medicinal–Mineral Resources

Among medicinal–mineral resources with considerable potential for development in Africa are spas, muds, and peat deposits. The word "spa" is derived from Latin abbreviations: S = salud, P = per, A = aqua, or "health through water." A spa originates at a location in the form of water from a spring or well. It is now well established that these waters, often warm (geothermal water) and charged with certain mineral constituents, are of therapeutic value with regard to a number of ailments. Associated with most spas is the use of muds (peloids) which either are found at the site or are imported from special locations. The muds are stored in tanks and "cured" for maximum benefit.

Muds can be classification into three categories:

1) Neutral – pure mineral (fango, mud)
2) Alkaline – mainly mineral (sea mud or liman)
3) Acidic – mainly vegetable peloid (moor, peat)

1.7.2 Balneotherapy

The word "balneology" is derived from the word "balneum," the Latin word for "bath." Balneology is the science of medical application of water, especially of curative springs and of peloids, mainly in spas. Waters from hot springs can contain Mg, Ca, Na, As, K, Mn, B, and I. The therapy is used to treat a variety of conditions, including:

- Extremely high blood pressure
- Fever
- Tumors or cancerous conditions
- Congestive heart failure.
- Liver, kidney, or circulatory conditions
- Anemia
- Risk of hemorrhaging in pregnancy

Raju et al. 2006 reported appreciable amounts of the elements K, Ca, Cr, Mn, Cu, and Zn in *Cassia auriculata, Eugenia jambolana, Gymnema sylvestre, Adhatoda vasica, Withania somnifera, Ocimum sanctum, Tinospora cordifolia*, and *Azadirachta indica*, which are responsible for potentiating insulin action.

The traditional medicinal system practiced in India for several centuries is well known as Ayurveda. According to this medicinal system, metal-based drugs known as "Bhasma" involve the conversion of a metal into its mixed oxides. During these transformations, the zero valent metal state gets converted into a form with higher oxidation state and the most important aspect of this synthesis (known traditionally as "bhasmikarana") is that the toxic nature (i.e. systemic toxicity causing nausea, vomiting, stomach pain, etc.) of the resulting metal oxide is completely destroyed while inducing the medicinal properties into it.

The drugs known as "Bhasmas" are well known in the traditional Indian Ayurveda and these are chemically mixed oxides of one or more metals (Wadekar et al. 2005). Their traditional preparation involves conversion of a pure metal into its oxide form following a typical procedure, available in the ancient literature of Ayurveda (Wadekar et al. 2006). Such a procedure is believed to eliminate the harmful properties of metal oxides while inducing the medicinal properties into it. The critical steps in the preparation of "Bhasma" are (i) trituration of a pure metal with various plant juices for several hours and (ii) repeated cycles of high temperature calcination in a sealed earthen pot (Lemberkovics et al. 2002; Ražić, et al. 2005).

The essential elements are involved in many metabolic processes in the human organism, especially as enzymes activators, e.g. Fe, Zn, and Mn (Razic et al. 2005). They can also interact with some organic compounds – for instance with flavonoids, introduced into the human organism, after using the plant drugs, influencing their biological activity. Taking this into consideration, not only the total level of the essential elements in herbs but also their interactions with the plant's constituents should be studied.

With growing importance of nanotechnology in medicine and healthcare (Turki 2015) the advent of engineered nanoparticles in therapeutics is alarming for their possible harmful effects (Palkhiwala and Bakshi 2014). The ancient application of nanomedicine in the form of Ayurvedic Bhasma throws a light on the safer usage of present nanomedicine for a living being and the environment (Kapoor 2010; Sarkar and Chaudhary 2010; Chaudhary 2011). The use of nanoparticulate metals in therapeutics has been a common practice in Ayurveda. The present attempt is to revisit the Ayurvedic Bhasma concept as organometallic ethnonanomedicine in the surging area of nanomedicine (Paul and Chugh 2011). Bhasma, literally meaning ash, is a unique Ayurvedic herbo-mineral-metallic compound in the size of nanodimensions (usually 5–50 nm, as established by modern microscopic and spectroscopic techniques). These are the products of classical Indian alchemy, the "Ayurveda Rasa Shastra," used for treating diverse chronic ailments (Kumar et al. 2006). The essence of metal/mineral-based drugs is that they function best when converted from their original metal/mineral state to oxide state. The manufacturing process is very systematic and elaborate, called "Bhasmikarana" which converts the metal from its zero valent state to a form with higher oxidation state and eliminates the toxic nature of metal and its oxide while rendering the metal oxide with high medicinal value (Wadekar et al. 2005). During bhasmikarana, metals/minerals are subjected to various processes of purification and incineration before internal administration, aimed to reduce the particle size (for instance, the particle size of Swarna Bhasma [gold calx] was found between 1 and 2 μ) and thus, converting them into biocompatible, bio-assimilable, absorbable, and suitable form for the human body. Bhasma nanoparticles are organo-metallic/organo-mineral complexes as they are integrated with biological molecules (of organic liquid media), having improved stability, functionality, absorption, assimilation, bioavailability, biocompatibility, targeted delivery of ingredient, and effectiveness (Pandit et al. 1999; Wadekar et al. 2006; Paul and Chug 2011, Valodkar et al. 2012; Palkhiwala and Bakshi 2014). It is also noteworthy that Ayurvedic Bhasma is considered very safe and economical in comparison to contemporary metal-based nanomedicines (Panda 2004; Palkhiwala and Bakshi 2014). All Bhasmas have some common properties such as Rasayana (immunomodulation and antiaging quality), Yogavahi (target drug delivery), Alpamatra (prescribed in minute doses, i.e. 15–250 mg/day), Rasibhava (readily absorbable, adaptable, assimilable, and nontoxic), Shigravyapi (spreads quickly and fast acting), and Agnideepana (increases metabolism at cellular level and acts as catalyst). Bhasma can be employed for selective/targeted/controlled drug delivery as they are biocompatible, nontoxic, and nonantigenic in nature (Kapoor 2010, Paul and Chug 2011). Bhasma is biologically produced nanoparticles with quick and targeted action, e.g. gold nanoparticles in Swarna Bhasma at 27 ± 3 nm size have been found effective in ameliorating symptoms of arthritis and at 4 nm size helped in increased apoptosis in B-chronic lymphocytic leukemia (Paul and Chug 2011). Few more examples of frequently used Bhasma include: Mukta Shukti Bhasma (pearl-oyster calx) at 22.52 ± 0.45 μm size in acidity, pyrexia, tuberculosis, cough, asthma, etc., and also to increase the bone mineral density; Mandura Bhasma (iron calx) in microcytic anemia and hemolytic jaundice, Tamra Bhasma (copper calx) as hepatoprotective and antioxidant, Yashad Bhasma (Zinc calx) in myopia, diabetes, leucorrhea, etc. (Pal et al. 2014). The benefits of nanomedicines are indubitable and unstoppable, nevertheless, and safety-related studies should also be carried out rigorously and planned in order to provide guidelines for safer manufacturing practices, keeping care of ecology and environment. Hence, Ayurvedic Bhasma may hold strong relevance in the emerging era of nanomedicine and can serve as an excellent template for the development of nanomedicine for an efficient therapeutic cure.

1.8 Use of Clay Minerals in Water Purification

Scrano et al. considered the use of organoclay complexes for water purification. These authors stated that ". . . the cleaning of contaminated water is one of the most difficult and expensive goals in environmental engineering. Clay minerals are the most important inorganic components in soil for sorption of contaminants, but without modification, are ineffective sorbents for contaminant compounds. The treatment with surfactants changes the surface properties from hydrophilic to hydrophobic and greatly increases the basal spacing of the layers. This new material, named organoclay, is used in some environmental applications, such as the removal of contaminants from polluted waters; offering a promising device in decontamination planning and environmental remediation."

Cr is an essential mineral for ruminants in tropical regions. Its *pods* contain high Cr concentration (up to 150 μg/L). Thus, supplements Cr in fodder for animals (goats) Cr requirement for animals is >0.1 mg/L, while toxic level was 1000 mg/L (Figure 1.14)

Prosopis juliflora pods are rich in carbohydrates and minerals. Goats consume the pods to meet their nutritional requirement, for example, died pods are powdered and used as animal feed.

Figure 1.14 *Prosopis juliflora* - (a) profuse flowering, (b) tender pods, (c) mature pods and, (d) goats grazing tender pods under the shade of *P. juliflora* (e) Harvested mature pods ready for grinding (f) Cyclotec mill for grinding pods to powder (g) Pod powder as cattle feed.

1.9 Bathing in Radioactive Monazite-Rich Sands

Monazite is a rare phosphate mineral with a chemical composition of Ce, La, Nd, Th (PO_4, SiO_4). It usually occurs in small isolated grains, as an accessory mineral in igneous and metamorphic rocks such as granite, pegmatite, schist, and gneiss. These grains are resistant to weathering and become concentrated in soils and sediments downslope from the host rock. When in high enough concentrations, they are mined for their rare earth and thorium content.

Monazite is produced in the offshore sands of India, Malaysia, Vietnam, and Brazil. Southern India and Sri Lanka have the most extensive offshore monazite resources known.

The term "geochemical diseases" is attributable to such cases where the etiology of a certain disease is almost directly and clearly related to a characteristic elemental composition of the immediate geological environment. Very recently, Korobova et al. (2020) produced an interesting paper entitled "Endemic diseases of geochemical origin and methodological approaches toward their prevention and elimination." Two genetically different types of endemic diseases of geochemical origin are distinguished, each having a specifically spatial structure.

Natural clays (see chapter 16 for details) and nanoscience plays a significant role in medical geology (Figure 1.15 and Table 1.9).

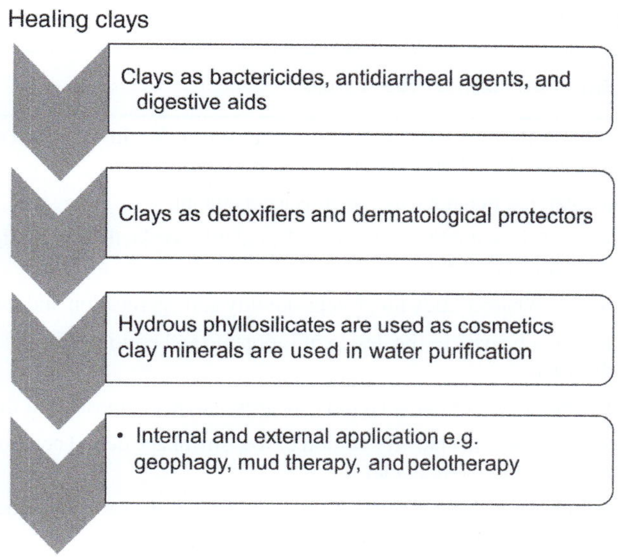

Figure 1.15 Natural clays with different healing functions.

Table 1.9 Nanoscience in medical geology.

Significant findings / observations	Author(s)/year
Ayurvedic Bhasma: the most ancient application of nanomedicine	Sarkar and Chaudhary (2010)
Assessing the role of Ayurvedic "Bhasmas" as ethno nanomedicine in the metalbased nanomedicine patent regime	Pauland and Chugh (2011)
Cytotoxicity evaluation and antimicrobial studies of starch capped water soluble copper nanoparticles	Valodkar M et al. (2012)
Bhasma: the ancient Indian nanomedicine	Pal et al. (2014)
Engineered nanoparticles: revisiting safety concerns in light of ethno medicine	Palkhiwala and Bakshi (2014)
The use of nanotechnology in medicine	Turki (2015)
Nanotechnology in medicine: leads from ayurveda	Sharma and Prajapati (2016)
Melanoma treatment via nonspecific adhesion of cancer cells using charged nanoclays in pre-clinical studies	Abduljauwad and Moy (2021)
Eco-friendly carboxymethyl cellulose hydrogels filled with nanocellulose or nanoclays for agriculture applications as soil conditioning and nutrient carrier and their impact on cucumber growing	Bauli et al. (2021)
Synthetic nanoclay gels do not cause skin irritation in healthy human volunteers	Bostan et al. (2021)
Composition of nanoclay supported silver nanoparticles in furtherance of mitigating cytotoxicity and genotoxicity	Chang et al. (2021)
Ayurvedic Bhasma: nanomedicine of ancient India – its global contemporary perspective	Chaudhary 2011
Synthesis of nanoclay composite material. In *Contemporary Nanomaterials in Material Engineering Applications*	Deheri and Kar (2021)
Nanoclayreinforced biomaterials for mending musculoskeletal tissue disorders	Erezuma et al. (2021)
3D printing unique nanoclay-incorporated double-network hydrogels for construction of complex tissue engineering scaffolds	Guo et al. (2021)

(Continued)

Table 1.9 (Continued)

Significant findings / observations	Author(s)/year
A stick-like intelligent multicolor nano-sensor for the detection of tetracycline: the integration of nanoclay and carbon dots	Jia et al. (2021)
bioinspired magnetic nanochains for medicine	Kralj and Marchesan (2021)
Fused filament fabrication of biodegradable polylactic acid reinforced by nanoclay as a potential biomedical material	Mihankhah et al. (2021)
Nanoclaybased study on protein stability and aggregation and its implication in human health	Parveen et al. (2021)
Recent advances in the design of inorganic and nano-clay particles for the treatment of brain disorders	Persano et al. (2021)
Emerging role of nanoclays in cancer research, diagnosis, and therapy	Peixoto et al. (2021)
Fabrication and characterization of low-shrinkage dental composites containing montmorillonite nanoclay	Ritto et al. (2021)
Production of starch-polyester bio-support for lipases immobilization: synergistic action of itaconic acid and nanoclay	Simões et al.(2021)
A new long-term composite drug delivery system based on thermo-responsive hydrogel and nanoclay	Tipa et al. (2021)
Nanoclay mineral-reinforced macroporous nanocomposite scaffolds for in situ bone regeneration: in *vitro* and in *vivo* studies	Wu et al. (2021)
Periodic charge matching driven immobilization of gentamicin in nanoclays for stable and long-term antibacterial coating	Yamaguchi et al. (2021)
Hypoxia-mimicking 3D bioglass-nanoclay scaffolds promote endogenous bone regeneration	Zheng et al. (2021)

Glossary

Aerosols Suspension of microscopic size solids, or tiny droplets of liquid, or gas in the air.

Al Eskan disease Recent term for pneumonia acquired from exposure to organic and inorganic substances embedded in dust storms, reported among deployed military personnel in the Middle East. Synonymous with Gulf War syndrome, Persian Gulf syndrome, and desert dust pneumonitis.

Alzheimer's disease A degenerative disease of the brain that results in progressive memory loss, impaired thinking, disorientation, and changes in personality and mood and is the most common form of dementia.

Amylophagy A form of geophagy expressed by compulsive tendency to eat raw starch.

Asbestosis A disease of the lungs caused by inhalation of asbestos fibers that is marked by thickening and scarring of lung tissue.

Balneotherapy The word "balneology" is derived from the word "balneum," the Latin word for "bath." Balneology is the science of medical application of water, especially of curative springs and of peloids, mainly in spas. Waters from hot springs can contain Mg, Ca, Na, As, K, Mn, B, and I.

Bioavailable A general term for a particular form of a chemical that can be absorbed in the body.

COPD Chronic obstructive pulmonary disease is also referred to as COPD. A pulmonary disease, such as emphysema or chronic.

Coccidioidomycosis A disease in humans and domestic animals that is caused by inhalation of spores from the fungi.

Cretinism A congenital condition marked by physical stunting and intellectual disability and caused by severe.

Crust The upper most layer of the earth, 5–40 km thick, relatively cold (compared to deeper layers) and comprising solid.

Dental caries Tooth decay marked by progressive destruction of bone or tooth.

Dental fluorosis An abnormal condition (such as mottling of the teeth) caused by fluorine or its compounds.

Dystrophy In general, any condition produced by faulty nutrition; the term is commonly used for a variety of other.

Fibrosis A condition marked by increase of interstitial fibrous tissues.

Gastroenteritis Inflammation of the lining membrane of the stomach and the intestines characterized especially by nausea.

Gem electuary A medicated paste prepared by pulverizing gem stones and mixing it with a sweetening substance, such as.

Geomedicine A branch of medicine that deals with geographic factors in medicine; common use of geospatial methods in spread of disease.

Geophagy (or geophagia), the habit of eating clay or earth.

Geopharmaceutical A broad term used for medical formulations in which geological materials are used as ingredients.

Germ theory A theory in medicine, emphasizing that infections, contagious diseases, and other conditions result from the action of microorganisms.

Gout A metabolic disease marked by painful inflammation of the joints, deposits of urates in and around the joints, commonly with an excessive amount of uric acid in the blood.

Hyperthyroidism Excessive functional activity of the thyroid gland that results in increased metabolic rate, enlargement of the thyroid gland, rapid heart rate, and high blood pressure.

Isopods Any species of a large order (Isopoda) of small immobile aquatic or terrestrial crustaceans.

Kaposi's sarcoma (KS) Kaposi's sarcoma is a mesenchymal tumor involving blood and lymphatic vessels (endemic to Africa).

Keratosis Area of skin marked by overgrowth of horny tissue.

Lassitude A condition of weariness or debility, accompanied by sickness and weight loss.

Lithotherapy An alternative system of medicine that uses the energy and colors of minerals and rocks for healing. The literal meaning of the Greek term is curing by using rocks.

Locosis A disease commonly in horses, mules, cattle, and sheep, caused by chronic poisoning with locoweeds that belong to any of several leguminous plants of the genera *Astragalus* and *Oxytropis*.

Magnitude, earthquake A measure of the energy output or power of the earthquake; the logarithmic magnitude scale ranges from 0 to 10, with each successive number indicating an order of magnitude difference.

Meningococcal meningitis A disease marked by inflammation of the membranes that enclose the brain and spinal cord (the meninges) that is either a relatively mild illness caused by a virus (such as various Coxsackieviruses) or a more severe, usually life-threatening illness, caused by the bacterium (*Neisseria meningitidis*).

Mesothelioma A type of cancer that develops from the thin layer of tissue that covers many of the internal organs (known as the mesothelium). The most common area affected is the lining of the lungs and chest wall. Prolonged exposure to some variety of asbestos family minerals can cause mesothelioma.

Miasmatic theory An old theory used to explain occurrence of disease due to inhalation of air.

Mseleni joint disease (MJD) Endemic osteoarthritis.

Mycosis disease Disease caused by a fungus (e.g. ring worm, etc.).

Neurodegenerative disease A general term for a range of conditions which primarily affect the neurons in the human brain, e.g.

Noncirrhotic portal A condition marked by increase of interstitial fibrous tissues in the veins that collect blood from one part of the body and distribute to another, e.g. from stomach to spleen, etc., but not into the liver.

Otitis media An inflammatory disease of the middle ear, common in children.

Pagophagy Compulsive desire to eat ice (a form of geophagy).

Parkinson's disease Neuropathy damage or dysfunction of one or more nerves that typically result in numbness, tingling, muscle weakness, and pain in the affected area.

Pelotherapy Application of peloids (=muds) is called pelotherapy. Cosmetic peloids.

Pica Medical term for craving to eat nonfood items such as soil, clay, ice, or raw starch.

Pneumoconiosis A disease of the lungs due to inhalation of dust, mineral particles, etc. characterized by inflammation, coughing, and fibrosis.

Podoconiosis It is an endemic nonfilarial elephantiasis, a geochemical disease in individuals exposed to red-clay soil derived from alkali volcanic rock.

Psammotherapy or arenotherapy naturotherapy Bathing in the naturally warm volcanic sands.

Pulmonary fibrosis A lung disease that causes the lung tissue to become damaged and thick. The thickened, stiff tissue makes it more difficult for the lung to work properly.

Rheumatism Any of various conditions characterized by inflammation or pain in muscles, joints, or fibrous tissue.

Sarcoidosis An inflammatory disease that affects multiple organs in the body, but mostly the lungs and lymph glands, resulting in the occurrence of abnormal masses of inflamed tissues in the body.

Silicosis Fibrosis of the lung, caused by the inhalation of dust containing silica.

Skeletal fluorosis A bone disease caused by excessive accumulation of fluoride in the bones. In advanced cases, it causes painful damage to bones and joints.

Slag wool A type of mineral wool manufactured from melts of metallurgical slags by converting them into glasslike fibers for thermal insulation.

SPA Derived from Latin abbreviations – s = salud, p = per, a = aqua, or "health through water."

Syncope Loss of consciousness resulting from insufficient blood flow to the brain.

Thalassotherapy The use of seawater in cosmetic and health treatment.

Therapeutic peloids Mud, or clay used therapeutically, as part of balneotherapy, or therapeutic bathing. Peloids consist of humus and minerals formed over many years by geological and biological, chemical and physical processes.

Urolithiasis A condition that is marked by the formation or presence of mineral salts (calculi) in the urinary tract.

Weathering A geologic process that operates at the earth's surface, resulting in gradual decomposition and disintegration of rocks into discrete mineral grains and constituent chemical elements and compounds, caused by a series of complex chemical, physical, and biochemical reactions.

Zoonotic disease Diseases caused by transfer of a virus from animals to humans, e.g. COVID-19, Ebola, zika, etc.

References

Abduljauwad, S.N. and Moy, V.T. (2021). Melanoma treatment via non-specific adhesion of cancer cells using charged nano-clays in pre-clinical studies. *Scientific Reports* 11 (1): 1–13.

Akbar, S., Ali, Z., Hussain, S. et al. (2021). Metal accumulation potential, human health risks, and yield attributes of hundred bread wheat genotypes on irrigation with municipal and remediated wastewater. *Environmental Science and Pollution Research* 28 (26): 1–15.

Avasarala, S. (2021). Techniques for assessing metal mobility in the environment: a geochemical perspective. In: *Practical Applications of Medical Geology* (ed. M. Siegel, O. Selinus and R. Finkelman), 139–167. Cham: Springer.

Aydar, E. and Diker, C. (2021). Carcinogen soil radon enrichment in a geothermal area: case of Güzelçamlı-Davutlar district of Aydın city, Western Turkey. *Ecotoxicology and Environmental Safety* 208: 111466.

Baba, A., Recepoglu, Y.K., and Yazdani, H. (2021). Naturally occurring arsenic and boron in geothermal systems and their health effects: a case study from Turkey. In: *Practical Applications of Medical Geology* (ed. M. Siegel, O. Selinus and R. Finkelman), 615–635. Cham: Springer.

Barn, P., Gombojav, E., Ochir, C. et al. (2018). The effect of portable HEPA filter air cleaners on indoor PM 2. 5 concentrations and second hand tobacco smoke exposure among pregnant women in Ulaanbaatar, Mongolia: the UGAAR randomized controlled trial. *Science of the Total Environment* 615: 1379–1389.

Bauli, C.R., Lima, G.F., de Souza, A.G. et al. (2021). Eco-friendly carboxymethyl cellulose hydrogels filled with nanocellulose or nanoclays for agriculture applications as soil conditioning and nutrient carrier and their impact on cucumber growing. *Colloids and Surfaces A: Physicochemical and Engineering Aspects* 623: 126771.

Bera, B., Bhattacharjee, S., Chamling, M. et al. (2021). Fluoride hazard and risk enumeration of hard rock unconfined aquifers in the extended part of Chhota Nagpur gneissic complex. *Journal of the Geological Society of India* 97 (2): 199–209.

Bjørklund, G., Christophersen, O.A., Chirumbolo, S. et al. (2017). Recent aspects of uranium toxicology in medical geology. *Environmental Research* 156: 526–533.

Bonotto, D.M. (2021). Hydrochemical and radiometric evaluation of fresh and thermal waters from Araxá city (Minas Gerais, Brazil). *Environmental Geochemistry and Health* 44 (7): 2163–2186.

Borah, S.N., Sen, S., Sarma, H., and Pakshirajan, K. (2021). Biological Remediation of Selenium in Soil and Water. In: *Handbook of Assisted and Amendment: Enhanced Sustainable Remediation Technology*, 403–421.

Bostan, L.E., Clarkin, C.E., Mousa, M. et al. (2021). Synthetic nanoclay gels do not cause skin irritation in healthy human volunteers. *ACS Biomaterials Science & Engineering* 7 (6): 2716–2722.

Bridle, T.G., Kumarathasan, P., and Gailer, J. (2021). Toxic metal species and 'endogenous' metalloproteins at the blood–organ interface: analytical and bioinorganic aspects. *Molecules* 26 (11): 3408.

Carretero, M.I. (2020a). Clays in pelotherapy: a review, part I – mineralogy, chemistry, physical and physicochemical properties. *Applied Clay Science* 189: https://doi.org/10.1016/j.clay.2020.105526.

Carretero, M.I. (2020b). Clays in pelotherapy: a review, part II – organic compounds, microbiology and medical application. *Applied Clay Science* 189: 105531.

Cave, M. and Wragg, J. (2021). Application of bioavailability measurements in medical geology. In: *Practical Applications of Medical Geology* (ed. M. Siegel, O. Selinus and R. Finkelman), 235–261. Cham: Springer.

Chabukdhara, M. and Nema, A.K. (2012). Assessment of heavy metal contamination in Hindon River sediments: a chemometric and geochemical approach. *Chemosphere* 87 (8): 945–953.

Chandler, D.J., Grijsen, M.L., and Fuller, L.C. (2021). With bare feet in the soil: podoconiosis, a neglected cause of tropical lymphoedema. *Dermatology* 237 (2): 236–247.

Chang, C.H., Lee, Y.H., Liao, Z.H. et al. (2021). Composition of nanoclay supported silver nanoparticles in furtherance of mitigating cytotoxicity and genotoxicity. *PLoS One* 16 (2): e0247531.

Chaudhary, A. (2011). Ayurvedic Bhasma: nanomedicine of ancient India – its global contemporary perspective. *Journal of Biomedical Nanotechnology* 7: 68–69.

Cheleschi, S., Gallo, I., and Tenti, S. (2020). A comprehensive analysis to understand the mechanism of action of balneotherapy: why, how, and where they can be used? Evidence from *in vitro* studies performed on human and animal samples. *International Journal of Biometeorology* (64): 1247–1261.

Chimeddorj, B., Munkhbat, D., Altanbaatar, B. et al. (2021). Hydrogeochemical characteristics and geothermometry of hot springs in the Mongolian Altai Region, Mongolia. *Geochemistry: Exploration, Environment, Analysis* 21 (4).

Churko, C., Asnakew Asfaw, M., Tunje, A. et al. (2021). Knowledge, attitude, practice and associated factors of health professionals towards podoconiosis in Gamo zone, Ethiopia, 2019. *Journal of Foot and Ankle Research* 14 (1): 1.

Çiftçi, A., Vural, A.V., and Ural, M.N. (2021). Analysis of some concepts related to the environment and health with the N-gram method. *Journal of International Health Sciences and Management* 7 (13): 47–54.

Cölfen, H., Griesshaber, E., and Schmahl, W.W. (2021). Biominerals: formation, function, properties. *Crystals* 11 (3): 299.

Crompton, T.R. (2020). *Determination of Metals and Anions in Soils, Sediments and Sludges*. CRC Press.

Cuney, M. (2021). Editorial for the special issue "development of W-Sn and rare-metal metallogenic systems in an orogenic belt". *Minerals* 11 (2): 183.

Davies, T.C. (2009). A medical geology curriculum for African geoscience institutions. *Scientific African* 6: e00131.

Davies, T.C. (2013). Geochemical variables as plausible aetiological cofactors in the incidence. *Journal of African Earth Sciences* 79: 24–49.

Deheri, P.K. and Kar, B. (2021). Synthesis of nanoclay composite material. In: *Contemporary Nanomaterials in Material Engineering Applications* (ed. N.M. Mubarak, M. Khalid, R. Walvekar and A. Numan), 69–103. Cham: Springer.

Dissanayake, C. (2005). Of stones and health: medical geology in Sri Lanka. *Science* 309 (5736): 883–885.

Duval, S., Branscomb, E., Trolard, F. et al. (2020). On the why's and how's of clay minerals importance in life's emergence. *Applied Clay Science* 195: 105737. https://doi.org/10.1016/j.clay.2020.105737.

Ekosse, G.I., Nkeng, G.E., Bukalo, N., and Oyebanjo, O. (2021). Geophagic clays from cameroon: provenance, metal contamination and health risk assessment. *International Journal of Environmental Research and Public Health* 18 (16): 8315.

Elbehri, A., Elliott, J., and Wheeler, T. (2015). Climate change, food security and trade: an overview of global assessments and policy insights. In: *Climate Change and Food Systems: Global Assessments and Implications for Food Security and Trade* (ed. A. Elbehri), 1–27. Food and Agriculture Organization of the United Nations.

Elizabeth, W., Elmansouri, A., Ross, R. et al. (2021). Ecological study of fractures in paediatric melanesian communities with varying endemic environmental fluoride exposure. *Osteology* 1 (3): 132–140.

Emmanuel, A. and Emmanuel, E. (2021). Chemical pollution of drinking water in Haiti: an important threat to public health. *Environmental Health* https://doi.org/10.5772/intechopen.97766.

Emmanuel, T., Lypaek, D., Johansen, C., and Iversen, L. (2020). Effect of dead sea climatotherapy on psoriasis: a prospective cohort study. *Frontiers in Medicine* https://doi.org/10.3389/fmed2020.00083.

van der Ent, A., Baker, A.J., Reeves, R.D. et al. (2015). Agromining: farming for metals in the future? *Environmental Science & Technology* 49 (8): 4773–4780.

Erezuma, I., Eufrasio-da-Silva, T., Golafshan, N. et al. (2021). Nanoclay reinforced biomaterials for mending musculoskeletal tissue disorders. *Advanced Healthcare Materials* 16: 2100217.

Farkhutdinov, I., Farkhutdinova, L., Zlobina, A. et al. (2020). Historical aspects of medical geology. *Earth Sciences History* 39 (1): 172–183.

Ferrell, R.E. (2008). Medicinal clay and spiritual healing. *Clays and Clay Minerals* 56 (6): 751–760.

Filippelli, G. and Finkelman, R.B. (2021). Exploring the intersections of environmental health and urban medical geology. In: *Practical Applications of Medical Geology* (ed. M. Siegel, O. Selinus and R. Finkelman), 721–748. Cham: Springer.

Fuoco, I., Apollaro, C., Criscuoli, A. et al. (2021). Fluoride polluted groundwaters in Calabria region (Southern Italy): natural source and remediation. *Water* 13 (12): 1626.

Garcia-Villén, F., Faccendini, A., Miele, D. et al. (2020). Wound healing activity of nanoclay/spring water hydrogels. *Pharmaceuticals*, MDPI 12: 467.

Gevera, P.K., Cave, M., Dowling, K. et al. (2021). A review on the occurrence of some potentially harmful elements in the natural environment and their health implications: examples of fluoride, iron and salinity in the South-Eastern Kenya region. In: *Practical Applications of Medical Geology* (ed. M. Siegel, O. Selinus and R. Finkelman), 637–670. Cham: Springer.

Giraldo-Osorio, A., Ruano-Ravina, A., Perez-Rios, M. et al. (2021). Residential radon in Manizales, Colombia: results of a pilot study. *International Journal of Environmental Research and Public Health* 18 (3): 1228.

Gomes, C. and Rautureau, M. (ed.) (2021). *Minerals Latu Sensu and Human Health: Benefits, Toxicity and Pathologies*, 668. Springer International Publishing AG.

Gomes, C.D.S.F. and Silva, J.B.P. (2007). Minerals and clay minerals in medical geology. *Applied Clay Science* 36 (1–3): 4–21.

Gomes, C., Silva, A.C., Marques, A.C. et al. (2020a). Biotechnology applied to cosmetics and aesthetic medicines. *Cosmetics* 7: 33.

Gomes, C.S.F., Gomes, J.H., and Silva, E.A.F. (2020b). Bacteriostatic and bactericidal clays: an overview. *Environmental Geochemistry and Health* 42 (11): 3507–3527.

Gomes, C.S., Rautureau, M., Gomes, J.H., and Silva, E.A. (2021). Interactions of clay and clay minerals with the human health. In: *Minerals Latu Sensu and Human Health* (ed. C. Gomes and M. Rautureau), 271–375. Cham: Springer.

Grant, W.B., Lahore, H., McDonnell, S.L. et al. (2020). Evidence that vitamin D supplementation could reduce risk of influenza and COVID-19 infections and deaths. *Nutrients* 12 (4): 988.

Grzebisz, W., Ciesla, L., Komisarek, J., and Potarzycki, J. (2002). Geochemical assessment of the heavy metals pollution of urban soils. *Polish Journal of Environmental Studies* 11 (5): 493–500.

Guo, Z., Dong, L., Xia, J. et al. (2021). 3D printing unique nanoclay-incorporated double-network hydrogels for construction of complex tissue engineering scaffolds. *Advanced Healthcare Materials* 10 (11): 2100036.

Gwenzi, W. (2021). Toxic geogenic contaminants in serpentinitic geological systems: occurrence, behavior, exposure pathways, and human health risks. *Geochemistry: Concepts and Applications* 1–21. https://doi.org/10.1002/9781119710134.ch1.

Gwenzi, W., Tagwireyi, C., Musiyiwa, K. et al. (2021). Occurrence, behavior, and human exposure and health risks of potentially toxic elements in edible mushrooms with focus on Africa. *Environmental Monitoring and Assessment* 193 (5): 1–30.

Haraguchi, H. (2004). Metallomics as integrated biometal science. *Journal of Analytical Atomic Spectrometry* 19 (1): 5–14.

Hasan, S.E., Finkelman, R.B., and Skinner, H.C. (2013). Geology and health: a brief history from the Pleistocene to today. *The Impact of the Geological Sciences on Society* 501: 155.

Hassina, M. (2021). Medical geology in Africa: an example of a successful educational and research initiative at the University of Johannesburg, South Africa. In: *Practical Applications of Medical Geology* (ed. M. Siegel, O. Selinus and R. Finkelman). Cham: Springer https://doi.org/10.1007/978-3-030-53893-4_27.

Hunt, A.J., Anderson, C.W., Bruce, N. et al. (2014). Phytoextraction as a tool for green chemistry. *Green Processing and Synthesis* 3 (1): 3–22.

Ibaraki, M., Mori, H., and (2017). *Progress in Medical Geology*. Cambridge Scholars Publishing.

Jia, L., Chen, R., Xu, J. et al. (2021). A stick-like intelligent multicolor nano-sensor for the detection of tetracycline: the integration of nano-clay and carbon dots. *Journal of Hazardous Materials* 413: 125296.

Jung, C. and Awad, J. (2021). The improvement of indoor air quality in residential buildings in Dubai, UAE. *Buildings* 11 (6): 250.

Kapoor, R.C. (2010). Some observations on the metal based preparations in the Indian system of medicine. *Indian Journal of Traditional Knowledge* 9: 562–575.

Kazapoe, R.W., Arhin, E., and Amuah, E.E.Y. (2021). Known and anticipated medical geology issues in Ghana. *Ecofeminism and Climate Change* 2: 169–184.

Korobova, E., Romanov, S., and Silenok, A. (2020). Endemic diseases of geochemical origin and methodological approaches toward their prevention and elimination. *Environmental Geochemistry and Health* 42: 2595–2608. https://doi.org/10.1007/s10653-019-00442-z.

Kralj, S. and Marchesan, S. (2021). Bioinspired magnetic nanochains for medicine. *Pharmaceutics* 13 (8): 1262.

Kravchenko, J. and Ruhl, L.S. (2021). Coal combustion residuals and health. In: *Practical Applications of Medical Geology* (ed. M. Siegel, O. Selinus and R. Finkelman), 429–474. Cham: Springer.

Krishnamachary, B., Pemiah, B., Krishnaswamy, S. et al. (2012). Elucidation of a core-shell model for lauha bhasma through physico-chemical characterization. *International Journal of Pharmacy and Pharmaceutical Sciences* 4: 644–9.12.

Kumar, A., Nair, A.G., Reddy, A.V., and Garg, A.N. (2006). Bhasmas: unique ayurvedic metallic-herbal preparations, chemical characterization. *Biological Trace Element Research* 109: 231–254.

Kumaresan, M. and Riyazuddin, P. (2001). Overview of speciation chemistry of arsenic. *Current Science* 80: 837–846.

Lemberkovics, É., Czinner, E., Szentmihályi, K. et al. (2002). Comparative evaluation of Helichrysi flos herbal extracts as dietary sources of plant polyphenols, and macro-and microelements. *Food Chemistry* 78 (1): 119–127.

Li, Y. and Chen, T. (2005). Concentrations of additive arsenic in Beijing pig feeds and the residues in pig manure resources. *Conservation and Recycling* 45: 356–367.

Lohar (2019). Medical geology in India
Longley, P.A., Goodchild, M.F., Maguire, D.J., and Rhind, D.W. (2005). *Geographic Information Systems and Science*, 2e, vol. 14, 317–319. Wiley.
Lopes, S.I., Nunes, L.J., and Curado, A. (2021). Designing an indoor radon risk exposure indicator (IRREI): an evaluation tool for risk management and communication in the IoT age. *International Journal of Environmental Research and Public Health* 18 (15): 7907.
Marks, P. (2014). Blood minerals. *New Scientist* 222 (2973): 22.
Magambo, I., Dikgang, J., Gelo, D., and Tregenna, F. (2021). *Gold-Mining Pollution Exposure, Health Effects and Private Healthcare Expenditure in Tanzania*. University Library LMU Munich.
Malamba-Lez, D., Tshala-Katumbay, D., Bito, V. et al. (2021). Concurrent heavy metal exposures and idiopathic dilated cardiomyopathy: a case-control study from the Katanga mining area of the Democratic Republic of Congo. *International Journal of Environmental Research and Public Health* 18 (9): 4956.
Manera, M. (2021). Perspectives on complexity, chaos and thermodynamics in environmental pathology. *International Journal of Environmental Research and Public Health* 18 (11): 5766.
Mihankhah, P., Azdast, T., Mohammadzadeh, H. et al. (2021). Fused filament fabrication of biodegradable polylactic acid reinforced by nanoclay as a potential biomedical material. *Journal of Thermoplastic Composite Materials* https://doi.org/10.1177/08927057211044185.
Mohaptra, S. and Jha, C.B. (2010). Physicochemical characterization of Ayurvedic Bhasma (Swarna makshika bhasma): an approach to standardization. *International Journal of Ayurveda Research* 1: 82–86.
Myburgh, J., McGowan, K., and Davis, A. (2021). Veterinary geology. In: *Practical Applications of Medical Geology* (ed. M. Siegel, O. Selinus and R. Finkelman), 833–851. Cham: Springer.
Nematollahi, M.J., Keshavarzi, B., Moore, F. et al. (2021). Hydrogeochemical and ecological risk assessments of trace elements in the coastal surface water of the Southern Caspian Sea. *Environmental Monitoring and Assessment* 193 (7): 1–21.
Nethaji, S., Kalaivanan, R., Viswam, A., and Jayaprakash, M. (2017). Geochemical assessment of heavy metals pollution in surface sediments of Vellar and Coleroon estuaries, Southeast coast of India. *Marine Pollution Bulletin* 115 (1–2): 469–479.
Nghargbu, K.T. (2021). Balneotherapy: an overview of healing with natural waters. *Practical Applications of Medical Geology* 697–719.
Ngigi, P.B., Laing, G.D., Masinde, P.W., and Lachat, C. (2020). Selenium deficiency risk in central Kenya highlands: an assessment from the soil to the body. *Environmental Geochemistry and Health* 42: 2233–2250.
Nichols, N.T.A. and Nelson, L. (2021). Health effects of exposure to specific geologic materials: summary of clinical findings, treatment, and prevention. In: *Practical Applications of Medical Geology* (ed. M. Siegel, O. Selinus and R. Finkelman), 525–563. Cham: Springer.
Nnorom, I.C., Igwe, J.C., and Oji-Nnorom, C.G. (2005). Trace metal contents of facial (make-up) cosmetics. *African Journal of Biotechnology* 4 (10): 1133–1138.
Ojok, W., Wanasolo, W., Wasswa, J. et al. (2021). Hydrochemistry and fluoride contamination in Ndali-Kasenda Crater Lakes, Albertine Graben: assessment based on multivariate statistical approach and human health risk. *Groundwater for Sustainable Development* 15: 100650.
Olumayowa Oluwasola, H., Oluoye, O., Mohammad Bashir, S. et al. (2021). Geochemical and health risk assessment of heavy metals concentration in soils around Oke-Ere mining area in Kogi State, Nigeria. *International Journal of Environmental Analytical Chemistry* 1–16. https://doi.org/10.1080/03067319.2020.1862817.
Ortega, J.L. (2021). Anthropocene: new geological era or new biophysical. *Preprints* doi: 10.20944/preprints202108.0008.v1.
Pal, D., Sahu, C.K., and Haldar, A. (2014). Bhasma: the ancient Indian nanomedicine. *Journal of Advanced Pharmaceutical Technology & Research* 5: 4–12.
Palkhiwala, S. and Bakshi, S.R. (2014). Engineered nanoparticles: revisiting safety concerns in light of ethno medicine. *Ayu* 35: 237–242.
Panda, H. (2004). *Handbook on Ayurvedic Medicines with Formulae, Processes and their Uses*, 1–582. India: National Institute of Industrial Research.
Pandit, S., Biswas, T.K., Debnath, P.K. et al. (1999). Chemical and pharmacological evaluation of different ayurvedic preparations of iron. *Journal of Ethnopharmacology* 65: 149–156.
Pardo, L., Domínguez-Maqueda, M., Cecilia, J.A. et al. (2020). Adsorption of Salmonella in clay minerals and clay-based materials. *Fortschritte der Mineralogie* 10: 130. https://doi.org/10.3390/min10020130.
Parivallal, A. (2012). Structural analysis of metallic medicines. *Ayurpharm International Journal of Ayurveda and Allied Sciences* 1: 83–89.
Parveen, R., Tarannum, Z., Ali, S., and Fatima, S. (2021). Nanoclay based study on protein stability and aggregation and its implication in human health. *International Journal of Biological Macromolecules* 166: 385–400.
Paul, S. and Chugh, A. (2011). Assessing the role of Ayurvedic 'bhasms' as ethno-nanomedicine in the metal based nanomedicine patent regime. *Journal of Intellectual Property Rights* 16: 509–515.

Pawlik-Sobecka, L., Górka-Dynysiewicz, J., and Kuciel-Lewandowska, J. (2021). Balneotherapy with the use of radon–sulfide water: the mechanisms of therapeutic effect. *Applied Sciences* 11 (6): 2849.

Peixoto, D., Pereira, I., Pereira-Silva, M. et al. (2021). Emerging role of nanoclays in cancer research, diagnosis, and therapy. *Coordination Chemistry Reviews* 440: 213956.

Pereg, L., Steffan, J.J., Gedeon, C. et al. (2021). Medical geology of soil ecology. In: *Practical Applications of Medical Geology* (ed. M. Siegel, O. Selinus and R. Finkelman), 343–401. Cham: Springer https://doi.org/10.1007/978-3-030-53893-4_12.

Persano, F., Batasheva, S., Fakhrullina, G. et al. (2021). Recent advances in the design of inorganic and nano-clay particles for the treatment of brain disorders. *Journal of Materials Chemistry B* 9 (12): 2756–2784.

Pinto, T., Dias, A.G., and Vasconcelos, C. (2021). Geology and environment: a problem-based learning study in higher education. *Geosciences* 11 (4): 173.

Pizzorno, P., Bühl, V., and Mañay, N. (2021). Human health risk assessment by exposure to arsenic through groundwater in a rural area of Canelones, Uruguay. *ISEE Conference Abstracts* 2021 (1).

Puccinelli, M., Pezzarossa, B., Pintimalli, L., and Malorgio, F. (2021). Selenium biofortification of three wild species, *Rumex acetosa* L., *Plantago coronopus* L., and *Portulaca oleracea* L., grown as microgreens. *Agronomy* 11 (6): 1155.

Radomska, D., Czarnomysy, R., Radomski, D. et al. (2021). Selenium as a bioactive micronutrient in the human diet and its cancer chemopreventive activity. *Nutrients* 13 (5): 1649.

Raju, G.N., Sarita, P., Murty, G.R. et al. (2006). Estimation of trace elements in some anti-diabetic medicinal plants using PIXE technique. *Applied Radiation and Isotopes* 64 (8): 893–900.

Ražić, S., Onjia, A., Đogo, S. et al. (2005). Determination of metal content in some herbal drugs – empirical and chemometric approach. *Talanta* 67 (1): 233–239.

Ríos-Reyes, C.A., Ríos-Gutiérrez, M.P., and Joya-Neira, S. (2021). The importance of minerals in medical geology: impacts of the environment on health. *Archivos de Medicina (Col)* 21 (1): 182–202.

Ritto, F.P., da Silva, E.M., Borges, A.L.S. et al. (2021). Fabrication and characterization of low-shrinkage dental composites containing montmorillonite nanoclay. *Odontology* 110 (1): 35–43.

Rodrigues, I., Torres, J., and Favas, P. (2016). Geomedicine and history of science: a contribution to scientific culture. In: *Geoscience Education* (ed. C. Vasconcelos), 207–223. Cham: Springer.

Röllin, H.B., Channa, K., Olutola, B., and Odland, J.Ø. (2021). Selenium status, its interaction with selected essential and toxic elements, and a possible sex-dependent response. *International Journal of Environmental Research and Public Health* 18 (16): 8344.

Romero-Mujalli, G., Roisenberg, A., Cordova-Gonzalez, A., and Stefano, P.H. (2021). Indoor radon concentration and a diffusion model in dwellings situated in a subalkaline granitoid area, Southern Brazil. *Environmental Earth Sciences* 80 (17): 1.

Sadeghfam, S., Abdi, M., Khatibi, R., and Nadiri, A.A. (2021). An investigation into uncertainties within human health risk assessment to gain an insight into plans to mitigate impacts of arsenic contamination. *Journal of Cleaner Production* 311: 127667.

Samaddar, S., Karp, D.S., Schmidt, R. et al. (2021). Role of soil in the regulation of human and plant pathogens: soils' contributions to people. *Philosophical Transactions of the Royal Society B* 376 (1834): 20200179.

Sandri, G., Faccendini, A., Longo, M. et al. (2020). Halloysite- and montmorillonite-loaded scaffolds as enhancers of chronic wound healing. *Pharmaceuticals*, MDPI 12: 179. https://doi.org/10.3390/pharmaceutics12020179.

Sanyaolu, O.M., Mouri, H., Selinus, O., and Odukoya, A. (2021). Sources, pathways, and health effects of iodine in the environment. In: *Practical Applications of Medical Geology* (ed. O. Selinus, M. Siegel and R. Finkelman), 565–613. Cham: Springer.

Sarkar, P.K. and Chaudhary, A.K. (2010). Ayurvedic Bhasma: the most ancient application of nanomedicine. *Journal of Scientific and Industrial Research* 69: 901–905.

Schwartz, G.G. and Williamson, M.R. (2021). Acid precipitation and the prevalence of Parkinson's disease: an ecologic study in US States. *Brain Sciences* 11 (6): 779.

Selinus, O., Finkelman, R.B., Nichols, N.T.A., and Walvoord, K. (2021). A guide to the medical geology literature. In: *Practical Applications of Medical Geology* (ed. M. Siegel, O. Selinus and R. Finkelman), 855–863. Cham: Springer.

Sharma, R. and Prajapati, P.K. (2016). Nanotechnology in medicine: leads from ayurveda. *Journal of Pharmacy & Bioallied Sciences* 8 (1): 80–81.

Siegel, M. (2021a). A framework for applied medical geology: part I. The environmental pathways analysis. In: *Practical Applications of Medical Geology* (ed. M. Siegel, O. Selinus and R. Finkelman), 3–50. Cham: Springer.

Siegel, M. (2021b). A framework for applied medical geology: part II. The biological impact analysis. In: *Practical Applications of Medical Geology* (ed. M. Siegel, O. Selinus and R. Finkelman), 51–111. Cham: Springer.

Siegel, M., Selinus, O., and Finkelman, R. (2021). *Practical Applications of Medical Geology*. Cham: Springer https://doi.org/10.1007/978-3-030-53893-4_12.

Silva, A.L.O.D., Barrocas, P.R., Jacob, S.D.C., and Moreira, J.C. (2005). Dietary intake and health effects of selected toxic elements. *Brazilian Journal of Plant Physiology* 17: 79–93.

Simões, B.M., Grossmann, M.V.E., Baron, A.M. et al. (2021). Production of starch-polyester bio-support for lipases immobilization: synergistic action of itaconic acid and nanoclay. *Preparative Biochemistry & Biotechnology* 51 (6): 580–588.

Sultani, H.M., Okoyo, C., Kanyi, H.M. et al. (2021). Mapping of podoconiosis cases and risk factors in Kenya: a nationwide cross-sectional study. *The American Journal of Tropical Medicine and Hygiene* 10.4269/ajtmh.21-0248.

Sundaray, S.K., Nayak, B.B., Lin, S., and Bhatta, D. (2011). Geochemical speciation and risk assessment of heavy metals in the river estuarine sediments – a case study: Mahanadi basin, India. *Journal of Hazardous Materials* 186 (2–3): 1837–1846.

Suwatvitayakorn, P., Ko, M.-S., Kim, K.-W., and Chanpiwat, P. (2020, 2020). Human health risk assessment of cadmium exposure through rice consumption in cadmium contaminated areas of the Mae Tao sub-district, Tak, Thailand. *Environmental Geochemistry and Health* 42: 2331–2344.

Tappe O. (2021). Artisanal, small-scale and large-scale mining in Lao PDR.

Tashakor, M., Modabberi, S., and Argyraki, A. (2022). Assessing the contamination level, sources and risk of potentially toxic elements in urban soil and dust of Iranian cities using secondary data of published literature. *Environmental Geochemistry and Health* 44: 645–675.

Thomson, B. (2021). Environmental contamination from uranium mining and milling in the Western US. In: *Practical Applications of Medical Geology* (ed. M. Siegel, O. Selinus and R. Finkelman), 475–523. Cham: Springer.

Thompson, A.K., Monti, M.M., and Gribble, M.O. (2021). Co-occurrence of metal contaminants in United States public water systems in 2013–2015. *International Journal of Environmental Research and Public Health* 18 (15): 7884.

Tipa, C., Cidade, M.T., Vieira, T. et al. (2021). A new long-term composite drug delivery system based on thermo-responsive hydrogel and nanoclay. *Nanomaterials* 11 (1): 25.

Tunakova, Y., Shagidullin, A., Valiev, V. et al. (2021). Method for determining regional reference values of metal content in biological substrates and their intake into the body via drinking water. *International Journal of Environmental Research and Public Health* 18 (18): 9903.

Turki, Y.A. (2015). The use of nano technology in medicine. *Saudi Journal of Medicine and Medical Sciences* 3: 184–184.

Valodkar, M., Rathore, P.S., Jadeja, R.N. et al. (2012). Cytotoxicity evaluation and antimicrobial studies of starch capped water soluble copper nanoparticles. *Journal of Hazardous Materials* 201–202: 244–249.

Venieri, D., Gounaki, I., Christidis, G.E. et al. (2020). Bridging the gaps: bole and terra sigillata as artefacts, as simples and as antibacterial clays. *Minerals*. MDPI 10: 348. https://doi.org/10.3390/min10040348.

Volfson, I.F., Farrakhov, E.G., Pronin, A.P. et al. (2010). Medical Geology in Russia and the NIS. In : Selinus, O., Finkelman, R., Centeno, J. (eds) *Medical Geology*. International Year of Planet Earth. Springer, Dordrecht. *Medical Geology*, 221–258.

Wadekar, M.P., Rode, C.V., Bendale, Y.N. et al. (2005). Preparation and characterization of a copper based Indian traditional drug: Tamra bhasma. *Journal of Pharmaceutical and Biomedical Analysis* 39 (5): 951–955.

Wadekar, M.P., Rode, C.V., Bendale, Y.N. et al. (2006). Effect of calcination cycles on the preparation of tin oxide based traditional drug: studies on its formation and characterization. *Journal of Pharmaceutical and Biomedical Analysis* 41: 1473–1478.

Wålinder, R. (2021). Occupational health and medical geology. In: *Practical Applications of Medical Geology* (ed. M. Siegel, O. Selinus and R. Finkelman), 749–768. Cham: Springer.

Wang, J., Liu, G., Lu, L. et al. (2015). Geochemical normalization and assessment of heavy metals (Cu, Pb, Zn, and Ni) in sediments from the Huaihe River, Anhui, China. *Catena* 129: 30–38.

Whitbread, J.S., Murphy, K.J., Clifton, P.M., and Keogh, J.B. (2021). Iodine excretion and intake in women of reproductive age in South Australia eating plant-based and omnivore diets: a pilot study. *International Journal of Environmental Research and Public Health* 18 (7): 3547.

Wigginton, K.R. and Boehm, A.B. (2020). Environmental engineers and scientists have important roles to play in stemming outbreaks and pandemics caused by enveloped viruses. *Environmental Science & Technology* 54: 3736–3739.

Williams, L.B. (2019). Natural antibacterial clays: historical uses and modern advances. *Clays and Clay Minerals* 67 (1): 7–24.

Wu, M., Chen, F., Wu, P. et al. (2021). Nanoclay mineral-reinforced macroporous nanocomposite scaffolds for in situ bone regeneration: *in vitro* and *in vivo* studies. *Materials & Design* 205: 109734.

Yadav, M., Singh, G., and Jadeja, R.N. (2021). Fluoride contamination in groundwater, impacts, and their potential remediation techniques. In: *Groundwater Geochemistry: Pollution and Remediation Methods*, 22–41. https://doi.org/10.1002/9781119709732.ch2.

Yamaguchi, T., Ko, S.J., Jung, J.S. et al. (2021). Periodic charge matching driven immobilization of gentamicin in nanoclays for stable and long term antibacterial coating. *Dalton Transactions* 50: 14216–14222.

Zakharovskyi, V. and Németh, K. (2021). Quantitative-qualitative method for quick assessment of geodiversity. *Land* 10 (9): 946.

Zheng, X., Zhang, X., Wang, Y. et al. (2021). Hypoxia-mimicking 3D bioglass-nanoclay scaffolds promote endogenous bone regeneration. *Bioactive Materials* 6 (10): 3485–3495.

Ziarati, P., Hochwimmer, B., Cruz-Rodriguez, L. et al. (2021). Lung cancer disease and arsenic: the medical geology as key. *Journal of Medical Discovery* 6 (1): 1–16.

2

Biogeochemistry

Essential Link Between Geosphere and Biosphere

Dilan Ranaweera[1]*, Udayagee Kumarasinghe*[2]*, Chathuri Senanayake*[2]*, Rohan S. Dassanayake*[2]*, Pasan Chinthana Bandara*[2]*, and Pabasari Arundathi Koliyabandara*[1]

[1] *Department of Civil and Environmental Technology, Faculty of Technology, University of Sri Jayewardenepura, Homagama, Sri Lanka*
[2] *Department of Biosystems Technology, Faculty of Technology, University of Sri Jayewardenepura, Homagama, Sri Lanka*

2.1 Introduction to Biogeochemistry

Biogeochemistry is broadly defined as a branch of geobiology that involves the mutually exclusive interactions between the biology and chemistry of the Earth system (Cutter 2005). The term biogeochemistry was first introduced in 1926 by a Russian and Ukrainian scientist, Vladimir Vernadsky, in his provocative book *The Biosphere*. Vladimir identified three main spheres: abiotic sphere, biosphere, and noosphere (Vernadskiĭ et al. 2007; Bianchi 2021). The abiotic sphere comprises all the nonliving things, including energy and materials; the biosphere is all living organisms and life processes; the noosphere includes the human cognitive process interconnected and interacting through communication (Svoboda and Nabert 1999; Vernadskiĭ et al. 2007). Eville Gorham also provided the first historical basis of how Biogeochemistry came into existence in one of the papers entitled *Biogeochemistry: Its Origins and Development* (Gorham 1991). Furthermore, William H. Schlesinger also described biogeochemistry as a premier scientific discipline that evaluates human impacts on the environment in his article on *Better Living Through Biogeochemistry* (Schlesinger 2004).

Since the early twentieth century, biogeochemistry has emerged as a multidisciplinary scientific field, encompassing the physical, chemical, biological, and geological processes and reactions that govern the composition and changes of our natural environment. Biogeochemical approaches help to understand the complex natural processes that occur in the biosphere, including oceans, forests, agricultural areas, coastal sediments, deserts, and grasslands. Moreover, biogeochemistry studies the migration, distribution, and interactions of chemical substances including elements and compounds, with other substances and organisms in the earth's system, hydrosphere (water and ice), biosphere (life), lithosphere (rock and soil), and atmosphere (air). Those processes of biogenic mass exchange are cyclic in nature. The pathway by which particular chemical substances transport and transform into various substances in the earth's biotic and abiotic environment is called a geochemical cycle. The global geochemical cycles include nutrients such as carbon (C), nitrogen (N), sulfur (S), and phosphorous (P), essential elements such as iron (Fe), zinc (Zn), selenium (Se), and toxic trace elements, including cadmium (Cd), mercury (Hg), and lead (Pb). Biogeochemical cycles play an imperative role in ecosystem functioning. Moreover, biogeochemical cycles also cover the effect of organisms on the chemistry of manmade chemicals, such as the biodegradation of polychlorinated biphenyls (PCBs) (Hellweger 2008).

Biogeochemistry is becoming an increasingly popular area among the scientific community as it provides insights into ecological processes and how ecosystems respond to natural events such as forest fires, landslides, floods, snowfall, and anthropogenic activities, including deforestation, land use, household, and industrial waste. Moreover, modern biogeochemistry involves the use of enzymatic, biochemical, physiological, and omics approaches, including proteomics, metabolomics, genomics, and transcriptomics adapting to global climate change, mitigating pollution, enhancing food production and security, developing renewable energy, drug development, and disease-resistant crops, subsequently providing the efficient management of ecosystems and discovering new products and technologies (Bashkin 2003).

This chapter discusses the two major spheres of the earth's system, namely geosphere and biosphere, natural and artificial biogeochemical cycles, soil biogeochemistry, and the impact of natural and anthropogenic activities on biogeochemical processes and future perspectives of biogeochemistry.

Medical Geology: En route to One Health, First Edition. Edited by Majeti Narasimha Vara Prasad and Meththika Vithanage.

2.2 Geosphere: Formation, Evolution, and Isotopes

Basically, the rock, minerals, and landforms of the earth's surface and interior can be referred to as the geosphere. In this context, this sphere includes all endogenous and exogenous activities that make up the crust and the core of the earth as well as it interacts strongly with the hydrosphere, atmosphere, biosphere, and anthrosphere, which covers everything natural and lifeless on the surface of the earth (Smith 1971; Thomas and Spicer 1987; Trenberth 2002; Sorokhtin et al. 2011). Mainly the composition of the geosphere consists of five major dynamic processes as fluvial and alluvial processes (the physical interaction of flowing water and the natural channels of rivers and streams), aeolian processes (involving erosion, transportation, and deposition of sand and dust by the wind), tectonic processes (dealing with the dynamics of Earth's outer shell such as folds, faults, and mountains), volcanic processes (related to the eruption of molten rocks, volcanoes, lava flows and magma, calderas, and volcanic deposits), and surface processes (including surface soil-related activities, landslides, weathering, and erosion process) (Smith 1971; Thomas and Spicer 1987; Trenberth 2002; Sorokhtin et al. 2011).

Earth is the fifth largest planet in our solar system, and the only known planet to have liquid water on the surface. The geosphere is a layered structure with a radius of 6371.3 km, the heart of the Earth is located in the inner core, a sphere of iron and nickel about 1221.4 km wide, and at 5426 °C temperature, but it is a solid form due to the enormous pressure. The inner core is 2253 km from the outer core surrounded by solid iron and nickel liquids, and beyond the outer core is the mantle, a layer of viscous molten rock 2897 km thick in the Earth's outer layer (Woodhouse et al. 1986; Vasco et al. 1999). The mantle is composed of rock and ranging in depth between 300 and 1890 km and the deeper inner mantle, though hot enough for the rock to be liquid under ordinary pressures is solid because of the enormous pressure to which it is subjected. At the top of the inner mantle, at a depth of 10–300 km, the outer mantle consists of hot molten magma (basaltic) (Nolet et al. 1994; Clayton and Comer 1983; Vasco et al. 1999). The dense lithosphere that floats on magma is composed of relatively strong rocks, ranges in thickness from a few kilometers to about 400 km and on average about 100 km (Clayton and Comer 1983; Nolet et al. 1994; Vasco et al. 1999). The transition layer between molten magma and the lithosphere is an asthenosphere composed of relatively weak and plastic hot rocks (Clayton and Comer 1983; Nolet et al. 1994; Vasco et al. 1999). The Lithosphere, the outermost layer of the earth's crust, which is about 40 km thick, but the land surface is about 30.5 km thick and the ocean surface is about 4.8 km thick (Phinney 1964; White 1988).

According to the theory of plate tectonics, the Earth's surface consists of massive lithospheric plates that hold continents and oceans and act as units. The Earth's crust is a dynamic system with its lithospheric plates moving about a few centimeters relative to each other on average, sometimes due to this movement, earthquakes are occurring (Phinney 1964; White 1988). On the other hand, due to the same process, magma comes to the surface along the plate boundaries with molten rock, ash, and gases as volcanoes. As a result of the process of the plate tectonic cycle (divergent plates, convergent plate boundaries, and transform fault boundaries) and the puzzles of geospheric phenomena, including the opening and widening of the ocean floor, the relative movement of continents, the formation of ridges, volcanic activity, and earthquakes are creating in the earth crust (Phinney 1964; White 1988).

There are about 2000 minerals on Earth, however, most of the rocks in the crust are made up of only 25 common minerals, mainly 49.5% oxygen and 25.7% silicon. The crust is largely composed of small amounts of aluminum, iron, carbon, sulfur, and other chemical compounds of these two elements (Yasuo and Kamitani 2006; US DOE 2010). Except for aluminum and iron, about 1.6% of the earth's surface is composed of metals, phosphorus, and rocks that serve as essential minerals for plant growth (Yasuo and Kamitani 2006; US DOE 2010). Part of the crust, which is crucial for the survival of humans and many other nonaquatic life forms, is a thin layer of weathered rock, semidecomposed organic matter, aerospace, and waterlogged soils that support plant life.

2.2.1 Evolution of the Geosphere

Focusing on geographical evolution, data on lithosphere evolution can be discussed, as the plates that lead to alteration and recycling of the lithosphere originate in the early stages of planetary differentiation into geospatial processes. Using the uranium-lead clock dating methods, in 1953, Claire C. Patterson of the California Institute of Technology established an age of 4.55 billion years for Earth and many of the meteorites that formed it. Accordingly, the main focus is on the pre-Cambrian period as the oldest evidence of the geosphere evolution as we as the modern ocean lithosphere is young, covering more than 85% of Earth's geological history over time and covering more than 70% of the continents (Michael and Rushmer 2005; Hamilton 2007; Hawkesworth et al. 2020). Knowledge of the lithosphere evolution is largely based on recent advances in the isotopic geochemistry and provides insights into the interactions of surface and deep processes, including the effects and manifestations of fundamental planetary heat loss, but there are some significant repercussions about interpretations, concepts, and understanding of how the early Earth operated (Dewey 2007; Hamilton 2007).

The Archean Eon, also known as the Archaean Eon, spelled out the two formal divisions of the Precambrian period (about 4.6–541 Gya) and the first period in which life appeared on Earth. Based on the minerals that do not recrystallize during later tectonic events and the evidence of the depth of >200 km of the subcrustal lithosphere, the age of the oldest continental lithosphere which is in the Archean Cratons is 3.0 Gya, as revealed by Sm–Nd and Re–Os isotopes studies in diamond inclusions (Carlson et al. 1994; Shirey and Richardson 2011). However, some petrological studies have shown that the entire lithospheric belt of the cratonic keel formed by a rapid tectono-magmatic event at relatively short intervals of at least 220–250 km, but isotope ages indicate the mantle lithosphere of the croutons and the crust were formed at the same time and have been connected ever since (Carlson et al. 1994; Pearson et al. 1995). Based on the fossil evidence and Archean greenstone-granite belts and many mineral deposits, including gold and silver, the earliest primitive life-forms (prokaryotic microbes) from the domain called Archaea and bacteria have appeared in rocks about 3.5–3.7 Gya also suggesting that life could have emerged sometime before 3.95 Gya (Carlson et al. 1994; Pearson et al. 1995). The oxygen content of the present atmosphere must have begun in Archean times and slowly accumulated over time, while the free oxygen anaerobic cyanobacteria in the Archaean atmosphere synthesized the organic photosynthesis of carbon dioxide (CO_2) and water (H_2O) (Carlson et al. 1994; Pearson et al. 1995).

Throughout the Archean Eon, the oceanic and arctic crust has been in semi-continuous production for 1.5 Gya. Thus, most Archean rocks are igneous and the oldest rocks on Earth are estimated to be 4.28 Gya (Artemievaa et al. 2002).

The Proterozoic Eon extended from 2.5 billion years ago and ended 542.0 million years ago and is the second part of the Precambrian time. It is subdivided into three geologic eras, the Paleoproterozoic (2.5–1.6 Gya), Mesoproterozoic (1.6–1 Gya), and Neoproterozoic (1 billion to 542.0 Mya) (Rogers and Santosh 2003). One of the most well-known events of this period was the transition to an oxygen-rich atmosphere during the Mesoproterozoic period. Several glaciers, including the hypothetical snowball Earth of the age of the Neoproterozoic period, and the Ediacaran period (635–542 Mya) are characterized by the evolution of abundant soft-bodied multicellular organisms (Rogers and Santosh 2003). The first known glaciations in the history of the Geosphere occurred during the Proterozoic epoch, one shortly after the beginning of the eon, and at least four during the Neoproterozoic period, climaxing with the Snowball Earth of the Sturtian and Marinoan glaciations (Young 2013). Though oxygen is believed to have been released by photosynthesis as far back as Archean times, the most important event in the aspect of biogeochemistry of the Proterozoic was the accumulation of oxygen in the Earth's atmosphere (Young 2013). Although it is believed that the release of oxygen by photosynthesis dates back as far as the Archean Age, the most important event in protozoanism was the accumulation of oxygen in the Earth's atmosphere (Young 2013). Moreover, until the chemical sinks-unoxidized sulfur and iron could not build up to any significant level at least approximately 2.3 Gya, oxygen could only be 1–2% of its current level (Young 2013).

The Phanerozoic Eon from 541 Mya and to the present is divided into three major spans of time largely based on the characteristic assemblages of life-forms, the Paleozoic (541–252 Mya), Mesozoic (252–66 Mya), and Cenozoic (66 Mya to the present) eras (Stanley 1989). The assembly of the Paleozoic-era supercontinent of Pangaea, roughly 200 million years ago, merged into a continental supercontinent, but since then the plate tectonics have slowly separated as Tethys Sea that eventually divided the landmass into two smaller supercontinents, Laurasia and Gondwana, creating the layout of the continents as today world (Torsvik et al. 2021). Pangaea is the grouping of a series of geographical events, including the island-arc accumulation of all the major continents, the continental conflict, and the closing of the ocean basin (Torsvik et al. 2021). Also climatically, there is evidence that sea levels rose and fell four times during the Paleozoic period, according to abundant oceanic sedimentary rocks such as fossil corals and iodized limestone. The most important aspect is the assembly of the pangia, leading to a supercontinental arid climate and even greater diversification and dominance of invertebrates and vertebrates, mainly adapting to terrestrial life during the Middle Paleozoic. Groups that grew up in this hot climate eventually became dinosaurs (Torsvik et al. 2021). Another group called synapses eventually evolved into mammals.

Pangaea began to break up in the Late Triassic about 210 Mya, during which time the South Atlantic Ocean opened up and South America seceded from Africa, with the opening of the Atlantic Ocean as the young Mid-Atlantic began to form seabed (Torsvik et al. 2021). This was accompanied by the separation of Greenland and Scandinavia. The Mesozoic era was dominated by reptiles and, more specifically, dinosaurs (Gradstein et al. 2004; Haq and Schutter 2008). The Triassic Permian saw extinct ecosystems that took more than 30 million years to fully re-emerge after the Great Extinction (Gradstein et al. 2004). It was at this point that the first appearance of many modern animal groups, which later prospered (Gradstein et al. 2004). The first flying vertebrate group, such as dinosaurs and mammals, the pterosaurs started small in the Triassic.

In Cenozoic, the Earth's plates moved to more familiar locations, the biggest difference being the closure of the Sea of Tethys with collisions such as the Alps, Sagittarius, the Himalayas Rockies, Black Hills, Wind River Range, Uinta Mountains, and the San Rafael Swell (Gradstein et al. 2004; Haq and Schutter 2008). Also, instead of descending directly into the mantle, the submergence plate shallows eastward beneath the continental shelf, affecting the continent hundreds of miles east of the continental boundary and forming high mountains. There are five original mammalian groups in the fossil record, based primarily on fossil teeth, the hardest bone in the vertebrate skeleton (Gradstein et al. 2004). For the purpose of this text, the two major groups of mammals of the Cretaceous period, the Marsupials (such as the Sinodelphys) and the Eupantotheres, which migrated to the placenta or eutherians (such as the Eomaia), were the most important group and later diversified into the Cenozoic (Gradstein et al. 2004). Hominids, the name for the human-like primates, originated in East Africa millions of years ago.

2.3 Biosphere

The term "biosphere" refers to the earth's life that has existed for approximately 4 billion years. Water, air, and nutrients (macro and micronutrients) are the essential components to form life (Dobretsov et al. 2008). Therefore, the biosphere lies in all the other spheres including, the atmosphere, hydrosphere, and lithosphere. Over the years, simple organisms present in aquatic bodies evolved into complex organisms. The first organisms were single-celled prokaryotes that lack membrane-bound nuclei. The early microorganisms were found in deep-sea hydrothermal vents, underground, and in rocks. The oldest microorganisms evolved around 4.4 billion years ago. The microfossils found in the bedded chart unit of the Archean Apex Basalt of Australia were morphologically similar to the existing cyanobacteria (Kutschera and Niklas 2004). The earliest organisms consumed the organic matter nearby and digested it through fermentation.

As the earth's population increased immensely, there was a great need to make food, which led to the evolution of photosynthesis. Photosynthetic cells use carbon dioxide (CO_2) and solar energy during photosynthesis and produce sugar and oxygen (O_2). These sugar molecules serve as the basis for more complex organic molecules. Chemosynthesis is also another energy-yielding reaction that involves the production of organic carbon from CO_2 in the dark (Enrich-Pr et al. 2014). Microorganisms that live in deep-sea hydrothermal vents acquire energy via chemosynthesis. Earth's life is believed to be originated at hydrothermal vents located in volcanic areas on the ocean floor. Therefore, it is more likely that chemosynthesis has evolved before photosynthesis. Later, eukaryotes with complex cell structures evolved.

2.3.1 Evolution of the Biosphere

The evolution of the biosphere was an aged-long process. The beginning of life was around 4.4 billion years ago, as soon as the cooling of the first ocean happened (Wilde et al. 2001). The extreme temperature (~300 °C) and the high acidity resulting from the serpentinization of pyroxenes at very high temperatures were not ideal for the emergence of life. Such extreme conditions cause destructive effects on organic molecules like RNA in living cells. However, the early ocean would have provided the essential elements that help onset life in the ocean. The electrochemical disequilibria resulting from hydrothermal convection within the thick, fractured, and permeable Hadean crust was the driving force for the emergence of life on the Earth. Russell and Ponce (2020) stated that olivine ($[Mg>Fe]_2SiO_4$), bridgmanite ($[Mg,Fe]SiO_3$), serpentine ($[Mg,Fe,]_{2\text{-}3}Si_2O_5[OH)]_4$), and pyrrhotite ($Fe_{(1-x)}S$) were the four most essential minerals to create such a disequilibrium on earth. Fougerite ($[Fe^{2+}{}_{6x}Fe^{3+}{}_{6(x-1)}O_{12}\mathbf{H}_{\mathbf{2}(7-3x)}]^{2+}\cdot[(CO^{2-})\cdot 3H_2O]^{2-}$) and mackinawite ($Fe[Ni]S$) were vital as initial "free energy" conductors and converters of such disequilibria (Russell and Ponce 2020).

Hydrogen (H_2) is considered one crucial fuel for the beginning of life on Earth, produced via oxidation of remaining native iron in the Hadean crust (Righter et al. 1997). Redox and proton gradients provided by H_2 were greatly responsible for the emergence of life on the Earth. Nitrogen dioxide (NO_2) was produced by cloud-to-cloud lightning, dissolved in the ocean, and generated nitrate and nitrite ions. Alkaline (pH ~11–11.5) springs with moderate temperatures bearing HS^- were generated due to the partial dissolution of pyrrhotite and the hydration and oxidation of olivine to serpentine in the oceanic crust through the convective circulation of ocean water. CO_2 degassed from the hot mantle was dissolved in the ocean water and produced a carbonic ocean with high pH (~5.5). CO_2 trapped in the oceanic crust was slowly converted to methane and formate, which was dissolved in the moderate temperature alkaline hydrothermal systems. The reduction of the ocean's temperature favored the reaction between H_2, which continuously comes out from Earth's interior, and CO_2 in the atmosphere and ocean. This reaction yielded acetate and water as the main products. Glycine (amino acetate), other amino acids, and traces of nucleic acids are the other products of this reaction which are the essential constituents of organic molecules (Huber and Wächtershäuser 1997). Energy from the acetate reaction and the redox and proton gradient provided by H_2 formed pyrophosphate, which had the energy to polymerize amino acids into peptides. RNA molecules were developed into a state where they could be transferred to the next generation.

Initial cell membranes and cell walls are composed of hydrophobic heterochiral peptides, including iron sulfides as stabilizers and electron transfer agents. Later lipids would have been generated from acetyl-coenzyme-A, one of the cellular structures' constituents. It was believed that the earliest form of bacteria would have emerged near the mound's surface at moderate temperature (40 °C) where chemical and electrochemical gradients were high. On the other hand, the earliest Archaea originally evolved at relatively high temperatures (~60 °C). However, the ability to live in high temperatures was passed back to the bacteria resulting in the evolution of thermophilic bacteria, which is considered the last common community (Gaucher et al. 2003). In ancient cells, interactions with the neighboring cells may have happened by gene swapping, providing nutrients, and removing waste. Proto-Archaea lived in the hydrothermal vents in the deep ocean and acquired energy via chemosynthesis. During chemosynthesis, the energy stored in the chemical bonds of organic sulfides (i.e. hydrogen sulfide-H_2S) was used to produce food energy from water CO_2. A small number of Archaea cells that could withstand elevated temperatures generated methane as waste during methanogenesis in the anaerobic environment. Those methanogenic cells differentiated from the parent population to become ancestors of Archaea.

As the Earth's population increased immensely, the oceanic sediments and crust could only provide nutrients to the earlier communities. This situation led to the evolution of photosynthesis about 4 billion years ago (Pratt 1993). Obduction and uplift of oceanic sediments and the hydrated basalt beneath moved a portion of bacteria to the photic zone on ocean island chains. Those cells were adapted to protect from harmful UV solar radiation by developing a UV pigment protector. These pigment protectors may have been modified later as electron transfer agents, photosynthetic reaction centers, and antenna proteins to facilitate photosynthesis. The first photosynthetic bacteria were green sulfur bacteria that used H_2S as an electron donor and generated elemental sulfur (S) as waste and gained electrons for the biosynthesis process (Eq. 2.1).

Photosynthetic reaction centers in green sulfur bacteria

$$H_2S + CO_2 + \text{light} \rightarrow (CH_2O)_{\text{life}} + H_2O + 2S^0 \quad (2.1)$$

Photosynthetic precursors of the heliobacteria were able to fix CO_2 by the oxidation of organic waste, and at the same time, they used a similar reaction as in Eq. (2.1) (Baymann et al. 2001). The evolution of photosystem 2 (PS2) and photosystem (PS1) resulted due to the hybridization of the two reaction centers. PS2, together with PS1, oxidizes two molecules of water to produce one molecule of oxygen while gaining four electrons and four protons for the fixation of CO_2. The emergence of oxygenic photosynthesis, 3.7 billion years ago, and the presence of atmospheric trace oxygen, 3.5 billion years ago was one of the leading causes of the evolution of multicellular eukaryotes (Ohmoto et al. 1993). The evolution of earth's life from single-celled organisms to multicellular eukaryotes was a prolonged process. Paleontological and biochemical evidence suggests that the appearance of the first eukaryotic cells happened about 2–1.5 billion years ago. The oldest multicellular eukaryote dated 1.2 billion years ago was *Bangiomorpha* (Butterfield 2015). The evolution of multicellular animals happened 300 million years ago when there was enough atmospheric oxygen in the environment. The oxygen produced by plants by photosynthesis is used for the respiration process by animals. Respiration, which resembles photosynthesis in reverse, is a chemical process that releases energy from glucose in the presence of oxygen.

2.3.2 Bacteria: The Most Primitive Organisms on Earth

Bacteria are usually single-celled organisms with an average diameter of 0.5–2.0 μm. They are the most abundant and diverse group of organisms that dominate in aquatic systems and extreme environments. All bacteria cells consist of simple structures such as a rigid cell wall, cell membrane, cytoplasm, nuclear (genetic material), and ribosome (Figure 2.1). At the same time, some bacteria cells consist of the flagellum, capsule, pilus, etc. The small size and the rigid cell wall enable bacteria to uptake only low-molecular-weight compounds via cell membrane by facilitated diffusion and active transport. The small size also results in a higher metabolic rate and faster generation time (15–30 minutes) than multicellular organisms.

The shape is one of the important characteristics of bacteria, and it evolved as a result of the adaption to environmental factors, and it is one of the major factors in contributing to their versatile behavior. Apart from basic shapes including; rod, spherical, comma, or helicoidal shapes, bacteria also possess other shapes such as stars, mustaches, serpentines, and branches (Young 2006). Like all living organisms, the metabolism of bacteria is composed of catabolism and anabolism. Catabolic reactions, also referred to as energy-yielding reactions, involve the breakdown of molecules into smaller components. On the other hand, anabolic reactions build larger molecules from smaller molecules that usually require energy. Similar to other complex eukaryotic cells, adenosine triphosphate (ATP) is used by bacterial cells as the primary energy source. Specific enzymes catalyze those metabolic reactions.

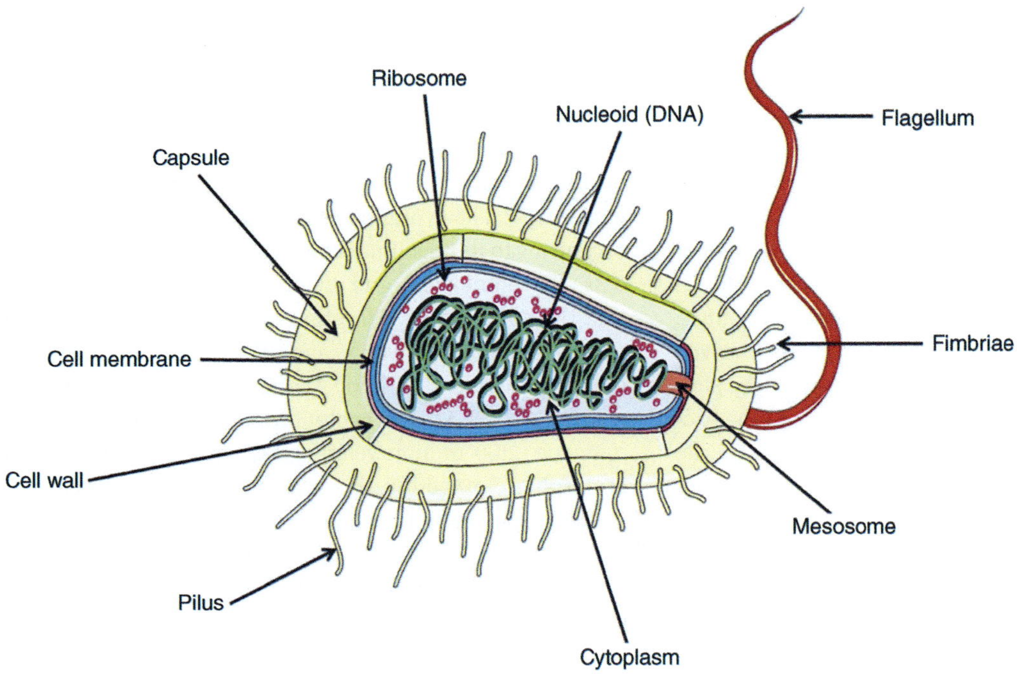

Figure 2.1 Structure of a bacterial cell. (*Source:* Jimenez-Jimenez et al. 2022 / MDPI / Public domain).

2.4 Natural Biogeochemistry Cycles (C, N, P, and S)

Material cycling among different spheres is important for both terrestrial and oceanic ecosystems. The most significant chemical elements: namely carbon (C), nitrogen (N), phosphorous (P), oxygen (O), and sulfur (S) prevail in dynamic fluxes between biotic and abiotic environments. These fluxes occur with the environment cyclically due to biological, physical, and chemical spontaneous reactions as well as due to interventions. Collectively, all these functions are considered biogeochemical cycles. These cycles are interconnected and primarily powered by solar energy. There can be exceptions due to localization. Oxidation and reduction reactions are mainly responsible for the transformations occurring within the cycles. These cycles are interconnected to one another and operative in both oxidative and reductive states.

2.4.1 Carbon Cycle

The carbon cycle elaborates the circulation of carbon in varied spheres namely the biosphere, geosphere, atmosphere, and hydrosphere. Photosynthesis, respiration, and microbial decay aids in the cycling of C. Atmospheric carbon dioxide is a result of respiration, decomposition, combustion, and mineralization. Cyclic processes of exchange of carbon mass are of particular importance for the global biosphere, both in terrestrial and oceanic ecosystems. Photosynthesis is the main process of fixing atmospheric CO_2. It is transferred to the chemical energy of the organisms and flows through the trophic levels. The carbon cycle consists of exogenic and endogenic processes. Carbon in its oxidized state can be observed in the dissolved form in aquatic systems and the atmosphere. C is present as carbonate minerals in soils and sediments while C is in its reduced form in biomass, sedimentary organic matter. A considerable portion of C is present in minerals as carbonates. The biogeochemical processes cause the formation of coal, liquid petroleum, and natural gas from the carbon present in minerals (Pulselli 2015).

Fixed atmospheric CO_2 is gradually released back to the environment either as CO_2 to air or to the sea as organic matter. The processes involved are respiration and decomposition. The agent of photosynthesis in oceans is phytoplankton. Decomposers on the oceanic floor use the organic sediments and produce CO_2, which is partially absorbed in the depths of the sea and partially released back into the atmosphere. Carbon can be observed in two major forms in ocean environments as a component of organic matter and as a component of exchangeable inorganic anion HCO_3^-, CO_3^{2-}, and CO_2 (Hannah 2021).

Anthropogenic activities: namely fossil fuel combustion, deforestation, livestock breeding, solid waste, and wastewater management activities directly affect the accumulation of CO_2 (Figure 2.2).

2.4.2 Nitrogen Cycle

Being the fourth most abundant element on the earth, N plays an important role in an organism's life cycle. The

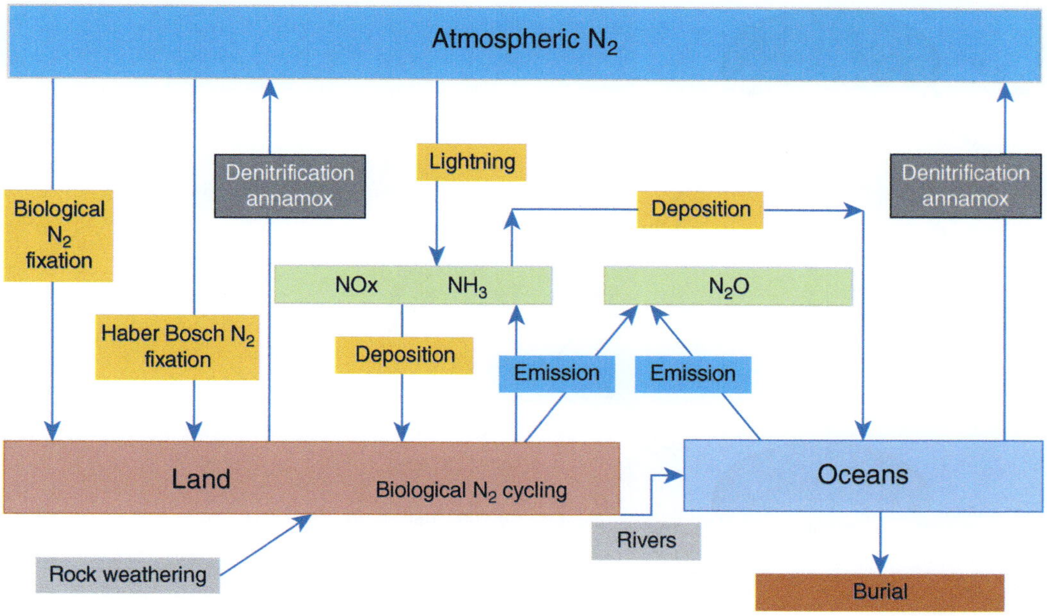

Figure 2.2 Schematic diagram of the carbon cycle.

amounts and types of N available for organisms are controlled by the N cycle. Different speciation can be observed as ammonium (NH_4^+) and nitrite (NO_2^-) may act as a reductant of oxygen, while nitrite (NO_2^-) and nitrate (NO_3^-) act as an oxidant for organic carbon participating in redox reactions. The oceanic environment accounts for the highest level of N content, while a significant amount of freshwater N is found in groundwaters (Hansen et al. 2012). Weathering, erosion, microbial denitrification, and leaching can cause loss of N, while the bioavailable N has increased due to the utilization of N-rich fertilizer, rise in legume cultivation, etc.

N input to the cycling process is by N_2 fixation contributed by lightening, anthropogenic activities transformation of N in the biosphere is significantly influenced by the microbial activity. Due to the high dependency on microbes, the N_2 cycle is also denoted as the microbial N cycle. Addition or removal of N is occurred by the processes namely nitrogen fixation, denitrification, and anammox. Spatial distribution and internal cycling of N species can occur via assimilation, mineralization, nitrification, and dissimilatory nitrate reduction. The additional microbial pathways termed anaerobic ammonium oxidation (anammox) and dissimilatory nitrate reduction to ammonium (DNAR) contribute to the N_2 cycle in soils. N_2 fixation occurs by the diazotrophs. These prokaryotes reduce N_2 to ammonia.

Nitrogen assimilation can be described as the process of producing organic nitrogen from external nitrogen species. Ammonia diffusion in a cell is followed by the addition of ammonia into amino acids. Nitrate assimilation can be stated as a function that is more complicated in terms of metabolic pathways because it has to be reduced to ammonia prior to assimilating its N into the biomolecules (Leigh and Dodsworth 2007). Mineralization is the process by which biomass-derived N is converted to an inorganic form of ammonia. Heterotrophs carry out mineralization mainly on oceanic floors, while in terrestrial environments the process is carried out mainly by fungi and earthworms. Nitrification is the process by which the fixed ammonium is oxidized to nitrates. Nitrosomonas (ammonia-oxidizing bacteria) and Achaea oxidize ammonia to nitrite. Nitrobacter (nitrite oxidizing bacteria oxidizes nitrite to nitrate in aerobic conditions. Complete ammonia oxidation (comammox) bacteria are capable of oxidizing ammonia to nitrate (Van Kessel et al. 2015). Dissimilatory nitrate reduction to ammonium (DNRA) is a process by which the nitrate is first reduced to nitrite and then to ammonium. This process recycles the fixed nitrogen within an ecosystem. In denitrification, the nitrates are converted back to N_2. Facultatively aerobic, heterotrophic bacteria are mostly found in denitrifying organisms. These bacteria who respire nitrate in the nonoxygenated environments consume organic matter and yield ammonium and N_2 as the final products (Zhang et al. 2020). The industrial Haber Bosch process is used for the conversion of N_2 and H_2 to produce NH_3 under elevated temperature and pressure conditions. The process accounts mostly for the similar quantity of fixed nitrogen to the biosphere considering the natural processes (e.g. 100 MT of nitrogen fertilizer each year) (Lehnert et al. 2021) (Figure 2.3).

2.4.3 Phosphorous Cycle

Phosphorous is a vital element of all forms of life. It is a major component in genetic materials (RNA, DNA),

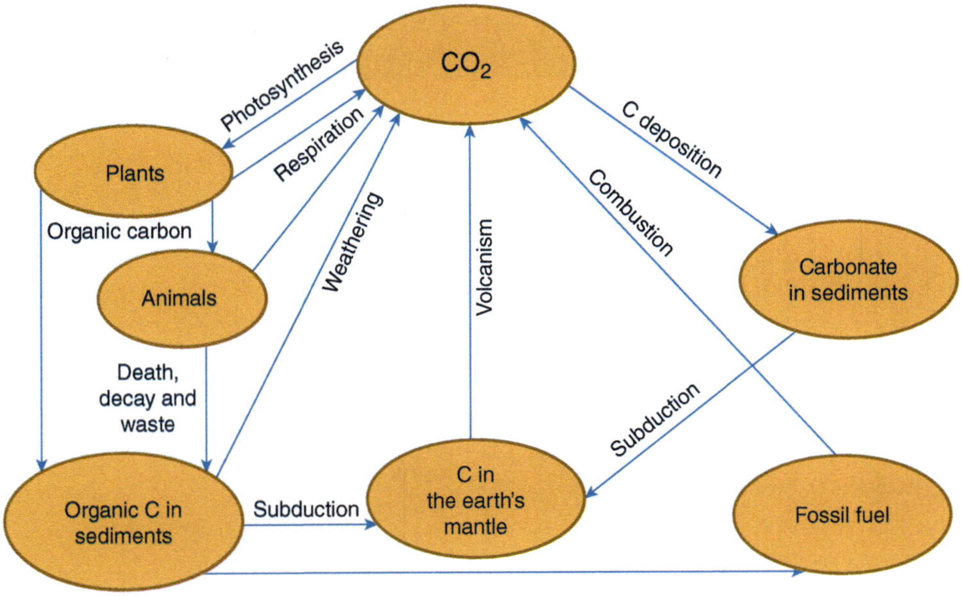

Figure 2.3 Schematic diagram of the global N_2 cycle.

phospholipids, bones, energy transfer molecules, etc. The global phosphorus cycle consists of microbial biomass-mediated mineralization, immobilization reactions, and the solubilization of the inorganic phosphorus-containing minerals (Gentry et al. 2021). Key areas can be identified in the phosphorus cycle as physical erosion of rocks, tectonic uplift, and exposure of P-bearing rocks to the forces of weathering, chemical weathering of rocks, sediments, riverine transport of P, and sedimentation of P.

P in soils, sediments are not directly up taken by the organisms but the dissolved form of orthophosphate. The P cycle aids in the process. Ocean sediments are often known as the largest reservoir for biogeochemical cycles. The sediments are built up as calcium phosphates (Ruttenberg 2001). Particulate and dissolved forms can be observed in riverine P. Anthropogenic activities namely deforestation often leads to increased suspended matter in riverine particulates. Rivers play a major role in transporting P to the oceans. $Al(OH)_3$ and clays are notable carriers of P (Karl and Björkman 2015). The particulate P is mostly held within the lattices, and they are not participating in the biogenic cycles. The terrestrial P cycle is majorly influenced by anthropogenic activities namely agriculture, deforestation, sewage, usage of detergents, urban and industrial waste disposal, etc. P stocks in plant matter are directly released to ash when burning occurs, and these are rapidly dissolved and leached to aquatic sources (Bharathi and Cycle 2008). P can be seen in combined forms in the environment due to its high productivity. Soluble phosphates are formed by microbial activity, and these are passed through the food webs. Ultimately, the P is returned to the environment upon death and decay. In terrestrial environments, P is available in soil, sediments, and biomass. Weathering of the bedrock is the source for plant uptake and returning to the soil by decay. Increased P contents in aquatic systems often lead to eutrophication, and the natural population declines due to hypoxic and anoxic conditions (Figure 2.4).

2.4.4 Sulfur Cycle

Being the 14th most abundant element on earth, sulfur plays an important role in geochemical cycling. Sulfur is available in different forms namely organic sulfur, sulfide, elemental S, thiosulfate, tetrathionate, sulfur dioxide, sulfite, sulfur trioxide, and sulfate. Hydrothermal reaction, weathering sulfide oxidation/sulfate reduction, causes the production of elemental sulfur. Oceanic compositions and the earth's redox balance are governed by sulfur transformation. A diverse nature can be observed regarding the varied oxidation states and the mode of transformation (bacteriologically or chemically). Eight significant oxidation/reduction reactions can be observed in the cycling process of S among the most reduced form H_2S (−2) most oxidized form SO_4^{2-} (+6) (Bharathi and Cycle 2008). Sources of atmospheric S can be stated as H_2S and SO_2 from active volcanoes, swaps like anaerobic systems, sulfate salts, and decaying, etc. S cycle is dispensed between different pools namely atmosphere, terrestrial systems, and aquatic systems. Organic matter degradation in coastal lake sediments is highly influenced by microbial-mediated functions. Sediments get enriched with sulfur due to the hydrolysis of sulfate esters, overlying water, and oxidation processes. SO_2 in the atmosphere contributes to the movement of S by dissolving into acid rains. Due to the ubiquitous distribution of S in marine

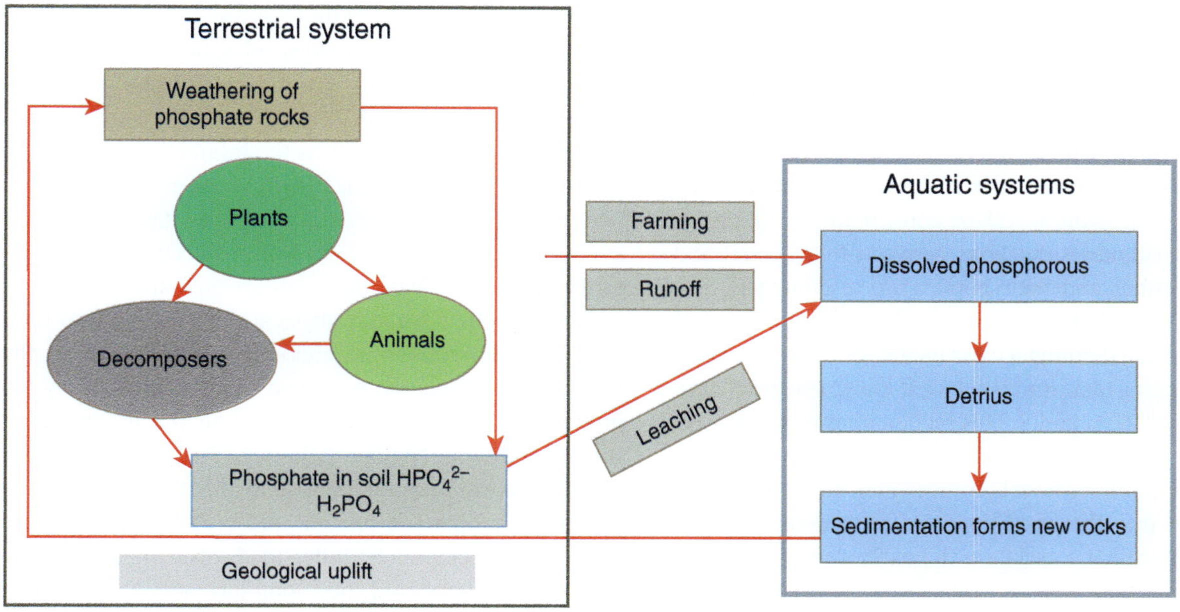

Figure 2.4 Global cycle of P.

environments, S cycling is prominent in such areas. Sulfate is reduced to H_2S by sulfate-reducing bacteria in anaerobic conditions. Eutrophic conditions favor sulfate reduction as of the presence of organic matter and sulfates. Reduction of S and reoxidation aid in S deposition gaseous sulfur compounds namely H_2S, dimethyl sulfoxide (DMSO), dimethyl sulfide (DMS), carbonyl sulfide (OCS), carbon disulfide (CS2), and dimethyl disulfide (DMDS) have an important role in S cycle. Nearly 75% of S entering the atmosphere and approximately 90% of biogenic S emissions from marine systems are contributed by DMS (Vincent et al. 2021). Anthropogenic activities namely mining and burning of fossil fuels alter the natural balance of the S cycle. Human combustion activities account for the significant portion of S present in the troposphere.

2.5 Artificial Biogeochemistry Cycles

In biogeochemistry, we essentially consider the geosphere as a chemical system, therefore, any links that appear between its elements and compounds with the biosphere are vital. In addition to the natural biogeochemistry cycles, there are many important human-induced biogeochemistry cycles in nature. The influence of anthropogenic activities on these novel chemicals is significant. Therefore, they are termed "artificial biogeochemistry cycles" and studied as a specialized branch in biogeochemistry.

Humans as the most evolved species on earth have invented and altered many naturally equilibrated processes as we know them now. Not only we have spent time utilizing natural resources for our basic needs but we have also invested in developing new materials for our specific needs. It may not be an exaggeration to say these were done at unsustainable rates over the last few decades. Hence, these approaches have allowed significant alteration and/or formation of biogeochemical dynamics of elements and chemical compounds that are not naturally linked with the biosphere. Trace elements such as Fe, Li, Se, Zn, Ni, Cu, Cd, Hg, Mo, and Pb, radioactive elements such as U and Rn, synthetic organic compounds, nanoparticles, and nanocomposites, and even nanoplastics and microplastics can be considered as novel entities that form their complex biogeochemistry cycles (Conley 2002; EPA 2021).

Surely the anthropogenic activities have increased their mobility through and between the geosphere and biosphere making them more bioavailable than they naturally are. Water naturally immobilizes ions and compounds; however, higher levels of anthropogenic releases of these entities can result in increased transport. This allows the transfer of materials between abiotic and biotic components of the environment. Air currents can also act as a medium for these entities, especially in terms of long-distance transport (De Santiago et al. 2014). In general, these transport mechanisms complicate the biogeochemistry cycles of entities with anthropogenic origins making the available to interact with the biosphere more often than naturally be. Ocean also acts as a major reservoir in many artificial biogeochemical cycles (Conley 2002). In some cases, entities interact to form complex transport phenomena making biogeochemical cycles alter. For example, plastic debris is known to aid hydrophobic compounds and microbial communities, altering the Hg biogeochemical cycle in freshwater and saltwater ecosystems (Bowman et al. 2021).

Due to the improperly managed industrial discharges, agricultural runoff, and vehicular emissions, many novel understudied entities are entering the biosphere at an unprecedented rate and may cause risks and chronic effects that are still unknown and unclear. Therefore, this is indeed a novel perspective in the field of biogeochemistry as many researchers around the world are contributing to filling the gap in understanding the links between these novel and emerging entities and their importance to biotic life. The complex nature of these entities and processes related to them have made defining a clear picture of biogeochemical cycles a daunting task, unlike the well-known biogeochemical cycles of C, N, P, and S. This section aims to summarize updated knowledge on such artificial biogeochemical cycles.

To start with, these novel entities enter the biosphere or get released into the environment due to many reasons. Their complexity and variation themselves make it harder to make subtle conclusions, however, some general trends can be highlighted. Deep earth mining operations such as oil and coal exploration can cause the release of elements such as Se, Hg, As, Si, and S at unprecedented concentrations into the ground and surface water bodies (Hedar and Budiyono 2018; Haddaway et al. 2019).

Similarly, radioactive elements are released during uranium mining in the subsequent processes. Not only does radioactive waste contaminate topsoil and water bodies, but it can also enter the atmosphere as a result of radioactive decay. This allows the bioavailability of radioactive gas radon not only to the miners but also to nearby communities as air currents can carry these to such populated areas (Debabrata et al. 2018). Furthermore, improperly disposed of electronic devices and e-waste serves as another route where toxic heavy metals can enter soil and water bodies. Particulate matter in the air also works as a carrier for heavy metals allowing their long-range transport.

Nanoplastics and microplastics, on the other hand, can be resulted from incomplete degradation and become bioaccumulative. The fate of nanoplastics and microplastics in the ocean and oceanic biosphere is a well-studied area. Microplastics can end up in the ocean, while some sink to the bottom and some beaches along the shorelines. Marine fish, seabirds, and aquatic mammals are shown to contain microplastics in their tissues in the stomach, intestines, and liver while larger plastics can appear in gastrointestinal tracts (Marte et al. 2021). Beached macroplastics due to oceanic currents can degrade to form nano and microplastics, and recent studies have shown an increasing trend in concentrations indicating their impact on the biosphere is becoming further significant. Atmospheric deposition of microplastics is also possible not only in populated areas but also in remote locations (Wright et al. 2020). However, studies relating to the biogeochemistry of nano and microplastics are limited and often challenged by geographical limitations. Current study findings are more localized and specific to regions and animal species, but their impact on the general environment is not yet quantified or understood clearly (Smith et al. 2018).

Fate and transport as well as the interaction of nanomaterials and nanocomposites in the real environment are poorly understood in general. However, there are many attempts to characterize novel nanomaterials at the laboratory level at least. Nanomaterials are well known for their remarkable chemical and physical properties due to their nanoscale dimensions. Therefore, their transport in the natural environment is governed and complicated by the interaction between nanoparticles and the properties of the environment. Nanoparticles are often known to form interactions with natural organic matter, hence considered in many nanoparticle-related studies. These interactions are yet to be completely understood through careful sampling and analysis, and their long-term effect on the biosphere is yet to be fully unrevealed (Batley et al. 2013).

However, quantification and modeling of the transportation and transformation of these novel entities in their biogeochemical cycles require ample specific data. Both intrinsic and extrinsic chemical and physical properties of these entities and transport mediums are important factors to be considered. For example, biogeochemical interactions between nanoparticles and their medium are well understood when properties such as nanoparticle concentration, speciation and complexation of nanoparticles, stability and degradability of nanoparticles, medium pH, and the effect of natural organic matter and other biological matters are considered in the account. The basic idea is, with these factors affecting, nanoparticles in the same transport media with varying conditions could behave differently therefore their effect on biogeochemical cycles can also be different.

2.6 Soil Biogeochemistry

2.6.1 Introduction to Soil Biogeochemistry

As signifies, the term soil biogeochemistry implies the scientific study of the complex and inseparable actions and interactions between soil biotic and abiotic constituents. Soil biogeochemistry overviews the interactions of climate, geology, and biology in the formation and evolution of soil and soil biota. The chemical and mineralogical nature of the soil is understandable and measurable; however, the involvement of soil biota, including micro and macro animals, in these chemical reactions leads to complicated relationships to be understood. However, the influence of human beings has made alterations in soil biogeochemistry, which needs the knowledge and understanding to implement the recovery actions and prevent the risk of man-made problems. This chapter focuses on understanding the basic concept of soil formation and evolution, the involvement of natural factors in soil formation and evolution, and human involvement in soil bio-geochemistry.

2.6.2 Soil Formation and Evolution

According to soil scientists or pedologists, the soil is a more or less loose or friable material resulting from weathering of the solid Earth's crust. Silicates dominate the soil on Earth's surface, yet the soil is a carbon-rich material involved in different biological processors. Various factors determine the composition of the soil. Soil scientists highlight the state factors of soil formation or the soil evolution; parent material, topography, climate, soil organisms, and time (van Breemen and Buurman 1998; Bernasconi 2008).

Soil parent material or solid matter-making soils can consist of consolidated rocks. Soil parent material may also consider river glacial tills, alluvium, volcanic ash, lake or marine sediments, and organic matter accumulating in swamps or bogs. The soil parent materials directly influence the texture, structure, stratification, composition, and chemical nature of the soil. Different processors of weathering influence the breakdown of the soil parent material; (i) physical weathering (temperature changes, gravity, and pressure differences), (ii) chemical weathering, and (iii) biological weathering (Hillel 2008).

Temperature and precipitation directly influence the speed of soil formation by involving chemical and biological reactions. Faster reactions occur in higher temperatures, whereas lower reaction rates occur at colder environments. Furthermore, physical weathering also boosts with temperature fluctuations. Precipitation is a vital component in rock weathering; water movement through rocks or water leaching, water evaporation, and evapotranspiration influence the chemical and biological reactions happening in the soil. The formation of different soil types, such as alluvial, colluvial, and aeolian soil, is evolved with water movement through gravity. Furthermore, the topography, slope length factors, and gravity determine the water and sediment moment in the soil particles leading to physiological alterations in the soils (Baldi 2021).

The chemical nature of Earth's crust is unique. It has an exceptional combination of chemical elements (O, Si, Al, Fe, Ca, Na, Mg, and K) inherent to Earth that is not expected in our solar system. Earth soils are dominated by Si and O. The chemical reactions over time changed the mineral composition of the topsoil; the weatherable primary minerals converted to the most resistant minerals such as quartz. Furthermore, quartz is becoming increasingly high in warm and humid conditions. This is boosted by the acidic rainwater passing through the soil's parent materials, which could dissolve the silicate minerals (Arce et al. 2019).

The formation of soil organic matter has a history of more than 400 years. The emergence of the land plants was found in the Silurian as considered the secondary process of soil formation and the starting of the organic matter cycle. The plant matter is dominated by C, H, O, N, and P, where the other elements are trace levels. Plants passively or actively take up nutrients dissolved in the soil water other than the C fixed during photosynthesis. The plants that grow in the land consist of roots below the ground level, directly incorporated into the soil when plants die. The above ground parts may enter the soil with biological degradations and leaching with water. Thus, soil property changes with the growth of biotic constituents. The presence of organic matter (plant or plant residues) could create an acidic microenvironment leading to parent material or rock weathering. Other than the plants, soil micro-organisms such as bacteria or fungi, burrowing insects, animals, and humans influence soil evolution. The organic matter may convert to leachable nutrients, peat, or humus by the involvement of soil biota or other animals. Figure 2.5 exemplifies the biotic and abiotic interactions happening in the soil in the presence of organic matter. The process of rock weathering illustrates that the formation or the evolution of soil is not an individual process, but is intimately linked to physical, chemical, and biological processes (Meng et al. 2022).

As soil develops over time, soil layers can be seen as soil profiles where the physical, chemical, and biological characteristics vary in different horizons. Especially, horizons are defined by the color and texture of the soil. A typical soil profile consists of an organic matter layer as the top horizon (O) followed by an organic matter mixed mineral matter mixed horizon (A), which almost obliterates the mineral matter. Most of the biological activities occur in the A horizon, thus mentioned as biomantle. Soil micro and macro-organisms are concentrated here, associated with plants roots. A well-developed mineral matter layer called the B horizon underlines a loose, poorly developed mineral matter layer (C). B horizon or the subsoil dominated with iron or the clay minerals, the presence of organic matter or the mineral matter is also possible resulting in leaching from A horizon. The biological activities are less in the B horizon, although the plant roots penetrate up to this layer. The bedrock or the parent material is the bottom layer of a soil profile. Other than these master horizons, some transitional horizons may carry combined properties of two adjacent master horizons (van Breemen and Buurman 1998; Hillel 2008).

2.6.3 Soil and Ecosystem Balancing

Soil is a mixture of minerals, soil water, and soil air which the composition depends on numerous factors. The porous structure of soil provides habitats to millions of species of microorganisms, macro-organisms, and plants. The health,

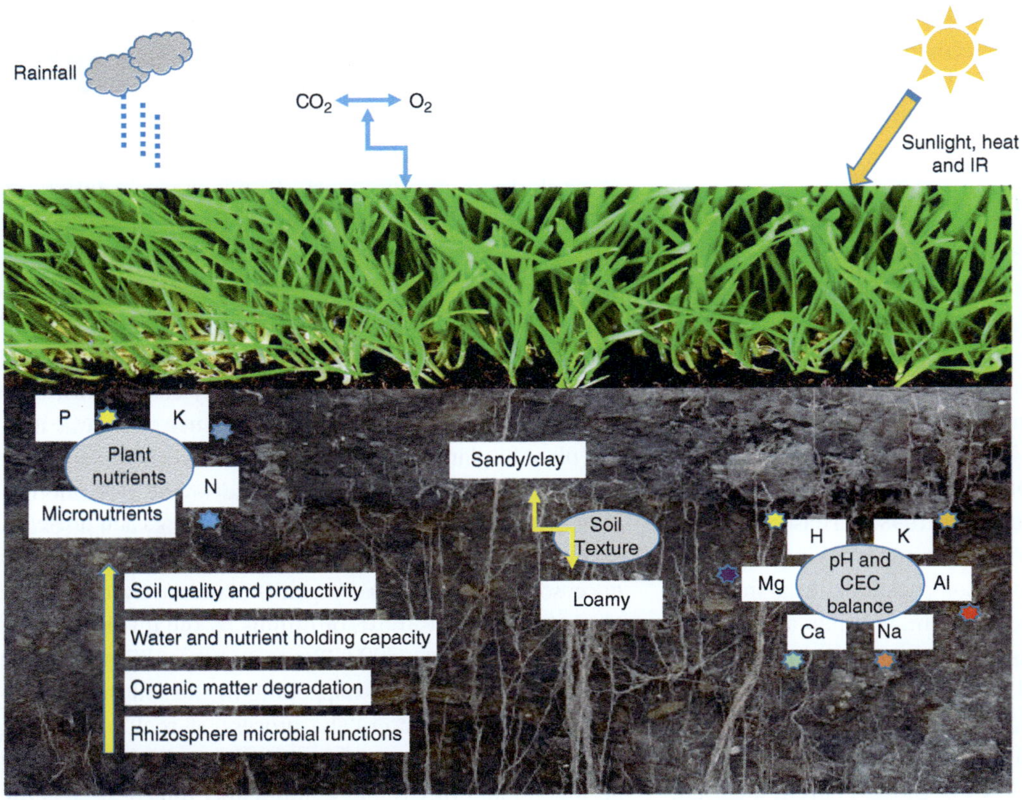

Figure 2.5 The biotic and abiotic interactions occur in the soil in the presence of organic matter.

productivity, and soil ecological balance highly depend on the interactions and wide range of activities with one another among these biotic and abiotic constituents. Soil provides minerals, water, heat, and a hospitable place for plants to grow, thus leading to enriching the soil with soil organic matter that will enter the nutrient cycles and provide habitats and substrates to soil organisms. The edible forms of plants are the primary source of food for the animals living on earth thus directly involved in ecosystem balancing. Furthermore, the spongiest nature of the soil and the soil water infiltration act as a major role in water and another nutrient cycling on earth. Moreover, the soil is a better filter for the water; as water passes through the soil, the contaminants might be trapped by the soil particles which the process is known as natural attenuation (Meng et al. 2022).

Soils health and climate change have always been closely related even though often neglected components of the climate system. Soil is the second largest carbon store on the Earth, being stored in plants that could become stable for thousands of years if undisturbed. Thus, healthy soils can thus mitigate climate change. However, climate change puts soil under pressure. The plant growth rates, mineralization, and organic matter decomposition increase with increasing temperature. The functions of soil microorganisms can boost up with the increasing atmospheric carbon dioxide, potentially releasing even more carbon dioxide by reducing soil carbon storage. Furthermore, the rising temperature directly correlates with the soil moisture content. Higher evaporation rates and unexpected rainfall events risk the whole soil ecosystem by changing the physical, chemical, and biological processes and their interactions (Yin et al. 2020).

2.7 Impact of Natural and Anthropogenic Activities on Biogeochemical Processes

The evolution of biogeochemical processes facing the stresses of natural and anthropogenic impacts drives modern ecosystems. Processes such as landslides, volcanic eruptions, intensive agriculture, industrial activities, ocean acidification, and climate change have a direct impact on biogeochemical cycles.

2.7.1 Landslide

Landslide is the movement of rock, debris, earth, or soil mass on a downhill slope, and landslides occur when gravity and other types of shear stresses on a slope exceed the shear strength of the materials that make up the slope (Meng 2022). Landslides can significantly affect flora and fauna, changes in surface and groundwater systems, affect species, destroy agriculture, and create artificial dams (Ewa et al. 2017). Landslides can erode large amounts of particulate carbon (POC) in a short period of time and can play an important role in the carbon cycle on a regional scale

(Ewa et al. 2017; Yang et al. 2016). Storms and earthquakes trigger landslides, linking climate and geological changes to biochemical cycles (Oli 2012). Their deposits protect the biosphere-mobilized POC from oxidation and act as short-term CO_2 zinc. However, these deposits are prone to further erosion, which can give the rivers POC (Oli 2012; Ewa et al. 2017). These POCs can then be transported to sedimentary basins such as lakes and oceans that act as long-term carbon–zinc (Oli 2012; Ewa et al. 2017). Therefore, it is important to understand the fate of POC post-landslides in order to predict the consequences of landslides on the carbon cycle. In contrast to the rock-derived sources of POC that act to separate CO_2 potential from the modern atmosphere, it is also important to limit the carbon source to biosphere POC only (Oli 2012; Ewa et al. 2017)

2.7.2 Volcanic Eruptions

Volcanic eruptions occur on the Earth's surface but begin at the Earth's core, where molten rock, hot rock fragments, and hot gases explode into the atmosphere. Magma that has erupted is called lava (Decker 2022). Volcanic eruptions provide sharp, transient, and relatively well-understood forces that cause short-term global surface cooling and, as a result, lower marginal warming (Robock and Mao 1995; Rothenberg et al. 2012). The process of volcanic eruptions causes a short-term global decrease in both surface temperature and precipitation, as well as a dynamic response in a climate system characterized by a biochemical response (Robock and Mao 1995; Rothenberg et al. 2012). This reaction is caused by the injection of large sulfate atmospheres into the periphery by explosions, which increase the amount of solar radiation that is reflected into space at the top of the atmosphere (Schneider et al. 2009). Although extreme spherical aerosols tend to mix relatively quickly in the atmosphere, tropical eruptions tend to have greater climatic effects than at higher latitudes, due to the longer duration of their aerosols as well as the apparent stronger dynamic response (Schneider et al. 2009).

2.7.3 Industrial Activities

Major impacts on the ecosystems are posed by anthropogenic activities irrespective of technological advancement. Various industrial activities have a direct toll on natural and artificial biogeochemistry cycles. Industries use various raw materials and convert them to value-added products using utilities such as fuels, electricity, thermal power, and water. Other than the commercial outputs, industries produce wastes and by-products can alter the balance of the biogeochemistry cycles. These imbalances are normalized in the long run, however, can cause localized and spontaneous changes, as well as significant changes in gaseous compositions, atmospheric temperature, rainfall, and habitat loss (Siudek and Frankowski 2017; Zang et al. 2021). The changes in biogeochemical cycles can be seen not only near the industrial zones but also in faraway regions due to the transport of elements via groundwater and surface water, ocean, air currents, rain, and snow, making its effect a complicated affair. Studies have shown chemicals such as heavy metals being blown away to remote forest areas and aquatic ecosystems due to the wind over industrialized or urbanized regions (Türtscher et al. 2017; Xing et al. 2017; Pandey et al. 2010). Such transportation results in the expansion of the span of biogeochemistry cycles of anthropogenic elements drawing previously unaffected areas and ecosystems to an unprecedented threat. In fact, this is a potential research prospective in biogeochemistry. The effect of industrialization on biogeochemistry cycles can be further evaluated by looking at many long-term processes such as intensive agriculture, greenhouse gas emissions, ocean acidification, and climate change. These processes and their results are interconnected and can cause integrated changes to various biogeochemistry cycles.

2.7.4 Intensive Agriculture

The most profound revolution on the Earth, farming, started about 10 000 years ago, interrupting the planetary evolution built-in resilience for several billion years. Then the boundary conditions were changed suddenly, affecting the direct interactions between the biosphere and geosphere (Sun et al. 2021; Amundson 2022). The impact caused by agriculture on soil systems and elemental balance is well noticed as the nutrient mass balance was in a steady state in the pre-agricultural Holocene. Fundamentally, the agricultural practices disrupt the natural steady state of the nutrient cycles leading long-term changes and depletions of C, N, P elements and eventually reducing the soil thickness (Amundson 2022). Thousands of years of agricultural practices removed nearly or entirely the biomass from the lands. High-yielding varieties associated with agrochemicals, fertilizers, heavy machinery, controlled water application methods, and new farming methods were introduced with the third agricultural revolution called the green revolution, exploring a new era (Sun et al. 2021). Today, we are using nearly all of the usable lands for agriculture. The nutrient cycles were artificially modified, borrowing additional natural resources to improve land productivity. However, the impact caused by intensive agriculture on the biogeochemistry of Earth cannot be negligible (Borrelli et al. 2017). Roughly half of the nitrogen-based fertilizers wash out from the farm fields altering the micro soil environment and expanding the "dead zones" affected by eutrophication in surface water resources (Vaiphei and Kurakalva 2021). The expanded use of pesticides imposes a negative impact on biodiversity. The hydrological cycles were changed, quality and quantity of soil and water were altered with the operation of irrigation schemes over the agricultural fields. The underground aquifers were depleted due to overdrafting, and the quality was lessened with the leaching down of the agrochemicals into deeper aquifers. Soil degradation was often observed; salting, waterlogging, compaction, contamination with agrochemicals, a decline in soil structure quality, loss of fertility, soil acidification, salinization, and erosion were accelerated with poorly controlled agricultural practices (Borrelli et al. 2017; Usero et al. 2021). The agriculture livestock production sector releases a large quantity of greenhouse gas while clearing up the forests to fulfill the requirements of pasturing and grazing (Wang et al. 2021). Agriculture has threatened ecosystems devastating the soil biochemical cycles, soil biodiversity, and its balances showing the importance of implementing sustainable approaches in agriculture.

2.7.5 Greenhouse Gas Emissions

The direct impact of the alarming levels of greenhouse gas emissions impacts the long-term shift of temperature, weather patterns, and biodiversity. Not only the carbon cycle that affects the climate change but also the other cycles namely the N cycle, water cycle, etc. Effects of increasing temperature of the biosphere can directly affect the precipitation variations, weather instability, changes in hydrological cycles, desertification, etc. The acidification of the upper oceanic waters shows the cascade effect of climate warming (Bijma et al. 2009; Chen et al. 2013). An increase in sea levels, stronger hurricanes, unusual patterns of flooding, reduction of freshwater, and ultimately an ecosystem change will not be reversible if the current trends of warming patterns continue (Caney 2015).

2.7.6 Ocean Acidification

Increasing levels of carbon dioxide in the atmosphere are coveted more carbon into oceans due to the human and natural processes, known as "ocean acidification." When carbon dioxide (CO_2) dissolves in seawater, it can react with water molecules to form carbonic acid (H_2CO_3), a weak acid that severely degrades the marine carbonate system, thereby lowering surface pH and reducing carbonate saturation, as acidification can be further explained (Rafferty 2022). Ocean acidification is not a climatic process, but ocean chemistry is directly affected by the fact that seawater absorbs CO_2 from the atmosphere (Pilcher et al. 2019; Rafferty 2022). It does, however, present a separate set of environmental changes affecting marine ecosystems, fisheries, and other marine resources (Rassmann et al. 2018; Pilcher et al. 2019; Rafferty 2022). It also directly affects the carbon and nitrogen cycles, reducing the ability of much marine life to form shells and altering organisms and ecosystems. Ocean acidification currently has a global pH increase of 0.1, equivalent to a ~30% increase in acidity since the beginning of the industrial revolution (Rassmann et al. 2018; Pilcher et al. 2019; Rafferty 2022).

However, the global ocean absorbs 41% of all human CO_2 emissions as a result of the burning of fossil fuels and industrial processes, thus reducing some of the warming associated with climate change (Khatiwala et al. 2009). It has been found that naturally cold, poorly protected waters and ocean mixing processes pose a high risk to ocean acidification, resulting in ocean acidification patterns that increase ocean circulation patterns and CO_2 gas solubility (Khatiwala et al. 2009). As a result, seawater is already undergoing an unexpected change in the oceanic regions of the high latitudes, where the concentration of carbonate ions is naturally low. In addition, ocean acidification is caused by changes in the pH of marine biological processes or by changes in the concentration of liquid CO_2, bicarbonate ions or carbonate ions, photosynthesis, respiration rate, growth rate, calcification, etc. Almost every major biological activity that responds to change can be different. It determines the main nutrient cycles in the ocean, including geological, chemical, physical, and biological processes, as well as the oceans' ability to absorb nutrients that support all marine life, as well as their ability to separate CO_2 from the atmosphere and ocean acidification can alter both chemical and biological processes affecting nutrient and carbon cycles, such as alteration of nitrogen fixation by certain marine algae (Khatiwala et al. 2009; Rassmann et al. 2018; Rafferty 2022).

2.8 Conclusion and Future Perspectives

Biogeochemistry is an interdisciplinary field that integrates numerous disciplines, including biology, chemistry, and geology. It investigates how the biological processes mediate the chemical and geological dynamics of the Earth's atmosphere, hydrosphere, and lithosphere. Biogeochemical cycles are pathways that involve the distribution of elements and chemicals through ecosystems and their physical environments. Biogeochemistry plays an imperative role in underrating global challenges related to water, soil, and air quality, climate change, effects of emissions on human and ecological health, and impact on regulatory activities. Knowledge of biogeochemistry enhances the ability to address global issues, including climate events, food production and security, energy crisis, environmental pollution, health issues, and improving quality of life. Therefore, the fundamentals of biogeochemistry, natural and artificial biogeochemical cycles, soil biogeochemistry, and the impact of natural and anthropogenic activities on biogeochemical processes have been extensively discussed in this chapter.

Understanding the interaction between the human population and ecosystems is essential with the increasing world population. Biogeochemistry has been an integral component of sustainability, addressing many global challenges and exploring new dimensions of sustainability via a diverse range of research. Key future trends in biogeochemistry include stable isotope biogeochemistry, human biogeochemistry, biogeochemical mapping and modeling, coastal biogeochemistry, and urban biogeochemistry.

Stable isotope biogeochemistry is a rapidly developing area of modern biogeochemistry and is employed in many fields such as agriculture, anthropology, astronomy, oceanography, biology, medicine, and geochemistry. Stable isotope biogeochemistry uses the stable isotopic compositions of elements of biological interest, including C, N, S, Fe, Mg, Cu, and Zn, that makeup minerals and organic matter to study biogeochemical cycles. Stable isotope biogeochemistry helps understand the biotic and abiotic cycling of elements and chemicals in anoxic environments, including freshwater, deep-sea sediments, and soil-aquifer systems. It also assists in reconstructing the ecosystem dynamics by evaluating the activity and regulation of microorganisms that govern the cycling of elements in biogeochemical cycles. In addition, human biogeochemistry has emerged as a novel branch of modern biogeochemistry dealing with the quantitative assessment of relationships between the migration of chemical species in food webs and human health.

Biogeochemical mapping and modeling have recently received significant attention as they provide abstract and simplified representations of biogeochemical processes. Biogeochemical maps and models are used to identify and simulate the interaction of abiotic and biotic variables via time and across space to determine rates of spatial and temporal biogeochemical fluxes. These maps and models provide vital information about climate, land use, water quality, vegetation, and environmental impacts. Biogeochemical mapping and modeling are also useful in the ecological management of the regional and global landscapes. Coastal biogeochemistry is another area of modern biogeochemistry where it provides a better understanding of the role of the coastal ocean in the global biogeochemical processes. Coastal biogeochemistry helps study coastal productivity, marine nitrogen fixation, coastal and open ocean exchange of carbon and minerals, hydrodynamics and biogeochemical interactions, mesoscale dynamics, interannual climate variability, benthic fluxes of elements, and other substances. Urban biogeochemistry is a novel approach in modern biogeochemistry that examines the effects of urban demographics, engineering, household, and industrial actions on biogeochemical cycles.

References

Amundson, R. (2022). Soil biogeochemistry and the global agricultural footprint. *Soil Security* 6: 100022. https://doi.org/10.1016/j.soisec.2021.100022.

Arce, M.I., Mendoza-Lera, C., Almagro, M. et al. (2019). A conceptual framework for understanding the biogeochemistry of dry riverbeds through the lens of soil science. *Earth-Science Reviews* 188 (2018): 441–453. https://doi.org/10.1016/j.earscirev.2018.12.001.

Artemievaa, I.M., Mooney, W.D., Perchuc, E., and Thybo, H. (2002). Processes of lithosphere evolution: new evidence on the structure of the continental crust and uppermost mantle. *Tectonophysics* 358 (2002): 1–15.

Baldi, E. (2021). Soil–plant interaction: effects on plant growth and soil biodiversity. *Agronomy* 11 (12): 10–12. https://doi.org/10.3390/agronomy11122378.

Bashkin, V.N. (2003). *Modern Biogeochemistry*, 1e. Dordrecht: Springer.

Batley, G.E., Kirby, J.K., and McLaughlin, M.J. (2013). Fate and risks of nanomaterials in aquatic and terrestrial environments. *Accounts of Chemical Research* 46 (3): 854–862. https://doi.org/10.1021/ar2003368.

Baymann, F., Brugna, M., Mühlenhoff, U., and Nitschke, W. (2001). Daddy, where did (PS)I come from? *Biochimica et Biophysica Acta* 1507: 291–310.

Bernasconi, S.M. (2008). Weathering, soil formation and initial ecosystem evolution on a glacier forefield: a case study from the Damma Glacier, Switzerland. *Mineralogical Magazine* 72 (1): 19–22. https://doi.org/10.1180/minmag.2008.072.1.19.

Bharathi, L. and Cycle, P.S. (2008). *Encyclopedia of Ecology*. Amsterdam: Elsevier.

Bianchi, T.S. (2021). The evolution of biogeochemistry: revisited. *Biogeochemistry* 154 (2): 41–181.

Bijma, J., Barange, M., Brander, L. et al. (2009). Impacts of ocean acidification. *ESF Science Policy Briefing* 37: 1–12.

Borrelli, P. et al. (2017). An assessment of the global impact of 21st century land use change on soil erosion. *Nature Communications* 8 (1): https://doi.org/10.1038/s41467-017-02142-7.

Bowman, K.L., Lamborg, C.H., Agather, A.M., and Hammerschmidt, C.R. (2021). The role of plastic debris in the biogeochemical cycle of mercury in Lake Erie and San Francisco Bay. *Marine Pollution Bulletin* 171: 112768. https://doi.org/10.1016/j.marpolbul.2021.112768.

van Breemen, N. and Buurman, P. (1998). *Soil Formation*. Dordrecht: Kluwer Academic Publishers https://doi.org/10.1007/978-0-585-31788-5.

Butterfield, N.J. (2015). Early evolution of the Eukaryota. *Frontiers in Palaeontology* 58 (1): 5–17. https://doi.org/10.1111/pala.12139.

Caney, S. (2015). Climate Change. In: *The Routledge Handbook of Global Ethics* (ed. D. Moellendorf and H. Widdows), 372–386. Routledge.

Carlson, R.W., Shirey, S.B., Pearson, D.G., and Boyd, F.R. (1994). *The Mantle Beneath Continents*, vol. 93, 109–117. Washington, DC: Carnegie Inst.

Chen, H., Zhu, Q., Peng, C. et al. (2013). The impacts of climate change and human activities on biogeochemical cycles on the Qinghai-Tibetan Plateau. *Global Change Biology* 19: 2940–2955.

Clayton, R.W. and Comer, R.P. (1983). A tomographic analysis of mantle heterogeneities from body wave travel times. *Eos Transactions Amercian Geophysical Union* 64: 776.

Conley, D.J. (2002). Terrestrial ecosystems and the global biogeochemical silica cycle. *Global Biogeochemical Cycles* 16 (4): 1121. https://doi.org/10.1029/2002GB001894.

Cutter, G.A. (2005). Biogeochemistry: now and into the future. In: *Geobiology: Objectives, Concepts, Perspectives* (ed. N. Noffke), 191–198. Amsterdam: Elsevier.

De Santiago, A., Longo, A.F., Ingall, E.D. et al. (2014). Characterization of selenium in ambient aerosols and primary emission sources. *Environmental Science and Technology* 48 (16): 8988–8994. https://doi.org/10.1021/es500379y.

Debabrata, P., Lala, S.B., Niharbala, D., and Sridhara, A. (2018). Geochemical cycle of radon and its bioremediation opportunity from water environment: a review. *Recent Patents on Biotechnology* 12 (4): https://doi.org/10.2174/2211550107666180501123246.

Decker, R.W. (2022). *Volcano*. Encyclopedia Britannica.

Dewey, J. F. (2007) The secular evolution of plate tectonics and the continental crust: an outline. In: Hatcher, R. D., Jr., Carlson, M. P., McBride, J. H., Martínez Catalán, Editors. 4-D Framework of Continental Crust. Geological Society of America Memoir 200: 1–7.

Dobretsov, N.L., Kolchanov, N.A., Rozanov, A., and Zavarzin, G. (2008). *Biosphere Origin and Evolution*. Springer Science ISBN-13:978-0-387-68655-4.

Enrich-Prast, A., Bastviken, D., Crill, P. et al. (2014). Chemosynthesis. In: *Reference Modules in Earth Systems and Environmental Sciences*, 211–225. Elsevier.

EPA (2022) Radioactive waste from uranium mining and milling. https://www.epa.gov/radtown/radioactive-waste-uranium-mining-and-milling#:~:text=Mining%3A%20When%20uranium%20is%20near,then%20removed%20through%20underground%20tunnels (accessed 2021).

Ewa, B., Lasota, J., Piaszczyk, W. et al. (2017). *Journal of Soils Sediments*, 1775–1774. Springer.

Gaucher, E.A., Thomson, J.M., Burgan, M., and Benner, S.A. (2003). Inferring the palaeoenvironment of ancient bacteria on the basis of resurrected proteins. *Nature* 425: 285–288. https://doi.org/10.1038/nature01977.

Gentry, T.J., Fuhrmann, J.J., and Zuberer, D.A. (2021). *Principles and Applications of Soil Microbiology*. Elsevier.

Gorham, E. (1991). Biogeochemistry: its origins and development. *Biogeochemistry* 13 (3): 199–239.

Gradstein, F.M., Ogg, J.G., and Smith, A.G.A. (2004). *Geologic Time scale*. Cambridge University Press.

Haddaway, N.R., Cooke, S.J., Lesser, P. et al. (2019). Evidence of the impacts of metal mining and the effectiveness of mining mitigation measures on social-ecological systems in Arctic and boreal regions: a systematic map protocol. *Environmental Evidence* 8 (9): https://doi.org/10.1186/s13750-019-0152-8.

Hamilton, W.B. (2007). Earth's first two billion years the era of internally mobile crust. *Geological Society of America Memoir* 200: 233–296.

Hannah, L. (2021). *Climate Change Biology*. Academic Press.

Hansen, B., Dalgaard, T., Thorling, L. et al. (2012). Regional analysis of groundwater nitrate concentrations and trends in Denmark in regard to agricultural influence. *Biogeosciences* 9: 3277–3286.

Haq, B.U. and Schutter, S.R. (2008). A chronology of Paleozoic sea-level changes. *Science* 322 (5898): 64–68. 13.

Hawkesworth, C.J., Cawood, P.A., and Dhuime, B. (2020). The evolution of the continental crust and the onset of plate tectonics. *Frontiers in Earth Science* 8 (2020): https://doi.org/10.3389/feart.2020.00326.

Hedar, Y. and Budiyono (2018). Pollution impact and alternative treatment for produced water. In: *E3S Web of Conferences*, vol. 31, 03004. https://doi.org/10.1051/e3sconf/20183103004.

Hellweger, F.L. (2008). Biogeochemical models. In: *Encyclopedia of Ecology* (ed. S.E. Jørgensen and B.D. Fath), 386–396. Oxford: Academic Press.

Hillel, D. (2008). *Soil Formation, Soil in the Environment*, vol. 15–26. https://doi.org/10.1016/b978-0-12-348536-6.50008-3.

Huber, C. and Wächtershäuser, G. (1997). Activated acetic acid by carbon fixation on (Fe, Ni)S under primordial conditions. *Science* 276: 245–247. https://doi.org/10.1126/science.276.5310.245.

Jimenez-Jimenez, C., Moreno, V.M., and Vallet-Regi, M. (2022). Bacteria-assisted transport of nanomaterials to improve drug delivery in cancer therapy. *Nanomaterials (Basel)* 12 (2): 288. https://doi.org/10.3390/nano12020288.

Karl, D.M. and Björkman, K.M. (2015). Dynamics of dissolved organic phosphorus. In: *Biogeochemistry of Marine Dissolved Organic Matter* (ed. D.A. Hansell and C.A. Carlson), 233–334. Elsevier.

Khatiwala, S., Primeau, F., and Hall, T. (2009). Reconstruction of the history of anthropogenic CO_2 concentrations in the ocean. *Nature* 462: 346–349.

Kutschera, U. and Niklas, K.J. (2004). The modern theory of biological evolution: an expanded synthesis. *The Science of Nature* 91: 255–276. https://doi.org/10.1007/s00114-004-0515-y.

Lehnert, N., Musselman, B.W., and Seefeldt, L.C. (2021). Grand challenges in the nitrogen cycle. *Chemical Society Reviews* 50: 3640–3646.

Leigh, J.A. and Dodsworth, J.A. (2007). Nitrogen regulation in bacteria and archaea. *Annual Review of Microbiology* 61: 349–377.

Marte, H., Alessio, G., Jürgen, S. et al. (2021). Documentation of microplastics in tissues of wild coastal animals. *Frontiers in Environmental Science* 9: https://doi.org/10.3389/fenvs.2021.575058.

Meng, X. (2022). *Landslide*. Encyclopedia Britannica.

Meng, X. et al. (2022). The current and future role of biota in soil-landscape evolution models. *Earth-Science Reviews* 226: 103945. https://doi.org/10.1016/j.earscirev.2022.103945.

Michael, B. and Rushmer, T. (2005). *Evolution and Differentiation of the Continental Crust*. Cambridge University Press.

Nolet, G., Grand, S.P., and Kennett, B.L.N. (1994). Seismic heterogeneity in the upper mantle. *Journal of Geophysical Research* 99: 23753–23766.

Ohmoto, H., Kakegawa, T., and Lowe, D.R. (1993). 3.4-billion-year-old biogenic pyrites from Barberton, South Africa: sulfur isotope evidence. *Science* 262: 555–557.

Oliver, K. (2012). Landslides: types, mechanisms and modeling. In: *Clague and Douglas Stead* (ed. J. John), 10–23. Cambridge University Press.

Pandey, P.J., Shubhashish, K., and Pandey, R. (2010). Heavy metal contamination of Ganga river at Varanasi in relation to atmospheric deposition. *Tropical Ecology* 51: 365–373.

Pearson, D.G., Carlson, R.W., Shirey, S.B. et al. (1995). Stabilization of Archaean lithospheric mantle; a Re-Os isotope study of peridotite xenoliths from the Kaapvaal Craton. *Earth and Planetary Science Letters* 134: 341–357.

Phinney, R.A. (1964). Structure of the earth's crust from spectral behavior of long-period body waves. *Journal of Geophysical Research* 69 (14): https://doi.org/10.1029/JZ069I014P02997.

Pilcher, D.J., Naiman, D.M., Cross, J.N. et al. (2019). Modeled effect of coastal biogeochemical processes, climate variability, and ocean acidification on aragonite saturation state in the Bering sea. *Journal of Front. Marine Science* 05: 508.

Pratt, J.M. (1993). Nature's design and use of catalysts based on Co and the macrocyclic corrin ligand: 4 × 109 years of coordination chemistry. *Pure and Applied Chemistry* 65: 1513–1520.

Pulselli, F.M. (2015). *Global Warming Potential and the Net Carbon Balance*. Environmental Science.

Rafferty, J.P. (2022). *Ocean Acidification*. Encyclopedia Britannica.

Rassmann, J., Lansard, B., Gazeau, F. et al. (2018). Impact of ocean acidification on the biogeochemistry and meiofaunal assemblage of carbonate-rich sediments: results from core incubations (Bay of Villefranche, NW Mediterranean Sea). *Marine Chemistry* 203: 102–119.

Righter, K., Drake, M.J., and Yaxley, G. (1997). Prediction of siderophile element metal-silicate partition coefficients to 20 GPa and 2,800°C: the effects of pressure, temperature, oxygen fugacity, and silicate and metallic melt compositions. *Physics of the Earth and Planetary Interiors* 100: 115–134. https://doi.org/10.1016/S0031-9201(96)03235-9.

Robock, A. and Mao, J. (1995). The volcanic signal in surface temperature observations. *Journal of Climate* 8: 1086–1103.

Rogers, J.J.W. and Santosh, M. (2003). Supercontinents in earth history. *Gondwana Research* 6 (3): 357–368.

Rothenberg, D., Mahowald, N., Lindsay, K. et al. (2012). Volcano impacts on climate and biogeochemistryin a coupled carbon–climate model. *Earth System Dynamics* 3: 121–136.

Russell, M.J. and Ponce, A. (2020). Six 'must-have' minerals for life's emergence: Olivine, Pyrrhotite, Bridgmanite, Serpentine, Fougerite and Mackinawite. *Life* 10: 291. https://doi.org/10.3390/life10110291.

Ruttenberg, K. (2001). Phosphorus cycle. *Encyclopedia of Ocean Sciences* 4: 2149–2162.

Schlesinger, W.H. (2004). Better living through biogeochemistry. *Ecology* 85 (9): 2402–2407.

Schneider, D.P., Ammann, C.M., Otto-Bliesner, B.L., and Kaufman, D.S. (2009). Climate response to large, high-latitude and low-latitude volcanic eruptions in the community climate system model. *Journal of Geophysical Research* 114: D15101. https://doi.org/10.1029/2008JD011222.

Shirey, S.B. and Richardson, S.H. (2011). Start of the wilson cycle at 3 Ga shown by diamonds from subcontinental mantle. *Science* 333: 434–436.

Siudek, P. and Frankowski, M. (2017). Atmospheric deposition of trace elements at urban and forest sites in central Poland – insight into seasonal variability and sources. *Atmospheric Research* 198: 123–131.

Smith, P.J. (1971). The climate system. In: *Understanding the Earth* (ed. I.G. Gass, P.J. Smith and R.C.L. Wilson), 407–586. Sussex: The Artemis Press, Horsham.

Smith, M., Love, D.C., Rochman, C.M. et al. (2018). Microplastics in seafood and the implications for human health. *Current Environmental Health Reports* 5: 375–386. https://doi.org/10.1007/s40572-018-0206-z.

Sorokhtin, O.G., Chilingar, G.V., and Sorokhtin, N.O. (2011). Origin of earth, its past and future, evolution of earth and its climate: birth, life and death of earth. *Developments in Earth and Environmental Sciences* 10: 61–111.

Stanley, S.M. (1989). *Earth and Life Through Time*, 2e. New York: W. H. Freeman.

Sun, Q., Qi, W., and Yu, X. (2021). Impacts of land use change on ecosystem services in the intensive agricultural area of North China based on Multi-scenario analysis. *Alexandria Engineering Journal* 60 (1): 1703–1716. https://doi.org/10.1016/j.aej.2020.11.020.

Svoboda, J. and Nabert, D. (1999). *Noosphere in Environmental Geology*, 438–440. Dordrecht: Springer.

Thomas, B. and Spicer, R.A. (1987). *The Evolution and Palaeobiology of Land Plants*. London: Croom Helm.

Torsvik, T.H., Domeier, M., and Cocks, L.R.M. (2021). Phanerozoic paleogeography and pangea. In: *Ancient Supercontinents and the Paleogeography of Earth*, 577–603. https://doi.org/10.1016/B978-0-12-818533-9.00003-5.

Trenberth, K. (2002). *Earth System Processes, Encyclopedia of Global Environmental Change*. Chichester: Wiley.

Türtscher, S. et al. (2017). Declining atmospheric deposition of heavy metals over the last three decades is reflected in soil and foliage of 97 beech (*Fagus sylvatica*) stands in the Vienna Woods. *Environmental Pollution* 230: 561–573. https://doi.org/10.1016/j.envpol.2017.06.080.

US DOE (2010). Critical material strategy. http://www.energy.gov (accessed 21 August 2011).

Usero, F.M. et al. (2021). Effects of soil microbial communities associated to different soil fertilization practices on tomato growth in intensive greenhouse agriculture. *Applied Soil Ecology* 162: https://doi.org/10.1016/j.apsoil.2021.103896.

Vaiphei, S.P. and Kurakalva, R.M. (2021). Hydrochemical characteristics and nitrate health risk assessment of groundwater through seasonal variations from an intensive agricultural region of upper Krishna River basin, Telangana, India. *Ecotoxicology and Environmental Safety* 213: 112073. https://doi.org/10.1016/j.ecoenv.2021.112073.

Van Kessel, M.A., Speth, D.R., Albertsen, M. et al. (2015). Complete nitrification by a single microorganism. *Nature* 528: 555–559.

Vasco, D.W., Johnson, L.R., and Marques, O. (1999). Global earth structure: inference and assessment. *Geophysics Journal* 137: 381–407.

Vernadskiĭ, V.I., Salisbury, F.B., and Yanshin, A. (2007). *Geochemistry and the Biosphere*. Synergetic Press.

Vincent, S.G.T., Jennerjahn, T.C., and Ramasamy, K. (2021). *Microbial Communities in Coastal Sediments: Structure and Functions*. Elsevier.

Wang, T. et al. (2021). Expanding grass-based agriculture on marginal land in the U.S. Great Plains: the role of management intensive grazing. *Land Use Policy* 104: 105155. https://doi.org/10.1016/j.landusepol.2020.105155.

White, R.S. (1988). The earth's crust and lithosphere. *Journal of Petrology* 1: 1–10.

Wilde, S.A., Valley, J.W., Peck, W.H., and Graham, C.M. (2001). Evidence from detrital zircons for the existence of continental crust and oceans on the Earth 4.4 Gyr ago. *Nature* 409: 175–178. https://doi.org/10.1038/35051550.

Woodhouse, J.H., Giardini, D., and Li, X.D. (1986). Evidence for inner core anisotropy from free oscillations. *Geophysical Research Letters* 13: 1549–1552.

Wright, S.L., Ulke, J., Font, A. et al. (2020). Atmospheric microplastic deposition in an urban environment and an evaluation of transport. *Environment International* 136: 105411. https://doi.org/10.1016/j.envint.2019.105411.

Xing, J. et al. (2017). Atmospheric wet deposition of dissolved trace elements to Jiaozhou Bay, North China: fluxes, sources and potential effects on aquatic environments. *Chemosphere* 174: 428–436. https://doi.org/10.1016/j.chemosphere.2017.02.004.

Yang, N., Zou, D., Yang, M., and Lin, Z. (2016). Variations in soil microbial biomass carbon and soil dissolved organic carbon in the re-vegetation of hilly slopes with purple soil. *PLoS One* 11 (12): e0166536.

Yasuo, K. and Kamitani, M. (2006). Rare earth minerals and resources in the world. *Journal of Alloys and Compounds* 408–412: 1339–1343.

Yin, S. et al. (2020). Carbon sequestration and emissions mitigation in paddy fields based on the DNDC model: a review. *Artificial Intelligence in Agriculture* 4: 140–149. https://doi.org/10.1016/j.aiia.2020.07.002.

Young, K.D. (2006). The selective value of bacterial shape. *Microbiology and Molecular Biology Reviews* 70: 660–700. https://doi.org/10.1128/MMBR.00001-06.

Young, G.M. (2013). Precambrian supercontinents, glaciations, atmospheric oxygenation, metazoan evolution and an impact that may have changed the second half of Earth history. *Geoscience Frontiers* 4 (3): 247–261.

Zang, F. et al. (2021). Atmospheric wet deposition of trace elements to forest ecosystem of the Qilian mountains, Northwest China. *Catena* 197: 104966. https://doi.org/10.1016/j.catena.2020.104966.

Zhang, X., Ward, B.B., and Sigman, D.M. (2020). Global nitrogen cycle: critical enzymes, organisms, and processes for nitrogen budgets and dynamics. *Chemical Reviews* 120: 5308–5351.

3

Geochemical Release and Environmental Interfaces

C. C. Kadigamuwa[1], Nuwan T. Perera[2], and B. S. S. Perera[1]

[1] Department of Chemistry, Faculty of Science, University of Kelaniya, Kelaniya, Sri Lanka
[2] Department of Chemistry and Physics, College of Arts and Sciences, Western Carolina University, Cullowhee, NC, USA

3.1 Introduction

Geochemistry is the study of principles of chemistry which explains the properties and chemical operations of major geological systems within and on the surface of the earth, both in the past and the present (Mather 2013).

The geochemical release is the process of chemical weathering of minerals from their parent materials. These minerals originate from naturally occurring solids, such as sediments, and rocks, which interact with the environment through their surfaces. Depending on different environmental conditions that affect these surfaces, they could release mineral and gases to the environment.

Weathering is a major process in releasing of minerals from natural solids. It is the alternation or breakdown of rocks and other naturally occurring solids in its original form at the earth surface into simpler forms, such as sediments, clay, gases, soils, and substances that dissolve in water (Saunders and Fookes 1970).

Weathering at the earth surface happens in two different pathways namely, physical weathering and chemical weathering as shown in Figure 3.1. Physical weathering is disintegration of rocks by mechanical processes such as rock fracturing, freezing, and thawing, or breakdown in transportation by rivers and glaciers (Saunders and Fookes 1970). Chemical weathering occurs through chemical reactions between minerals and external agents such as air and water (Chung et al. 2020).

Geochemical release is the end product of chemical weathering of many geological systems. It can be separated into two different types, namely mineral release and gas release.

Figure 3.1 Picture of both physical and chemical weathering. *Source:* U.S. Department of the Interior / Public Domain.

3.1.1 Mineral Release

As mentioned earlier, release of minerals from rocks happens due to both chemical and physical weathering. The weathering of rocks occurs due to various factors, such as (Patel 2017):

- Properties of the parent rock: the minerology and the structure affect the susceptibility of the weathering process.
- Climate: high temperatures and rainfalls increase the rate of weathering
- Soil: moisture, pH, organisms, and host vegetation in soil form an environment that promotes the weathering of rocks.
- Length of exposure: longer the exposure, higher the rate of weathering.
- Topography: lower the topographic elevations, more susceptible to the weathering.

Medical Geology: En route to One Health, First Edition. Edited by Majeti Narasimha Vara Prasad and Meththika Vithanage.

Minerals that weathered chemically from naturally occurring solids are due to many chemical reactions with air and water. These processes may include dissolution of minerals or sometimes combination with other atmospheric components in order to form new minerals, such as clay and oxides. Chemical weathering (Figure 3.2) occurs due to three main reactions, such as (Viers et al. 2007):

3.1.1.1 Dissolution

Carbonic acid forms from carbon dioxide (CO_2) that released from decaying organic matter and the atmosphere, dissolved in rainwater:

$$CO_2 + H_2O \rightarrow H_2CO_3 \rightarrow H^+ + (HCO_3)^-$$

This carbonic acid reacts with limestones, dissolving carbonates yielding an aqueous solution of bicarbonate and calcium ions:

$$\underset{\text{Limestone}}{CaCO_3} + \underset{\text{carbonic acid}}{H_2CO_3} \rightarrow \underset{\text{calcium ions}}{Ca^{2+}} + \underset{\text{bicarbonate ions}}{2(HCO_3)^-}$$

Most of the underground cavern systems are created by the dissolution of limestones.

3.1.1.2 Oxidation and Hydration

Oxidation produces iron oxide minerals (hematite and limonite) in well-aerated soils, usually in the presence of water.

High iron content substances, such as magnetite, pyrite, pyroxene, amphibole, and olivine, are more prone to the oxidation.

These minerals combine with oxygen and water to form hydrated iron oxides as follows:

$$\underset{\text{Iron-rich minerals}}{4Fe^{++}} + \underset{\text{Oxygen}}{3O_2} + \underset{\text{Water}}{6H_2O} \rightarrow \underset{\text{Limonite}}{2(Fe_2O_3.3H_2O)}$$

3.1.1.3 Hydrolysis

This reaction is responsible for the formation of clays, the most important mineral in soils. A typical hydrolytic reaction occurs when orthoclase feldspar reacts with slightly acidic water to form clay minerals, potassium ions, and silica in solution:

$$\underset{\text{Orthoclase feldspar}}{2KAlSi_3O_8} + \underset{\text{acid}}{2H^+} + \underset{\text{water}}{9H_2O} \rightarrow \underset{\text{clay mineral}}{Al_2Si_2O_5(OH)_4} + \underset{\text{potassium ion}}{2K^+} + \underset{\text{soluble silica}}{4H_4SiO_4}$$

The ions released from silicate minerals in the weathering process are sodium, potassium, calcium, iron, and magnesium ions. They are carried away by rain and river waters and become important soil nutrients. (Viers et al. 2007)

Figure 3.2 Chemical weathering. *Source:* Bonnie Moreland / Flickr / Public domain. Accessed from: https://www.geographyrealm.com/physical-chemical-of-weathering-of-rocks/

Physical weathering is a mechanical process that fragments minerals and rocks, and this may happen due to certain reasons. Natural zone of weakness is one of the main reasons. Some of the most common processes are bedding planes in sedimentary rocks, joints in massive igneous rocks, and exfoliation in metamorphic rocks. Through uplift and erosion, rocks rise slowly to the earth's surface, where they are free from the weight of overlying rock; thus, their fractures will open slightly. This allows chemical and physical weathering to widen the cracks. Living organisms in the rocks may also promote weathering as shown in Figure 3.3. For instance, tree root may occupy and amplify cracks in rocks (Macheyeki et al. 2020).

Hot and cold cycles can also occupy in weathering of minerals by thermal expansion and contraction depending on the region. Mineral crystallization of aqueous solutions also contributes to the fractures to expand.

The weathered minerals are then transported through erosion. Erosion is a natural process that transports weathered or unweathered materials. It loosens, removes, and

Figure 3.3 Root weathering. *Source:* National Park Service / Jim Peaco / Public Domain.

Figure 3.4 Erosion by wind. *Source:* Vastram/Adobe Stock, EdNurg/Adobe stock, maxos_dim / 55 images/Pixabay. Accessed from: https://www.geo.fu-berlin.de/en/v/geolearning/mountain_building/weathering/Erosion/index.html

transports materials like, sand, sediment, soil, mud, and other weathered minerals. Erosion uses mediums, such as wind (Figure 3.4), underground water as well as running water, waves, glaciers, and gravity (Moses et al. 2014).

3.1.2 Gas Release

When it comes to geochemical release of substances, it is very crucial to gain a knowledge about the gases that are released from geological systems, mainly from earth surfaces and also from the deep earth crust. It is a well-known fact that some minerals, such as quartz and pumice rocks, have cavities that contain gases like, CO_2, hydrogen, and nitrogen. Gases that release from natural disasters, mainly from volcanic eruptions, earthquakes, and anthropogenic breakage of rocks, also contribute to gas release. Emission of noble gases from geological resources also contributes widely to the geochemical releases.

There are three main hypotheses to explain the states in which the gases exist in rocks. The simplest of all is that the gases in cavities and pores have been entrapped in the rocks during the process of solidification of magma. It is supported with microscopic studies of minerals like quartz and topaz that observe numerous small gas bubbles (Chamberlin 1909).

The gases that present in many minerals are usually CO_2 and hydrogen found in hints enclosed in the cavities of the rock. Many gases are released to the environment with the heat reactions of nongaseous substances of rocks. Hydrogen liberates with temperatures higher than

500 °C. CO_2 is produced when carbonates are heated sufficiently in the presences of iron. CO is formed from the CO_2 when heated. Methane is also produced by decomposition of organic matter and the reaction between CO_2 and water. Likewise, many gases can be obtained with thermal reaction of nongaseous substances of rocks (Chamberlin 1909).

Shale gas (Figure 3.5) is another type of natural gas, which is trapped in the fine-grained sedimentary rocks which act as a reservoir as well as source. Shale is a combination of mud (clay and slit), clay, carbonates (dolomite and calcite), silica, and organic materials. The gas is normally stored in three distinctive ways: (i) free gas, gas within the rock cavity; (ii) adsorbed gas, gas that is absorbed onto the organic materials; and finally, (iii) dissolved gas, gas that is dissolved in the organic materials. These gases are obtained through horizontal drilling and hydraulic fracturing (Zendehboudi and Bahadori 2017).

Figure 3.5 A gas shale bedrock with the layered structure. *Source:* Zhang et al. 2020/John Wiley & Sons, Inc.

When considering geochemical releasing of gases, volcanic gases get an immense place. Relatively, volcanic gases are substances that releases from volcanos or volcanically active areas. Volatile substances are chemical elements that dissolved in magma that forms gases when placed toward low pressure and temperature.

Volcanic gases (Figure 3.6) mainly constitute about 90% of water vapors and 10–40% of CO_2. It also contains sulfur compounds such as sulfur dioxide and hydrogen sulfide (Aiuppa et al. 2005). It also consists of chemical elements like F, Cl, and B and is also a composition of H_2, CO, CH_4, S_2, N_2, NH_3, and O_2. Metals such as Pb, Zn, Cd, Hg, Cu, Bi, Na, and K also present at trace levels (Aiuppa et al. 2007).

The three main reservoirs that contribute to the formation of volcanic gases are as follows (Lin et al. 2010):

- The mantle
- The crust
- The atmosphere

Many volcanos in the continental regions are highly explosive, and the gases that release are more versatile and variable due to their composition. These volcanos contain higher amount of water and thick magma, which expand 1000 times more during an eruption. The water comes from two main sources. Many volcanos carry hydrated minerals beneath the earth crust which releases water under high pressure. Another source of water is meteoric water that derived from precipitation (snow and rain). Since this water contains other dissolved substances such as O_2, the volcanic gas composition is localized to its region.

Another fact worth knowing is that there are many places that release volcanic gases without a magma degassing (Figure 3.7). These types of gases are diffused through

Figure 3.6 Volcanic gases in Hawaii are rich in carbon dioxide and sulfur dioxide. Flank of Kilauea between Pu'u O'o and coast. *Source:* Tada Images/Adobe Stock. Accessed from: https://www.sandatlas.org/volcanic-gases/

Figure 3.7 Steam and volcanic gases escape from a crack in the roadway in the Leilani Estates neighborhood in the aftermath of eruptions from the Kilauea volcano on Hawaii's Big Island on 9 May 2018 in Pahoa, Hawaii. *Source:* U.S. Geological Survey/NASA Earth Observatory.

the ground or may be released through cracks of the earth surface. This is mainly due to the ground water heated by nearby volcanos. These gas vapors mainly constitute water vapor but may contain nonreactive species like CO_2 and He^3. Volcanic gases: Overview (sandatlas.org).

Another geochemical gas release to be noticed is the gases that are released due to breakages of rocks (Figure 3.8). Slight amount of helium and argon gases released from rocks and earth crust under stress, which can be used to forecast natural disasters such as earthquakes before they happen. Rocks contain slight amounts of geogenic gases like radon, helium and argon that form over many years when radioactive compounds decays. The composition of these gases can be considered as a function of mineralogical composition, rock matrix, and the depositional, thermal, erosional, and tectonic history of the formation. Hence, the release of these noble gases gives a precursory signal before the deformation or degradation of rocks (Byerlee 1978).

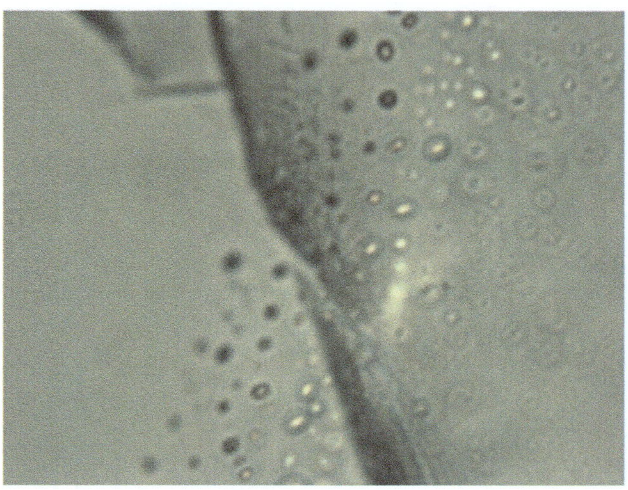

Figure 3.8 A magnified image of tiny gas-filled pores in the rock, with a stress-induced crack. *Source:* Gas released from rocks can predict impending breakage – GeoSpace – AGU Blogosphere. Accessed from: https://blogs.agu.org/geospace/2016/12/30/gas-released-from-rocks-can-predict-impending-breakage/

3.2 Environmental Interfaces

Environmental interfaces are broadly defined as any surface in equilibrium with its surrounding environment. There are various types of environmental interfaces, including geochemical, atmospheric aerosols, nanomaterials, and indoor surfaces. Thus, investigating the connection between geochemical release and these varied and complex environmental interfaces is necessary to understand both beneficial and adverse effects on living organisms.

3.2.1 Atmospheric Aerosol Interface

Atmospheric aerosols are tiny particles of solids and liquids. An atmospheric aerosol interface is an equilibrium of these particles with the surrounding environment. Aerosol particles play an important role in the atmosphere. By absorbing and scattering solar radiation, aerosols reduce the amount of energy reaching the earth's surface, while at the same time they enhance the greenhouse effect by absorbing and emitting long-wave radiation (Pöschl 2005).

The aerosol particles are meticulously related with chemical reactions of gas and liquid phases at the gas–aerosol particle interface. The gas–aerosol interface is a localized region where both inorganic and organic species are enriched with. The reactions take place according to chemical and physical properties of the species. For instance, hydrophobic organic species are favorably absorbing on to the gas–particle interface through the particle phase. There are three main reactions that the gas–aerosol interface experiences, such as gas phase diffusion, interfacial transport, and particle phase diffusion, to achieve the gas–particle equilibrium (Qian et al. 2019).

Aerosols are sites that certain chemical reactions occur in. One of the most abundant reactions is the destruction of the stratospheric ozone. During winter, aerosols form polar stratospheric clouds. These clouds act as a site for the chemical reactions to happen. Volcanic aerosols preferentially contribute for destroying the stratospheric ozone.

There are three types of aerosols that mainly effect the climate. Among them, volcanic aerosols impact enormously on the climate. After a massive volcanic eruption, a large, wide layer of volcanic gases form in the stratosphere. The gases are highly enriched with sulfur dioxide ions.

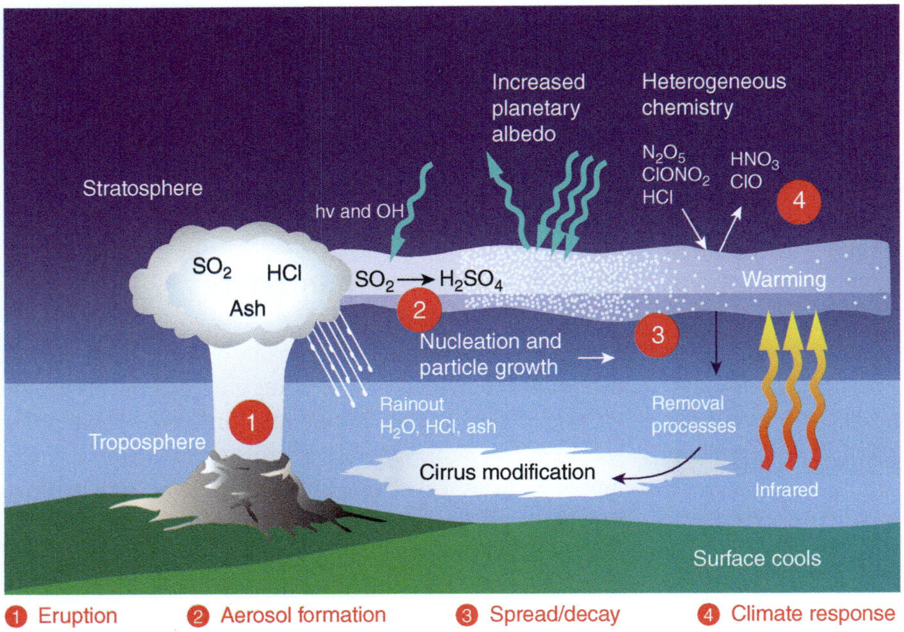

Figure 3.9 Illustration of production and effect of volcanic aerosols, including SO_2. *Source:* Windows to the universe staff (Lisa Gardiner) volcanoes influence climate – windows to the universe (windows2universe.org).

Sulfur dioxide is a major substance that contributes to the depletion of ozone. Sulfur dioxide gases formed are converted to sulfuric acid, which condenses to produce aerosol particles that remain in the atmosphere for a vast period of time (Figure 3.9). Volcanic gases also contain hydrochloric (HCl) acid which condenses with water vapor and rained out of volcanic cloud formation. Volcanic aerosols increase the echelons of chlorine (Cl_2) gas that reacts with N_2 in the atmosphere, which is a key donor toward the ozone annihilations (Kirk-Davidoff 2018).

Desert dust is an another method that effect the climate. Desert dust is a composition of many minerals which are suspended in the atmosphere (Figure 3.10). The particles are relatively large for atmospheric aerosols and would not retain in the atmosphere for a long time, if they were not launched by intense storms to higher altitudes. Since the dust is made up of minerals, the particles absorb sunlight and disperse it. By the absorbing sunlight, the dust particles tend to heat the atmosphere, which in turn inhibits the formation of storm clouds. By suppressing the storm clouds and their rain, it is believed that the veil of dust will continue the expansion of the desert (Mahowald et al. 2014).

The third type of aerosol comes from human activities. While, much of the man-made aerosols come in the form of smoke from the forest fire by burning coal and oil. The concentration of artificial sulfate aerosols in the atmosphere has increased since the beginning of the industrial revolution. At current production level, man-made sulfate particles are believed to be outnumbered the naturally produced particles. Sulfate aerosols do not absorb sunlight but reflect it, which decrease the amount of sunlight reaching the earth.

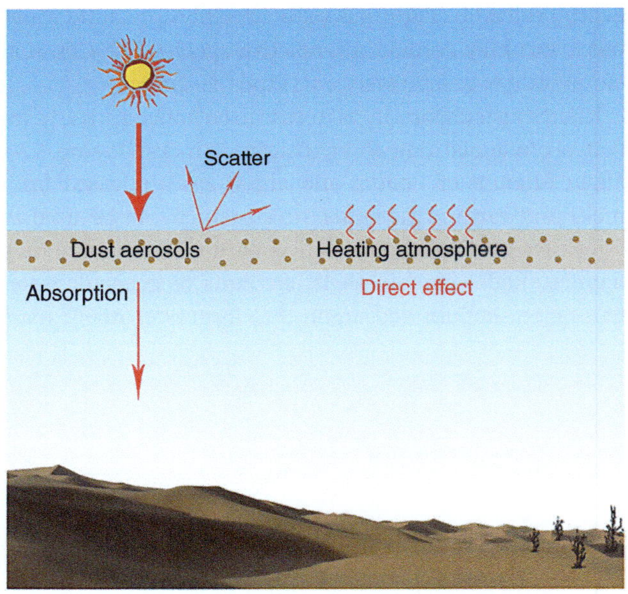

Figure 3.10 A schematic depiction of the direct radiative effect of dust aerosols. *Source:* Huang et al. (2014) / John Wiley & Sons.

3.2.2 Nanomaterial Interfaces

Nanoparticles (NPs) are of 1–100 nm in size and composed of organic matter, carbon, metal, metal oxide, or aerosol particles. In this size range, increase in surface area, change of electronic state, or deficient or development of new lattice structures result in the appearance of unique magnetic, biological, optical, and mechanical properties. The extremely high surface area specifies the importance of the surface of a NP and distinctiveness compared to its core which in various cases is linked to its final application. Therefore, surface functionalization is critical in displaying its distinct properties which in turn would weaken upon aggregation (Perera et al. 2021).

Nanomaterials can be divided into two basic sections, namely natural nanomaterials and synthetic nanomaterials (Ghaemi et al. 2017). Natural nanomaterials are the substances that occur naturally in the earth crust. These particles are generally formed by various geological processes, such as volcanic eruptions, lightning flash pyrolysis, and biochemical processes as well. Thousands of different nanomaterials are added to the atmosphere by each and every active volcano. These materials can spread through a large area with the help of the wind containing elements like Ni, Cd, Pb, S, Fe, Ag, Ca, Si, and Cu. Synthetic nanomaterials are substances that are made through anthropogenic activities, both purposely and inadvertently. Synthetic nanomaterials are classified as engineered nanomaterials and incident nanomaterials (Barhoum et al. 2022).

Nanomaterials from volcanic ash, lightning, forest fire, and even from outer space are naturally present in the atmosphere. For instance, multiwalled carbon nanotubes produced from soot are present in the air in the size range of 15–70 nm (Griffin et al. 2017). Nanomaterials of silicon dioxide are suspended in the atmosphere due to volcanic eruptions (Michel et al. 2013).

Synthetic nanomaterials, both engineered and incident nanomaterials, are dissolved in the atmosphere. They released to the environment by the sources of productions. Gases that release from the road vehicles are considered to be the prime supplier for these engineered nanomaterials to be suspended in the air. Incidental nanomaterials like titanium oxide, that were used in many industries, can be found in water sources contaminated with waste waters from different industries, such as paint industry (Sani-Kast et al. 2015). Titanium oxide nanomaterials mostly accumulated in sludge-treated soils than by landfill and sediments. Zinc oxide (ZnO) is another incidental nanomaterial that accumulates in earth by disposal of electronics, medicines, and cosmetics such as sunscreens containing ZnO (Durenkamp et al. 2016). Carbon NPs, both incidental and engineered nanomaterials, are released mainly due to the emission of carbon from vehicles and industrial wastes, which may be waste water or waste gas. When these nanomaterials suspended in the atmosphere, they will have a direct effect on all geological systems.

3.2.3 Effect of Geochemical Release on Environmental Interfaces

3.2.3.1 Adverse Effects of Geochemical Release on Environmental Interfaces

The release of minerals and gases from the geochemical systems sometimes effects negatively to the whole atmosphere from earth crust to upper atmosphere. This includes air pollution and land pollution, as well as adverse effects on the human health.

One of the main pollutants that cause adverse effects is the release of polluted gases from natural and anthropogenic causes. This may include volcanic gases, desert dusts, carbon emission from road vehicles and industries, smoking, forest fires, etc.

Volcanic gases and dust that are blown to the atmosphere during an eruption have an impact on environment and human health. Despite the distinct locations of the volcanos situated, their effect can be widely spread around the atmosphere. Volcanic gases mainly consist of sulfur dioxide, CO_2, hydrogen sulfides, and hydrogen halides which would affect negatively not only on atmosphere but also on human health, animals, properties, and vegetation. These compounds are potent greenhouse gases. SO_2 is a greenhouse gas having both warming and cooling effects, and it reacts with water to form sulfuric acid which is the key source of acid rain. Acid rain causes many environmental effects like deforestation, corroding buildings, contaminating water ways, and aquatic systems (Tricker and Tricker 1999). CO_2 is another gas that is harmless in small quantities, but when in high quantities, it causes global warming that effects badly on the living organisms. It also has an effect on vegetation and human health as well. At higher concentrations, it may cause coma and sometimes death as well. For instance, the incident in Nyos Lake event killed almost more than 1000 of people in Cameroon (Tchindjang 2018). Fumes that come from volcanos are acidic not only with sulfuric but also with HCL and HF as well (Aiuppa 2009). These also contribute to the acid rain significantly.

Another major adverse effect of geochemical release is an effect due to mineral release through the weathering process, due to human activities like mechanical weathering. Mechanical weathering, if done incorrectly, causes landslides, deformation of rocks, and also the formation of mountains as well.

Release of heavy metals and metalloids such as arsenic, chromium, and lead can cause severe health problems. For example, exposure to arsenic is known to cause health effects from skin tumors, cancers, and children's cognitive developments. Inorganic arsenic has been identified as a carcinogen that can induce cancers in human body (Hong et al. 2014). Additionally, arsenic is known to effect on major organs such as the dermal system, cardiovascular system, renal system, nervous system, hepatic system, endocrine system, and hematological system (Abdul et al. 2015). Additionally, the release of nonmetallic ions such as fluoride can also cause adverse health effects. Excess amount of fluoride in drinking water can cause dental fluorosis. It is reported that more than 200 million people have dental fluorosis that caused primarily by the excess amount of fluoride in drinking water (Jagadeshan et al. 2015).

3.2.4 Benefits of Geochemical Releases on Environment

Even though there are some of negative effects of geochemical releases, they also have positive effects toward the environment.

The most important benefit is the release of plant nutrients into the environment. These nutrients are including macronutrients such as K, Ca, Mg, P, and S, macronutrients such as B, Cl, Cu, Fe, Mn, Mo, Ni, and Zn, and some other nutrients that are essential for plant life under particular environmental conditions (Co, Na, and Si). These nutrients are present in rocks, and they are slowly released to the environment during weathering of these rocks. It is suggested that ground rock acts as a slowly release fertilizer that provides the nutrients top soil that can be utilized by plants.

As mentioned earlier, volcanic gases contain SO_2 substances that cause the greenhouse effect but also have a cooling effect to the atmosphere, when SO_2 reacts with water vapor to form sulfuric acid which has a property of cooling the stratosphere by back-scattering the sunlight radiation, making the atmosphere cool. (Burtraw et al. 2005)

Volcanic eruptions are also beneficial. Volcanic rocks and substances are weathered to form most fertile soils in the earth, which can be used in cultivations and forest civilizations. Some of the young volcanos can be used as harnesses to produce geothermal energy. Many volcanic materials are sources to the metallic minerals such as Ag, Au, Cu, Pb, and Zn.

The release of noble gases such as He and Ar from the breakage of rocks are also useful in finding when and where the natural disasters like earthquakes, take place, hence, preventing the loss due to natural disasters before they happen. (Bauer et al. 2016)

References

Abdul, K.S.M., Jayasinghe, S.S., Chandana, E.P. et al. (2015). Arsenic and human health effects: a review. *Environ. Toxicol. Pharmacol.* 40 (3): 828–846.

Aiuppa, A. (2009). Degassing of halogens from basaltic volcanism: insights from volcanic gas observations. *Chem. Geol.* 263 (1): 99–109. https://www.sciencedirect.com/science/article/pii/S0009254108003781 (accessed 12 April 2022).

Aiuppa, A., Inguaggiato, S., McGonigle, A.J.S. et al. (2005). H2S fluxes from Mt. Etna, Stromboli, and Vulcano (Italy) and implications for the sulfur budget at volcanoes. *Geochim. Cosmochim. Acta* 69 (7): 1861–1871. https://www.sciencedirect.com/science/article/pii/S0016703704007574 (accessed 12 April 2022).

Aiuppa, A., Moretti, R., Federico, C. et al. (2007). Forecasting Etna eruptions by real-time observation of volcanic gas composition. *Geology* 35 (12): 1115–1118. https://doi.org/10.1130/G24149A.1.

Barhoum, A., García-Betancourt, M.L., Jeevanandam, J. et al. (2022). Review on natural, incidental, bioinspired, and engineered nanomaterials: history, definitions, classifications, synthesis, properties, market, toxicities, risks, and regulations. *Nano* 12: 177.

Bauer, S.J., Gardner, W.P., and Lee, H. (2016). Release of radiogenic noble gases as a new signal of rock deformation. *Geophys. Lett.* 43: 10,688–10,694.

Burtraw, D., Evans, D.A., Krupnick, A. et al. (2005). Economics of pollution trading for SO_2 and NO_x. *Annu. Rev. Environ. Resour.* 30 (1): 253–289. https://doi.org/10.1146/annurev.energy.30.081804.121028.

Byerlee, J. (1978). A review of rock mechanics studies in the United States pertinent to earthquake prediction BT – rock friction and earthquake prediction. In: *Current Research in Geophysics (CCRG)* (ed. J.D. Byerlee and M. Wyss), 586–602. Basel: Birkhäuser Basel https://doi.org/10.1007/978-3-0348-7182-2_2.

Chamberlin, R. (1909). The gases in rocks. *J. Geol.* 17 (06): 534–568. https://www.jstor.org/stable/30058558 (accessed 11 April 2022).

Chung, H.Y., Jung, J., Lee, D.H. et al. (2020). Chemical weathering of granite in ice and its implication for weathering in polar regions. *Fortschr. Mineral.* 10: https://doi.org/10.3390/min10020185.

Durenkamp, M., Pawlett, M., Ritz, K. et al. (2016). Nanoparticles within WWTP sludges have minimal impact on leachate quality and soil microbial community structure and function. *Environ. Pollut.* 211: 399–405.

Ghaemi, F., Abdullah, L.C., and Tahir, P.M. (2017). Natural and synthetics nanomaterials: comparative study on their mechanical and thermal properties as nanofiller in polymer composite. *J. Phys. Conf. Ser.* [Internet] 914 (1): 12014. Available from: https://doi.org/10.1088/1742-6596/914/1/012014.

Griffin, S., Masood, M.I., Nasim, M.J. et al. (2017). Natural nanoparticles: a particular matter inspired by nature. *Antioxidants (Basel)* 7 (1): 3. https://pubmed.ncbi.nlm.nih.gov/29286304 (accessed 12 April 2022).

Hong, Y.S., Song, K.H., and Chung, J.Y. (2014). Health effects of chronic arsenic exposure. *J. Prev. Med. Public Health* 47 (5): 245–252. https://doi.org/10.3961/jpmph.14.035. Epub 2014 Sep 11. PMID: 25284195; PMCID: PMC4186552.

Huang, J., Wang, T., Wang, W. et al. (2014), Climate effects of dust aerosols over East Asian arid and semiarid regions, *J. Geophys. Res. Atmos* 119, 11,398–11,416. https://doi.org/10.1002/2014JD021796

Jagadeshan, G., Kalpana, L. & Elango, L. Major ion signatures for identification of geochemical reactions responsible for release of fluoride from geogenic sources to groundwater and associated risk in Vaniyar River basin, Dharmapuri district, Tamil Nadu, India *Environ. Earth Sci.* 74, 2439–2450 (2015). https://doi.org/10.1007/s12665-015-4250-9.

Kirk-Davidoff, D. (2018). Chapter 3.4 – the greenhouse effect, aerosols, and climate change. In: *Green Chemistry* (ed. B. Török and T. Dransfield), 211–234. Elsevier https://www.sciencedirect.com/science/article/pii/B9780128092705000091 (accessed 12 April 2022).

Lin, Y., Qiquan, R. and Yongle, H. (2010). A new method to describe the formation framework of the volcanic gas reservoir hierarchically. *International Oil and Gas Conference and Exhibition in China*. p. SPE-131932-MS. https://doi.org/10.2118/131932-MS.

Macheyeki, A.S., Li, X., Kafumu, D.P., and Yuan, F. (ed.) (2020). Chapter 1 – elements of exploration geochemistry. In: *Applied Geochemistry Advances in Mineral Exploration Techniques*, 1–43. Elsevier https://www.sciencedirect.com/science/article/pii/B9780128194959000013 (accessed 12 April 2022).

Mahowald, N., Albani, S., Kok, J.F. et al. (2014). The size distribution of desert dust aerosols and its impact on the Earth system. *Aeolian Res.* 15: 53–71. https://www.sciencedirect.com/science/article/pii/S1875963713000736 (accessed 11 May 2022).

Mather, T.A. (2013). Geochemistry. In: *Reference Module in Earth Systems and Environmental Sciences*. Elsevier https://www.sciencedirect.com/science/article/pii/B9780124095489059297 (accessed 11 April 2022).

Michel, K., Scheel, J., Karsten, S. et al. (2013). Risk assessment of amorphous silicon dioxide nanoparticles in a glass cleaner formulation. *Nanotoxicology* 7 (5): 974–988: https://doi.org/10.3109/17435390.2012.689881.

Moses, C., Robinson, D., and Barlow, J. (2014). Methods for measuring rock surface weathering and erosion: a critical review. *Earth-Sci. Rev.* 135: 141–161. https://www.sciencedirect.com/science/article/pii/S0012825214000749 (accessed 11 April 2022).

Patel, A. (2017). Classification of weathering in rocks and its engineering implications. *Civ. Eng. Res. J.* 2 (3): 20–23.

Perera, S., Wijesekara, D., Thiripuranathar, G., and Menaa, F. (2021 Jul). The use of nanoparticles to enhance performance in the textile industry – a concise review. *Curr. Nanosci.* 15: 17.

Pöschl, U. (2005). Atmospheric chemistry atmospheric aerosols: composition, transformation, climate and health effects. *Angew. Chem.* 44 (46): 7520–7540.

Qian, Y., Deng, G., Lapp, J., and Rao, Y. (2019). Interfaces of gas – aerosol particles: relative humidity and salt concentration effects. *J. Phys. Chem. A* 123: 6304–6312.

Sani-Kast, N., Scheringer, M., Slomberg, D. et al. (2015). Addressing the complexity of water chemistry in environmental fate modeling for engineered nanoparticles. *Sci. Total Environ.* 535: 150–159. https://www.sciencedirect.com/science/article/pii/S0048969714017276 (accessed 12 April 2022).

Saunders, M.K. and Fookes, P.G. (1970). A review of the relationship of rock weathering and climate and its significance to foundation engineering. *Eng. Geol.* 4 (4): 289–325. https://www.sciencedirect.com/science/article/pii/0013795270900219 (accessed 11 April 2022).

Tchindjang, M. (2018). Lake Nyos, a multirisk and vulnerability appraisal. *Geosciences* 8 (9): https://www.mdpi.com/2076-3263/8/9/312 (accessed 11 May 2022).

Tricker, R. and Tricker, S. (ed.) (1999). Pollutants and contaminants. In: *Environmental Requirements for Electromechanical and Electronic Equipment*, 158–194. Oxford: Newnes https://www.sciencedirect.com/science/article/pii/B9780750639026500103 (accessed 11 May 2022).

Viers, J., Oliva, P., Dandurand, J.-L. et al. (2007). 5.20 – chemical weathering rates, CO_2 consumption, and control parameters deduced from the chemical composition of rivers. In: *Treatise on Geochemistry* (ed. H.D. Holland and K.K. Turekian), 1–25. Oxford: Pergamon https://www.sciencedirect.com/science/article/pii/B9780080437514002492 (accessed 11 April 2022).

Zendehboudi, S. and Bahadori, A. (ed.) (2017). Chapter one – shale gas: introduction, basics, and definitions. In: *Shale Oil and Gas Handbook Theory, Technologies, and Challenges*, 1–26. Gulf Professional Publishing https://www.sciencedirect.com/science/article/pii/B9780128021002000010 (accessed 11 May 2022).

Zhang, B., Li, X., Zhao, Y. et al. (2020). A review of gas flow and its mathematical models in shale gas reservoirs. Wei W *Geofluids*. 2020;2020:8877777. https://doi.org/10.1155/2020/8877777

Section 2

Dust Storms and Health

4

Minerogenic Dust and Human Health

Majeti Narasimha Vara Prasad

School of Life Sciences, University of Hyderabad (an Institution of Eminence), Hyderabad, Telangana, India

> *"Tree bark pockets as pollution time capsules for historical monitoring of minerogenic dust"*
>
> – K. Satake 2001

Graphic abstract

a) Air pollution from abandoned mines with dense particulates of toxic minerals, Sao Domingo, Portugal
b) Phosphogypsum transport and dust, Russia
c) Bare hill tops after iron ore mining, China
d) Volcano eruption plume and volcanic activity, Iceland
e) Uranium mine waste subjected to wind erosion, Brazil
f) Open cast copper mine, USA
g) Open cast zinc mine, Thailand
h) Open-cast manganese mine, India
i) Open-cast magnesite mining, India
j) Zinc silicate rock from zinc mine, Thailand
k) Sedimentation pond of zinc mine smelter, Thailand
l) Artisanal and small-scale gold mining waste released mercury vapors and health risks, Nigeria
m) Lignite burn fly ash, Poland
n) Bauxite spent pot line (hazardous) containing cyanide and fluoride. Packed in polybags and stored in barracks, India

Medical Geology: En route to One Health, First Edition. Edited by Majeti Narasimha Vara Prasad and Meththika Vithanage.

4.1 Introduction

Titanium is the ninth most abundant element, approximately 0.7% of the earth crust. It is used worldwide in large quantities for various applications. The IARC includes TiO_2 as possibly carcinogenic to humans, suggesting that pathological effects correlate with particle size and shape. Natural TiO_2 particles are released during mining activity involving meta-basalt and shale lithologies. The airborne concentrations in meta-basalt were 4.21 and 23.94 ff/L in shale. In both cases, the concentration never exceeds the limits established by various organizations for workers health protection. Nevertheless, TiO_2-elongated particles, recognized as rutile, showed the dimensional characteristic of fibers, as reported by WHO. These fibers deserve particular attention because they can reach the alveolar space and trigger inflammation and chronic diseases. The results indicate that monitoring the TiO_2 in both working environments and Ti-rich geological formations associated with epidemiological studies may represent a useful tool to determine the exposure risk of workers and the general population (Derbyshire 2007; La Maestra et al. 2021). Silicosis, asbestosis, and talcosis due to volcanic and industrial gas emissions can be toxic to humanity. Minerogenic dust in the environment includes radionuclides that can be toxic to humanity (Figure 4.1) (Hu et al. 2010; Heldal et al. 2021; Pourcelot et al. 2021; Yan et al. 2021).

Figure 4.1 Dust and pollution in selected countries (a–c) India. (a) Dust and atmospheric pollution, (b) clean environment without atmospheric pollution, and (d–f) Portugal, dust and atmospheric pollution fur to forest fires, etc. *Source:* Anushree Fadnavis/Reuters, mrallen/Adobe Stock, Tebyan Cultural and Information Institute, inforse.org.

The Republic of Kazakhstan (northeastern part) consists of 470 nuclear test sites comprising an area of 430 km^2 conducted during 1949–1989; 26 of the sites were surface, 87 air, and 357 underground. An area of about 430 km^2 is polluted with radionuclides. The main contribution to radioactive pollution was made by aboveground nuclear tests conducted until 1963. After 1963, radioactive pollution was caused by radionuclide splinters of nuclear fuel fission, unreacted nuclear fuel, and radionuclide products of explosive devices activated by neutrons. Nuclear decay of the underground nuclear tests and the subsequent decay products eventually accumulated in these test sites. Dangerous radionuclides such as $^{239,240}Pu$, ^{90}Sr, and ^{137}Cs (^{137}Cs-has a half-life of 30 years and ^{90}Sr has a half-life of 28 years) leached to surface waters and groundwaters. ^{137}Cs acts similarly to potassium and ^{90}Sr behaves similarly to calcium. Vertical and horizontal migration of these decay products of radionuclides enters plants and reaches humans through the food chain and is thus a health hazard. Unauthorized activities of local people to recover cables and radioactive scrap further aggravate the situation. Presently, this contaminated territory is a grazing land and is partly mined for coal and gold. Strategies for rehabilitation of ecosystems destroyed by nuclear tests have not been planned, although radio-phytoremediation is gaining considerable progress in other parts of the world (Sagyndyk et al. 2007, 2008).

Semipalatinsk Test Site facilities (three research reactors) of the Republic of Kazakhstan, which is involved in civilian activities and conversion of the site to nondefense uses (Figure 4.2a, b). The Semipalatinsk test range covered an area of 18 000 km^2. Between 1949 and 1989, 470 nuclear tests, including 357 underground and 113 atmospheric tests, were conducted at Semipalatinsk Test Site facilities (Sagyndyk et al. 2008).

Radioecological investigations of the territory, migration of radionuclides from underground water into soil and into the roots of plants, uptake and translocation to aerial parts, the transfer factor, and concomitant effects on the food chain have not been investigated (Figure 4.3a, b).

Minerogenic dust released into the ambient air and health implications are summarized in Tables 4.1 and 4.2.

4.2 Tree "Bark Pockets" as Pollution Time Capsules

Tree "bark pockets" are designated as pollution time capsules for historical monitoring (Satake et al. 1996; Satake 2001). Chemical weapons and weapons of mass destruction when used, the residual chemicals, some of them are often called "forever chemicals," get concentrated in the bark pockets (Zhao et al. 2021). Likewise, atmospheric pollution monitoring for trace element concentrations helped to assess mortality and morbidity of urban population in highly industrialized mega cities. Airborne toxic pollutants such as uranium, polybrominated diphenyl ethers (PBDEs), road traffic emissions viz., heavy metals (Janta et al. 2016) and steel plant emissions (Kousehlar et al. 2021) are successfully monitored with the help of tree bark pockets (Figures 4.4 and 4.5) (Bellis et al. 2001; Chen et al. 2020; Drava et al. 2020; dos Santos et al. 2021; Flett et al. 2021). Accumulation of lead in tree trunk bark pockets, bioindication of atmospheric heavy metal pollution, and particulates of a source of c8–c12 perfluoroalkyl carboxylates and 10 : 2 fluorotelomer have been successfully

Figure 4.2 (a, b) Kazakhstan's Semipalatinsk nuclear facility. *Source:* Sagyndyk et al. (2008), John Wiley & Sons.

Figure 4.3 (a, b) Morphological changes exhibited by plants growing around the area (a) *Potentilla strigosa* Pall.ex Pursch. and (b) *Artemisia sieversiana* Wild. *Source:* Figures 4.1 and 4.2 courtesy Dr Sagyndyk.

Table 4.1 Minerogenic dust.

Significant findings / observations	Author(s)/Year
Minerals in the air: an environmental perspective.	Buseck et al. (2000)
Natural sources of metals in the environment.	Garret (2000)
Aerosol particles in the troposphere: a mineralogical introduction	Posfai and Molnar (2000)
African dust in America.	Prospero (2001)
Analysis of potentially toxic metals in airborne cement dust around Sagamu, Southwestern Nigeria.	Gbadebo and Bankole (2007)
Natural minerogenic dust and human health.	Derbyshire (2007)
Super volcanoes and their super eruptions.	Miller and Wark (2008)
Health effects of construction materials and construction products.	van Thienen and Spee (2008)
Impact of direct soil exposures from airborne dust and geophagy on human health.	Sing and Sing (2010)
Mine wastes: past, present, future.	Hudson-Edwards et al. (2011)
Geochemistry and mineralogy of solid mine waste: essential knowledge for predicting environmental impact.	Jamieson (2011)
Waste streams of mined oil sands: characteristics and remediation.	Kasperski and Mikula (2011)
Mine waters: acidic to circumneutral.	Nordstrom (2011)
Mine wastes and human health.	Plumlee and Morman (2011)
Recycling, reuse, and rehabilitation of mine wastes.	Lottermoser (2011)
The impact of the atmosphere on the Eyjafjallajökull 2010 eruption plume.	Petersen et al. (2012)
Minerals in the air.	Gieré and Vaughan (2013)
Health effects associated with inhalation of airborne arsenic arising from mining operations.	Martin et al. (2014)
Calcium phosphate biominerals.	Edmonds et al. 2018
Causes and effects of sand and dust storms: what has past research taught us? A survey.	Opp et al. (2021)
Silicon in the soil–plant continuum: intricate feedback mechanisms within ecosystems.	Katz et al. (2021)
Associations between natural resource extraction and incidence of acute and chronic health conditions: evidence from Tanzania.	Lyatuu et al. (2021)

Table 4.2 Recent research (2021) on dust storms.

Significant findings / observations	Author(s)/Year
Investigation of the 2018 Shiraz dust event: potential sources of metals, rare earth elements, and radionuclides; health assessment	Abbasi et al. (2021a)
Dust storm source detection using ANP and WRF models in Southwest of Iran. *Arabian Journal of Geosciences* 14(15): 1–15. ANP and the WRF models.	Abbasi et al. (2021b)
Origin and transport pathway of dust storm and its contribution to particulate air pollution in northeast edge of Taklimakan Desert, China.	Aili et al. (2021)
Synoptic analysis of dust storm in Iraq. *Environment Asia*	Al-Khalidi et al. (2021)
Dust storm satellite images. In atlas of fallen dust in Kuwait, 1–46.	Al-Obed et al. (2021)
Dust transport from Inland Australia and its impact on air quality and health on the eastern coast of Australia during the February 2019 dust storm.	Aragnou et al. (2021)
Land degradation modeling of dust storm sources using MODIS and meteorological time series data.	Bakhtiari et al. (2021)
Meteorological characteristics of dust storm events in Turkey.	Baltaci (2021)
The unusual severe dust storm of May 2018 over Northern India: genesis, propagation, and associated conditions.	Banerjee et al. (2021)
The mars dust activity database (MDAD): a comprehensive statistical study of dust storm sequences.	Battalio and Wang (2021)
Revisiting Andhi in Northern India: a case study of severe dust storm over the urban megacity of New Delhi.	Chakravarty et al. (2021)
Holocene dust storm variations over Northern China: transition from a natural forcing to an anthropogenic forcing.	Chen et al. (2021a)
Tracheitis hospital admissions are associated with Asia dust storm.	Chen et al. (2021b)
Climatology of dust storm frequency and its association with temperature and precipitation patterns over Pakistan.	Dar et al. (2021)
Evaluation of machine learning models for predicting the temporal variations of dust storm index in arid regions of Iran.	Ebrahimi-Khusfi et al. (2021)
The economic effects of dust storm.	Faridi et al. (2021)
Air pollution in the Gobi desert region: analysis of dust storm events. *Quarterly Journal of the Royal Meteorological Society*.	Filonchyk et al. (2021)
Mixing timescales of dustiness and some associated effects at middle atmosphere during the 2018 global dust storm.	Guha and Panda (2021)
Associations between short term exposure to ambient particulate matter from dust storm and anthropogenic sources and inflammatory biomarkers in healthy young adults	Jaafari et al. (2021)
Analysis of sand and dust storm events over Saudi Arabia in relation with meteorological parameters and ENSO.	Labban and Butt (2021)
Revealing the dust transport processes of the 2021 mega dust storm event in northern China	Liang et al. (2021)
Increases in the formation of water-soluble organic nitrogen during Asian dust storm episodes.	Liu et al. (2021)
A method to select sites for sand and dust storm source mitigation: case study in the Sistan region of Southeast Iran.	Maleki et al. (2021)
An investigation into climatic and terrestrial drivers of dust storms in the Sistan region of Iran in the early twenty-first century	Miri et al. (2021)
Dust storm frequency change in relation to climate drivers	Modarres (2021)
The effect of the Martian 2018 global dust storm on HDO as predicted by a mars global climate model	Rossi et al. (2021b)
A new approach for dust storm detection using MODIS data.	Sarikhani et al. (2021)
Asymmetric impacts on Mars' polar vortices from an equinoctial global dust storm.	Streeter et al. (2021)
Detection of a dust storm in 2020 by a multi-observation platform over the Northwest China.	Yang et al. (2021)
Biological crust in sand and dust storm source areas of Asia and its impact on dust emission.	Zhang et al. (2021a)
Dust storms in Northern China during the last 500 years.	Zhang et al. (2021b)
Representations of dynamics size distributions of mineral dust over East Asia by a regional sand and dust storm model.	Zhou et al. (2021)

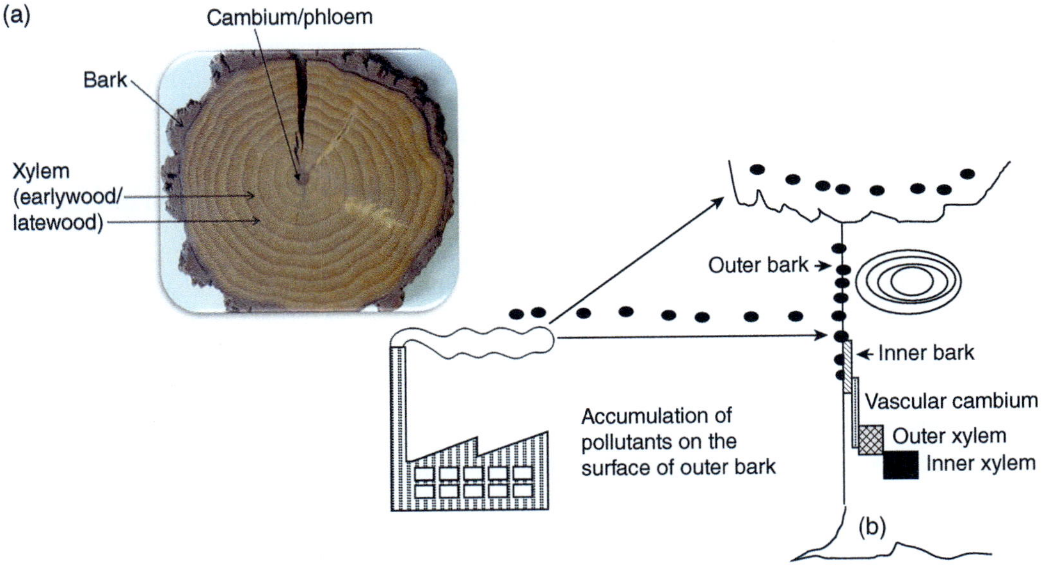

Figure 4.4 Capture of minerogenic dust by tree bark. (a) Cross section of mature tree trunk showing different tissues (xylem and phloem) and outer bark. (b) Industrial particulate pollutants get deposited on the surface of the tree bark.

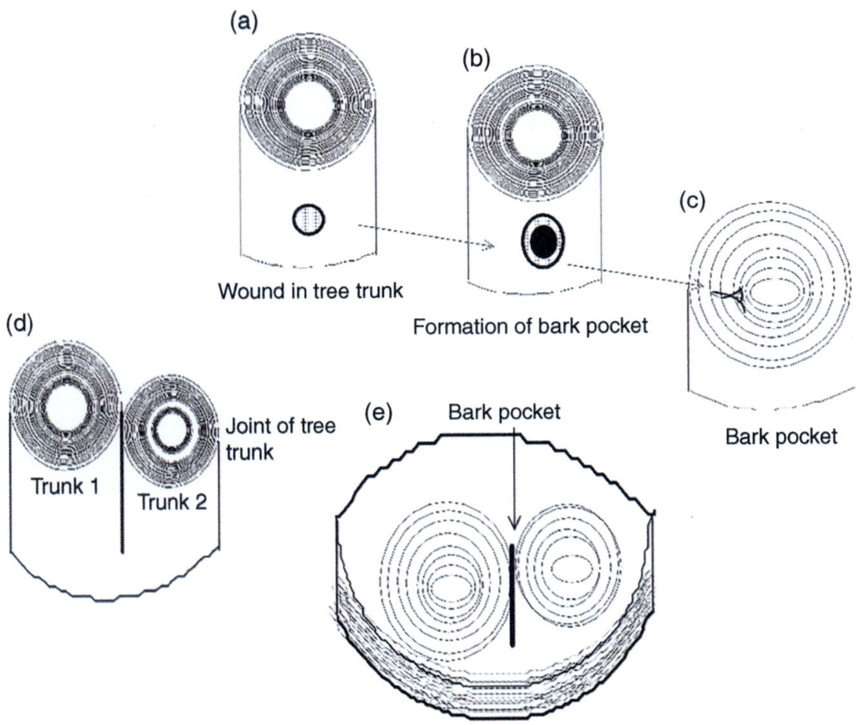

Figure 4.5 Tree trunk wounds and bark pockets ad minerogenis dust collectors. (a) wound in tree trunk, (b) formation of bark pocket, (c) bark pocket, (d) joint pf tree trunk and (e) formation of bark pocket at the junction of two trunks.

monitored for protecting human health and evolving environmental regulations (Patel et al. 2020; Sut-Lohmann et al. 2020; Yousaf et al. 2020)

Tree bark and tree trunk wounds serve as dust pollution time capsules. Trees growing in industrial belts in the vicinity of urban environment, in/nearby war field areas are known to accumulate particulates and the finger print analysis aid in deciphering the type of weapons used (Figures 4.4 and 4.5)

Magnesite (refractory material) mining (Küçüker 2021; Štofejová et al. 2021) produced dust is hazardous to human (Figure 4.6)

Figure 4.6 Magnesite mining dust. (a) Transport of Magnesite ore for and (b) production of refractory material and occupational hazard for workers exposed to dust.

4.3 Asbestosis

Asbestos is the naturally occurring silicate mineral excavated mainly from Asia and Middle East countries of the world. In spite of its toxic effect on human (asbestosis, lung cancer, mesothelioma, and other malignant tumor and pleural diseases) and environment pollution problems (airborne disease and metal enrichment in water and soil), its production is still in progress (Figure 4.7). It is mainly used by cement, ceilings and flooring, textiles, electrical,

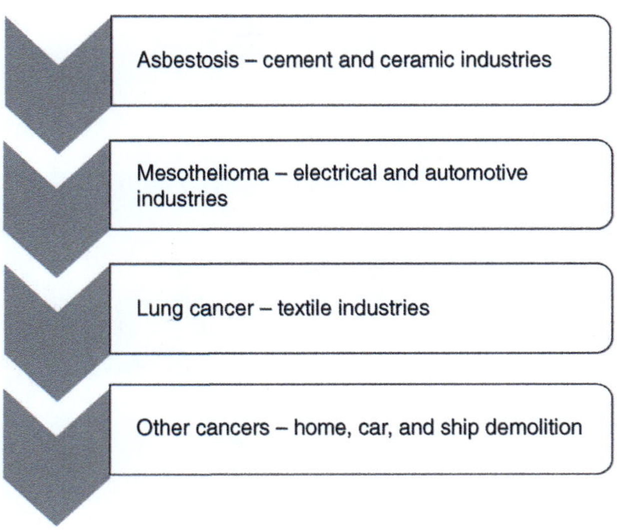

Figure 4.7 Occupational hazards of asbestos.

and automotive industries. However, huge mine waste generated carrying tons of asbestos fibers could be utilized in different industrial sectors such as carbon nanotubes, semiconductors, and thermally modified ceramic formulations. Bioremediation/phytoremediation is eco-friendly and cost-effective technology which can dilute and stabilize the loose asbestos waste, reduce the environmental and health problems, and can generate good economy for the local people.

Active and abandoned asbestos mine/waste causes severe environmental pollution to the surroundings. In spite of its toxic nature, its production and consumption are still increasing. Asbestos has average tensile strength and resistance to heat, fire, electrical, and chemical damage. Because of its numerous vital properties, it became the first choice for almost all sectors of industries (textile, chemical, electrical, automotive, roofing, ceramic, cement, and paper). Because of its lethal effect, it has been banned in some developed countries. These wastes contain minute small size asbestos fibers which can easily transfer from one place to other by air and cause huge impact on environment (soil, air, and water) and human health. Asbestos is highly carcinogenic, and their exposure may result in lung cancer, mesothelioma, pneumoconiosis, and other respiratory problems. In addition, leaching of heavy metals from asbestos waste in water bodies may destroy the aquatic ecosystem, whereas high metal contamination and lack of nutrients in soil result in vegetation and habitat loss. Heavy run-off during monsoon causes deposition of

Table 4.3 Hazards of asbestos.

Asbestosis	Author(s)/Year
Risk of asbestosis, mesothelioma, other lung disease or death among motor vehicle mechanics: a 45-year Danish cohort study.	Thomsen et al. (2021)
Asbestosis leading to the diagnosis of malignant peritoneal mesothelioma.	Martínez et al. (2021)
Asbestosis mimicking metastatic lung cancer: case report.	An et al. (2021)
Incidence of mesothelioma and asbestosis by occupation in a diverse workforce.	DeBono et al. (2021)
Underreporting of mesothelioma, asbestosis, and pleural plaques in Brazil.	Santana et al. (2021)
Pleural effusion–a radiological review.	Bhat (2021)
Costs of asbestosis and silicosis hospitalization in Italy (2001–2018)	Ferrante (2021)
Asbestos-free aluminum dross brake pad: a mini review.	Dirisu et al. (2021)

thick layer of asbestos waste in nearby agricultural fields, fallow lands, and vegetative areas. Pleural effusion could be due to a benign (empyema, asbestosis, and hemothorax) or a malignant process "Fibrothorax." This process of pleural thickening is gross and generalized and may involve parenchymal fibrosis and lung damage (Table 4.3).

4.4 Phosphogypsum Dust (Anthropogenic Radioactivity)

Literature reported that the risks caused by dust emissions coming from the gypsum not only by the dust itself but also by the radon which it emits. Stacked gypsum does not give off dust profusely. However, it is a matter of concern about the radioactive risks of radon given off by phosphogypsum dust (Figures 4.8 and 4.9; Table 4.4).

Radon is a rare gas which can accumulate in air-tight cavities. If radon is ingested, the radon emitted will remain blocked in the organism and will exercise its effects in totality. On the other hand, radon in a gypsum stack is not harmful, since it is not enclosed but open to the air. On the other hand, that of a phosphate rock deposit, where a proportion 50% higher than in the gypsum can be found, can sometimes be more dangerous if houses are constructed on the surface. In fact, the emissions of radon reach the surface through fissures in the land and can accumulate in non-ventilated rooms in the house. This is why a ventilated empty space is recommended under houses constructed on land from which radon is emitted.

From an esthetic point of view and also to protect against wind and rain erosion, generally revegetation of gypsum stacks is recommended. If the stack is not too acid, the revegetation occurs naturally due to seeds which are deposited by the wind and birds.

Figure 4.8 (a–f) Transport and use of phosphogypsum in road making causing dust pollution.

Figure 4.9 (a–c) Piledup phosphogypsum stack using heavy machinery.

Table 4.4 Hazards of phosphogypsum.

Significant findings / observations	Author(s)/Year
Biogeochemical aspects of uranium mineralization, mining, milling, and remediation.	Campbell et al. (2015)
Optimized and validated wavelength-dispersive X-ray fluorescence spectrometry (WDXRF) method for the determination of uranium in phosphogypsum.	Guitouni (2015)
Uranium (^{234}U, ^{235}U, and ^{238}U) contamination of the environment surrounding phosphogypsum waste heap in in wiślinka (Northern Poland).	Olszewski et al. (2015)
Redox chemistry of sulfate and uranium in a phosphogypsum tailings dump.	Papanicolaou et al. (2010)
Radioactive materials in phosphogypsum.	Bolívar et al. (1998)
Sources of anthropogenic radionuclides in the environment: a review.	Hu et al. (2010)
Using the activity of naturally occurring radionuclides to identify the contribution of the Al industry to the atmosphere and plants.	Pourcelot et al. (2021)
Distribution of natural and anthropogenic radionuclides in sediments from the Vefsnfjord, Norway.	Heldal et al. (2021)
Phytoremediation of radionuclides in soil, sediments, and water.	Yan et al. (2021)

There has been experience of the plant at Engis in Belgium (Belboom (2015)). Trials of planting directly on the gypsum stack have given negative results, since the plants have not found a medium favorable to their establishment. On the other hand, a plantation of birch, alder, and acacia at the periphery of the stack was almost completely successful and, after a few years, there was a natural seeding of the stack itself by alder and birch; it can be deduced from this that the seeds brought in by the wind and the birds find in the gypsum stack growth conditions more favorable than plants which were more developed at the start. Evidently, the rate of natural revegetation is fairly slow; it can be accelerated (by a factor of 10) by deposing some tens of centimeters of earth and planting suitable species such as alder and birch. On very acid stacks, the literature (Belboom (2015)) gives evidence of successful revegetation but with very intensive and costly measures, i.e. the installation of an impermeable sheet covered by a 45 cm layer of earth on which grass, bushes, and trees are planted.

Phosphosypsum is considered as a naturally occurring radioactive material (NORM). Phosphogypsum consists primarily of calcium sulfate dihydrate with small amounts of silica, usually as quartz and unreacted phosphate rock and traces of radioactive material (like radium and uranium).

During sulfuric acid processing of phosphate raw materials with radioactive isotopes, there goes an uneven transition to liquid and solid phases with uranium (^{238}U) enrichment of phosphoric acid, and thorium (^{230}Th, ^{232}Th) and radium (^{226}Ra) enrichment of phosphogypsum. For non-radioactive phosphate raw materials like Kola apatite concentrate, its derivatives are also are non-radioactive.

Sand of volcanic nature are of medical geological importance. Sand baths also can already take place in bath-houses/balnearies, being the sand heated inside the spa by artificial heating. There are also examples of sand bathing or psammotherapy practiced in certain countries, as is the case of places in southern Japan where the sand of volcanic nature is naturally heated by exposure to the volcanic heat, and the case of certain beaches of Guarapari, in Victória do Espirito Santo district, Brazil, where health benefits are assumed to be related to the radon emitted from the radioactive monazite-rich sands.

Hazardous materials (HMs) or disaster materials (DMs), such as volcanic ash, volcanic gas, volcanic mud, mine wastes and tailings, acid mine drainage, debris flow, muddy flood, dust plumes, and smoke plumes, are after released into the environment by various natural and man-made disasters, and the toxicological characteristics of those HM or DM.

Chemical elements, metals, and metalloids present in natural gas emissions (volcanic emissions) of potentially hazardous materials produced by natural disasters such as volcanic eruptions, volcanic gases, volcanic ashes, landslides and debris flows, hurricanes, floods, wildfires, and mine wastes and mill tailing impoundment failures. The human and animal health impacts of pandemics, climate changes, volcanic eruptions, earthquakes, landslides, tsunamis, and hurricanes are actual and promising issues of Medical Geology.

Gayo Pliny Segundo, known as Pliny the Elder (23–79 A.C.) and author of the book *Historia Naturalis*, described the use of clays, some of volcanic origin with occurrences near Naples, especially as medicines for the stomach and intestines.

On natural sources of airborne minerals, such as desert dusts, volcanic dusts, and volcanic gases, and health hazards, Posfai and Molnar (2000), Garret (2000), Buseck et al. (2000), Prospero (2001), Gieré and Vaughan (2013), Miller and Wark (2008), Williams-Jones and Rymer (2015), and Edmonds et al. (2018) are excellent references. Regarding the anthropogenic sources of airborne minerals and health hazards, Martin et al. (2014), van Thienen and Spee (2008), and Gbadebo and Bankole (2007) are excellent references too.

4.5 Silicosis

Occupational exposures to dust with elevated levels of respirable crystalline silica in artificial stone (AS) increase workers' risk for silicosis. Silicosis is an incurable but preventable disease caused by inhalation of respirable crystalline silica dust. (Adamcakova and Mokra 2021; Burki 2021; Ehrlich et al. 2021; Hoy 2021; Hoy et al. 2021; Reynolds and Jerome 2021; Tan and Chen 2021). Based on outbreaks in other countries, the Occupational Safety and Health Administration (OSHA) along with the National Institute for Occupational Safety and Health (NIOSH) warned about the silica health hazards for workers involved in manufacturing, finishing, and installing manufactured stone countertop products.

Manufactured, engineered, and AS describe the commercial product that combines silica with a binding agent to produce a stone-like countertop, commonly called "quartz." The silica percentages in natural stone vary, but AS surfaces can be over 93% silica. Grinding, cutting, drilling, and polishing stone and AS countertops release hazardous silica dust into the air, which creates a widely known and established occupational health hazard. The rapid increase of stone and AS surface imports over the last decade has placed increasingly large numbers of workers at risk of silicosis in the United States and globally.

4.6 Volcanic Ash

The use of radioactivity for 100 years is added to the natural inventories. The ban of aboveground testing of nuclear weapons reduced the amounts and also due to the shorter half-lives of many of the nuclides, it has seen a marked decrease. Silicon is released from silicates during weathering processes, and it is also released under water during volcanic activity (Williams-Jones and Rymer 2015; Delmelle et al. 2021; Rossi et al. 2021a). Very rarely does silicon occur in free form in nature, and when such happens in aqueous solutions it occurs as orthosilicic acid, H_4SiO_4 or $Si(OH)_4$.

The main natural sources of mercury are volcanic eruptions, while the main anthropogenic sources of mercury are the combustion of fossil energy resources, coal, oil, and natural gas, as well as pharmaceuticals, paper and pulp preservatives, agriculture industry, and chlorine and caustic soda production industry (Table 4.5). Classic example of

Table 4.5 Toxic gases due to volcanic eruption.

Significant findings / observations	Author(s)/Year
Hazards of volcanic gases	Williams-Jones and Rymer (2015)
Eruptive style controls the formation of silicon hexafluoride salts on volcanic ash: the case of the 2010 eruption of Eyjafjallajökull volcano, Iceland.	Delmelle et al. (2021)
The fate of volcanic ash: premature or delayed sedimentation?	Rossi et al. (2021a)

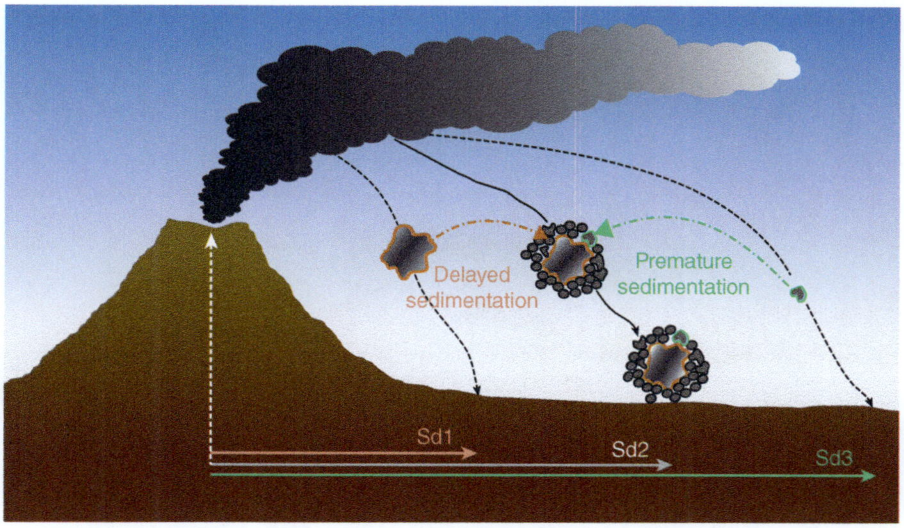

Figure 4.10 Premature and delayed sedimentation. Simplified sketch showing both premature and delayed sedimentation of single particles due to aggregation. Premature sedimentation of a single particle (typically <63 microns) occurs when it falls closer to the volcano (at distance Sd2) than it would do if not part of an aggregate (distance Sd3). On the contrary, a particle that acts as an aggregate core in a cored cluster could be rafted and sediment at larger distance from vent (Sd2) than if it was falling individually (distance Sd1). *Source:* Rossi et al. (2021a).

eruption volcano is Eyjafjallajökull in 2010 (near ice land) (see Figure 4.10)

Volcanic activity is the main source of Se, and concentrations of several 100 of ppm may occur in soils of volcanic territories.

4.7 Dust and Gases from Volcanic Eruptions

Volcanic eruption leads to the propulsion of an enormous content of aerosols and fine particles into the atmosphere with sizes within the range from micrometers to several nanometers.

According to Taylor (2002), a single volcanic eruption can release up to 30×106 tons of Nano particles (NPs) in the form of ash into the atmosphere. The released NPs spread throughout the world and settle in the stratosphere and the troposphere which are the lowest atmospheric layers. The effect of NPs will be significant in areas of 10 km approximately from the volcano, and the particulate materials will affect human, animal, and plant activities by blocking and scattering the sunlight.

The volcanically erupted particles may possess heavy metals that are toxic to humans. The short-term effects of such particles include nose, throat, eye, and skin irritations and bronchial symptoms, while the long-term effects include diseases such as podoconiosis and the Kaposi's sarcoma.

As aforesaid, volcanic eruptions are important sources not only of mineral dust emitted into the atmosphere but also of toxic gases which may cause health conditions. Finely particulate eruption of volcanic matter may be ejected and reach high altitudes and long distances from the volcanic vent.

Gieré and Vaughan (2013), in an article entitled "Minerals in the Air," report in a synthetic form the impacts of the presence of minerals in the atmosphere, either of natural origin or of anthropogenic origin, on environment and humans. In the authors' opinion, it is only during super eruptions (potentially long-lasting, minimum weeks) that gigantic clouds of volcanic ash are produced and that humans become aware of the potential impacts of airborne minerals on health. A recent example of these gigantic clouds of volcanic ash was the super eruption of the Icelandic volcano Eyjafjallajokull. An old example mentioned by the authors is the devastating and asphyxiating clouds of ash produced by the Plinian eruption of Mount Vesuvius.

4.8 Artisanal and Small-Scale Gold Mining Activities in Nigeria

Artisanal and small-scale mining (ASM) is one of the most important economic activities in many of the sub-Saharan Africa rural communities. Widespread artisanal alongside small-scale mining operations are currently increasing in intensity in Nigeria. These activities are causing immeasurable damage to the environment and populations that live in the vicinity of these mine fields. The discharges of potentially harmful elements (PHEs) from the exposed mine-out/mineral processing sites and their subsequent remobilization into the soils and natural water bodies constitute serious human health problems. Artisanal mining in Nigeria and especially in its southwestern sector is reported to be at best unplanned and haphazard with attendant negative impact on the environment

In 2008, traditional farmers began small-scale gold mining near Bagega (Town in Zamfara State, Nigeria). When gold prices skyrocketed during this time due to the Great Recession, gold mining for the farmers became lucrative. People from other villages converged on Bagega to mine.

Figure 4.11 Artisanal and small-scale gold mining and exposure to dust and toxic fumes, Mararaba, Kebbi State, Nigeria. (a) Artisanal and small-scale gold mining workers exposed to hazardous potentially harmful toxic elements; (b) risks from inhalation of toxic mine dust as the mine workers stay temporarily on the site for prolonged period; (c) mine pits prone to erosion and dust pollution; (d) mine waste with hazardous dust; (e) extraction of gold remains; (f) different stages of gold purification including production of pure gold nuggests; and (g) processed gold samples.

Even when only small particles of gold were recovered, it still paid well for these small-scale miners. The mines in this area contained lead which is not typical in most gold mining areas. Exposure to lead (Pb) can cause learning and behavior problems or even death. In addition to being exposed to lead, these artisanal miners used mercury (Hg) to extract gold particles. Mercury has its own hazardous properties. It is estimated that the global release of mercury is about 2503 tons/year from anthropogenic sources (mainly biomass burning and artisanal gold mining production).

Artisanal and small-scale gold mining (ASGM) has long been in practice not only in Nigeria but also in Africa and some other parts of the world. ASGM is the major gold mining practice in northern part of Nigeria and is usually managed locally. Skyrocketing gold price, lack of viable alternative livelihoods, and availability of mercury at affordable cost have been a resurgence of ASGM activities in Northern Nigeria in recent years. ASM refers to informal mining activities undertaken by individuals and groups who rely heavily on manual labor, using simple implements and methods. However, because of the large number of people involved in ASM operations, significant eco-toxicological risks that could be created within a very short period of time are being faced by the labor involved in ASM.

Gold mining in Kebbi area (Figure 4.11) has been in practice since the colonial period. It has been done in a crude manner using traditional and crude methods and implements. It was observed that unprofessional practices such as lack of use of gas mask while working in the dust-filled mine and at the mills could expose workers to risks of inhalation of crystalline silica as well as exposure to radon gas posing respiratory disorders. Hazards such as drowning by animals and humans, erosion, soil degradation, alteration of fluvial systems, and clearing of vegetation at the abandoned sites are the normal scenarios. Socioeconomic and law and order problems are prevalent in communities where ASGM is taking place. The artisanal miners are exposed to dust and toxic fumes.

Acknowledgments

The author would like to thank for the license for the reuse of Figures 4.10 (Springer Nature) and 4.11 (Elsevier). The author also would like to thank Dr. Boris Levin, Phosagro, Russia (Leninsky prospect, Moscow) for providing illustrations Figure 4.8 d–f and Figure 4.9 a–c on phosphogypsum and its use in road construction.

References

Abbasi, E., Etemadi, H., Smoak, J.M. et al. (2021a). Dust storm source detection using ANP and WRF models in Southwest of Iran. *Arabian Journal of Geosciences* 14 (15): 1–15.

Abbasi, S., Rezaei, M., Keshavarzi, B. et al. (2021b). Investigation of the 2018 Shiraz dust event: potential sources of metals, rare earth elements, and radionuclides; health assessment. *Chemosphere* 279: 130533.

Adamcakova, J. and Mokra, D. (2021). New insights into pathomechanisms and treatment possibilities for lung silicosis. *International Journal of Molecular Sciences* 22 (8): 4162.

Aili, A., Xu, H., Kasim, T., and Abulikemu, A. (2021). Origin and transport pathway of dust storm and its contribution to particulate air pollution in northeast edge of Taklimakan Desert, China. *Atmosphere* 12 (1): 113.

Al-Khalidi, J., Bakr, D., and Abdullah, A.A. (2021). Synoptic analysis of dust storm in Iraq. *Environment Asia* 14 (1): 13–22.

Al-Obed, M., Uddin, S., and Ramadhan, A. (2021). Dust storm satellite images. In: *Atlas of Fallen Dust In Kuwait* (ed. A. Al-Dousari), 1–46. Cham: Springer.

An, J., Song, M., and Chang, B. (2021). Asbestosis mimicking metastatic lung cancer: case report. *Medicina* 57 (5): 402.

Aragnou, E., Watt, S., Duc, H.N. et al. (2021). Dust transport from Inland Australia and its impact on air quality and health on the Eastern coast of Australia during the February 2019 dust storm. *Atmosphere* 12 (2): 141.

Bakhtiari, M., Boloorani, A.D., Kakroodi, A.A. et al. (2021). Land degradation modeling of dust storm sources using MODIS and meteorological time series data. *Journal of Arid Environments* 190: 104507.

Baltaci, H. (2021). Meteorological characteristics of dust storm events in Turkey. *Aeolian Research* 50: 100673.

Banerjee, P., Satheesh, S.K., and Moorthy, K.K. (2021). The unusual severe dust storm of May 2018 over Northern India: genesis, propagation, and associated conditions. *Journal of Geophysical Research-Atmospheres* 126 (7): e2020JD032369.

Battalio, M. and Wang, H. (2021). The mars dust activity database (MDAD): a comprehensive statistical study of dust storm sequences. *Icarus* 354: 114059.

Belboom, S., Szöcs, C. and Léonard, A., 2015. Environmental impacts of phosphoric acid production using di-hemihydrate process: a Belgian case study. *Journal of Cleaner Production*, 108: 978–986.

Bellis, D., Ma, R., Bramall, N. et al. (2001). Airborne uranium contamination – as revealed through elemental and isotopic analysis of tree bark. *Environmental Pollution* 114 (3): 383–387.

Bhat, R. (2021). Pleural effusion–a radiological review. *APIK Journal of Internal Medicine* 9 (1): 58–61.

Burki, T.K. (2021). Silicosis and the countertop industry. *The Lancet Respiratory Medicine* https://doi.org/10.1016/S2213-2600(21)00452-5.

Buseck, P.R., Jacob, D.J., Pósfai, M. et al. (2000). Minerals in the air: an environmental perspective. *International Geology Review* 42: 577–593.

Campbell, K.M., Gallegos, T.J., and Landa, E.R. (2015). Biogeochemical aspects of uranium mineralization, mining, milling, and remediation. *Applied Geochemistry* 57: 206–235.

Chakravarty, K., Vincent, V., Vellore, R. et al. (2021). Revisiting Andhi in Northern India: a case study of severe dust-storm over the urban megacity of New Delhi. *Urban Climate* 37: 100825.

Chen, Y., Zhang, A., Li, H. et al. (2020). Concentrations and distributions of polybrominated diphenyl ethers (PBDEs) in surface soils and tree bark in Inner Mongolia, Northern China, and the risks posed to humans. *Chemosphere* 247: 125950.

Chen, C.S., Chan, Y.S., and Liu, T.C. (2021a). Tracheitis hospital admissions are associated with Asia dust storm. *International Journal of Environmental Health Research* 32 (6): 1337–1343.

Chen, S., Liu, J., Wang, X. et al. (2021b). Holocene dust storm variations over Northern China: transition from a natural forcing to an anthropogenic forcing. *Science Bulletin* https://doi.org/10.1016/j.scib.2021.08.008.

Dar, M.A., Ahmed, R., Latif, M. and Azam, M., 2021. Climatology of dust storm frequency and its association with temperature and precipitation patterns over Pakistan. *Natural Hazards*, pp.1–23. https://doi.org/10.1007/s11069-021-04962-9.

DeBono, N.L., Warden, H., Logar-Henderson, C. et al. (2021). Incidence of mesothelioma and asbestosis by occupation in a diverse workforce. *American Journal of Industrial Medicine* 64 (6): 476–487.

Delmelle, P., Maters, E.C., Calkins, J.A. et al. (2021). Eruptive style controls the formation of silicon hexafluoride salts on volcanic ash: the case of the 2010 eruption of Eyjafjallajökull volcano, Iceland. *Chemical Geology* 579: 120327.

Derbyshire, E. (2007). Natural minerogenic dust and human health. *Ambio: A Journal of the Human Environment* 36 (1): 73–77.

Dirisu, J.O., Fayomi, O.S.I., Oyedepo, S.O., and Udoye, N.E. (2021). Asbestos-free aluminium dross brake pad: a mini review. In: *IOP Conference Series: Materials Science and Engineering*, vol. 1107, No. 1, 012034. IOP Publishing https://doi.org/10.1088/1757-899X/1107/1/012034.

Drava, G., Ailuno, G., and Minganti, V. (2020). Trace element concentrations measured in a biomonitor (tree bark) for assessing mortality and morbidity of urban population: a new promising approach for exploiting the potential of public health data. *Atmosphere* 11 (8): 783.

Ebrahimi-Khusfi, Z., Taghizadeh-Mehrjardi, R., and Mirakbari, M. (2021). Evaluation of machine learning models for predicting the temporal variations of dust storm index in arid regions of Iran. *Atmospheric Pollution Research* 12 (1): 134–147.

Edmonds, M., Grattan, J., and Michnowicz, S. (2018). Volcanic gases: silent killers. In: *Advances in Volcanology* (ed. C.J. Fearnley, D.K. Bird, K. Haynes, et al.), 65–83. Springer.

Ehrlich, R., Akugizibwe, P., Siegfried, N., and Rees, D. (2021). The association between silica exposure, silicosis and tuberculosis: a systematic review and meta-analysis. *BMC Public Health* 21 (1): 1–18.

Faridi, S., Rahmani, S., Hashemi, N. et al. (2021). The economic effects of dust storm. *Journal of Health* 11 (5): 699–713.

Ferrante, P. (2021). Costs of asbestosis and silicosis hospitalization in Italy (2001–2018). *International Archives of Occupational and Environmental Health* 94 (4): 763–771.

Filonchyk, M., Peterson, M., and Hurynovich, V. (2021). Air pollution in the Gobi desert region: analysis of dust-storm events. *Quarterly Journal of the Royal Meteorological Society* 147 (735): 1097–1111.

Flett, L., McLeod, C.L., McCarty, J.L. et al. (2021). Monitoring uranium mine pollution on Native American lands: insights from tree bark particulate matter on the Spokane Reservation, Washington, USA. *Environmental Research* 194: 110619.

Garret, R.G. (2000). Natural sources of metals in the environment. *Human and Ecological Risk Assessment* 6 (6): 954–963.

Gbadebo, A.M. and Bankole, O.D. (2007). Analysis of potentially toxic metals in airborne cement dust around Sagamu, Southwestern Nigeria. *Journal of Applied Sciences* 7: 35–40. https://doi.org/10.3923/jas.2007.

Gieré, R. and Vaughan, D.J. (2013). Minerals in the air. *Elements* 9: 410–411.

Guha, B.K. and Panda, J. (2021). Mixing time scales of dustiness and some associated effects at middle atmosphere during the 2018 global dust storm. *Advances in Space Research* https://doi.org/10.1016/j.asr.2021.05.030.

Guitouni, M. (2015). Optimized and validated wavelength-dispersive X-ray fluorescence spectrometry (WDXRF) method for the determination of uranium in phosphogypsum. *Journal of Radioanalytical and Nuclear Chemistry* 303 (3): 1649–1657.

Heldal, H.E., Helvik, L., Haanes, H. et al. (2021). Distribution of natural and anthropogenic radionuclides in sediments from the Vefsnfjord, Norway. *Marine Pollution Bulletin* 172: 112822.

Hoy, R.F. (2021). Artificial stone silicosis. *Current Opinion in Allergy and Clinical Immunology* 21 (2): 114–120.

Hoy, R.F., Glass, D.C., Dimitriadis, C. et al. (2021). Identification of early-stage silicosis through health screening of stone benchtop industry workers in Victoria, Australia. *Occupational and Environmental Medicine* 78 (4): 296–302.

Hu, Q.H., Weng, J.Q., and Wang, J.S. (2010). Sources of anthropogenic radionuclides in the environment: a review. *Journal of Environmental Radioactivity* 101 (6): 426–437.

Hudson-Edwards, K.A., Jamieson, H.E., and Lottermoser, B.G. (2011). Mine wastes: past, present, future. *Elements* 7 (6): 375–380.

Jaafari, J., Naddafi, K., Yunesian, M. et al. (2021). Associations between short term exposure to ambient particulate matter from dust storm and anthropogenic sources and inflammatory biomarkers in healthy young adults. *Science of the Total Environment* 761: 144503.

Jamieson, H.E. (2011). Geochemistry and mineralogy of solid mine waste: essential knowledge for predicting environmental impact. *Elements* 7 (6): 381–386.

Janta, R., Chantara, S., Inta, A. et al. (2016). Levels of road traffic heavy metals in tree bark layers of Cassia fistula tree. *International Journal of Environmental Science and Development* 7 (5): 385.

Kasperski, K.L. and Mikula, R.J. (2011). Waste streams of mined oil sands: characteristics and remediation. *Elements* 7 (6): 387–392.

Katz, O., Puppe, D., Kaczorek, D. et al. (2021). Silicon in the soil–plant continuum: intricate feedback mechanisms within ecosystems. *Plants* 10 (4): 652.

Kousehlar, M., Widom, E., and Kuentz, D. (2021). Osmium isotope geochemistry of steel plant emissions using tree bark biomonitoring. *Environmental Pollution* 272: 115976.

Küçüker, M.A. (2021). Characterization of asbestos in ambient air during refractory material production from magnesite ore. *Journal of Advanced Research in Natural and Applied Sciences* 7 (3): 423–436.

La Maestra, S., D'Agostini, F., Sanguineti, E. et al. (2021). Dispersion of natural airborne TiO_2 fibres in excavation activity as a potential environmental and human health risk. *International Journal of Environmental Research and Public Health* 18 (12): 6587.

Labban, A.H. and Butt, M.J. (2021). Analysis of sand and dust storm events over Saudi Arabia in relation with meteorological parameters and ENSO. *Arabian Journal of Geosciences* 14 (1): 1–12.

Liang, P., Chen, B., Yang, X. et al. (2021). Revealing the dust transport processes of the 2021 mega dust storm event in Northern China. *Science Bulletin* 67: 21–24.

Lottermoser, B.G. (2011). Recycling, reuse and rehabilitation of mine wastes. *Elements* 7 (6): 405–410.

Lyatuu, I., Loss, G., Farnham, A. et al. (2021). Associations between natural resource extraction and incidence of acute and chronic health conditions: evidence from Tanzania. *International Journal of Environmental Research and Public Health* 18 (11): 6052.

Maleki, S., Miri, A., Rahdari, V., and Dragovich, D. (2021). A method to select sites for sand and dust storm source mitigation: case study in the Sistan region of Southeast Iran. *Journal of Environmental Planning and Management* 64 (12): 2192–2213.

Martin, R., Dowling, K., Pearce, D. et al. (2014). Health effects associated with inhalation of airborne arsenic arising from mining operations. *Geosciences* 4: 128–175. https://doi.org/10.3390/geosciences4030128.

Martínez, A.G. Sr., Guerrero, I.P., and Gonzaga, L.C. (2021). Omental cake in a patient with asbestosis leading to the diagnosis of malignant peritoneal mesothelioma. *Cureus* 13 (5): https://doi.org/10.7759/cureus.15116.

Miller, C.F. and Wark, D.A. (2008). Super volcanoes and their super eruptions. *Elements* 4: 11–16.

Miri, A., Maleki, S., and Middleton, N. (2021). An investigation into climatic and terrestrial drivers of dust storms in the Sistan region of Iran in the early twenty-first century. *Science of the Total Environment* 757: 143952.

Modarres, R. (2021). Dust storm frequency change in relation to climate drivers. *International Journal of Climatology* 41: E187–E199.

Nordstrom, D.K. (2011). Mine waters: acidic to circmneutral. *Elements* 7 (6): 393–398.

Olszewski, G., Boryło, A., and Skwarzec, B. (2015). Uranium (^{234}U, ^{235}U and ^{238}U) contamination of the environment surrounding phosphogypsumwaste heap in wiślinka (Northern Poland). *Journal of Environmental Radioactivity* 146: 56–66.

Opp, C., Groll, M., Abbasi, H., and Foroushani, M.A. (2021). Causes and effects of sand and dust storms: what has past research taught us? A survey. *Journal of Risk and Financial Management* 14 (7): 326.

Papanicolaou, F., Antoniou, S., and Pashalidis, I. (2010). Redox chemistry of sulphate and uranium in a phosphogypsum tailings dump. *Journal of Environmental Radioactivity* 101: 601–605.

Patel, K.S., Yadav, A., Sahu, Y.K. et al. (2020). Tree bark as a bioindicator for arsenic and heavy metal air pollution in Rajnandgaon district, Chhattisgarh, India. *Journal of Hazardous, Toxic, and Radioactive Waste* 24 (1): 05019006.

Petersen, G.N., Bjornsson, H., and Arason, P. (2012). The impact of the atmosphere on the Eyjafjallajökull 2010 eruption plume. *Journal of Geophysical Research-Atmospheres* 117 (D20): https://doi.org/10.1029/2011JD016762.

Plumlee, G.S. and Morman, S.A. (2011). Mine wastes and human health. *Elements* 7 (6): 399–404.

Posfai, M. and Molnar, A. (2000). Aerosol particles in the troposphere: a mineralogical introduction. In: *EMU Notes in Mineralogy 2, Enviromental Mineralogy, Chapter 10* (ed. D.J. Vaughan and R.A. Wogelius). Budapest: Eotvos University Press https://doi.org/10.1180/EMU-notes.2.6.

Pourcelot, L., Masson, O., Beguin-Leprieur, M. et al. (2021). Using the activity of naturally occurring radionuclides to identify the contribution of the Al industry to the atmosphere and plants. *Applied Geochemistry* 131: 105033.

Prospero, J.M. (2001). African dust in America. *Geotimes* 46 (1): 24–27.

Reynolds, K. and Jerome, J. (2021). Silicosis. *Workplace Health & Safety* 69 (1): 51–51.

Rossi, E., Bagheri, G., Beckett, F., and Bonadonna, C. (2021a). The fate of volcanic ash: premature or delayed sedimentation? *Nature Communications* 12 (1): 1–9.

Rossi, L., Vals, M., Montmessin, F. et al. (2021b). The effect of the Martian 2018 global dust storm on HDO as predicted by a mars global climate model. *Geophysical Research Letters* 48 (7): e2020GL090962.

Sagyndyk, K.S., Aidossova, S.S., and Prasad, M.N.V. (2007). Grasses tolerant to radionuclides growing in Kazakhstan nuclear test sites exhibit structural and ultrastructural changes – implications for phytoremedaition and involved risks. *Terrestrial and Aquatic Ecotoxicology* 1: 70–77.

Sagyndyk, K.S., Aidossova, S.S., and Prasad, M.N.V. (2008). Floristic composition at Kazakhstan's Semipalatinsk nuclear test site: relevance to the containment of radionuclides to safeguard ecosystems and human health. In: *Trace Elements as Contaminants and Nutrients: Consequences in Ecosystems and Human Health* (ed. M.N.V. Prasad), 281–294. New York: Wiley.

Santana, V.S., Salvi, L., Cavalcante, F. et al. (2021). Underreporting of mesothelioma, asbestosis and pleural plaques in Brazil. *Occupational Medicine* 71 (4–5): 223–230.

dos Santos, E.C., Maihara, V.A., Genezine, F.A., and Saiki, M. (2021). A study on tree bark samples for atmospheric pollution monitoring. *Brazilian Journal of Radiation Sciences* 9 (1A).

Sarikhani, A., Dehghani, M., Karimi-Jashni, A., and Saadat, S. (2021). A new approach for dust storm detection using MODIS data. *Iranian Journal of Science and Technology, Transactions of Civil Engineering* 45 (2): 963–969.

Satake, K. (2001). Tree bark: tree "bark pockets" as pollution time capsules for historical monitoring, analysis by biodiversity. In: *Metals in the Environment* (ed. M.N.V. Prasad), 373–386. CRC Press p. 487.

Satake, K.I., Tanaka, A., and Kimura, K. (1996). Accumulation of lead in tree trunk bark pockets as pollution time capsules. *Science of the Total Environment* 181 (1): 25–30.

Sing, D. and Sing, C.F. (2010). Impact of direct soil exposures from airborne dust and geophagy on human health. *International Journal of Environmental Research and Public Health* 7 (3): 1205–1223.

Štofejová, L., Fazekaš, J., and Fazekašová, D. (2021). Analysis of heavy metal content in soil and plants in the dumping ground of magnesite mining factory Jelšava-Lubeník (Slovakia). *Sustainability* 13: 4508.

Streeter, P.M., Lewis, S.R., Patel, M.R. et al. (2021). Asymmetric impacts on Mars polar vortices from an equinoctial global dust storm. *Journal of Geophysical Research, Planets* 126 (5): e2020JE006774.

Sut-Lohmann, M., Jonczak, J., Parzych, A. et al. (2020). Accumulation of airborne potentially toxic elements in *Pinus sylvestris* L. bark collected in three Central European medium-sized cities. *Ecotoxicology and Environmental Safety* 200: 110758.

Tan, S. and Chen, S. (2021). Macrophage autophagy and silicosis: current perspective and latest insights. *International Journal of Molecular Sciences* 22 (1): 453.

Taylor, D.A. (2002). Dust in the wind. *Environmental health perspectives*, 110(2), pp. A80–A87.

van Thienen, G. and Spee, T. (2008). Health effects of construction materials and construction products. *Journal of Occupational Science* 14 (13): 2–23.

Thomsen, R.W., Riis, A.H., Flachs, E.M. et al. (2021). Risk of asbestosis, mesothelioma, other lung disease or death among motor vehicle mechanics: a 45-year Danish cohort study. *Thorax* 77 (5): 477–485.

Williams-Jones, G. and Rymer, H. (2015). Hazards of volcanic gases. In: *The Encyclopedia of Volcanoes*, 2e (ed. M.N.V. Prasad, P.J.C. Favas and S.K. Maiti), 985–992.

Yan, L., Van Le, Q., Sonne, C. et al. (2021). Phytoremediation of radionuclides in soil, sediments and water. *Journal of Hazardous Materials* 407: 124771.

Yang, L., Hu, Z., Huang, Z. et al. (2021). Detection of a dust storm in 2020 by a multi-observation platform over the Northwest China. *Remote Sensing* 13 (6): 1056.

Yousaf, M., Mandiwana, K.L., Baig, K.S., and Lu, J. (2020). Evaluation of acer rubrum tree bark as a bioindicator of atmospheric heavy metal pollution in Toronto, Canada. *Water, Air, & Soil Pollution* 231 (8): 1–9.

Zhang, J., Zhou, C.H., Gui, H.L., and Zhang, X.Y. (2021a). Biological crust in sand and dust storm source areas of Asia and its impact on dust emission. *Advances in Climate Change Research* 12 (3): 395–408.

Zhang, S., Xu, H., Lan, J. et al. (2021b). Dust storms in Northern China during the last 500 years. *Science China Earth Sciences* 64 (5): 813–824.

Zhao, N., Zhao, M., Liu, W., and Jin, H. (2021). Atmospheric particulate represents a source of C8–C12 perfluoroalkyl carboxylates and 10:2 fluorotelomer alcohol in tree bark. *Envi-ronmental Pollution* 273: 116475.

Zhou, C., Zhang, X., Zhang, J., and Zhang, X. (2021). Representations of dynamics size distributions of mineral dust over East Asia by a regional sand and dust storm model. *Atmospheric Research* 250: 105403.

5

Silicosis and Asbestosis

Mapa S. T. Mapa[1] *and Rangika S. Hikkaduwa Koralege*[2]

[1] *Department of Biosystems Technology, Faculty of Technology, University of Sri Jayewardenepura, Homagama, Sri Lanka*
[2] *Department of Chemistry and Physics, College of Arts and Sciences, Western Carolina University, Cullowhee, NC, USA*

5.1 Introduction

Occupational lung diseases resulting from inhalation of inorganic dusts in the working environment are one of the common employment-related health hazardous conditions. Long-term exposure to many different airborne dust particles leads to chronic inflammatory processes in lower respiratory tract. Chronic exposure to quartz or silica leads to silicosis, and asbestos leads to asbestosis. Therefore, asbestosis and silicosis are commonly observed among overexposed workers in mining, construction, manufacturing, and building maintenance industries. Both fibrotic diseases deserve special attention due to similarities in the causative factors which occur years after initial exposure and the incurable end up impact. Therefore, early recognition and supportive interventions are important to reduce the risk of these chronic nonmalignant lung diseases.

5.2 Silicosis

5.2.1 Structure and Properties of Silica

Silica, an oxide of silicon, is a group IV metal oxide. Silica naturally occurs in both crystalline and amorphous forms. In crystalline silica, the silicon and oxygen atoms are arranged in a fixed geometric pattern. In comparison, in amorphous silica, no spatial arrangement of atoms is present. The basic unit of all crystalline and amorphous forms of silica is silicon–oxygen tetrahedra (SiO_4). Stishovite is an exception where under extreme pressure conditions, silicon is forced to bind to six oxygen atoms in an octahedral coordination. In each silicon–oxygen tetrahedron, each silicon atom is surrounded by four oxygen atoms and each oxygen atom is shared by two tetrahedra. The three-dimensional structure of crystalline silicas is determined by the regular arrangement of the tetrahedra, and the differences in the orientation and position of the tetrahedra create various polymorphs of silica. The amorphous silica forms are also composed of tetrahedra sharing their oxygen atoms. However, the orientation of the bonds in these silicas is random and lacks any long-range periodicity (Greenberg et al. 2007; Mizutani et al. 2016; Pollard 2016; Rice 2000). The classification and nomenclature of silica forms are shown in Table 5.1 (Lyon 2012).

The most common form of crystalline silica is α-quartz, and it is thermodynamically stable at ambient conditions. Although α-quartz is the only silica phase stable under ambient conditions, other silica polymorphs, namely α-tridymite, α-cristobalite, coesite, and stishovite, exist with metastability at the earth's surface. In nomenclature, α is used for a lower-temperature phase, and β is used for a higher-temperature phase. The different arrangements of tetrahedra in the various polymorphs and the presence of octahedrally coordinated silicon in stishovite imply notable differences in the distance between the silicon and oxygen atoms and the density (Lyon 2012; Pollard 2016; Rice 2000; Wagner 1997). A correlation between different physical parameters of silica and their biological responses have been reported in literature. Different crystal forms of silica affected lysis of human red blood cells and induced pulmonary inflammation and fibrosis in mice. Consequently, the physical parameters of these silicas play an important role in their biological responses which will be discussed in detail in a later section of this chapter. Silica has poor water solubility, and it is higher for the amorphous form than for the crystalline morphologies. Solubility increases with temperature and pH and is affected by the presence of trace metals (Callis et al. 1985; Greenberg et al. 2007).

Medical Geology: En route to One Health, First Edition. Edited by Majeti Narasimha Vara Prasad and Meththika Vithanage.

Table 5.1 Classification and nomenclature of silica forms.

Chemical name: Silicon dioxide	
Structure: Crystalline, amorphous, or cryptocrystalline	
Origin: Mineral, biogenic. or synthetic	
Classification	
Crystalline forms	*Natural* – α, β quartz; α, β_1, β_2 tridymite; α, β cristobalite; coesite; stishovite; moganite *Synthetic* – keatite; silica W; porosils (zeosils and clathrasils)
Amorphous forms	*Natural* – opal; biogenic silica; diatomaceous earths; silica fibers (biogenic); vitreous silica *Synthetic* – fused silica; pyrogenic or fumed silica; precipitated silica; colloidal silica; silica gel
Silica rocks (>90% SiO_2)	Quartzite, quartz arenite, diatomite, porcellanite, radiolarite, chert, geyserite
Varietal names	
Crystalline forms	*Natural* – α-quartz: agate; chalcedony; chert; flint; jasper; novaculite; quartzite; sandstone; silica sand; tripoli
Amorphous forms	Natural – diatomaceous earths: diatomite, kieselguhr, tripolite

5.2.2 Environmental Occurrence of Silica

Quartz, cristobalite, and tridymite are the most commonly occurring polymorphs of silica, which are found in rocks and soil. Keatite, coesite, stishovite, and moganite are rarely found in nature. Quartz, cristobalite, and tridymite can be released to the environment via both natural and anthropogenic sources. Some examples include foundry processes, brick and ceramics manufacturing, silicon carbide production, burning of agricultural waste or products, or calcining of diatomaceous earth. Quartz, being a hard, inert, and insoluble mineral, tolerates through the various weathering processes. Quartz is a major component of soils, composing 90–95% of all sand and silt fractions in a soil. α-Quartz is also found in trace to major amounts in most rock types, including igneous, sedimentary, metamorphic, and argillaceous (Fubini et al. 1995; Lyon 2012).

Crystalline tridymite and cristobalite are found in acid volcanic rocks. Cristobalite also occurs in some bentonite clays and as traces in diatomite. Coesite and stishovite are rarely found in nature and have been found in rocks that equilibrated in short-lived high-pressure environments such as meteoritic impact craters. Keatite has been found in high-altitude atmospheric dusts, which are believed to originate from volcanic sources (Fubini et al. 1995; Lyon 2012).

5.2.3 Industrial Applications and Human Exposure to Silica

Owing to its extensive natural occurrence in the earth's crust and the wide variety of uses of the material, workers may be exposed to crystalline silica in a large variety of industries and occupations. Table 5.2 summarizes the main industries and activities in which workers could be exposed to crystalline silica. The general population (non-occupationally exposed) is exposed to silica mainly due to the use of commercial products containing quartz. These products include cleansers, cosmetics, art clays and glazes, pet litter, talcum powder, caulk, putty, paint, and mortar. The general population may also be exposed via ingestion of potable water containing quartz particles (Rice 2000).

The primary exposure route of silica is by inhalation. Crystalline silica is considered respirable when particles are <10 μm in diameter. However, these particles are often described by a size range. "Coarse" particles are usually described as particles with a diameter more than 2 μm; "fine" particles are those with diameters in the range of 0.1–2.0 μm; and "ultrafine" particles are described as particles with a diameter less than 0.1 μm. In humans, inhalation of "respirable" particles defined as exposure to the particles in a mineral dust that are able to penetrate into the alveolar spaces of the lungs. Generally, particles have an aerodynamic diameter of <3–4 μm, while most particles larger than 5 μm may be deposited in the tracheobronchial airways and thus not reach the alveolar region. Respirable quartz levels exceeding 0.1 mg/m^3 have been reported in many industries worldwide. These levels are most frequently found: in metal, nonmetal, and coal mines and mills; in granite quarrying and processing, crushed stone and related industries; in foundries; in the ceramics industry; and in construction and sandblasting operations.

5.2.4 Silicosis and Its Pathologic Mechanisms

Silicosis is the oldest known occupational lung disease. Ancient Greeks (Hippocrates) who worked in quarries described a condition of "breathlessness" in miners. In 1690, Lohneiss noted that when "the dust and stones fall upon the lungs, the men have lung disease, breathe with difficulty (Greenberg et al. 2007). This disease was known for centuries, and in 1997, the International Agency for Research on Cancer (IARC) classified crystalline silica as a group 1 (known) human carcinogen (Lyon 2012). Silicosis is characterized by chronic inflammation and scarring in the upper lobes of the lungs. It can be categorized based on the quantity inhaled and length of exposure (Pollard 2016).

1) Chronic simple silicosis: it occurs after 15–20 years of moderate to low exposures to respirable crystalline silica.

Table 5.2 Main activities in which workers may be exposed to crystalline silica.

Industry/activity	Specific operation/task	Source material
Agriculture	Ploughing, harvesting, and use of machinery	Soil
Mining and related milling operations	Most occupations (underground, surface, and mill) and mines (metal and nonmetal, and coal)	Ores and associated rock
Quarrying and related milling operations	Crushing stone, sand and gravel processing, monumental stone cutting and abrasive blasting, slate work, and diatomite calcination	Sandstone, granite, flint, sand, gravel, slate, and diatomaceous earth
Construction	Abrasive blasting of structures buildings	Sand, concrete
	Highway and tunnel construction excavation and earth-moving masonry, concrete work, and demolition	Rock
		Soil and rock
		Concrete, mortar, and plaster
Glass, including fiberglass	Raw material processing	Sand and crushed quartz
	Refractory installation and repair	Refractory materials
Cement	Raw materials processing	Clay, sand, limestone, and diatomaceous earth
Abrasives	Silicon carbide production	Sand
	Abrasive products fabrication	Tripoli and sandstone
Ceramics, including bricks, tiles, sanitary ware, porcelain, pottery, refractories, and vitreous enamels	Mixing, molding, glaze, or enamel spraying, and finishing	Clay, shale, flint, sand, quartzite, and diatomaceous earth
Iron and steel mills	Refractory preparation and furnace repair	Refractory material
Silicon and ferro-silicon	Raw materials handling	Sand
Foundries (ferrous and nonferrous)	Casting, shaking out	Sand
	Abrasive blasting, fettling	Sand
	Furnace installation and repair	Refractory material
Metal products including structural metal, machinery, and transportation equipment	Abrasive blasting	Sand
Shipbuilding and repair	Abrasive blasting	Sand
Rubber and plastics	Raw material handling	Fillers (tripoli and diatomaceous earth)
Paint	Raw materials handling	Fillers (tripoli, diatomaceous earth, and silica flour)
Soaps and cosmetics	Abrasive soaps and scouring powders	Silica flour
Asphalt and roofing felt	Filling and granule application	Sand and aggregate, and diatomaceous earth
Agricultural chemicals	Raw material crushing and handling	Phosphate ores and rock
Jewelry	Cutting, grinding, polishing, and buffing	Semiprecious gems or stones, and abrasives
Dental material	Sandblasting and polishing	Sand and abrasives
Automobile repair	Abrasive blasting	Sand
Boiler scaling	Coal-fired boilers	Ash and concretions

Source: Lyon (2012).

2) Accelerated form of silicosis: it occurs after 5–10 years of high exposures to respirable crystalline silica.
3) Acute silicosis, or silicoproteinosis: it occurs after a few months or as long as five years following exposures to extremely high concentrations of respirable crystalline silica. This is the most severe form of silicosis.

The main pathophysiological event occurs in silicosis involves deposition of tiny silica particles into alveoli where they cannot be cleared. Upon particle deposition, macrophages present in alveoli initiates an inflammatory response, which stimulates fibroblasts to proliferate and produce collagen. The deposited silica particles will be enclosed by collagen leading to fibrosis and nodular lesions that are characteristic of the disease. The collagen containing silicotic nodules are a specific response to crystalline silica. However, the cellular and molecular components responsible for the inflammatory and fibrotic responses are not the same.

There are numerous animal experimental studies done on silicosis and silica-induced inflammatory responses. Warheit et al. observed a persistent pulmonary inflammatory response and impairment of alveolar macrophage clearance functions in rats after a short-term inhalation bioassay of the pulmonary toxicity of aerosolized quartz particles with an aerodynamic diameter of 3.7 μm. They have observed progressive lesions after one month after a three-day (6 h/day) aerosolized quartz exposure of 100 mg/m^3. The lesions had developed into a multifocal, granulomatous-type pneumonitis after two months (Warheit et al. 1997). Cassel et al. has reported that inflammatory response and development of pulmonary fibrosis after inhalation of silica is dependent on the Nalp3 inflammasome (Cassel et al. 2008). Activation of inflammasome occurs following a cascade of reactions, including uptake of silica by macrophage scavenger receptors, lysosomal rupture, and release of cathepsin B accompanied by production of reactive oxygen species (ROS) and potassium efflux. Although scavenger receptors play a significant role in clearing silica, the binding of silica to scavenger receptors mediate apoptosis of macrophages. Pro-inflammatory cytokines released during apoptosis contribute to lung inflammation and fibrosis (Cassel et al. 2008; Pollard 2016).

Acute and chronic inflammation and fibrosis have different molecular and cellular requirements as shown in Figure 5.1 (Pollard 2016). As shown in Figure 5.1, inflammation and fibrosis occur independently of T, B, NKT, and NK cells (T lymphocyte; T, B lymphocyte; B, natural killer T cell; NKT, interleukin; NK) (Fubini et al. 1995). Acute inflammation, but not chronic inflammation or fibrosis, requires IL-17; conversely, chronic inflammation, but not acute inflammation or fibrosis, requires type I IFN and IRF7 (Bagchi 1992; Heppleston and Styles 1967; Suratt et al. 1977). Further studies have shown that presence of innate immune response components (particularly IL-1 receptor, IL-1, ASC, NALP3, IL-18 receptor, IL-33 receptor, TRIF, and TLR2, 3, and 4) are not required for the accumulation of collagen in the lung. However, inflammation, neutrophil accumulation, IL-1β release, and granuloma formation require MyD88 (Suratt et al. 1977).

Acute silicosis involves a variety of different mechanisms compared to chronic silicosis. Electron microscopy of images of lungs affected by acute silicosis have shown hypertrophic type II pneumocytes lining the alveoli. Hypertrophic pneumocytes could produce excessive

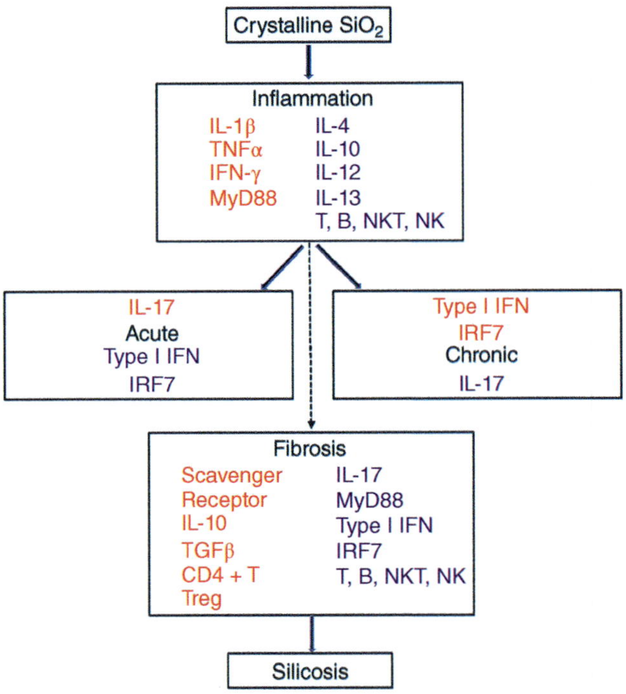

Figure 5.1 Molecular and cellular components are involved in silica-induced inflammation and fibrosis. Components required for acute or chronic inflammation and fibrosis are highlighted in red, while those not essential have been highlighted in green. (TNF, tumor necrosis factor; IFN, interferon; MyD88, myeloid differentiation primary response gene 88; IL, interleukin; NK, natural killer; NKT, natural killer T cell; IRF, interferon regulatory factor; TGF, transforming growth factor, regulatory T cell; Treg, T lymphocytes; CD4). *Source:* Pollard (2016) / Frontiers Media S.A / CC BY 4.0.

amounts of proteinaceous material and surfactant proteins. This may lead healthy alveoli become filled with protein-containing material (Hoffmann et al. 1973). Another factor that contributes the development of silicotic lung disease is excessive free-radical formation. Generation of free radicals is dependent on surface morphology of silica particles. Freshly fractured silica contains more cleaved particle surfaces than intact silica that could aid the formation of surface ROS such as peroxides and hydroperoxides. These free radicals are responsible for cell/ or DNA damage (Pollard 2016).

Chronic silicosis is associated with chronic inflammatory changes within the alveoli, and this condition will ultimately result in the development of pulmonary fibrosis. The onset of chronic silicosis occurs due to phagocytosis of silica particles by alveolar macrophages in an attempt to clear them from the lung. If silica particles are not cleared efficiently, these alveolar macrophages get damaged and release ROS, reactive nitrogen species, and excess free radicals. This triggers the release of transcription factors which will then trigger the production and release of inflammatory cytokines (TNF-, IL-1, and IL-6), proteases, and arachidonic acid metabolites (leukotriene-B4 and prostaglandin E2). Released silica particles from dying alveolar macrophages can be re-engulfed by other alveolar macrophages, and this induces the injury cycle once more. Repetitive injury cycles will lead to further inflammatory changes in the lung. It has been shown that a constant production of fibrotic factors contributes to the evolution of silicotic lesions and production of large amounts of fibronectin and collagen. At this stage, pulmonary architecture becomes permanently changed due to scar tissues (Dubois et al. 1989; Fubini et al. 1995; Greenberg et al. 2007).

The diagnosis of silicosis is primarily based on a clearly documented history of substantial silica exposure, usually in the occupational setting. However, in combination with excessive exposure to silica, medical findings such as chest radiographs confirming the presence of nodular opacities of the lung are also important. Chronic silicosis can be distinguished from acute silicosis using radiography by the presence of large upper lobe opacities in conjunction with small, diffuse nodular lesions. Acute silicosis could be differentiated from chronic silicosis on chest X-rays due to the ground-glass appearance of the lung due to acute alveolar filling. Apart from acute and chronic silicosis, epidemiological studies show an association between lung cancer and inhaled crystalline silica (α-quartz and cristobalite) from occupational exposure (Greenberg et al. 2007; Lahiri et al. 2005).

5.2.5 Prevention and Treatment of Silicosis

Silica is a major component of sand, granite, quartz, and most stone. Exposure to silica occurs in numerous industries, including construction, mining, mineral processing, foundries, pottery, and glass. Silica dust is produced in a variety of industrial processes as discussed in Table 5.2. Silicosis is a debilitating and often fatal lung disease that is completely preventable. Silicosis can progress even after the person has been removed from the dusty environment, because scar tissue in the lungs continues to grow in response to the silica particles embedded in alveolar tissues. Silicosis also makes people vulnerable to other disease conditions such as tuberculosis. Interventions to control silica varies from industry to industry, and it is dependent on the process that silica been used. Depending on the process, a combination of interventions could also be used. Types of interventions used include substitution for silica sand, wet method, ventilation systems, worker training, and personal protective equipment (respirators). The best way to prevent silicosis is to remove silica from the entire operation, and various other substitutes for silica are available, including specular hematite, coal, copper and nickel slags, and aluminum oxide. However, when substitution is not an option, providing adequate knowledge of the hazard and training on handling and working with silica dust are the first steps to prevent exposure. Additionally, silica dust could be controlled through wetting the surface. Local exhaust ventilation (LEV) is another method to control silica dust. It has been shown that the exposure levels were reduced from 20 mg of respirable dust per cubic centimeter to less than 0.2 mg/cc with the use of a ventilated grinder. The use of respirators is also common to prevent silica exposure. However, for silicosis prevention in high-exposure jobs, respirators are insufficient (Greenberg et al. 2007; Health 2002).

Several treatment methods that have been developed aiming at decreasing the pulmonary inflammatory response to silica are available. Corticosteroids and aluminum citrate have been used in pharmacologic treatment protocols. But they have shown varying success for silicosis. One study demonstrated an improvement in both inflammatory bronchoalveolar lavage and pulmonary function in chronic silicosis after the administration of corticosteroids. Another study reported that acute silicosis was reversed by corticosteroid therapy. Some other studies with administration of aluminum citrate powder by inhalation have shown both positive and negative effects on silicosis. Polyvinylpyridine-*N*-oxide (PVPNO) has been shown to slow down the course of disease by concentrating silica particles inside of cells. PVPNO has shown to decrease the production of ROS, acting as a hydrogen acceptor, and has been capable of coating the surface of silica particles and thereby decrease the potential for silica toxicity in both in vitro and in vivo models. Unfortunately, none of the current therapies have clearly reduced the mortality associated with silicosis. Therefore, it is very important that other disease conditions arising due to silicosis, such as mycobacterial infections, tuberculosis, and other pulmonary infections, should be identified and treated promptly in all silicosis patients (Greenberg et al. 2007).

5.3 Asbestosis

5.3.1 Structure of Asbestos

Naturally occurring mineral silicate fibers of serpentine (90%) and amphibole series (10%) are identified as asbestos. The silicate tetrahedra which is the building block of asbestos arrangement is leading to these categories as sheets $(Si_4O_{10})^{-4}$ in serpentine and double chains $(Si_4O_{11})^{-6}$ in the amphiboles (Virta 2002).

Chrysotile is the only fibrous sheet silicate mineral type of serpentine asbestiform, and five types of chain structure amphibole asbestiform are belong to this category. They are crocidolite (riebeckite asbestos), amosite (grunerite asbestos), anthophyllite, tremolite, and actinolite, as shown in Figure 5.2 (Gualtieri et al. 2018; Kusiorowski et al. 2012).

Crystal structure of chrysotile consists of wrapped sheets composed of layers of silicate tetrahedra in a pseudo-hexagonal network joined to layers of $Mg_6O_4(OH)_8)^{-4}$ octahedral brucite layer, in which on one side, two out of every three hydroxyls are replaced by apical oxygens of the silicate tetrahedra (Bernstein et al. 2013; Cressey and Whittaker 1993; Virta 2002). The silica layer has the smaller dimensions, and roughly each layer is about 7.3 Å. There is no chemical bonding identified in between the layers. The different dimensions of these two components cause the structural mismatch and curl, concentrically or spirally layers in the chrysotile. 12–20 of these layers with some mechanical interlocking build the fiber walls. Finally, this structural arrangement creates very long, smooth, silky chrysotile fibers which are typically around 200 Å in diameter (Bernstein et al. 2013; Cressey and Whittaker 1993).

In contrast to the chrysotile, fiber amphibole fibers appear as individually, rather than in fiber bundles. The amphibole asbestiform types mainly consist of a double chain of corner-sharing tetrahedra and a strip of edge-sharing octahedra (Bernstein et al. 2013; Cressey and Whittaker 1993; Virta 2002). The layer of anionic valency sites formed in this double-chain arrangement is

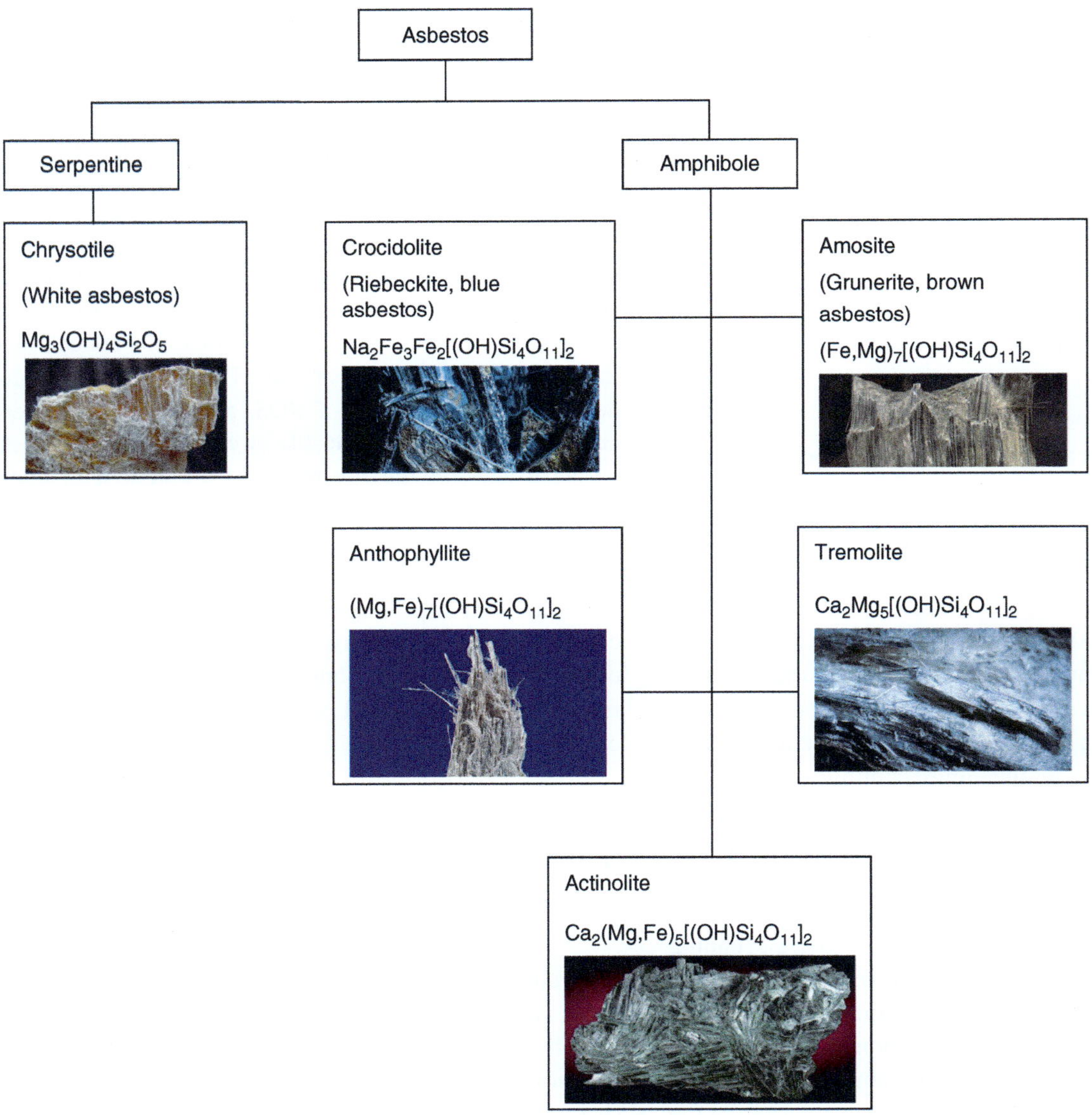

Figure 5.2 Classification of asbestose minerals. *Source:* Asbestos.com.

neutralized by the metal cations in 16 sites of 4 different types. The complex chemical composition is represented as:

$$A_{0-1},B_2C_5T_8O_{22}\left(OH,O,F,Cl\right)_2$$

where

A = Na, K
B = Na, Ca, Mg, Fe^{+2}, Mn, Li
C = Al, Fe^{+2}, Fe^{+3}, Ti, Mg, Mn, Cr
T = Si, Al

A, B, and C each represent cationic sites within the crystal structure and can host a large variety of metal cations without substantial disruption of the lattice. This mixture of different ions accommodation in the space between the silicate chains is determined by the host rock. This double chain of tetrahedral silicate with the silica on the outside of the fiber makes amphibole varieties as very strong and durable structure. Based on this generic representation, same structure contains amphibole asbestos divided into five different varieties as shown in Figure 5.2.

5.3.2 Properties of Asbestos

Chemically inert or nearly inert asbestos fibers possess several physical and chemical properties. Flexibility, high-tensile strength, wear and friction characteristics, large surface area, incombustibility, thermal, electrical, and acoustic insulation, abrasion resistance, adsorption capacity, and resistance to chemical, thermal, and biological degradation are the general characteristic features of asbestos (Guidotti et al. 2004; Virta 2002).

Each type of asbestos has different physical characteristics due to the mineralogical composition and their structure which determine the industrial applications. Table 5.3 summarizes physical and chemical properties of the six types of asbestos fibers.

5.3.3 Sources of Asbestos Fiber

Asbestos occurs naturally on every continent in the world, and they are referred as naturally occurring asbestos (NOA) (Lee et al. 2008). These fibrous mineral NOA can be found in certain rocks or soil as a result of natural geological processes. Serpentinites, altered ultramafic rocks, and some mafic rocks are the most commonly found rocks. In most of the areas, asbestos minerals are found in very low quantities, and it is not sufficient to explore in commercial scale. Natural weathering condition and human activities may disrupt these NOA and release fibers into the air. These airborne asbestos fibers can contaminate the nearby water sources.

Table 5.3 The physical and chemical properties of the six types of asbestos fibers.

Property \ Name	Chrysotile	Crocidolite	Amosite	Anthophyllite	Tremolite	Actinolite
Physical state	Solid	Solid	Solid	Solid	Solid	Solid
Color	White, gray, green, and yellowish	Lavender, blue, and green	Brown, gray, and greenish	Gray, white, brown-gray, and green	White to pale green with increasing iron content	White to pale green with increasing iron content
Density (g/cm^3)	2.55	3.37	3.43	2.85–3.1	2.9–3.2	3.0–3.2
Flexibility	Good	Good	Fair	Fair to brittle	Brittle	Fair to brittle
Decomposition temperature (°C)	600–850	400–600	600–800	600–850	950–1040	620–960
Flammability limits	Nonflammable	Nonflammable	Nonflammable	Nonflammable	Nonflammable	Nonflammable
Tensile strength (10^3 kg/cm^2)	31	35	17	(<7)	5	5
Solubility: water	Insoluble	Insoluble	Insoluble	Insoluble	Insoluble	Insoluble
Organic solvents	Insoluble	Insoluble	Insoluble	Insoluble	Insoluble	Insoluble
Resistance to acids	Fibers degrade in dilute acid	Good	Fibers degrade in dilute acid	Very good	Very good	Fibers degrade in dilute acid
Resistance to alkaline	Very good	Good	Good	Very good	Good	Good
Electrical charge at neutral pH	Positive	Negative	Negative	Negative		

5.4 Industrial Application

As described above, asbestos has several chemical and physical properties which make it ideal raw material for many industries. Early asbestos uses were initiated with its thermal behavior and tensile strength in between 2000 and 3000 BCE. Egyptian used asbestos containing fabrics to wrap the embalmed bodies of pharaohs to protect the bodies from deterioration. Anthophyllite contain clay utensils, and pottery was discovered in Finland which was dating back to approximately 2500 BCE. (Virta 2002; Wyers 1949). In the nineteenth century, the commercial usage of asbestos fibers was begun in Italy with the development of asbestos textiles. By the time of the end of the nineteenth century, significant number of asbestos deposits were discovered throughout the world, and their exploitation had begun mainly in Canada, South Africa, and the Union of Soviet Socialist Republics. Later, the demand for asbestos fibers grew dramatically for numerous industrial applications from the beginning of the twentieth century. Due to the physio-chemical properties, manufacturers utilized asbestos fibers as raw material for their numerous products (Bowles 1946; Guidotti et al. 2004; Virta 2002). In 1980s, Johns Manville, National Gypsum Company, Armstrong World Industries, and Raybestos-Manhattan Co. were some of the distinguished leading asbestos manufacturing companies.

The most commonly used form of asbestos is chrysotile or white asbestos, and it can be found in the roofs, ceilings, walls, and floors of homes and businesses. Not only that but also chrysotile asbestos are used in automobile brake linings, gaskets and boiler seals, and insulation for pipes, ducts and appliances. These fibers have been used to strengthen the plastic products made from PVC, phenolics, polypropylene, nylon, etc. (Virta 2002). The two most important amphibole asbestos minerals are amosite and crocidolite, and both are hydrated silicates of iron, magnesium, and sodium (crocidolite only). Brown asbestos or amosite is most frequently used in cement sheets and pipe insulation. It can also be found in insulating board, ceiling tiles, and thermal insulation products. Blue asbestos or Crocidolite is commonly used to insulate steam engines and, in some spray, on coatings, in pipe insulation, in plastics, and in cement products. Anthophyllite is used in limited quantities for insulation products and construction materials. Tremolite and actinolite contain similar chemical components, and they are not used commercially. But both of them and anthophyllite can be found as contaminants in chrysotile asbestos, vermiculite, and talc.

Due to this huge demand in the consumption, world production of asbestos peaked at about 4.8×10^6 tons in 1975. In the first half of the century, Canada was the main producer, then Soviet Union had become the leading manufacture with some other countries such as Brazil, China, South Africa, Zimbabwe, and Kazakhstan by supplying more than 90% of the requirement (Dummer and Gotay 2015; Virta 2002; Vogel 2005). Heavy utilization of asbestos consumption was identified in the industrialized countries like Russia, China, Brazil, India, etc. In recent years, asbestos production and consumption were declined due to the identification of health disasters with rising labor complain against the usage. With this, most of the countries introduced the strict regulations for asbestos and/or outright bans on exposure. Asbestos is banned in 67 countries according to the International Ban Asbestos Secretariat, but China, Russia, the United States, Canada, and India are not included in this list (Guidotti et al. 2004). Even with that according to the latest available consumption data, worldwide 2 030 000 tons consumed annually (Furuya et al. 2018). This indicates that asbestos is widely used material up to date.

5.5 Exposure to Mineral Fiber

Microscopic asbestos fibers can be remained in the air for hours and become airborne toxic mineral. In an ideal environment with little disturbances, asbestos fibers may take 48–72 hours to settle. But it can easily become aerial with a little disturbance, since it is so light-weighted material. Therefore, these airborne asbestos fibers place anyone in danger by inhaling or ingesting. After the long-term inhalation or ingestion of asbestos, the fibers cannot be removed from the body. The fibers may embed into the tissue and cause irritation and scarring overtime by leading to the health defects.

Most people are exposed to asbestos fibers through their occupation, and it is a No. 1 cause for work-related deaths in the world (Furuya et al. 2018). Miners and millers of asbestos, workers in the manufacturing of asbestos containing material and laborers, plumbers, carpenters, personal in the construction places, shipyard and aerospace, etc., are subjected to substantial direct exposure of asbestos fibers (Wagner 1997). In the early 1900s, researchers recognized an unusual fibrosis of the lungs from the factory inspectors in the United Kingdom and France (Murray 1990; Virta 2002; Wyers 1949). It was estimated that 27.5 million US workers were potentially exposed to the asbestos from 1940 to 1979, and in the 1990s approximately 6000 incident annual US lung cancer cases are assessed related to this (Steenland and Stayner 2014). Even though most of the companies stopped using asbestos in the 1980s, remaining millions of older buildings worldwide contain asbestos-containing materials. Therefore, not only workers who are going to expose during the original installation but also in the maintenance, repair, and removal of asbestos-containing material release considerable quantities of asbestos fibers (Guidotti et al. 2004).

The most recent incident of tons of crushed asbestos fibers releasement was 9/11 attack on the World Trade Center in the New York city, USA in 2001. This caused a sudden and very serious exposure difficulty for rescue, recovery, and cleanup workers who remained at the site for months. In 2006, study was conducted on them and identified about 70% of them are suffered new or worsened respiratory problems, and about 28% had abnormal lung function tests (Herbert et al. 2006). This provides clear evidence for occupational health effect on direct exposure to asbestos fibers. Other than direct exposure, especially women and children may indirectly expose to this mainly by an asbestos worker who unintentionally brings asbestos fibers to the home with their work clothes, hair, and skin. Similarly, professionals like electricians, masons, and painters can be witness to this second form of asbestos exposure. The third and most common form of asbestos exposure is the general community exposure, such as the use of asbestos for road surfaces, playground material, landfills, and chemical paints. Direct exposure has high risk of asbestosis compared to the other forms of exposure, since asbestosis is a dose-dependent disease.

5.6 Disease Description and Mechanisms of Action

Recurring exposure of human tissues to the asbestos fibers may result in the accumulation, inflammation, and DNA damage. Lung tissues are the major exposure area of the human body, since asbestos fibers enter into the human body through the nose and mouth via inhalation and drinking. Later heart, diaphragm, and intestine tissues may get the asbestos fibers Figure 5.3). There are differences in the mineralogical characteristics, patterns of translocation after inhalation, and biological persistence of various forms of asbestos fibers (Sayers and Dreessen 1938; Wyers 1949). Practically, commercial usage of asbestos involves various mineral species. Therefore, exposures tend to be mixed fiber types. But, overtime of exposure to all forms of asbestos can cause those cellular changes that lead to cancer and other associated diseases (Guidotti et al. 2004; Liu et al. 2013; Mossman and Churg 1998). Four different types of cancers can develop due to the asbestos fiber exposure. They are mesothelioma, lung cancer, ovarian cancer, and laryngeal cancer. Asbestosis, pleural plaques, pleural effusion, diffuse pleural thickening, pleurisy, and atelectasis are some of the several pulmonary conditions caused by the asbestos fiber.

Bilateral diffuse interstitial fibrosis of the lungs caused by the inhalation of asbestos fibers is defined as asbestosis (Guidotti et al. 2004; Mossman and Churg 1998;

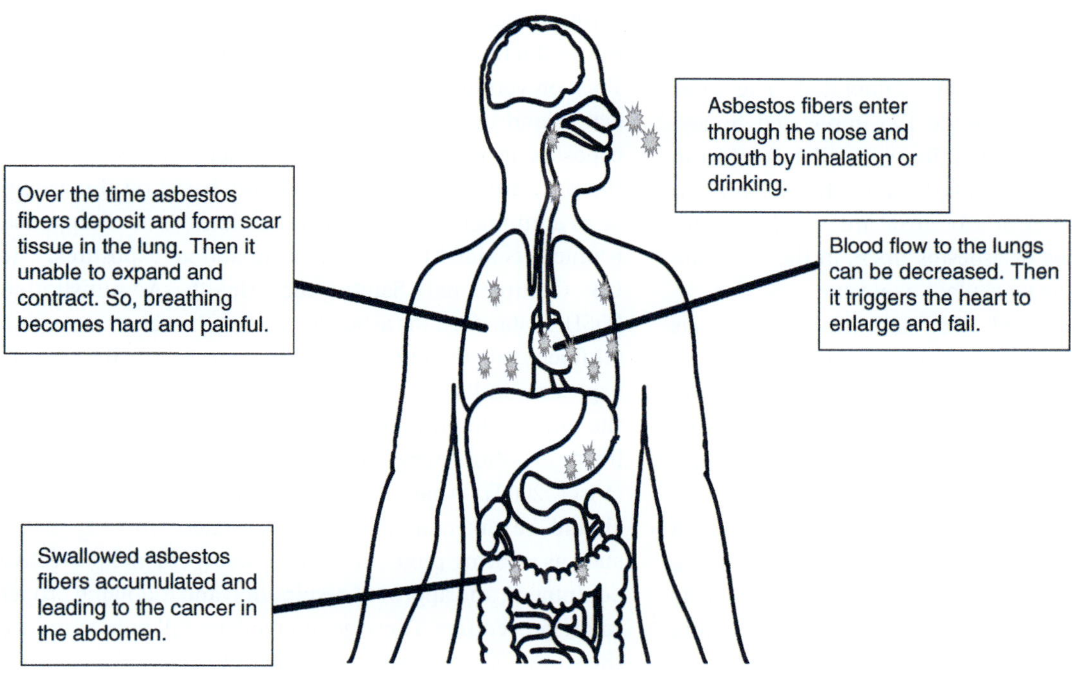

Figure 5.3 The effect of asbestos fiber exposure.

Wagner 1997). So, infected scarred lungs could not expand and relax normally. Asbestosis can develop in 5–10 years after initial exposure of asbestos fiber exceeding 5 million particles per cubic foot (m.p.p.c.f.) (Sayers and Dreessen 1938). The common symptoms of asbestosis are shortness of breath due to decreasing lung volumes, persistence cough, wheezing, fatigue, chest or shoulder pain, weight loss due to loss of appetite, dry or crackling sound in the lungs while breathing, and swollen fingertips and toes in sever stage (Wagner 1997). Most patients with this gradually progressive disease may be free of symptoms, and it is first noticeable only on physical exertion (chest X-ray or CT scan of the lungs) or during the surveillance activities. But asbestosis is more prevalent and more complex in the smokers and ex-smokers who have worked in dusty environments. Even though the explanation is not clear yet, this may be due to that the cigarette smoke exposure increases the retention of short fibers much more than the retention of long fiber (Guidotti et al. 2004; Mossman and Churg 1998).

Asbestosis is characteristically most advanced and appears earliest in the lower lung fields. But rarely massive bilateral upper lobe fibrosis observed as well characterized condition. When considering the severity of asbestosis, the grade I is the mildest form, and it involves the alveolated walls of respiratory bronchioles and the alveolar ducts. Alveolar ducts and/or two or more layers of adjacent alveoli involving acinus move to more severe grade II fibrosis. At this stage, normal lung remains in a zone between adjacent bronchioles. When entire acinus involves by advanced and coalescent fibrosis, all lung between at least two adjacent bronchioles is affected in grade III asbestosis and some alveoli are completely obliterated. The most severe grade of asbestosis (grade IV) implicated honeycomb cysts, areas of confluent fibrosis, and large (up to 1 cm) dilated spaces grossly visible in parenchyma (Guidotti et al. 2004). These grades of asbestosis are correlated with counts and frequencies of asbestos fibers in the lung and estimates of cumulative workplace exposure.

Molecular mechanism of fiber dose-driven asbestosis disease is not fully understood yet, and scientists believe that multiple mode of actions is associated here (Gwinn 2011). Alveolar epithelial cell (AEC) apoptosis and inflammatory cells involvement modulated by endoplasmic reticulum (ER) are the crucial early events in asbestosis. The accumulation of alveolar macrophage with fibroblast proliferation and collagen deposition in the walls of the respiratory bronchioles and alveolar ducts where fiber deposited were observed. Highly ROS were generated by these deposited asbestos fibers because they contain <2% of total weight iron and are wrapped with mucopolysaccharides and iron-rich proteins (e.g. ferritin and hemosiderin). ROS production can be occurred on fiber surface, from alveolar macrophage and mitochondria of AEC and other target cells. These generated ROS alter the mitochondrial membrane potential and release cytochrome *c* from the mitochondria into the cytosol. Then caspase-9 and -3 can be activated and lead to the apoptosis (Kamp and Weitzman 1999; Liu et al. 2013). However, future studies with transgenic animal models and humans are essential to understand more details of asbestosis molecular mechanism.

5.7 Prevention and Treatment Plans

According to latest information, asbestos causes approximately 255 000 deaths annually, and out of that 233 000 deaths are responsible for work-related exposures (Furuya et al. 2018). Predominantly, construction workers such as carpenters and joiners, pipe fitters, laggers, laborers, painters, and shop fitters are included in this number (Wagner 1997; Walters et al. 2018). There is no effective treatment to reverse the course of asbestosis. Only early identification of this irreversible asbestosis followed by abandonment from asbestos exposure may lead to more favorable long-term results. At that time, supportive care such as pulmonary rehabilitation, oxygen therapy, and use of inhaler with mild symptoms are the only available option to treat these patients. Therefore, elimination of hazardous exposure conditions is prime importance to prevent this long-term lung disease.

Introducing wide range of laws and regulations to asbestos production and usage is the major prevention method. This prevention method covers worker's compensation claims and injury litigation (EPA 2014). Since cumulative exposure to respirable hazardous asbestos dust increases the risk, introducing exposure limits based on the weight or as a number of fibers-per-unit-volume in working environment is aided to regulate occupational exposures. The US Occupational Safety and Health Administration (OSHA) standard for asbestos recently has been revised to 0.1 fibers/mL (Steenland and Stayner 2014). In general, exposure limits vary considerably among countries with the range being from 0.1 fibers per cubic centimeter (f/cc) for 4- or 8-hour time-weighted-average exposure to 2 f/cc (Virta 2002). Apart from these, necessary precautions should be taken in the asbestos-related working places such as wearing protective equipment (personal protective equipment, and approved respirators) and following proper safety procedures. These good practices will help to reduce the worker's asbestosis risk in the future.

Regardless of the usefulness of asbestos, present efforts are to eliminate this man-made problem by considering the harmful effects of asbestos on health. Therefore, applying programs like the International Labor Organization/World Health Organization Joint Program for the Elimination of Asbestos-Related Diseases and policies like banning of new asbestos use and tight control and management of existing structures containing asbestos are essential (Virta 2002). Currently, 67 countries have completely banned with exemptions for the use of asbestos and, in some cases, asbestos products. So, overtime consumption can be declined, and alternative sustainable products will enter the market. That will be the best option to decline this lung disease. However, in future, asbestos probably will remain in specialized applications particularly for matrix-based products.

References

U.S. Environmental Protection Agency (2014). *Occupational Safety and Health (Control of Asbestos at Work) Regulations*. (EPA), U. S. E. P. A.

Bagchi, N. (1992). What makes silica toxic? *British Journal of Industrial Medicine* 49 (3): 163–166.

Bernstein, D., Dunnigan, J., Hesterberg, T. et al. (2013). Health risk of chrysotile revisited. *Critical Reviews in Toxicology* 43 (2): 154–183.

Bowles, O. (1946). *Asbestos – The Silk of the Mineral Kingdom*. New York: The Ruberoid Co.

Callis, A.H., Sohnle, P.G., Mandel, G.S. et al. (1985). Kinetics of inflammatory and fibrotic pulmonary changes in a murine model of silicosis. *The Journal of Laboratory and Clinical Medicine* 105 (5): 547–553.

Cassel, S.L., Eisenbarth, S.C., Iyer, S.S. et al. (2008). The Nalp3 inflammasome is essential for the development of silicosis. *Proceedings of the National Academy of Sciences of the United States of America* 105 (26): 9035–9040.

Cressey, B.A. and Whittaker, E.J.W. (1993). Five-fold symmetry in chrysotile asbestos revealed by transmission electron microscopy. *Mineralogical Magazine* 57 (389): 729–732.

Department of Health and Human Services (DHHS), The National Institute for Occupational Safety and Health (NIOSH) (2002). *Health Effects of Occupational Exposure to Respirable Crystalline Silica*, 129. DHHS (NIOSH).

Dubois, C.M., Bissonnette, E., and Rola-Pleszczynski, M. (1989). Asbestos fibers and silica particles stimulate rat alveolar macrophages to release tumor necrosis factor. Autoregulatory role of leukotriene B4. *The American Review of Respiratory Disease* 139 (5): 1257–1264.

Dummer, T. and Gotay, C. (2015). Asbestos in Canada: time to change our legacy. *Canadian Medical Association Journal* 187 (10): E315–E316.

Fubini, B., Bolis, V., Cavenago, A., and Volante, M. (1995). Physicochemical properties of crystalline silica dusts and their possible implication in various biological responses. *Scandinavian Journal of Work, Environment & Health* 21 (2): 9–14.

Furuya, S., Chimed-Ochir, O., Takahashi, K. et al. (2018). Global asbestos disaster. *International Journal of Environmental Research and Public Health* 15 (5): 1000.

Greenberg, M.I., Waksman, J., and Curtis, J. (2007). Silicosis: a review. *Disease-a-Month* 53 (8): 394–416.

Gualtieri, A.F., Pollastri, S., Gandolfi, N.B., and Gualtieri, M.L. (2018). In vitro acellular dissolution of mineral fibres: a comparative study. *Scientific Reports* 8 (1): 1–12.

Guidotti, T.L., Miller, A., Christiani, D. et al. (2004). Diagnosis and initial management of nonmalignant diseases related to asbestos. *American Journal of Respiratory and Critical Care Medicine* 170 (6): 691–715.

Gwinn, M.R. (2011). Multiple modes of action of asbestos and related mineral fibers. *Journal of Toxicology and Environmental Health – Part B: Critical Reviews* 14 (1–4): 1–2.

Heppleston, A.G. and Styles, J.A. (1967). Activity of a macrophage factor in collagen formation by silica. *Nature* 214 (5087): 521–522.

Herbert, R., Moline, J., Skloot, G. et al. (2006). The world trade center disaster and the health of workers: five-year assessment of a unique medical screening program. *Environmental Health Perspectives* 114 (12): 1853–1858.

Hoffmann, E.O., Lamberty, J., Pizzolato, P., and Coover, J. (1973). The ultrastructure of acute silicosis. *Archives of Pathology & Laboratory Medicine* 96 (2): 104–107.

Kamp, D.W. and Weitzman, S.A. (1999). The molecular basis of asbestos induced lung injury. *Thorax* 54 (7): 638–652.

Kusiorowski, R., Zaremba, T., Piotrowski, J., and Adamek, J. (2012). Thermal decomposition of different types of asbestos. *Journal of Thermal Analysis and Calorimetry* 109 (2): 693–704.

Lahiri, S., Levenstein, C., Nelson, D.I., and Rosenberg, B.J. (2005). The cost effectiveness of occupational health interventions: prevention of silicosis. *American Journal of Industrial Medicine* 48 (6): 503–514.

Lee, R.J., Strohmeier, B.R., Bunker, K.L., and Van Orden, D.R. (2008). Naturally occurring asbestos-a recurring public policy challenge. *Journal of Hazardous Materials* 153 (1–2): 1–21.

Liu, G., Cheresh, P., and Kamp, D.W. (2013). Molecular basis of asbestos-induced lung disease. *Annual Review of Pathology: Mechanisms of Disease* 8: 161–187.

Lyon, F.R. (2012). Arsenic, Metals, Fibres, and Dusts. In: *International Agency for Research on Cancer. Monographs on the Evaluation of Carcinogenic Risks to Humans*, vol. 100C, 11–465. IARC Working Group on the Evaluation of Carcinogenic Risks to Human.

Mizutani, R.F., Lombardi, E.M.S., de Paula Santos, U., and Terra-Filho, M. (2016). Silica exposure, silicosis, autoimmune diseases, tuberculosis and nontuberculous pulmonary mycobacterial disease. *European Respiratory Journal* 48 (suppl. 60): PA1171.

Mossman, B.T. and Churg, A. (1998). Mechanisms in the pathogenesis of asbestosis and silicosis. *American Journal of Respiratory and Critical Care Medicine* 157 (5 PART I): 1666–1680.

Murray, R. (1990). Asbestos: a chronology of its origins and health effects. *British Journal of Industrial Medicine* 47 (6): 361–365.

Pollard, K.M. (2016). Silica, silicosis, and autoimmunity. *Frontiers in Immunology* 7: 97.

Rice, F. (2000). *Crystalline Silica, Quartz*. Geneva: World Health Organization.

Sayers, R.R. and Dreessen, W.C. (1938). Asbestosis. *American Journal of Public Health* 29: 205–214.

Steenland, K. and Stayner, L. (2014). Silica. *Cancer Causes & Control* 8 (3): 491–503.

Suratt, P.M., Winn, W.C., Brody, A.R. et al. (1977). Acute silicosis in tombstone sandblasters. *The American Review of Respiratory Disease* 115 (3): 521–529.

Virta, R.L. (2002). *Asbestos: Geology, Mineralogy, Mining, and Uses*, 1–28. U.S. Department of the Interior – U.S. Geological Survey.

Vogel, L. (2005). Asbestos in the world. In: *European Asbestos Conference* (ed. L. Kazan-Allen), 7–21. HESA Newsletter.

Wagner, G.R. (1997). Asbestosis and silicosis. *Lancet* 349 (9061): 1311–1315.

Walters, G.I., Robertson, A.S., Bhomra, P.S., and Burge, P.S. (2018). Asbestosis is prevalent in a variety of construction industry trades. *NPJ Primary Care Respiratory Medicine* 28 (1): 2017–2019.

Warheit, D.B., Hansen, J.F., Yuen, I.S. et al. (1997). Inhalation of high concentrations of low toxicity dusts in rats results in impaired pulmonary clearance mechanisms and persistent inflammation. *Toxicology and Applied Pharmacology* 145 (1): 10–22.

Wyers, H. (1949). Asbestosis. *Postgraduate Medical Journal* 25 (290): 631–638.

6

Radon and Health

Dinusha Peramune[1], Niwanthi Dissanayake[2], Vidura D. Thalangamaarachchige[2], Mohamed N. M. Farhath[3], and Rohan S. Dassanayake[1]

[1] *Department of Biosystems Technology, Faculty of Technology, University of Sri Jayewardenepura, Homagama, Sri Lanka*
[2] *Department of Chemistry and Physics, McNeese State University, Lake Charles, LA, USA*
[3] *Department of Chemical Sciences, Faculty of Applied Sciences, South Eastern University of Sri Lanka, Sammanthurai, Sri Lanka*

6.1 Introduction

The radioactivity was accidentally discovered in 1896 by the French Physicist Henry Becquerel while investigating the properties of fluorescent materials that emit X-rays upon stimulated by light. Becquerel observed the emission of penetration radiation by potassium uranyl sulfate crystals that can be registered on a photographic plate in the absence of sunlight. Further studies proved that the uranium (U) salt emitted spontaneous radiation owing to its inherent radioactive properties and that radiation differed from X-rays, deflecting by both electric and magnetic fields. In the next years, Pierre and Marie Curie found that thorium (Th) was radioactive and discovered two new radioactive elements, polonium (Po) and radium (Ra). They also reported that the radioactive radiation could be separated into three components, namely, alpha (α), beta (β), and gamma (γ) rays (Radvanyi and Villain 2017). Becquerel and the Curies won the Nobel Prize in 1903 for their work on radioactivity. The third radioactive element, actinium (Ac), was later identified by André Debierne (Radvanyi and Villain 2017).

Marie and Pierre Curies were the first to observe that the decay product of radium is radioactive; however, the gaseous form of the product was not known. The decay product of radium was first discovered by the German Chemist Friedrich Ernst Dorn in 1900 while working on the radioactive materials emitted by thorium (Nagaratnam 1994; Degu Belete and Alemu Anteneh 2021). Then also found out that radon and thorium both emitted the same radioactive gaseous substance. This decay product was originally named niton after the Latin word for shining, "nitens." It has been called radon since 1923 (Nagaratnam 1994).

Radon is a radionuclide (^{222}Rn) produced by the decay of ^{238}U naturally present in the terrestrial crust, including rock and soil. The half-life of ^{222}Rn, the amount of time it requires for half of the radioactive radon atoms in a sample to decay into more stable forms, is 3.82 days (Al-Zoughool and Krewski 2009). The transient disintegrated radionuclides of radon, also known as radon progeny, radon daughters, or radon decay products (RDPs), include polonium (Po), bismuth (Bi), and lead (Pb). During the spontaneous transformation, the energy is released in the form of particles such as α, β, and or photons including γ- and X-rays. Part of the emitted radon escapes to the soil surface and enters the air, while some retain below the soil surface and enter the groundwater via continuous diffusion from earth crust and rocks (Al-Zoughool and Krewski 2009). Rocks such as phosphorites, uranium-bearing granites, glauconite sandstones, metamorphic, and carbonaceous shales exhibit high Rn emanation characteristics (Nagda 1994). However, the radon emanation potential depends on the type of the rock and the soil features, including porosity, moisture content, stability, and gas barrier properties (Nagda 1994). Radon (^{222}Rn) is a highly water-soluble noble gas and can be found in groundwater resources where uranium (^{238}U) and radon (^{222}Rn) are present in the ground. Therefore, drinking water obtained from groundwater resources, including wells, springs, and boreholes, has a high radon concentration.

Radon concentration is normally reported in the SI unit of Bq (Becquerel)/m^3. Curie (Ci) is also used as a measurement unit of radioactivity. One Bq equals one disintegration per second, whereas one Ci is 3.7×10^{10} disintegrations per second. Radon (^{222}Rn) can exist in the outdoor and indoor environment. Generally, in the open air, the

Medical Geology: En route to One Health, First Edition. Edited by Majeti Narasimha Vara Prasad and Meththika Vithanage.

concentration of radon gas is significantly low and hence does not cause a considerable health concern. However, the outdoor radon level may be high during temperature inversions at higher altitudes over the regions with high uranium and radon in the ground (Segovia et al. 2007). Temperature inversion is a natural meteorological phenomenon that occurs when cool air becomes trapped in the ground under a layer of warm air. For instance, high outdoor radon levels were reported in Norway, Serbia, and Spain (Sáinz et al. 2007; Zunic et al. 2007; European Commission and Joint Research Centre 2020). Radon (^{222}Rn) can be found indoors in low concentrations. However, building materials can contribute to high indoor radon levels as major parts of the Earth's crust consist of uranium. For instance, building materials, including brick, concrete, tiles, and wallboard, are manufactured using rock types that contain radon (^{222}Rn), which may contribute to the indoor radon (Ackers et al. 1985; Man et al. 1989). Radon penetrates buildings through structural defects such as pores in hollow-block walls, sumps, construction joints, drains, cavity walls, cracks in the floors and walls, gaps in timber floors, and suspended concrete. Moreover, radon concentrations are typically higher in confined spaces such as basements, living spaces, and cellars in contact with the ground, and the air exchange is minimal. Radon accounts for the leading source of the background radiation dose of over 55% received by the environment and is found in different concentrations depending on the location (Chalmers et al. 2001; Ravikumar and Somashekar 2013). The health effects have long been known.

The standard radon safe limits vary depending on the regulating agency, including Miners Safety and Health Act (MSHA), Occupational Safety and Health Act (OSHA, International Commission on Radiological Protection (ICRP), United States Environmental Protection Agency (US EPA), and World Health Organization (WHO) (Degu Belete and Alemu Anteneh 2021). Several techniques and instruments are available for the detection of radon and its progeny. The most commonly used radon detection techniques include solid-state nuclear track detectors, activated carbon method, electret ion chambers (EICs), electronic integrating devices (EIDs), and continuous radon monitors (CRMs) (Organization 2007). Radon exposure has been associated with genetic and cytogenetic effects in human and animal studies (Fowler et al. 2020; Shergill et al. 2021). For instance, numerous epidemiological cohort and case–control studies have been reported linking indoor radon exposure and lung cancer. In 2005, WHO established the International Radon Project to study effective strategies for minimizing the health impact of long-term occupational and residential radon exposure and raising general public awareness with over 30 countries worldwide (Organization 2007). The current chapter provides comprehensive coverage on the chemistry, safety levels, detection techniques, and health effects of radon.

6.2 Radon Chemistry

Radon is a radioactive gas, and there are three main natural isotopes of radon; ^{219}Rn, ^{220}Rn, and ^{222}Rn, with half-life values ($T_{1/2}$) of 3.96 seconds, 56.6 seconds and 3.823 days, respectively. Regarding the $T_{1/2}$ values, ^{222}Rn isotope (Radon-222) is the most stable isotope, and it is the greatest concern as a threat to human health. Radon is naturally formed as a product of the uranium-238 (^{238}U) decay series, see Figure 6.1.

Radon-222 is the eighty-sixth (86) element of the periodic table. It is a colorless, odorless, and tasteless monoatomic gas, making it impossible to detect by human senses alone. Interestingly, radon exhibits phosphorescence below its freezing point, becoming yellow and red at lower temperatures. Under standard temperature and pressure, radon gas has a density of 9.73 kg/m^3, making it the heaviest gas in nature. The melting and boiling point of radon are at 202 K and 211.5 K, respectively.

Due to its closed-shell atomic structure, radon is categorized as a noble gas. The electronic configuration of neutral radon is [Xe] $4f^{14}$ $5d^{10}$ $6s^2$ $6p^6$. This electronic structure of radon suggests limited chemical properties. However, radon's low first ionization energy (1037.1 kJ/mol) compared to the other noble gases in the group allows it to undergo a limited number of chemical reactions. However, the radioactivity of Rn-222 makes the synthesis and identification of radon compounds more difficult than the other noble gas compounds. Radon was earlier considered a metalloid due to its similarities with boron (B), germanium (Ge), polonium (Po), and antimony (Sb). It is reported that the fluorides of radon, RnF_2 and RnF_4, are produced upon the reaction of radon with liquid halogen fluorides such as ClF, ClF_3, ClF_5, BrF_3, and IF_7 (Nabiev et al. 2014). The physical and chemical properties of radon are summarized in Table 6.1.

6.3 Sources of Radon

Groundwater is sourced from the geological formations known as aquifers. Uranium can occur in the groundwater as a result of the natural dissolution of uranium-bearing minerals such as granite aquifers. Since radon is a daughter

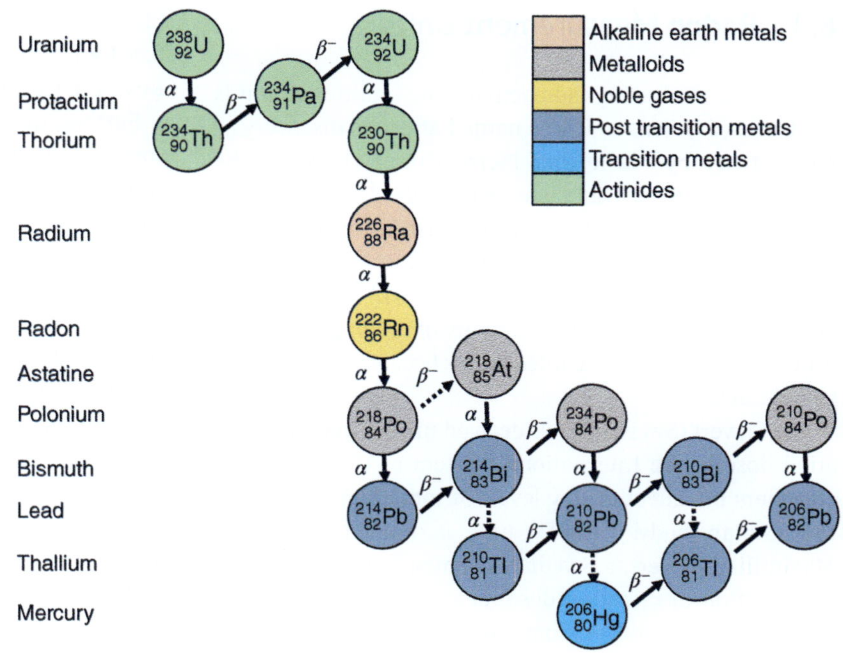

Figure 6.1 Decay series of uranium (^{238}U).

Table 6.1 Physical and chemical properties of radon-222.

Property	Description
Physical properties	
Phase at STP	Gaseous
Melting point	202 K (−71 °C)
Boiling point	211.5 K (−67.5 °C)
Density	9.73 kg/m^3
Heat of fusion	3.247 kg/mol
Heat of vaporization	18.10 kg/mol
Molar heat capacity	20.786 J/mol/K
Chemical properties	
Atomic number	86
Atomic mass	222 g/mol
Oxidation states	+2, +4
Electronegativity	Paulin scale: 2.2
First ionization energy	1037.1 kJ/mol
Covalent radius	150 ppm
Van der Waals radius	220 ppm
Electron configuration	[Xe] $4f^{14}\ 5d^{10}\ 6s^2\ 6p^6$

element of radium, both Ra-238 and Rn-222 find high groundwater concentrations in Jurassic granite terrain worldwide (Reddy et al. 2017; Telahigue et al. 2018; Cho and Choo 2019; Kelly et al. 2019; Alaboodi et al. 2020; Hwang and Kim 2021; Shu'aibu et al. 2021). Rn-222 is a highly water-soluble gas, and the groundwater containing Rn-222 can be mixed with spring water on the Earth's surface causing drinking water contamination. Kawabatha and co-workers analyzed the post and pre-seismic Rn-222 concentrations in spring water before and after the 2016 Kumamoto earthquake. The authors found increased Rn-222 concentration after the earthquake and a gradual decrease in Rn-222 in spring water systems near the Futagawa-Hinagu fault zone. This is due to the groundwater mixing with spring water, causing Rn-222 contamination. They also reported the earthquake caused the change of the saturation ratio of the aquifer, increased rock surface area, and the changes in the porosity of the aquifer, causing the rise of the Rn-222 concentration in the surface water (Kawabata et al. 2020). Shale gas formation is another area of concern associated with radon production. Since shale contains high concentrations of uranium and radium, radon production with shale gas formation is a possible cause (Tian et al. 2019).

Technologically enhanced naturally occurring radioactive material (TENORM) wastes such as coal ash, red mud, phosphogypsum, and sulfur polymer composites are known for their extensive use in the construction field. The radiation exposure from these construction materials is divided into external and internal. The external radiation exposure is mainly due to the U-238 decay series and the radioactive isotope of potassium (K-40) existing in building materials. Therefore, Rn-222 exhalation is a possible effect of TENORM. Among many building materials, phosphogypsum-based materials have the highest radon exhalation rate of over 9 Bq/m^2/h, according to Szajerski and co-workers (2019).

6.4 Radon Measurement Units

In terms of measuring radioactivity, the traditional unit is the Curie (Ci). This unit was named after the discovery of radioactivity by Marie and Pierre Curie. However, the Curie is a non-SI unit, and scientists used it during the early days of radioactivity analysis. It is referred to as the activity of 1 g of radium (Ra). Since Curie is a unit with a large magnitude, picocuries per liter (pCi/L) is generally employed to express the radioactivity of radon gas. The pCi unit is only used in the United States because it is required by federal law.

The Sievert (Sv) is another derived unit of ionizing radiation dose in the International System of Units (SI) that considers the effect of low levels of ionizing radiation on the human body. The Sv unit is named after Rolf Maximilian Sievert, a Swedish scientist renowned for his contribution to radiation dose measurement and investigating the effects of ionizing radiations on biological systems (Sekiya and Yamasaki 2016). Still, the Sv is too large to examine the impacts of ionizing radiation. As a result, the radon dose is generally expressed in milliSv (10^{-3} mSv), microSv (10^{-6} μSv), or even nanoSv (10^{-9} nSv).

Nevertheless, many countries adhere to the SI unit, named the Becquerel (Bq). The Becquerel is another radioactivity measuring unit named after Henri Becquerel. This unit is much smaller than the Curie, and radon gas concentration in the air is commonly expressed in Becquerels per cubic meter (Bq/m^3). One Becquerel equals one radioactive disintegration per second. One pCi/L is equal to 37 Bq/m^3. Incidentally, it is worth noting that the units pCi/L and Bq/m^3 refer to radon concentration in the air, while unit mSv per year indicates the health effects of the Rn concentration in the human body (Board on Radiation Effects Research National Research Council 1999).

As it is not possible to quantify the activities of individual radon progeny in the air and inside the body, a unit called the "working level (WL)" is used to measure the cumulative radiation dose from RDPs. One WL is referred to as the amount of short-lived RDPs in one liter of air that will ultimately decay by releasing 1.3×10^5 MeV of alpha (α) energy. Once the decay chain approaches 100% equilibrium, WL equals 100 pCi/L of Rn^{222}. However, approaching indoor equilibrium is extremely rare due to constant changes in Rn and RDPs concentration as ventilation gets rid of some of the Rn and its decay products. Also, it requires some time for the incoming Rn to produce its decay products. An equilibrium factor (F) is applied for unit conversion in such a scenario. Hence, the National Research Council Committee on Health Risks of Exposure to Radon (BEIR VI) assumes equilibrium between radon and RDPs indoors is 40% ($F = 0.4$). In that case, one pCi/L (37 Bq/m^3) of Rn^{222} in the air is roughly equivalent to 0.004 WL (Board on Radiation Effects Research National Research Council 1999).

Further, due to the highly electrostatic nature of RDPs, some of them tend to attach to physical objects such as walls, floors, ceilings, furniture, or even clothing of persons, and this process is known as "plating out" (Vogiannis and Nikolopoulos 2015). Consequently, all RDPs produced are not circulating in the air and are not breathable or measurable. Thus, the equilibrium ratio (ER) is the more practical unit to estimate the RDPs level in the air. By definition, ER is the concentration of RDPs in the air (measured RDPs) relative to the total RDPs produced by Rn (when the RDPs were in equilibrium with the Rn gas) (Board on Radiation Effects Research National Research Council 1999). If the RDPs were not plated out and still in the air, the ER would be 1 (or 100%). Generally, ER varies between 0.3 and 0.7 due to ventilation, plating out, and environmental conditions. Therefore, an ER of 0.5 is commonly assumed as an average (Board on Radiation Effects Research National Research Council 1999).

Human exposure to RDPs is generally recorded in working level months (WLM) units. When this unit was initially developed, it was believed that miners would work for 170 h/month. Thus, a WLM is defined as the equivalent of breathing one (01) WL for 170 working hours. Residential exposure times are much greater than 170 h/month (Board on Radiation Effects Research National Research Council 1999). The SI unit for WLM is J h/m^3; 1 WLM = 3.6×10^{-3} J h/m^3.

6.5 Safe Radon Levels

Generally, outdoors radon is not a cause of concern since it rapidly dilutes to low concentrations. The average outdoor radon level ranges from 5 to 15 Bq/m^3 (Khan et al. 2019). Nevertheless, indoor radon concentrations, especially in areas with minimal ventilation, are relatively higher than outdoors. The highest levels have been found in mines, caves, and water treatment facilities.

While there is no internationally accepted level designated as a safe radon level, different countries recommend their own acceptable radon level, which is considered low risk and does not cause health concerns. The action and target level are the two radon-level classifications established by the UK Health Security Agency (UKHSA). The action level refers to the annual average concentration in a home. The radon target level is ideal for remediation works in existing buildings and protective measures in new buildings. If the radon level is high, then the "target level" is the radon concentration level that mitigation measures should help you reach. According to the UKHSA and Public Health England (PHE), radon's action level and target levels are 200 Bq/m^3 and 100 Bq/m^3, respectively (Robertson et al. 2013; UK Health Security Agency 2022). Generally, the indoor radon level should be at the target level or lower. When the radon level is equal to or above the action level, necessary action needs to be taken to reduce the radon level in the house to minimize the health risk. If the radon level is between the target and action level, necessary measures should be taken to reduce the radon level.

By righteously identifying the health challenges of radon gas and its progeny, the United States (US) Congress passed "Radon Act 51," which set the natural outdoor radon gas level of 0.4 pCi/L as the target radon level for indoor. The US EPA prepared practical guidelines and recommendations to achieve in response to the Radon Act 51 and declared an action level of 4 pCi/L (148 Bq/m^3) (Robertson et al. 2013; U.S. Environmental Protection Agency 2016, 2022). If the indoor radon level is at or above the action level, the EPA guidance recommends taking remedial measures to decrease exposure to radon gas. However, this does not mean that the radon level under 4.0 pCi/L is considered safe or acceptable, as stated in the BEIR VI study (National Research Council 2006; U.S. Environmental Protection Agency 2016). As US EPA stated correctly, "Any radon exposure has some risk of causing lung cancer." It has been estimated that a decrease of radon levels to under 2 pCi/L most probably diminishes the annual lung cancer deaths accredited to radon by 50% (U.S. Environmental Protection Agency 2016). Even with the action level of 2.0 pCi/L, the cancer risk caused by radon gas is a hundred times larger than the risks associated with carcinogens in food and water.

Meanwhile, the WHO recognized that radon exposure is a key and rising public health threat based on the thorough research on radon gas and published a handbook on Indoor Radon in 2009 (Organization 2009). Thus, the WHO established a commonly accepted action level of 100 Bq/m^3 (2.7 pCi/L) for indoor radon (Organization 2009). If the radon level is higher than the WHO action level, homeowners are advised to take corrective action to reduce the radon levels. The WHO further suggested that if this level cannot be achieved because of country-specific location and conditions, the action level should not exceed 300 Bq (8 pCi/L) (Organization 2009).

Numerous comprehensive studies on radon and RDPs in confined areas such as caves, mines, or tunnels, revealed significantly higher radon levels mainly due to low ventilation. To protect the worker in such confined environments, International Atomic Energy Agency (IAEA) established an action level of 1000 Bq/m^3 for radon concentration (IAEA 2003). However, due to modern technology and efficient ventilation systems installed, the average radon and RDPs concentration in mines are lower than IAEA's action level of 1000 Bq/m^3 (IAEA 2003). However, due to modern technology and efficient ventilation systems installed, the average radon and RDPs concentration in mines are lower than IAEA's action level of 1000 Bq/m^3 (Margineanu 2019). Many countries consider a value between 200 and 400 Bq/m^3 as their action level for indoor radon. This somewhat varies from the action level for radon in drinking water. For example, the USA has set their action level of radon for drinking water at 150 Bq/L, while the European Union has recommended a 1000 Bq/L of the action level. Table 6.2 shows the average of indoor radon

Table 6.2 Comparison of average of indoor radon and indoor radon action level enforced by different countries and organizations.

Country	Average radon concentration in homes (Bq/m^3)	Action level (Bq/m^3)
South Korea	102	200
Australia	[a]	200
United States	46	150
Finland	123	400
Canada	[a]	200
China	[a]	100–200
Greece	[a]	200–400
Belgium	[a]	400
United Kingdom	20	200
Norway	51–60	200
Israel	[a]	200
Poland	[a]	400
Germany	50	250
Ireland	60	200
Sweden	108	400
Switzerland	70	1000
Lithuania	37	100
Luxembourg	[a]	250
EU	200	300
WHO	[a]	100 (300 if 100 cannot be achieved
ICRP	[a]	300

[a] Not specified.

EU, European Union; ICRP, International Commission on Radiological Protection; USEPA, United States Environmental Protection Agency; WHO, World Health Organization.

Source: Adapted from Degu Belete and Alemu Anteneh (2021), Ruano-Ravina et al. (2017).

and their indoor radon action level enforced by different countries and organizations (Bowie and Bowie 1991; Zeeb et al. 2009; Benghida 2016).

An ideal safe radon level for indoors should be zero, but unfortunately, this is impossible to achieve. However, the risk of radon-related health problems can be significantly lowered by reducing indoor radon levels. Therefore, installing at least a couple of radon detectors at the indoor property is always advisable. This allows residents and building owners to act promptly in case of high radon levels.

6.6 Radon Detection Methods

It is important to detect indoor radon levels to identify the houses to mitigate the safety level. Moreover, the detection of underground radon levels is also used for earthquake prediction as the earthquake-preparation process relates to physical parameters of Earth's crust, including electric resistivity, geoelectric potentials, or concentration of radon gas (Studnička et al. 2019). The detection of α, β, and γ radiations emitted by radon and radon progeny is complex and needs to utilize expensive techniques to detect and analyze radon progeny compared to radon itself. Different techniques have been developed to measure α-particles emitted by radon or a combination of radiations of RDPs, which range from 5.5 to 7.7 MeV (Bem et al. 2020).

Generally, simple instruments are used to measure the short-term exposure of radon from a few days to several months. In contrast, detecting long-term exposure of radon from several years to several decades requires complex instruments. The short-term radon measurement can be unreliable due to the high temperature-dependent radon levels. Therefore, long-term measurements are much preferred when calculating the annual average of radon levels. Radon and its decay particles detection methods can be divided into grab and instantaneous, integrating, and continuous. Sampling devices to measure radon levels in soil, gas, and groundwater can be active or passive, and it depends on the requirement of electric power to collect the samples. Active devices require electricity, whereas passive devices do not require electrical power or a pump to work in the sampling setting. The selection of the specific detection method is based on the application, duration, type of information, and cost. The sensitivity of the instrument is also a key factor in selecting the detection method to measure radon. The selection of suitable device depends on the measuring of radon or its decay products. For example, radon detection is usually based on determining the existence of α-particles emitted by radon. However, α, β, and γ radiations emitted by its decay products are also utilized to detect radon.

6.7 Detection of Radon and Radon Decay Products by Grab Sampling Method

Grab sampling methods are the most convenient methods for measuring indoor radon levels. These methods are also useful in calibrating other methods and cross-checking the data obtained. When using these methods, sample collection is performed over a short period. The most commonly used grab sampling methods for detecting radon include ionization chambers and the scintillation cell method. The analysis of radon progeny is more challenging and expensive compared to radon. To determine the radon progeny, alpha radiation is collected in a given volume for a specific time (2–10 minutes) in a filter media. In this method, α-particles counting is performed for a specified duration. Grab sampling methods employed for radon progeny are gross alpha counting and alpha spectrometry.

6.7.1 Ionization Chambers

Ionization chambers are the preferred instruments to analyze grab samples, including outdoor air containing low radon concentrations. The sample collection is performed in the field using different containers or trappings on activated carbon. These samples are then transferred to the ionization chamber, constructed of electropolished stainless steel. Nowadays, a sealed chamber is used to collect the measurements (Studnička et al. 2019). First, water vapor and oxygen in the chamber are removed to assist the fast pulse counting. After that, the sample is directed to a chamber filled with 85% nitrogen and 15% hydrogen gases and a platinum catalyst. During the catalytic reaction, water is generated and then removed by a drierite. The lower detection limit for a 2 L sealed ionization chamber is 0.7 mBq of radon for 17 hours (George 1990).

Ionization chambers provide highly accurate measurements of samples collected in grab sampling that contain low radon concentrations. However, this method involves complex and expensive instruments. Further, to operate these devices, trained operators are required.

6.7.2 Scintillation Cell Method

This method is one of the most famous and oldest radon detection methods performed via grab sampling in the field or laboratory (Lucas 1957; Geor 1976; Abdalla et al. 2021). These scintillation cells are generally 0.09–3.0 L range in size. They are made up of metal, glass, or plastic containers, and the internal surface is coated with silver (Ag) activated zinc sulfide (ZnS) powder (Abdalla et al. 2021). The scintillation cell is coupled with a photomultiplier tube assembly system to count the light photons. The radiation emitted by radon interact with ZnS-Ag phosphor, which leads to the formation of light photons eventually counted. For the scintillation cells, sizes ranging from 0.1 to 3.0 L acquire sensitivities of range from 0.8 to 16 counts per hour (CPH) per Bq/m^3 (George 1990). Scintillation cells are also used to analyze radon progeny using the gross alpha counting method. In this method, the alpha particles from ^{218}Po and ^{214}Po are counted by a system containing a photomultiplier tube and ZnS-Ag scintillator.

Scintillation cells are helpful for radon source detection and remediation and study the impact of mitigation measures on reducing radon levels. This method is comparatively intense and eliminates the transferring of the sample before counting. The proper maintenance of scintillation cells allows them to use for years. However, there are several disadvantages of this method. One disadvantage is that this may give erroneous readings due to improper calibration. Moreover, leaking of the cells is also another issue with this method. At times the photomultiplier tube in this system malfunctions giving inaccurate readings. Therefore, scintillation cell-photomultiplier tube calibrated for one system should not be used in another, except both systems are cross-calibrated.

6.7.3 Liquid Scintillation Counting (LSC)

Liquid scintillation counting (LSC) is the most widely popular laboratory method to measure radon in water bodies. It can be used as an alternative method to count Rn-222 in soil gas (Bem et al. 2020). The high solubility of radon in organic solvents is used in this technique to extract radon from aqueous samples into a preloaded scintillation cocktail that contains organic solvents, including toluene, xylene, and mineral oil (Schubert et al. 2014). Therefore, the radon in water partitions in the aqueous/organic phases and the air space inside the scintillation vial. The photons emitted from the scintillation fluid will be counted to quantify the radon activity of the added water sample.

6.7.4 Gross Alpha Counting

This is the most popular radon progeny analysis method (Lopes et al. 2005; Ha et al. 2011). In this method, two distinct energies are counted for alpha particles from ^{218}Po and ^{214}Po. The counting system is equipped with a photomultiplier tube and ZnS-Ag scintillator, and they are placed in close proximity to the filter. Various counting methods are used to identify individual activity concentrations of 218Po, ^{214}Pb, and ^{214}Bi and their contribution to the overall potential alpha energy concentration (PAEC) (Raabe and Wrenn 1969; Thomas 1972). The concentrations of three radon progeny represent three independent variables and three measurement periods. Therefore, three independent simultaneous equations are used to collect their concentrations in air (Thomas 1972). The lower limit of detection for ^{218}Po, ^{214}Pb, ^{214}Bi, and PAEC for 5 minutes sample at a flow rate of 10 L/min is 25 Bq/m^3, 7 Bq/m^3, 22 Bq/m^3, and 50 nJ/m^3, respectively (George 1990).

In gross alpha counting, higher flow rates can be used to determine the environmental levels of radon progeny. This method is easier and provides more sensitive measurements than alpha spectrometry. Alpha spectrometry is a more sophisticated and expensive method. It is used widely in the laboratory and occasionally in specific field research. During this method, false readings may be caused by improper air sample volume determinations, low-quality filter media, poor counting geometry, and counting efficiency. In addition, the random nature of radioactive decay can cause counting errors. The improper calibration of the air metering device, air pump with lower stability, issues with filter holder, and inaccurate timing may result in inaccuracy in flow rate measurements.

6.7.5 Alpha Spectrometry

This is a more precise laboratory method to analyze individual radon progeny (Ha et al. 2011). Surface barrier or diffuse junction detectors are generally used. The detectors provide higher resolution spectra of alpha particles collected on the filter. Alpha spectrometry is the best method when other interfering alpha emitters are also present. This technique provides the measurements of radon progeny concentration of $<20\ Bq/m^3$ at a flow rate of 10 L/min (Nazaroff 1984).

6.8 Detection of Radon and Radon Decay Products by Integrated Measurement Methods

This is the most popular method to measure radon and its progeny over a period. Early integrated instruments consisted of passive environmental monitors (PERM) and radon progeny integrated sampling units (RPISU) (Schiager 1974). However, current commercial devices are simpler, portable, and can be deployed inside the building by the owners. Integrated methods for radon include solid-state nuclear track detectors, activated carbon method, electret-ionization chamber (EIC) method, and solid-state detection monitors. These devices are sensitive to measuring indoor radon levels, and they average the short-term flocculation of radon levels.

6.8.1 Solid-State Nuclear Track Detector

In a solid-state nuclear track detector, α particles are captured by a solid-state material incorporated into the detector. This solid-state material comprises cellulose nitrate and allyl diglycol carbonate plastic films (Alter and Fleischer 1981; Gillm and Wertheim 2018). The plastic detector is placed in a small container, and only radon can diffuse through the filter. Inside the container, α-particles from radon and radon progeny irradiate the plastic detector, producing damage tracks. The visibility of these tracks is enhanced by chemical or electrochemical etching using NaOH solution. Electrochemical etching generates tracks with a large diameter which simplifies the track counting. The measurements for track density are obtained through manual counting and scanning. Additionally, automated optical scanning systems are also used. Depending on the size of the scanned area of the detector, the lower limit of detection for a three-month exposure ranges from 2 to 8 Bq/m^3.

The solid-state nuclear track (alpha track) method has a low sensitivity. Hence, more extended exposure periods are required under normal environmental conditions. One drawback is the analysis is not cost effective and is typically conducted in large laboratories. Moreover, product quality control, including frequent calibrations and duplicate exposures, is an essential step in the alpha track method to minimize the effect of the different variabilities, such as the thickness of solid-state nuclear track plastics, acquired background, and temperature during preparation.

6.8.2 Activated Carbon Method

The activated carbon collector is another popular monitoring device for assessing indoor radon levels for integrated periods of up to 7 days (George 1984; Sohrabi and Ebrahiminezhad 2018; Wakabayashi et al. 2019). Radon is adsorbed onto activated carbon through passive diffusion. The emitted α, β, and γ radiations are accumulated in the collector and, after sealing the collector, sent to a laboratory for further analysis. A minimum exposure of three hours is needed to enable the equilibration between radon and its progeny before counting. Simple detection techniques, including gamma counting on a sodium iodide (NaI(T1)) system or liquid scintillation system, are used to perceive α, β, and γ radiations resulting from the radon progeny (Chittaporn et al. 1981; Messier and Serre 2017; Maier et al. 2021). The impact of water vapor in gamma counting collectors is minimized by having carbon (25–200 g) with an open-faced system or with a diffusion barrier. The average radon concentration during the sampling is calculated by integrating the peaks (area under the peaks) from ^{214}Pb and ^{214}Bi (Maier et al. 2021). In the liquid scintillation method, only 1–2 g of activated carbon is used in collectors. The adsorbed radon onto carbon is transferred into a scintillant blend. The sensitivity increases as the collectors include more carbon. Depending on collector size, the lower detection limit ranges from 3.7 to 7.5 Bq/m^3 of radon for a four-day exposure period (George 1990). Novel approaches have been included to develop active charcoal adsorption detectors more effectively by combining active charcoal with techniques such as polycarbonate track detecting (Sohrabi and Ebrahiminezhad 2018).

The activated carbon method is a simple, straightforward, and cost-effective analysis method that can detect α, β, and γ radiations. The devices and equipment utilized in this method are found in most radiological laboratories. Additionally, this technique is the most practical method for accurate and precise measurements of the average indoor radon concentrations if a proper quality assurance program is in place.

6.8.3 Electret-Ionization Chamber (EIC) Method

EIC is a passive radon detecting system widely used to measure radon in both air and water. It can measure the average radon concentration during the period of measurement. An EIC contains a chamber with an electric field generator and a filtered inlet. Radon gas enters the chamber through the inlet via passive diffusion, and the radiation from radon and its progeny ionizes the air inside the chamber. The positive electret of the detector collects the formed negative ions inside the chamber. The electret discharge over a known time interval is related to the radon concentration. EICs come in two types: short-term and long-term detectors. The short-term EICs can measure radon over 2–15 days, while long-term detectors function for 3–12 months. In this method, the lowest detection limit is 7 Bq/m^3 for the short-term EIC during a three-day exposure period (George 1990).

The main advantage of the EIC method is the instantaneous availability of the results instantly after the exposure. Moreover, these devices have no considerable effect from the other factors, such as humidity and air temperature. The main disadvantage of this method is that it responds to natural gamma radiation, which can be different in concentrations from site to site.

6.8.4 Solid-State Detection Monitors

These are sophisticated radon detection systems that store and analyze data using microcomputer technology. A solid-state silicon detector identifies the α radiation produced from radon. Several radon progeny monitor versions have been developed, following the same basic principle. The air flows through it at a constant flow rate through the filter-detector assembly in the sampling head. Multiple sensing devices are used such as thermoluminescent dosimeter (TLD) chips, alpha scintillation phosphors, solid-state nuclear track (track etch), surface barrier or diffused junction, and electret ion chamber detectors (Gillm and Wertheim 2018). Generally, the detector is placed close to the filter to irradiate the detector by the alpha particles emitted by the radon progeny. The lowest detection limit in this method for two-day exposure is 4 Bq/m^3 (George 1990).

Solid-state detection monitors are helpful devices to collect pre- and post-mitigation diagnosis data. TLD chips used in this method are labor intensive and require highly trained employees to operate. The filters can also be clogged if the samples are collected in a highly dusty environment. Solid-state nuclear track monitoring devices are smaller than TLD chips and can be sent for exposure and returned by mail for analysis to a central laboratory. One of the main disadvantages of the method requires exceptional quality control, particularly when solid-state nuclear track detectors are used. In addition, trained operators are needed to install the total gross alpha detector systems; hence, using these devices for a single measurement is relatively expensive.

6.9 Detection of Radon and Radon Decay Products by Continuous Monitors

In the continuous method, sampling and analysis of radon and radon progeny occur simultaneously; hence, it provides real-time measurements. This is an important method for situations in which radon concentrations change significantly and rapidly. Furthermore, radon measurement with continuous monitors is easier than that of radon progeny measurement.

6.9.1 Continuous Scintillation Cell Monitor

One of the main features of this device is the two-port scintillation cell, which is coated with ZnS-Ag phosphor. The underlying principle in this method mimics grab sampling. The only exception is that air is collected via the cell in this method. The radon and its decay particles are deposited in the internal surface scintillation cell, producing α radiations identified by a photomultiplier tube. The lowest detection limit lies in the range of 1–37 Bq/m^3 for 30 minutes intervals. However, this value varies based on the size of the scintillation cell.

6.9.2 Passive Continuous Radon Monitor

Passive diffusion continuous radon monitor is one of the devices for continuous measurements for radon. This is equivalent to a scintillation cell, and the data collection mechanism in this method is exactly similar to scintillation cell monitors. The only difference of passive diffusion continuous radon monitor is the lack of the air pump. Instead, the air diffuses into the scintillation cell through 1 cm thick foam after removing radon progeny and other airborne particles. The lower detection limit for one hour counting interval is about 37 Bq/m^3.

Portable pulse ion chambers are also used as passive devices to collect continuous measurements of radon. In this device, air diffuses through the ionization chamber, and an electrometer detects radon and decay products. The lower detection limit of this device is 18 Bq/m^3 for an hourly exposure. The diffusion radon only monitors detect only radon (Chittaporn et al. 1981). The sensing device is a modified scintillation cell connected to a charged electret. The function of electret is to remove radon progeny from the collecting sample before depositing on

ZnS-Ag, detecting phosphor. This only detects alpha pulses from radon. This device has a lower detection limit of around 10 Bq/m^3 over 15-minutes intervals.

6.9.3 Continuous Radon Monitors for Radon Progeny

In this method, air sampling is performed continuously using an air pump and detection via a filter detector. In order to analyze samples, integration of the alpha radioactivity is carried out at several minutes to an hour. This method has an automation system that allows for frequent sampling and counting. The microprocessor stores data such as the number of counts and the elapsed time. In this method, solid state or diffused junction-type detectors collect α-particles from 2 to 8 MeV emitted by ^{218}Po and ^{214}Po. The rate of counting is directly correlated to the number of alpha particles emitted during the measurements. This technique helps analyze different sampling methods through customized programming due to connected computer interphase.

Continuous radon and radon progeny monitors enable the instant to obtain instant real-time radon concentration data. However, these are costly and complex devices that trained workers must operate. One feature in these monitors is count rate does not alter rapidly with sudden variations in radon concentration. The portable microprocessor in a portable pulse ion chamber allows real-time data collection. Screening and follow-up radon measurements are convenient with this device and provide real-time monitoring of fluctuations of radon concentrations, along with hourly and on-command printouts. Diffusion radon only monitor is a common device in research and mitigation work. Computer interphase capability in continuous radon progeny monitors facilitates customized programming for different sampling protocols. These devices have higher sensitivity and require less labor. However, these instruments are relatively expensive and need a trained operator to collect the data.

6.10 Health Effects of Radon

Radon (^{222}Rn) is a relatively inert noble gas that does not readily interact chemically with other elements and can exist indoors and outdoors. However, radon is a radioactive element, contributing to ionizing radiation. It is electrically charged transient progeny can readily attach to dust and aerosol particles. If someone inhaled or ingested radon progeny, they tend to deposit in different regions of the respiratory or digestive system, exposing to the α-radiation emitted by two main radon daughters, ^{218}Po and ^{214}Po. These α-particles are highly ionizing due to their double positive charge and large mass. They can impart their energies to the walls of the human respiratory and digestive systems, interacting with the sensitive cells in the organs such as the lung and stomach (Chalmers et al. 2001; Messier and Serre 2017). Moreover, the radioactive radon progeny can possibly transport to other regions of the human body via blood circulation (Degu Belete and Alemu Anteneh 2021).

The human body contains around 60% of water, and ionizing radiation, including α-emission, undergoes radiolysis of water. Radiolysis of water generates hydrated electron (e_{aq}^-), which further react with water to produce highly reactive oxygen species, including hydrogen radicals ($^{\cdot}H$), hydroxyl radicals ($^{\cdot}OH$), superoxide ($O_2^{-\cdot}$), and hydrogen peroxide (H_2O_2) (Dassanayake et al. 2013, 2015). Equation (6.1)–(6.4) show the key reactions involved in the radiolysis of water.

$$\text{Ionizing radiation } H_2O \rightarrow H_2O^+ + e^- \tag{6.1}$$

$$\text{Ionizing radiation } H_2O + e^- \rightarrow H_2O^- \tag{6.2}$$

$$\text{Ionizing radiation } H_2O + e^- \rightarrow H_2O^+ + 2e^- \tag{6.3}$$

$$\text{Ionizing radiation } H_2O^- \rightarrow {}^{\cdot}H + OH^- \tag{6.4}$$

$$\text{Ionizing radiation } H_2O^+ \rightarrow {}^{\cdot}HO + {}^{\cdot}H \tag{6.5}$$

Exposure to ionization radiation affects the cells by generating highly reactive free radicals or damaging cellular generic materials, including DNA. The reaction between free radicals and DNA may cause cellular damage, DNA breakage, accurate/inaccurate repair, apoptosis, gene mutations, and genetic instability, leading to cancer. A number of scientific researches have been reported on the effect of radon inhalation on lung cancer; however, the effect of radon on other diseases is yet to be fully explored. Several studies have been conducted to identify a relationship between other cancerous and noncancerous diseases and radon exposure. The evidence for the risk of skin cancer, leukemia, and noncancerous diseases such as circulatory system diseases (CSDs) associated with radon exposure has been reported.

6.10.1 Lung Cancer

Lung cancer is one of the most common causes of death from cancer worldwide. In 2018, lung cancer accounted for over 2.1 million new cases and 1.8 million deaths, representing 11.6 and 18.4% of total cancer-related cases (Li et al. 2020). Indoor radon exposure has been identified as the second most source of human lung carcinogenicity after smoking tobacco (Cousins et al. 2013; Degu Belete and Alemu Anteneh 2021). Radon is considered lung carcinogen as adverse respiratory mutagenesis has been reported in laboratory animals exposed to radon and humans under occupational conditions (Cross 1988; Samet and Hornung 1990). In the early and mid-twentieth century, miners from numerous countries, including the USA, Canada, Sweden, France, China, Germany, and the Czech Republic, were diagnosed with lung cancer (Lubin et al. 1995; Hornung 2001). Therefore, epidemiologic cohort studies of radon-exposed miners were conducted to investigate the association between radon exposure and lung cancer. All these cohort studies showed a significant correlation between radon exposure in mines and the risk of lung cancer. Therefore, the first evidence of radon lung carcinogenicity came from those early occupational studies on uranium miners who worked with high underground radon levels and particulate matter (Lundin et al. 1969;Archer et al. 1973; Gottlieb and Husen 1982; Whittemore and McMillan 1983; Ackers et al. 1985; Man et al. 1989; Samet and Hornung 1990). Moreover, radon (^{222}Rn) has been identified as an environmental carcinogen by global health agencies, including WHO, the International Agency for Research on Cancer, and the US EPA (Pawel and Puskin 2004; Li et al. 2020). In 2014, the special report released by the International Committee on Radiological Protection (ICRP) on "Radiological Protection against Radon Exposure" (ICRP 126) also emphasized that public exposure to indoor radon should not be disregarded (Lecomte et al. 2014).

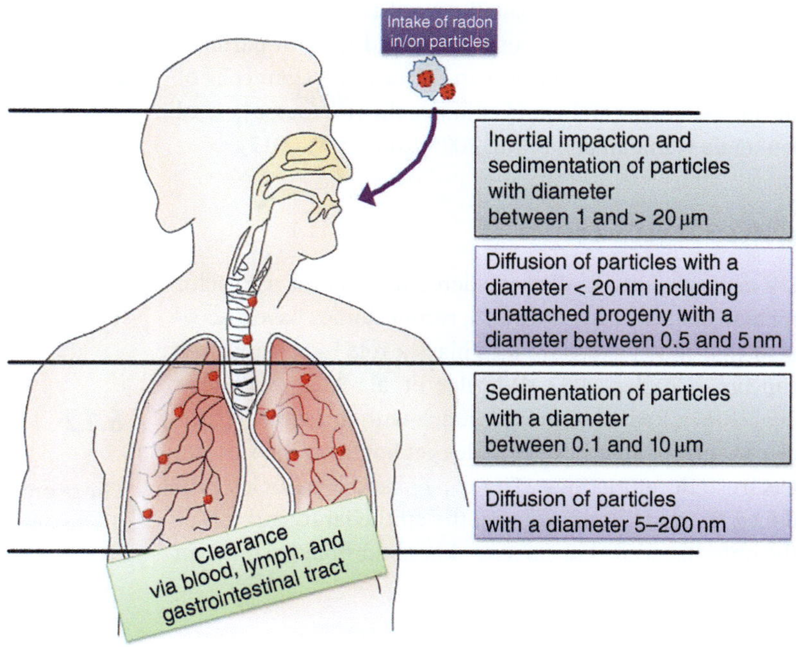

Figure 6.2 Different deposition mechanisms of radon containing aerosols with different particle sizes. *Source:* Drawing was taken from OpenClipart-Vectors on Pixabay under Creative Commons Attribution 4.0 International.

The risk of developing lung cancer results from the amount of radon delivered to the bronchial epithelium cells via the inhalation of aerosols containing radon and its progeny (Degu Belete and Alemu Anteneh 2021). The dose of radon retained in the main regions of the lung, including bronchial (BB), bronchiolar (bb), and alveolar intestinal (AI), depends on the deposition and clearance mechanisms of the short-lived radon progeny (Winkler-Heil and Hofmann 2002). Since the radon dose on the respiratory tract cannot be determined directly, several models have been proposed. Based on the model calculations, three radon deposition mechanisms have been proposed: inertial impaction, sedimentation, and diffusion (Maier et al. 2021). The extent to which each mechanism engages in the radon deposition relies on the physical characteristics of aerosols such as size and shape, breathing pattern, and the lung geometry of the respiratory tract (Maier et al. 2021).

Due to the inertial momentum, radon-containing aerosol particles with 2–20 μm diameter deposit in the upper part of the respiratory tract, including trachea, nasopharynx, and upper bronchi (Chittaporn et al. 1981). This process is called inertial impaction (Chittaporn et al. 1981). The sedimentation mechanism involves depositing smaller particles with 0.1–50 μm diameter in the upper respiratory tract, bronchioles, and alveoli (Chittaporn et al. 1981). The sedimentation process is governed by gravity. Radon progeny with a diameter less than 1 μm undergoes Brownian diffusion in the gas exchanging regions of the lung and subsequently depositing on respiratory surfaces (Chittaporn et al. 1981). Due to the smaller size of the radon progeny, both sedimentation and diffusion mechanisms predominate over inertial impaction.

After the deposition of the radon and its progeny containing aerosols in the different regions of the respiratory tract, elimination occurs via three main routes: bloodstream, lymphatic drainage system, or gastrointestinal tract (Carvalho et al. 2011). In the trachea or bronchial tubes, particles eliminate by phagocytosis or mucociliary transport by pulmonary alveolar macrophages. Particles trapped in the gas exchange regions, transportation and elimination occur through the bloodstream, and phagocytosis or lymphatic channels (Carvalho et al. 2011). Figure 6.2 shows the three deposition mechanisms for aerosol particles of different sizes.

6.10.2 Leukemia

The α-particles emitted by radon daughter atoms may deliver to the human tissues other than the respiratory system and cause damage. For example, a long-time scientist believed that α-radiation could damage the red bone marrow and cause leukemia in people. In the early twenties, this idea was abandoned since the research did not provide enough scientific evidence of a relationship between radon exposure and leukemia (Laurier et al. 2001). However, recent studies reveal that there may be an influence of radon exposure on leukemia (Tong et al. 2012; Schwartz and Klug 2016; Lu et al. 2020). The long-term inhaled radon can be deposited in the respiratory tract and further decay inside the body. A small part of the decay products can reach bone marrow and brain, and these α-particles emitted by the radon atoms can damage the stem cells of the bone marrow and could enhance the risk factor toward leukemia (Kendall and Smith 2005; Tong et al. 2012).

6.10.3 Skin Cancer

UV exposure is typically considered the highest risk factor for skin cancer. Other than that, recent studies have discovered that radon exposure also plays a role in causing skin cancers (Charles 2007; Wheeler et al. 2012; Vienneau et al. 2017). Radon atoms can attach with the aerosol particles in the air, and these particles get bound to the skin via electrostatic attractions. The α-, β-, and γ-particles produced by radon decay damage the epithelial tissue cells of the outer layer of skin causing skin abnormalities that can lead to skin cancer (Charles 2007).

6.10.4 Circulatory System Diseases (CSDs)

Exposure to highly ionizing radiation is known to cause CSDs and cancerous diseases. Few cohort studies using cohorts of uranium miners have been conducted to explore a relationship between radon exposure and CSD. They have found a weak association of radon exposure in cerebrovascular diseases (Nusinovici et al. 2010; Drubay et al. 2015). His relationship is suggested that the long-term exposure to minor α-radiation by radon gas can damage the arterial walls, which could cause atherosclerosis and, consequently, CSDs (Nusinovici et al. 2010).

6.11 Prevention and Mitigation of Radon in Indoor Settings

The common steps of prevention and mitigation of indoor radon levels include:

- increasing underfloor ventilation system
- installing well tested and durable indoor radon detectors in homes and places where the public is likely to be exposed for extended periods, including schools, government buildings, and lodging facilities.
- preventing the passage of radon from the basement into living spaces
- sealing floors and walls with plaster, caulk, or other materials
- improving the ventilation of the building
- stopping smoking and discourage smoking in your home
- asking for radon-resistant construction techniques when buying a new house
- introducing radon prevention standards and regulations, including building codes for new dwelling construction
- implementing radon mitigation standards and requirements for remediation of existing dwellings
- implementing appropriate training and certification of building professionals to ensure the efficiency of prevention and mitigation actions.

6.12 Conclusion

The overall goal of this chapter is to provide an overview of the major aspects of radon, safety levels, detection methods, and health effects. Radon (Rn) is a highly radioactive noble gas that does not chemically react with other elements. Among all radon radioactive isotopes, only ^{220}Rn (thoron) and ^{222}Rn (radon) are major radioactive isotopes found in significant quantities in nature. These radon isotopes are the intermediate products of long decay chains commencing with isotopes of ^{232}Th and ^{238}U, respectively, and terminating with stable lead (Pb). All the intermediates between radon and stable Pb are radon daughters or RDPs. Radon (Rn) and its RDPs emit α, β, and γ particles. The health risk from radon does not come solely from radon itself but its RDPs as well. Due to its natural occurrence, radioactivity, and health risk, radon has been subjected to many recent studies. The WHO recommended indoor radon safety level is 100 Bq/m3 (2.7 pCi/L); however, acceptable radon levels vary depending on the country or region. Numerous methods are available to detect radon and its progeny, both indoor and outdoor. Most widely used radon detection methods include alpha-track detectors, activated carbon method, electret ion chambers, electronic integrating devices, and continuous radon monitors. Humans are exposed to radon and its RDPs through inhalation, ingestion, and even via skin. Many regulatory agencies, including WHO and the International Agency for Research on Cancer, have identified Rn as a category one human carcinogen due to extensive studies on radon exposure. Further, the report published by the National Academy of Sciences (NAS) in 1999 determined that indoor air Rn is the second primary reason for lung cancer in the USA after cigarette smoking. It has also been reported that radon can cause leukemia, skin cancer, and noncancerous diseases, including CSDs. Several actions for radon prevention and mitigation can be implemented to achieve an overall risk reduction.

References

Abdalla, A.M., Ismail, A.M., and Al Madiy, A.A. (2021). Soil radon detection using active scintillation cell. *Journal of Radiation Research and Applied Science* 14 (1): 16–22.

Ackers, J.G., Den Boer, J.F., De Jong, P. et al. (1985). Radioactivity and radon exhalation rates of building materials in The Netherlands. *Science of the Total Environment* 45: 151–156.

UK Health Security Agency (2022). UKradon. https://www.ukradon.org/information/whatisradon (accessed 27 February 2022).

U.S. Environmental Protection Agency 2016. A citizen's guide to radon: the guide to protecting yourself and your family from radon. https://www.epa.gov/sites/default/files/2016-12/documents/2016_a_citizens_guide_to_radon.pdf (accessed 5 March 2022).

U.S. Environmental Protection Agency (2022). EPA facts about radon. https://semspub.epa.gov/work/HQ/176336.pdf (accessed 27 February 2022).

Alaboodi, A.S., Kadhim, N.A., Abojassim, A.A. et al. (2020). Radiological hazards due to natural radioactivity and radon concentrations in water samples at Al-Hurrah city, Iraq. *International Journal of Radiation Research* 18 (1): 1–11.

Alter, H.W. and Fleischer, R.L. (1981). Passive integrating radon monitor for environmental monitoring. *Health Physics* 40 (5): 693–702.

Al-Zoughool, M. and Krewski, D. (2009). Health effects of radon: a review of the literature. *International Journal of Radiation Biology* 85 (1): 57–69.

Archer, V.E., Wagoner, J.K., and Lundin, F.E. (1973). Lung cancer among uranium miners in the United States. *Health Physics* 25 (4): 351–371.

Bem, H., Gasiorowski, A., and Szajerski, P. (2020). A fast method for the simultaneous determination of soil radon (^{222}Rn) and thoron (^{220}Rn) concentrations by liquid scintillation counting. *Science of the Total Environment* 709: 136127.

Benghida, D. (2016). Indoor radon mitigation in South Korea. *International Journal of Applied Engineering Research* 11 (15): 8521–8523.

Board on Radiation Effects Research National Research Council (1999). *Evaluation of Guidelines for Exposures to Technologically Enhanced Naturally Occurring Radioactive Materials.* Indoor-Radon Guidelines and Recommendations https://www.ncbi.nlm.nih.gov/books/NBK230646/ (accessed 5 March 2022).

Bowie, C. and Bowie, S.H. (1991). Radon and health. *Lancet* 337: 409–413.

Carvalho, T.C., Peters, J.I., and Williams, R.O. (2011). Influence of particle size on regional lung deposition – what evidence is there? *International Journal of Pharmaceutics* 406 (1): 1–10.

Chalmers, B., Mangiaterra, V., and Porter, R. (2001). WHO principles of perinatal care: the essential antenatal, perinatal, and postpartum care course. *Birth* 28 (3): 202–207.

Charles, M.W. (2007). Radon exposure of the skin: II. Estimation of the attributable risk for skin cancer incidence. *Journal of Radiological Protection* 27 (3): 253–274.

Chittaporn, P., Eisenbud, M., and Harley, N.H. (1981). A continuous monitor for the measurement of environmental radon. *Health Physics* 41 (2): 405–410.

Cho, B.W. and Choo, C.O. (2019). Geochemical behavior of uranium and radon in groundwater of Jurassic granite area, Icheon, Middle Korea. *Water* 11 (6): 1278.

National Research Council (2006). *Health Risks from Exposure to Low Levels of Ionizing Radiation: BEIR VII Phase 2.* The National Academies Press.

Cousins, C., Miller, D.L., Bernardi, G. et al. (2013). ICRP publication 120: radiological protection in cardiology. *Annals of the ICRP* 42 (1): 1–125.

Cross, F.T. (1988). Radon inhalation studies in animals. *Radiation Protection Dosimetry* 24 (1–4): 463–466.

Dassanayake, R.S., Cabelli, D.E., and Brasch, N.E. (2013). Pulse radiolysis studies on the reaction of the reduced vitamin B_{12} complex Cob(II)alamin with superoxide. *ChemBioChem: A European Journal of Chemical Biology* 14 (9): 1081–1083.

Dassanayake, R.S., Cabelli, D.E., and Brasch, N.E. (2015). Pulse radiolysis studies of the reactions of nitrogen dioxide with the vitamin B12 complexes cob(II)alamin and nitrocobalamin. *Journal of Inorganic Biochemistry* 142: 54–58.

Degu Belete, G. and Alemu Anteneh, Y. (2021). General overview of radon studies in health hazard perspectives. *Journal of Oncology* 2021: 6659795.

Drubay, D., Caër-Lorho, S., Laroche, P. et al. (2015). Mortality from circulatory system diseases among french uranium miners: a nested case-control study. *Radiation Research* 183 (5): 550–562.

European Commission and Joint Research Centre (2020). *European Atlas of Natural Radiation.* Publications Office.

Fowler, B.L., Johannes, C.M., O'Connor, A. et al. (2020). Ecological level analysis of primary lung tumors in dogs and cats and environmental radon activity. *Journal of Veterinary Internal Medicine* 34 (6): 2660–2670.

George, A.C. (1984). Passive, integrated measurement of indoor radon using activated carbon. *Health Physics* 46 (4): 867–872.

George, A.C. (1990). An overview of instrumentation for measuring environmental radon and radon progeny. *IEEE Transactions on Nuclear Science* 37 (2): 892–901.

George, A.C. (2018). Scintillation flasks for the determination of low level concentrations of radon. In: *Operational Health Physics: Proceedings*, 112–115. New York, USA: Energy Research and Development Administration.

Gillmore, G. and Wertheim, D. (2018). Solid state nuclear track detectors for radon exposure-dose measurements and the application of confocal imaging for 3D visualisation of tracks. *Conference Paper, 20th EGU General Assembly*, EGU2018 (April), Vienna, Austria. 17480.

Gottlieb, L.S. and Husen, L.A. (1982). Lung cancer among Navajo uranium miners. *Chest* 81 (4): 449–452.

Hamzah, Z., Saat, A., Alias, M. et al. (2011). Measurement of gross alpha, gross beta, radon and radium activity concentrations in aqueous samples using liquid scintillation technique. *Journal of Nuclear and Related Technologies* 8 (2): 85–93.

Hornung, R.W. (2001). Health effects in underground uranium miners. *Occupational Medicine* 16 (2): 331–344.

Hwang, J. and Kim, S.W.K. (2021). Occurrence of uranopolycrase in pegmatite related to a Jurassic granite pluton in Cheongwon, South Korea: mineral chemistry and texture of a potential source of uranium and radon in a granite aquifer. *Environment and Earth Science* 14: 1–11.

IAEA (2003). Radiation protection against radon in workplaces other than mines. https://www-pub.iaea.org/MTCD/Publications/PDF/Pub1168_web.pdf (accessed 5 February 2022). I.A.E.A. Safety Reports Series No 33.

Kawabata, K., Sato, T., Takahashi, H.A. et al. (2020). Changes in groundwater radon concentrations caused by the 2016 Kumamoto earthquake. *Journal of Hydrology* 584: 124712.

Kelly, J.L., Dulai, H., Glenn, C.R. et al. (2019). Integration of aerial infrared thermography and in situ radon-222 to investigate submarine groundwater discharge to Pearl Harbor, Hawaii, USA. *Limnology and Oceanography* 64 (1): 238–257.

Kendall, G.M. and Smith, T.J. (2005). Doses from radon and its decay products to children. *Journal of Radiological Protection* 25 (3): 241–256.

Khan, S.M., Gomes, J., and Krewski, D.R. (2019). Radon interventions around the globe: a systematic review. *Heliyon* 5 (5): e01737.

Laurier, D., Valenty, M., and Tirmarche, M. (2001). Radon exposure and the risk of leukemia: a review of epidemiological studies. *Health Physics* 81 (3): 272–288.

Lecomte, J.-F., Solomon, S., Takala, J. et al. (2014). ICRP publication 126: radiological protection against radon exposure. *Annals of the ICRP* 43: 5–73.

Li, C., Wang, C., Yu, J. et al. (2020). Residential radon and histological types of lung cancer: a meta-analysis of case–control studies. *International Journal of Environmental Research and Public Health* 17 (4): 1457.

Lopes, I., Madruga, M., and Carvalho, F. (2005). Application of liquid scintillation counting techniques to gross alpha, gross beta, radon and radium measurement in Portuguese waters. *NORM IV-Naturally Occurring Radioactive Materials* 37 (10): 357–367.

Lu, Y., Liu, L., Chen, Q. et al. (2020). Domestic radon exposure and risk of childhood leukemia: a meta-analysis. *Journal of B.U.ON.: Official Journal of the Balkan Union of Oncology* 25: 1035–1041.

Lubin, J.H., Boice, J.D. Jr., Edling, C. et al. (1995). Lung cancer in radon-exposed miners and estimation of risk from indoor exposure. *JNCI: Journal of the National Cancer Institute* 87 (11): 817–827.

Lucas, H.F. (1957). Improved low-level alpha-scintillation counter for radon. *Review of Scientific Instruments* 28 (9): 680–683.

Lundin, F.E.J., Lloyd, J.W., and Smith, E.M. (1969). Mortality of uranium miners in relation to radiation exposure, hard-rock mining and cigarette smoking-1950 through September 1967. *Health Physics* 16 (5): 571–578.

Maier, A., Jones, J., Sternkopf, S. et al. (2021). Radon adsorption in charcoal. *International Journal of Environmental Research and Public Health* 18 (9): 4454.

Maier, A., Wiedemann, J., Rapp, F. et al. (2021). Radon exposure – therapeutic effect and cancer risk. *International Journal of Molecular Sciences* 22 (1): 316.

Man, C.K., Lau, S.Y., Au, S.C. et al. (1989). Radionuclide contents in building materials used in Hong Kong. *Health Physics* 57 (3): 397–401.

Margineanu, R.M. (2019). Radon measurements in underground mines and caves from several European countries. In: *AIP Conference Proceedings*, vol. 2076, 050004. AIP Publishing LLC.

Messier, K.P. and Serre, M.L. (2017). Lung and stomach cancer associations with groundwater radon in North Carolina, USA. *International Journal of Epidemiology* 46 (2): 676–685.

Nabiev, S.S., Sokolov, V.B., and Chaivanov, B.B. (2014). Molecular and crystal structures of noble gas compounds. *Russian Chemical Reviews* 83 (12): 1135–1180.

Nagaratnam, A. (1994). Radon: a historical overview. *Bulletin of Radiation Protection* 17 (3–4): 1–9.

Nazaroff, W.W. (1984). Optimizing the total-alpha three-count technique for measuring concentrations of radon progeny in residences. *Health Physics* 46 (2): 395–405.

Nusinovici, S., Vacquier, B., Leuraud, K. et al. (2010). Mortality from circulatory system diseases and low-level radon exposure in the French cohort study of uranium miners, 1946–1999. *Scandinavian Journal of Work, Environment & Health* 36 (5): 373–383.

Organization, W.H. (2007). *International Radon Project Survey on Radon Guidelines, Programmes and Activities*. Geneva: WHO.

Organization, W.H. (2009). *WHO Handbook on Indoor Radon: A Public Health Perspective*. World Health Organization.

Pawel, D.J. and Puskin, J.S. (2004). The U.S. Environmental Protection Agency's assessment of risks from indoor radon. *Health Physics* 87 (1): 68–74.

Raabe, O.G. and Wrenn, M.E. (1969). Analysis of the activity of radon daughter samples by weighted least squares. *Health Physics* 17 (4): 593–605.

Nagda, N.L. (1994). *Radon: Prevalence, Measurements, Health Risks and Control*, ASTM Manual Series: MNL, 15e. Philadelphia: ASTM International.

Radvanyi, P. and Villain, J. (2017). The discovery of radioactivity. *Comptes Rendus Physique* 18 (9): 544–550.

Ravikumar, P. and Somashekar, R.K. (2013). Estimates of the dose of radon and its progeny inhaled inside buildings. *European Journal of Environmental Sciences* 3: 88–95.

Reddy, K.U., Ningappa, C., Sannappa, J. et al. (2017). Concentration of radon and physicochemical parameters in ground water around Kolar Gold Fields, Karnataka State, India. *Journal of Radioanalytical and Nuclear Chemistry* 314: 907–915.

Robertson, A., Allen, J., Laney, R. et al. (2013). The cellular and molecular carcinogenic effects of radon exposure: a review. *International Journal of Molecular Sciences* 14 (7): 14024–14063.

Ruano-Ravina, A., Kelsey, K.T., Fernández-Villa, A. et al. (2017). Action levels for indoor radon: different risks for the same lung carcinogen? *European Respiratory Journal* 50 (5): 1701609.

Sáinz, C., Quindós, L.S., Fernández, P.L. et al. (2007). High background radiation areas: the case of Villar de la Yegua village (Spain). *Radiation Protection Dosimetry* 125 (1–4): 565–567.

Samet, J.M. and Hornung, R.W. (1990). Review of radon and lung cancer risk. *Risk Analysis* 10 (1): 65–75.

Schiager, K.J. (1974). Integrating radon progeny air sampler. *American Industrial Hygiene Association Journal* 35 (3): 165–174.

Schubert, M., Kopitz, J., and Chałupnik, S. (2014). Sample volume optimization for radon-in-water detection by liquid scintillation counting. *Journal of Environmental Radioactivity* 134: 109–113.

Schwartz, G.G. and Klug, M.G. (2016). Incidence rates of chronic lymphocytic leukemia in US states are associated with residential radon levels. *Future Oncology* 12 (2): 165–174.

Segovia, N., Gaso, M.I., and Armienta, M.A. (2007). Environmental radon studies in Mexico. *Environmental Geochemistry and Health* 29 (2): 143–153.

Sekiya, M. and Yamasaki, M. (2016). Rolf Maximilian Sievert (1896–1966): father of radiation protection. *Radiological Physics and Technology* 9 (1): 1–5.

Shergill, S., Forsman-Phillips, L., and Nicol, A.-M. (2021). Radon in schools: a review of radon testing efforts in canadian schools. *International Journal of Environmental Research and Public Health* 18 (10): 5469.

Shu'aibu, H.K., Khandaker, M.U., Baballe, A. et al. (2021). Determination of radon concentration in groundwater of Gadau, Bauchi State, Nigeria and estimation of effective dose. *Radiation Physics and Chemistry* 178: 108934.

Sohrabi, M. and Ebrahiminezhad, F. (2018). Concentration and duration dependence of a new prototype polycarbonate/activated-carbon fabric individual and environmental radon monitor. *Journal of Instrumentation* 13: P11012–P11012.

Studnička, F., Štěpán, J., and Šlégr, J. (2019). Low-cost radon detector with low-voltage air-ionization chamber. *Sensors (Basel, Switzerland)* 19 (17): 3721.

Szajerski, P., Celinska, J., Bem, H. et al. (2019). Radium content and radon exhalation rate from sulfur polymer composites (SPC) based on mineral fillers. *Construction and Building Materials* 198: 390–398.

Telahigue, F., Agoubi, B., Souid, H. et al. (2018). Groundwater chemistry and radon-222 distribution in Jerba Island, Tunisia. *Journal of Environmental Radioactivity* 182: 74–84.

Thomas, J.W. (1972). Measurement of radon daughters in air. *Health Physics* 23 (6): 783–789.

Tian, W. et al. (2019). Investigating effects of pore size distribution and pore shape on radon production in marcellus shale gas formation. *Energy & Fuels* 33 (2): 700–707.

Tong, J., Qin, C., Cao, Y. et al. (2012). Environmental radon exposure and childhood leukemia. *Journal of Toxicology and Environmental Health, Part B* 15 (5): 332–347.

Vienneau, D., De Hoogh, K., Hauri, D. et al. (2017). Effects of radon and UV exposure on skin cancer mortality in Switzerland. *Environmental Health Perspectives* 125 (6): 067009.

Vogiannis, E.G. and Nikolopoulos, D. (2015). Radon sources and associated risk in terms of exposure and dose. *Frontiers in Public Health* 2: 207–207.

Wakabayashi, A., Yasuoka, Y., Janik, M. et al. (2019). Derivation and validity evaluation of calibration factors for activated-charcoal radon collectors. *Radioisotopes* 68 (5): 317–329.

Wheeler, B.W., Allen, J., Depledge, M.H. et al. (2012). Radon and skin cancer in southwest England: an ecologic study. *Epidemiology* 23 (1): 44–52.

Whittemore, A.S. and McMillan, A. (1983). Lung cancer mortality among U.S. uranium miners: a reappraisal. *JNCI: Journal of the National Cancer Institute* 71 (3): 489–499.

Winkler-Heil, R. and Hofmann, W. (2002). Comparison of modeling concepts for radon progeny lung dosimetry. *International Congress Series* 1225: 169–177.

Zeeb, H., Shannoun, F., and Health, O.W. (2009). *WHO Handbook on Indoor Radon: A Public Health Perspective*. World Health Organization.

Zunic, Z.S., Yarmoshenko, I.V., Birovljev, A. et al. (2007). Radon survey in the high natural radiation region of Niska Banja, Serbia. *Journal of Environmental Radioactivity* 92 (3): 165–174.

Section 3

Medical Geology of the Hydrosphere

7

Water–Rock Interactions

Mineral Dissolution

Huaming Guo[1,2], *Zhipeng Gao*[1,2], *and Shiping Xing*[1,2]

[1] *MOE Key Laboratory of Groundwater Circulation and Environmental Evolution & School of Water Resources and Environment, China University of Geosciences (Beijing), Beijing, China*
[2] *State Key Laboratory of Biogeology and Environmental Geology, China University of Geosciences (Beijing), Beijing, China*

7.1 Introduction

Mineral dissolution is an important geochemical process to introduce chemical compositions into groundwater during water–rock interaction. Mineral dissolution can be classified into three groups: congruent (simple) dissolution, incongruent dissolution, and reductive dissolution. The congruent dissolution of a mineral produces only dissolved components, while incongruent dissolution produces both dissolved components and undissolved solids. The latter is also called as mineral weathering. However, reductive dissolution of solids, especially Fe/Mn oxides, generally occurs under reducing conditions. The congruent dissolution is thermodynamically controlled, while the incongruent dissolution and the reductive dissolution are mainly controlled by kinetic rates of the reactions. Although the dissolution of individual minerals in aquifer is congruent, the dissolution of one mineral leads to the precipitation of another due to different mineral solubilities. The congruent dissolution of different minerals leading to precipitation of other minerals, which is also called as incongruent dissolution. This incongruent dissolution leads to decreases in concentrations of components in precipitated minerals and increases in concentrations of components in dissolved minerals, which considerably alters groundwater chemistry (Brantley et al. 2008). Since most of incongruent dissolutions consume protons in groundwater, it gradually increases groundwater pH. This would partly explain high pH values of deep groundwater with long residence time (Xing et al. 2022). The high pH would lead to desportion of anions and/or oxyanions, such as F^- and As. Recent studies have shown that weathering of silicates is one of the major processes promoting mobility of F^- and As in aquifers via desorption (Ali et al. 2019; Wang et al. 2021; Knappett et al. 2022; Xing et al. 2022).

Reductive dissolution of Fe/Mn oxides releases not only Fe and/or Mn but also the adsorbed trace elements into groundwater (Guo et al. 2014). During the reductive dissolution, reductants (or electron donors) are the key factor in delivering electrons to Fe/Mn oxides, which include organic matter, NH_4^+, S^0, S^{2-}, CH_4, and H_2 (Kocar et al. 2010; Yang et al. 2012; Xiu et al. 2020; Glodowska et al. 2021). The reduction of Fe/Mn oxides coupled to oxidation of the above-mentioned reductants is usually mediated by microbes (Lovely 1991; Islam et al. 2004; Li et al. 2014), which affect metabolism pathways of elemental cycling. Organic matter acts as both carbon source and electron donor, and therefore biodegradable NOM is crucial for the reductive dissolution of Fe/Mn oxides (Guo et al. 2019; Qiao et al. 2020). One of the most important consequences is that the reduction dissolution of Fe(III) oxides results in As enrichment in groundwater (Smedley and Kinniburgh 2002).

Therefore, the objectives of this chapter are to introduce concepts of three kinds of dissolutions and review causes and consequences of those dissolutions in aquifer system.

7.2 Congruent (Simple) Dissolution

7.2.1 Simple Dissolution of Minerals in Groundwater System

Generally, an increase in total dissolved solid (TDS) occurs along groundwater flow path mainly due to mineral dissolution. The solubility of the mineral is crucial in determining concentrations of mineral-containing components in

Medical Geology: En route to One Health, First Edition. Edited by Majeti Narasimha Vara Prasad and Meththika Vithanage.

groundwater. In the recharge area with low TDS groundwater, minerals with low solubility would be dissolved in aquifer systems, while minerals with moderate solubility would be dissolved in the transition area with moderate TDS groundwater. Groundwater in discharge areas generally has high TDS, where minerals with relatively high solubility would be dissolved. Therefore, mineral dissolution not only regulates groundwater chemical composition but also is controlled by groundwater chemicals.

If the groundwater does not initially contain any of the components of the mineral, disequilibrium occurs between the groundwater and the mineral and the mineral will partly be dissolved to input its components into the groundwater. The concentrations of the components in the groundwater may increase continuously and finally keep constant when the equilibrium between dissolution and precipitation arrives. However, if the groundwater has already contained the components of the mineral, the solubility of the mineral would be reduced because less would be required to dissolve to obtain equilibrium. Therefore, mineral dissolution would be also controlled by the initial concentration of the mineral components.

7.2.2 Saturation Index

Saturation index (SI) of groundwater with respect to the minerals can be used to reflect the possible dissolution of the mineral in the groundwater, which can be calculated according to Eq. (7.1).

$$SI = \log\left(IAP / K_{mineral}\right) \tag{7.1}$$

where $K_{mineral}$ is the solubility product; IAP is the ion activity product, which is the product of the activities of mineral components derived from water sample analyses. Taking anhydrite dissolution as an example, the IAP is the product of the activities of Ca^{2+} and SO_4^{2-} derived from water sample analyses, and $K_{mineral}$ is the solubility product of the mineral.

For SI = 0, there is equilibrium between the mineral and the solution; $SI < 0$ reflects that the groundwater is undersaturated with respect to the mineral, and $SI > 0$ shows that the groundwater is oversaturated with the components of mineral.

Since there are uncertainties during the calculation of saturation indices (including the accuracy of the chemical analyses, the method of ion activity calculation, and mineral equilibrium constant), a range of SI values near zero is commonly believed to be within the equilibrium for a mineral. In various studies, ranges of $SI = 0 \pm 0.5$ and $SI = 0 \pm 0.1$ have been used (Paces 1975; Appelo and Postma 2005).

SI calculations for groundwater samples can be used to predict the existence of reactive minerals in the groundwater system and predict the reactivity of minerals. The minerals being normally in equilibrium with groundwater are believed to be reactive in groundwater environments, which significantly control concentrations of dissolved components. The reactive mineral with SI near zero in groundwater likely exists in the aquifer environments and regulates concentrations of dissolved components. In addition, the SI calculated from groundwater data can be used to predict the reactive minerals in aquifer solids. Although the presence of a reactive mineral can be validated by analyzing the mineralogy of the solid phase, it cannot be identified in the solid due to its poor crystallinity or low content below the detection limit.

Groundwater is undersaturated with respect to a mineral with the SI less than zero. It indicates that the mineral, if present, should dissolve into groundwater to reach equilibrium. Since the groundwater residence times, being usually years to decades or even longer, are longer than equilibrium time of a reactive mineral dissolution, the reactive mineral may not be kept undersaturated in groundwater. The most plausible explanation of $SI < 0$ is that either the mineral is not reactive in the aquifer or the mineral is reactive but not present in the aquifer.

In case that SI of a mineral is >0 in groundwater, the mineral is generally believed to be not reactive. For example, saturation indices of hematite are often greater than 10 in oxic groundwater, but those of ferrihydrite are generally near 0. The reason is that ferrihydrite is the reactive mineral in aquifer solid, but hematite is not reactive. Hematite would precipitate very slowly, even on the scale of groundwater residence time, which cannot limit concentration of dissolved iron (Deutsch 1997).

7.2.3 Chemical Evolution of Groundwater Controlled by Congruent Dissolution

Chemical components of groundwater can be controlled by congruent dissolution. Due to dissolution of calcite, groundwater in a karst system is usually dominated by Ca^{2+} and HCO_3^-. Generally, along the groundwater flow path, Ca^{2+} concentration increases, and SI of calcite gradually increases and approaches to zero, which mainly results from dissolution of calcite in aquifers. Near the recharge area with low TDS, groundwater is undersaturated with respect to calcite. It indicates that groundwater may not equilibrate with calcite due to the rapid groundwater flow rate, or it is flushed out the aquifer solid by the fresh recharge water. Near the recharge area with high TDS, groundwaters are mostly in equilibrium with calcite, suggesting the presence of calcite in the aquifer.

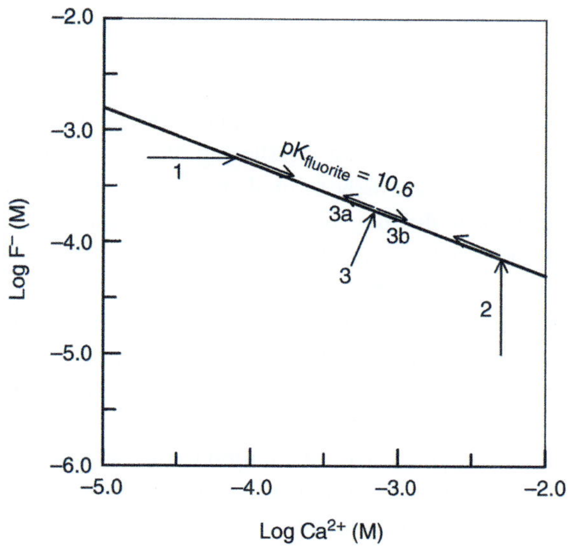

Figure 7.1 Fluorite equilibrium and evolutions of Ca^{2+} and F^- activities in groundwater along the flow path.

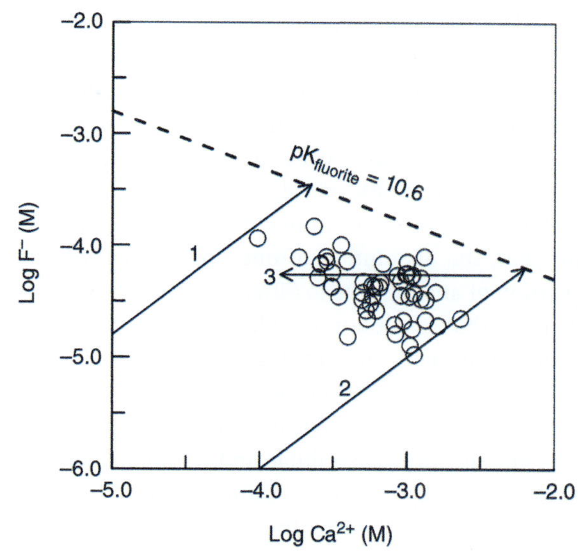

Figure 7.2 Variation of F^- concentration with Ca^{2+} concentration of groundwater samples from the Hetao basin (Line 1 shows dissolution of fluorite; Line 2 shows dissolution of calcite and fluorite at a molar ratio of 200; Line 3 indicates Ca mineral precipitation or cation exchange).

Equilibrium plot can be used to evaluate equilibrium condition of a mineral in an aquifer. As shown in Figure 7.1, groundwater samples plotted along the line of $pK_{fluorite} = 10.6$ have the activities of Ca^{2+} and F^- being in equilibrium with fluorite. Samples being located to the left of the $pK_{fluorite}$ line are undersaturated, while those to the right are oversaturated with respect to fluorite. In addition, it is useful to evaluating chemical evolution of groundwater in aquifers. As shown in Figure 7.1, groundwater samples taken along the flow path are located along the line 1, indicating that there is a source of Ca^{2+} to groundwater (such as calcite), but the absence of F^- source because of F^- activity keeps constant until it reaches the fluorite equilibrium line. If Ca^{2+} is continuously introduced into groundwater, precipitation of fluorite occurs, which lowers F^- activity. In spite of fluorite precipitation, Ca^{2+} activity continuously increases along the line 1. Groundwater samples along the line 2 of Figure 7.1 indicate that there is additional source for groundwater F^-, but no source for Ca^{2+}. Line 3 may show that sources of both F^- and Ca^{2+} are present along the flow path. Their activities gradually increase until fluoride equilibrium is reached. If F^- being introduced to groundwater are higher than two times of Ca^{2+}, the precipitation of fluorite lowers the activity of dissolved Ca^{2+}, while F^- activity continuously increases (Line 3a). If the input flux of Ca^{2+} is greater than the half of the input flux of F^-, the evolution of F^- and Ca^{2+} will follow the Line 3b of Figure 7.1.

Another example is genesis of high F^- groundwater in aquifer system (Guo et al. 2012). Generally, a negative correlation was observed between Ca^{2+} and F^- in groundwater from the Hetao basin, suggesting that fluorite solubility may limit F^- concentration ($K_{fluorite} = 10^{-10.6}$ from Parkhurst and Appelo 1999). This is supported by the fact that groundwater samples are undersaturated with respect to fluorite (Figure 7.2). When fluorite congruently dissolves, activities of Ca^{2+} and F^- follow line 1 of Figure 7.2. However, the Ca^{2+} and F^- activities of most samples are located in the right side of the line 1, which indicates that calcium is introduced into the groundwater from other sources.

Other Ca minerals free of F^-, being present in the Hetao basin sediments such as calcite (3.6~25%) and dolomite (<1.0~10%) (Guo et al. 2008), could dissolve to increase Ca^{2+} concentration. Considering that calcite coexists with fluorite in the aquifers and they dissolve at a molar ratio of 200 (calcite:fluorite), the activities of Ca^{2+} and F^- are predicted to follow line 2 in Figure 7.2. Groundwater samples are mostly plotted between lines 1 and 2 of Figure 7.2, demonstrating that F^- concentration is limited by Ca^{2+} originating from fluorite dissolution and calcite dissolution. The activity of Ca^{2+} may be decreased by mineral precipitation or cation exchange (line 3 of Figure 7.2). Saturation indices calculated with PHREEQC (Parkhurst and Appelo 1999) show that all groundwater samples are oversaturated with respect to calcite. Precipitation of calcite would lower the dissolved Ca^{2+} concentration and favor dissolution of fluorite.

7.3 Incongruent Dissolution

7.3.1 Incongruent Dissolution in Aquifer Systems

Incongruent dissolution of minerals is the most common water–rock interaction in groundwater system, which provides various solutes and precipitation of secondary clays, oxides and other minerals, enhancing element mobility (White et al. 2001). In addition to dissolved species, incongruent dissolution produces one or more minerals or amorphous solid substances, such as the weathering of silicate producing clay minerals and Fe-oxides minerals (Appelo and Postma 2004; Bonte et al. 2013). In some cases, although the dissolution of individual minerals in aquifer is congruent, the dissolution of one mineral leads to the precipitation of another due to different mineral solubility, for example, the dissolution of calcite leading to the precipitation of fluorite (Kumar et al. 2020), which is also considered as incongruent dissolution.

The weathering of silicates, one kind of incongruent dissolution, not only provides soluble and insoluble materials to groundwater (Appelo and Postma 2004; White and Buss 2014), but also changes the physical and chemical characteristics of groundwater system (such as pH) (Xing et al. 2022). The weathering of silicates into secondary minerals is controlled by many factors, including acidic and alkaline conditions, temperature, hydrologic permeability, and solubility of minerals (White et al. 2001; Gruber et al. 2016).

7.3.2 Weathering of Silicates

The weathering of silicates is significantly related to geochemistry of groundwater and sediment formation (Correns 1961). Silicates are weathered into several different components, which generally causes the formation of secondary minerals such as kaolinite, gibbsite, montmorillonite, and Fe-oxides (Figure 7.3).

The reaction involving the incongruent dissolution of feldspars (e.g. albite, anorthite, and microcline) to kaolinite, montmorillonite, and gibbsite can be represented by Eqs. (7.2–7.5), Eq. (7.6) and Eqs. (7.7) to (7.8), respectively (Table 7.1). During silicate weathering, the formation of montmorillonite is mostly favored in dry climate, where the rate of soil flushing is relatively slow (Appelo and Postma 2004; Halim et al. 2010). Generally, the enhanced degree of silicate weathering decreases Si content in secondary minerals, from montmorillonite over kaolinite to gibbsite. When albite alters to montmorillonite, 89% of the Si is preserved in the secondary mineral, which decreases to 33% for weathering to kaolinite and to 0% to gibbsite (Appelo and Postma 2004). In addition to the formation of secondary aluminosilicate minerals, the weathering products of silicates can be Fe-oxide minerals (Figure 7.3), and the reaction involving the incongruent dissolution of pyroxene (e.g. augite) to Fe-oxide can be described as Eq. (7.9) (Table 7.1).

The weathering rate of silicate determines the formation of secondary minerals, and transformation of silicates into different secondary minerals under different hydrological conditions is dependent on the weathering rate of silicate. The precipitation of secondary minerals enhances the weathering rate of silicates, which can explain that laboratory dissolution experiments predict slower rates than those observed in the field (Ganor et al. 2007). Different weathering rates of primary minerals contribute to different secondary minerals. Feldspars are more susceptible for weathering and alteration than quartz in silicate rocks. Different weathering rate of silicates is commonly observed

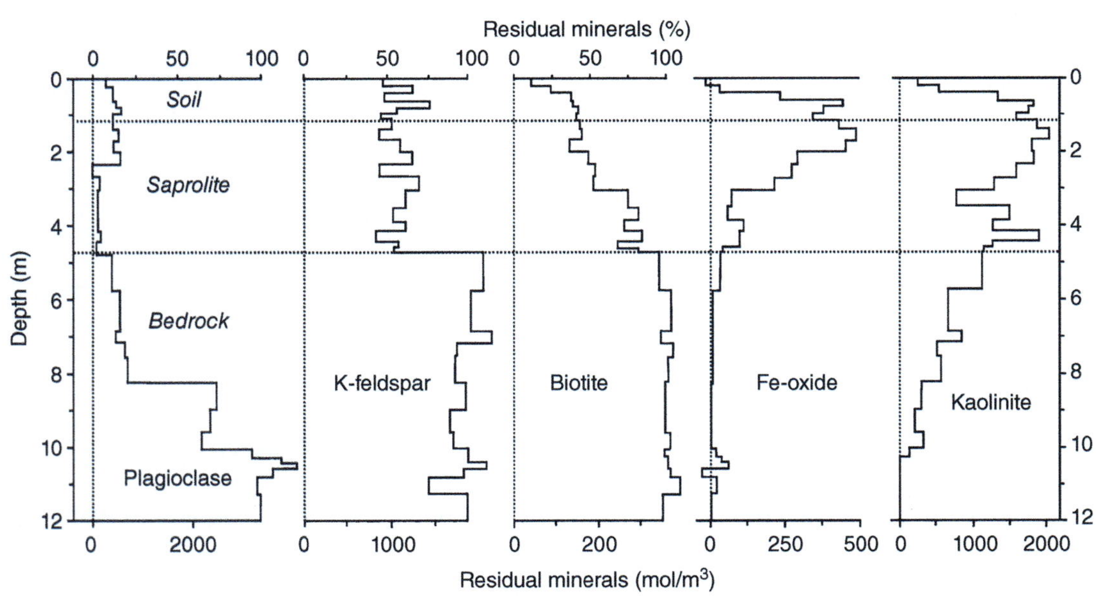

Figure 7.3 Residual mineral developed on granodiorite distributions with depth (White et al. 2001).

Table 7.1 Weathering reactions for different silicate minerals to the clay mineral (e.g. kaolinite, montmorillonite, and gibbsite) and Fe-oxide minerals.

Sodic Plagioclase (Albite)	Kaolinite	Eq.
$2Na(AlSi_3)O_8 + 2H^+ + 9H_2O$	$\rightarrow Al_2Si_2O_5(OH)_4 + 2Na^+ + 4H_4SiO_4$	(7.2)
Calcic Plagioclase (Anorthite)	Kaolinite	
$Ca(Al_2Si_2)O_8 + 2H^+ + H_2O$	$\rightarrow Al_2Si_2O_5(OH)_4 + Ca^{2+}$	(7.3)
K-feldspar (Microcline)	Kaolinite	
$2K(AlSi_3)O_8 + 2H^+ + 9H_2O$	$\rightarrow Al_2Si_2O_5(OH)_4 + 2K^+ + 4H_4SiO_4$	(7.4)
Pyroxene (Augite)	Kaolinite	
$(Mg_{0.7}CaAl_{0.3})(Al_{0.3}Si_{1.7})O_6 + 3.4H^+ + 1.1H_2O$	$\rightarrow 0.3Al_2Si_2O_5(OH)_4 + Ca^{2+} + 0.7Mg^{2+} + 1.1H_4SiO_4$	(7.5)
Sodic Plagioclase (Albite)	Montmorillonite	
$2NaAlSi_3O_8 + Mg^{2+} + 4H_2O$	$\rightarrow 2Na_{0.5}Al_{1.5}Mg_{0.5}Si_4O_{10}(OH)_2 + 2Na^+ + H_4SiO_4$	(7.6)
Sodic Plagioclase (Albite)	Gibbsite	
$Na(AlSi_3)O_8 + H^+ + 7H_2O$	$\rightarrow Al(OH)_3 + Na^+ + 3H_4SiO_4$	(7.7)
Calcic Plagioclase (Anorthite)	Gibbsite	
$CaAl_2Si_2O_8 + 2H^+ + 6H_2O$	$\rightarrow 2Al(OH)_3 + Ca^{2+} + 2H_4SiO_4$	(7.8)
Mica (Biotite)	Kaolinite and Fe-oxide	
$2K(Mg_2Fe)(AlSi_3)O_{10}(OH)_2 + 10H^+ + 0.5O_2 + 7H_2O$	$\rightarrow Al_2Si_2O_5(OH)_4 + 2K^+ + 4Mg^{2+} + 2Fe(OH)_3 + 4H_4SiO_4$	(7.9)

in the natural environment, which are dependent on solubility, chemical kinetics, and hydrologic permeability.

Normally, the weathering rate of silicates is a function of pH, showing U shape (Figure 7.4). The pH-dependent weathering rate of silicates (e.g. albite) is generally interpreted as: (i) proton promoted, which dominates the reaction rate under acidic condition; (ii) hydroxyl promoted, which dominates the reaction rate under alkaline condition, (iii) water promoted, which dominates the reaction rate under near neutral conditions. Accordingly, these three independent reaction mechanisms can be written as Eq. (7.10) (Hellmann 1994; Blum and Stillings 1995).

$$\text{Rate}_{\text{wea}} = k'_{H^+} \cdot a_{H^+}^{n_{H^+}} + k'_{H_2O} + k'_{OH^-} \cdot a_{OH^-}^{n_{OH^-}} \tag{7.10}$$

where k_i' is a rate coefficient (mol/m²/s) of reaction mechanism i; a_i is the activity of aqueous species i, and n_i is the reaction order with respect to species i; i refers to species of H^+ or H_2O, or OH^-. Furthermore, according to the relationship between weathering rate of silicate and pH (Figure 7.4), there is only one of these reaction mechanisms dominating the overall rate. Gruber et al (2016). reported that weathering rate of albite was under far-from-equilibrium conditions ($-58\,\text{kJ/mol} > \Delta G_r$) at pH 5 and at 3.6 and at temperature 25 and 50 °C.

The rate of silicate weathering is controlled by temperature (Figure 7.4). Temperature effects on the weathering rate of silicates are observed in many well-characterized environments (Bonte et al. 2014; Xing et al. 2022). Bonte et al. (2013) reported column experiment results with temperature ranging from 5 to 60 °C, showing that temperature induced chemical effects are believed to be caused by the incongruent dissolution of silicates. According to the Arrhenius expression, the weathering rate of silicate generally increases exponentially with temperature, which can be expressed as the ratio of reaction rates R/R_0 at different temperatures T and T_0 (K) in Eq. (7.11).

$$\frac{R}{R_0} = \exp\left[\frac{E_\alpha}{R'}\left(\frac{1}{T_0} - \frac{1}{T}\right)\right] \tag{7.11}$$

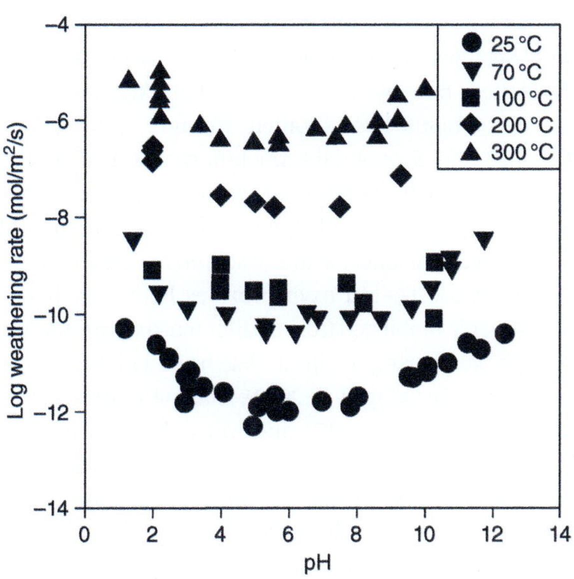

Figure 7.4 The change in log albite weathering rate (mol/m²/s) as a function of pH at various temperatures (25, 70, 100, 200, and 300 °C) (Gruber et al. 2016).

The activation energy $E\alpha$ (kJ/mol/K) for weathering of most common silicate minerals varies between 50 and 120 kJ. There is about an order of magnitude increase in silicate weathering rate when temperature increases from 0 to 25 °C, which is the most common temperature in the natural environment (White and Buss 2014).

Temperature not only controls silicate weathering kinetics but also has effect on the thermodynamic saturation state of a groundwater–rock system. Especially in aquifer environments with low permeability, kinetic reaction rate of silicate weathering is more dependent on solubility equilibrium (White et al. 2001). Silicate weathering rates are related to both the activation energy and the enthalpy of dissolution, which can be expressed via the van't Hoff equation (Eq. 7.12).

$$\frac{\mathrm{dln}K}{\mathrm{d}T} = \frac{\Delta H^0}{RT^2} \tag{7.12}$$

where K is the reaction equilibrium constant, T is the temperature (K), ΔH^0 is the reaction enthalpy (kJ/mol), and R is the gas constant (8.314 J K/mol). Gruber et al. (2016) proposed a rate law predicting that silicate weathering is temperature dependent, and silicate weathering rate is strengthened under high temperature.

Primary and secondary hydraulic conductivities (i.e. hydrologic permeability), which determine fluid volumes passing through the aquifer pore and the thermodynamic saturation of the silicates, control the weathering rate of silicate. The secondary to primary hydraulic conductivity ratio in the Panola bedrock kinetically controls plagioclase dissolution, but thermodynamically inhibits K-feldspar reaction (White et al. 2001). The initial weathering rate of silicates, when meteoric wetting front reaches a specific depth, is expected to be extremely slow and limited by fluid conductivities. Continued weathering of silicates produces secondary minerals (e.g. kaolinite), replacing the primary minerals and leading to mass loss, which increases the aquifer permeability. In addition, the increased permeability accelerates weathering rate of silicate in turn, which produces further secondary permeability (White et al. 2001).

Silicate weathering can be traced by isotopic techniques. The increased Sr isotope ratios in groundwater have been linked to the enhanced weathering of silicates since Sr isotope ratios are most probably controlled by exchange on secondary minerals (e.g. clays) deriving from weathering of the silicate minerals, which is promoted by the low groundwater flow rates (Cartwright et al. 2007). In addition, Sr isotope ratios have been used to examine mineral dissolution in silicate-dominated aquifers, and the high Sr isotope ratios are ultimately controlled by weathering of silicates particularly when weathering of silicate is believed to significantly contribute to dissolved solute compositions (Harrington and Herczeg 2003). However, Sr isotope ratios are poor tracer of silicate weathering when isotopic signature of carbonates is partially derived from co-existing silicate minerals due to diffusional exchange during regional metamorphism (Palmer and Edmond 1989). Especially at Himalaya, carbonates and calc-silicates have unusually radiogenic Sr isotope ratios (Quade et al. 2003), and thus lithium isotopes are another tracer for weathering of silicates (Kisakurek et al. 2005). In addition to strontium isotopes and lithium isotopes, boron isotopes in groundwater are also good tracer for silicates weathering (Negrel et al. 2012). The chemical weathering of silicates from bedrock may control dissolved boron in groundwater, and groundwater interacting with silicates generally has low δ^{11}B values from −10 to +10‰, which is proved by numerous previous studies (Negrel et al. 2002; Millot et al. 2007).

7.3.3 Consequence of Incongruent Dissolution of Silicates

In addition to secondary solid phase, incongruent dissolution (weathering) of silicates contributes greatly to elevated concentrations of groundwater solutes, including trace elements (e.g., As and F^-). Weathering of silicates can not only release trace elements bound in silicate directly into groundwater but also promotes the enrichment of those trace elements in groundwater by changing the characteristics of groundwater environment (Alam et al. 2014; Knappett et al. 2020; Xing et al. 2022).

Weathering of silicates releases silicate-bound As into groundwater, leading to elevated concentration of dissolved As. Arsenic can substitute for Si, Al, Fe, and Ti in many silicate mineral structures and is therefore present in many silicates (Smedley and Kinnibugh 2002). Feldspar normally has As content ranging from <0.1 to 2.1 mg/kg, and phyllosilicate such as biotite has As content around 1.4 mg/kg (Baur and Onishi 1969). Sequential extraction results turn out that silicates are the main As reservoir in the aquifer sediments/sandstone (Alam et al. 2014; Xing et al. 2022). Alam et al. (2014) reported that silicates in the sediment contained 75% of total As, while Fe–Mn oxyhydroxides hold only 16% of total As. Arsenic mobilized from aquifer matrix to groundwater through weathering of silicate has been widely reported in the Guide basin (Xing et al. 2022), shallow aquifer of West Bengal (Desbarats et al. 2017), and Antsirabe in the Central Highlands of Madagascar (Sracek et al. 2019).

In addition to As, weathering of silicates may also release F^- into groundwater directly. High F^- in groundwater is generally from geological sources, and the median content of fluorine with respect to natural geology environment is as follows: metamorphic rocks ≥ granitoids ≥ complex rocks ≥ volcanic rocks ≥ sedimentary rocks (Chae et al. 2007).

Furthermore, fluorine in groundwater is reported originating from silicates presenting in the granitic and gneissic rocks (Kumar et al. 2015), especially in micas (Rafique et al. 2015) and biotite (Chae et al. 2006). Biotite may contain fluorine at the OH- sites of the octahedral sheet, and positive correlations between F^- and SiO_2 and between F^- and Na^+ support the silicate origin of fluoride (Chae et al. 2006). Results of batch dissolution experiments showed that finer-sized biotite (<63 μm) leached out F^- up to 6 mg/L within 200 hours and F^- leached from coarser-sized biotite (125~180 and 250~600 μm) progressively increased within the reaction time (Chae et al. 2007).

Additionally, weathering of silicates can contribute to the accumulation of As and F^- in groundwater indirectly via changing hydrogeochemical characteristics of groundwater system. The weathering leads to the formation of clay minerals (e.g. kaolinite), which increases the abundance of cation exchange sites, and depletes the concentration of dissolved Ca^{2+} and releases more Na^+ in groundwater (Rango et al. 2013; Knappett et al. 2020). High concentration of Na^+ is beneficial to As enrichment by decreasing As adsorption (Smith et al. 2002; Scanlon et al. 2009), although groundwater with high concentration of Ca^{2+} is not in favor of As accumulation due to a divalent cation effect. The decreasing concentration of Ca^{2+} in groundwater due to cation exchange between Ca^{2+} and Na^+ favors the dissolution of fluorine-bearing minerals, such as fluorite and fluorapatite, which promotes F^- enrichment. This is consistent with positive correlations between As and Na^+/Ca^{2+} (Bondu et al. 2017) and between F^- and Na^+/Ca^{2+} (Hossain et al. 2016; Wang et al. 2021) in the groundwater systems. Weathering of silicate may increase concentration of HCO_3^- and raise pH in groundwater (Rango et al. 2013; Hossain et al. 2016; Knappett et al. 2020). High concentration of HCO_3^- in groundwater is conducive to As enrichment mainly via competitive adsorption sites (Gao et al. 2020) and/or inhibiting the formation of ternary goethite-As(III)-Fe(II) complex (Stolze et al. 2019). Previous studies also reported the competitive adsorption between HCO_3^- and F^- (Hossain et al. 2016). In the alkaline environment, binding sites of surface area in solid phases are readily changed from positively to negatively charged, which is beneficial for desorption of As and/or F^- (Guo et al. 2014; Rafique et al. 2015).

7.4 Reductive Dissolution of Fe(III) Oxides

7.4.1 Fe(III) Oxide Mineral and Reductive Dissolution

Iron (Fe) is ranked as the fourth most abundant element (around 5%) in earth's crust, which commonly occurs as two redox forms (e.g. ferrous iron, Fe(II) and ferric ion, Fe(III)) in environment (Wu et al. 2019). Among them, Fe(III) is poorly soluble at near-neutral pH and usually occurs as Fe(III) oxides, hydroxides, or oxide-hydroxides (hereafter referred as Fe(III) oxides) in sediments (Cornell and Schwertmann 2003). Ferrihydrite, lepidocrocite, goethite, and hematite are the most common Fe(III) oxides, with the former two as amorphous phases and the latter two as more crystalline ones (Bonneville et al. 2009). However, Fe(II) is relatively more soluble and is a common constituent of anoxic groundwater (Appelo and Postma 2005). In sediments, it usually presents as siderite ($FeCO_3$), mackaniwite (FeS), pyrite (FeS_2), and vivanite ($Fe_3(PO4)_2$). Green rust ($[Fe^{II}{}_{(6-x)}Fe^{III}(OH)_{12}]^{x+}[A^{2-}]_{x/2}\cdot yH_2O]^{x-}$) and magnetite ($Fe^{II}Fe^{III}{}_2O_4$) are the two commonly occurring Fe(II)-Fe(III) mixed minerals. Transformation between the two Fe phases is one of the most important biogeochemical processes in environments, which poses crucial impacts on fates of nutrients (e.g. C, N, P, and S) (Wu et al. 2019) and toxic metals (e.g. As, Cr, and U) (Borch et al. 2010).

Iron(III) oxides in environments are generally formed from aerobic weathering of magmatic rocks and redistributed from transportation and deposition (Cornell and Schwertmann 2003). When exposure to natural oxidants (such as O_2, NO_3^-, and birnessite), oxidation of reduced Fe species (such as dissolved Fe(II), siderite, mackinawite, and pyrite) occurs to form Fe(III) oxides in environments.

When environments are favorable (e.g. O_2 and NO_3-N concentrations lower than 0.5 mg/L), Fe(III) oxide reduction may occur (McMahon and Chapelle 2008). The reductive dissolution of Fe(III) oxides is an abiotic or biotic kinetic process being controlled by their intrinsic properties, such as crystallinity, specific surface area, particle size, and solubility (Bonneville et al. 2009). Collectively, these variables result in a wide range of reactivity of Fe(III) oxides, which can be explained by a reactivity model ($J/m_0 = k'\cdot(m/m_0)^\gamma$) (Postma 1993) or a rate equation ($R = m_0\cdot k\cdot(m/m_0)^\gamma\cdot$(1-SR(FeOOH))) (Postma et al. 2016; Gao et al. 2020). The reactivity model has been widely used to characterize reactivity or reduction kinetics of synthetic Fe(III) oxides (Pedersen et al. 2006; Davranche et al. 2013) or Fe(III) oxide minerals in sediments (Postma et al. 2010; Seddique et al. 2011; Chen et al. 2013; Zhu et al. 2014; Shen et al. 2018). Based on the model, the reactivity or reduction kinetics of Fe(III) oxides decreases in the order of ferrihydrite > lepidocrocite > goethite > hematite regarding both abiotic (Larsen and Postma 2001) and microbial reduction of Fe(III) oxides (Bonneville et al. 2004; Xiu et al. 2020). Therefore, the reactivity of Fe(III) oxides can be qualitatively evaluated by their crystallinity. Our knowledge on compositions of Fe oxide minerals in sediments highly relies on bulk sediment analysis (e.g. X-ray fluorescence), wet-sediment geochemical extractions (Keon et al. 2001; Wenzel et al. 2001; Postma et al. 2012), X-ray diffraction patterns, microscopic (such as transmission electron microscopy (TEM) and scanning electron microscopy (SEM)), and spectroscopic techniques (Mössbauer spectroscopy, X-ray absorption spectroscopy, X-ray absorption near edge structure (XANES) spectra, and extended X-ray adsorption fine structure (EXAFS) spectroscopy) (Kappler et al. 2021). These results can provide information on Fe oxide mineral identity and crystallinity (Kappler et al. 2021) and can be used to evaluate extent of Fe(III) oxide reduction and mobility and availability of elements in environments (e.g. Postma et al. 2012; Nguyen et al. 2014; Stuckey et al. 2015; Desbarats et al. 2017; Gillispie et al. 2019).

Extrinsic factors influencing Fe(III) oxide reduction include electron donors and their availability (see below), abundances of functional microorganisms that can mediate Fe(III) oxide reduction (Lovely 1991 and references therein), and other terminal electron accepting processes (TEAPs) competing for electrons (Rolle et al. 2008).

Iron(III) oxide reduction is tightly connected to microbial transformations of carbon, nitrogen, phosphorus, and sulfur and dynamics of toxic metals and organic pollutants (Wu et al. 2019; Kappler et al. 2021). One of the most important consequences is that extensive Fe(III) oxide reduction may result in high-As groundwater (Smedley and Kinniburgh 2002). Here, we present a short review to introduce electron donors for Fe(III) oxide reduction and its relevant environmental consequences.

7.4.2 Cause of Reductive Dissolution of Fe(III) Oxides

Various kinds of electron donors may be used for the reductive dissolution of Fe(III) oxides under reducing conditions, including natural organic matter (NOM), NH_4^+, H_2S, CH_4, and H_2 (Figure 7.5; Table 7.2). The reductive dissolution of Fe(III) oxides triggered by biodegradable NOM degradation is a microbial-mediated process (Qiao et al. 2020). In most cases, these biodegradable NOM (e.g. carbohydrates and aliphatic/proteins) is decomposed by fermenting bacteria and further degraded to acetate, formate, and H_2 (Appelo and Postma 2005), which are thereafter utilized by Fe-reducing bacteria, such as *Geobacter* spp. (Lovely and Phillips 1988), *Shewanela* spp. (Myers and Nealson 1990), and *Geothrix fermentans* (Coates et al. 1999). However, the process may also be controlled by the accessibility of these Fe(III)-reducing bacteria to Fe(III) oxides (Kappler et al. 2021). One proposed mechanism is that some recalcitrant NOM, such as humic substances, may serve as electron shuttles for Fe(III) oxide reduction (Mladenov et al. 2015). During NOM degradation, oxidized

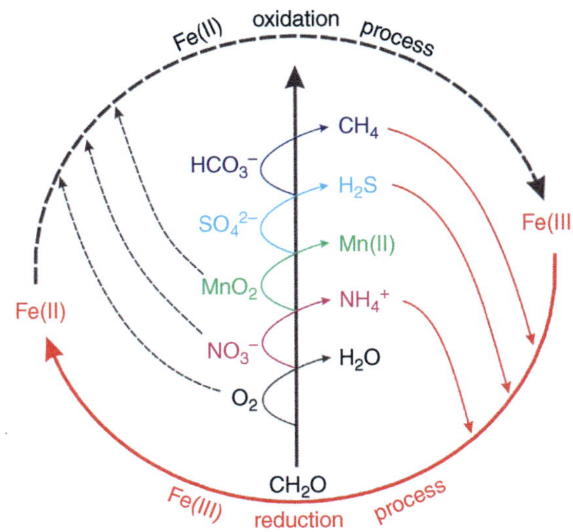

Figure 7.5 Overview of geochemical processes that are involved in Fe(II) oxidation and Fe(III) reduction. Solid lines indicate reduction process, while dotted lines represent oxidation processes.

humic substances are first reduced, and then quinone moieties in the reduced humic substances may shuttle electrons to accelerate microbial Fe(III) oxide reduction (Lovely et al. 1996).

Furthermore, anaerobic NH_4^+ oxidation coupled to Fe(III) oxide reduction (termed as Feammox) has been increasingly proposed as a Fe(III) reduction pathway (Yang et al. 2012; Ding et al. 2014; Bao and Li 2017; Xiu et al. 2020; Gao et al. 2021). The occurrence of the Feammox process can be mediated by Fe(III) reducing bacteria, such as *Geobacter* spp. and *Shewanela* spp. (Kappler et al. 2021). Furthermore, the *Acidimicrobiaceae* family has also been identified as an important microbe that is involved in Feammox process (Huang et al. 2016). During the process, pH values, Fe(III) oxide contents, and NH_4^+ concentrations are the primary controlling factors (Huang et al. 2016). NOM degradation and dissimilatory NO_3^- reduction to NH_4^+ (DNRA) are two important natural sources of dissolved NH_4^+ in groundwater under anoxic conditions (Gao et al. 2021). High abundances of Fe(III) oxides and NH_4^+ and acidic pH may be favorable for the Feammox process (Huang et al. 2016). Nevertheless, reduction kinetics of the process are still unavailable now.

The reductive dissolution of Fe(III) oxides stimulated by reduced sulfur species (e.g. H_2S, S^0, FeS_2, and FeS) is thermodynamically favorable (Table 7.2). Because the ΔG^0 of SO_4^{2-} (−744.53 kJ/mol) is very close to that of Fe(III) oxides with −708.5 kJ/mol, reductions of Fe(III) and SO_4^{2-} usually occur simultaneously, leading to the coexistence of reduced sulfur species and Fe(III) oxides. In this case, chemical H_2S oxidation coupled to Fe(III)

Table 7.2 Reaction formulas and related ΔG^0 values of Fe(III) oxide reduction (using ferrihydrite, $Fe(OH)_3$ as examples) being coupled to NOM, CH_4, NH_4^+, HS^-, S^0, FeS_2, and FeS.

Number	Reaction formulas	ΔG^0(kJ/mol)	Reference
C-Fe	$CH_3COO^- + 8Fe(OH)_3 + 15H^+ = 2HCO_3^- + 8Fe^{2+} + 20H_2O$	−612	Kocar and Fendorf (2009)
	$C_3H_5O_3^- + 4Fe(OH)_3 + 7H^+ = CH_3COO^- + HCO_3^- + 4Fe^{2+} + 10H_2O$	−347.3	Kocar and Fendorf (2009)
	$CH_2O + 7CO_2 + 4Fe(OH)_3 = 4Fe^{2+} + 8HCO_3^- + 3H_2O$	−114	Postma and Jakobsen (1996)
	$CH_4 + 8Fe(OH)_3 + 15H^+ = HCO_3^- + 8Fe^{2+} + 21H_2O$	−270.3	Beal et al. (2009)
N-Fe	$3Fe(OH)_3 + 5H^+ + NH_4^+ = 3Fe^{2+} + 9H_2O + 0.5N_2$	−245	Yang et al. (2012)
S-Fe	$2Fe(OH)_3 + HS^- + 5H^+ = 2Fe^{2+} + S^0 + 6H_2O$	−77.5	Poulton et al. (2004)
	$10S^0 + 4Fe(OH)_3 = 4Fe^{2+} + 3SO_4^{2-} + 7HS^- + 5H^+$	−92	Thamdrup et al. (1993)
	$FeS_2 + 14Fe(OH)_3 + 15HCO_3^- + 11H+ = 15FeCO_3 + 2SO_4^{2-} + 34H_2O$	−806	Schippers and Jørgensen (2002)
	$FeS + 8Fe(OH)_3 + 9HCO_3^- + 7H^+ = FeCO_3 + SO_4^{2-} + 20H_2O$	−515	Schippers and Jørgensen (2002)

oxide reduction may occur (Poulton et al. 2004; Kocar et al. 2010). The process is kinetically controlled by Fe(III) oxide contents and H_2S concentrations with a rate equation of $R = k \cdot m_{Fe(III)} \cdot (H_2S)^{0.5}$ (Poulton et al. 2004; Gao et al. 2020). Our previous study utilizing reactive transport model has calculated that H_2S oxidation may be responsible for around 7.7–25% of Fe(III) oxide reduction in the Hetao Basin, China (Gao et al. 2020). Furthermore, the elemental sulfur (S^0), as an intermediate of SO_4^{2-} reduction, can react with Fe(III) oxides via a bacterial disproportion process and thus result in Fe(III) oxide reduction (Guo et al. 2016). Although FeS_2 and FeS oxidation coupled to Fe(III) oxide reduction is thermodynamically favorable (Table 7.2), these reactions should be very limited because of their low contact to Fe(III) oxides (Schippers and Jørgensen 2001). Dissolved Fe(III) may serve as oxidant if it is adsorbed onto the surface of FeS_2 or FeS (Schippers and Jørgensen 2001).

Anaerobic oxidation of methane (CH_4) coupled to Fe(III) oxide reduction (Fe-AOM) has been recently proposed as an important CH_4 sink in environment (Beal et al. 2009; Yang et al. 2021). Many studies have provided geochemical and isotopic evidence of the AOM in marine sediments (Beal et al. 2009), porewater of lake sediment (Sivan et al. 2011), and groundwater (Glodowska et al. 2021; Pienkowska et al. 2021; Stopelli et al. 2021). The process is primarily mediated by anaerobic methane-oxidizing (ANME) archaea, which may result in an increase in ^{13}C-CH_4 but a decrease in ^{13}C-DIC, since lighter $^{12}CH_4$ is preferentially oxidized into HCO_3^- during the process (Stopelli et al. 2021). The calculated AOM rates were 1.2 and 6 μmol/year/$cm^3_{sediment}$ for lake (Sivan et al. 2011) and marine sediments (Beal et al. 2009), respectively. Nevertheless, relative importance of the Fe-AOM process in Fe(III) oxide reduction is unclear in CH_4-rich aquifers.

7.4.3 Consequence of Reductive Dissolution of Fe(III) Oxides

The reductive dissolution of Fe(III) oxides may result in the release of associated trace metals (e.g. As), organics, and ligands (e.g. phosphate) (Poulton et al. 2004). The process has been accepted as the primary mechanism of As enrichment in groundwater from anoxic aquifers (Fendorf et al. 2010; Guo et al. 2014), being mainly due to the fact that As bound in Fe(III) oxides represents a large proportion of sedimentary As (Figure 7.6). In general, Fe(III) oxide-binding sedimentary As can be structurally incorporated in layers of Fe(III) oxides or adsorbed on surface of Fe(III) oxides (Neidhardt et al. 2021). These two kinds of

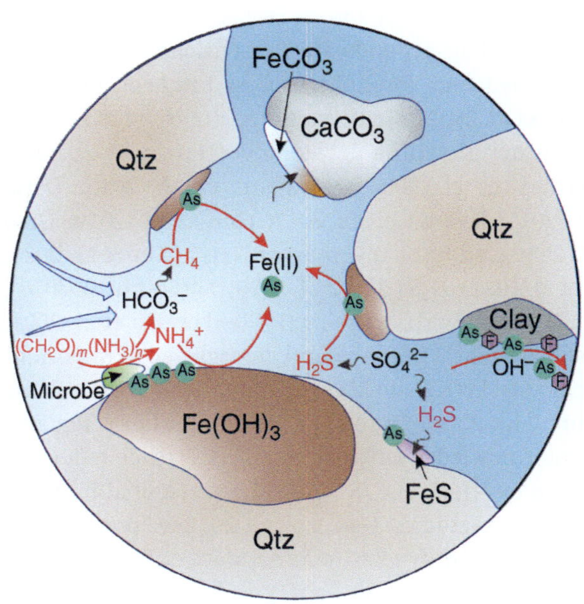

Figure 7.6 Schematic diagram illustrating biogeochemical processes of Fe(III) oxide reduction and As release. *Source:* Modified from Steefel et al. (2005).

Fe(III) oxide-binding As represent the most labile As pools in sediments, which are quite sensitive to changes in groundwater environments and redox conditions, especially the process of reductive dissolution of Fe(III) oxides. Fe(III) oxide reduction may also result in decreases in the number of surface sites on Fe(III) oxides (Jung et al. 2012), and thus both the Fe(III) oxide bound and adsorbed As are released during the process (Gao et al. 2020). Therefore, As contents in Fe(III) oxides (or As/Fe molar ratios) should be the most crucial parameter determining extent of As mobilization during the Fe(III) oxide reduction. Methods quantifying As contents in Fe(III) oxides include various selective extractions (Keon et al. 2001; Wenzel et al. 2001; Stuckey et al. 2015; Sø et al. 2018; Gao et al. 2020) and Fe EXAFS in combination with As XANES (Stuckey et al. 2015; Schaefer et al. 2017; Sun et al. 2021). In fact, aquifers with higher As/Fe molar ratios are typically related to higher As groundwaters (Sø et al. 2018; Gao et al. 2020).

Environmental variables that may influence the process of Fe(III) oxide reduction are also interfering factors controlling groundwater As enrichment. The crystallinity of Fe(III) oxides is quite important in controlling groundwater As mobility. Previous studies have indicated that high abundance of amorphous Fe(III) oxides in sediments may be related to high-As groundwater, since this kind of Fe(III) oxides is relatively more reducible than other crystalline ones (Kocar and Fendorf 2009). Another crucial variable is the availability and sources of electron donors. As mentioned above, reduced substances, such as NOM, NH_4^+, H_2S, S^0, CH_4, and H_2, are potential triggers of Fe(III) oxide reduction and As release. The predominant role of NOM in groundwater As enrichment has long been studied ever since the process has been proposed (Harvey et al. 2002). However, sources of NOM in high-As groundwater are still a debate now (Harvey et al. 2002; McArthur et al. 2004; Neumann et al. 2010; Datta et al. 2011; Postma et al. 2012; Guo et al. 2019; Qiao et al. 2020). Additionally, roles of NH_4^+ (Xiu et al. 2020; Gao et al. 2021), H_2S (Gao et al. 2020), S^0 (Guo et al. 2016), and CH_4 (Glodowska et al. 2021; Pienkowska et al. 2021; Stopelli et al. 2021) were recently recognized as important drivers of Fe(III) oxide reduction and As release (Figure 7.6), although their relative importance remains unclear. Knowledge on kinetics of Fe(III) oxide reduction triggered by NH_4^+, H_2S, S^0, and CH_4 is essential for these quantification studies. Gao et al. (2020) have provided the first trial to quantify relative importance of H_2S oxidation and NOM degradation in Fe(III) oxide reduction and As release and found that NOM degradation was the primary electron donors (>75%). Future studies are required to investigate relative contributions of NOM, NH_4^+, H_2S, S^0, and CH_4 in Fe(III) oxide reduction and As release, since they may coexist in whole or in part in high-As groundwater systems.

7.5 Conclusion Remarks

Mineral dissolution introduces dissolved components into groundwater. As an important process, congruent dissolution mainly contributes to the increase in groundwater TDS along the flow path. In the recharge area, minerals with low solubility would be congruently dissolved in aquifer systems, while congruent dissolution of minerals with high solubility mainly occurs in the discharge area, usually with relatively high TDS. SI can be used to reflect the possible dissolution of the mineral in the groundwater. Minerals with $SI < 0$ have the potential to be dissolved, while those with $SI > 0$ have the potential to be precipitated in groundwater. Although the dissolution of individual minerals in aquifer is congruent, the dissolution of one mineral leads to the precipitation of another due to different mineral solubilities. Incongruent dissolution of minerals not only produces dissolved components but also forms undissolved solids. In addition to directly introducing dissolved components into groundwater, incongruent dissolution or weathering of minerals indirectly promotes the mobility of trace elements in aquifer systems by elevating groundwater pH and HCO_3^-. Those processes contribute to As and/or F^- enrichment in groundwater, especially old groundwater with long residence time. The reductive dissolution of Fe/Mn oxides not only introduces Fe and/or Mn into groundwater but also mobilizes the bound trace elements (such as As), which is dependent on electron donors and their availability, abundances of functional microorganisms and other TEAPs. One of the most important consequences is that the reduction dissolution of Fe(III) oxides results in As enrichment in groundwater.

Acknowledgments

The study was financially supported by the National Natural Science Foundation of China (grant nos. 42130509 and 42102285), 111 project (No. B20010), and the Fundamental Research Funds for the Central Universities (grant Nos. 2652020623 and 2652019298).

References

Alam, M.S., Wu, Y., and Cheng, T. (2014). Silicate minerals as a source of arsenic contamination in groundwater. *Water Air and Soil Pollution* 225: 1–15.

Ali, W., Mushtaq, N., Javed, T. et al. (2019). Vertical mixing with return irrigation water the cause of arsenic enrichment in groundwater of district Larkana Sindh. Pakistan. *Environmental Pollution* 245: 77–88.

Appelo, C.A.J. and Postma, D. (2004). *Geochemistry, Groundwater and Pollution*. CRC Press.

Appelo, C.A.J. and Postma, D. (2005). *Geochemistry, Groundwater and Pollution*. Amsterdam: A.A. Balkema Publishers.

Bao, P. and Li, G.X. (2017). Sulfur-driven iron reduction coupled to anaerobic ammonium oxidation. *Environmental Science & Technology* 51: 6691–6698.

Baur, W.H. and Onishi, B.M.H. (1969). Arsenic. In: *Handbook of Geochemistry* (ed. K.H. Wedepohl), 33-A-1–33-0-5. Berlin: Springer-Verlag.

Beal, E.J., House, C.H., and Orphan, V.J. (2009). Manganese- and iron-dependent marine methane oxidation. *Science* 325: 184–187.

Blum, A. and Stillings, L.L. (1995). Feldspar dissolution kinetics. In: *Chemical Weathering Rates of Silicate Minerals* (ed. A.F. White and S.L. Brantley). Washington: Mineralogical Society of America.

Bondu, R., Cloutier, V., Rosa, E., and Benzaazoua, M. (2017). Mobility and speciation of geogenic arsenic in bedrock groundwater from the Canadian Shield in Western Quebec, Canada. *Science of the Total Environment* 574: 509–519.

Bonneville, S., Van Cappellen, P., and Behrends, T. (2004). Microbial reduction of iron(III) oxyhydroxides: effects of mineral solubility and availability. *Chemical Geology* 212: 255–268.

Bonneville, S., Behrends, T., and Van Cappellen, P. (2009). Solubility and dissimilatory reduction kinetics of iron(III) oxyhydroxides: a linear free energy relationship. *Geochimica et Cosmochimica Acta* 73: 5273–5282.

Bonte, M., van Breukelen, B.M., and Stuyfzand, P.J. (2013). Temperature-induced impacts on groundwater quality and arsenic mobility in anoxic aquifer sediments used for both drinking water and shallow geothermal energy production. *Water Research* 47 (14): 5088–5100.

Bonte, M., Stuyfzand, P.J., and Breukelen, B.M. (2014). Reactive transport modeling of thermal column experiments to investigate the impacts of aquifer thermal energy storage on groundwater quality. *Environmental Science & Technology* 48 (20): 12099–12107.

Borch, T., Kretzschmar, R., Kappler, A. et al. (2010). Biogeochemical redox processes and their impact on contaminant dynamics. *Environmental Science & Technology* 44: 15–23.

Brantley, S.L., White, A.F., and Kubicki, J.D. (2008). *Kinetics of Water-Rock Interaction*. New York: Springer.

Cartwright, I., Weaver, T., and Petrides, B. (2007). Controls on 87Sr/86Sr ratios of groundwater in silicate-dominated aquifers: SE Murray Basin, Australia. *Chemical Geology* 246 (1–2): 107–123.

Chae, G.T., Yun, S.T., Kwon, M.J. et al. (2006). Batch dissolution of granite and biotite in water: implication for fluorine geochemistry in groundwater. *Geochemical Journal* 40 (1): 95–102.

Chae, G.T., Yun, S.T., Mayer, B. et al. (2007). Fluorine geochemistry in bedrock groundwater of South Korea. *Science of the Total Environment* 385 (1–3): 272–283.

Chen, L.J., Zhu, M.X., Yang, G.P., and Huang, X.L. (2013). Reductive reactivity of iron(III) oxides in the East China Sea sediments: characterization by selective extraction and kinetic dissolution. *PloS One* 8: 1–12.

Coates, J.D., Ellis, D.J., Gaw, C.V., and Lovley, D.R. (1999). *Geothrix fermentans* gen. nov., sp. nov., a novel Fe(III)-reducing bacterium from a hydrocarbon-contaminated aquifer. *International Journal of Systematic and Evolutionary Microbiology* 49: 1615–1622.

Cornell, R.M. and Schwertmann, U. (2003). *The Iron Oxides*, 2e. Druckhaus Darmstadt: Wiley- VCH.

Correns, C.W. (1961). The experimental chemical weathering of silicates. *Clay Minerals Bulletin* 4 (26): 249–265.

Datta, S., Neal, A.W., Mohajerin, T.J. et al. (2011). Perennial ponds are not an important source of water or dissolved organic matter to groundwaters with high arsenic concentrations in West Bengal, India. *Geophysical Research Letters* 38: L20404.

Davranche, M., Dia, A., Fakih, M. et al. (2013). Organic matter control on the reactivity of Fe(III)-oxyhydroxides and associated as in wetland soils: a kinetic modeling study. *Chemical Geology* 335: 24–35.

Desbarats, A.J., Pal, T., Mukhejee, P.K., and Beckie, R.D. (2017). Geochemical evolution of groundwater flowing through arsenic source sediments in an aquifer system of West Bengal, India. *Water Resources Research* 54: 8715–8735.

Deutsch, W.J. (1997). *Groundwater Geochemistry: Fundamentals and Applications to Contamination*. New York: CRC Press.

Ding, L.J., An, X.L., Li, S. et al. (2014). Nitrogen loss through anaerobic ammonium oxidation coupled to iron reduction from paddy soils in a chronosequence. *Environmental Science & Technology* 48: 10641–10647.

Fendorf, S., Michael, H.A., and van Geen, A. (2010). Spatial and temporal variations of groundwater arsenic in South and SouthEast Asia. *Science* 328: 1123–1127.

Ganor, J., Peng, L., Zheng, Z., and Chen, Z. (2007). Bridging the gap between laboratory measurements and field estimations of silicate weathering using simple calculations. *Environmental Geology* 53 (3): 599–610.

Gao, Z.P., Jia, Y.F., Guo, H.M. et al. (2020). Quantifying geochemical processes of arsenic mobility in groundwater from an inland basin using a reactive transport model. *Water Resources Research* 56: e2019WR025492.

Gao, Z.P., Weng, H.C., and Guo, H.M. (2021). Unraveling influences of nitrogen cycling on arsenic enrichment in groundwater from the Hetao Basin using geochemical and multi-isotopic approaches. *Journal of Hydrology* 595: 125981.

Gillispie, E.C., Matteson, A.R., Duckworth, O.W. et al. (2019). Chemical variability of sediment and groundwater in a Pleistocene aquifer of Cambodia: implications for arsenic pollution potential. *Geochimica et Cosmochimica Acta* 245: 441–458.

Glodowska, M., Stopelli, E., Schneider, M. et al. (2021). Arsenic mobilization by anaerobic iron-dependent methane oxidation. *Water Research* 4: 42.

Gruber, C., Kutuzov, I., and Ganor, J. (2016). The combined effect of temperature and pH on albite dissolution rate under far-from-equilibrium conditions. *Geochimica et Cosmochimica Acta* 186: 154–167.

Guo, H.M., Yang, S.Z., Tang, X.H. et al. (2008). Groundwater geochemistry and its implications for arsenic mobilization in shallow aquifers of the Hetao basin, Inner Mongolia. *Science of the Total Environment* 393: 131–144.

Guo, H.M., Zhang, Y., Xing, L.N., and Jia, Y.F. (2012). Spatial variation in arsenic and fluoride concentrations of shallow groundwater from the Shahai town of the Hetao basin, Inner Mongolia. *Applied Geochemistry* 27: 2187–2196.

Guo, H.M., Wen, D.G., Liu, Z.Y. et al. (2014). A review of high arsenic groundwater in Mainland and Taiwan, China: distribution, characteristics and geochemical processes. *Applied Geochemistry* 41: 196–217.

Guo, H.M., Zhou, Y.Z., Jia, Y.F. et al. (2016). Sulfur cycling-related biogeochemical processes of arsenic mobilization in the Western Hetao Basin, China: evidence from multiple isotope approaches. *Environmental Science & Technology* 50 (23): 12650–12659.

Guo, H.M., Li, X.M., Xiu, W. et al. (2019). Controls of organic matter bioreactivity on arsenic mobility in shallow aquifers of the Hetao Basin, P.R. China. *Journal of Hydrology* 2019: 448–459.

Halim, M.A., Majumder, R.K., Nessa, S.A. et al. (2010). Evaluation of processes controlling the geochemical constituents in deep groundwater in Bangladesh: spatial variability on arsenic and boron enrichment. *Journal of Hazardous Materials* 180 (1–3): 50–62.

Harrington, G.A. and Herczeg, A.L. (2003). The importance of silicate weathering of a sedimentary aquifer in arid Central Australia indicated by very high $^{87}Sr/^{86}Sr$ ratios. *Chemical Geology* 199 (3–4): 281–292.

Harvey, C.F., Swartz, C.H., Badruzzaman, A.B.M. et al. (2002). Arsenic mobility and groundwater extraction in Bangladesh. *Science* 298 (5598): 1602–1606.

Hellmann, R. (1994). The albite-water system: part I. The kinetics of dissolution as a function of pH at 100, 200 and 300°C. *Geochimica et Cosmochimica Acta* 58 (2): 595–611.

Hossain, S., Hosono, T., Yang, H., and Shimada, J. (2016). Geochemical processes controlling fluoride enrichment in groundwater at the Western part of Kumamoto area, Japan. *Water, Air, & Soil Pollution* 227 (10): 1–14.

Huang, S., Chen, C., Peng, X.C., and Jaffé, P.R. (2016). Environmental factors affecting the presence of *Acidimicrobiaceae* and ammonium removal under iron-reducing conditions in soil environments. *Soil Biology and Biogeochemistry* 98: 148–158.

Islam, F.S., Gault, A.G., Boothman, C. et al. (2004). Role of metal reducing bacteria in arsenic release from Bengal delta sediments. *Nature* 430: 68–71.

Jung, H.B., Bostick, B.C., and Zheng, Y. (2012). Field, experimental, and modeling study of arsenic partitioning across a redox transition in a Bangladesh aquifer. *Environmental Science & Technology* 46: 1388–1395.

Kappler, A., Bryce, C., Mansor, M. et al. (2021). An evolving view on biogeochemical cycling of iron. *Nature Reviews Microbiology* 19: 360–374.

Keon, N.E., Swartz, C.H., Brabander, D.J. et al. (2001). Validation of an arsenic sequential extraction method for evaluating mobility in sediments. *Environmental Science & Technology* 35 (13): 2778.

Kisakurek, B., James, R.H., and Harris, N.B.W. (2005). Li and δ^7Li in Himalayan rivers: proxies for silicate weathering? *Earth and Planetary Science Letters* 237 (3–4): 387–401.

Knappett, P.S.K., Li, Y., Loza, I. et al. (2020). Rising arsenic concentrations from dewatering a geothermally influenced aquifer in central Mexico. *Water Research* 185: 116257. https://doi.org/10.1016/j.watres.2020.116257.

Knappett, P.S.K., Farias, P., Miller, G.R. et al. (2022). A systems approach to remediating human exposure to arsenic and fluoride from overexploited aquifers. *GeoHealth* 6: e2022GH000592.

Kocar, B.D. and Fendorf, S. (2009). Thermodynamic constraints on reductive reactions influencing the biogeochemistry of arsenic in soils and sediments. *Environmental Science & Technology* 43: 4871–4877.

Kocar, B.D., Borch, T., and Fendorf, S. (2010). Arsenic repartitioning during biogenic sulfidization and transformation of ferrihydrite. *Geochimica et Cosmochimica Acta* 74: 980–994.

Kumar, P.J.S., Jegathambal, P., Nair, S., and James, E.J. (2015). Temperature and pH dependent geochemical modeling of fluoride mobilization in the groundwater of a crystalline aquifer in Southern India. *Journal of Geochemical Exploration* 156: 1–9.

Kumar, M., Goswami, R., Patel, A.K. et al. (2020). Scenario, perspectives and mechanism of arsenic and fluoride co-occurrence in the groundwater: a review. *Chemosphere* 249: 126126.

Larsen, O. and Postma, D. (2001). Kinetics of reductive bulk dissolution of lepidocrocite, ferrihydrite, and goethite. *Geochimica et Cosmochimica Acta* 65: 1367–1379.

Li, Y., Guo, H.M., and Hao, C.B. (2014). Arsenic release from shallow aquifers of the Hetao Basin, Inner Mongolia: evidence from bacterial community in aquifer sediments and groundwater. *Ecotoxicology* 23: 1900–1914.

Lovely, D.R. (1991). Dissimilatory Fe(III) and Mn(IV) reduction. *Microbiological Reviews* 55: 259–287.

Lovely, D.R. and Phillips, E.J.P. (1988). Novel mode of microbial energy metabolism: organic carbon oxidation coupled to dissimilatory reduction of iron or manganese. *Applied and Environmental Microbiology* 54: 1472–1480.

Lovley, D.R., Coates, J.D., Blunt-Harris, E.L. et al. (1996). Humic substances as electron acceptors for microbial respiration. *Nature* 382: 445–448.

McArthur, J.M., Banerjee, D.M., Hudson-Edwards, K.A. et al. (2004). Natural organic matter in sedimentary basins and its relation to arsenic in anoxic ground water: the example of West Bengal and its worldwide implications. *Applied Geochemistry* 19 (8): 1255–1293.

McMahon, P.B. and Chapelle, F.H. (2008). Redox processes and water quality of selected principal aquifer systems. *Groundwater* 46 (2): 259–271.

Millot, R., Negrel, P., and Petelet-Giraud, E. (2007). Multi-isotopic (Li, B, Sr, Nd) approach for geothermal reservoir characterization in the Limagne Basin (massif Central, France). *Applied Geochemistry* 22 (11): 2307–2325.

Mladenov, N., Zheng, Y., Simone, B. et al. (2015). Dissolved organic matter quality in a shallow aquifer of Bangladesh: implications for arsenic mobility. *Environmental Science & Technology* 49: 10815–10824.

Myers, C.R. and Nealson, K.H. (1990). Respiration-linked proton translocation coupled to anaerobic reduction of manganese(IV) and iron(III) in *Shewanella putrefaciens* MR-1. *Journal of Bacteriology* 172: 6232–6238.

Negrel, P., Petelet-Giraud, E., Kloppmann, W., and Casanova, J. (2002). Boron isotope signatures in the coastal groundwaters of French Guiana. *Water Resources Research* 38 (11): 44-1–44-5.

Negrel, P., Millot, R., Guerrot, C. et al. (2012). Heterogeneities and interconnections in groundwaters: coupled B, Li and stable-isotope variations in a large aquifer system (Eocene Sand aquifer, SouthWestern France). *Chemical Geology* 296: 83–95.

Neidhardt, H., Rudischer, S., Eiche, E. et al. (2021). Phosphate immobilisation dynamics and interaction with arsenic sorption at redox transition zones in floodplain aquifers: insights from the Red River Delta, Vietnam. *Journal of Hazardous Materials* 411: 125128.

Neumann, R.B., Ashfaque, K.N., Badruzzaman, A.B.M. et al. (2010). Anthropogenic influence on groundwater arsenic concentrations in Bangladesh. *Nature Geoscience* 3: 46–52.

Nguyen, T.H.M., Postma, D., Pham, T.K.T. et al. (2014). Adsorption and desorption of arsenic to aquifer sediment on the Red River floodplain at Nam Du, Vietnam. *Geochimica et Cosmochimica Acta* 142: 587–600.

Paces, T. (1975). *Kinetics of Natural Water System.* Vienna: IAEA.

Palmer, M.R. and Edmond, J.M. (1989). The strontium isotope budget of the modern ocean. *Earth and Planetary Science Letters* 92 (1): 11–26.

Parkhurst, D.L. and Appelo, C.A.J. (1999). User's guide to PHREEQC (version 2)-a computer program for speciation, batch-reaction, one dimensional transport and inverse geochemical calculation. *USGS Water Resources Investigation Report 99-4259*. Denver: U.S. Geological Survey; 312 pp.

Pedersen, H.D., Postma, D., and Jakobsen, R. (2006). Release of arsenic associated with the reduction and transformation of iron oxides. *Geochimica et Cosmochimica Acta* 70: 4116–4129.

Pienkowska, A., Glodowska, M., Mansor, M. et al. (2021). Isotopic labeling reveals microbial methane oxidation coupled to Fe(III) mineral reduction in sediments from an As-contaminated aquifer. *Environmental Science & Technology Letters* 8: 832–837.

Postma, D. (1993). The reactivity of iron oxides in sediments: a kinetic approach. *Geochimica et Cosmochimica Acta* 57: 5027–5034.

Postma, D. and Jakobsen, R. (1996). Redox zonation: equilibrium constraints on the Fe(III)/SO_4-reduction interface. *Geochimica et Cosmochimica Acta* 60: 3169–3175.

Postma, D., Jessen, S., Nguyen Thi Minh, H. et al. (2010). Mobilization of arsenic and iron from Red River floodplain sediments, Vietnam. *Geochimica et Cosmochimica Acta* 74: 3367–3381.

Postma, D., Larsen, F., Nguyen, T.T. et al. (2012). Groundwater arsenic concentration in Vietnam controlled by sediment age. *Nature Geoscience* 5: 656–661.

Postma, D., Pham, T.K.T., Sø, H.U. et al. (2016). A model for the evolution in water chemistry of an arsenic contaminated aquifer over the last 6000 years, Red River floodplain, Vietnam. *Geochimica et Cosmochimica Acta* 195: 277–292.

Poulton, S.W., Krom, M.D., and Raiswell, R. (2004). A revised scheme for the reactivity of iron (oxhydr)oxide minerals towards dissolved sulfide. *Geochimica et Cosmochimica Acta* 68: 3703–3715.

Qiao, W., Guo, H.M., He, C. et al. (2020). Molecular evidence of arsenic mobility linked to biodegradable organic matter. *Environmental Science & Technology* 54 (12): 7280–7290.

Quade, J., English, N., and DeCelles, P.G. (2003). Silicate versus carbonate weathering in the Himalaya: a comparison of the Arun and Seti River watersheds. *Chemical Geology* 202 (3–4): 275–296.

Rafique, T., Naseem, S., Ozsvath, D. et al. (2015). Geochemical controls of high fluoride groundwater in Umarkot sub-district, Thar Desert, Pakistan. *Science of the Total Environment* 530: 271–278.

Rango, T., Vengosh, A., Dwyer, G., and Bianchini, G. (2013). Mobilization of arsenic and other naturally occurring contaminants in groundwater of the Main Ethiopian Rift aquifers. *Water Research* 47 (15): 5801–5818.

Rolle, M., Clement, T.P., Sethi, R., and Molfetta, A.D. (2008). A kinetic approach for simulating redox-controlled fringe and core biodegradation processes in groundwater: model development and application to a landfill site in Piedmont, Italy. *Hydrological Processes* 22: 4905–4921.

Scanlon, B.R., Nicot, J.P., Reedy, R.C. et al. (2009). Elevated naturally occurring arsenic in a semiarid oxidizing system, Southern High Plains aquifer, Texas, USA. *Applied Geochemistry* 24 (11): 2061–2071.

Schaefer, M.V., Guo, X., Gan, Y. et al. (2017). Redox controls on arsenic enrichment and release from aquifer sediments in central Yangtze River Basin. *Geochimica et Cosmochimica Acta* 204: 104–119.

Schippers, A. and Jørgensen, B.B. (2001). Oxidation of pyrite and iron sulfide by manganese dioxide in marine sediments. *Geochimica et Cosmochimica Acta* 65: 915–922.

Schippers, A. and Jørgensen, B.B. (2002). Biogeochemistry of pyrite and iron sulfide oxidation in marine sediments. *Geochimica et Cosmochimica Acta* 66: 85–92.

Seddique, A.A., Masuda, H., Mitamura, M. et al. (2011). Mineralogy and geochemistry of shallow sediments of Sonargaon, Bangladesh and implications for arsenic dynamics: focusing on the role of organic matter. *Applied Geochemistry* 26: 587–599.

Shen, M., Guo, H., Jia, Y. et al. (2018). Partitioning and reactivity of iron oxide minerals in aquifer sediments hosting high arsenic groundwater from the Hetao basin, P. R. China. *Applied Geochemistry* 89: 190–201.

Sivan, O., Adler, M., Pearson, A. et al. (2011). Geochemical evidence for iron-mediated anaerobic oxidation of methane. *Limnology and Oceanography* 56 (4): 1536–1544.

Smedley, P.L. and Kinniburgh, D.G. (2002). A review of the source, behaviour and distribution of arsenic in natural waters. *Applied Geochemistry* 17: 517–568.

Smith, E., Naidu, R., and Alston, A.M. (2002). Chemistry of inorganic arsenic in soils: II. effect of phosphorus, sodium, and calcium on arsenic sorption. *Journal of Environmental Quality* 31 (2): 557–563.

Sø, H.U., Postma, D., Lan, V.M. et al. (2018). Arsenic in Holocene aquifers of the Red River floodplain, Vietnam: effects of sediment-water interactions, sediment burial age and groundwater residence time. *Geochimica et Cosmochimica Acta* 225: 192–209.

Sracek, O., Rahobisoa, J.J., Trubač, J. et al. (2019). Geochemistry of thermal waters and arsenic enrichment at Antsirabe, Central Highlands of Madagascar. *Journal of Hydrology* 577: 123895.

Steefel, C.I., Depaolo, D.J., and Lichtner, P.C. (2005). Reactive transport modeling: an essential tool and a new research approach for the Earth sciences. *Earth and Planetary Science Letters* 240: 539–558.

Stolze, L., Zhang, D., Guo, H., and Rolle, M. (2019). Surface complexation modeling of arsenic mobilization from goethite: interpretation of an in-situ experiment. *Geochimica et Cosmochimica Acta* 248: 274–288.

Stopelli, E., Duyen, V.T., Prommer, H. et al. (2021). Carbon and methane cycling in arsenic-contaminated aquifers. *Water Research* 200: 117300.

Stuckey, J.W., Schaefer, M.V., Benner, S.G., and Fendorf, S. (2015). Reactivity and speciation of mineral-associated arsenic in seasonal and permanent wetlands of the Mekong Delta. *Geochimica et Cosmochimica Acta* 171: 143–155.

Sun, Y.Q., Sun, J., Nghiem, A.A. et al. (2021). Reduction of iron (hydr)oxide-bound arsenate: evidence from high depth resolution sampling of a reducing aquifer in Yinchuan Plain, China. *Journal of Hazardous Materials* 406: 124615.

Thamdrup, B., Finster, K., Hansen, W., and Bak, F. (1993). Bacterial disproportionation of elemental sulfur coupled to chemical reduction of iron and manganese. *Applied and Environmental Microbiology* 59: 101–108.

Wang, Z., Guo, H.M., Xing, S.P., and Liu, H.Y. (2021). Hydrogeochemical and geothermal controls on the formation of high fluoride groundwater in the Guide basin, NorthWestern China. *Journal of Hydrology* 598: 126372.

Wenzel, W.W., Kirchbaumer, N., Prohaska, T. et al. (2001). Arsenic fractionation in soils using an improved sequential extraction procedure. *Analytica Chimica Acta* 436: 309–323.

White, A.F. and Buss, H.L. (2014). *Natural Weathering Rates of Silicate Minerals. Treatise on Geochemistry*, 2e, 115–155. Oxford: Elsevier.

White, A.F., Bullen, T.D., Schulz, M.S. et al. (2001). Differential rates of feldspar weathering in granitic regoliths. *Geochimica et Cosmochimica Acta* 65 (6): 847–869.

Wu, S., Vymazal, J., and Brix, H. (2019). Critical review: biogeochemical networking of iron in constructed wetlands for wastewater treatment. *Environmental Science & Technology* 53: 7930–7944.

Xing, S., Guo, H., Zhang, L. et al. (2022). Silicate weathering contributed to arsenic enrichment in geotherm-affected groundwater in Pliocene aquifers of the Guide basin, China. *Journal of Hydrology* 127444.

Xiu, W., Lloyd, J., Guo, H.M. et al. (2020). Linking microbial community composition to hydrogeochemistry in the Western Hetao Basin: potential importance of ammonium as an electron donor during arsenic mobilization. *Environment International* 136: 105489.

Xiu, W., Yuan, W.J., Polya, D.A. et al. (2021). A critical review of abiotic and microbially-mediated chemical reduction rates of Fe(III) (oxyhydr)oxides using a reactivity model. *Applied Geochemistry*, 126, 104895.

Yang, W.H., Weber, K.A., and Silver, W.L. (2012). Nitrogen loss from soil through anaerobic ammonium oxidation coupled to iron reduction. *Nature Geoscience* 5: 538–541.

Yang, H., Yu, S., and Lu, H. (2021). Iron-coupled anaerobic oxidation of methane in marine sediments: a review. *Journal of Marine Science and Engineering* 9: 875.

Zhu, M.X., Chen, L.J., Yang, G.P. et al. (2014). Kinetic characterization on reductive reactivity of iron(III) oxides in surface sediments of the East China Sea and the influence of repeated redox cycles: implications for microbial iron reduction. *Applied Geochemistry* 42: 16–26.

8

Water Hardness and Health

W. P. R. T. Perera

Department of Indigenous Medical Resources, Faculty of Indigenous Health Sciences and Technology, Gampaha Wickramarachchi University of Indigenous Medicine, Yakkala, Sri Lanka

8.1 Water Hardness – Overview

Improving global access to safe drinking water is one of the most cost-effective and life-saving ways to improve public health. Owing to the lack of having safe and clean water, there will be argumentative health effects and long-term exposure to unsafe water including hard water. In addition to waterborne, water-based, and water-related diseases, a long-term revelation that water contains a higher amount of ion content also causes long-term health concerns (Hasan et al. 2019). Meanwhile, hard water plays a significant role in generating unsuitable conditions for drinking water in different regions of the world. It is caused by a variety of dissolved polyvalent metallic ions, pretty much exclusively calcium and magnesium cations, though other cations such as aluminum barium, iron, manganese, zinc, and strontium are also contributing. Actually, water hardness is the traditional measure of the capacity of water to react with soap. It describes water's ability to bind soap to form a lather, which is a chemical reaction detrimental to the washing process (Rubenowitz-lundin and Hiscock 2013). In layman's terms, when hands still feel slimy after washing with soap and water, or when drinking glasses at home become less than crystal clear, water hardness may occur. Calcium and magnesium were always primarily found in natural water bound to bicarbonate, sulfate, or chloride. Bicarbonate converts to carbonate and precipitates with Ca^{2+} to form calcium carbonate ($CaCO_3$) when hard water evaporates or is heated above 60°C (Lesnic et al. 2021).

Carbonate hardness includes the total hardness that is equivalent to bicarbonate and carbonate, and it means alkalinity. If the hardness exceeds the alkalinity, the excess is denoted as noncarbonate hardness, and it is a measure of calcium and magnesium sulfates. Carbonate hardness is called temporary hardness, and it can be mitigated upon precipitating as $CaCO_3$ by heating. However, noncarbonate hardness leads to a permanent hardness in water (Rubenowitz-lundin and Hiscock 2013).

Carbonate hardness compounds; $CaCO_3$, $MgCO_3$, $Ca(HCO_3)_2$, $Mg(HCO_3)_2$, $Ca(OH)_2$, and $Mg(OH)_2$

Non-carbonate hardness compounds; $CaSO_4$, $MgSO_4$, $CaCl_2$, and $MgCl_2$

When considering the water hardness as a numerical value, if it enhances beyond 150 mg/L, that water called hard water and if the water hardness value reported less than 60 mg/L it is called soft water (Table 8.1) (World Health Organization 2011; Bouhoun et al. 2021).

Despite having a negative impression on the water hardness, hard water, on the other hand, can have some advantages. Humans require minerals to stay healthy, and the World Health Organization (WHO) states that drinking water may be a source of calcium and magnesium in the diet, which may be especially important for those who are deficient in calcium and magnesium. Calcium and magnesium are both essential minerals that are beneficial to human health in a range of ways. However, inadequate intake of either nutrient can have adverse health effects (World Health Organization 2011).

Water hardness is a concern for groundwater systems. As water moves through soil and rock that contains magnesium and calcium-bearing minerals, it dissolves small amounts of naturally occurring minerals and carries them into the groundwater supply (Jawadi et al. 2020). Since water is an excellent solvent for calcium and magnesium, hard water may be delivered to homes if the minerals are present in the soil around a water supply well.

Many studies have been conducted to investigate the potential health benefits of drinking water hardness, and most of these were ecological epidemiological studies that

Medical Geology: En route to One Health, First Edition. Edited by Majeti Narasimha Vara Prasad and Meththika Vithanage.

Table 8.1 Further classification of water based on the water hardness.

Water type	Dissolved calcium and magnesium (mg/L)
Soft	0–60
Moderate	61–120
Hard	121–180
Very hard	More than 180

found an inverse relationship between water hardness and cardiovascular mortality (Bagheri et al. 2021). In addition to that, exposure to inadequate calcium and magnesium levels facilitated different kinds of illnesses including cancers (Kozisek 2020).

8.2 Origin of Water Hardness

Dissolved polyvalent metallic ions from sedimentary rocks, as well as seepage and runoff from soils, are the primary natural sources of hardness in water. Biochemical and organic sediments, as well as sedimentary rocks, are formed by living organisms. After an organism dies, remains can be accumulated into sedimentary rocks concentrating a significant amount of calcium and magnesium, e.g. calcite ($CaCO_3$) is precipitated by organisms to form a shell or other skeletal structure. The accumulation of these skeletal remains results in limestone. Apart from that, dolostones are limestones that have been chemically modified by Mg-rich fluids flowing through the rock. $CaCO_3$ recrystallizes to form the new mineral dolomite $Ca.Mg(CO_3)_2$. Hence, those types of calcium and magnesium-rich sedimentary rocks facilitate augmenting calcium, magnesium, and other divalent metal ions in drinking water sources (Ram et al. 2021).

When atmospheric CO_2 dissolves in the water, acid rain may occur, and it will be the main reason for enhancing the carbonic acid in water resources. Figure 8.1 is pointing out the whole process of generating the water hardness and upon increasing the pH value of the water, up to around 7.6, carbonate (CO_3^{-2}) generates, and it reacts with calcium or magnesium ions to produce $CaCO_3/MgCO_3$. Meanwhile, the application of fertilizers containing a high amount of calcium, magnesium causes enhancement of relevant metal ions in the soil. Due to the acid rain followed by surface runoff, it may enter the surface water. Hence,

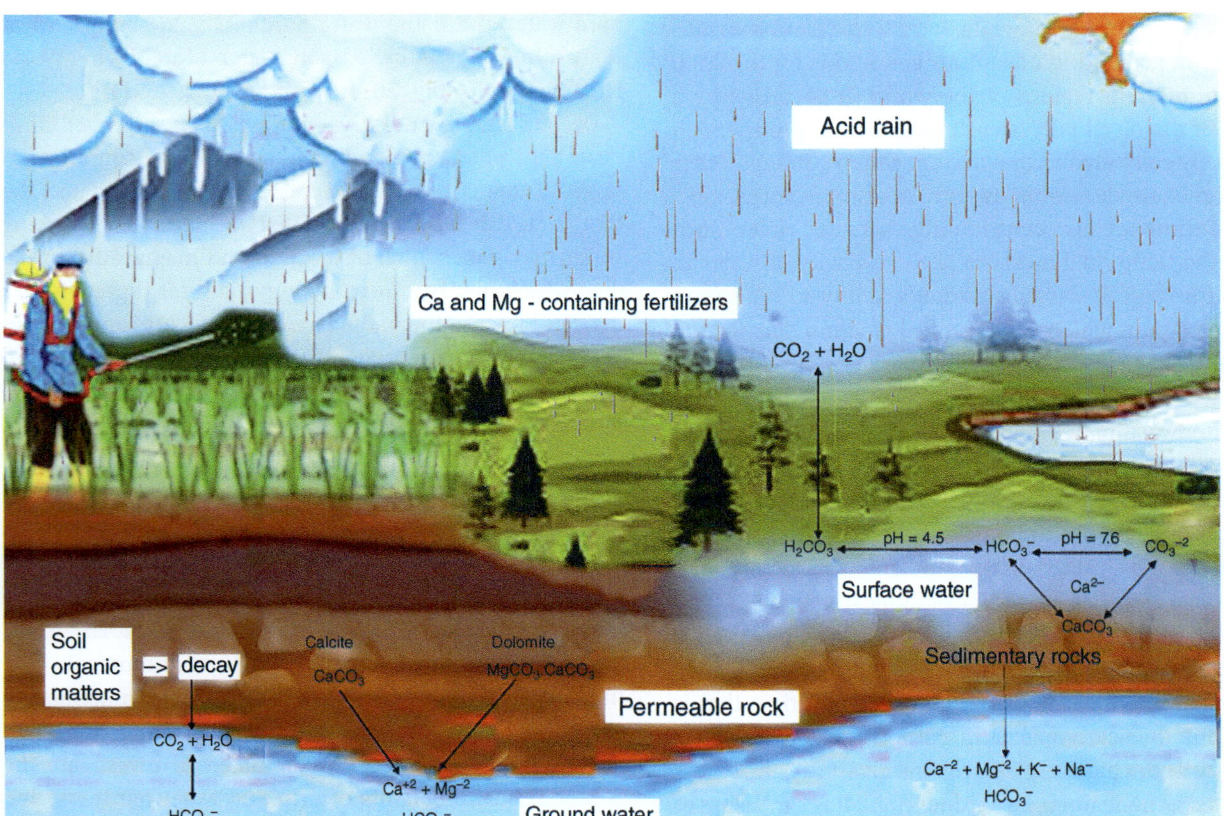

Figure 8.1 Overall illustration of the origin of water hardness based on the combination of anthropogenic and natural routes.

contents of calcium, magnesium, and some other relevant metal ions (sodium and potassium) are enriched in the surface water (Li et al. 2018).

Groundwater hardness mostly depends on the geology and hydrology of the river catchments. If the particular catchment is underlain by the impermeable rock, surface water is considered soft water due to the resistance to the erosion of the rocks reducing surface runoff for weathering. Contrariwise, catchment areas underlain by the permeable rocks allow water to infiltrate beneath the ground surface, so that groundwater in contact with the rock mass promotes weathering, potentially leading to the formation of fissures and conduits. Hence, the groundwater achieves a high concentration of dissolved materials, and it is called hard water. The fact that the releasing of elements into the water is determined by the geochemistry of the rock weathering. Apart from that, owing to the decay process of the soil organic matter CO_2 concentration of the soil is further enhanced, and it is dissolved in infiltrate water as carbonic acid (Rubenowitz-lundin and Hiscock 2013). This weak acid dissolves soluble calcium and magnesium carbonates such as calcite and dolomite within the soil and unsaturated zone of sedimentary rocks, yielding high concentrations of calcium, magnesium, and bicarbonate (Figure 8.1). In addition to that, carbonic acid reacts with igneous rocks and sedimentary rocks, and as a result, concentrations of calcium, magnesium, potassium, and sodium are released into infiltrating soil water or groundwater, as well as bicarbonate also enhance in the groundwater (Li et al. 2021).

Thus, it is depicted that both anthropogenic and natural processes contribute to enhancing the water hardness level in groundwater as well as surface water. Surface runoff is the major course of contamination of surface water with elements that contribute to water hardness, and acid rains accelerate the surface runoff and augment the carbonate content in water bodies enhancing the water hardness. Permeable rocks also contribute to the contamination of calcium and magnesium in surface water (Figure 8.1).

8.3 Water Hardness and Health Influence – Background

There are several research studies that have been focused on the interrelationship of water hardness and health effects in different dimensions. Hard water's health impacts are predominantly due to the effects of the salts dissolved in it, primarily calcium and magnesium. Meanwhile, very low or very high calcium (Ca) and magnesium (Mg) concentrations, or total hardness in drinking water have been empirically identified as a cause of health issues or water taste problems. Since 1960, epidemiologic studies completed in foreign nations by various research teams have consistently reported chronic health effects of naturally occurring low mineral water or artificially softened water. Most recent studies have proved the inverse association between cardiovascular diseases (CVDs) and Mg concentration as well as elevated calcium in drinking water. Not only CVD, but researchers have also focused on the protective effect of hard water against other diseases, including cancer development (Rapant et al. 2019; Yousefi et al. 2019; Bagheri et al. 2021). Based on the previous research investigations on the health impact of water hardness, it can be concluded that intake of sufficient calcium and magnesium from hard water prevents a wide spectrum of diseases including cancers and CVD. Despite there being many more supportive pieces of evidence on the health benefits of water hardness, some significant studies discovered that highly excessive amounts of calcium and magnesium levels may negatively affect the healthy life of humans, especially for children (Rahman et al. 2021).

8.3.1 Cardiovascular Diseases and Water Hardness

In Western countries, CVD is the leading cause of death (Meier et al. 2019; Feigin et al. 2016). The trend is being distributed in Asian countries also due to several reasons, including fast foods and unhealthy life patterns. However, there are several studies that have focused on the interconnection between the CVD and risk of CVDs in the past few years. Along with it, the first study was published indicating an inverse relationship between drinking water hardness and cardiovascular mortality in Japan and the United States (Catling et al. 2008). After that, based on the findings, several studies have been published in different countries indicating the association between CVD and water hardness and later magnesium and calcium concentrations (Wodschow et al. 2021). Several epidemiological studies conducted over the last 50 years have found a link between the risk of CVD development and the hardness of drinking water or its Mg and Ca content (Rylander 2008). Table 8.2 recaps some of the important research studies indicating the beneficial side of water hardness for CVD.

8.3.2 Prevention Mechanism of CVD by Water Hardness

Since there are enough research findings on the inverse relationship between CVD mortality and consumption of hard water (Table 8.2), it is beneficial to identify the preventive mechanism. Most of the studies have focused on magnesium and CVD preventive mechanisms (Pickering et al. 2021).

Table 8.2 Analysis of the relationship between the CVD mortality and consumption of hard water based on statistical analysis in some research studies.

Year and country	Objective of study	Method (brief)	Outcome of the study	Reference
2019, Hungarian	To determine the relationship of water hardness and socioeconomic deprivation with mortality rates of CVD	Hardness of water and CVD records were collected in selected areas. And for the calculations, IBM SPSS ver. 24 program (IBM Corp, Armonk, NY, USA) was used.	The regents reported high total water hardness such as Szekszárd-Villány, Balaton, and Eger fewer CVD cases were observed.	Nagy et al. (2019)
2017, Iran	To study mortality zoning of CVDs and water hardness by GIS, as well as statistical relationships between the variables mentioned during the years 2013–2015 in Isfahan.	Statistical analysis was done using the SPSS software (Version 20, IBM Corporation, Armonk, NY, USA) and results were compared to the results of the GIS zoning map.	A nonsignificant inverse relationship was observed between the water hardness and CVDs	Hossienifar et al. (2019)
2013, Slovak Republic	To determine which chemical elements in groundwater are most closely related to CVD mortality and recommend limit concentrations for which CVD mortality is as low as possible.	Correlation analysis was done between elemental composition found in groundwater and CVD mortality	Relative mortality of cardiovascular disease is significantly lower when Ca greater than 89.4 mg/L, Mg between 42 and 78.1 mg/L, and Ca + Mg between 4.4 and 7.6 mmol/L. Those values were 2–3 times higher than Slovakia's drinking water standards.	Rapant et al. (2015)
2010, Bosnia and Herzegovina	To evaluate the correlation between the water hardness and CVD based on the data collection and statistical analysis in Mostar city	N = 1021 the study group was divided into two: soft water consumers (603 people) and hard water consumers (618 people), and statistical analysis (SPSS 16.0) was done to find the correlation between CVD and water hardness	The prevalence of CVD was 21.3 % in the study area where soft water was consumed, while the prevalence of CVD was 13.7 % in the study area where hard water was consumed	Knezović et al. (2014)
2010, Iran	To investigate the relationship between calcium and magnesium concentrations in drinking water and CVD in urban and rural areas of an Iranian city	Data collection; water hardness, Ca and Mg contents of drinking water in selected areas, and data on CVD patients admitted into the city hospital (in the years 2010 and 2011) SPSS software was used to conduct the analysis (SPSS, Chicago, IL., USA, version 20.0)	In 2010, water calcium content greater than 72 mg/L was linked to a lower number of CVDs in 1000 people. Increased Mg hardness levels above 31 mg/L in 2010 and above 26 mg/L in 2011 were linked with a lower number of CVDs per 1000 people.	Momeni et al. (2014)

2009, UK	To investigate associations between cardiovascular mortality and levels of water hardness, calcium, or magnesium using time-series data from populations experiencing a change in drinking water	The drinking water regulator in England and Wales identified 37 areas where drinking water hardness changed between 1981 and 2005. Using the CVD mortality data correlation analysis was done using STATA S.E. 8.0	There was no evidence of such an association found, though the lack of large population areas and a reasonable change in magnesium level prevents a particular conclusion about the impact of this cation	Lake et al. (2010)
2008, Finland	To investigate the relationship between calcium, magnesium, aluminum, copper, iron, zinc, and nitrates in local groundwater and the spatial variation in the incidence of acute myocardial infarction (AMI) in men and women in rural Finland from 1991 to 2003 (people aged 35–74 years were studied)	Bayesian modeling was used for the statistical analysis. To divide the study area, 10 × 10 km grid cells were used instead of administrative boundaries.	A 1 mg/L increase in Mg level in local groundwater was associated with a 2% reduction in AMI incidence in the rural population. Finally, a high AMI incidence in eastern Finland is associated with soft groundwater that is low in Mg	Kousa et al. (2008)
2004, Finland	To explore the relation between spatial variation in acute myocardial infarction (AMI) incidence and potential environmental determinants in groundwater such as total water hardness, calcium, magnesium, fluoride, iron, copper, zinc, nitrate, and aluminum.	Bayesian modeling was used in a small area study, with geo-referenced data aggregated into 10 km × 10 km cells.	The findings of this study, which used a more specific Bayesian statistical analysis, confirm previous observations of an inverse relationship between water hardness and coronary heart disease	Kousa et al. (2004)

Magnesium is acting as a major contributor to water hardness together with calcium. On the biochemical and cellular levels, magnesium plays a lot of functions in the pathogenesis of CVDs. Because principally magnesium activates the adenosine triphosphate (ATPase), which is essential for cellular function, providing more energy to the Na^+–K+ pump (Purvis and Movahed 1992). Alone with it, some previous research studies have ascertained that magnesium deficiency reduces the activity of the Na^+–K^+ pump and because of intracellular sodium accumulation, magnesium deficiency caused the membrane potential to be less polarized, implying Na^+–K^+ pump inhibition (Kazama 2021). Additionally, intracellular magnesium deficiency could also lead to an increase in intracellular sodium and calcium, which can lead to arterial vasospasm, increased catecholamine release, as well as an increase in fatty acids and lipids (Dinicolantonio et al. 2018). According to those findings, a solid conclusion can be made that magnesium contributes essentially to preventing CVD, and hard water contributes to a higher intake of magnesium via drinking water. However, there may be adverse health impacts to humans due to the overdose of magnesium other than preventing CVD. Hence, without enough scientific evidence, we cannot comment on whether the daily consumption of hard water is beneficial or safe. On the other hand, one of the pooled data analyses from several large-scale, population-based, prospective cohort studies revealed a nonsignificant inverse relationship between dietary calcium intake and incident CAD and stroke (Wang et al. 2012). One of the meta-analysis reports based on the calcium intake and cardiovascular disease risk concluded that in generally healthy adults, calcium intake within tolerable upper intake levels (2000–2500 mg/day) is not associated with CVD risk (Chung et al. 2016).

Nevertheless, some research studies proved that magnesium and calcium cause reduced CVD. But profound studies need to be carried out to comment on the overall health effect and hard water consumption and the safety of consuming hard water. Most of the studies have focused on kidney disease and water hardness. In most cases, excess fluoride and hardness (commonly $[Ca^{2+}] > [Mg^{2+}]$ with major ions ($[HCO_3^-] > [Cl^-] > [SO_4^{-2}]$) in water appear to augment the prevalence of CKDu (chronic kidney disease of unknown etiology) (Cooray et al. 2019). Hence, the association between kidney disease and water hardness also needs to be scrutinized.

8.3.3 Kidney Disease and Water Hardness

When considering the contribution of water hardness for kidney disease, concern on the prevalence of CKDu in the world, including South Asia regent, is also very significant. In spite of previous research studies that have focused on water hardness and CKDu prevalence (Balasooriya et al. 2020), a clear-cut correlation between water hardness and CKDu prevalence could not be found. However, a recent study commented on the correlation between CKDu prevalence and water hardness. According to the study, all groundwater (collected from the CKDu prone areas in Sri Lanka) can be characterized primarily by Ca–Mg–CO_3 predominant geochemistry and undesirable contents (>maximum allowable limits for drinking) of alkalinity, hardness, and turbidity. A statistical comparison of groundwater quality between high-risk and low-risk regions revealed high hardness due to elevated magnesium and high ionicity due to major dissolved ions related to the occurrence of CKDu (Imbulana et al. 2020). In addition to that, another recent research study from Sri Lanka revealed that elevated fluoride levels, in combination with water hardness associated with higher Mg^{2+} levels, may be linked to CKDu and influence disease progression (Liyanage et al. 2022). Even though recent research studies commented on the positive link between water hardness and prevalence of CKDu, the direct influence of calcium and magnesium on kidney health needs to be focused on. On the other hand, renal tubular damages can occur due to the calcium phosphate crystals in the renal tubular fluid. When phosphate intake is excessive, FGF23 hormone will be circulated, and it suppresses the phosphate reabsorption in renal tubules, resulting in an increase in phosphate concentration in tubule fluid and phosphate together with calcium augment the risk of the crystal formation causing renal tubular damages (Shiizaki et al. 2021). More than 3/4 of kidney stones are composed of calcium salt, which usually manifests as calcium oxalate and, less frequently, as calcium phosphate (Sengupta 2013). Hence, consumption of an overdose of calcium may generate a risk for tubular damages. However, several researchers have discovered no link between water hardness and the occurrence of urinary stones (Sengupta 2013). On the other hand, a recent study carried out based on serum magnesium, mortality, and disease progression in chronic kidney disease concluded that serum magnesium levels in CKD patients are not associated with a difference in the rate of eGFR decline (Azem et al. 2020). When concerning recent research studies together, the correlation between consumption of excessively hard water and renal damages is still equivocate.

8.3.4 Protective Competence of the Hard Water Against Cancer Development

There so many studies based on water consumption patterns and cancer effects depicted that negative correlation between hard water consumption and risk of cancers and a

Table 8.3 Analysis of the protective effect of hard water consumption against the development of various types of cancers based on research studies conducted by Taiwan research group.

Country and year of the study	Cancer type	Major finding (protective effect of hard water)	Source
Taiwan, 2009	Pancreatic cancer	Trihalomethane exposure enhanced the risk of pancreatic cancer and enhanced magnesium level in drinking water positively influences the relationship between trihalomethane and development of pancreatic cancers	Chiu et al. (2010)
Taiwan, 2002	Esophageal cancer	Higher magnesium levels significantly reduced the risk of developing esophageal cancer when compared to those with magnesium levels less than 7.0 mg/L. (The adjusted odd ratios were 0.75 (0.62–0.91.)	
Taiwan, 2000	Prostate cancer	Magnesium intake from drinking water and other dietary sources may have a significant protective effect against the development of prostate cancer. 0.73 (95% CI = 0.51–1.03) and 0.64 (95% CI = 0.43–0.96). However, there was no statistical evidence that association between calcium and protective effect against the development of prostate cancer	Yang et al. (2002)
Taiwan, 2000	Colon cancer	There is a protective effect of calcium intake from drinking water against colon cancer (adjusted odds ratio for 95% confidence intervals were 0.58 (0.47 + 0.73) for the group with calcium levels 42.4 mg/L or more) but not magnesium	Yang et al. (1997)
Taiwan, 1999	Rectal cancer	There is a significant inverse relationship reported between drinking water hardness and rectal cancer mortality (the odds ratio were 1.24 (1.01–1.55), 1.38 (1.10–1.73), and 95% confidence interval)	Yang et al. (1999b)
Taiwan, 1998	Esophageal cancer	The use of soft water is associated with a 42% increased risk of esophageal cancer mortality (adjusted odds ratio for 95% confidence intervals was 1.42 (1.22 + 1.66)).	Yang et al. (1999a)
Taiwan, 1997	Gastric cancer	The findings indicate a significant inverse relationship between drinking water hardness and gastric cancer mortality. The odds ratioswere 1.16 (1.07–1.26) and 1.65 (1.52–1.79).	Yang et al. (1997)

Taiwan research group has studied the cancer mortality and consumption of hard water based on various types of cancers and they reported solid shreds of evidence the protective effect of hard water against the development of cancers (Table 8.3) (Yin et al. 2020). However, Taiwan research groups have focused on the protective effect of water hardness, magnesium, or calcium against cancers (Table 8.3), and clear-cut research-based shreds of evidence can be found from those studies.

8.3.5 Calcium and Magnesium Intake and Other Health Effects

According to the WHO report 2009, intake of dietary calcium and magnesium is described as a beneficial habit through an investigation into the public health implications of calcium and magnesium in drinking water.

The report summarizes the study findings, which show that increasing magnesium levels in drinking water reduces the risk of cardiovascular disease mortality. It also discusses the protective role of magnesium in hypertension, cardiac arrhythmias, pre-eclampsia (hypertension after 20 weeks of pregnancy), atherosclerosis, and diabetes mellitus. In terms of calcium, the survey indicates that a higher calcium intake may positively affect osteoporosis by increasing bone mass (World Health Organization 2009).

Owing to the enzyme-regulating function, magnesium plays a significant role in preventing diabetes. Because all kinases and other ATP-related enzymes and channels that regulate insulin action are magnesium dependent, some clinical studies revealed that sun-clinical Mg deficiency might cause the diabetic state, and Mg^{2+} influences constructively insulin secretion via effects on K^+-ATP channels (Xiong et al. 2019).

The influence of the excessive intake of calcium on the male reproductive system has been investigated using male healthy rats. According to the results, excessive dietary calcium appears to increase the generation of free radicals, resulting in structural and functional disruption of male reproduction (Chandra et al. 2012). The health influence of magnesium on the male reproductive system also has been investigated using male sexually matured Wistar rats, and the study reported that excessive magnesium does not cause oxidative stress in the testis at the concentrations used as well as magnesium positively affected the overall reproductive health by enhancing the androgenic P-450 enzyme activity, serum testosterone, and relative sperm count (Chandra et al. 2012).

Meanwhile, several research studies suggest that dietary calcium has an inverse relationship with blood pressure (Morvaridzadeh et al. 2020). Apart from that, dietary calcium consumption could reduce the risk of cardiovascular events like stroke, which are commonly associated with hypertension (Sengupta 2013). High calcium and magnesium intakes are most beneficial for bone mineral health too. To confirm that statement, one of the investigations was carried out in Italy. According to that study, women aged 30–70 years living in Sangemini, a region of central Italy, who consumed the local high calcium water (318 mg/L) had significantly higher spine mineral density than women in the same region who consumed low calcium water (60 mg/L) (Knezović et al. 2014). Increased dietary Ca intake has also been shown to improve bone mineral density (BMD) in a short period of time. Increased Ca supply during the growth period improves bone mineral accretion by 1–5%, according to Cashman and Flynn. BMD can catch up during the bone consolidation phase to compensate for the impaired BMA caused by insufficient Ca intake during the growth spurt. Nonetheless, this study found that people with low calcium intake may not reach their target metabolic panel. This phenomenon also occurs when long-term exposure to soft water. Apart from that, magnesium is required for bone formation and mineralization. It increases the activity of osteoblasts and phosphatase enzymes, which are involved in the bone formation process. In contrast, elevated Mg levels may also be detrimental to bone stiffness and strength. Increased Mg intake in postmenopausal women was found to be associated with an increase in wrist fractures. These findings are consistent with previous research indicating that a higher Mg concentration may have a negative impact on bone metabolism and parathyroid function, resulting in poor mineralization (Zofková et al. 2013; Ciosek et al. 2021).

Exposure to high water hardness is not suitable especially for children. To prove that statement, one of the Japanese studies can be described. Accordingly, atopic eczema (AE) prevalence was significantly higher in the highest water hardness category than in the lowest, and water hardness may increase the risk of atopic dermatitis (AD) in Japanese elementary school children (Miyake et al. 2004).

Along with the previous studies, a recent investigation has been done based on one region in Iran (Slovak municipalities) based on two groups: soft water consuming and hard water consuming. Based on the statistical analysis, they investigated four major causes of death: cardiovascular disease, cancer, gastrointestinal disease, and respiratory disease. According to their findings, the relative mortality for CVDs, oncological diseases, digestive tract diseases, and respiratory diseases was 56, 62, 128, and 121% higher in the soft drinking water population compared to the hard water population, respectively. Apart from that, it has been concluded that life expectancy is also higher in an enhanced hard water-consuming group. Finally, it has been emphasized that observation backs up previous research findings on the relationship between water hardness and human health (Rapant et al. 2021).

8.3.6 Physiological Significance of Magnesium

Magnesium is a very critical element for the biochemical functions of the body, and it is acting as a protective element preventing most noncommunicable diseases including CVD, diabetes, bone diseases, etc. It is significant for activating ATP demanding enzymatic reactions at the cellular level. Magnesium contributes to activating the Na^+–K^+ pump, hereafter the ion transport process also activates in the cardiac tissues reducing the risk of CVD. Not only that, insulin secretion is also augmented. Apart from that, magnesium triggers various types of enzymatic reactions enhancing the metabolism in the body (Ajith 2020; Dubey et al. 2020; Barbagallo et al. 2021), and it will be more beneficial for reproductive health and enhance the insulin level. Serum magnesium is very constructive for the bone health of humans (Menzel et al. 2021) (Figure 8.2). In addition to that, magnesium is reported as a very advantageous element for mental health. Some research findings depicted that low magnesium levels have been linked to attention-deficit hyperactivity disorder, and lower magnesium levels have been found in autism spectrum disorder patients (Botturi et al. 2020). The above-mentioned research findings depict the significance of magnesium as an element for balancing the health of humans.

Intake of magnesium; weathering process of the rock promotes the magnesium contents in the water. Magnesium-containing igneous rocks are olivine, pyroxene, amphiboles, and dark-colored mica; sedimentary rocks are magnesite and dolomite; magnesian minerals are chlorite, serpentine, and montmorillonite. It has been proposed that the quantitative contribution of water magnesium to body magnesium status may be critical for those who have a low dietary intake and use water with high magnesium levels. In spite of having food as the main reason for the magnesium source, water magnesium contributes considerably. Because magnesium in

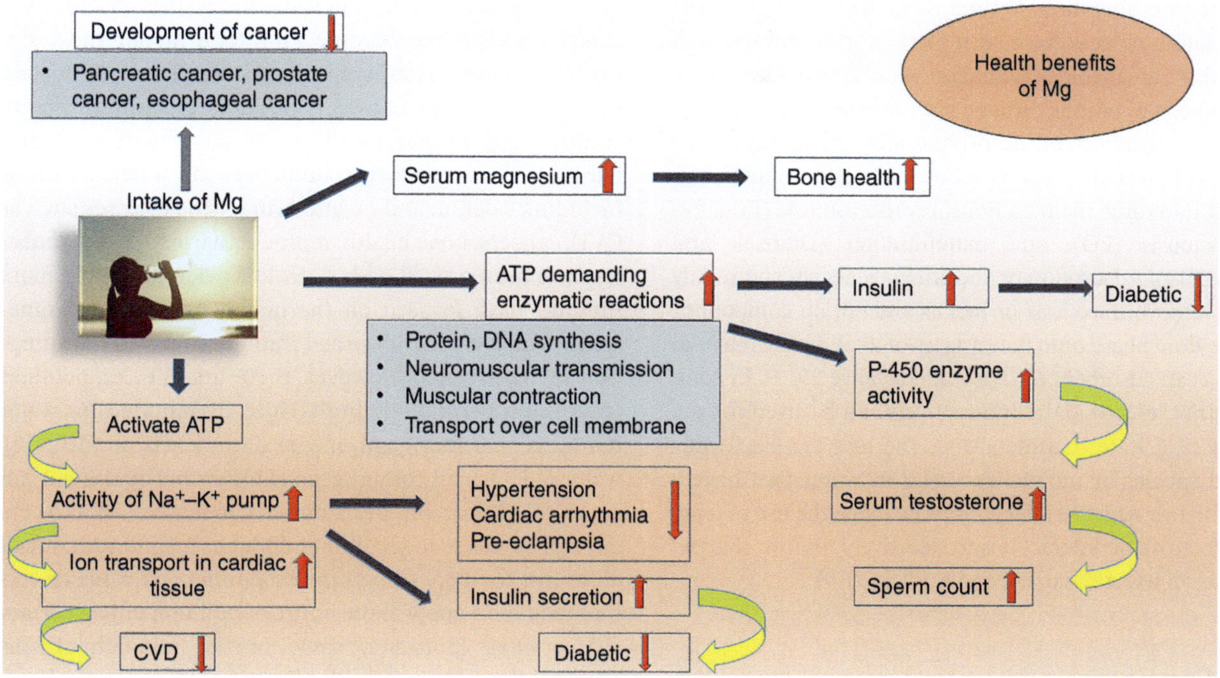

Figure 8.2 Key health benefits of magnesium intake based on evidence extracted from different clinical and statistical research studies (↑ indicates the effects or concentration will be enhanced and ↓ indicates effects or concentration will be reduced). *Source:* Antonioguillem / Adobe Stock.

water, as hydrated ions, is more bioavailable than magnesium in food (Rubenowitz-lundin and Hiscock 2013).

8.3.7 Health Drawbacks of Water Hardness

Although water hardness is associated with various kinds of health benefits for humans, there is a harmful influence of excess water hardness, especially among children. Accordingly, one of the studies suggested that domestic water hardness is a significant risk factor for AD development and skin barrier dysfunction in the first three months of life, particularly in genetically predisposed infants (Perkin et al. 2016). The interaction of hard water and surfactants contributes to the development of AD and early water softener installation may be able to prevent the development of AD (Danby et al. 2018). Another recent study found that there was a link found between living in a hard water area (range: 76–> 350 mg/L $CaCO_3$) and AE in children (Jabbar-Lopez et al. 2021). In addition to it, some of the research studies describe exposure to extremely hard water may have a negative impact on kidney stone formation (Esmail et al. 2020). Meanwhile, Sri Lankan studies on the distribution of CKDu and the water hardness reported that the presence of hardness may alter the chemical forms of fluoride and trace elements in water, acting as a secondary factor in disease etiology (Wasana et al. 2016). Hence, long-term exposure to very hard water may generate dangerous health effects, and remediation technologies need to be developed to mitigate the excessive hardness of the water.

8.4 Mitigation of Water Hardness

Hardness is a serious challenge for groundwater, which is promised to be a source of water for a variety of purposes. It causes scaling in boilers, washing machines, and pipes, difficult lathering in industries, as well as causing some negative health influences especially among children (Konale et al. 2020). Therefore, excess water hardness mitigation or water softening process is critical in terms of water purification.

In general, there are five main principles that can be applied for water softening; desalination, thermal process, chemical process, physical process, and physico-chemical process (Dudziak and Kudlek 2019).

Evaporation is a thermal desalination technique. The process is carried out by converting the source water (liquid) into vapor (gas) or ice (solid) (Humplik et al. 2011). In addition to the liquid–vapor phase transition, crystallization is also used in phase-change desalination. The phase change allows water to be converted from a liquid to a solid state (Xie et al. 2018). In terms of energy consumption, crystallization utilizes less energy than evaporation. On the other hand, ion transport and chelating process have been focused in recent research studies that can enable desalination via selective transport of water molecules or ions through a semipermeable barrier, or via chelating agents or ion adsorption/absorption on a porous material (Zhou et al. 2018). Selective ion transport can be categorized into three main sections: reverse osmosis (RO), forward osmosis (FO), and nanofiltration (Dudziak and Kudlek 2019). Adsorption or chelating agents are commonly used to remove trace ions or metals and enrich compounds from the fluid phase onto the surfaces of a solid-state material known as an adsorbent (Aragaw and Ayalwe 2019). In addition to that, electrostatic interactions can be used for the mitigation of water hardness since the ions are electrically charged species or molecules found in water that have a net positive or negative charge. Ions facilitate the use of long-range electrostatic interactions to selectively remove charged species from water (Dudziak and Kudlek 2019).

8.5 Conclusions

Natural processes, as well as anthropogenic processes, are contributing to the hardness of the groundwater and surface water. Owing to the unpleasant taste of the water and other difficulties such as detrimental to the washing process, hard water has become a hot topic among the scientific community. However, research studies are reporting the health benefits of water hardness including calcium and magnesium that are the major contributors of water hardness. The protective influences of water hardness including calcium and magnesium for the diseases such as CVD, cancers, bone health, reproductive health, and diabetes have been proved with statistical evidence. Many more findings have focused on the health benefits of magnesium, which can be absorbed into the body via consumption of hard water. However, there are WHO guidelines (maximum permissible limits) already reported for water hardness, calcium, and magnesium levels in drinking water. WHO or any other responsible authority still has not come to a clear-cut decision or conclusion on is there enough evidence to recommend higher magnesium or calcium intake, if so, should the recommended value or recommendation apply to the entire population or just certain risk groups. Moreover, some of the research studies reported the negative health effect of excessive water hardness on AD development and skin barrier dysfunction in children. Hence, much more studies need to be focused on the influence of water hardness on human health to acquire a precise conclusion.

References

Ajith, T.A. (2020). Possible therapeutic effect of magnesium in ocular diseases. *Journal of Basic and Clinical Physiology and Pharmacology* 31 (2): https://doi.org/10.1515/jbcpp-2019-0107.

Aragaw, T.A., & Ayalew, A.A. (2019). Removal of water hardness using zeolite synthesized from Ethiopian kaolin by hydrothermal method. *Water Practice and Technology*, 14(1): 145–159.

Azem, R., Daou, R., Bassil, E. et al. (2020). Serum magnesium, mortality and disease progression in chronic kidney disease. *BMC Nephrology* 21 (1): 1–10.

Bagheri, A., Naghshi, S., Sadeghi, O. et al. (2021). Total, dietary, and supplemental magnesium intakes and risk of all-cause, cardiovascular, and cancer mortality: a systematic review and dose–response meta-analysis of prospective cohort studies. *Advances in Nutrition* 12 (4): 1196–1210.

Balasooriya, S., Munasinghe, H., Herath, A.T. et al. (2020). Possible links between groundwater geochemistry and chronic kidney disease of unknown etiology (CKDu): an investigation from the Ginnoruwa region in Sri Lanka. *Exposure and Health* 12 (4): 823–834.

Barbagallo, M., Veronese, N., and Dominguez, L.J. (2021). Magnesium in aging, health and diseases. *Nutrients* 13: 463. https://doi.org/10.3390/nu13020463.

Botturi, A., Ciappolino, V., Delvecchio, G. et al. (2020). The role and the effect of magnesium in mental disorders: a systematic review. *Nutrients* 12 (6): 1661.

Bouhoun, M.L., Blondeau, P., Louafi, Y., and Andrade, F.J. (2021). A paper-based potentiometric platform for determination of water hardness. *Chemosensors* 9 (5): 96.

Catling, L.A., Abubakar, I., Lake, I.R. et al. (2008). A systematic review of analytical observational studies investigating the association between cardiovascular disease and drinking water hardness. *Journal of Water and Health* 6 (4): 433–442.

Chandra, K., Sengupta, P., Goswami, H., and Sarkar, M. (2012). Excessive dietary calcium in the disruption of structural and functional status of adult male reproductive system in rat with possible mechanism. *Molecular and Cellular Biochemistry* 364 (1): 181–191.

Chiu, H.F., Tsai, S.S., Wu, T.N., and Yang, C.Y. (2010). Effect modification of the association between trihalomethanes and pancreatic cancer by drinking water hardness: evidence from an ecological study. *Environmental Research* 110 (5): 513–518.

Chung, M., Tang, A.M., Fu, Z. et al. (2016). Calcium intake and cardiovascular disease risk: an updated systematic review and meta-analysis. *Annals of Internal Medicine* 165 (12): 856–866.

Ciosek, Ż., Kot, K., Kosik-Bogacka, D. et al. (2021). The effects of calcium, magnesium, phosphorus, fluoride, and lead on bone tissue. *Biomolecules*, 11(4): 506.

Cooray, T., Wei, Y., Zhong, H. et al. (2019). Assessment of groundwater quality in CKDu affected areas of Sri Lanka: implications for drinking water treatment. *International Journal of Environmental Research and Public Health* 16 (10): 1698.

Danby, S.G., Brown, K., Wigley, A.M. et al. (2018). The effect of water hardness on surfactant deposition after washing and subsequent skin irritation in atopic dermatitis patients and healthy control subjects. *Journal of Investigative Dermatology* 138 (1): 68–77.

DiNicolantonio, J.J., Liu, J., and O'Keefe, J.H. (2018). Magnesium for the prevention and treatment of cardiovascular disease. *Open Heart* 5 (2): e000775.

Dubey, P., Thakur, V., and Chattopadhyay, M. (2020). Role of minerals and trace elements in diabetes and insulin resistance. *Nutrients* 12 (6): 1864.

Dudziak, M. and Kudlek, E. (2019). Removal of hardness in wastewater effluent using membrane filtration. *Architecture Civil Engineering Environment* 12 (2): 141–147.

Esmail, A.O., Qadir, B.A., and Hamad, H.Q. (2020). Effect of drinking water hardness on kidney stones formation in Ranya District. *Cihan University-Erbil Scientific Journal* 4 (1): 1–6. https://doi.org/10.24086/cues.v4n1y2020.pp1-6.

Feigin, V.L., Roth, G.A., Naghavi, M. et al. (2016). Global burden of stroke and risk factors in 188 countries, during 1990–2013: a systematic analysis for the Global Burden of Disease Study 2013. *The Lancet Neurology* 15 (9): 913–924.

Hasan, K., Shahriar, A., and Ullah, K. (2019). Heliyon water pollution in Bangladesh and its impact on public health. *Heliyon* 5 (July): e02145. https://doi.org/10.1016/j.heliyon.2019.e02145.

Hossienifar, F., Entezari, M., and Hosseini, S. (2019). Water hardness zoning of Isfahan Province, Iran, and its relationship with cardiovascular mortality, 2013–2015. *ARYA Atherosclerosis* 15 (6): 275.

Humplik, T., Lee, J., O'hern, S.C. et al. (2011). Nanostructured materials for water desalination. *Nanotechnology* 22 (29): 292001.

Imbulana, S., Oguma, K., and Takizawa, S. (2020). Evaluation of groundwater quality and reverse osmosis water treatment plants in the endemic areas of Chronic Kidney Disease of Unknown Etiology (CKDu) in Sri Lanka. *Science of the Total Environment* 745: 140716.

Jabbar-Lopez, Z.K., Ung, C.Y., Alexander, H. et al. (2021). The effect of water hardness on atopic eczema, skin barrier function: a systematic review, meta-analysis. *Clinical & Experimental Allergy* 51 (3): 430–451.

Jawadi, H.A., Sagin, J., and Snow, D.D. (2020). A detailed assessment of groundwater quality in the Kabul basin, Afghanistan, and suitability for future development. *Water* (Switzerland) 12 (10): 1–19. https://doi.org/10.3390/w12102890.

Kazama, I. (2021). High-magnesium exposure to bullfrog heart causes ST segment elevation. *Journal of Veterinary Medical Science* 83 (4): 675–679.

Knezović, N.J., Memić, M., Mabić, M. et al. (2014). Correlation between water hardness and cardiovascular diseases in Mostar city, Bosnia and Herzegovina. *Journal of Water and Health* 12 (4): 817–823.

Konale, R.A., Mahale, N.K., and Ingle, S.T. (2020). Nano-zeolite-graphene oxide composite for calcium hardness removal: isotherm and kinetic study. *Water Practice and Technology* 15 (4): 1011–1031.

Kousa, A., Moltchanova, E., Viik-Kajander, M. et al. (2004). Geochemistry of ground water and the incidence of acute myocardial infarction in Finland. *Journal of Epidemiology & Community Health* 58 (2): 136–139.

Kousa, A., Havulinna, A.S., Moltchanova, E. et al. (2008). Magnesium in well water and the spatial variation of acute myocardial infarction incidence in rural Finland. *Applied Geochemistry* 23 (4): 632–640.

Kozisek, F. (2020). Regulations for calcium, magnesium or hardness in drinking water in the European Union member states. *Regulatory Toxicology and Pharmacology* 112: 104589.

Lake, I.R., Swift, L., Catling, L.A. et al. (2010). Effect of water hardness on cardiovascular mortality: an ecological time series approach. *Journal of Public Health* 32 (4): 479–487.

Lesnic, M., Ştefan, D.S., Moga, I.C., and Ştefan, M. (2021). Uncompresive requirements for drinking water hardness: conservation indicators. *International Journal of Conservation Science* 12 (2): 641–646.

Li, Y., Li, J., Gao, L., and Tian, Y. (2018). Irrigation has more influence than fertilization on leaching water quality and the potential environmental risk in excessively fertilized vegetable soils. *PloS one* 13 (9): e0204570.

Li, P., Karunanidhi, D., Subramani, T., and Srinivasamoorthy, K. (2021). Sources and consequences of groundwater contamination. *Archives of Environmental Contamination and Toxicology* 80 (1): 1–10.

Liyanage, D.N.D., Diyabalanage, S., Dunuweera, S.P. et al. (2022). Significance of Mg-hardness and fluoride in drinking water on chronic kidney disease of unknown etiology in Monaragala. *Sri Lanka. Environmental Research* 203: 111779.

Meier, T., Gräfe, K., Senn, F. et al. (2019). Cardiovascular mortality attributable to dietary risk factors in 51 countries in the WHO European Region from 1990 to 2016: a systematic analysis of the Global Burden of Disease Study. *European Journal of Epidemiology* 34 (1): 37–55.

Menzel, J., Abraham, K., Stangl, G.I. et al. (2021). Vegan diet and bone health – results from the cross-sectional RBVD study. *Nutrients* 13 (2) 685: 1–16.

Miyake, Y., Yokoyama, T., Yura, A. et al. (2004). Ecological association of water hardness with prevalence of childhood atopic dermatitis in a Japanese urban area. *Environmental Research* 94 (1): 33–37.

Momeni, M., Gharedaghi, Z., Amin, M.M. et al. (2014). Does water hardness have preventive effect on cardiovascular disease? *International Journal of Preventive Medicine* 5 (2): 159.

Morvaridzadeh, M., Sepidarkish, M., Fazelian, S. et al. (2020). Effect of calcium and vitamin D co-supplementation on blood pressure: a systematic review and meta-analysis. *Clinical Therapeutics* 42 (3): e45–e63.

Nagy, J., Sipka, S., Kocsis, J., and Horváth, Z. (2019). The hardness of drinking water negatively while socio-economic deprivation positively correlate with the age-adjusted mortality rates due to cardiovascular diseases in Hungarian wine regions. *International Journal of Environmental Research and Public Health* 16 (18): 3437.

Perkin, M.R., Craven, J., Logan, K. et al. (2016). Association between domestic water hardness, chlorine, and atopic dermatitis risk in early life: a population-based cross-sectional study. *Journal of Allergy and Clinical Immunology* 138 (2): 509–516.

Pickering, R.T., Bradlee, M.L., Singer, M.R., and Moore, L.L. (2021). Higher intakes of potassium and magnesium, but not lower sodium, reduce cardiovascular risk in the Framingham Offspring Study. *Nutrients* 13 (1): 269.

Purvis, J.R. and Movahed, A. (1992). Magnesium disorders and cardiovascular diseases. *Clinical Cardiology* 15 (8): 556–568.

Rahman, M., Kumar, S., Lamb, D., and Rahman, M.M. (2021). Health risk assessment of arsenic, manganese, and iron from drinking water for high school children. *Water, Air, & Soil Pollution* 232 (7): 1–13.

Ram, A., Tiwari, S.K., Pandey, H.K. et al. (2021). Groundwater quality assessment using water quality index (WQI) under GIS framework. *Applied Water Science* 11 (2): 1–20.

Rapant, S., Fajčíková, K., Cvečková, V. et al. (2015). Chemical composition of groundwater and relative mortality for cardiovascular diseases in the Slovak Republic. *Environmental Geochemistry and Health* 37 (4): 745–756.

Rapant, S., Cvečková, V., Fajčíková, K. et al. (2019). Hard water, more elastic arteries: a case study from Krupina district, Slovakia. *International Journal of Environmental Research and Public Health* 16 (9): 1521.

Rapant, S., Letkovičová, A., Jurkovičová, D. et al. (2021). Differences in health status of Slovak municipalities supplied with drinking water of different hardness values. *Environmental Geochemistry and Health* 43 (7): 2665–2677.

Rubenowitz-Lundin, E. and Hiscock, K.M. (2013). Water hardness and health effects. In: *Essentials of Medical Geology* (ed. O. Selinus), 337–350. Dordrecht: Springer.

Rylander, R. (2008). Drinking water constituents and disease. *The Journal of Nutrition* 138 (2): 423S–425S.

Sengupta, P. (2013). Potential health impacts of hard water. *International Journal of Preventive Medicine* 4 (8): 866.

Shiizaki, K., Tsubouchi, A., Miura, Y. et al. (2021). Calcium phosphate microcrystals in the renal tubular fluid accelerate chronic kidney disease progression. *The Journal of Clinical Investigation* 131 (16).

Wang, L., Manson, J.E., and Sesso, H.D. (2012). Calcium intake and risk of cardiovascular disease. *American Journal of Cardiovascular Drugs* 12 (2): 105–116.

Wasana, H., Aluthpatabendi, D., Kularatne, W.M.T.D. et al. (2016). Drinking water quality and chronic kidney disease of unknown etiology (CKDu): synergic effects of fluoride, cadmium and hardness of water. *Environmental Geochemistry and Health* 38 (1): 157–168.

Wodschow, K., Villanueva, C.M., Larsen, M.L. et al. (2021). Association between magnesium in drinking water and atrial fibrillation incidence: a nationwide population-based cohort study, 2002–2015. *Environmental Health* 20 (1): 1–13.

World Health Organization (2009). *WHO Library Cataloguing-in-Publication Data Calcium and Magnesium in Drinking-Water: Public Health Significance*. World Health Organization.

World Health Organization (2011). *Hardness in Drinking-water, Background Document for Development of WHO Guidelines for Drinking-water Quality*. World Health Organization.

Xie, C., Zhang, L., Liu, Y. et al. (2018). A direct contact type ice generator for seawater freezing desalination using LNG cold energy. *Desalination* 435: 293–300.

Xiong, J., He, T., Wang, M. et al. (2019). Serum magnesium, mortality, and cardiovascular disease in chronic kidney disease and end-stage renal disease patients: a systematic review and meta-analysis. *Journal of nephrology*, 32(5): 791–802.

Yang, C.Y., Chiu, H.F., Chiu, J.F. et al. (1997). Gastric cancer mortality and drinking water qualities in Taiwan. *Archives of Environmental Contamination and Toxicology* 33 (3): 336–340.

Yang, C.Y., Chiu, H.F., Cheng, M.F. et al. (1999a). Esophageal cancer mortality and total hardness levels in Taiwan's drinking water. *Environmental Research* 81 (4): 302–308.

Yang, C.Y., Tsai, S.S., Lai, T.C. et al. (1999b). Rectal cancer mortality and total hardness levels in Taiwan's drinking water. *Environmental Research* 80 (4): 311–316.

Yang, C.Y., Chiu, H.F., Tsai, S.S. et al. (2002). Magnesium and calcium in drinking water and the risk of death from esophageal cancer. *Magnesium Research* 15 (3–4): 215–222.

Yin, J., Wu, X., Li, S. et al. (2020). Impact of environmental factors on gastric cancer: a review of the scientific evidence, human prevention and adaptation. *Journal of Environmental Sciences* 89: 65–79.

Yousefi, M., Najafi Saleh, H., Yaseri, M. et al. (2019). Association of consumption of excess hard water, body mass index and waist circumference with risk of hypertension in individuals living in hard and soft water areas. *Environmental Geochemistry and Health* 41 (3): 1213–1221.

Zhou, X., Essawy, H.A., Mohamed, M.F. et al. (2018). Grafting polymerization of acrylic acid onto chitosan-cellulose hybrid and application of the graft as highly efficient ligand for elimination of water hardness: adsorption isotherms, kinetic modeling and regeneration. *Journal of Environmental Chemical Engineering* 6 (2): 2137–2147.

Zofková, I., Nemcikova, P., and Matucha, P. (2013). Trace elements and bone health. *Clinical Chemistry and Laboratory Medicine (CCLM)* 51 (8): 1555–1561.

9

Geochemistry of Fluoride in the Environment and Human Health

Rohana Chandrajith[1,2], *C.B. Dissanayake*[1], *and Johannes A.C. Barth*[2]

[1] *Department of Geology, Faculty of Science, University of Peradeniya, Peradeniya, Sri Lanka*
[2] *GeoZentrum Nordbayern, Department Geographie und Geowissenschaften, Friedrich-Alexander-Universität Erlangen-Nürnberg (FAU), Erlangen, Germany*

9.1 Introduction

The close relationship between human health and the geochemical environment is more apparent in tropical and subtropical terrains where climatic and geoenvironmental factors are unique and communities mostly depend on their immediate natural environment for drinking water and food (Dissanayake and Chandrajith 1999). Among many reported diseases, the geochemistry of fluoride in drinking water in areas with dental and skeletal fluorosis is a well-established link in medical geology. It was estimated that over 200 million people in the world are suffering from dental fluorosis and many other fluoride-related health problems. Such problems are common in tropical regions of Africa (e.g. Syria, Egypt, Kenya, Ghana, Tanzania, etc.), Asia (India, Sri Lanka, Pakistan, Bangladesh Thailand, and China), North America (USA), and South America (Mexico and Argentina), in addition to Japan and Australia (Figure 9.1).

Fluoride is considered an essential element but becomes toxic for humans and animals when ingested in higher doses. Although the essentiality of fluoride for human health is still under debate, excessive intake may cause dental and skeletal fluorosis in addition to some other chronic diseases including neurological complications in humans as well as among animals (Choubisa 2014; Chandrajith et al. 2020). In the case of fluoride, which is considered a required element for the teeth, the essentiality and toxicity are determined by a narrow range (1.0 mg/L) as indicated by the WHO (WHO 2011). Food is the principal source of many trace elements that are essential for the physiological functions of the human body, but fluoride enters the body mainly through drinking water. Therefore, the geochemistry of fluoride in drinking water is of major interest in many parts of the world. In addition, recent studies indicated that crops and vegetables that are grown on fluoride-contaminated soils or irrigated with fluoride-rich groundwater adversely affect their physio-biological parameters and also decline the productivity (Banerjee and Roychoudhury 2019). Although contamination of environmental media by fluoride is primarily geogenic, anthropogenic activities also made a considerable contribution. Weathering and leaching from fluoride-bearing minerals and volcanic eruptions are the main geogenic sources. Coal burning, fertilizer application, glass, ceramic, and steel manufacturing industries, and phosphate ore processing are considered as the anthropogenic sources of fluoride (Prasad and Mondal 2007; Brindha and Elango 2011; Vithanage and Bhattacharya 2015). World Health Organization has provided a safe limit for fluoride in drinking water as 1.5 mg/L (WHO 1996, 2011), but in many tropical counties, the limit is not applicable, and the toxic effect of chronic fluoride ingestion appeared even in much lower concentrations (Chandrajith et al. 2020).

9.2 Geochemistry of Fluoride

Fluoride is ubiquitous in the environment and is also the 13th most abundant element in the earth's crust (Edmunds and Smedley 2013). The crustal abundance of fluoride is 557 mg/kg (Rudnick et al. 2003). It is the most electronegative element and, therefore, highly reactive. Due to it is reactivity, fluoride is more common in many geological and environmental materials. Fluoride is mostly found in silicate minerals, particularly those of volcanic origin (Edmunds and Smedley 2013; Vithanage and Bhattacharya 2015). Fluorite (49%F), cryolite (54%F), apatite (3.8%F), gagarinite (33–36%F), topaz (13–20%F), biotite (3.5%F), hornblende (2.9%F), and sphene (1.36%F) are common fluoride-bearing

Medical Geology: En route to One Health, First Edition. Edited by Majeti Narasimha Vara Prasad and Meththika Vithanage.

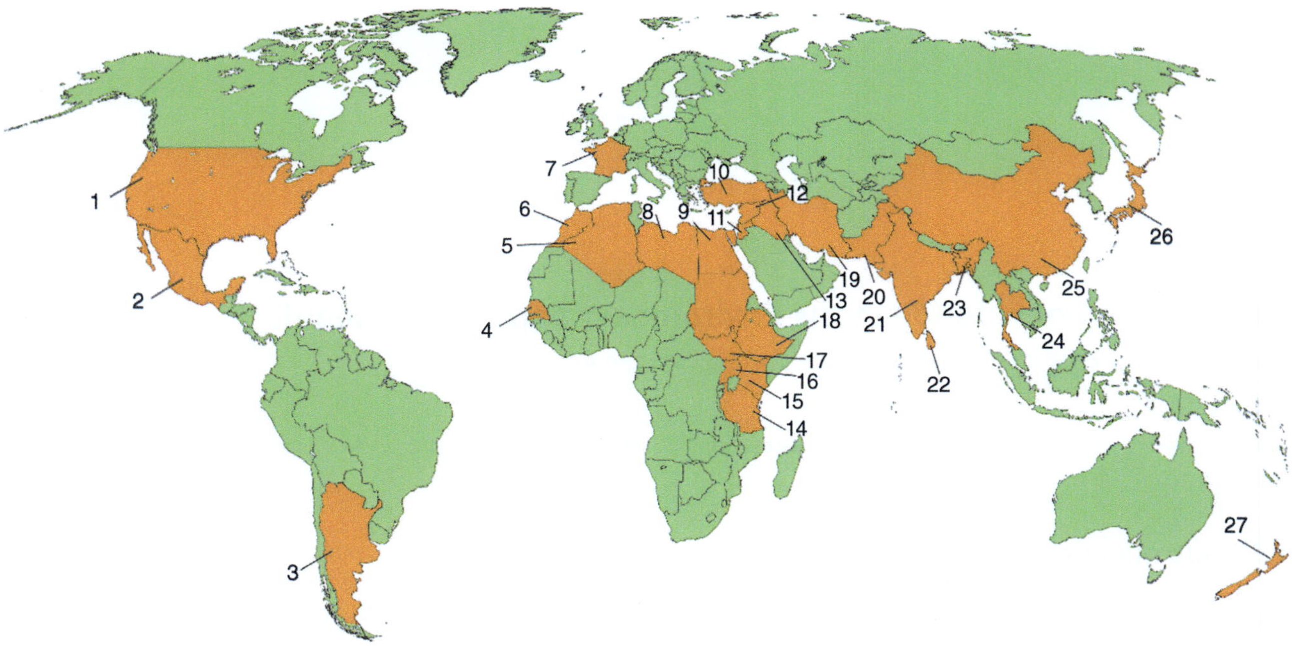

Figure 9.1 Countries with dental and skeletal fluorosis due to excessive ingestion of fluoride mainly from drinking water. *Source:* Adapted from UNICEF.

minerals among many others. Fluoride (ionic radius 136 pm) can be easily exchanged with hydroxyl ion (ionic radius 140 pm) due to their remarkable similarity in size and charge. Therefore, many hydrous minerals such as mica and amphibole groups of minerals largely incorporate fluoride in their structure. For instance, hydroxylapatite [$Ca_5(PO_4)_3OH$], which is the main mineral phase of teeth and bones, and fluorapatite [$Ca_5(PO_4)_3F$] are isomorphic end members in the solid solution series of apatite.

9.3 Fluoride in Rocks

Among rocks, fluoride is more abundant in igneous suites such as granites (750 mg/kg), calc-alkaline igneous rocks (950 mg/kg), and volcanic ash (750 mg/kg) (Aswathanarayana et al. 1985) (Table 9.1). Since residual magma is often rich with volatile fluoride and is also a ubiquitous constituent, later stage magmatic minerals such as topaz, fluorite, and tourmaline and rocks (e.g. pegmatites) incorporate fluoride (Brunt et al. 2004). Due to the higher content of hydrous minerals such as clay and phosphate minerals (e.g. francolite), sedimentary rocks also contain elevated fluoride levels. For instance, sedimentary limestone generally contains up

Table 9.1 Fluoride content in different rocks and other environmental materials.

Rock type	Fluoride range (mg/kg)	Mean (mg/kg)
Basalts	20–1 060	360
Granite and gneisses	20–2 700	870
Shales	10–7 600	800
Limestone	Upto 1 200	220
Sandsotne/greywacke	10–1 100	200
Phosphate/phosphorite	24 000–41 500	31 000
Coal ash	40–480	80
Dolomite	110–400	260
Volcanic ash	100–2 900	750
Marine sediments	100–1 600	730
Pegmatite	365–22 680	7 550
Rapakivi granite	20 000–40 000	na
Meta-igneous rocks	383–2685	1014
Meta-sedimentary rocks	167–1207	542

Source: Adapted from Fleischer and Robinson (1963), Fuge and Andrews (1988), Lahermo et al. (1991).

to 1200 mg/kg of fluoride, and shale shows values up to 7600 mg/kg. Excessive contacts of fluoride can also be found in sedimentary phosphorites (2.4–4.2%). Metamorphic rocks particularly high-grade varieties have a fluoride content of up to 100 mg/kg, while contact metamorphic rocks are enriched with fluoride with a content of 5000 mg/kg (Frencken 1992).

9.4 Fluoride in Soil

The fluoride content in soils varied from 20 to 500 mg/kg, which is mostly influenced by pH and oxides of Al and Ca that make complexes in soil (Kabata-Pendias and Pendias 1984). Fluoride is incorporated in clays and oxy-hydroxides that are more common constituents in noncalcareous soils, displacing hydroxide groups; therefore, fluoride in soils is immobile and mostly insoluble (Hong et al. 2016). Calcareous soils also fix fluoride as CaF_2. In addition, anthropogenic activities such as the application of phosphate fertilizer and coal ash fallouts from power plants contaminate soils by adding additional fluoride. Therefore, fluoride content in soils can vary considerably. For instance, fluoride content in soils from England varied from 321 to 2412 mg/kg, while the highest content was found in soils overlying siltstone (Mikkonen et al. 2018). Jayawardana et al. (2012) noted that soils in Sri Lanka vary from 4 to 411 mg/kg, while remarkable enrichment was noted in soils from dryer areas compared to wet regions with higher annual rainfall. The retention of fluoride in soils depends on the rate of water percolation that is preliminarily controlled by the permeability of the soil horizon (Abugri and Pelig-Ba 2011).

9.5 Fluoride in Plants

Fluoride is not considered an essential component for the growth of plants. In general, fluoride is immobile in soils, and usually, the content increases toward the C-horizon. Due to the immobility of fluoride in soil, it is not easily available for uptake by plants. When the clay minerals, organic matter, and other oxy-hydroxides are lower in soil, fluoride bioavailability can be increased and leads to uptake via plant roots (Anshumali 2014). Alkaline conditions in soil also enhance the bioavailability due to the repulsion of fluoride from excess negative chargers in the soil surfaces (Banerjee and Roychoudhury 2019). Therefore, the bioavailability of fluoride heavily depends on the pH and mineralogical properties of soils (Figure 9.2). Fluoride that absorbs through the

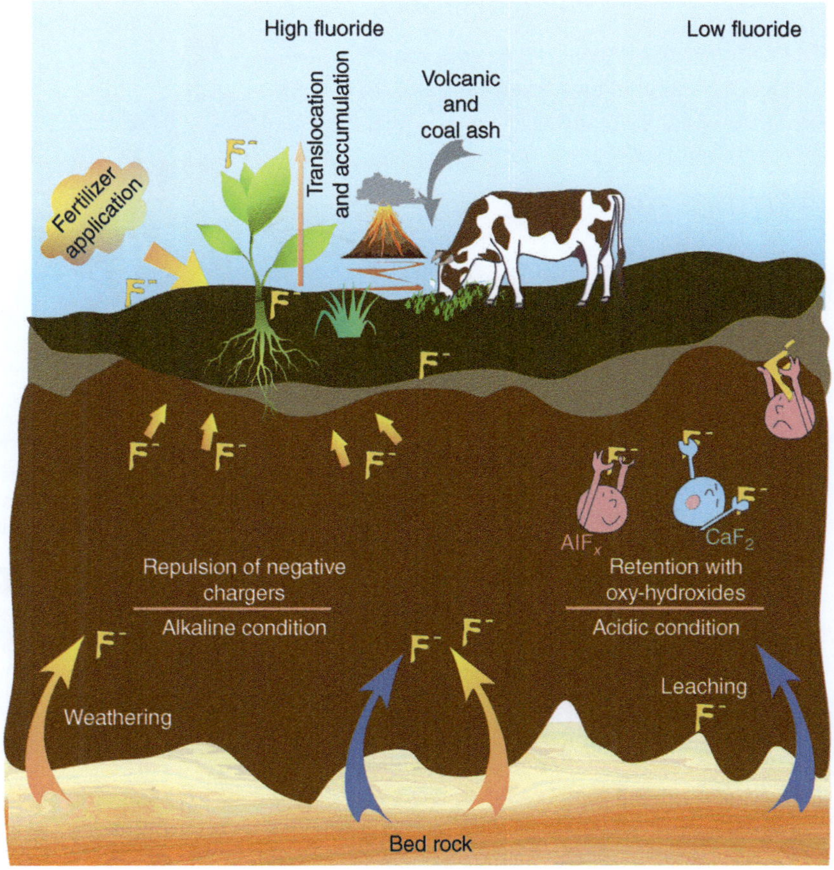

Figure 9.2 Behavior of fluoride under different pH conditions.

roots of plants then transports by the symplastic or apoplastic pathways via xylematic flow to the transpiratory organs (Elloumi et al. 2005; Singh et al. 2018). Plants grown in fluoride-contaminated soil mainly due to natural or anthropogenic activities may have enriched levels compared to uncontaminated sites.

Tea (*Camellia sinensis*) is a well-known plant that accumulates fluoride at higher levels. For instance, 214 mg/kg of fluoride was observed in tea leaves that are grown in contaminated soils (Baunthiyal and Ranghar 2015). During the preparation of tea liquor, a substantial amount of fluoride can be leached from tea leaves to the liquid (Chandrajith et al. 2007, 2022). In tea leaves, fluoride is associated with aluminum and then released into the infusion as free fluoride (Huang et al. 2020). Under laboratory experimental conditions, rice (*Oryza sativa* L.) cultivars showed a higher accumulation of fluoride when irrigated with fluoride-contaminated water, in addition to the adverse impact on growth parameters of the plant such as lipid peroxidation, root ion leakage, and catalase activity (Chakrabarti et al. 2013; Chakrabarti and Patra 2015; Mondal 2017). Fluoride accumulation in rice plants follows the order of soil > root > shoot > grain (Chakrabarti et al. 2013). Cabbage (*Brassica oleracea*), Spinach (*Spinacea oleracea*), and Celery (*Apium graveolens*) are also well-known edible plants among many others that can accumulate fluoride (Hong et al. 2016).

9.6 Fluoride in Natural Water

When considering the environmental components, groundwater is the main source of fluoride for human exposure. The main sources of fluoride in natural water are the chemical weathering of rocks and minerals. Particularly in tropical terrains, rock weathering is more intense, and fluoride can be easily leached out into the aqueous medium. When rainwater is seeping through soils and rocks, fluoride can easily be dissolved, contaminating groundwater. Therefore, higher content of fluoride can be observed in groundwater compared to surface water (Chandrajith et al. 2012). Resident time is one of the most significant factors in controlling the fluoride in groundwater. Ion exchange, dissolution, and precipitation are the main reactions that control groundwater fluoride levels (Figure 9.3). The alkalinity of water is one of the important factors that control the leachability of fluoride (Saxena and Ahmed 2003), while the availability of alkaline earth metals, mainly Ca and Mg, in groundwater acts as a good sink for fluoride (Dissanayake 1996, 2010; Jacks et al. 2005).

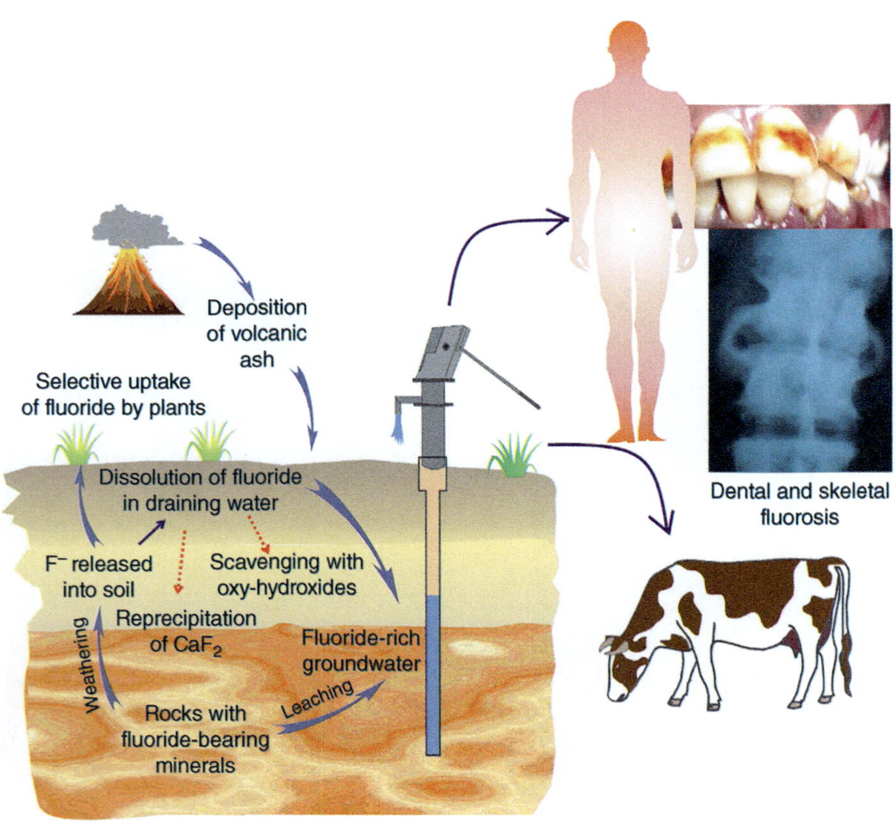

Figure 9.3 Schematic diagram showing pathways of fluoride in the natural environment.

Weathering and leachability of fluoride in terrains are, therefore, the significant factors that influence the fluoride content in groundwater. Heavy rainfall and subsequent evapotranspiration, which is pronounced in arid and semi-arid regions, tend to increase the fluoride in groundwater (Dissanayake 1996). Jacks et al. (2005) showed that evapotranspiration under tropical conditions causes precipitation of Ca and results in an increase in Na/Ca ratios allowing an increase of fluoride in groundwater. Excessively higher fluoride levels were reported in groundwater from the African continent, mainly along with the countries in the East African Rift Valley, for instance, Kenya (0.5–75 mg/L), Tanzania (10.5–46 mg/L) and Ethiopia (1.5–68 mg/L) (Onipe et al. 2020). In India, groundwater fluoride levels as high as 48 mg/L were reported (Susheela 1999) However, uncontaminated surface water mostly contains lower levels (<1.0 mg/L) due to its minimum interaction with rock (Edmunds and Smedley 2013). However, some regions showed excessively higher levels, which are mostly draining through regions with active volcanos (Table 9.2). In addition, deep bore wells showed higher fluoride levels compared to shallow dug wells (Chandrajith et al. 2012).

Table 9.2 Fluoride levels in selected surface water bodies in some selected countries in the world.

Surface water	Fluoride (range) in mg/L	Source
Telangana river (India)	0.37–3.48	More et al. (2021)
Yongding river (China)	0.30–1.50	Wang et al. (2020)
Chasico' lake (Argentine)	6.74–8.50	Puntoriero et al. (2014)
Chinnaeru river (India)	0.90–3.50	Machender et al. (2014)
Nalgonda reservoir (India)	0.50–1.00	Machender et al. (2014)
Fal river (England)	0.25–0.40	Fuge and Andrews (1988)
Maijia river (Chain)	0.08–9.0 (Summer); 0.16–5.22 (Winter)	Wei et al. (2021)
Rivers in Africa	10.70–39.0 (mean 1034)	Malago et al. (2017)
Lakes in Africa	1.30–250 (mean 130)	Malago et al. (2017)
Pampa Plain/rivers (Argentina)	0.19–1.64	Rosso et al. (2011)
Uyyakondan channel (India)	0.30–0.70	Brindha and Kavitha (2015)
Lake water (Norway)	<0.005–0.56 mg/L	Skjelkvåle (1994)
River and lake water (Nigeria)	0.03–2.07 mg/L	Waziri et al. (2012)

Another important feature observed in terrains with high-grade metamorphic rocks is the significant variation of the fluoride content within short distances. It was noted that high fluoride wells (>2 mg/L) are located within an average distance of 500 m from low fluoride wells (<0.5 mg/L) after analyzing 6107 wells. The minimum distance observed between high and low fluoride wells was 42 m and the maximum distance was 9 km (Ranasinghe et al. 2019). Therefore, it is necessary to investigate individual wells for their fluoride content for proper management of the problem.

9.7 Fluoride and Human Health

In the past few decades, numerous studies have been carried out on fluoride toxicity due to different toxicological impacts on humans, animals, and plants. It is widely believed that lower fluoride in drinking water (<0.5 mg/L) causes tooth decay. Therefore, fluoride is also known as a therapeutic material to treat dental caries. Higher concentrations of fluoride in drinking water (>1.5 mg/L) lead to detrimental health effects such as dental fluorosis (mottling of teeth) (Figure 9.4), skeletal fluorosis, and many other diseases (Chandrajith et al. 2020; Edmunds and Smedley 2013). Both the beneficial and harmful effects of fluoride are falling within a very narrow change. Dental and skeletal fluorosis in human and livestock occur

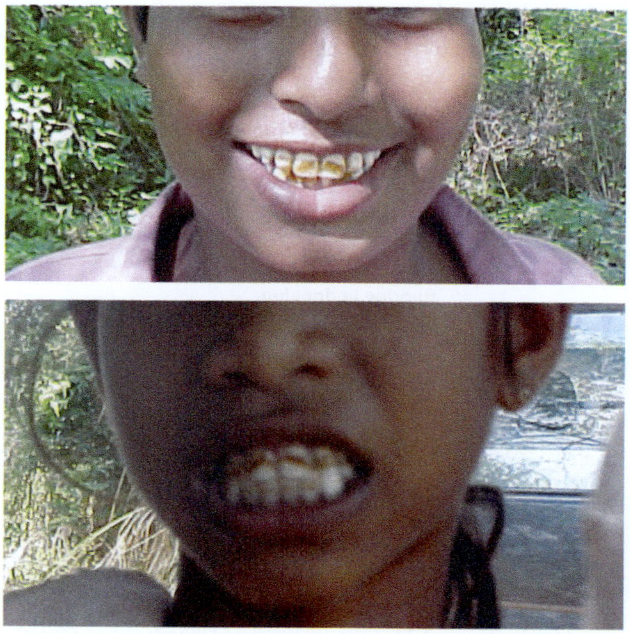

Figure 9.4 Typical case of dental fluorosis due to intake of high fluoride groundwater.

mainly due to the extreme affinity of fluoride onto calcified tissues, which predominantly consist of mineral hydroxylapatite (Choubisa 2014; Choubisa and Choubisa 2016). Therefore, fluorosis poses an important public health issue in many parts of the world, mainly due to the drinking of fluoride-contaminated water. According to the WHO (WHO 2011), drinking water with fluoride levels higher than 1.0 mg/L could lead to dental fluorosis, particularly in tropical countries. Drinking water with over 4 mg/L could cause skeletal fluorosis, while much higher contents would result in severe osteal deformities such as crippling fluorosis and genu-varum (Choubisa and Choubisa 2019). However, in the tropical regions of Sri Lanka, fluorosis appears at fluoride levels higher than 0.6 mg/L (Warnakulasuriya et al. 1992), while Khan et al. (2004) observed a mild degree of fluorosis in Pakistan children who drink water with 0.35 mg/L fluoride. Therefore, countries particularly in the tropical terrain adapted their own recommended standards for fluoride in drinking water (Table 9.3).

It is well-known fact that fluoride affects calcified tissues notably dental enamel, vertebral column, pelvic girdle, and ribs. Due to high fluoride ingestion, the fusion of vertebral bodies can occur resulting in neurological symptoms and stiff necks (Ghosh et al. 2013). Dental fluorosis is the discoloration and weakening of teeth due to hypomineralization that usually occurs with chronic ingestion of fluoride mainly through drinking water (Dissanayake 1991, 2005; Gupta et al. 2007; Daesslé et al. 2009; Edmunds and Smedley 2013; Vithanage and Bhattacharya 2015). This particularly affects growing teeth hence more pronounced among young children (Figure 9.3). Although dental fluorosis is not a deadly chronic disease, it makes an esthetic impact due to staining of the teeth. It was noted that psychosocial impact emanates among adolescents with dental fluorosis and their smiles are destroyed (Ranasinghe and Kanthi 2016).

Table 9.3 Countries having their own recommended standard for fluoride in water.

Country	Accepted level of fluoride in drinking water (mg/L)	Source
Botswana	<0.50	Zietsman (1991)
Sudan	<0.50	Ibrahim et al. (1995)
Senegal	0.60	Brouwer et al. (1988)
Chile	0.5–0.6	Locker (1999)
Sri Lanka	0.60	Warnakulasuriya et al. (1992)
Pakistan	0.39	Khan et al. (2004)
Western India	0.34	Chandra et al. (1980)
WHO value	0.50–1.00	WHO (2011)

Skeletal fluorosis is the most common and severe manifestation of fluorosis among adult populations who ingest higher doses of fluoride (>4 mg/L) through drinking water or other sources such as drinking brick tea and using coal (Zhang and Cao 1996; Liu et al. 2007; Choubisa and Choubisa 2019). Lumbar and leg joint pain, rigidity in joints, deformation of bones, and compression of the spinal cord and nerve roots are the clinical presentations of skeletal fluorosis (Choubisa and Choubisa 2019). X-ray images of skeletal fluorosis show increased bone density and fibrous ostephours and classification of tendons (Choubisa 2014; Yang et al. 2017; Sellami et al. 2020) (Figure 9.5). More advanced stages of skeletal fluorosis lead to the status of crippling fluorosis while both are incurable. Crippling fluorosis was also observed in the population who consume even lower fluoride (2.8 mg/L) drinking water, and extreme ingestion of fluoride causes kyphosis (curvature in the spin) and genu varum (bow legs) (Choubisa and Choubisa 2016, 2019).

Although excessive fluoride intake and its health effects mainly on hard tissues are well documented, chronic fluoride exposure may also lead to severe negative impacts on soft tissues. Due to its oxidizing nature, fluoride can cause the inhabitation of enzymes in the human body

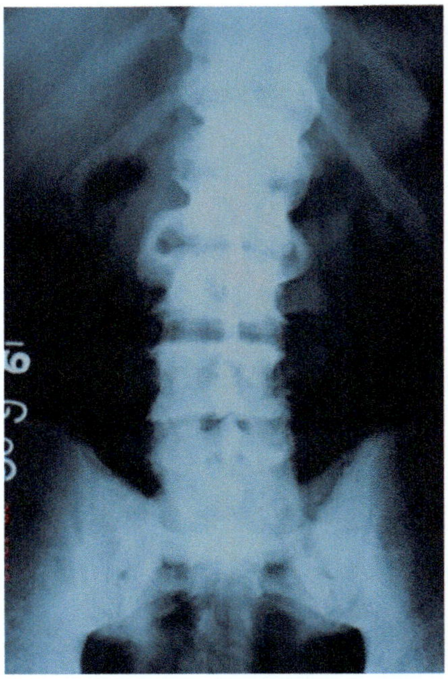

Figure 9.5 X-ray image of the pelvic bone with spines of a subject with severe skeletal fluorosis. Pelvic bones appear dense, and bridging osteophytes and calcification in spines or vertebrae can be visible with irregular new bone formation.

(García-Montalvo et al. 2009). Detrimental impacts of fluoride on the liver, kidney, heart, lungs, brain, and nervous and reproductive systems were reported in many early studies (Ozsvath 2009; Kumari and Kumar 2011; Nakamoto and Rawls 2018; Zuo et al. 2018; Dissanayake and Chandrajith 2019; Gulegoda et al. 2022). Recently, synergetic impact of fluoride and hard water was suggested as a causative factor for kidney damage and end-stage renal failures in tropical terrains (Balasooriya et al. 2019; Dissanayake and Chandrajith 2019; Nanayakkara et al. 2020; Liyanage et al. 2022). Studies have also shown that fluoride can damage mitochondria that increase the release of cytochrome-c (Cyt-c) and induce caspase-dependent apoptosis (Choi et al. 2012; Tan et al. 2018; Araujo et al. 2019).

In addition, fluoride alone with calcium can accumulate in the pineal gland, which produces melatonin hormone leading to calcification of the gland that may result in melatonin deficiency (Chlubek and Sikora 2020). Drinking water with high fluoride is believed to decrease insulin mRNA and affects the secretion of β-cells, and fluoride exposure possibly impaired glucose homeostasis, leading to type-2 diabetes mellitus (García-Montalvo et al. 2009). In addition, drinking high fluoride water, particularly during childhood, lowers the Intelligence Quotient (IQ) level among children (Xiang et al. 2003; Anna et al. 2012; Kundu et al. 2015). Even a high fluoride diet during pregnancy was also found to significantly affect the IQ levels of children (Kundu et al. 2015).

9.8 Conclusions

Although fluoride is considered an essential element for humans, excess fluoride in the environment poses severe heath impact mainly on humans but also impacts animals and plants. Fluoride toxicity mainly causes dental and skeletal fluorosis in humans as well as in animals, and it also leads to adverse effects on soft tissues in the human body leading to many other health issues. In recent years, many studies are focused on fluoride-related health issues in humans and animals. This chapter reviews the present status of adverse effects of fluoride intoxication on various organs in humans as well as other biological components aiming to summarize the understanding of fluoride toxicity due to natural ingestion, particularly through drinking water.

Acknowledgments

RC gratefully acknowledges the Alexander von Humboldt Foundation, Germany, for awarding the Georg Forster Research Award during the writing stages of this paper.

References

Abugri, D.A. and Pelig-Ba, K.B. (2011). Assessment of fluoride content in tropical surface soils used for crop cultivation. *African Journal of Environmental Science and Technology* 5 (9): 653–660.

Anna, L.C., Guifan, S., Ying, Z., and Philippe, G. (2012). Developmental fluoride neurotoxicity: a systematic review and meta-analysis. *Environmental Health Perspectives* 120 (10): 1362–1368. https://doi.org/10.1289/ehp.1104912.

Anshumali, B.K. (2014). Fluoride in agricultural soil: a review on its sources and toxicity to plants. *Global Sustainability Transitions: Impacts and Innovations* 3: 29–37.

Araujo, T.T., Barbosa Silva Pereira, H.A., Dionizio, A. et al. (2019). Changes in energy metabolism induced by fluoride: insights from inside the mitochondria. *Chemosphere* 236: 124357. https://doi.org/10.1016/j.chemosphere.2019.124357.

Aswathanarayana, U., Lahermo, P., Malisa, E., and Nanyaro, J.T. (1985). High fluoride waters in an endemic fluorosis area in Northern Tanzania. In: *Proceedings of the 1st International Symposium on Geochemistry and Health*, Monograph Series – Environmental Geochemistry and Health (ed. I. Thornton), 243–249. London: Royal Society of London.

Balasooriya, S., Munasinghe, H., Herath, A.T. et al. (2019). Possible links between groundwater geochemistry and chronic kidney disease of unknown etiology (CKDu): an investigation from the Ginnoruwa region in Sri Lanka. *Exposure and Health* https://doi.org/10.1007/s12403-019-00340-w.

Banerjee, A. and Roychoudhury, A. (2019). Fluorine: a biohazardous agent for plants and phytoremediation strategies for its removal from the environment. *Biologia Plantarum* 63 (1): 104–112. https://doi.org/10.32615/bp.2019.013.

Baunthiyal, M. and Ranghar, S. (2015). Accumulation of fluoride by plants: potential for phytoremediation. *CLEAN – Soil, Air, Water* 43 (1): 127–132. https://doi.org/10.1002/clen.201300353.

Brindha, K. and Elango, L. (2011). Fluoride in groundwater: causes, implications and mitigation measures. In: *Fluoride Properties, Applications and Environmental Management*, vol. 1 (ed. S.D. Monroy), 111–136. NY: Nova Science Publisher.

Brindha, K. and Kavitha, R. (2015). Hydrochemical assessment of surface water and groundwater quality along Uyyakondan channel, South India. *Environmental Earth Sciences* 73 (9): 5383–5393. https://doi.org/10.1007/s12665-014-3793-5.

Brouwer, I.D., De Bruin, A., Dirks, O.B., and Hautvast, J.G.A.J. (1988). Unsuitability of world health organisation guidelines for fluoride concentrations in drinking water in Senegal. *The Lancet* 331 (8579): 223–225. https://doi.org/10.1016/S0140-6736(88)91073-2.

Brunt, R., Vasak, L., and Griffioen, J. (2004). *Fluoride in Groundwater: Probability of Occurrence of Excessive*. Utrecht: International Groundwater Resources Assessment Centre.

Chakrabarti, S. and Patra, P.K. (2015). Biochemical and antioxidant responses of paddy (*Oryza sativa* L.) to fluoride stress. *Fluoride* 48 (1): 56–61.

Chakrabarti, S., Patra, P.K., and Mondal, B. (2013). Uptake of fluoride by two paddy (*Oryza sativa* L.) varieties treated with fluoride-contaminated water. *Paddy and Water Environment* 11 (1): 619–623. https://doi.org/10.1007/s10333-012-0342-2.

Chandra, S., Sharma, R., Thergaonkar, V.P., and Chaturvedi, S.K. (1980). Determination of optimal fluoride concentration in drinking water in an area in India with dental fluorosis. *Community Dentistry and Oral Epidemiology* 8 (2): 92–96. https://doi.org/10.1111/j.1600-0528.1980.tb01264.x.

Chandrajith, R., Abeypala, U., Dissanayake, C.B., and Tobschall, H.J. (2007). Fluoride in Ceylon tea and its implications to dental health. *Environmental Geochemistry and Health* 29 (5): 429–434. https://doi.org/10.1007/s10653-007-9087-z.

Chandrajith, R., Padmasiri, J.P., Dissanayake, C.B., and Prematilaka, K.M. (2012). Spatial distribution of fluoride in groundwater of Sri Lanka. *Journal of the National Science Foundation of Sri Lanka* 40 (4): 303–309.

Chandrajith, R., Diyabalanage, S., and Dissanayake, C.B. (2020). Geogenic fluoride and arsenic in groundwater of Sri Lanka and its implications to community health. *Groundwater for Sustainable Development* 10: 100359. https://doi.org/10.1016/j.gsd.2020.100359.

Chandrajith, R., Bhagya, S., Diyabalanage, S. et al. (2022). Exposure sssessment of fluoride intake through commercially available black tea (*Camellia sinensis* L.) from areas with high incidences of chronic kidney disease with undetermined origin (CKDu) in Sri Lanka. *Biological Trace Element Research* 200 (2): 526–534. https://doi.org/10.1007/s12011-021-02694-2.

Chlubek, D. and Sikora, M. (2020). Fluoride and pineal gland. *Applied Sciences* 10 (8): 2885.

Choi, A.L., Sun, G., Zhang, Y., and Grandjean, P. (2012). Developmental fluoride neurotoxicity: a systematic review and meta-analysis. *Environmental Health Perspectives* 120 (10): 1362–1368. https://doi.org/10.1289/ehp.1104912.

Choubisa, S.L. (2014). Bovine calves as ideal bio-indicators for fluoridated drinking water and endemic osteo-dental fluorosis. *Environmental Monitoring and Assessment* 186 (7): 4493–4498. https://doi.org/10.1007/s10661-014-3713-x.

Choubisa, S.L. and Choubisa, D. (2016). Status of industrial fluoride pollution and its diverse adverse health effects in man and domestic animals in India. *Environmental Science and Pollution Research* 23 (8): 7244–7254. https://doi.org/10.1007/s11356-016-6319-8.

Choubisa, S.L. and Choubisa, D. (2019). Genu-valgum (knock-knee) syndrome in fluorosis-endemic Rajasthan and its current status in India. *Fluoride* 52 (2): 161–168.

Daesslé, L.W., Ruiz-Montoya, L., Tobschall, H.J. et al. (2009). Fluoride, nitrate and water hardness in groundwater supplied to the rural communities of Ensenada County, Baja California, Mexico. *Environmental Geology* 58 (2): 419–429.

Dissanayake, C.B. (1991). The fluoride problem in the ground water of Sri Lanka – environmental management and health. *International Journal of Environmental Studies* 38 (2–3): 137–155. https://doi.org/10.1080/00207239108710658.

Dissanayake, C.B. (1996). *Water Quality and Dental Health in the Dry Zone of Sri Lanka*, vol. 113 (1), 131–140. London: Geological Society https://doi.org/10.1144/GSL.SP.1996.113.01.10.

Dissanayake, C. (2005). Of stones and health: medical geology in Sri Lanka. *Science* 309 (5736): 883–885.

Dissanayake, C.B. (2010). Water quality in the dry zone of Sri Lanka-some interesting health aspects. *Journal of the National Science Foundation of Sri Lanka* 33 (3): 161–168.

Dissanayake, C.B. and Chandrajith, R. (1999). Medical geochemistry of tropical environments. *Earth-Science Reviews* 47 (3): 219–258. https://doi.org/10.1016/S0012-8252(99)00033-1.

Dissanayake, C.B. and Chandrajith, R. (2019). Fluoride and hardness in groundwater of tropical regions-review of recent evidence indicating tissue calcification and calcium phosphate nanoparticle formation in kidney tubules. *Ceylon Journal of Science* 48 (3): 197–207. https://doi.org/10.4038/cjs.v48i3.7643.

Edmunds, W.M. and Smedley, P.L. (2013). Fluoride in natural waters. In: *Essentials of Medical Geology*, 311–336. Springer https://doi.org/10.1007/978-94-007-4375-5_13.

Elloumi, N., Abdallah, F.B., Mezghani, I. et al. (2005). Effect of fluoride on almond seedlings in culture solution. *Fluoride* 38 (3): 193–198.

Fleischer, M. and Robinson, W.O. (1963). Some problems of the geochemistry of fluorine. In: *Studies in Analytical Geochemistry*, vol. 6 (ed. D.M. Shaw), 58–75. Toronto: University of Toronto Press.

Frencken, J.E. (1992). *Endemic Fluorosis in Developing Countries, Causes, Effects and Possible Solutions*. Leiden: TNO Institute for Preventive Health Care. NIPG-TNO.

Fuge, R. and Andrews, M.J. (1988). Fluorine in the UK environment. *Environmental Geochemistry and Health* 10 (3): 96–104. https://doi.org/10.1007/bf01758677.

García-Montalvo, E.A., Reyes-Pérez, H., and Del Razo, L.M. (2009). Fluoride exposure impairs glucose tolerance via decreased insulin expression and oxidative stress. *Toxicology* 263 (2): 75–83. https://doi.org/10.1016/j.tox.2009.06.008.

Ghosh, A., Mukherjee, K., Ghosh, S.L., and Saha, B. (2013). Sources and toxicity of fluoride in the environment. *Research on Chemical Intermediates* 39 (7): 2881–2915. https://doi.org/10.1007/s11164-012-0841-1.

Gulegoda, C.R., Dissanayake, C.B., Amarasekara, D.S. et al. (2022). Impact of fluoride exposure on male reproductive parameters: a pilot case–control study in Sri Lanka. *Exposure and Health* https://doi.org/10.1007/s12403-022-00465-5.

Gupta, R., Kumar, A.N., Bandhu, S., and Gupta, S. (2007). Skeletal fluorosis mimicking seronegative arthritis. *Scandinavian Journal of Rheumatology* 36 (2): 154–155. https://doi.org/10.1080/03009740600759845.

Hong, B.D., Joo, R.N., Lee, K.S. et al. (2016). Fluoride in soil and plant. *Korean Journal of Agricultural Science* 43 (4): 522–536. https://doi.org/10.7744/kjoas.20160054.

Huang, C., Zhang, H., Zeng, W. et al. (2020). Enhanced fluoride adsorption of aluminum humate and its resistance on fluoride accumulation in tea leaves. *Environmental Technology* 41 (3): 329–338. https://doi.org/10.1080/09593330.2018.1498135.

Ibrahim, Y.E., Abuaffan, A.H., and Bjorvatn, K. (1995). Prevalence of dental fluorosis in Sudanese children from two villages with 0.25 and 2.56 ppm fluoride in the drinking water. *International Journal of Paediatric Dentistry* 5 (4): 223–229. https://doi.org/10.1111/j.1365-263X.1995.tb00183.x.

Jacks, G., Bhattacharya, P., Chaudhary, V., and Singh, K.P. (2005). Controls on the genesis of some high-fluoride groundwaters in India. *Applied Geochemistry* 20 (2): 221–228. https://doi.org/10.1016/j.apgeochem.2004.07.002.

Jayawardana, D.T., Pitawala, H.M.T.G.A., and Ishiga, H. (2012). Geochemical assessment of soils in districts of fluoride-rich and fluoride-poor groundwater, North-Central Sri Lanka. *Journal of Geochemical Exploration* 114: 118–125. https://doi.org/10.1016/j.gexplo.2012.01.004.

Kabata-Pendias, A. and Pendias, H. (1984). *Trace Elements in the Soil and Plants*. Boca Raton, FL: CRC Press.

Khan, A.A., Whelton, H., and O'Mullane, D. (2004). Determining the optimal concentration of fluoride in drinking water in Pakistan. *Community Dentistry and Oral Epidemiology* 32 (3): 166–172. https://doi.org/10.1111/j.1600-0528.2004.00152.x.

Kumari, S. and Kumar, A. (2011). Fluoride toxicity enhances phagocytic activity of macrophages in spleen of rats. *Asian Journal of Experimental Biological Sciences* 2 (2): 283–287.

Kundu, H., Basavaraj, P., Singla, A. et al. (2015). Effect of fluoride in drinking water on children's intelligence in high and low fluoride areas of Delhi. *Journal of Indian Association of Public Health Dentistry* 13 (2): 116–121. https://doi.org/10.4103/2319-5932.159043.

Lahermo, P., Sandström, H., and Malisa, E. (1991). The occurrence and geochemistry of fluorides in natural waters in Finland and East Africa with reference to their geomedical implications. *Journal of Geochemical Exploration* 41 (1): 65–79. https://doi.org/10.1016/0375-6742(91)90075-6.

Liu, G., Zheng, L., Qi, C., and Zhang, Y. (2007). Environmental geochemistry and health of fluorine in Chinese coals. *Environmental Geology* 52 (7): 1307–1313. https://doi.org/10.1007/s00254-006-0569-6.

Liyanage, D.N.D., Diyabalanage, S., Dunuweera, S.P. et al. (2022). Significance of Mg-hardness and fluoride in drinking water on chronic kidney disease of unknown etiology in Monaragala, Sri Lanka. *Environmental Research* 203: 111779. https://doi.org/10.1016/j.envres.2021.111779.

Locker, D. (1999). *Benefits and Risks of Water Fluoridation*. Toronto: Ontario Ministry of Health.

Machender, G., Dhakate, R., and Reddy, M.N. (2014). Hydrochemistry of groundwater (GW) and surface water (SW) for assessment of fluoride in Chinnaeru river basin, Nalgonda district, (AP) India. *Environmental Earth Sciences* 72 (10): 4017–4034. https://doi.org/10.1007/s12665-014-3291-9.

Malago, J., Makoba, E., and Muzuka, A.N. (2017). Fluoride levels in surface and groundwater in Africa: a review. *American Journal of Water Science and Engineering* 3 (1): 1–17. https://doi.org/10.11648/j.ajwse.20170301.11.

Mikkonen, H.G., van de Graaff, R., Mikkonen, A.T. et al. (2018). Environmental and anthropogenic influences on ambient background concentrations of fluoride in soil. *Environmental Pollution* 242: 1838–1849. https://doi.org/10.1016/j.envpol.2018.07.083.

Mondal, N.K. (2017). Effect of fluoride on photosynthesis, growth and accumulation of four widely cultivated rice

(*Oryza sativa* L.) varieties in India. *Ecotoxicology and Environmental Safety* 144: 36–44. https://doi.org/10.1016/j.ecoenv.2017.06.009.

More, S., Dhakate, R., Ratnalu, G.V., and Machender, G. (2021). Hydrogeochemistry and Health Risk Assessment of groundwater and surface water in fluoride affected area of Yadadri-Bhuvanagiri District, Telangana State, India. *Environmental Earth Sciences* 80 (7): 262. https://doi.org/10.1007/s12665-021-09544-3.

Nakamoto, T. and Rawls, H.R. (2018). Fluoride exposure in early life as the possible root cause of disease In later life. *Journal of Clinical Pediatric Dentistry* 42 (5): 325–330. https://doi.org/10.17796/1053-4625-42.5.1.

Nanayakkara, S., Senevirathna, S.T.M.L.D., Harada, K.H. et al. (2020). The Influence of fluoride on chronic kidney disease of uncertain aetiology (CKDu) in Sri Lanka. *Chemosphere* 257: 127186. https://doi.org/10.1016/j.chemosphere.2020.127186.

Onipe, T., Edokpayi, J.N., and Odiyo, J.O. (2020). A review on the potential sources and health implications of fluoride in groundwater of Sub-Saharan Africa. *Journal of Environmental Science and Health, Part A* 55 (9): 1078–1093. https://doi.org/10.1080/10934529.2020.1770516.

Ozsvath, D.L. (2009). Fluoride and environmental health: a review. *Reviews in Environmental Science and Bio/Technology* 8 (1): 59–79. https://doi.org/10.1007/s11157-008-9136-9.

Prasad, B. and Mondal, K. (2007). Leaching characteristics of Fluoride from coal ash. *Asian Journal of Water, Environment and Pollution* 4 (2): 17–21.

Puntoriero, M.L., Volpedo, A.V., and Fernández Cirelli, A. (2014). Arsenic, Fluoride, and Vanadium in surface water (Chasicó Lake, Argentina). *Frontiers in Environmental Science* 2: https://doi.org/10.3389/fenvs.2014.00023.

Ranasinghe, N. and Kanthi, R.D.F.C. (2016). Psychosocial impact of dental appearance and associated factors among 15 year old school children in anuradhapura municipal council area, in Sri Lanka. *Asian Pacific Journal of Health Science* 3 (4S): 50–58. https://doi.org/10.21276/apjhs.2016.3.4S.7.

Ranasinghe, N., Kruger, E., Chandrajith, R., and Tennant, M. (2019). The heterogeneous nature of water well fluoride levels in Sri Lanka: an opportunity to mitigate the dental fluorosis. *Community Dentistry and Oral Epidemiology* 47 (3): 236–242. https://doi.org/10.1111/cdoe.12449.

Rosso, J.J., Puntoriero, M.L., Troncoso, J.J. et al. (2011). Occurrence of fluoride in arsenic-rich surface waters: a case study in the Pampa Plain, Argentina. *Bulletin of Environmental Contamination and Toxicology* 87 (4): 409. https://doi.org/10.1007/s00128-011-0358-0.

Rudnick, R.L., Gao, S., Holland, H.D., and Turekian, K.K. (2003). Composition of the continental crust. *The Crust* 3: 1–64.

Saxena, V. and Ahmed, S. (2003). Inferring the chemical parameters for the dissolution of fluoride in groundwater. *Environmental Geology* 43 (6): 731–736. https://doi.org/10.1007/s00254-002-0672-2.

Sellami, M., Riahi, H., Maatallah, K. et al. (2020). Skeletal fluorosis: don't miss the diagnosis! *Skeletal Radiology* 49 (3): 345–357. https://doi.org/10.1007/s00256-019-03302-0.

Singh, G., Kumari, B., Sinam, G. et al. (2018). Fluoride distribution and contamination in the water, soil and plants continuum and its remedial technologies, an Indian perspective – a review. *Environmental Pollution* 239: 95–108. https://doi.org/10.1016/j.envpol.2018.04.002.

Skjelkvåle, B.L. (1994). Factors influencing fluoride concentrations in Norwegian lakes. *Water, Air, and Soil Pollution* 77 (1): 151–167. https://doi.org/10.1007/bf00483055.

Susheela, A.K. (1999). Fluorosis management programme in India. *Current Science* 77 (10): 1250–1256.

Tan, P.-p., Zhou, B.-h., Zhao, W.-p. et al. (2018). Mitochondria-mediated pathway regulates C2C12 cell apoptosis induced by fluoride. *Biological Trace Element Research* 185 (2): 440–447. https://doi.org/10.1007/s12011-018-1265-6.

Vithanage, M. and Bhattacharya, P. (2015). Fluoride in the environment: sources, distribution and defluoridation. *Environmental Chemistry Letters* 13 (2): 131–147. https://doi.org/10.1007/s10311-015-0496-4.

Wang, T., Shao, Z., Yu, H., and Bah, H. (2020). Distribution of fluoride in surface water and a health risk assessment in the upper reaches of the Yongding River. *Journal of Geographical Sciences* 30 (6): 908–920. https://doi.org/10.1007/s11442-020-1761-4.

Warnakulasuriya, K.A.A.S., Balasuriya, S., Perera, P.A.J., and Peiris, L.C.L. (1992). Determining optimal levels of fluoride in drinking water for hot, dry climates – a case study in Sri Lanka. *Community Dentistry and Oral Epidemiology* 20 (6): 364–367. https://doi.org/10.1111/j.1600-0528.1992.tb00700.x.

Waziri, M., Musa, U., and Hati, S.S. (2012). Assessment of fluoride concentrations in surface waters and groundwater sources in Northeastern Nigeria. *Resources and Environment* 2 (2): 67–72. https://doi.org/10.5923/j.re.20120202.10.

Wei, H.W., Hassan, M., Che, Y. et al. (2021). Spatio-temporal characteristics and source apportionment of water pollutants in upper reaches of Maotiao River, Southwest of China, from 2003 to 2015. *Journal of Environmental Informatics* 37 (2): 93–106.

WHO (1996). *World Health Organization, Guidelines for Drinking Water Quality Recommendations*. Geneva: World Health Organization.

WHO (2011). *Guidelines for Drinking-Water Quality*, 4e. Geneva: World Health Organization.

Xiang, Q., Liang, Y., Chen, L. et al. (2003). Effect of fluoride in drinking water on children's intelligence. *Fluoride* 36 (2): 84–94.

Yang, C., Wang, Y., and Xu, H. (2017). Treatment and prevention of skeletal fluorosis. *Biomedical and Environmental Sciences* 30 (2): 147–149. https://doi.org/10.3967/bes2017.020.

Zhang, Y. and Cao, S. (1996). Coal burning induced endemic fluorosis in China. *Fluoride* 29 (4): 207–211.

Zietsman, S. (1991). Spatial variation of fluorosis and fluoride content of water in an endemic area in Bophuthatswana. *The Journal of the Dental Association of South Africa* 46 (1): 11–15.

Zuo, H., Chen, L., Kong, M. et al. (2018). Toxic effects of fluoride on organisms. *Life Sciences* 198: 18–24. https://doi.org/10.1016/j.lfs.2018.02.001.

10

Iodine Essentiality for Human Health: Sources, Toxicity, Biogeochemistry, and Strategies for Alleviation of Iodine Deficiency Disorders

Majeti Narasimha Vara Prasad

School of Life Sciences, University of Hyderabad (an Institution of Eminence), Hyderabad, Telangana, India

10.1 Introduction

Several trace elements viz. Ca, Cl, Co, Cr, Cu, F, Fe, I, K, Ca, Mg, Mn, Mo, N, P, S, Se, Si, and Zn are required for the metabolism of human system (see Chapter 10 for additional information). The trace elements are essential not only for life and life quality (Gomes and Rautureau 2021). A holistic approach to mineral nutrition and human health was appraised by Gomes and Rautureau (2021) by elucidating the general and specific interactions – a broad concept of mineral, mineral *latu sensu* (mineral *l.s.*). The concept of minerals *strictu sensu* (mineral *s.s.*) or mineral itself – natural, inorganic, solid, and crystalline product which is an important component of rocks and soils. For instance, in earth sciences, soil sciences, and materials science; and the concept of mineral – chemical element of natural and inorganic origin, as a rule metal and metalloid, which is an important component of the human body, food (solid and liquid), and pharmaceuticals – adopted as so, for instance, in medical sciences, pharmaceutical sciences, and nutritional sciences. It is exemplified with a modern endemic disease expressed by the increasing number of people affected by thyroid cancer caused by the fallout of radioactive iodine isotope ^{131}I. Specific chemical are concentrated in particular areas/parts of the human body. Iodine, for instance, is concentrated by the thyroid gland. Some thyroid cancers can be identified and then treated using radioactive iodine^{131}I. (De la Vieja and Santisteban 2018).

Water and diet are the important sources of iodine. Once ingested, over 90% is absorbed in the stomach and duodenum through the sodium–iodide symporter. The expression of sodium–iodide symporter is downregulated when the amount of iodine ingested is high, and vice versa. It is speculated that iodine may well have a major role as a potential therapy for unexplained infertility. In several countries, the bioavailability of iodine via natural resources is insufficient, and the people suffer from iodine deficiency disorders (IDD). In order to circumvent the IDD in those countries, different strategies in alleviation of iodine deficiency are adopted. Excess of iodine causes toxicity.

In this chapter, biogeochemistry of iodine is discussed, and iodine essentiality for human health, toxicity, sources, deficiency disorders, strategies for alleviation, and biogeochemical pathways of iodine are covered (Pennington 1990).

Iodine is one of the mineral nutrients required by the body in trace or minute quantities. The human body contains about 50 mg of iodine with 70–80% found in the thyroid gland. A person requires only about a teaspoonful of iodine throughout his or her lifetime, a minimum daily requirement of about 15–20 mg, but it is a vital raw material required by the thyroid gland for the production of its product called thyroid hormones (Bamidele et al. 2021). Prevalence of iodine deficiency is regulated by iodized salt (Hatch-McChesney et al. 2022). Thyroid hormones that stimulate the body is responsible for the proper development of the brain. Thus, when the diet lacks iodine, these functions are impaired, and prolonged chronic deficiency may result in any of the definite diseases called IDD (Bamidele et al. 2021).

Medical Geology: En route to One Health, First Edition. Edited by Majeti Narasimha Vara Prasad and Meththika Vithanage.

10.2 Iodine Essentiality for Human Health

Iodine is essential for human health as a micronutrient. At the same time, excess iodine is toxic. Iodine essentiality and toxicity are briefed in Table 10.1. Table 10.2 highlights the countries reported for excessive iodine intake. Iodine is essential for the production of thyroid hormone in the thyroid gland. Thyroid hormones viz. thyroxine (T4) and triiodothyronine (T3) contain 65 and 59% of iodine, respectively. These hormones regulate the key processes of metabolism, brain development, and growth. The role of iodine in human health is elaborated later in this chapter (Turianica and Rothová 2021).

10.3 Role of Iodine in Thyroid Function

Iodine plays a pivotal role in thyroid functioning which itself regulates the health of human (Lee and Pearce 2022). Therefore, optimum dose of iodine availability in nutrition is important to avoid thyroiditis and thyroid cancer and to balance thyroid autoimmunity (Apaydin et al. 2021; Atmis et al. 2021; Chaboki et al. 2021; Çolak et al. 2021; Delange and Hetzel 2008; Gholami et al 2021; Habib et al. 2021; Hashemipour et al. 2021; Kart et al. 2021; Koukkou et al. 2017; Laurberg et al. 2000; Lazarus 2014; Lee 2021; Lee et al 2021; Masilamani et al. 2021; Nettore et al. 2018; Ning et al. 2020; Olivieri et al. 2021; Ruggeri and Trimarchi 2021; Yordam et al. 1999; Zimmermann and Galetti 2015). Thyroid is a small organ, normally weighing about 25–30 g found within the front part of the neck (Figure 10.1). It is made up of two parts, called flaps, that lie along the windpipe (trachea) and are joined together by a contract band of thyroid tissue, known as the isthmus. The thyroid is arranged fair underneath the “Adams apple” or larynx. The function of the thyroid is to acquire iodine found in numerous nourishments and convert it into thyroid hormones: T4 and T3. Thyroid cells only can assimilate iodine. These cells combine iodine and the amino acid tyrosine to form T3 and T4. T3 is discharged into the blood stream and is transported all through the body where they control metabolism (conversion of oxygen and calories to energy). Every cell within the body depends upon thyroid hormones for their functions. The typical thyroid organ produces around 80% of T4 and approximately 20% of T3.

Table 10.1 Iodine essentiality and toxicity.

Salient findings	Author(s)/year
Initial evaluation and treatment of iodine oral poisonings	Larsen and Cummings (1998)
Thyroid disorders in mild iodine deficiency	Laurberg et al. (2000)
The re-emergence of iodine deficiency in New Zealand	Mann and Aitken (2003)
Iodine supplementation improves cognition in iodine-deficient school children	Zimmermann et al.(2006)
Iodine deficiency in pregnancy	Zimmermann (2009b)
Iodine is a chemical messenger through which the brain regulates and controls important body functions	Bamidele et al. (2021)
The role of iodine in human growth and development	Zimmermann (2011)
Iodine intake as a risk factor for thyroid cancer	Zimmermann and Galetti (2015)
Iodine deficiency may impair fertility	Kuehn (2018)
Iodine-induced retinopathy	Kobat et al. (2020)
Iodine nutrition: disorders, monitoring, and policies	Bertinato (2021)
Monitoring iodine nutritional status: adults or school children	Lazarus (2021)
Maternal iodine supplementation during lactation may improve child's cognitive development at 3 years of age in a mild-to-moderately iodine-deficient population	Lee (2021)
Assessment of iodine nutritional status in school children concerning autoimmune thyroiditis in Tamil Nadu, India	Masilamani et al. (2021)
Prevalence of iodine deficiency disorders (IDD) among school-going children in rural areas of Bangalore	Anil et al. (2021)
Excess iodine is toxic	Southern and Jwayyed (2021)
Essential minerals: nutritional requirements, dietary sources, and deficiencies	Tako(2022)
Iodine, an effective substance against the COVID-19 pandemic	Rezaei Ahvanooei (2022)

Table 10.2 Countries reported for excessive iodine intake based on global scorecard of iodine nutrition in 2021 in the general population based on SAC.

Country or territory	Median UIC (µg/L)	Date of survey	Data type	Population surveyed
Benin	318	2011	N	SAC (6–12)
Cameroon	>300	2014, 2017, 2018	S	SAC
Costa Rica	314	2008–2009	N	SAC
Djibouti	335	2015	N	SAC
Equatorial Guinea	564	2007	S	SAC
Honduras	356	2005	S	SAC
Korea, Democratic People's Republic of	449	2013–2015	N	SAC, adolescents (6–19)
Liechtenstein	96	2010	N	SAC (6–12)
Lithuania*	75	1995	N	SAC
Madagascar	46	2015	N	WRA (15–49)
Qatar	341	2014	N	SAC (6–12)
Solomon Islands	328	2007–2010	S	SAC (6–12)
Trinidad and Tobago	311	2018	N	SAC
Uganda	464	2005	N	SAC (6–12)

Source: 2021, Iodine Global Network.
N, Nationally representative survey; NPNLW, Nonpregnant, nonlactating women; P, Date published; S, Subnational survey (all administrative levels); SAC, School-age children; UIC, Urinary iodine concentration; WRA, Women of reproductive age.
*Countries in which the most recent available survey is older than 15 years (i.e., conducted prior to 2005), are marked with an asterisk (*)

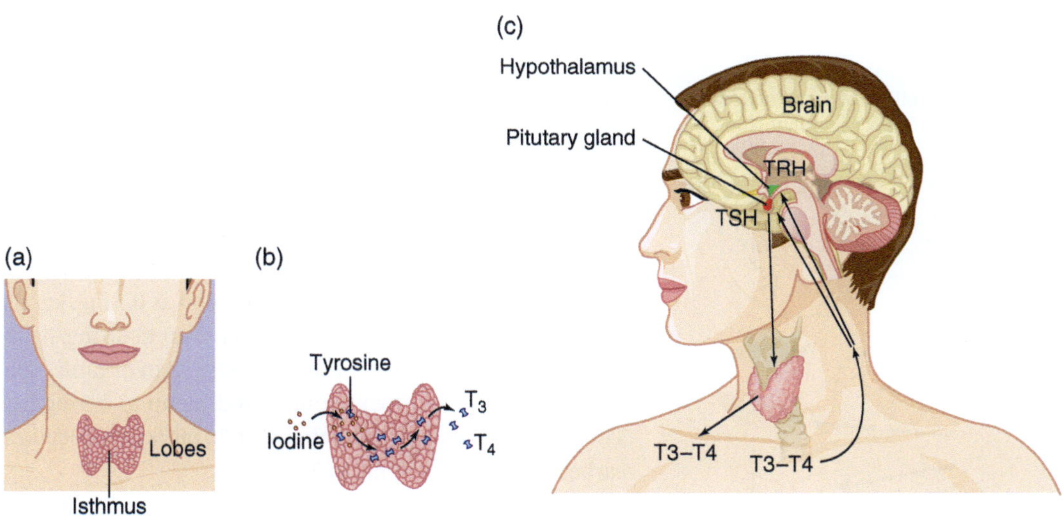

Figure 10.1 Thyroid organ and its parts – role in iodine metabolism. (a) Isthumus and thyroid gland lobes (b) Thyroid gland lobes secreting tyrosine and thyroxine (T4) and triiodothyronine (T3) (c) The hypothalamus is a part of the brain that produces TSH (thyroid fortifying hormone) discharging hormone (TRH) which tells the pituitary organ to stimulate the thyroid gland.

The thyroid organ is under the control of the pituitary organ, which is of the size of a peanut at the base of the brain. When the level of thyroid hormones (T3 and T4) drops, the pituitary organ produces thyroid-stimulating hormone (TSH) which fortifies the thyroid organ to deliver more hormones. When the level of thyroid hormones (T3 and T4) drops too low, the pituitary gland produces TSH which stimulates the thyroid gland to produce more hormones. Under the influence of TSH, the thyroid will secrete T3 and T4, thereby raising the blood levels. The pituitary senses this and responds by decreasing its TSH production. One can imagine the thyroid gland as a furnace and the pituitary gland as the thermostat. Thyroid hormones are like heat. When the heat gets back to the thermostat, it turns the thermostat off. As the room cools (the thyroid hormone levels drop), the thermostat turns back on (TSH increases), and the furnace produces more heat (thyroid hormones).

The pituitary organ itself is directed by another organ, known as the hypothalamus. The hypothalamus is a portion of the brain and produces thyroid-releasing hormone which tells the pituitary organ to stimulate the thyroid gland (release TSH). One might envision the hypothalamus as the individual who controls the thermostat since it tells the pituitary gland at what level the thyroid ought to be set.

Thyroid enlargement (goiter) is a typical visible sign of iodine deficiency and can occur across all age groups. It is a physiological adaptation to inadequate iodine status; under the negative feedback system, TSH secretion increases to maximize iodine uptake by the thyroid, while also stimulating thyroid hypertrophy (Figure 10.2).

The decrease in thyroid hormone leads to an increase in the TSH, resulting in the development of goiter, the disease that affects 740 million people worldwide, accounting for 13% of the world's total population. Cretinism, a neurological disorder, is another disease caused by iodine deficiency. Vitti et al. (2001) provided information on IDD in Europe (97 million people are reported to have goiter due to iodine insufficiency).

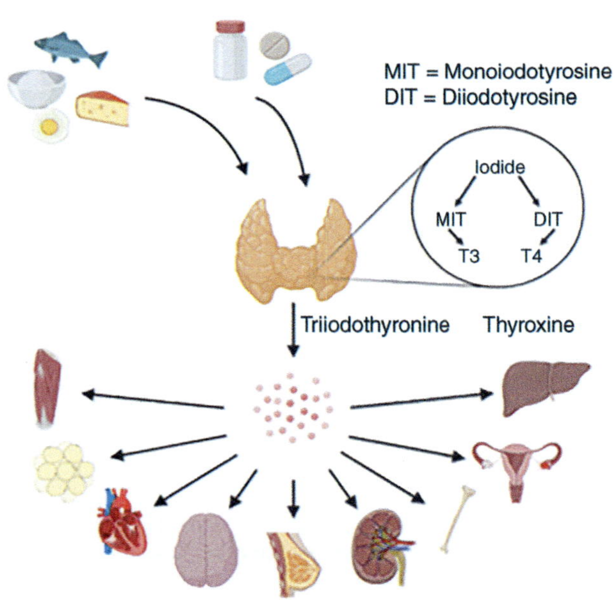

Figure 10.2 Iodine can be ingested through diet and dietary supplements. The primary sources of iodine are salt, seafood, fish, algae, milk, and dairy. Iodine is taken up by the thyroid gland and used for the synthesis of monoiodotyrosine (MIT) and diiodotyrosine (DIT), which are used for the biosynthesis of thyroid hormones (T3 and T4). Receptors of thyroid hormones are on the surface of different organs; thus, thyroid hormones travel in the blood and may affect many systems such as reproductive system, skeletal system, or muscular system. *Source:* Krela-Kazmierczak et al. (2021) / MDPI / Public domain.

Iodine is a reactive trace element, not occurring in its pure form (SCF 2002). It is essential for the production of thyroid hormone production in the thyroid gland (Walther et al. 2018). T4 and T3 contain 65 and 59% of iodine, respectively (SCF 2002). These hormones regulate key processes of metabolism, brain development (Delange 2000; Grau et al. 2015; SCF 2002). Moreover, iodine might also be important due to its immunomodulatory effects on human immunity (Iannaccone et al. 2019). Therefore, iodine is very important for maintaining physiological processes (Flachowsky et al. 2014). If these demands are not fulfilled, the insufficient iodine status may result in the deficiency of thyroid hormones and subsequently in the deficiency symptoms including cretinism, brain damage, irreversible mental retardation, deaf-mutism, and goiter (Troshina et al. 2021 and Walther et al. 2018). In this respect, it is estimated that more than 2 billion people worldwide (Zimmermann 2009a) and 30% of the European countries have insufficient iodine intake and are at risk of iodine deficiency (Lazarus 2014). This has been confirmed by Johner et al. (2012), who found insufficient iodine intake in 30% of the participants of a German study.

10.4 Iodine Sources in Biogeosphere and Hydrosphere

Iodine is considered an essential micronutrient in the environment as its deficiencies are associated with health risks for both humans and wildlife (Edmunds and Smedley 1996; Jooste et al. 1882; Milewski and Diamond 2000; Neal et al. 2007; Phaneuf et al. 1999; Sack et al. 2000). Iodine content of the Earth's crust is extremely variable; however, its crustal average abundance is estimated as 0.5 mg/kg. Its highest content is in sedimentary rocks: up to 38 mg/kg in argillaceous sediments and up to 17 mg/kg in organic-rich shales. Igneous acid rocks contain iodine up to 0.5 mg/kg. The high average iodine contents of 30 mg/kg in deep-sea carbonates (Kabata-Pendias 2011).

Among micronutrients, undernutrition of iron, vitamin A, iodine, and zinc remains a significant problem in developing country populations (Kennedy et al. 2003; SCN 2004). Biofortification addresses specific micronutrient deficiencies attributable to inadequate intake; thus, it can help populations at risk for these deficiencies when other options are not available.

Seaweeds and kelps, also known as macroalgae, have long been common in the East-Asian diet. During recent years, macroalgae have entered the global food market, and a variety of macroalgae products are now available for consumers. Some macroalgae species are known to be particularly rich in iodine, but little data regarding the iodine content of macroalgae-containing foods exist. Figure 10.3 summarizes iodine in some common rock types, Figure 10.4 appraises iodine in various types of soils, and Figure 10.5 shows iodine in different types of vegetation. Oceans are the major source of iodine which volatilizes and gets redeposited in soil. However, the levels of iodine in soil vary globally due to glaciation, flooding, high levels of rain fall, and erosion (Figures 10.4 and 10.5).

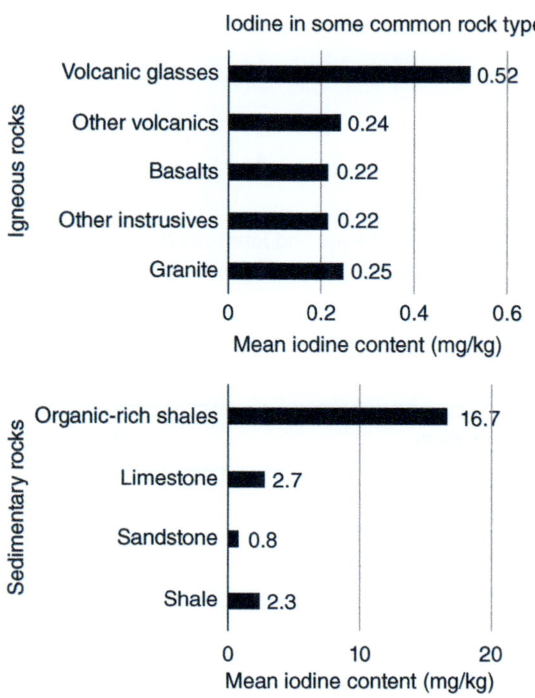

Figure 10.3 Iodine in some common rock types. *Source:* Redrawn from the data of table 1 of Fuge and Johnson (2015).

10.5 Iodine in Diets

The bulk of iodine required for human consumption is derived from food (80%) and water (20%) (Pehrsson et al. 2021, 2022; Sprague et al. 2022; Stevenson et al. 2018; Szymandera-Buszka et al. 2021). Sea foods have the highest iodine content, in which it can be as high as 3 mg/kg. Ocean water contains about 50 μg of iodine per liter and is the primary sources of iodine in the world (Koutras et al. 1985). Fish and seafood are naturally rich sources due to bioaccumulation of iodine from seawater. The other main dietary sources of iodine are milk (Arrizabalaga et al. 2015), egg, cereal, and meat. The average iodine content of some of the foods (fresh and dry basis) is reported by Koutras et al. (1985). According to Koutras et al. (1985), these average iodine values in foods cannot be used universally for estimating iodine intake as the content varies with geographical location and environmental media. Reference may be made to Table 10.3 for iodine concentrations of selected foods and feeds, Table 10.4 for recommended daily iodine requirements, Table 10.5 for cow's milk iodine concentration in several European countries, Table 10.6 for recommended dietary allowances, and Table 10.7 for iodine content in selected foods and feeds. Halophytes such as *Allenrolfea occidentalis* are capable of facultative or obligatory growth on arid soils and can contain a high level of iodine (Soudek et al. 2006). The range of iodine content μg/g DW of examples of the

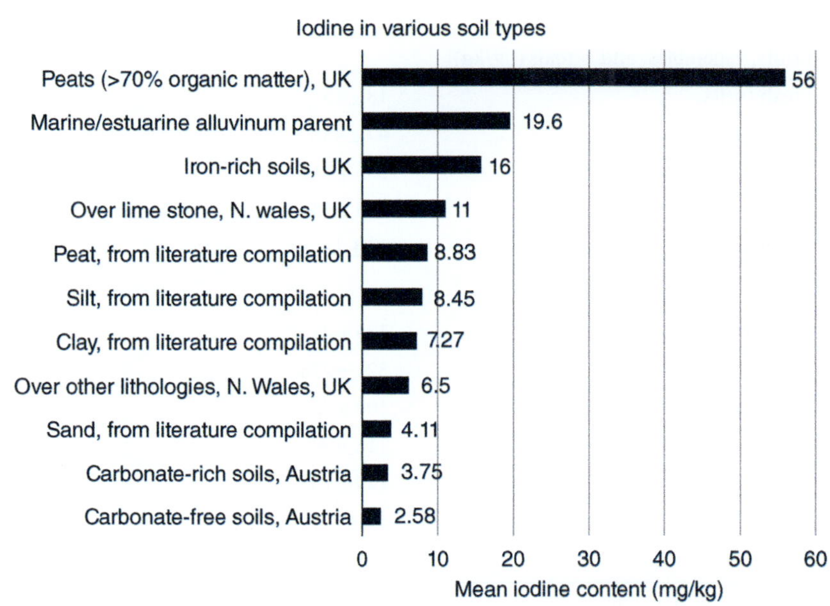

Figure 10.4 Iodine in various soil types. *Source:* Redrawn from the data of table 1 of Fuge and Johnson (2015).

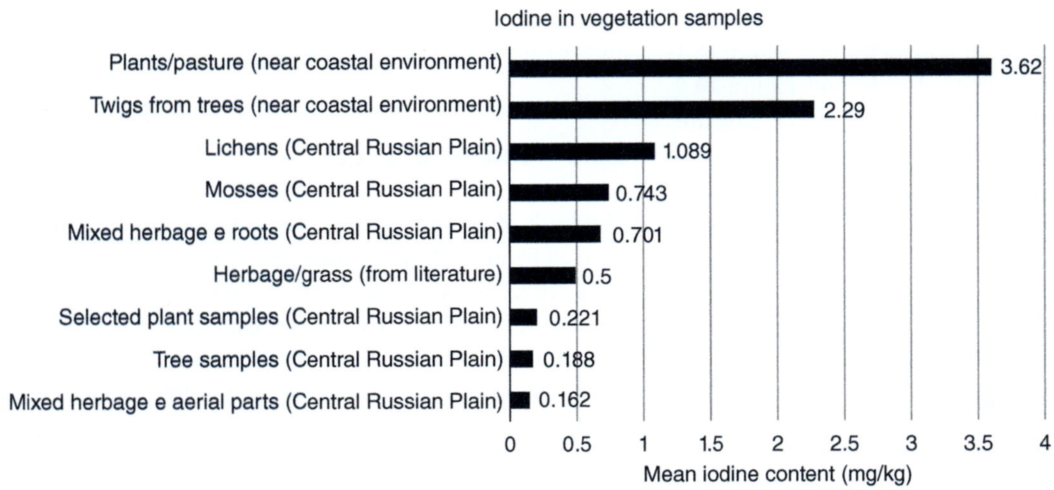

Figure 10.5 Iodine in different types of vegetation. *Source:* Redrawn from the data of table 1 of Fuge and Johnson (2015).

Table 10.3 Iodine concentrations of selected foods and feeds.

Food item	Iodine concentration		
	Mean	Median	Min. – max.
Dairy products (μg/kg)			
Cow milk	251±110	—	147–605
Cow cheese	473±289	396	146–1323
Goat cheese	700	—	250–1040
Sheep cheese	861	—	—
Yoghurt	670±313	556	347–1239
Meat, protein sources (μg/kg)			
Egg	360	—	340–380
Meat	40	—	2–430
Fish, marine	2112±1713	1440	387–6926
Fish, freshwater	375±436	205	11–1571
Fruits, vegetables, and cereals (μg/kg)			
Fresh fruits	18±15	15	2–75
Fresh vegetables	47±43	33	9–203
Bread	393±213	392	25–1032
Rice	333±341	250	11–934
Potatoes	16±11	18	4–26
Feeds (μg/kg dry matter)			
Corn silage	125±22	94	29–479
Grass silage	321±278	176	105–949
Maize silage	109±42	102	54–197
Hay	112±94	78	23–523
Pasture	149±105	119	27–555
Grains	55±7	40	28–270
Soy products	101±21	62	29–320

Source: Mikláš et al. (2021).

Table 10.4 Recommended daily iodine requirements and upper tolerable iodine intake (μg/day) for the chosen age categories and physiological states.

Age/physiological stage	Iodine requirements	Tolerable upper intake level of iodine (SCF 2002)
0–5 years	90	—
1–3 years	—	200
4–6 years	—	250
6–12 years	120	—
7–10 years	—	300
11–14 years	—	450
>12 years	150	—
15–17 years	—	500
Adults	150	600
Pregnant and lactating women	2500	60

Source: Mikláš et al. (2021).

Table 10.5 Impact of organic and conventional farming systems on cow's milk iodine concentration in several European countries.

	Milk iodine (μg/L)		
Country	Organic	Conventional	Notes
Czechia	174	463	Raw
Czechia	119	136	Raw
Denmark	17	27	—
Germany	58	112	Pasteurized, UHT
Spain	78	157	Raw
Spain	56	163	UHT
Switzerland	55	93	Raw
Switzerland	71	111	UHT
United Kingdom	399	578	Pasteurized
United Kingdom pasteurized	343	474	UHT

Source: Mikláš et al. (2021).
UHT, ultra-heat treatment.

Table 10.6 Recommended dietary allowances (RDA) and adequate intake for iodine.

Group RDA	RDA (μg)	Adequate intake (AI) (μg)
0–6 months		110
7–12 months		130
1–3 years	90	
4–8 years	90	
9–13 years	120	
14–18 years	150	
≥19 years	150	
Pregnancy	220	
Breastfeeding women	290	

Source: Krela-Kazmierczak et al. (2021) / MDPI / Public domain.

Table 10.7 Iodine content in the selected food products.

Product	Iodine content (μg/100 g of product)
Codfish, fresh	110
Salmon, fresh	0.7–44
Pike, fresh	8
Kefir, 2% of fat	7.5
Skimmed milk	19.5–21
Cheese, full-fat	7.7–30
Chicken eggs	9.5–57.6
Oat flakes	0.5–5.9
Vegetable	1–31
Nuts	4–9
Salt, iodized	2293.4

Source: Krela-Kazmierczak et al. (2021) / MDPI / Public domain.

three main classes of edible seaweed are summarized in Table 10.8 (Smyth 2021).

Iodized salt remains an important and recommended source of iodine (Lopes-Pereira et al. 2022). In the middle of last century, iodine status improved markedly in New Zealand by post-mandatory iodization of table salt. In 2002, 84.6% of parents reported using iodized salt at home. Similarly, in the 2008–2009 Adult Nutrition Survey, 85.7% of New Zealand adults aged 15 years and over who used salt at home reported using the iodized variety. It is difficult to accurately quantify the amount of salt dispensed from a shaker, and salt is often used in recipes "to taste." Because of this, there is little data about the proportional contribution salt makes to iodine intake in the diet. Currently, where discretionary iodized salt use is reported, a bolus amount of 1 g iodized salt containing 48 μg iodine is added to the total intake. This estimate is derived from the amount of salt consumed in processed foods, as well as an average of the iodine concentration in iodized table salts in New Zealand.

Table 10.8 Range of iodine content μg/g DW of examples of the three main classes of edible seaweeds.

Classification and species	Common names	Iodine content, μg/g DW (range)	Grams of seaweed required to achieve daily RNI of 150 μg/day
Green algae (Chlorophyta)			
Undaria pinnatifida	Wakame	30–185	2.0–6.0
Ulva lactuca	Sea lettuce		
Ulva intestinalis			
Red algae (Rhodophyta)			
Palmaria	Dulse, dillisk	20–200	0.3–3.0
Porphyra	Nori		
Alaria	Irish wakame		
Chondrus crispus	Irish moss		
Brown algae (Phaeophyceae)			
Laminaria	Kombu	2500–10 000	0.01–0.04
Ascophyllum			
Fucoids			

Source: Smyth (2021) / Karger Publishers.

10.6 Iodine in Watersheds

Halogens in drinking water sources such as bromine (Br) and iodine (I) pose no direct health risk, but are critical precursors in the formation of cyto- and genotoxic brominated and iodinated (Br^-/I^-) disinfection by-product (DBPs). Iodine species are reported in drinking water sources (groundwater and surface water). Potable water is contaminated with iodide ion uptake inhibitors, and their determination is necessary (Cengiz et al. 2022).

Milk and dairy products, as important sources of iodine (Arrizabalaga et al. 2015; van der Reijden et al. 2019), might help to alleviate this problem, in case of proper supplementation of mineral premixes and licks to dairy animals (Schone and Rajendram 2009), which can bring a positive outcome not only for human but also for animal nutrition (Flachowsky et al. 2007).

Iodine is an essential element for mammals including humans (Flachowsky et al. 2014). However, it was not proved indispensable for plants. Nevertheless, in kelp and seaweeds, iodine exists (Aakre et al. 2021) in the form of organic compounds, such as 3,5-di-iodine tyrosine.

The long-lived radioisotope I^{129} is a uranium fission product, an environmental contaminant of the nuclear age (Herod et al. 2016). Iodine concentration in seawater averages around 50 μg/L, and the concentrations in rainfall and river water are usually an order of magnitude lower (Moran et al. 2002). However, much higher concentrations can occur in brines associated with oil fields where concentrations can average 30 000–40 000 μg/L (Greenwood and Ernshaw 2005). A major proportion of iodine in surface freshwater has significant sources: oceanic cyclic iodine that is delivered to rover watersheds atmospherically and iodine weathered from soils and rocks.

The transfer of iodine from atmosphere is a complex process involving both inorganic and organic components, and the processes include sea spray and biogenic production of organo-iodine (Radlinger and Huemann 2000), whereas iodine is concentrated in soils, especially organic rich ones, and can be released back to solution under reducing conditions (Muramatsu and Yoshida 1999). Other minor sources of iodine in freshwaters volatilized from plants or derived from upwelling, deeply sourced ground waters. Because different sources have unique I^{129}/I ratios, this ratio is widely used as an indicator to trace sources and examine mixing (Moran et al. 2002). Iodine exists in a number of oxidation states with −1 (iodide) and +5 (iodate) being the most common ones in natural waters. Iodine can also be present as organo-iodine in waters and fixed onto humic substances, making it a precursor for the formation of DBP during disinfection process (Neal et al. 2007). With organo-iodine and iodide being favored over iodate likely makes it less available for plant uptake as land plants do not concentrate iodine as efficiently as marine organisms. Iodine concentrations in the water depend on the source and fluctuate with seasons. Spring freshet meltwater is observed to have the highest concentration of iodine followed by a decline over the summer (Herod et al. 2016). The spatial variability in iodine concentrations and I^{129}/I ratios reveal information about the sources of iodine and I^{129} in the water bodies. In watersheds, this concentration is dependent on evapotranspiration, input of iodine from soil strata, thermodynamic conditions, and land use practices.

Iodine can also be found in drinking water industry as an effective disinfectant. The traces of iodine usage for water treatment can be found in early 1920s when it was added to prevent goiter (Hanke and Boland 1971). Iodine was used for disinfection during the First and Second World Wars by France and the Unites States, respectively. More recently, NASA started using it for disinfection in space flights (Punyani et al. 2006). Iodine is known to be an effective broad-spectrum disinfectant with small doses with no efficiency impacts due to impurities and can eliminate the chances of iodine deficiency diseases when used as a disinfectant in drinking water treatment process. In adult populations, when used up to 20 ppm as pure crystalline iodine, no prolonged psychological impacts were detected, although consideration needs to be given when used with more vulnerable populations as seniors and children. The feasibility of iodine for disinfection largely depends on the ability to use it to kill or inactivate pathogens on contact without releasing iodine into the drinking water. Aerotherapy (***baths of iodine-rich*** air and aerosol) is a special type of thalassotherapy comprises surfbathing, by definition is the use of seawater, sea products (sand, mud, salt, algae, aerosols, etc.), and seaside climate as a form of therapy. This kind of treatment should have a medical prescription issued by a competent medical practitioner, reporting the clinical state of the patient and the prescribed treatments, acquainted with treatment and precautions (Gomes and Rautureau 2021).

10.7 Iodine Deficiency Disorders

Iodine is essential for proper functioning of thyroid (see Section 10.3). The consequences of inadequate thyroid hormone production due to iodine deficiency range from infant mortality and cretinism in the neonate to impaired mental function and physical development in adolescents and adults (Zeru et al. 2021). Collectively, these effects are termed IDD. The World Health Organization (WHO) estimates iodine deficiency as the single most preventable cause of brain damage worldwide. Essential and toxic elements in seaweeds and their consumption can disturb iodine metabolism in the thyroid gland (Desideri et al. 2016; Nanri et al. 2017). Global iodine availability status and insufficient iodine bioavailable regions of the world are shown in Figures 10.6 and 10.7, respectively.

Plant and animal tissues contain inorganic iodide which is readily absorbed, whereas iodate used in fortification is reduced in the gut and absorbed as iodide (Zimmermann 2011). Iodide is released into the

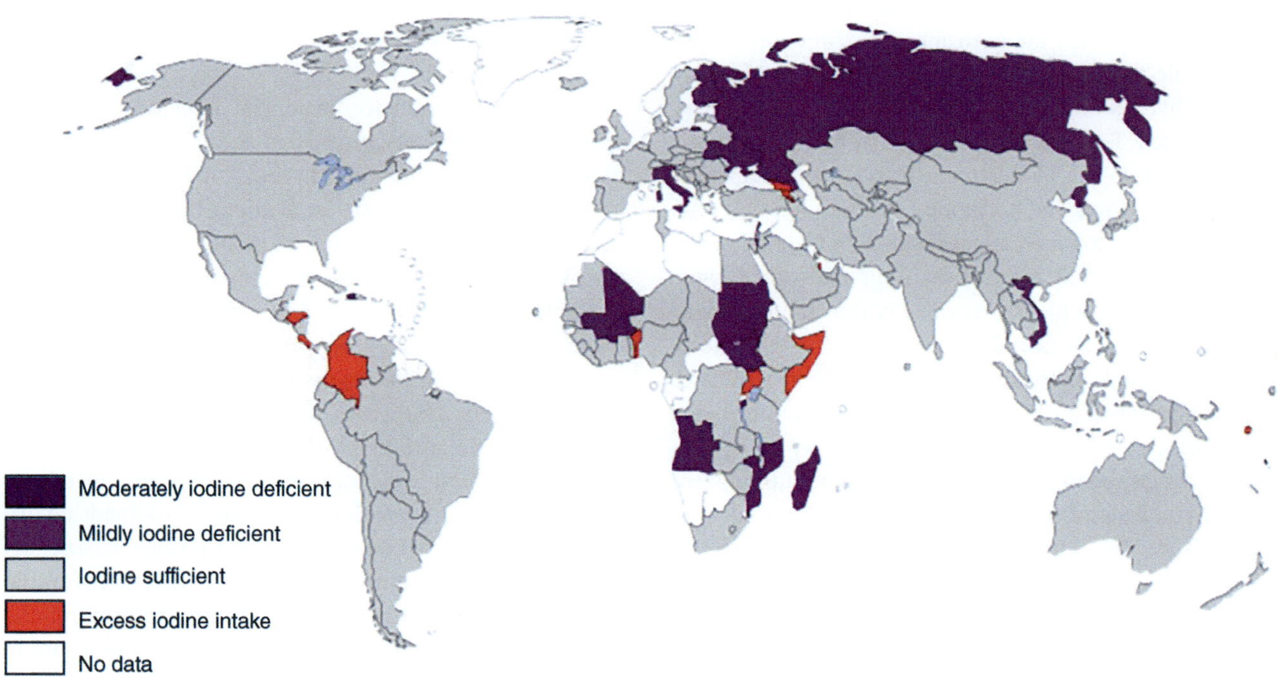

Figure 10.6 Global iodine availability status. *Source:* Accessed from: https://www.ncbi.nlm.nih.gov/books/NBK285556.

Figure 10.7 Insufficient iodine intake is observed in many regions, mainly African and Asian countries. Optimal intake of iodine occurs in the United States, Canada, France, UK, and some countries in Europe. However, intake of iodine is not adequate in many European regions. *Source:* Krela-Kaźmierczak et al. (2021) / MDPI / Public domain.

bloodstream and taken up by the thyroid gland through active transport. The activity of the sodium–iodide symporter is 20–50 times greater in thyroid tissue compared to elsewhere in the body, so iodide is preferentially absorbed from the blood into the thyroid (Zimmermann 2011). Iodide ions are oxidized to form iodine molecules and attached to tyrosyl residues in thyroglobulin to produce the hormone precursors iodotyrosine and diiodotyrosine. These couple to form T3 and T4, which are stored within iodinated thyroglobulin in the follicular cells of thyrocytes. When required, T3 and T4 are removed from thyroglobulin and released into the bloodstream through the monocarboxylate transporter. The hormones attach to binding proteins and migrate to target tissues where T4 is deiodinated to T3, the metabolically active form. Degradation of T3 and T4 releases iodine, which is recycled within the thyrocyte or excreted by the kidney (Zimmermann 2011). Iodine is also concentrated in the mammary glands, where it is secreted into breastmilk during lactation (Zimmermann et al. 2006).

Iodine is required in the body in minute amounts, mainly in the thyroid gland for the production of thyroid hormones, T4 and T3. These hormones are essential for the normal development and functions of nervous system and body metabolism (Dunn and Van der Haar 1990). The normal thyroid contains between 2 and 20 mg of iodine. Under normal circumstances, the thyroid takes about 20% of the available iodine, which is adequate for normal physiological life processes. When the intake is less than the physiological requirements of the body, thyroid function impairs and is unable to synthesize sufficient amounts of thyroid hormone; as a result, a series of functional and developmental abnormalities occurs, which includes endemic goiter and cretinism, mental retardation, decreased fertility rate, and increased prenatal death and infant mortality, which are referred to as IDD (Mathews et al. 2021).

In addition, when too much is ingested, the major epidemiological consequence is iodine-induced hyperthyroidism, whereby the thyroid may shut down or, under certain circumstances, it might become overactive (Delange 2000; Todd et al. 1995; Stanbury et al. 1998).

It is reported that iodine is lacking in several regions of the Himalayas. Soils in mountain ranges and in areas subjected either to flooding or to intense and prolonged glaciations are particularly likely to be deficient in iodine. Evidence for the lack of iodine in Bhutan is supported circumstantially by the occurrence of goiter and cretinism, which were common in the people of the region prior to the widespread use of iodized salt in 1985. Among the many control programmes, iodine fortification and distribution of salt fortified with iodine has been successful in combating and eliminating the prevalence of IDD in the country (Abdelrahman et al. 2020; Apaydin et al. 2021; Ashabi et al. 2021; Olivieri et al. 2021) (Figure 10.8).

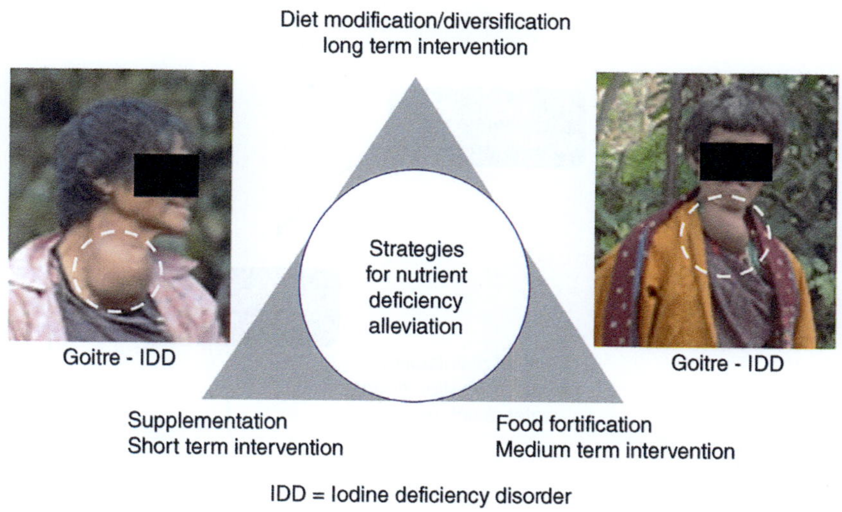

Figure 10.8 Strategies for iodine deficiency alleviation. Efficacy and effectiveness of the measures are being researched for cost-effectiveness and acceptability.

Iodine supplementation for pregnant women was advocated by various investigators (Aktaş et al. 2021; Apaydin et al. 2021; Delshad et al 2021; Gholami et al 2021; Grau et al. 2015; Pearce 2018; Pearce et al. 2016; Pinheiro et al 2021; Vural et al. 2021; Zimmermann 2009b).

10.8 Biogeochemical Cycling of Iodine

Biogeochemical cycling of iodine in hydrological cycle and gaseous cycle is shown in Figures 10.9–10.11. Halogens in drinking water sources such as bromine (Br) and iodine (I) pose no direct health risk but are critical precursors in the formation of cyto- and genotoxic brominated and iodinated (Br^-/I^-) DBPs. Iodine species are reported in drinking water sources (groundwater and surface water).

Iodine (I) in drinking water poses no direct health risk but is a critical precursor in the formation of cyto- and genotoxic iodinated (I^-) DBPs. Occurrence and speciation of iodine in groundwater and surface water serve as the source for drinking water treatment plants (DWTPs) (Naushita Sharma 2021). Sharma et al. (2019) conducted the first national iodine species survey in the United States. About >9000 samples of nondrinking water were collected

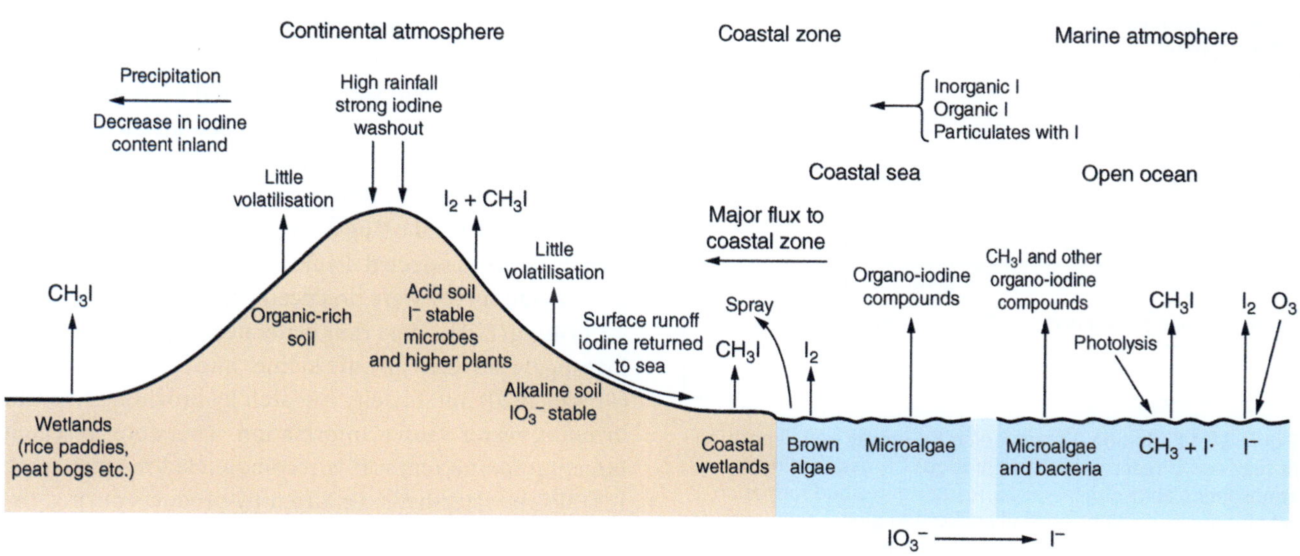

Figure 10.9 Biogeochemical cycling of iodine – role of environmental geochemistry and diet. *Source:* Fuge and Johnson (2015) / With permission from Elsevier.

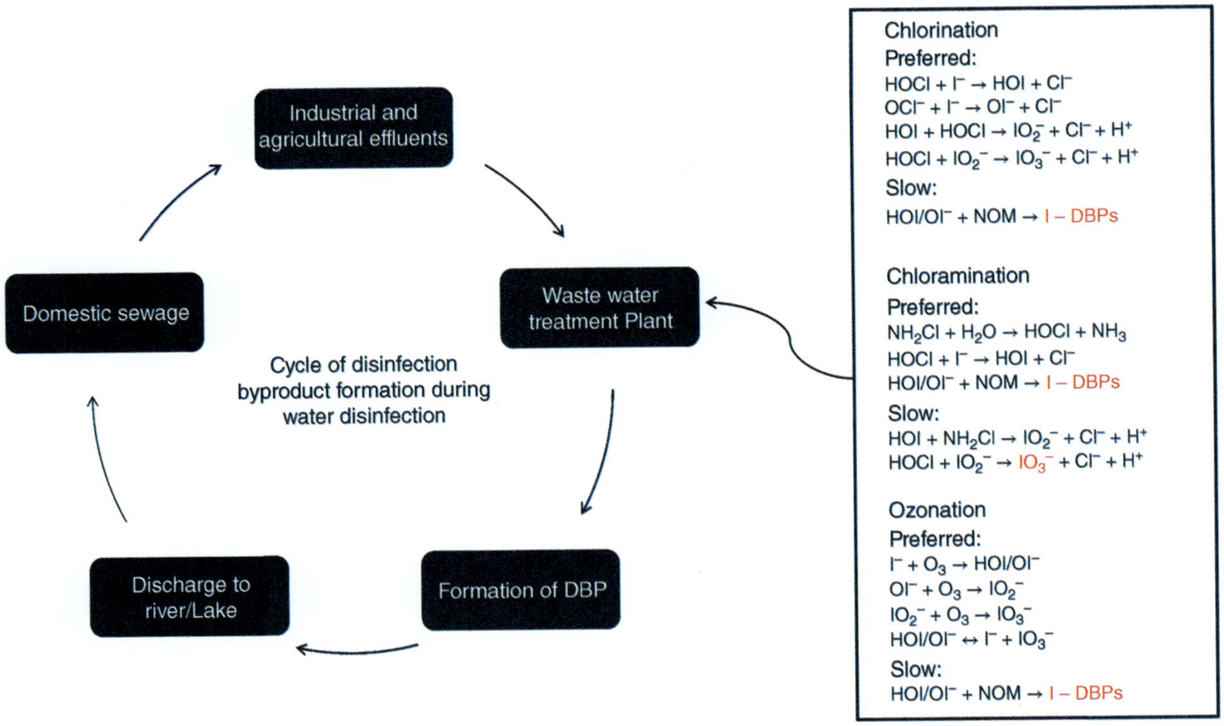

Figure 10.10 Cycle of disinfection by-product formation during water disinfection via chlorination, chloramination, and ozonation. By-products in bold are the preferred products, and those highlighted in red are undesirable products of disinfection.

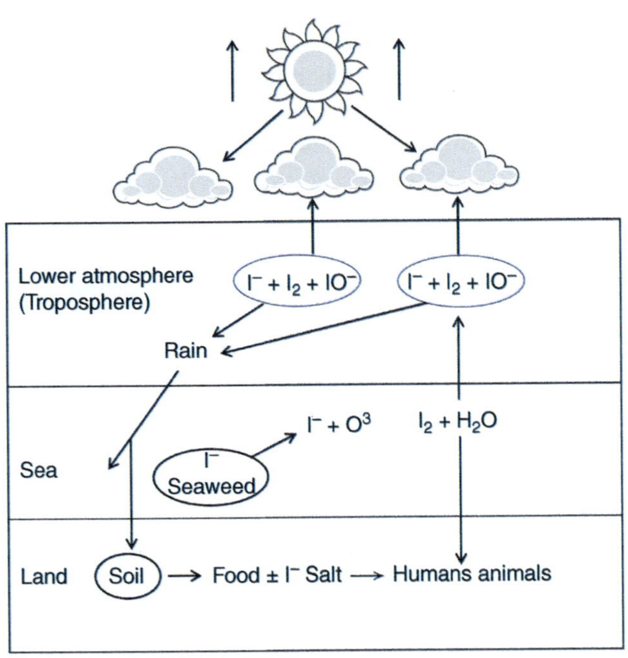

Figure 10.11 Iodide (I^-) and iodine (I_2) uptake and deposition. In seaweed, I^- reacts with O_3 to produce I^2I_2 gas, leading to atmospheric aerosol generation, resulting in cloud formation which reflects sunlight, partially offsetting global compound can be deposited in rain, and gaseous I^{2-} can be taken up by respiration. *Source:* Smyth (2021).

from the United States and analyzed. Iodine concentrations were 12 and 13 μg/L in surface and groundwater, respectively. Although these locations were not drinking water sources, this as the first of its kind analysis provides potential bounds for iodine to focus specifically on DWTP sources.

Survey of >250 drinking water sources conducted between 2018 and 2020 (iodide and iodate ions) reported the occurrence of organic iodine. Iodine concentrations ranged from 1 to 250 μg/L. To quantify iodine species, tandem ion chromatography and inductively coupled plasma mass spectrometry were applied for the first time in drinking water sources Sharma et al. (2019). Iodine and IO_3^- peaks were well resolved and have minimum detection limit of 0.4 and 0.7 μg/L, respectively. Organic iodine (Org-I) peaks in selected drinking water samples from the nationwide survey were practically resolved ranging from <5 to 40 μg/L. The data thus generated would be useful to investigate further about iodine and DBP formation (Prasad 2020) and toxicity research by providing relevant drinking water source information. Therefore, research targeting iodine removal is recommended for managing I^- DBPs in watersheds. Oxidized by-products during disinfection are due to the presence of iodide in drinking water sources.

Another nondietary role of iodine is the involvement of seaweed-generated iodine gas (I_2) in influencing weather and global warming/cooling (Smyth 2021). The origin of atmospheric gaseous iodine (I_2) is believed to be the reaction of inorganic iodide (I^-) present in seaweed with ozone present in the lower atmosphere. This reaction is believed to be catalyzed by haloperoxidases and gives rise to iodine oxides (IO) and elemental iodine (I_2). When the tide recedes, seaweed is exposed to the atmosphere and desiccated and becomes stressed. This stress involves contact with atmospheric O_3, leading to the following reactions (iodide (I^-) and iodine (I_2) uptake and deposition):

$$O_3 + H + I^- \rightarrow HOI + O_2$$

$$H^+ + HOI + I^- \rightarrow I_2 + H_2O.$$

I^- in seaweed reacts with O_3 to produce I_2 gas, leading to atmospheric aerosol generation, resulting in cloud formation, which reflects sunlight partially offsetting the global compound which can be deposited in rain, and gaseous I_2 can be taken up by respiration.

This is a simplified version of the reactions involving iodide and other iodine oxides. O_3 in the lower atmosphere (troposphere) is the product of the reaction of chemical pollutants with sunlight and is a strong reactive oxygen species whose removal by seaweed is beneficial to human respiratory health. Different seaweeds grow at different tidal levels with the result that some will be uncovered by seawater for longer or shorter times which will also be determined by tidal conditions, high or low tides. Both living and decaying seaweeds release iodine not only into the atmosphere but also into the surrounding seawater, thus contributing to global iodine fluxes.

Coastal seaweed-rich areas are responsible for most of the iodine although oceanic phytoplankton also plays a role. Iodine vapor condenses to form nuclear clusters which can coalesce to produce new particles leading to aerosol and cloud formation, affecting climate change. The clouds so formed provide a heat shield, contributing to global cooling by partially offsetting global warming due to greenhouse gases. Gaseous iodine formed over a seaweed bed is often short lived and subject to photolysis by sunlight and dispersion by wind. Thus, the highest concentration of gaseous iodine is found over or close to the seaweed bed after dark on a calm night. In addition to iodine released from seaweeds, it appears that iodine deposition as rain has increased in recent times. An increase in atmospheric iodine over the North Atlantic has been recently reported to have a threefold increase in iodine deposition in Europe since 1950. Increasing emissions of pollutants O_3 and NO_2 react with seaweed-derived I^- to produce I_2 and iodine oxides including $IONO_2$ which is very water soluble and thus deposited in rain. The combination of atmospheric and deposited iodine will make a contribution to iodine flux.

10.9 Conclusions

According to the WHO, iodine deficiency disorders (IDD) have been identified in 130 countries, and 2.2 billion people, or approximately 38% of the world's population, live in iodine-deficient regions (Carlsen et al. 2018). The body of a healthy person (weighing about 70 kg) contains 15–20 mg of iodine, the minimum daily requirement is 100–150 μg, and almost all of it is taken up by the thyroid gland and used to synthetize the thyroid hormone, while the remaining iodine (about 100–200 μg/L) is excreted in the urine (de Oliveira Campos et al. 2021; Iitaka 2004; Zou et al. 2021).

Adequate iodine status is crucial from the first trimester through to the second year of life. Endemic cretinism due to severe maternal iodine deficiency can result in mental retardation, spastic diplegia, and deaf-mutism. Poor iodine intake during infancy and in childhood may also affect neurodevelopment, somatic growth, and cognitive and motor function.

The WHO, UNICEF, and the International Council for Control of Iodine Deficiency Disorders are actively engaged in the eradication of IDD; since iodine deficiency can be prevented by iodine supplementation (Çelmeli et al. 2020; Ershow et al. 2018; Lopes-Pereira et al. 2022). Commitment and motivation of countries' government are essential to eliminate IDD. Benefits, barriers, and harms of a prevention program for IDD have to be carefully analyzed (Schaffner et al. 2021a, b).

The One-Health approach in seaweed food production has a principle in which human, animal, and environmental health outcomes are connected (Bizzaro et al. 2022). In algae and seaweeds, the iodine concentration can reach up to 624.5 μg/g DW (e.g. *Laminaria digitata*). Iodine dose of 150 μg/day has been suggested to avoid thyroid impairment (Desideri et al. 2016; Nitschke 2016). High concentration of essential metals and iodine in the environment, their bioaccumulation in edible food, and feed may derive optimum benefits of iodine (Figure 10.12).

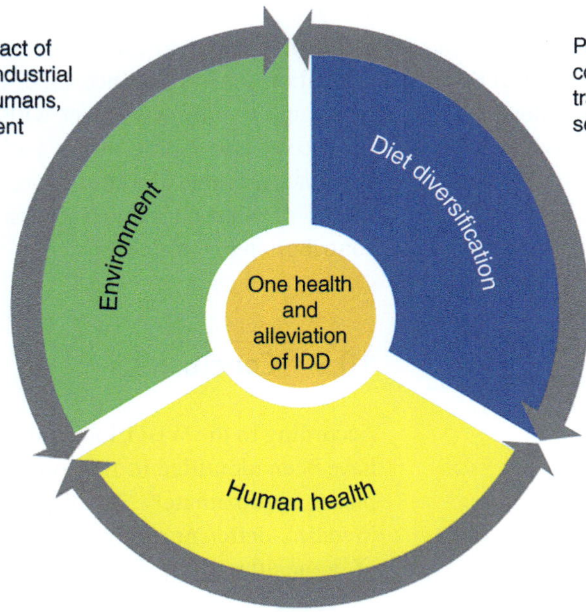

Figure 10.12 The One-Health approach in seaweed food production relevant to iodine essentiality for human health: sources, toxicity, and strategies for the alleviation of IDD.

Table 10.9 The iodine deficiency disorders, by age group.

Age groups	Health consequences of iodine deficiency
	Goiter, including toxic nodular goiter; increased occurrence of hypothyroidism in moderate-to-severe iodine deficiency; decreased occurrence of hypothyroidism in mild-to-moderate iodine deficiency; increased susceptibility of the thyroid gland to damage and thyroid cancer from iodine radioisotopes (e.g. from radioactive fall-out)
Fetus	Abortion, stillbirth, congenital anomalies, perinatal mortality
Neonate	Infant mortality, endemic cretinism
Child and adolescent	Impaired mental function, delayed physical development
Adults	Impaired mental function, iodine-induced hyperthyroidism
	Overall, moderate-to-severe iodine deficiency causes subtle but widespread adverse effects in a population secondary to hypothyroidism, including decreased educability, apathy, and reduced work productivity, resulting in impaired social and economic development

Source: Zimmermann (2008).

Table 10.10 Countries reported for insufficient iodine intake based on global scorecard of iodine nutrition in 2021 in the general population based on school-age children (SAC).

Country or territory	Median UIC (μg/L)	Date of survey	Data type	Population surveyed
Central African Republic*	21	1993	S	All
Estonia*	65	1995	N	SAC (8–10)
Finland	96	2017	N	Adults (25–74)
Germany	89	2014–2017	N	SAC, adolescents (3–17)

Table 10.10 (Continued)

Country or territory	Median UIC (μg/L)	Date of survey	Data type	Population surveyed
Haiti	77	2018	N	WRA (15–49)
Iraq	84	2011–2012	N	SAC (6–12)
Israel	83	2016	N	SAC
Korea, Democratic People's Republic of	97	2009–2010	N	SAC (6–12)
Lebanon	66	2013	N	SAC
Liechtenstein	96	2010	N	SAC (6–12)
Lithuania*	75	1995	N	SAC
Madagascar	46	2015	N	WRA (15–49)
Mali	69	2005	N	SAC (6–12)
Mozambique	97	2011–2012	N	WRA (15–49)
Nicaragua	90	2018	N	NPNLW (14–75)
Norway	75	2017–2018	S	WRA (18–30)
Russian Federation	<100	2008–2020	S	SAC
Samoa	88	2013	N	WRA (18–45)
South Sudan	94	2006	S	SAC (6–12)
Tajikistan	75	2016	N	WRA (15–49)
Ukraine*	90	2002	N	SAC (6–11)
Vanuatu	77	2017	N	WRA (15–49)
Viet Nam	84	2013–2014	N	SAC (8–10)

Source: Adapted from The Iodine Global Network, 2021.
N, Nationally representative survey; NPNLW, Nonpregnant, nonlactating women; P, Date published; S, Subnational survey (all administrative levels); SAC, School-age children; WRA, Women of reproductive age.
*Countries in which the most recent available survey is older than 15 years (i.e. conducted prior to 2005), are marked with an asterisk (*)

Table 10.11 Prevalence of iodine deficiency, as total number (millions) and percentages, in general population (all age groups) and in school-age children (6–12 year) in 2007 and the percentage of households with access to iodized salt.

WHO regions[a]	Population with urinary iodine 100 mg/L[b]		% of households with access to iodized salt[c]
	General population	School-age children	
Africa	312.9 (41.5%)	57.7 (40.8%)	66.6
	98.6 (11.0%)	11.6 (10.6%)	86.8
Americas			
Eastern Mediterranean	259.3 (47.2%)	43.3 (48.8%)	47.3
Europe	459.7 (52.0%)	38.7 (52.4%)	49.2
Southeast Asia	503.6 (30.0%)	73.1 (30.3%)	61.0
Western Pacific	374.7 (21.2%)	41.6 (22.7%)	89.5
Total	2000.0 (30.6%)	263.7 (31.5%)	70.0

[a] 193 WHO Member States.
[b] Based on the population estimates for 2006 (United Nations, Population Division, World Population Prospects: the 2004 revision).
[c] These figures do not include data for non-United Nations International Children's Emergency Fund (UNICEF) countries (e.g. the United States and Western Europe). *Source:* Zimmermann (2008).

Acknowledgments

The author thankfully acknowledge for according license for reuse of Figures 10.1, 10.4, and 10.5 (Elsevier); Figure 10.5; Tables 10.6–10.8 (Dr Helena Čermáková); Tables 10.9 and 10.10 (MDPI), and Table 10.11 (Karger Publishers).

References

Aakre, I., Solli, D.D., Markhus, M.W. et al. (2021). Commercially available kelp and seaweed products – valuable iodine source or risk of excess intake? *Food & Nutrition Research 65*: 7584. https://doi.org/10.29219/fnr.v65.7584.

Abdelrahman, A., Salih, L.M.A., and Saeed, E. (2020). Knowledge, attitude, and practice of iodized salt use in Al-Riyadh and Al-Ozozab areas, Khartoum, Sudan. *Sudanese Journal of Paediatrics* 20 (1): 26–33.

Aboyede, O.J. (2016). *National Bulletin on Main Nutrients Deficiencies Control in Nigeria*, 34–40. Aramed Pub. Co.

Aktaş, K., Gökçay, A., Canpolat, A. et al. (2021). Intensifying iodine deficiency throughout trimesters of pregnancy in a borderline iodine-sufficient urban area, Ankara, Turkey. *Biological Trace Element Research* 1: 1–6.

Anil, N.S., Deepthi, R., Pratibha, K. et al. (2021). Prevalence of iodine deficiency disorders (IDD) among school going children in rural areas of Bangalore. *The Indian Journal of Nutrition and Dietetics* 58 (2): 213–221.

Apaydin, M., Demirci, T., Başer, Ö.Ö. et al. (2021). The effects of salt consumption habits on iodine status and thyroid functions during pregnancy. *Turkish Journal of Medical Sciences* 51 (2): 766–771.

Arrizabalaga, J.J., Jalon, M., Espada, M. et al. (2015). Iodine concentration in ultra-high temperature pasteurized cow's milk. Applications in clinical practice and in community nutrition. *Medicina Clínica* 145 (2): 55.

Ashabi, A., Heydarieh, S.A., Monjazeb Marvdashti, L. et al. (2021). Iodized salt: assessment of nutritional status, iodine intake and iodine exposure. *Journal of Chemical Health Risks* 11 (3): 245–253.

Atmis, V., Bülbül, B., Bahşi, R. et al. (2021). Iodine concentration and prevalence of thyroid disease in older people after salt iodization in Turkey. *Eastern Mediterranean Health Journal* 27 (2): 151–158.

Bamidele, H.A., Hussaina, T. and Adegboye, Y., 2021. Prevention and control of iodine deficiency disorders (IDD) among the people of Nasarawa local government of Kano State. *Bakolori Journal of General Studies* 11(1): 1–2.

Bamidele, H.A., Tanko Hussaina, O., and Adegboye, Y. (2021). Prevention and control of iodine deficiency disorders (IDD) among the people of Nasarawa local government of Kano state. *Bakolori Journal of General Studies* 11 (1): 2999–3005.

Bertinato, J. (2021). Iodine nutrition: disorders, monitoring and policies. *Advances in Food and Nutrition Research* 96: 365–415.

Bizzaro, G., Vatland, A.K., and Pampanin, D.M. (2022). The one-health approach in seaweed food production. *Environment International* 158: 106948.

Carlsen, M.H., Andersen, L.F., Dahl, L. et al. (2018). New iodine food composition database and updated calculations of iodine intake among Norwegians. *Nutrients* 10 (7): 930.

Çelmeli, G., Çürek, Y., Küçükçetin, İ.Ö. et al. (2020). The results of 16 years of iodization: assessment of iodine deficiency among school-age children in Antalya, Turkey. *Journal of Clinical Research in Pediatric Endocrinology* 12 (3): 256.

Cengiz, M.F., Sen, F., Bilgin, A.K., and Boyaci-Gunduz, C.P. (2022). Determination of exposure to major iodide ion uptake inhibitors through drinking waters. *Environmental Research* 204: 112345.

Chaboki, B.G., Tabrizi, M., Meymeh, M.H. et al. (2021). Mapping the relative risk of congenital hypothyroidism incidence via spatial zero-inflated poisson model in Guilan province, Iran. *International Journal of Preventive Medicine* 12 (1): 53.

Çolak, E., Özkan, B., Genç, S., and Polat, B. (2021). Ultrasonographic determination of thyroid volume in infants and children from Aegean region of Turkey and comparison with national and international references. *Journal of Pediatric Endocrinology and Metabolism* 34 (4): 457–464.

De la Vieja, A. and Santisteban, P. (2018). Role of iodide metabolism in physiology and cancer. *Endocrine-Related Cancer* 5 (4): R225–R245.

Delange, F. (2000). The role of iodine in brain development. *The Proceedings of the Nutrition Society* 59 (1): 75–79.

Delange, F. and Hetzel, B. (2008). The iodine deficiency disorders. In: *The Thyroid and Its Diseases* (ed. L.E. DeGroot and G. Hannemann). http://www.thyroidmanager.org/S (accessed 18 January 2008).

Desideri, D., Cantaluppi, C., Ceccotto, F. et al. (2016). Essential and toxic elements in seaweeds for human consumption. *Journal of Toxicology and Environmental Health, Part A 79* (3): 112–122.

Dunn, J.T. and Van der Haar, F. 1990. A practical guide to the correction of iodine deficiency. Report of the International Council for the Control of Iodine Deficiency Disorders, Netherlands, 62.

Edmunds, W.M. and Smedley, P.L. (1996). Groundwater geochemistry and health: an overview. In: *Environmental Geochemistry and Health*, vol. 113 (ed. J.D. Appleton, R. Fuge and G.J.H. McCall), 91–105. Geological Society (London) Special Publications.

Ershow, A.G., Skeaff, S.A., Merkel, J.M., and Pehrsson, P.R. (2018). Development of data bases on iodine in foods and dietary supplements. *Nutrients* 10 (1): 100.

Flachowsky, G., Schone, F., Leiterer, M. et al. (2007). Influence of an iodine depletion period and teat dipping in the iodine concentration of serum and milk of cows. *Journal of Animal and Feed Sciences* 16(1): 18.

Flachowsky, G., Franke, K., Meyer, U., Leiterer, M. and Schöne, F. (2014). Influencing factors on iodine content of cow milk. *European journal of nutrition*, 53(2): 351–365.

Fuge, R. and Johnson, C.C. (2015). Iodine and human health, the role of environmental geochemistry and diet, a review. *Applied Geochemistry 63*: 282–302.

Gomes, C. and Rautureau, M. (2021). *Minerals Latu Sensu and Human Health: Benefits, Toxicity and Pathologies*. Springer International Publishing AG.

Gorstein, J.L., Bagriansky, J., Pearce, E.N. et al. (2020). Estimating the health and economic benefits of universal salt iodization programs to correct iodine deficiency disorders. *Thyroid* 30 (12): 1802–1809.

Grau, G., Aguayo, A., Vela, A. et al. (2015). Normal intellectual development in children born from women with hypothyroxinemia during their pregnancy. *Journal of Trace Elements in Medicine and Biology* 1 (31): 18–24.

Greenwood, N.N. and Ernshaw, A. (2005). *Chemistry of the Elements*. Amsterdam: Elsevier 1342 pp.

Habib, A., Shojazadeh, A., Molayemat, M. et al. (2021). Prevalence and predictive factors of transient and permanent congenital hypothyroidism in Fars province, Iran. *BMC Pediatrics* 21 (1): 1–9.

Hanke, S.H. and Boland, J.J. (1971). Water requirements or water demands? *American Water Works Association* 63 (11): 677–681.

Hashemipour, M., Kelishadi, R., Amin, M.M. et al. (2021). The association between familial and environmental factors and prevalence of congenital hypothyroidism in center of Iran. *Environmental Science and Pollution Research* 28 (7): 8434–8441.

Hatch-McChesney, A. and Lieberman, H.R. (2022). Iodine and Iodine Deficiency: A Comprehensive Review of a Re-Emerging Issue. *Nutrients* 14(17): 3474.

Herod, M.N., Li, T., Pellerin, A. et al. (2016). The seasonal fluctuations and accumulation of iodine-129 in relation to the hydrogeochemistry of the Wolf Creek Research Basin, a discontinuous permafrost watershed. *Science of the Total Environment* 559: 1212–1223.

Iannaccone, M., Ianni, A., Elgendy, R. et al. (2019). Iodine supplemented diet positively affect immune response and dairy product quality in fresian cow. *Animals*, 9(11): 866.

Iitaka, M. (2004). Iodine deficiency disorder and clinical practice. *Japan Medical Association Journal* 47(8): 371–375.

Johner, S.A., von Nida, K., Jahreis, G. and Remer, T. (2012). Time trends and seasonal variation of iodine content in German cow's milk--investigations from Northrhine-Westfalia. *Berliner und Munchener Tierarztliche Wochenschrift*, 125(1–2): 76–82.

Jooste, P.L., Badenhorst, C.J., and Truesdale, V.W. (1882). Determination of total iodine and iodate-iodine in natural freshwaters. *Analyst* 107: 1417–1424.

Kabata-Pendias, A. (2011). *Trace Elements in Soils and Plants*, 4e, 548. Boca Raton, FL: CRC Press is an imprint of Taylor & Francis Group, an Informa Business.

Kart, P.Ö., Türkmen, M.K., Anık, A. et al. (2021). The association of lactating mothers' urinary and breast milk iodine levels with iodine nutrition status and thyroid hormone levels of newborns. *Turkish Archives of Pediatrics* 56 (3): 207.

Kennedy, G., Nantel, G., and Shetty, P. (2003). The scourge of 'Hidden hunger': global dimensions of micronutrient deficiencies. In: *Food, Nutrition and Agriculture*, 8–14. Rome: FAO.

Kobat, S.G., Gül, F.C., and Turgut, B. (2020). Iodine-induced retinopathy: a case report. *Turkish Journal of Ophthalmology* 50 (4): 255.

Koukkou, E.G., Roupas, N.D., and Markou, K.B. (2017). Effect of excess iodine intake on thyroid on human health. *Minerva Medica* 108 (2): 136–146.

Koutras, DA, Matovinovic, J, Vought, R. (1985). The ecology of iodine. In: Stanbury J.B., Hetzel B.S. (eds). Endemic goitre and cretinism, iodine nutrition in health and disease. *Wiley Eastern Limited New York*: 185–95.

Krela-Kaźmierczak, I., Czarnywojtek, A., Skoracka, K. et al. (2021). Is there an ideal diet to protect against iodine deficiency? *Nutrients* 13 (2): 513.

Kuehn, B. (2018). Iodine deficiency may impair fertility. *JAMA* 319 (8): 760–760.

Larsen, L.C. and Cummings, D.M. (1998). Oral poisonings: guidelines for initial evaluation and treatment. *American Family Physician* 57 (1): 85–92.

Laurberg, P., Nøhr, S.B., Pedersen, K.M. et al. (2000). Thyroid disorders in mild iodine deficiency. *Thyroid* 10 (11): 951–963.

Lazarus, J. (2014). Iodine status in Europe in 2014. *European Thyroid Journal* 3 (1): 3–6.

Lazarus, J.H. (2021). Monitoring iodine nutritional status: adults or school children? *Journal of Endocrinological Investigation* 44: 383–385.

Lee, S.Y. (2021). Maternal iodine supplementation during lactation may improve child's cognitive development at 3 years of age in a mild-to-moderately iodine-deficient population. *Clinical Thyroidology* 33 (7): 302–304.

Lee, S.Y. and Pearce, E.N. (2022). Assessment and treatment of thyroid disorders in pregnancy and the postpartum period. *Nature Reviews Endocrinology* 1–14. https://doi.org/10.1038/s41574-021-00604-z.

Lopes-Pereira, M., Quialheiro, A., Costa, P. et al. (2022). Iodine supplementation: compliance and association with adverse obstetric and neonatal outcomes. *European Thyroid Journal* 11 (1): https://doi.org/10.1530/ETJ-21-0035.

Mann, J. and Aitken, E. (2003). The re-emergence of iodine deficiency in New Zealand? *New Zealand Medical Journal* 14: 116.

Masilamani, D., Arasan, D., Pyarejan, K., Venkatasamy, S., Kalyanasundaram, K., Jeganathan, S., Krishnasamy, A. and Manjunathan, R., 2021. Assessment of iodine nutritional status in school children concerning autoimmune thyroiditis in Tamil Nadu, India. *Indian Journal of Child Health*. https://doi.org/10.32677/IJCH.2021.v08.i02.006 *8*(2), pp.90–94.

Mathews, D.M., Johnson, N.P., Sim, R.G. et al. (2021). Iodine and fertility: do we know enough? *Human Reproduction* 36 (2): 265–274.

Mikláš, Š., Tančin, V., Toman, R., and Trávníček, J. (2021). Iodine concentration in milk and human nutrition: a review. *Czech Journal of Animal Science* 66: 189.

Milewski, A.V. and Diamond, R.E. (2000). Why are very large herbivores absent from Australia? A new theory of micronutrients. *Journal of Biogeography* 27: 957–978.

Moran, J.E., Oktay, S.D., and Santschi, P.H. (2002). Sources of iodine and Iodine 129 in rivers. *Water Resources Research* 38 (8): 1149–1159.

Muramatsu, Y. and Yoshida, S. (1999). Effects of micro-organisms on the fate of iodine in the soil environment. *Geomicrobiology Journal* 16: 85–93.

Nanri, A., Mizoue, T., Shimazu, T. et al. (2017). Dietary patterns and all-cause, cancer, and cardiovascular disease mortality in Japanese men and women: the Japan public health center-based prospective study. *PLoS One 12* (4): e0174848.

Neal, C., Neal, M., Wickham, H. et al. (2007). Dissolved iodine in rainfall, cloud, stream and groundwater in the Plynlimon area of mid-Wales. *Hydrology and Earth System Sciences* 11 (1): 283–293.

Nettore, I.C., Colao, A., and Macchia, P.E. (2018). Nutritional and environmental factors in thyroid carcinogenesis. *International Journal of Environmental Research and Public Health* 15 (8): 1735. (PMC free article: PMC6121258).

Ning, P., Ren, Q., Teng, D. et al. (2020). Current iodine nutrition status and prevalence of thyroid disorders in Tibetan adults in an oxygen-deficient plateau, Tibet, China: a population-based study. *Thyroid* 30 (5): 759–766.

Nitschke, U. (2016). Quantification of iodine loss in edible Irish seaweeds during processing. *Journal of Applied Phycology* 28 (6): 3527–3533.

de Oliveira Campos, R., Lima, S.C.R., de Souza Braga Filho, J. et al. (2021). Association of salt iodization and urine iodine concentration in schoolchildren from public schools in Northeast of Brazil. *Biological Trace Element Research* 199 (12): 4423–4429.

de Oliveira Campos, R., Lima, S.C.R., de Souza Braga Filho, J.et al. (2021). Association of Salt Iodization and Urine Iodine Concentration in Schoolchildren from Public Schools in Northeast of Brazil. *Biological Trace Element Research*, 199(12): 4423–4429.

Olivieri, A., Angelis, S.D., Moleti, M., and Vermiglio, F. (2021). Iodine deficiency and thyroid function. In: *Thyroid, Obesity and Metabolism* (ed. L. Luzi), 3–20. Cham: Springer.

Pearce, E.N. (2018). Iodine nutrition: recent research and unanswered questions. *European Journal of Clinical Nutrition* 72 (9): 1226–1228.

Pearce, E.N., Lazarus, J.H., Moreno-Reyes, R., and Zimmermann, M.B. (2016). Consequences of iodine deficiency and excess in pregnant women: an overview of current knowns and unknowns. *The American Journal of Clinical Nutrition* 104 (Suppl 3): 918S–923S.

Pehrsson, P., Spungen, J., Ershow, A. et al. (2021). USDA, NIH and FDA iodine database of US foods for estimating iodine intakes. Current developments. *Nutrition* 5 (Supplement_2): 175–175.

Pehrsson, P.R., Roseland, J.M., Patterson, K.Y. et al. (2022). Iodine in foods and dietary supplements: a collaborative database developed by NIH, FDA and USDA. *Journal of Food Composition and Analysis* 109: 104369.

Pennington, J.A. (1990). A review of iodine toxicity reports. *Journal of the American Dietetic Association* 90 (11): 1571–1581.

Phaneuf, D., Cote, I., Dumas, P. et al. (1999). Evaluation of the contamination of marine algae (seaweed) from the St. Lawrence River and likely to be consumed by humans. *Environmental Research* 20: S175–S182.

Prasad, M.N.V. ed., (2020). Disinfection By-products in Drinking Water: Detection and Treatment. Butterworth-Heinemann. 464.

Punyani, S., Narayan, P., Singh, H., and Vasudevan, P. (2006). Iodine based water disinfection: a review. *Journal of Scientific and Industrial Research* 65: 116–120.

Radlinger, G. and Huemann, K.G. (2000). Transformation of iodide in natural and wastewater systems by fixation on humic substances. *Environmental Science & Technology* 34: 3932–3936.

van der Reijden, O.L., Galetti, V., Herter-Aeberli, I. et al. (2019). Effects of feed iodine concentrations and milk processing on iodine concentrations of cows' milk and dairy products, and potential impact on iodine intake in Swiss adults. *The British Journal of Nutrition* 122 (2): 172–185.

Rezaei Ahvanooei, M.R. (2022). Iodine, an effective substance against the COVID-19 pandemic. *Austin Journal of Veterinary Science & Animal Husbandry* 9 (1): 1086.

Ruggeri, R.M. and Trimarchi, F. (2021). Iodine nutrition optimization: are there risks for thyroid autoimmunity? *Journal of Endocrinological Investigation* 1–9. https://doi.org/10.1007/s40618-021-01548-x.

Sack, J., Kaiserman, I., Tulchinski, T. et al. (2000). Geographic variation in groundwater iodine and iodine deficiency in Isreal, the West bank and Gaza. *Journal of Pediatric Endocrinology and Metabolism* 13: 185–190.

SCF – Scientific Committee on Food (2002). *Opinion of the Scientific Committee on Food on the Tolerable Upper Intake Level of Iodine*. Brussels, Belgium: European Commision – Health and Consumer Protection Directorate General 25 p.

Schaffner, M., Mühlberger, N., Conrads-Frank, A. et al. (2021a). Benefits and harms of a prevention program for iodine deficiency disorders: predictions of the decision-analytic EUthyroid model. *Thyroid* 31 (3): 494–508.

Schaffner, M., Rochau, U., Stojkov, I. et al. (2021b). Barriers against prevention programs for iodine deficiency disorders in Europe: a Delphi study. *Thyroid* 31 (4): 649–657.

Schone, F. and Rajendram, R. (2009). Iodine in farm animals. In: *Comprehensive Handbook of Iodine: Nutritional, Biochemical, Pathological and Therapeutic Aspects* (ed. V.R. Preedy, G.N. Burrow and R. Watson), 151–170. Amsterdam: Academic Press.

SCN (2004). Nutrition to improve development outcomes. *5th report on the World Nutrition situation. UN standing committee on Nutrition*, Geneva Standing Committee on Nutrition (SCN).

Sharma, N., Karanfil, T. and Westerhoff, P. (2019). Historical and future needs for geospatial iodide occurrence in surface and groundwaters of the United States of America. *Environmental Science & Technology Letters* 6(7): 379–388.

Smyth, P.P.A. (2021). Iodine, seaweed, and the thyroid. *European Thyroid Journal* 10: 101–108.

Sorokman, T., Bachu, M., and Sokolhyk, I. (2021). Frequency of iodine deficiency disorders among children living in the iodine deficiency region. *Romanian Journal of Pediatrics* 70 (1): 44–52.

Soudek, P., Tykva, R., Vaňková, R., and Vaněk, T. (2006). Accumulation of radioiodine from aqueous solution by hydroponically cultivated sunflower (*Helianthus annuus* L.). *Environmental and Experimental Botany* 57 (3): 220–225.

Southern, A.P. and Jwayyed, S. (2021). Iodine toxicity. *StatPearls (Internet)*.

Sprague, M., Chau, T.C., and Givens, D.I. (2022). Iodine content of wild and farmed seafood and its estimated contribution to UK dietary iodine intake. *Nutrients* 14 (1): 195.

Stanbury, J.B., Ermans, A.E., Bourdoux, P., Todd, C., Oken, E., Tonglet, R. (1998). Iodine-induced hyperthyroidism: occurrence and epidemiology. *Thyroid* 8: 83–100.

Stevenson, M.C., Drake, C., and Givens, D.I. (2018). Further studies on the iodine concentration of conventional, organic and UHT semi-skimmed milk at retail in the UK. *Food Chemistry* 15 (239): 551–555.

Szymandera-Buszka, K., Waszkowiak, K., Kaczmarek, A., and Zaremba, A. (2021). Wheat dietary fibre and soy protein as new carriers of iodine compounds for food fortification – the effect of storage conditions on the stability of potassium iodide and potassium iodate. *Lebensmittel-Wissenschaft & Technologie* 137: 110424.

Tako, E. (2022). Essential minerals: nutritional requirements, dietary sources, and deficiencies. In: *Nutrition Guide for Physicians and Related Healthcare Professions*, 365–376. Cham: Humana.

Todd, C.H., Allain, T., Gomo, Z.A. et al. (1995). Increase in thyrotoxicosis associated with iodine supplements in Zimbabwe. *Lancet* 346: 1563–1564.

Turianica, I. and Rothová, M. (2021). Symptoms and prevention of the iodine deficiency. *Journal of Microbiology, Biotechnology and Food Sciences* 2021: 803–812.

Vitti, P., Rago, T., Aghini-Lombardi, F. and Pinchera, A. (2001). Iodine deficiency disorders in Europe. *Public Health Nutrition* 4(2b): 529–535.

Vural, M., Koc, E., Evliyaoglu, O. et al. (2021). Iodine status of Turkish pregnant women and their offspring: a national cross-sectional survey. *Journal of Trace Elements in Medicine and Biology* 63: 126664.

Walther, B., Wechsler, D., Schlegel, P., and Haldimann, M. (2018). Iodine in Swiss milk depending on production (conventional versus organic) and on processing (raw versus UHT) and the contribution of milk to the human iodine supply. *Journal of Trace Elements in Medicine and Biology* 1 (46): 138–143.

Yordam, N., Özön, A., Alikaşifoğlu, A. et al. (1999). Iodine deficiency in Turkey. *European Journal of Pediatrics* 158 (6): 501–505.

Zeru, A.B., Muluneh, M.A., Giorgis, K.K.H. et al. (2021). Iodine deficiency disorder and knowledge about benefit and food source of iodine among adolescent girls in the North Shewa zone of Amhara region. *Journal of Nutrition and Metabolism* 2021: https://doi.org/10.1155/2021/8892180.

Zimmermann, M.B. (2008). Iodine requirements and the risks and benefits of correcting iodine deficiency in populations. *Journal of Trace Elements in Medicine and Biology* 22(2): 81–92.

Zimmermann, M.B. (2009a). Iodine deficiency. *Endocrine Reviews* 30 (4): 376–408.

Zimmermann, M.B. (2009b). Iodine deficiency in pregnancy and the effects of maternal iodine supplementation on the offspring: a review. *The American Journal of Clinical Nutrition* 89 (2 Suppl): 668S–672S.

Zimmermann, M.B. (2011). The role of iodine in human growth and development. *Seminars in Cell & Developmental Biology* 22 (6): 645–652.

Zimmermann, M.B. and Galetti, V. (2015). Iodine intake as a risk factor for thyroid cancer: a comprehensive review of animal and human studies. *Thyroid Research* 8: 8. (PMC free article: PMC4490680).

Zimmermann, M.B., Connolly, K., Bozo, M. et al. (2006). Iodine supplementation improves cognition in iodine-deficient schoolchildren in Albania: a randomized, controlled, double-blind study. *The American Journal of Clinical Nutrition* 83 (1): 108, 14.

Zou, Y., Li, H., Pang, J. et al. (2021). An evaluation of urine and serum iodine status in the population of Tibet, China: no longer an iodine-deficient region. *Nutrition* 82: 111033.

11

Understanding Nexus Between Hydrogeochemical Cycling and Medical Geology of Arsenic

Muhammad Mahroz Hussain[1], Natasha[2], Irshad Bibi[1], Hamna Bashir[1], Muhammad Shahid[2], and Nabeel Khan Niazi[1]

[1] *Institute of Soil and Environmental Sciences, University of Agriculture Faisalabad, Faisalabad, Pakistan*
[2] *Department of Environmental Sciences, COMSATS University Islamabad, Vehari Campus, Vehari, Pakistan*

11.1 Introduction

Arsenic (As) is an abundant element in soils, sediments, waters, and the environment (Bhattacharya et al. 2007; Hussain et al. 2019a). Contamination of groundwater with As is a global environmental issue due to its distinct toxicity and carcinogenicity. It is a Class I human carcinogen and it belongs to Group V of the periodic table (Crabbe et al. 2017). Arsenic is the 20th most abundant element found in the earth's crust, and its concentration in natural water is found from 0.5 to 5000 mg/L. In freshwater, its concentration ranges from <1 to <10 μg/L (Smedley 2008). Arsenite (As(III)) and arsenate (As(V)) are inorganic, toxic, and more mobile than organic species of As (e.g. monomethyl As(V) (MMAA)) (Shakoor et al. 2016). Inorganic As is mainly present in the form of As(III) and As(V), and its abundance depends on pH and redox potential, iron, and manganese oxides (Wang et al. 2018). Arsenic predominates at high pH (>7.5), while As(V) at low pH (Niazi et al. 2011).

Arsenic is debated whether it is essential, or beneficial to human health, but widely known for its toxicity and carcinogenicity (Yang et al. 2016). Bangladesh has the "worst population poisoning" in Asia, according to the World Health Organization (WHO) (Singh et al. 2015). Before the 1990s, the book Atlas of Endemic Diseases in the People's Republic of China and its Environment was the most thorough study of endemic disease prevalence and its link to environmental circumstances in China (Li et al. 2012). Due to this endemic, Chinese government and researcher increasingly concerned about endemic illnesses caused by metal(liods) and have developed a number of efficient treatments and remedation strategies.

Furthermore, daily intake of As-contaminated drinking water can have acute and chronic health effects. As in the case of Southeast Asian countries, groundwater As contamination in Pakistan (<5–916 μg/L) (Nickson et al. 2005; Shakoor et al. 2018), India (200 μg/L) (Bhattacharya et al. 2013), Bangladesh (300 μg/L) (Rahman et al. 2014), Vietnam (3050 μg/L) (Berg et al. 2007), Cambodia (159 μg/L) (Berg et al. 2007), the United States (1300 μg/L) (Robertson 1989), and Nepal is said to endanger human health (Azizullah et al. 2011) (Table 11.1). Low levels of As exposure can cause gastric irritation, reduced red and white blood cells, arrhythmia, vascular injury, and both hands "needling" influenza (Rosen and Liu 2009). Due to increasing As concentration in the terrestrial and aquatic environments, chronic exposures may lead to cancer, cardiovascular diseases, foot diseases, hyperkeratosis, melanosis, and neuropathy (Chakraborti et al. 2003; Fatmi et al. 2009; Ali et al. 2018). The main purpose of this chapter is to develop the research understanding, and progress of the human health impacts of As, particularly from the perspective of medical geology.

11.2 What Is Medical Geology?

For better understanding of rock formation and googology, soil, and groundwater in human and animal health, geoscientists, mineralogists, and health experts have to collaborate to explore the geogenic contamination potential of geogenic contaminants. The study is thus required to develop the link between natural geological cycling and human health, as well as the impact of environmental elements on the geographic distribution of health and

Medical Geology: En route to One Health, First Edition. Edited by Majeti Narasimha Vara Prasad and Meththika Vithanage.

Table 11.1 Elevated arsenic concentration effects on human and plants.

Humans					
Acute	**Reference**	**Chronic**	**Reference**	**Plants**	**Reference**
Headache	Mohan and Pittman (2007)	Human carcinogen	Cardoso et al. (2018), Zhou and Xi (2018)	Disturb enzyme function	Mishra et al. (2016)
Malaise	Pathak and Pathak (2018)	Skin cancer	Hoque et al. (2017)	Loss of turgor	Chandrakar et al. (2018)
Abdominal pain	Zhu et al. (2017)	Lung cancer	Doll (2017)	Wilting	Jasrotia et al. (2017)
Nausea	Tan et al. (2015)	Bladder cancer	Koutros et al. (2018)	Alteration in membrane integrity	Kharroubi et al. (2017)
Vomiting	Tan et al. (2015)	Liver cancer	Sadaf et al. (2018)	Membrane degradation	Boehme et al. (2016)
Acute hemolysis	Guo et al. (2019)	Prostate cancer	Wang et al. (2017)	Chlorosis	Wang et al. (2018)
Thickening of the skin	Tanga et al. (2016)	Diabetes	Muñoz et al. (2018)	Affects phosphorylation	Li (2017)
Dark skin	Mohammed et al. (2017)	Neurological disorders	Brown and Sen (2017)	Defoliation	Kofroňová et al. (2018)
Abdominal pain	Mohan and Pittman (2007)	Cardiac disorders	Emi (2017)	Guard cells, tissue death, and fruit drop-off	Xue and Yi (2017)
Numbness	Sharma et al. (2016)	Reproductive organs failure	Khatun et al. (2018)	Reduced yield	Iqbal et al. (2016)
>200 enzymes inactivation	Ratnaike (2003)	Eyes burning	Bhattacharya et al. (2015)	Stunted growth	Ali et al. (2016)
Muscle cramping	Zhang (2016)	Cough	Ergün et al. (2017)	Necrosis	Gonçalves et al. (2017)
Weakness	Sharma et al. (2016)	Heamoptysis	Patir et al. (2016)	Inhibit phosphatase enzymes	Chen and Lee (2016)
dizziness	Ishii et al. (2018)	Dyspnea	Sanchez et al. (2018)	Affects metabolic processes	Xiao et al. (2016)
Muscle and bone pain	Mohan and Pittman (2007)	Paresthesia	Mohan and Pittman (2007)	Cause oxidative damage	Singh and Singh (2018)
Insomnia	Niño et al. (2021)	Leukopenia	Ally et al. (2016)	Sterility of the florates/ spikelets in rice	Li et al. (2017)
Rigors	Haghi et al. (2022)	Proteinuria	Cheng et al. (2020)	Damages the chloroplast membrane	Gupta et al. (2016)
Loss of appetite	Hasanato and Almomen (2015)	Inflammation of respiratory mucosa	Torjussen (2017)	Disorganizes the functions of integral photosynthetic process	Chandrakar et al. (2016)
Skin lesions	Niedzwiecki et al. (2018)	Short-term memory	Hasanato and Almomen (2015)	Decrease in pigment synthesis	Anjum et al. (2017)
Slurred speech capability	Kühn et al. (2016)	Tremors	Alvarez et al. (2018)	Oxidative stress	Singh and Singh (2018)

disease, and is referred to as medical geology (Carretero et al. 2013). As a result, medical geology is a large and complicated subject that need interdisciplinary contributions from a variety of scientific disciplines in order to comprehend, lessen, or resolve problems (Goovaerts and Glass 2014).

The realization that small amounts of inorganic elements are necessary for human health, but that excess can be harmful, caught modernist attention in the nineteenth century. The nineteenth-century "Fowler's solution" contained As, and excessive use caused skin ulcers and cancer: Black feet disease from As-contaminated drinking water in Taiwan has been reported to have major impact on human population (Ziarati et al. 2021). Itai-itai disease in Japan was linked to Cd contamination of food and water in the mid-twentieth century, while methylmercury poisoning in Japan is known as Minamata disease. It was soon discovered that microorganisms and dissolved organic matter can convert mercury into fat-soluble methylmercury, which is found in higher concentrations in the food chain (Ziarati et al. 2021).

In recent years, there has been a lot of discussion about the occurrence and distribution of high levels of natural As throughout groundwater in Southeast Asia, particularly in heavily populated river deltas (Herath et al. 2016). Bangladesh, West Bengal (India), China, Vietnam, Taiwan, Thailand, and Nepal are among the nations in the region where As-related disorders are prevalent (Smedley 2008). Natural pollution, such as As, can be found in the Ganges–Megna–Brahmaputra Basin in South Asia. Groundwater is a major source of inorganic As intake by humans in Southeast Asia. More than 100 million people in the region are considered at risk of As poisoning in groundwater, and more than 700 000 people are affected by As-related diseases (Herath et al. 2016). In rural areas of the region, most groundwater used for drinking water has been found to be contaminated with As in concentrations that exceed the World Health Organization's drinking water limit of 10 g/L. In Bangladesh, groundwater As concentrations were higher in aquifers covered by finer sediments and lower in aquifers with permeable sand material or a thin silt/clay layer overlying them.

Within shallow layers (15 m), this near-surface lithology determines the geographical distribution of groundwater As. Significant amounts of As (>10 g/L, WHO recommended guideline) were found in groundwater in 14 out of 19 districts in West Bengal, including the Kolkata metropolitan area (India) (Shakoor et al. 2018). In West Bengal, more than 9.2 million people drink contaminated groundwater (>10 g/L). In a study of 4780 tubular well water samples from three districts of Uttar Pradesh (Ballia, Varanasi and Ghazipur in the upper and middle Ganges of India), 46.5% of the samples contained As concentrations above 10 and 50 g/L in 26.7% and 300 g/L in 10% of the samples (maximum As content, 3192 g/L).

11.3 Arsenic Release Mechanisms

Millions of people are at As toxicity stake due to limitation of clean water availability and lack of infrastructure to remediate As-contaminated groundwater. It is a need of an hour to understand the key geo-environmental resources which are contaminating and releasing the As and influencing its mobility into the groundwater. It is very complex to study the geochemistry of As because of its variable oxidation state, speciation, and redox potential. The release of As mechanism in groundwater rocks or sediments is related to the magnitude of As in the groundwater system (Ormachea Muñoz et al. 2016; Shakoor et al. 2017; Shahid et al. 2018).

A large number of population is at As peak of As contamination risk due to the usage of the As-contaminated surface water, and reports indicate that the concentration is crossing the WHO limits (As >10 μg/L) (Table 11.2). Usually in the earth crust, As average concentration ranges between 1 and 2 mg/kg which is very low (Hussain et al. 2020). However, in life-threatening environments, the main processes which include sorption, redox reactions, and sulfide mineral oxidation can be the availability of high release of As in the groundwater (Figure 11.1). Therefore, these geochemical processes not only cause the As release in the aquifer but also enhance the As concentration in the groundwater (Saifullah et al. 2018). The presence of As in the groundwater and deep wells is ultimately the source of accumulation of As in the respective soil. Iron oxides reductive dissolution and the oxidation of the sulfide in water body deposits are considered as the most important triggers for the release of As introduced the geochemically into the water (Ali et al. 2018). Best-known geochemical process is adsorption which is responsible for the transport and migration of the As into water bodies. It also controls the release and concentration of As in the water body. Its activity depends on many factors which include surface, pH, E_h, As concentration, common ions, and the morphology of As in the medium. A major source of As in groundwater may be the mineral oxides surfaces of Fe, Al, and Mn which are active in redox reactions. These oxides' adsorption properties are generally the result of the surfaces which contain hydroxyl groups. The adsorption properties of these oxides are generally the result of surface groups containing OH groups (Hussain et al. 2020, 2021).

Table 11.2 Arsenic concentrations in different countries of the world.

Country	Areas	Source of water	Depth (m)	Source	Arsenic content (µg/L)	References
Bangladesh	Rajshahi District	GW	100	Deltaic and flood plains	10–2 110	Chakraborti et al. (2010)
	Khulna	GW	100	Deposits of sediments	375–3 145	
	Barisal	GW	100	NG	10–1 170	
	Sylhet	GW	100	NG	100–300	
	Dhaka	GW	100	NG	224–1750	
	Chittagong	GW	100	NG	210–4730	
	Ganges–Brahmaputra–Meghna river basin, coastal plains	GW	50	Sedimentary aquifers	2.5–846 239	Ahmed et al. (2004)
	Matlab	SHTW	24–48	Deltaic deposits	2–900	Hossain et al. (2014)
India	West Bengal, Delta plain (BDP)	SW	50	Alluvial deposits	1–1 300	Biswas et al. (2012)
	West Bengal, Gangetic plain (Barasat)	GW100	100	Alluvial deposits	10–3 200	Rahman et al. (2003), Rahman et al. (2014), Shankar et al. (2014)
	Uttar Pradesh, Bihar	SHTW 24–48	48	Alluvial deposits	10–538	Chakraborti et al. (2003, Ahamed et al. 2006)
	Kaudikasa, Rajnandgaon district	GWW	100	Igneous territory	Above 250	Shukla et al. (2010), Das et al. (2015)
Thailand	Ron Phibun	SGW	50	Mining and digging	01–5 000	Kim et al. (2011)
Taiwan	Chianan plains	GW	277		1.3–575	Maity et al. (2011)
	Lanyang plains	GW	277	Well-grained fluvial pledges	2.5–348	Maity et al. (2011)
	Coastal zones	GW	50	Black shales	10–1 820	Nordstrom (2002)
	Southern Choushui river alluvial fan	SGW	50	Aquifers and aquitards	=590	Liu et al. (2013)
Pakistan	Jamshoro and Sindh	SW, GW	100	Deposits	3.0–106	Baig et al. (2009)
Cambodia	Mekong river (Southern Vietnam)	GW	100	Iron phases in sediment aquifers	1–1 610	Berg et al. (2007)
Vietnam	Red river delta	GW	100	-do-	01–3 050	Berg et al. (2007)
Afghanistan	Ghazni metropolitan area	W 65	100	Aquifer deposits	10–500	Smedley (2008)
Canada	Nova Scotia	GW	100	NG	1.5–738	Nriagu et al. (2007)
	Northern Saskatchewan	SW	50	Uranium pits	9.6–71	Donahue and Hendry (2003)
Chile	—	GW	100	Volcanogenic deposits	100–1 000	Nordstrom (2002), Nriagu et al. (2007)

Mexico	Zimapan and la Lagunéra region	GW	100	Thermal springs, mining, volcanic deposits, and mining	08–620	Armienta et al. (1997), Rodriguez et al. (2004), Boochs et al. (2014)
Brazil	Minas Gerais	SW	50	Mining of gold	0.4–350	Mukherjee et al. (2006)
UK	Southwest England	GW	100	Mining	01–80	Nriagu et al. (2007)
Argentina	Monte Quemado of Cordoba province	DW	100	Volcanic eruption ash	Above 100	Bundschuh et al. (2012), Nicolli et al. (2012a), Smedley (2008), Bejarano Sifuentes and Nordberg (2003), Bundschuh et al. (2004), Smedley and Kinniburgh (2002), Smedley et al. (2005), Bhattacharya et al. (2006), Nicolli et al. (2012b), Aullón Alcaine et al. (2012), Panigatti et al. (2014)
	North western Argentina	GWW	100	Thermal springs	52–1045	
	Carcarana River	GW	100	NG	10–720	
	Cordoba	GW	100	NG	100–2000	
Lao People's Democratic Republic	Laos	TW	100	Alluvial and deltaic aquifers	Almost 277.8	Nordstrom (2002)
New Zealand	Taupo Volcanic Zone	SW	100	Volcanic eruption	=21	Robinson et al. (2006)
Japan	Fukuoka	GWW	100	Industrial trash consist of As	01–293	Mukherjee et al. (2006)
	Sendai,	GWW	100	-do-	01–35	
	Takatsuki	GWW	100	-do-	03–60	
Germany		GW	100	Mineralized sandstone	10–150	Nordstrom (2002)
Slovakia	Banska Bystrica and Nitra counties	GW	100	Aquifer sediments	37–39	Leonardi et al. (2012)
Eastern Croatia	Osijek area	SGW	50	Aquifer sediments	27	Romić et al. (2011)
Spain	NG	GW	100	Alluvial sediments	<1–100	Nordstrom (2002)
Ghana	NG	GW	100	Gold mining	<1–175	Nordstrom (2002)
Greece	NG	GW	100	Mine tailings	10000	Donahue and Hendry (2003)
Italy	Vulcano Island and Phlegrean Fields	GW	100	Volcanic aquifers	0.1–6940	Aiuppa et al. (2003)
Romania and Hungary	Pannonian Basin	GWW	50	Aquifer sediments	0.5–240	Rowland et al. (2011)
Australia	Perth	SGW	50	Pyritic sediments	7000	Appleyard et al. (2006)
	Holocene barrier, Northern, New South Wales	SW	50	Al hydroxides and Fe oxyhydroxides	52–337	Smith et al. (2003)
	Victoria	GW	100	Gold mining	<1–300000	Mukherjee et al. (2006)

DW, Drinking water; GW, Ground water; GWW, Ground water wells; NG, Not given, SGWW, Shallow ground water well; SHTW, Shallow hand tube wells; SW, Shallow well; SW, Surface water; TW, Tube wells.

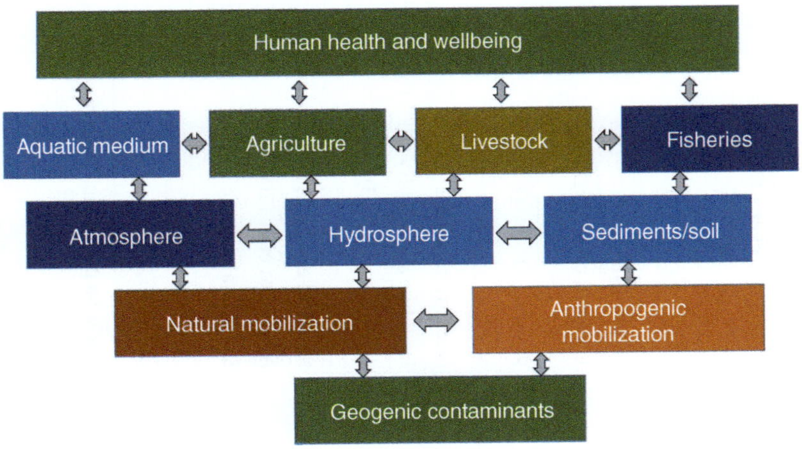

Figure 11.1 A schematic diagram showing the impact of arsenic hydrogeochemistry on human health in medical geology.

When ions of metal are exposed to the water-bound surface of the oxide, the hydroxyl group can complete their coordination shell, and it depends on the pH, and these can fix or discharge hydroxyl groups, which are usually associated with the control of specific adsorption properties. Surface charge. Adsorption mechanism. It has been found that As(V) adsorption occurs at the net release of hydroxyl ions with a pH range of 4.6–9.2, while As(III) adsorption results in a net release of OH at pH 9 2, and a net release of H^+ at pH 4.6 (Rinklebe et al. 2016; El-Naggar et al. 2019).

The process of As release can occur in a huge number of geochemical surroundings, but severe As events occur in recent alluvial basins near the active mountain belt. Recent alluvial and delta sediment deposits, particularly in Pakistan (Punjab and Sindh), Bangladesh, and India (West Bengal), are quite similar in geochemistry, and it is estimated that mobilization of As process is almost similar. Pakistan As levels in groundwater is lower than in India and Bangladesh due to redox situations and the existence of unrestricted water bodies (Hussain et al. 2019b). Several researchers have revealed that different regions of Punjab and Sindh are strongly influenced by the anthropogenic and natural sources of As (Nickson et al. 2005; Niazi et al. 2011).

The peoples living around the river Indus are under swearing risk of As (10 μg/L) exposure in their drinking water. About 13 million peoples are more vulnerable to As exposure out of 40 million people in 27 areas. In 2013, Pakistan projected that the total population of 182 million people are at As toxicity stake (Rabbani et al. 2017). In addition, a local survey found that 9% of all the studied water sources in Pakistan had an increased value of As > 10 μg/L. A high level of As in the 2517 groundwater sample of the Khair Pur (mirs) district of Sindh has also been reported (Das et al. 2004; Rabbani et al. 2017). The designated average concentration in As samples were 14.98 μg/L, and the maximum As concentration was noted at 350 μg/L. As a result, most of the people residing near these locations were found severely influenced by As contaminated groundwater. Except for sulfide oxidation, all these processes are fundamentally independent of anthropogenic activities in the subterranean areas (Abdul et al. 2015).

11.4 Exposure and Effects of As on Humans and Plants

Many studies have shown that exposure can cause cancer in addition to the pathological effects on humans. Animal studies have shown that chronic As exposure can have serious neurological effects and lead to endocrine disruptors (Sracek et al. 2004). Exposure to inorganic and organic As compounds is affecting millions of people's health across the world. Exposure to As is associated with cancer-related and non-cancer-related effects in almost all organs of the body, and there are indications that the health effects of lower As exposure are greater than previously thought (Carlin et al. 2016).

Continuous and long exposure to As from contaminated drinking water cause various health problems, such as nerves, skin, reproductive as well as pregnancy effects, cardiovascular effects, diabetes, respiratory and gastrointestinal diseases, and cancer, usually bladder, skin, liver, and lung. Almost 4.5% of the 8000 children in affected villages in the affected state have mild to moderate skin lesions. First survey results showed that above 10 000 patients get registered with various As-related disease signs, and above 100 000 were studied from affected states (Mukherjee et al. 2006; Chakrabarti et al. 2013).

Arsenic can present a threat to birds in these habitats, affecting reproduction, hatching ability, chick survival, and neurological behavior (n = 34) (Tsipoura et al. 2011). Arsenic is normally found in a variety of environments, occupations, and dietary settings, and they have a common neurotoxic effect that is the gradual onset and may be irreversible. Prolonged exposure at low concentrations may therefore pose a toxicity risk and cause oxidative stress, interference with neurotransmitters, and effects on the hematopoietic system (Andrade et al. 2015).

Chronic As exposure to humans has many biological effects which favor oxidative stress, DNA damage, and production of reactive oxygen species. It also leads to cancer because it induces genetic instability, epigenetic modification in DNA, immune regulation, and inflammation. High exposure is epidemiologically associated with bladder, skin, lung, liver, pancreatic, and kidney cancers as well as cardiovascular, neuronal, and other diseases (Rao et al. 2017). Arsenic is also an effective chemotherapeutic agent against acute promyelocytic leukemia, which may increase the cytotoxicity of chemotherapeutic DNA damaging drugs such as cisplatin. Arsenic has long been associated with inhibition of DNA repair, cell cycle disruption, and ubiquitination dysregulation, all of which adversely affect the response to DNA damage and may lead to the carcinogenic and chemotactic potential of As (Muenyi and Ljungman 2015). This provides evidence that the effects of As on the risk of cardiovascular disease vary from population to population and must be taken into account in the risk assessment (Ameer et al. 2015). In connection with the persuasion of various complications in the organ system of the body: rind, nerves, respiration, cardiovascular, hematopoietic, immunological, endocrine, liver, kidney, reproductive system, and developmental disorders, as this epigenetic changes in the body (in the uterus) and genetic mutations (the main cause of cancer). Arsenic metabolites can favor the methylation process in DNA and modifications in the base of p53 that play a role in the toxicity of As (Wen et al. 2020).

11.5 Conclusions and Outlooks

Arsenic toxicity is an eminent threat to the human population due to its geogenic release in to the aquifers. Arsenic hydrogeochemical cycling is complex under different redox and pH conditions; thus, better understanding is required to fulfill the research gap and open new gateway of research. Improved communication across the many disciplines dealing with illnesses caused by geological factors that impact human and animal health will be critical in the future. There is also a need to develop educational materials for schools and public and private organizations interested in medical geology, to demonstrate the impact of geological factors on human and animal health, and to hold joint technical meetings where geoscientists and others address medical geologists discuss problems. Geological survey offices, universities, and geomedical societies are playing a greater role in providing information on medical-geological conditions and encouraging the establishment of multidisciplinary medical-geological expert working groups at the local level. Research is encouraged to develop more efficient methods of studying geological problems in environmental medicine and provide recommendations to mitigate the consequences of natural and man-made hazardous geochemical situations. Therefore, medical geology is an interdisciplinary subject that will become even better known in the future.

Acknowledgments

The authors are thankful to Higher Education Commission of Pakistan for the financial support.

References

Abdul, K.S.M., Jayasinghe, S.S., Chandana, E.P. et al. (2015). Arsenic and human health effects: a review. *Environmental Toxicology and Pharmacology* 40: 828–846.

Ahamed, S., Sengupta, M.K., Mukherjee, A. et al. (2006). Arsenic groundwater contamination and its health effects in the state of Uttar Pradesh (UP) in upper and middle Ganga plain, India: a severe danger. *Science of the Total Environment* 370 (2–3): 310–322.

Ahmed, K.M., Bhattacharya, P., Hasan, M.A. et al. (2004). Arsenic enrichment in groundwater of the alluvial aquifers in Bangladesh: an overview. *Applied Geochemistry* 19: 181–200.

Aiuppa, A., D'Alessandro, W., Federico, C. et al. (2003). The aquatic geochemistry of arsenic in volcanic ground waters from Southern Italy. *Applied Geochemistry* 18: 1283–1296.

Ali, A., Hira Zafar, M.Z., ul Haq, I., Phull, A.R., Ali, J.S., Hussain, A. 2016. Synthesis, characterization, applications, and challenges of iron oxide nanoparticles. *Nanotechnology, Science and Applications*, 9, 49.

Ali, W., Rasool, A., Junaid, M., and Zhang, H. (2018). A comprehensive review on current status, mechanism, and possible sources of arsenic contamination in groundwater: a global perspective with prominence of Pakistan scenario. *Environmental Geochemistry and Health* 41: 737–760.

Ally, M.S., Ransohoff, K., Sarin, K. et al. (2016). Effects of combined treatment with arsenic trioxide and itraconazole in patients with refractory metastatic basal cell carcinoma. *JAMA Dermatology* 152 (4): 452–456.

Alvarez, C., Contreras-Kallens, P., Terraza, C. et al. (2018). Mesenchymal stem cells and their immunosuppressive role in transplantation tolerance. *Annals of the New York Academy of Sciences* 1417 (1): 35–56.

Ameer, S.S., Engström, K., Harari, F. et al. (2015). The effects of arsenic exposure on blood pressure and early risk markers of cardiovascular disease: evidence for population differences. *Environmental Research* 140: 32–36.

Andrade, V., Mateus, M., Batoréu, M. et al. (2015). Lead, arsenic, and manganese metal mixture exposures: focus on biomarkers of effect. *Biological Trace Element Research* 166: 13–23.

Anjum, S.A., Ashraf, U., Tanveer, M. et al. (2017). Drought induced changes in growth, osmolyte accumulation and antioxidant metabolism of three maize hybrids. *Frontiers in Plant Science* 8: 69.

Appleyard, S.J., Angeloni, J., and Watkins, R. (2006). Arsenic-rich groundwater in an urban area experiencing drought and increasing population density, Perth, Australia. *Applied Geochemistry* 21: 83–97.

Armienta, M.A., Rodriguez, R., Aguayo, A. et al. (1997). Arsenic contamination of groundwater at Zimapán, Mexico. *Hydrogeology Journal* 5: 39–46.

Aullón Alcaine, A., Sandhi, A., Bhattacharya, P. et al. (2012). Distribution and mobility of geogenic arsenic in the shallow aquifers of the Northeast of La Pampa, Argentina. In: *Understanding the Geological and Medical Interface of Arsenic As 2012*, Interdisciplinary Book Series: Arsenic in the Environment – Proceedings. Series editors: J. Bundschuh and P. Bhattacharya (ed. J.C. Ng, B.N. Noller, R. Naidu, et al.), 132–134. CRC Press/Taylor and Francis (ISBN-13: 978-0-415-63763-3).

Azizullah, A., Khattak, M.N.K., Richter, P., and Häder, D.-P. (2011). Water pollution in Pakistan and its impact on public health – a review. *Environment International* 37: 479–497.

Baig, J.A., Kazi, T.G., Arain, M.B. et al. (2009). Evaluation of arsenic and other physico-chemical parameters of surface and ground water of Jamshoro, Pakistan. *Journal of Hazardous Materials* 166: 662–669.

Bejarano Sifuentes, G. and Nordberg, E. (2003). Mobilisation of arsenic in the Rio Dulce alluvial cone, Santiago del Estero province, Argentina. TRITA-LWR Master thesis. Department of Land and Water Resources Engineering, KTH Royal Institute of Technology, Stockholm, Sweden. 75p.

Berg, M., Stengel, C., Trang, P.T.K. et al. (2007). Magnitude of arsenic pollution in the Mekong and Red River deltas – Cambodia and Vietnam. *Science of the Total Environment* 372: 413–425.

Bhattacharya, P., Claesson, M., Bundschuh, J. et al. (2006). Distribution and mobility of arsenic in the Río Dulce alluvial aquifers in Santiago del Estero province, Argentina. *Science of the Total Environment* 358 (1–3): 97–120.

Bhattacharya, P., Welch, A.H., Stollenwerk, K.G. et al. (2007). Arsenic in the environment: biology and chemistry. *Science of the Total Environment* 379 (2–3): 109–120.

Bhattacharya, P., Samal, A.C., Majumdar, J. et al. (2013). In vitro assessment on the impact of soil arsenic in the eight rice varieties of West Bengal, India. *Journal of Hazardous Materials* 262: 1091–1097.

Bhattacharya, P. and Bundschuh, J. (2015). Groundwater for sustainable development: cross cutting the UN sustainable development goals. *Groundwater for Sustainable Development* 1: 155–157.

Biswas, A., Nath, B., Bhattacharya, P. et al. (2012). Hydrogeochemical contrast between brown and grey sand aquifers in shallow depth of Bengal Basin: consequences for sustainable drinking water supply. *Science of the Total Environment* 431: 402–412.

Boehme, K.A., Zaborski, J.J., Riester, R. et al. (2016). Targeting hedgehog signalling by arsenic trioxide reduces cell growth and induces apoptosis in rhabdomyosarcoma. *International Journal of Oncology* 48 (2): 801–812.

Boochs, P.W., Billib, M., Gutiérrez, C., and Aparicio, J. (2014). Groundwater contamination with arsenic, Región Lagunera,México. In: *Once Century of the Discovery of Arsenicosis in Latin America (1914–2014) As 2014*, Interdisciplinary Book Series: Arsenic in the Environment – Proceedings. Series editors: J. Bundschuh and P. Bhattacharya (ed. M.I. Litter, H.B. Nicolli, M. Meichtry, et al.), 132–134. CRC Press/Taylor and Francis (ISBN 978-1-138-00141-1).

Brown, I. and Sen, A. (2017). Neurological disorders. In: *Textbook of Occupational Medicine Practice*, vol. 2, 287.

Bundschuh, J., Farias, B., Martin, R. et al. (2004). Grounwater arsenic in the Chaco-Pampean plain, Argentina: case study from Robles County, Santiago del Estero Province. *Applied Geochemistry* 19 (2): 231–243.

Bundschuh, J., Litter, M.I., Parvez, F. et al. (2012). One century of arsenic exposure in Latin America: a review of history and occurrence from 14 countries. *Science of the Total Environment* 429: 2–35.

Cardoso, A.P.F., Al-Eryani, L., and States, J.C. (2018). Arsenic-induced carcinogenesis: the impact of miRNA dysregulation. *Toxicological Sciences* 165 (2): 284–290.

Carlin, D.J., Naujokas, M.F., Bradham, K.D. et al. (2016). Arsenic and environmental health: state of the science and future research opportunities. *Environmental Health Perspectives* 124: 890–899.

Carretero, M.I., Gomes, C.S.F., and Tateo, F. (2013). Clays, drugs, and human health. In: *Developments in Clay Science*, vol. 5, 711–764. Elsevier.

Chakrabarti, D., Singh, S.K., Rashid, M.H., and Rahman, M.M. (2013). Arsenic: occurrence in groundwater. *Encyclopedia of Environmental Health 2*: 1e17.

Chakraborti, D., Mukherjee, S.C., Pati, S. et al. (2003). Arsenic groundwater contamination in middle ganga plain, Bihar, India: a future danger? *Environmental Health Perspectives* 111 (9): 1194–1201.

Chakraborti, D., Rahman, M.M., Das, B. et al. (2010). Status of groundwater arsenic contamination in Bangladesh: a 14-year study report. *Water Research* 44: 5789–5802.

Chandrakar, V., Naithani, S.C., and Keshavkant, S. (2016). Arsenic-induced metabolic disturbances and their mitigation mechanisms in crop plants: a review. *Biologia* 71 (4): 367–377.

Chandrakar, V., Pandey, N., and Keshavkant, S. (2018). Plant responses to arsenic toxicity: morphology and physiology. In: *Mechanisms of Arsenic Toxicity and Tolerance in Plants*, 27–48. Singapore: Springer.

Chen, B.-H. and Lee, W.-D. (2016). Comparison of extraction efficiency of tanshinones from S. miltiorrhiza by solvent and supercritical carbon dioxide. *International Journal of Food Processing Technology* 3: 1–10.

Cheng, C., Nie, Z.W., He, L.Y., and Sheng, X.F. (2020). Rice-derived facultative endophytic Serratia liquefaciens F2 decreases rice grain arsenic accumulation in arsenic-polluted soil. *Environmental Pollution 259*: 113832.

Crabbe, H., Fletcher, T., Close, R. et al. (2017). Hazard ranking method for populations exposed to arsenic in private water supplies: relation to bedrock geology. *International Journal of Environmental Research and Public Health* 14: 1490.

Das, H., Mitra, A.K., Sengupta, P. et al. (2004). Arsenic concentrations in rice, vegetables, and fish in Bangladesh: a preliminary study. *Environment International* 30: 383–387.

Das, N., Patel, A.K., Deka, G. et al. (2015). Geochemical controls and future perspective of arsenic mobilization for sustainable groundwater management: a study from Northeast India. *Groundwater for Sustainable Development* 1: 92–104.

Doll, M., Zoeller, E., Kleemann, J. et al. (2017). Arsenic trioxide decreases lymphangiogenesis by inducing intrinsic apoptotic pathways and inhibition of important lymphatic endothelial cell receptors. In: *Journal der Deutschen Dermatologischen Gesellschaft*, vol. 15, 20–20. Wiley.

Donahue, R. and Hendry, M. (2003). Geochemistry of arsenic in uranium mine mill tailings, Saskatchewan, Canada. *Applied Geochemistry* 18: 1733–1750.

El-Naggar, A., Shaheen, S.M., Hseu, Z.-Y. et al. (2019). Release dynamics of As, Co, and Mo in a biochar treated soil under pre-definite redox conditions. *Science of the Total Environment* 657: 686–695.

Emi, N. (2017). Arsenic trioxide: clinical pharmacology and therapeutic results. In: *Chemotherapy for Leukemia*, 221–238. Singapore: Springer.

Ergün, R., Evcik, E., Ergün, D. et al. (2017). High-resolution computed tomography and pulmonary function findings of occupational arsenic exposure in workers. *Balkan Medical Journal* 34 (3): 263–268.

Fatmi, Z., Azam, I., Ahmed, F. et al. (2009). Health burden of skin lesions at low arsenic exposure through groundwater in Pakistan. Is river the source? *Environmental Research* 109: 575–581.

Gonçalves, A.L., Pires, J.C., and Simões, M. (2017). A review on the use of microalgal consortia for wastewater treatment. *Algal Research* 24: 403–415.

Goovaerts, P. and Glass, G. (2014). Geostatistical modeling of the spatial distribution of surface soil arsenic around a smelter. *Journal of the Japanese Society of Soil Physics* 128: 5–10.

Guo, J., Yan, C., Luo, Z. et al. (2019). Synthesis of a novel ternary HA/Fe-Mn oxides-loaded biochar composite and its application in cadmium (II) and arsenic (V) adsorption. *Journal of Environmental Sciences* 85: 168–176.

Gupta, A., Verma, N.C., Khan, S., and Nandi, C.K. (2016). Carbon dots for naked eye colorimetric ultrasensitive arsenic and glutathione detection. *Biosensors and Bioelectronics* **81**: 465–472.

Haghi, V., Namdjoyan, S., and Soorki, A.A. (2022). Interactive effects of exogenous melatonin and hydrogen sulfide in alleviating lead toxicity in safflower seedlings. *Industrial Crops and Products* 187: 115523.

Hasanato, R.M. and Almomen, A.M. (2015). Unusual presentation of arsenic poisoning in a case of celiac disease. *Annals of Saudi Medicine* 35 (2): 165–167.

Herath, I., Vithanage, M., Bundschuh, J. et al. (2016). Natural arsenic in global groundwaters: distribution and geochemical triggers for mobilization. *Current Pollution Reports* 2: 68–89.

Hoque, M.A., Burgess, W.G., and Ahmed, K.M. (2017). Integration of aquifer geology, groundwater flow and arsenic distribution in deltaic aquifers–A unifying concept. *Hydrological Processes* 31 (11): 2095–2109.

Hossain, M., Bhattacharya, P., Frape, S.K. et al. (2014). Sediment color tool for targeting arsenicsafe aquifers for the installation of shallow drinking water tubewells. *Science of the Total Environment* 493: 615–625.

Hussain, M.M., Bibi, I., Shahid, M. et al. (2019a). Biogeochemical cycling, speciation and transformation pathways of arsenic. In: *Aquatic Environments with the*

Emphasis on Algae (ed. A.C. Duarte and V. Reis), 15–51. Amsterdam, Netherlands: Elsevier.

Hussain, M.M., Bibi, I., Shahid, M. et al. (2019b). Chapter two – biogeochemical cycling, speciation and transformation pathways of arsenic in aquatic environments with the emphasis on algae. In: *Comprehensive Analytical Chemistry* (ed. A.C. Duarte and V. Reis), 15–51. Elsevier.

Hussain, M.M., Wang, J., Bibi, I. et al. (2020). Arsenic speciation and biotransformation pathways in the aquatic ecosystem: the significance of algae. *Journal of Hazardous Materials* 403: 124027.

Hussain, M.M., Bibi, I., Niazi, N.K. et al. (2021). Arsenic biogeochemical cycling in paddy soil-rice system: interaction with various factors, amendments and mineral nutrients. *Science of the Total Environment* 773: 145040.

Iqbal, M.O.S.U.D., GKMM, R., GM, P. et al. (2016). Response of yield and yield attributes of different rice genotypes to soil arsenic. *Research on Crop Ecophysiology* 11: 1–17.

Ishii, N., Mochizuki, H., Ebihara, Y. et al. (2018). Clinical symptoms, neurological signs, and electrophysiological findings in surviving residents with probable arsenic exposure in Toroku, Japan. *Archives of Environmental Contamination and Toxicology* 75 (4): 521–529.

Jasrotia, S., Kansal, A., and Mehra, A. (2017). Performance of aquatic plant species for phytoremediation of arsenic-contaminated water. *Applied Water Science* 7 (2): 889–896.

Kharroubi, W., Ahmed, S.H., Nury, T. et al. (2017). Mitochondrial dysfunction, oxidative stress and apoptotic induction in microglial BV-2 cells treated with sodium arsenate. *Journal of Environmental Sciences* 51: 44–51.

Khatun, S., Maity, M., Perveen, H. et al. (2018). Spirulina platensis ameliorates arsenic-mediated uterine damage and ovarian steroidogenic disorder. *Facets* 3 (1): 736–753.

Kim, K.-W., Chanpiwat, P., Hanh, H. et al. (2011). Arsenic geochemistry of groundwater in Southeast Asia. *Frontiers in Medicine* 5: 420–433.

Kofroňová, M., Mašková, P., and Lipavská, H. (2018). Two facets of world arsenic problem solution: crop poisoning restriction and enforcement of phytoremediation. *Planta* 248 (1): 19–35.

Koutros, S., Baris, D., Waddell, R. et al. (2018). Potential effect modifiers of the arsenic–bladder cancer risk relationship. *International Journal of Cancer* 143 (11): 2640–2646.

Kühn, M., Sammartin, K., Nabergoj, M., and Vianello, F. (2016). Severe acute axonal neuropathy following treatment with arsenic trioxide for acute promyelocytic leukemia: a case report. *Mediterranean Journal of Hematology and Infectious Diseases* 8 (1).

Leonardi, G., Vahter, M., Clemens, F. et al. (2012). Inorganic arsenic and basal cell carcinoma in areas of Hungary, Romania, and Slovakia: a case–control study. *Environmental Health Perspectives* 120: 721–726.

Li, S., Xiao, T., and Zheng, B. (2012). Medical geology of arsenic, selenium and thallium in China. *Science of the Total Environment* 421: 31–40.

Li, F., Guo, H., Zhou, X. et al. (2017). Impact of natural organic matter on arsenic removal by modified granular natural siderite: evidence of ternary complex formation by HPSEC-UV-ICP-MS. *Chemosphere* 168: 777–785.

Liu, C.-C., Maity, J.P., Jean, J.-S. et al. (2013). The geochemical characteristics of the mud liquids in the Wushanting and Hsiaokunshui mud volcano region in Southern Taiwan: implications of humic substances for binding and mobilization of arsenic. *Journal of Geochemical Exploration* 128: 62–71.

Maity, J.P., Nath, B., Chen, C.-Y. et al. (2011). Arsenic-enriched groundwaters of India, Bangladesh and Taiwan – comparison of hydrochemical characteristics and mobility constraints. *Journal of Environmental Science and Health* 46: 1163–1176.

Mishra, D. S. and Verma, N. (2016). Carbon Bead-supported Hollow Carbon Nanofibers Synthesized via Templating Method for the Removal of Hexavalent Chromium. *Journal of Industrial and Engineering Chemistry* **36**: 346–354.

Mohammed, T., Mohammed, E., and Bascombe, S. (2017). The evaluation of total mercury and arsenic in skin bleaching creams commonly used in Trinidad and Tobago and their potential risk to the people of the Caribbean. *Journal of Public Health Research* 6 (3).

Mohan, D. and Pittman, C.U. Jr. (2007). Arsenic removal from water/wastewater using adsorbents—a critical review. *Journal of Hazardous Materials* 142 (1-2): 1–53.

Muenyi, C.S. and Ljungman, M. (2015). Arsenic disruption of DNA damage responses – potential role in carcinogenesis and chemotherapy. *Biomolecules* 5: 2184–2193.

Mukherjee, A., Sengupta, M.K., Hossain, M.A. et al. (2006). Arsenic contamination in groundwater: a global perspective with emphasis on the Asian scenario. *Journal of Health, Population, and Nutrition* 24: 142–163.

Muñoz, M.P., Valdés, M., Muñoz-Quezada, M.T. et al. (2018). Urinary inorganic arsenic concentration and gestational diabetes mellitus in pregnant women from Arica, Chile. *International Journal of Environmental Research and Public Health* 15 (7): 1418.

Niazi, N.K., Singh, B., and Shah, P. (2011). Arsenic speciation and phytoavailability in contaminated soils using a sequential extraction procedure and XANES spectroscopy. *Environmental Science & Technology* 45: 7135–7142.

Nickson, R., McArthur, J., Shrestha, B. et al. (2005). Arsenic and other drinking water quality issues, Muzaffargarh District, Pakistan. *Applied Geochemistry* 20: 55–68.

Nicolli, H.B., Bundschuh, J., Blanco, M.C. et al. (2012a). Arsenic and associated trace-elements in groundwater from the Chaco-Pampean plain, Argentina: results from 100 years of research. *Science of the Total Environment* 429: 36–56.

Nicolli, H.B., García, J.W., Falcón, C.M., and Smedley, P.L. (2012b). Mobilization of arsenic and other trace elements of health concern in groundwater from the Salí River basin, Tucumán Province, Argentina. *Environmental Geochemistry and Health* 34: 251–262.

Niedzwiecki, M.M., Liu, X., Zhu, H. et al. (2018). Serum homocysteine, arsenic methylation, and arsenic-induced skin lesion incidence in Bangladesh: a one-carbon metabolism candidate gene study. *Environment International* 113: 133–142.

Niño, S.A., Vázquez-Hernández, N., Arevalo-Villalobos, J. et al. (2021). Cortical synaptic reorganization under chronic arsenic exposure. *Neurotoxicity Research* 39 (6): 1970–1980.

Nordstrom, D.K. (2002). Worldwide occurrences of arsenic in ground water. *Science* 296: 2143–2145.

Nriagu, J., Bhattacharya, P., Mukherjee, A. et al. (2007). Arsenic in soil and groundwater: an overview. *Trace Metals Contamination Environment* 9: 3–60.

Ormachea Muñoz, M., Garcia Aróstegui, J.L., Bhattacharya, P. et al. (2016). Geochemistry of naturally occurring arsenic in groundwater and surface-water in the southern part of the Poopó Lake basin, Bolivian Altiplano. *Groundwater for Sustainable Development* 2–3: 104–116.

Panigatti, M.C., Boglione, R.M., Griffa, C.A., and Schierano, M.C. (2014). Groundwater arsenic in the Central-west of the Santa Fe Province, Argentine. In: *Once Century of the Discovery of Arsenicosis in Latin America (1914–2014) As 2014*, Interdisciplinary Book Series: Arsenic in the Environment – Proceedings. Series editors: J. Bundschuh and P. Bhattacharya (ed. M.I. Litter, H.B. Nicolli, M. Meichtry, et al.), 159–161. CRC Press/Taylor and Francis (ISBN 978-1-138- 00141-1).

Pathak, M. and Pathak, R., 2018. Arsenic Attenuation Through Phytochemistry.

Patir, P., Oytun, M., Ozluk, A.A. et al. (2016). Bilateral sixth nerve palsies possibly associated with arsenic trioxide in acute promyelocytic leukaemia. *International Journal of Hematology and Oncology* 31 (4): 176–178.

Rabbani, U., Mahar, G., Siddique, A., and Fatmi, Z. (2017). Risk assessment for arsenic-contaminated groundwater along river Indus in Pakistan. *Environmental Geochemistry and Health* 39: 179–190.

Rahman, M.M., Mandal, B.K., Roychowdhury, T. et al. (2003). Arsenic groundwater contamination and sufferings of people in North 24-Parganas, one of the nine arsenic affected districts of West Bengal, India. *Journal of Environmental Science and Health – Part A Environmental Science* 38: 25–59.

Rahman, M.M., Mondal, D., Das, B. et al. (2014). Status of groundwater arsenic contamination in all 17 blocks of Nadia district in the state of West Bengal, India: a 23-year study report. *Journal of Hydrology* 518: 363–372.

Rao, C.V., Pal, S., Mohammed, A. et al. (2017). Biological effects and epidemiological consequences of arsenic exposure, and reagents that can ameliorate arsenic damage in vivo. *Oncotarget* 8: 57605.

Ratnaike, R.N. (2003). Acute and chronic arsenic toxicity. *Postgraduate Medical Journal* 79 (933): 391–396.

Rinklebe, J., Shaheen, S.M., and Yu, K. (2016). Release of As, Ba, Cd, Cu, Pb, and Sr under pre-definite redox conditions in different rice paddy soils originating from the U.S.A. and Asia. *Geoderma* 270: 21–32.

Robertson, F.N. (1989). Arsenic in ground-water under oxidizing conditions, South-west United States. *Environmental Geochemistry and Health* 11: 171–185.

Robinson, B., Kim, N., Marchetti, M. et al. (2006). Arsenic hyperaccumulation by aquatic macrophytes in the Taupo volcanic zone, New Zealand. *Environmental and Experimental Botany* 58: 206–215.

Rodriguez, R., Ramos, J.A., and Armienta, A. (2004). Groundwater arsenic variations: the role of local geology and rainfall. *Applied Geochemistry* 19 (2): 245–250.

Romić, Ž., Habuda-Stanić, M., Kalajdžić, B., and Kuleš, M. (2011). Arsenic distribution, concentration and speciation in groundwater of the Osijek area, Eastern Croatia. *Applied Geochemistry* 26: 37–44.

Rosen, B.P. and Liu, Z. (2009). Transport pathways for arsenic and selenium: a minireview. *Environment International* 35: 512–515.

Rowland, H.A., Omoregie, E.O., Millot, R. et al. (2011). Geochemistry and arsenic behaviour in groundwater resources of the Pannonian Basin (Hungary and Romania). *Applied Geochemistry* 26: 1–17.

Sadaf, N., Kumar, N., Ali, M. et al. (2018). Arsenic trioxide induces apoptosis and inhibits the growth of human liver cancer cells. *Life Sciences* 205: 9–17.

Saifullah, S., Dahlawi, A., Naeem, M. et al. (2018). Opportunities and challenges in the use of mineral nutrition for minimizing arsenic toxicity and accumulation in rice: A critical review. *Chemosphere* 194: 171–188.

Sanchez, T.R., Powers, M., Perzanowski, M. et al. (2018). A meta-analysis of arsenic exposure and lung function: is there evidence of restrictive or obstructive lung disease? *Current Environmental Health Reports* 5 (2): 244–254.

Shahid, M., Niazi, N.K., Dumat, C. et al. (2018). A meta-analysis of the distribution, sources and health risks of arsenic-contaminated groundwater in Pakistan. *Environmental Pollution* 242: 307–319.

Shakoor, M.B., Niazi, N.K., Bibi, I. et al. (2016). Remediation of arsenic-contaminated water using agricultural wastes as biosorbents. *Critical Reviews in Environmental Science and Technology* 46 (5): 467–499.

Shakoor, M.B., Nawaz, R., Hussain, F. et al. (2017). Human health implications, risk assessment and remediation of As-contaminated water: a critical review. *Science of the Total Environment* 601–602: 756–769.

Shakoor, M.B., Bibi, I., Niazi, N.K. et al. (2018). The evaluation of arsenic contamination potential, speciation and hydrogeochemical behaviour in aquifers of Punjab, Pakistan. *Chemosphere* 199: 737–746.

Shankar, S., Shanker, U., and Shikha. (2014). Arsenic contamination of groundwater: a review of sources, prevalence, health risks, and strategies for mitigation. *Scientific World Journal.*

Sharma, S., Kaur, J., Nagpal, A.K., and Kaur, I. (2016). Quantitative assessment of possible human health risk associated with consumption of arsenic contaminated groundwater and wheat grains from Ropar Wetand and its environs. *Environmental Monitoring and Assessment* 188 (9): 1–14.

Shukla, D.P., Dubey, C., Singh, N.P. et al. (2010). Sources and controls of arsenic contamination in groundwater of Rajnandgaon and Kanker District, Chattisgarh Central India. *Journal of Hydrology* 395: 49–66.

Singh, A.L. and Singh, V.K. (2018). Assessment of groundwater quality of Ballia district, Uttar Pradesh, India, with reference to arsenic contamination using multivariate statistical analysis. *Applied Water Science* 8 (3): 1–18.

Singh, R., Singh, S., Parihar, P. et al. (2015). Arsenic contamination, consequences and remediation techniques: a review. *Ecotoxicology and Environmental Safety* 112: 247–270.

Smedley, P.L. (2008). Sources and distribution of arsenic in groundwater and aquifers. In: *Arsenic in Groundwater: A World Problem* (ed. T. Appelo), 4–32. Utrecht: International Association of Hydrogeologists Publication.

Smedley, P. and Kinniburgh, D. (2002). A review of the source, behaviour and distribution of arsenic in natural waters. *Applied Geochemistry* 17 (5): 517–568.

Smedley, P.L., Kinniburgh, D.G., Macdonald, D.M.J. et al. (2005). Arsenic associations in sediments from the loess aquifer of La Pampa, Argentina. *Applied Geochemistry 20* (5): 989–1016.

Smith, J.V.S., Jankowski, J., and Sammut, J. (2003). Vertical distribution of As(III) and As(V) in a coastal sandy aquifer: factors controlling the concentration and speciation of arsenic in the Stuarts point groundwater system, New Northern New South Wales, Australia. *Applied Geochemistry* 18: 1479–1496.

Sracek, O., Bhattacharya, P., Jacks, G. et al. (2004). Behavior of arsenic and geochemical modeling of arsenic enrichment in aqueous environments. *Applied Geochemistry* 19: 169–180.

Tan, X., Liu, Y., Zeng, G. et al. (2015). Application of biochar for the removal of pollutants from aqueous solutions. *Chemosphere* 125: 70–85.

Tanga, S., Goel, V.R., and Patil, S. (2016). Chronic arsenic poisoning leading to skin malignancy in a community. *Clinical Skin Cancer* 1 (1): 15–19.

Torjussen, W. (2017). Nasal cancer in nickel workers. Histopathological findings and nickel concentrations in the nasal mucosa of nickel workers, and a short review of chromium and arsenic. *Nasal Tumors in Animals and Man* 33–54.

Tsipoura, N., Burger, J., Newhouse, M. et al. (2011). Lead, mercury, cadmium, chromium, and arsenic levels in eggs, feathers, and tissues of Canada geese of the New Jersey Meadowlands. *Environmental Research* 111: 775–784.

Wang, N., Xue, X.M., Juhasz, A.L. et al. (2017). Biochar increases arsenic release from an anaerobic paddy soil due to enhanced microbial reduction of iron and arsenic. *Environmental Pollution* 220: 514–522.

Wang, D., Gilliland, S.E. III, Yi, X. et al. (2018). An iron mesh-based metal organic framework filter for efficient arsenic removal. *Environmental Science and Technology* 52 (7): 4275–4284.

Wen, Z., Lu, J., Zhang, Y. et al. (2020). Facile inverse micelle fabrication of magnetic ordered mesoporous iron cerium bimetal oxides with excellent performance for arsenic removal from water. *Journal of Hazardous Materials* 383: 121172.

Xiao, K.Q., Li, L.G., Ma, L.P. et al. (2016). Metagenomic analysis revealed highly diverse microbial arsenic metabolism genes in paddy soils with low-arsenic contents. *Environmental Pollution* 211: 1–8.

Xue, M. and Yi, H. (2017). Nitric oxide signaling is involved in arsenic-induced guard cell death in Vicia faba L.(Fabaceae). *Brazilian Journal of Botany* 40 (3): 635–642.

Yang, Y., Deng, Y., and Wang, Y. (2016). Major geogenic factors controlling geographical clustering of urolithiasis in China. *Science of the Total Environment* 571: 1164–1171.

Zhang, Y., Wu, X., Li, Y. et al. (2016). Endothelial to mesenchymal transition contributes to arsenic-trioxide-induced cardiac fibrosis. *Scientific Reports* 6 (1): 1–12.

Zhou, Q. and S. Xi. (2018). A review on arsenic carcinogenesis: Epidemiology, metabolism, genotoxicity and epigenetic changes. *Regulatory Toxicology and Pharmacology* 99: 78–88.

Zhu, Y.G., Xue, X.M., Kappler, A. et al. (2017). Linking genes to microbial biogeochemical cycling: lessons from arsenic. *Environmental Science & Technology* 51 (13): 7326–7339.

Ziarati, P., Hochwimmer, B., Cruz-Rodriguez, L. et al. (2021). Lung cancer disease and arsenic: the medical geology as key. *Journal of Medical Discovery* 6 (1): 1–16.

12

Potentially Toxic Metals and Health

Pelin Yapıcıoğlu[1], Mehmet İrfan Yeşilnacar[1], and Engin Tutkun[2]

[1] *Environmental Engineering Department, Faculty of Engineering, Harran University, Sanliurfa, Turkey*
[2] *Department of Public Health, Faculty of Medicine, Bozok University, Yozgat, Turkey*

12.1 Introduction

Public health is highly dependent on environmental conditions (Komatina 2004). Toxic composition of natural resources is the major factor that creates the effective environmental conditions on public health (Selinus et al. 2010, 2016). Toxic materials originating from natural resources such as arsenic, lead, and the other toxic elements and compounds lead to soil and groundwater pollution have been defined as the historical beginnings of the medical geology. Trace elements present at minimum levels could have a serious effect on vital functions based on their concentrations (Selinus et al. 2016). According to the previous studies, the total concentrations of chemical compounds cannot ensure knowledge about their mobility, bioavailability, and possible effect on flora and fauna. Only information of the chemical kind of a compound could obtain knowledge considering eventual chemical and biochemical reactions, and it leads to large information about their toxicology. Toxicology obviously plays a crucial role at medical geology (Selinus et al. 2016). Toxicology is a measure of damage originated from toxic materials on human health. The presence of toxic metals at water resources and agricultural fields has an unfavorable impact on human health. Bioaccumulation of trace elements could lead to serious health problems. The toxic substances that are arsenic (As), cadmium (Cd), chromium (Cr), and lead (Pb) considered as trace elements due to their essentiality and very limited amount at human bodies (Stern 2010; Tchounwou et al. 2012).

Toxic metals have been regarded as trace elements (Arruti et al. 2010; He et al. 2005; Tchounwou et al. 2012). The industrial, municipal, agricultural, and medical implementations of heavy metals lead to extensive distribution in the ecosystem, and this matter could increase their potential impacts on human health and the flora and fauna (Adimalla et al. 2020; Bradl 2002; Kaur et al. 2020). There are many resources of heavy metals contain erosion, weathering process, mining process, industrial waste and wastewater discharges, urban runoff, sewage discharge, pesticides, and insecticides applications to crops etc. (Jaishankar et al. 2014). Their toxicity based on many factors containing the dose and chemical content of the heavy metal and the age, gender, and exposure frequency of exposed people (Tchounwou et al. 2012). Heavy metal toxicity is a major threat for public health, and there are various health risks associated with it (Ali et al. 2019; Baki et al. 2018; Corrias et al. 2020; Gu et al. 2018; Guzzi et al. 2021; Hao et al. 2019).

Around the world, water and soil have been polluted due to municipal and anthropogenic activities, rapid industrialization, and agricultural activities and natural resources such as rocks, minerals, and atmospheric deposition (Kaur et al. 2018, 2020). Many types of pollutant, such as heavy metals are emitted into the environment (Zibret et al. 2018). According to the scientists, the heavy metals among several pollutants play a crucial role and considered as the most hazardous contaminants because of their nondegradable composition, environmental risks on human health, toxicity, and environmental persistence (Kaur et al. 2020). They could lead to carcinogenic and noncarcinogenic impacts with skin contact, inhalation, and ingestion from water and soil and may cause various serious health impacts on human with relentless consequences (Zibret et al. 2018). Bioaccumulation of heavy metals could lead to serious health problems. Bioaccumulation can be described as the process by which living organisms absorb chemicals or elements directly from the environment (Metcalf and Eddy 2014). For bioaccumulation, the concentration and the exposure dose to heavy metals is decisive factor.

Medical Geology: En route to One Health, First Edition. Edited by Majeti Narasimha Vara Prasad and Meththika Vithanage.

Bioaccumulation of toxic elements, which are especially lead, cadmium, etc., disrupts the blood cells in human bodies. The bioaccumulation of the trace elements concerned is an important step in the food chain (Liyanage et al. 2022). There are many routes through which people could be exposed to these elements. The routes contain inhalation and absorption through the skin, the ingestion of contaminated food, and potable water. Heavy metals are transported to human bodies through plants and animals with food chain. Both plants and animals could bioaccumulate heavy metals from the environment (Liyanage et al. 2022). When plants and animals are used as food resources, so they could provide to the uptake of heavy metals by people. They reach to plants and animals with soil and water ecosystems. Some prevention policies were developed in order to take under control of heavy metal pollution. Treatment is one of these applications. Especially, drinking water treatment is a wide-spread control strategy for the protection of human health. Especially, groundwater has a significant heavy metal content due to its nature (Bilgili et al. 2018; Ilhan et al. 2021; Yapıcıoğlu et al. 2020; Yenigün et al. 2021; Yeşilnacar and Güllüoğlu 2008; Yeşilnacar and Yenigün 2011; Yetiş et al. 2021). Several conventional and innovative treatment methods have been carried out for toxic metal removal from water and soil. Heavy metal removal from soil has a vital importance due to the plant growth and the people consume plants from soil directly (Puga et al. 2020; Zhou et al. 2015). Biochar generated using various organic substances is a by-product of thermochemical processes could be applied for greenhouse gas emissions minimization, soil amendment, hazardous waste management, and potable water and sewage treatment (Alhashimi and Aktas 2017; Awasthi et al. 2016; Kończak and Oleszczuk 2018; Lehmann and Joseph 2015; Lehmann et al. 2011; Zhang et al. 2019). Biochar adsorption is a new innovative treatment technique for toxic elements removal from both soil and drinking water (Ahmad et al. 2014; Lehmann et al. 2011; Rajapaksha et al. 2016; Vithanage et al. 2017, 2014; Yapıcıoğlu and Yeşilnacar 2022; Yuan et al. 2017). Biochar could not only be used as soil amendment but also it has heavy metal adsorption capacity. Also, it could be drinking water treatment method for heavy metal removal in recent years (Lehmann et al. 2011; Qambrani et al. 2017; Yapıcıoğlu and Yeşilnacar 2022).

In this context, this chapter contains an overview of the concentrations and possible resources of toxic metals released by gelogenic processes, which may threat the health of several people. This chapter presented the resources of potential toxic metals, which were arsenic (As), cadmium (Cd), chromium (Cr), lead (Pb), and their impacts on public health. This study majorly concentrated on these metals and their origins and their effects on human health. Also, this chapter mainly aimed to investigate the biochar adsorption process among treatment methods of toxic metals from potable water. Biochar is an inexpensive biomass type has presented as an effective toxic metal removal method from potable water.

12.2 Toxic Metals and Their Resources

Rising energy demand, urbanization, mining processes, agricultural activities, and excess discharge of industrial wastewater have led to water, air, and soil contamination. Among these contaminants, heavy metals and metalloids are nonbiodegradable and have accumulated rapidly water, soil and plants while threatening human health (Dökmeci 2020; Jaishankar et al. 2014; Krishnamoorthy et al. 2019; Medina-Estévez et al. 2020; Patniak 2010). Mining process, industrial waste and wastewater discharges, urban runoff, sewage discharge, pesticides, and insecticides applications to crops have been regarded as the major heavy metals source in the ecosystem (Jaishankar et al. 2014; Komatina 2004). Due to industrialization, mining processes, pesticide addition and rising energy demands, and heavy metals have been regarded as the main reason of occupational diseases, and they have damaged the balance of ecosystems in the result of water, air, and soil pollution. Besides, anthropogenic activities such as mining and smelting have caused to them becoming concentrated in the earth, in several locations achieving potentially detrimental levels. Other sources of heavy metals are vehicle emissions, industrial wastewater, and manures also trigger to the stock of trace metals in the soil, atmosphere, and potable water resources (Komatina 2004; Masindi and Muedi 2018; Selinus et al. 2016; USEPA 2019). Heavy metal toxicity resulted from several anthropogenic, and geological activities poses a global threat to the human health (Bhattacharya 2020; Jyothi and Farook 2020; Kim et al. 2019; Selinus et al. 2016; Tchounwou et al. 2012). Environmental pollution could also form due to metal corrosion, soil erosions, atmospheric deposition, and leaching process of heavy metals from water supplies to soil and groundwater (Tchounwou et al. 2012).

Natural geological events as weathering and volcanic eruptions trigger the toxic metal contamination (Dissanayake and Chandrajith 2009). Industrial resources of heavy metals contain metal processing in refineries, coal burning, combustion plants, plastics, wood, and paper textiles processing industries (Dissanayake and Chandrajith 2009; Emslay 2011; Jyothi and Farook 2020; Suvarapu and Baek 2017). Among these toxic and trace elements, arsenic (As), cadmium (Cd), chromium (Cr), and lead (Pb) and their impacts on public health were investigated in this chapter.

In this section, toxic metals and their resources and environmental impacts were defined:

12.2.1 Arsenic (As)

Inorganic arsenic is a trace element located naturally in the earth's crust (Abbas et al. 2018; Bayhan et al. 2020; Dökmeci 2020; Mohebbrad et al. 2019; Yapıcıoğlu and Yeşilnacar 2022). As is a toxic element that occurs in various oxidation states and different forms such as arsenate (As(V)) and arsenide (As(III)). (Bilgin Simsek et al. 2018; Jamali-Behnam et al. 2018; Niazi et al. 2018; Oyem et al. 2015; Parviainen et al. 2015; Yapıcıoğlu and Yeşilnacar 2022; Yapıcıoğlu et al. 2020). Arsenic could not be easily disposed and could only be converted into several forms or insoluble composites (Agrafioti et al. 2014; Lata and Samadder 2016; Mohan and Pittman 2007; Vithanage et al. 2017).

Pure arsenic is rare at natural conditions and is not very toxic, besides it could be oxidized easily and convert to toxic forms. The micro level of arsenic is effective on human health. Hence, the allowable value is lower while comparing to other toxic metals (10 ppb) according to World Health Organization (WHO) (Vithanage et al. 2017). Humans are exposed to various degrees of inorganic arsenic with polluted potable water, using polluted water in meal and consuming of contaminated crops and smoking (WHO 2014).

Long-term exposure to inorganic arsenic has been resulted from potable water and food leading to chronic poisoning. Seafood is the other source of arsenic, besides exposure from these foods is usually less than exposure through polluted groundwater (WHO 2014). The biggest danger for human health from arsenic has resulted from polluted groundwater. Inorganic arsenic is naturally forming at high concentrations in the groundwater in several numbers of countries (WHO 2014). Arsenic is used in the glass, pigments, textiles, paper, and metal adhesives processing industries. Arsenic is also used in the in pesticides (especially insecticides) for agricultural activities and pharmaceuticals. People who smoke could be exposed to the natural inorganic arsenic content (WHO 2014).

12.2.2 Cadmium (Cd)

Cadmium is a toxic metal that could be found in the environment either in its elemental form or in many salts. Cadmium has located in air in the form of particles, where cadmium oxide is potentially a significant component (Friberg et al. 1986; Jazini et al. 2018; Li et al. 2018; Sdiri 2018). Cadmium is generated majorly as a by-product from mining, smelting, and refining (Satarug 2019; Zhu et al. 2020). Cadmium is emitted to the environment from not only natural but also anthropogenic supplies. The main natural resources of cadmium are volcanoes and weathering of rocks (WHO 2014). Waste incineration process is an anthropogenic resource of cadmium. Tobacco is a significant supply of cadmium exposure for smoking people when tobacco plants take up cadmium from the soil. A considerable part of cadmium is related to agricultural products (Genchi et al. 2020). Potable water contamination might form because of the presence of cadmium in galvanized pipes, heaters, coolers, or taps. A maximum acceptable concentration of 0.003 mg/L is reported for total cadmium in potable water (WHO 2011). Crops are the major source of cadmium uptake for nonoccupationally exposed human (Genchi et al. 2020).

12.2.3 Chromium (Cr)

Chromium is a natural trace metal in rocks, plants, animal soil, volcanic dust, and gases. It is widely in the form of chromium(III), which is a significant mineral, and chromium(VI) (Hexavalent Chromium) most generated by industrial processes is very toxic (Guo et al. 2020; Zhang et al. 2017; Zhao et al. 2018). Recently, toxic metal contamination has been playing a crucial role for human health due to its cumulative and nondegradable characterization (Guo et al. 2020; Zhou et al. 2016). Chrome electroplating, leather tanning, porcelain and ceramics manufacturing, wood preservatives, paints/pigments, and effluents from chemical plants are major anthropogenic sources that release chromium into the environment (Herath et al. 2021; Zhitkovich 2011). Chromium is a trace metal that majorly presents in the form of Cr(VI) and Cr(III) and has been defined as a top-priority pollutant by the U.S. Environmental Protection Agency due to its severe teratogenicity (Guo et al. 2020; Hajizadeh et al. 2020; Ma et al. 2014; Naghdali et al. 2020; Ravi and Anuradha Jabasingh 2018; Rudakiya et al. 2018). Chromium is generally present in nature as hexavalent Cr(VI) (Herath et al. 2021; Zhitkovich 2011). WHO has recommended a maximum allowable concentration of chromium in potable water as 0.05 ppm (WHO 2017).

12.2.4 Lead (Pb)

Lead is a cumulative toxic element that infused body systems and is harmful to children. Lead is a natural toxic element located in the earth's crust. Lead use has caused water, air, and soil pollution and serious health problems in several areas of the world (Kwak et al. 2019; Liu and Zhang 2009; Shen et al. 2019). Significant resources of pollution contain mining, smelting, production, and recycling processes, and, in several areas, the continued use of lead containing paint and lead-containing fuel (Abdullah et al. 2018; Han et al. 2017; Jianfui et al. 2019; Wu et al. 2021). Potable water can be contaminated with lead pipes. Another supply of exposure is the use of medicines and cosmetics (Wang et al. 2015; Zhang et al. 2018).

Several people could be exposed to lead with industrial and environmental resources. This exposure due to inhalation of lead particles produced by burning substances including lead and ingestion of lead-contaminated dust, lead-contaminated water, and food (Kwak et al. 2019). According to drinking water standards, maximum allowable concentration of lead is 10 ppb (WHO 2011).

12.3 The Effects of Toxic Metals on Human Health

Heavy metals have unfavorable impacts on public health, and exposure to these toxic substances has been raised due to industrial, agricultural, and anthropogenic activities and industrialization (Balali-Mood et al. 2021; Coen et al. 2001; Rehman et al. 2018). Pollution of water and air resulted from toxic metals is an environmental challenge for many people who have been affected in the world (Cooper et al. 2020; Sarkar et al. 2008). The toxic impacts of these metals, even though they have no biological role, recalcitrant in form harmful for the human health. These metals generally play as a pseudo element of the body when they might interfere with metabolic processes (Jaishankar et al. 2014). Drinking water and air pollution are the main influence method to human health of heavy metal. Acute or chronic poisonings might be resulted from heavy metal exposure by the way of water, air, and food. Bioaccumulation of the heavy metals has led to a diversity of toxic impacts on several body tissues and organs (Balali-Mood et al. 2021). Many acute and chronic toxic impacts of heavy metals impose various organs of human. Gastrointestinal and kidney dysfunction, skin lesions and irritations, vascular damage, nervous system disorders, immune system dysfunction, birth defects, and cancer are well-known complications of toxic metals (Gazwi et al. 2020; Lee et al. 2016; Kim et al. 2020, 2015). Simultaneous exposure to one more heavy metal might have synergistic impacts (Balali-Mood et al. 2021).

Heavy metal toxicity has lower energy levels and disrupt the functioning of the brain, lungs, kidney, liver, blood composition, and other significant human organs (Kim et al. 2015; Rahman and Singh 2019; Yolay et al. 2021). Long-term exposure to heavy metals could cause to physical, muscular, cardiovascular, and neurological disruption processes, which triggers disorders such as multiple sclerosis, Alzheimer's disease, Parkinson's disease, and muscular dystrophy (Jaishankar et al. 2014; Yolay et al. 2021).

In this section, the effects of arsenic, cadmium, chromium, and lead on human health were presented. Table 12.1 shows the health effects of these heavy metals in details.

There are several strategies in order to protect human health from toxic metal exposures. According to the findings, heavy metal-contaminated drinking water is the main source could be taken under control. Especially groundwater resources contain high amount of toxic metal due to their natural composition and geogenic structure. Potable water treatment is a prevention for heavy metal exposure. Groundwater is regarded as the primary ssupply of potable water in several countries all over the world. In this chapter, drinking water protection strategies were considered. There are several conventional and innovative water treatment methods to remove heavy metals in recent years. In the next section, biochar adsorption process as an innovative technology for drinking water treatment will be defined for heavy metal removal from groundwater resources. In this context, arsenic, lead, cadmium, and chromium removal from groundwater supplies using biochar application was investigated.

12.4 Toxic Metal Removal with Biochar Adsorption

The exposure of toxic metals has led to severe health problems such as cancer. From this point of view, toxic metal pollution should be removed from groundwater using proper and adequate treatment methods. Groundwater supplies have considerable toxic metal concentrations due to its natural composition. There are several conventional and innovative treatment techniques of toxic metals removal from groundwater (Gupt et al. 2021; Luo et al. 2021; Song et al. 2021; Yapıcıoğlu et al. 2020). Coagulation and flocculation, adsorption processes, membrane processes, advanced oxidation processes (AOPs), electrocoagulation, and biochar application (Abbasi et al. 2021; Agrawal et al. 2021; Asif et al. 2021; Eslami et al. 2022; Kushawaha and Aithani 2021; Pal et al. 2021; Siddiq et al. 2021; Song et al. 2021). The difficulty and high operational cost by conventional treatment systems for toxic metal removal have made several researchers to study on efficient, eco-friendly, and cheaper groundwater treatment methods (Siddiq et al. 2021). Among these innovative treatment techniques, biochar has been globally applied as an adsorbent for the groundwater treatment.

Table 12.1 The effects of toxic metals on human health.

Toxic metal	The effects on human health	References
Arsenic (As)	Inorganic arsenic is regarded as carcinogen and is the most important pollutant in potable water all over the world. Arsenic may also be in an organic form. Inorganic arsenic composites are highly toxic, while organic arsenic composites are less harmful to human health. The early symptoms of acute arsenic poisoning contain vomiting, abdominal pain, and diarrhea. The other acute effects are numbness and tingling of the extremities, muscle cramping, and death, in extreme events. The symptoms of long-term exposure are usually monitored in the skin and contain pigmentation changes, skin lesions, and hard patches on the palms and hyperkeratosis. Apart from skin cancer, long-term exposure to arsenic might also lead to lungs and bladder cancers. The International Agency for Research on Cancer (IARC) has categorized arsenic as carcinogenic. Diabetes, pulmonary disease, and cardiovascular disease are the other disorders in the result of long-term exposure to arsenic. In China (Province of Taiwan), arsenic exposure has been named as "Blackfoot disease," which is a serious disorder of blood vessels causing to gangrene. Arsenic is also correlated with adverse pregnancy outcomes and infant mortality. It has led to mortality in young adults due to lung disease, heart attacks, and kidney failure in early childhood. Many studies have showed unfavorable effects of arsenic exposure on intelligence and memory.	Jaishankar et al. (2014), Zaynab et al. (2022), Yolay et al. (2021), Radfard et al. (2018), Sheikhi et al. (2021), Berkani et al. (2021), Varol et al. (2021), Alsubih et al. (2021), Rahman et al. (2021), Long et al. (2021), Ahmed et al. (2021), Bibi et al. (2021), Sheng et al. (2021), Das et al. (2021), Mushtaq et al. (2021), Ali et al. (2021), Singh et al. (2021), Biswas et al. (2021), Sarwar et al. (2021), Liu et al. (2021), Patel et al. (2021), Sathe et al. (2021)
Cadmium (Cd)	Cadmium is classified as carcinogenic for humans by IARC. It might be regarded as lung carcinogen, also inducer of prostatic or renal cancers. The severe damage is that cadmium could disrupt testosterone generation. Cadmium influenced the resorption function of the proximal tubules, the first symptom being an increase in the urinary excretion, and it could lead to the prostate lesions. Cadmium damages people with inhalation and parenteral routes of exposure. With chronic oral exposure, the kidney is the most sensitive organ. Cadmium has affected blood, hair, and nails on long-term exposure. Cadmium has been stored in body for many times and leads to the mutations and chromosomal deletions.	Jaishankar et al. (2014), Zaynab et al. (2022), Rahimzadeh et al. (2017), Sheng et al. (2021), Gujre et al. (2021), Cheng et al. (2021), Ma et al. (2021), Yang et al. (2021), Mahajan et al. (2021), Cheng et al. (2021), Wijeyaratne and Kumari (2021), Wang et al. (2021), Wang et al. (2022), Gu et al. (2021), Fu et al. (2021), Gómez et al. (2021)
Chromium (Cr)	Chromium(III) has been described as a trace element, which is essential for living organisms. However, chromium(VI) compounds are toxic and carcinogenic. The bronchial tree is the main part of the body affected by carcinogenic Cr(VI) compounds, and cancer forms with inhalation exposure, uptake in the respiratory organs. IARC has declared that chromium is under the group 1 carcinogenic for humans. Ulcers, acute irritative dermatitis, and allergic dermatitis are reported in the result of chromium(VI) exposure as the main disorders. Many epidemiological case studies have been performed on the human exposure to chromium compounds and the risk of cancer, especially lung cancer, but only less measurement of exposure was reported.	Christou et al. (2021), Karunanidhi et al. (2021), Infante et al. (2021), Manoj et al. (2021), Ratnalu and Dhakate (2021), Wijeyaratne and Kumari (2021), Kazemi et al. (2022), Ceballos et al. (2021), Ugulu et al. (2021), Vijayakumar et al. (2021), Chandio et al. (2021), Oginawati et al. (2021)

(Continued)

Table 12.1 (Continued)

Toxic metal	The effects on human health	References
Lead (Pb)	Lead has been emitted to the brain, liver, kidney, and bone, and it has been stocked in the teeth and bones in the human bodies. Lead exposure has been evaluated with the measurement of lead in blood. Young children are especially vulnerable to lead poisoning among people. In addition, innate curiosity and their age-appropriate hand-to-mouth behavior result of children could cause to exposure. Young children in Nigeria, Senegal, and other countries have been affected by lead poisoning. When lead affects the body, it is damaged to organs such as the brain, kidneys, liver, and bones. The body stocks lead in the teeth and bones. Lead storage in bone might be emitted into the blood during pregnancy, so exposing the fetus. Young children and fetus are at highest risk of lead exposure. Lead exposure could have important effects for children health. At higher stages of exposure, lead affects the brain and central nervous system, leading to coma, convulsions, and death. Children who expose to serious lead poisoning might have intellectual disability and behavioral diseases. Also, lead could damage children's brain development, leading to decreased intelligence quotient (IQ), behavioral changes such as decreased attention span and raised antisocial behavior. Furthermore, anemia, hypertension, renal impairment, and immunotoxicity could be resulted in exposure to lead. The neurological and behavioral impacts of lead have been observed.	Ali et al. (2021), Wijeyaratne and Kumari (2021), Nag and Cummins (2022), Gómez et al. (2021), Lai et al. (2021), Amjad et al. (2021), Kumar et al. (2021), Shi et al. (2021), Wen et al. (2021), Zaynab et al. (2022), Boudebbouz et al. (2021), Sanaei et al. (2021), Widyantoro et al. (2021), Binh et al. (2021), Gupta et al. (2021), Javorac et al. (2022), Habibian et al. (2021), Hou et al. (2021), Yan et al. (2022), Sarlak et al. (2021), Słota et al. (2022), Maitlo et al. (2021), Ochoa-Martínez et al. (2021), Shawahna (2021), Kubicz et al. (2021), Galiciolli et al. (2022), Almerud et al. (2021), Paltseva et al. (2022), Rogers et al. (2021), Jallad et al. (2021), Ahmed et al. (2021), Mehouel and Fowler (2021)

Biochar is generated by pyrolysis of biomass such as agricultural crops, fertilizers, and solid wastes (Cuong et al. 2022; Paz-Ferreiro et al. 2018; Sadhu et al. 2021; Yapıcıoğlu and Yeşilanacar 2022; Yuan et al. 2017). Traditionally, biochar is carried out for carbon sequestration due to its adsorption capacity in soils and wastewaters so it could be a greenhouse gas mitigation technique (Cayuela et al. 2014; Chi et al. 2021; Demir and Yapıcıoğlu 2018; Qambrani et al. 2017; Xu et al. 2021). Biochar has been widely applied to enhance nutrient availability in soils and plant growth so it could be used as soil amendment (Cheng et al. 2021; Kamali et al. 2022; Yuan et al. 2021, 2017). Biochar is recently focused on adsorption of pollutants from groundwater by several researchers (Cheng et al. 2021; Yapıcıoğlu and Yeşilnacar 2022). Biochar is considered as a new technique to toxic metal removal from groundwater (Yapıcıoğlu and Yeşilnacar 2022).

There are limited studies related to this topic. Yapıcıoğlu and Yeşilanar (2022) investigated arsenic removal from groundwater supplies in Harran Plain. They reported arsenic removal efficiency nearly 60% using agricultural waste derived biochar. Niazi et al. (2018) used perilla leaf biochar for arsenic removal biochar. They found 37–39% of removal efficiency. Cuong et al. (2022) investigated arsenic removal from groundwater using biochar. They reported high efficiency (94%) of redox transformation of As(III) to As(V). Tabassum et al. (2019) reported that all the biosorbents and biochar efficiently removed (50–100%) As from groundwater. Another study was performed by He et al. (2018). They reported 70% of arsenic removal efficiency using biochar. Chen et al. (2021) reported a study including chromium could be removed from groundwater using biochar. Song et al. (2021) performed a study on chromium removal from groundwater using magnetic biochar. They underlined an effective Cr(VI) adsorbent, which could remove toxic metal from groundwater. Sarkar et al. (2008) investigated chromium removal using rice-husk-derived biochar. Abid et al. (2017) investigated cadmium removal from groundwater. They confirmed that cotton-derived biochar could adsorb cadmium from groundwater.

12.5 Conclusions and Recommendations

Arsenic, lead, cadmium, and chromium are the effective carcinogenic toxic metals for human health. Drinking water and food consumption are the main resources of toxic metals. Groundwater resources have a high heavy metal content due to its natural composition. Proper and adequate treatment of groundwater resources is a control and prevention strategy of heavy metal exposure. Biochar adsorption process is a new technique to toxic metal removal from groundwater. Further research is required to determine the effective mechanisms in biochar application to remove heavy metals from groundwater.

References

Abbas, Y., Majeed, S., Ali, S. et al. (2018). Design of gravel-sand filter for arsenic removal: a case study of Muzaffargarh district in Pakistan. *Water Environment Research* 90 (12): 2106–2113.

Abbasi, M., Safari, E., Baghdadi, M., and Janmohammadi, M. (2021). Enhanced adsorption of heavy metals in groundwater using sand columns enriched with graphene oxide: lab-scale experiments and process modeling. *Journal of Water Process Engineering* 40: 101961.

Abdullah, F., Mohd Yusoff, A.R., Wan Abu Bakar, W.A. et al. (2018). Preparation, characterization, and lead removal appraisal of zinc aluminate prepared at different calcination temperatures. *Journal of the Chinese Chemical Society* 65 (10): 1199–1209.

Abid, M., Danish, S., Zafar-ul-Hye, M. et al. (2017). Biochar increased photosynthetic and accessory pigments in tomato (*Solanum lycopersicum* L.) plants by reducing cadmium concentration under various irrigation waters. *Environmental Science and Pollution Research* 24 (27): 22111–22118.

Adimalla, N., Qian, H., and Nandan, M.J. (2020). Groundwater chemistry integrating the pollution index of groundwater and evaluation of potential human health risk: a case study from hard rock terrain of South India. *Ecotoxicology and Environmental Safety* 206: 111217.

Agrafioti, E., Kalderis, D., and Diamadopoulos, E. (2014). Arsenic and chromium removal from water using biochars derived from rice husk, organic solid wastes and sewage sludge. *Journal of Environmental Management* 133: 309–314.

Agrawal, P.R., Singhal, S., and Sharma, R. (2021). Heavy metal contamination in groundwater sources. In: *Groundwater Geochemistry: Pollution and Remediation Methods*, 57–78. https://doi.org/10.1002/9781119709732.ch4.

Ahmad, M., Rajapaksha, A.U., Lim, J.E. et al. (2014). Biochar as a sorbent for contaminant management in soil and water: a review. *Chemosphere* 99: 19–33.

Ahmed, M.F., Mokhtar, M.B., and Alam, L. (2021). Carcinogenic and non-carcinogenic health risk of arsenic ingestion via drinking water in Langat River basin,

Malaysia. *Environmental Geochemistry and Health* 43 (2): 897–914.

Alhashimi, H.A. and Aktas, C.B. (2017). Life cycle environmental and economic performance of biochar compared with activated carbon: a meta-analysis. *Resources, Conservation and Recycling* 118: 13–26.

Ali, H., Khan, E., and Ilahi, I. (2019). Environmental chemistry and ecotoxicology of hazardous heavy metals: environmental persistence, toxicity, and bioaccumulation. *Journal of Chemistry* 2019: 1–14.

Ali, N., Eqani, S.A.M.A.S., Nazar, E. et al. (2021). Arsenic and lead in the indoor residential settings of different socio-economic status; assessment of human health risk via dust exposure. *Environmental Science and Pollution Research* 28 (11): 13288–13299.

Almerud, P., Zamaratskaia, G., Lindroos, A.K. et al. (2021). Cadmium, total mercury, and lead in blood and associations with diet, sociodemographic factors, and smoking in Swedish adolescents. *Environmental Research* 197: 110991.

Alsubih, M., El Morabet, R., Khan, R.A. et al. (2021). Occurrence and health risk assessment of arsenic and heavy metals in groundwater of three industrial areas in Delhi, India. *Environmental Science and Pollution Research* 28 (44): 63017–63031.

Amjad, M., Iqbal, M.M., Abbas, G. et al. (2021). Assessment of cadmium and lead tolerance potential of quinoa (*Chenopodium quinoa Willd*) and its implications for phytoremediation and human health. *Environmental Geochemistry and Health* 1–14. https://doi.org/10.1007/s10653-021-00826-0.

Arruti, A., Fernández-Olmo, I., and Irabien, A. (2010). Evaluation of the contribution of local sources to trace metals levels in urban PM2.5 and PM10 in the Cantabria region (Northern Spain). *Journal of Environmental Monitoring* 12: 1451–1458.

Asif, M.B., Price, W.E., Fida, Z. et al. (2021). Acid mine drainage and sewage impacted groundwater treatment by membrane distillation: organic micropollutant and metal removal and membrane fouling. *Journal of Environmental Management* 291: 112708.

Awasthi, M.K., Wang, Q., Ren, X. et al. (2016). Role of biochar amendment in mitigation of nitrogen loss and greenhouse gas emission during sewage sludge composting. *Bioresource Technology* 219: 270–280.

Baki, M.A., Hossain, M.M., Akter, J. et al. (2018). Concentration of heavy metals in seafood (fishes, shrimp, lobster and crabs) and human health assessment in Saint Martin Island, Bangladesh. *Ecotoxicology and Environmental Safety* 159: 153–163.

Balali-Mood, M., Naseri, K., Tahergorabi, Z. et al. (2021). Toxic mechanisms of five heavy metals: Mercury, Lead, Chromium, Cadmium, and Arsenic. *Frontiers in Pharmacology* 12: 643972.

Bayhan, İ., Yeşilnacar, M.İ., Yetiş, A.D., and Tutkun, E. (2020). An evaluation of drinking-usage water quality in terms of environmental health: a case study of Siverek (Sanliurfa), Turkey. *Turkish Bulletin of Hygiene & Experimental Biology/ Türk Hijyen ve Deneysel Biyoloji* 77: 107–120.

Berkani, M., Vasseghian, Y., Dragoi, E.N., and Khaneghah, A.M. (2021). The Fenton-like reaction for arsenic removal from groundwater: health risk assessment. *Environmental Research* 202: 111698.

Bhattacharya, S. (2020). The role of Spirulina (Arthrospira) in the mitigation of heavy-metal toxicity: an appraisal. *Journal of Environmental Pathology, Toxicology and Oncology* 39 (2): 149–157.

Bibi, I., Shahid, M., Niazi, N.K. et al. (2021). Hydrogeochemical and health risk evaluation of arsenic in shallow and deep aquifers along the different floodplains of Punjab, Pakistan. *Journal of Hazardous Materials* 402: 124074.

Bilgili, A.V., Yeşilnacar, İ., Akihiko, K. et al. (2018). Post-irrigation degradation of land and environmental resources in the harran plain, Southeastern Turkey. *Environmental Monitoring and Assessment* 190 (11): 1–14.

Bilgin Simsek, E., Novak, I., Berek, D., and Beker, U. (2018). Novel composite sorbents based on carbon fibers decorated with ferric hydroxides – arsenic removal. *Asia-Pacific Journal of Chemical Engineering* 13 (5): e2237.

Binh, N.T.L., Hoang, N.T., Truc, N.T.T. et al. (2021). Estimating the possibility of lead contamination in soil surface due to Lead deposition in atmosphere. *Journal of Nanomaterials* 2021: https://doi.org/10.1155/2021/5586951.

Biswas, B., Chakraborty, A., Chatterjee, D. et al. (2021). Arsenic exposure from drinking water and staple food (rice): a field scale study in rural Bengal for assessment of human health risk. *Ecotoxicology and Environmental Safety* 228: 113012.

Boudebbouz, A., Boudalia, S., Bousbia, A. et al. (2021). Heavy metals levels in raw cow milk and health risk assessment across the globe: a systematic review. *Science of the Total Environment* 751: 141830.

Bradl, H. (2002). *Heavy Metals in the Environment: Origin, Interaction and Remediation*, vol. 6. London: Academic.

Cayuela, M.L., van Zwieten, L., Singh, B.P. et al. (2014). Biochar's role in mitigating soil nitrous oxide emissions: a review and meta-analysis. *Agriculture, Ecosystems and Environment* 191: 5–16.

Ceballos, E., Dubny, S., Othax, N. et al. (2021). Assessment of human health risk of chromium and nitrate pollution in groundwater and soil of the Matanza-Riachuelo River basin, Argentina. *Exposure and Health* 13: 1–14.

Chandio, T.A., Khan, M.N., Muhammad, M.T. et al. (2021). Health risk assessment of chromium contamination in the nearby population of mining plants, situated at Balochistan, Pakistan. *Environmental Science and Pollution Research* 28 (13): 16458–16469.

Cheng, N., Wang, B., Wu, P. et al. (2021). Adsorption of emerging contaminants from water and wastewater by modified biochar: a review. *Environmental Pollution* 273: 116448.

Chi, N.T.L., Anto, S., Ahamed, T.S. et al. (2021). A review on biochar production techniques and biochar based catalyst for biofuel production from algae. *Fuel* 287: 119411.

Christou, A., Georgiadou, E.C., Zissimos, A.M. et al. (2021). Uptake of hexavalent chromium by *Lactuca sativa and Triticum* aestivum plants and mediated effects on their performance, linked with associated public health risks. *Chemosphere* 267: 128912.

Coen, N., Mothersill, C., Kadhim, M., and Wright, E.G. (2001). Heavy metals of relevance to human health induce genomic instability. *The Journal of Pathology: A Journal of the Pathological Society of Great Britain and Ireland* 195 (3): 293–299.

Cooper, A.M., Felix, D., Alcantara, F. et al. (2020). Monitoring and mitigation of toxic heavy metals and arsenic accumulation in food crops: a case study of an urban community garden. *Plant Direct* 4 (1): e00198.

Corrias, F., Atzei, A., Addis, P. et al. (2020). Integrated environmental evaluation of heavy metals and metalloids bioaccumulation in invertebrates and seaweeds from different marine coastal areas of Sardinia, Mediterranean Sea. *Environmental Pollution* 266: 115048.

Cuong, D.V., Wu, P.C., Liou, S.Y.H., and Hou, C.H. (2022). An integrated active biochar filter and capacitive deionization system for high-performance removal of arsenic from groundwater. *Journal of Hazardous Materials* 423: 127084.

Das, A., Joardar, M., De, A. et al. (2021). Pollution index and health risk assessment of arsenic through different groundwater sources and its load on soil-paddy-rice system in a part of Murshidabad district of West Bengal, India. *Groundwater for Sustainable Development* 15: 100652.

Demir, Ö. and Yapıcıoğlu, P., (2018). Sludge-derived biochar application for the reduction of greenhouse gas emissions: a reviw. *7th Global Conference on Global Warming (GCGW-2018),* Izmir, Turkey(24–28 June 2018). pp. 362–366.

Dissanayake, C.B. and Chandrajith, R. (2009). *Introduction to Medical Geology*. Springer science & business media.

Dökmeci, A.H. (2020). Environmental impacts of heavy metals and their bioremediation. In: *Heavy Metals-their Environmental Impacts and Mitigation*. IntechOpen https://doi.org/10.5772/intechopen.95103.

Emslay, J. (2011). *Nature's Building Blocks*. Oxford: Oxford University Press ISBN: 978-0-19-960563-7.

Eslami, H., Esmaeili, A., Razaeian, M. et al. (2022). Potentially toxic metal concentration, spatial distribution, and health risk assessment in drinking groundwater resources of Southeast Iran. *Geoscience Frontiers* 13 (1): 101276.

Friberg, L., Nordberg, G.F., and Vouk, V.B. (ed.) (1986). *Handbook of the Toxicology of Metals*, vol. II, 130–184. Amsterdam: Elsevier.

Fu, T., Zhao, R., Hu, B. et al. (2021). Novel framework for modelling the cadmium balance and accumulation in farmland soil in Zhejiang Province, East China: sensitivity analysis, parameter optimisation, and forecast for 2050. *Journal of Cleaner Production* 279: 123674.

Galiciolli, M.E.A., Lima, L.S., da Costa, N.D.S. et al. (2022). IQ alteration induced by lead in developed and underdeveloped/developing countries: a systematic review and a meta-analysis. *Environmental Pollution* 292: 118316.

Gazwi, H.S., Yassien, E.E., and Hassan, H.M. (2020). Mitigation of lead neurotoxicity by the ethanolic extract of Laurus leaf in rats. *Ecotoxicology and Environmental Safety* 192: 110297.

Genchi, G., Sinicropi, M.S., Lauria, G. et al. (2020). The effects of cadmium toxicity. *International Journal of Environmental Research and Public Health* 17 (11): 3782.

Gómez, S.N., Marchisio, P.F., Balverdi, M.P., and Sales, A.M. (2021). Lead, cadmium and arsenic levels and health risk evaluation of marketed chia seeds in Tucumán, Argentina. *Journal of Food Composition and Analysis* 103: 104093.

Gu, X., Xu, L., Wang, Z. et al. (2021). Assessment of cadmium pollution and subsequent ecological and health risks in Jiaozhou Bay of the Yellow Sea. *Science of the Total Environment* 774: 145016.

Gu, Y.G., Ning, J.J., Ke, C.L., and Huang, H.H. (2018). Bioaccessibility and human health implications of heavy metals in different trophic level marine organisms: a case study of the South China Sea. *Ecotoxicology and Environmental Safety* 163: 551–557.

Gujre, N., Rangan, L., and Mitra, S. (2021). Occurrence, geochemical fraction, ecological and health risk assessment of cadmium, copper and nickel in soils contaminated with municipal solid wastes. *Chemosphere* 271: 129573.

Guo, X., Liu, A., Lu, J. et al. (2020). Adsorption mechanism of hexavalent chromium on biochar: kinetic, thermodynamic, and characterization studies. *ACS Omega* 5 (42): 27323–27331.

Gupt, C.B., Kushwaha, A., Prakash, A. et al. (2021). Mitigation of groundwater pollution: heavy metal retention characteristics of fly ash-based liner materials. In: *Fate and Transport of Subsurface Pollutants*, 79–104. Singapore: Springer https://doi.org/10.1007/978-981-15-6564-9_5.

Guzzi, G., Ronchi, A., and Pigatto, P. (2021). Toxic effects of mercury in humans and mammals. *Chemosphere* 263: 127990.

Habibian, A., Abyadeh, M., Abyareh, M. et al. (2021). Association of maternal lead exposure with the risk of preterm: a meta-analysis. *The Journal of Maternal-Fetal & Neonatal Medicine* 1–9. https://doi.org/10.1080/14767058.2021.1946780.

Hajizadeh, Z., Valadi, K., Taheri-Ledari, R., and Maleki, A. (2020). Convenient Cr(VI) removal from aqueous samples: executed by a promising clay-based catalytic system, magnetized by Fe_3O_4 nanoparticles and functionalized with humic acid. *ChemistrySelect* 5 (8): 2441–2448.

Han, L., Qian, L., Liu, R. et al. (2017). Lead adsorption by biochar under the elevated competition of cadmium and aluminum. *Scientific Reports* 7 (1): 1–11.

Hao, Z., Chen, L., Wang, C. et al. (2019). Heavy metal distribution and bioaccumulation ability in marine organisms from coastal regions of Hainan and Zhoushan, China. *Chemosphere* 226: 340–350.

He, R., Peng, Z., Lyu, H. et al. (2018). Synthesis and characterization of an iron-impregnated biochar for aqueous arsenic removal. *Science of the Total Environment* 612: 1177–1186.

He, Z.L., Yang, X.E., and Stoffella, P.J. (2005). Trace elements in agroecosystems and impacts on the environment. *Journal of Trace Elements in Medicine and Biology* 19 (2–3): 125–140.

Herath, A., Reid, C., Perez, F. et al. (2021). Biochar-supported polyaniline hybrid for aqueous chromium and nitrate adsorption. *Journal of Environmental Management* 296: 113186.

Hou, S., Dong, H., Du, X., and Feng, L. (2021). Early warning on risk development in compound lead and cadmium contaminated sites. *Journal of Hazardous Materials* 416: 126174.

Ilhan, N., Demir Yetiş, A., Yeşilnacar, M.İ., and Atasoy, A.D.S. (2021). Predictive modelling and seasonal analysis of water quality indicators: three different basins of Sanliurfa, Turkey. *Environment, Development and Sustainability* 1–35. https://doi.org/10.1007/s10668-021-01566-y.

Infante, E.F., Dulfo, C.P., Dicen, G.P. et al. (2021). Bioaccumulation and human health risk assessment of chromium and nickel in paddy rice grown in serpentine soils. *Environmental Science and Pollution Research* 28 (14): 17146–17157.

Jaishankar, M., Tseten, T., Anbalagan, N. et al. (2014). Toxicity mechanism and health effects of some heavy metals. *Interdisciplinary Toxicology* 7 (2): 60–72.

Jallad, R., Rao, M.S., and Rahman, A. (2021). Prevalence of lead toxicity in adolescents in Kuwait. *BMC Public Health* 21 (1): 1–10.

Jamali-Behnam, F., Najafpoor, A.A., Davoudi, M. et al. (2018). Adsorptive removal of arsenic from aqueous solutions using magnetite nanoparticles and silica-coated magnetite nanoparticles. *Environmental Progress & Sustainable Energy* 37 (3): 951–960.

Javorac, D., Tatović, S., Anđelković, M. et al. (2022). Low-lead doses induce oxidative damage in cardiac tissue: subacute toxicity study in Wistar rats and benchmark dose modelling. *Food and Chemical Toxicology* 161: 112825.

Jazini, R., Soleimani, M., and Mirghaffari, N. (2018). Characterization of barley straw biochar produced in various temperatures and its effect on lead and cadmium removal from aqueous solutions. *Water Environment Journal* 32 (1): 125–133.

Jianfui, C., Sarjadi, M.S., Musta, B. et al. (2019). Synthesis of sawdust-based Poly (amidoxime) ligand for heavy metals removal from wastewater. *ChemistrySelect* 4 (11): 2991–3001.

Jyothi, N.R. and Farook, N.A. (2020). *Heavy Metal Toxicity in Public Health*. UK: Intech Open.

Kamali, M., Sweygers, N., Al-Salem, S. et al. (2022). Biochar for soil applications-sustainability aspects, challenges and future prospects. *Chemical Engineering Journal* 428: 131189.

Karunanidhi, D., Aravinthasamy, P., Subramani, T. et al. (2021). Chromium contamination in groundwater and Sobol sensitivity model based human health risk evaluation from leather tanin industrial region of South India. *Environmental Research* 199: 111238.

Kaur, M., Kumar, A., Mehra, R., and Kaur, I. (2020). Quantitative assessment of exposure of heavy metals in groundwater and soil on human health in Reasi district, Jammu and Kashmir. *Environmental Geochemistry and Health* 42 (1): 77–94.

Kaur, M., Kumar, A., Mehra, R., and Mishra, R. (2018). Human health risk assessment from exposure of heavy metals in soil samples of Jammu district of Jammu and Kashmir, India. *Arabian Journal of Geosciences* 11 (15): 1–15.

Kazemi, A., Esmaeilbeigi, M., Sahebi, Z., and Ansari, A. (2022). Health risk assessment of total chromium in the qanat as historical drinking water supplying system. *Science of the Total Environment* 807: 150795.

Kim, H.S., Kim, Y.J., and Seo, Y.R. (2015). An overview of carcinogenic heavy metal: molecular toxicity mechanism and prevention. *Journal of Cancer Prevention* 20 (4): 232.

Kim, J.J., Kim, Y.S., and Kumar, V. (2019). Heavy metal toxicity: an update of chelating therapeutic strategies. *Journal of Trace Elements in Medicine and Biology* 54: 226–231.

Kim, T.H., Kim, J.H., Le Kim, M.D. et al. (2020). Exposure assessment and safe intake guidelines for heavy metals in consumed fishery products in the Republic of Korea. *Environmental Science and Pollution Research* 27 (26): 33042–33051.

Komatina, M.M. (2004). *Medical Geology: Effects of Geological Environments on Human Health*. Elsevier.

Kończak, M. and Oleszczuk, P. (2018). Application of biochar to sewage sludge reduces toxicity and improve organisms growth in sewage sludge-amended soil in long term field experiment. *Science of the Total Environment* 625: 8–15.

Krishnamoorthy, R., Venkatramanan, V., Senthilkumar, M. et al. (2019). Management of heavy metal polluted soils: perspective of arbuscular mycorrhizal fungi. In: *Sustainable Green Technologies for Environmental Management*, 67–85. Singapore: Springer https://doi.org/10.1007/978-981-13-2772-8_4.

Kubicz, J., Lochyński, P., Pawełczyk, A., and Karczewski, M. (2021). Effects of drought on environmental health risk posed by groundwater contamination. *Chemosphere* 263: 128145.

Kumar, S., Rahman, A., Islam, R. et al. (2021). Lead and other elements-based pollution in soil, crops and water near a lead-acid battery recycling factory in Bangladesh. *Chemosphere* 290: 133288.

Kushawaha, J. and Aithani, D. (2021). Geogenic pollutants in groundwater and their removal techniques. In: *Groundwater Geochemistry: Pollution and Remediation Methods*, 1–21. https://doi.org/10.1002/9781119709732.ch1.

Kwak, J.H., Islam, M.S., Wang, S. et al. (2019). Biochar properties and lead (II) adsorption capacity depend on feedstock type, pyrolysis temperature, and steam activation. *Chemosphere* 231: 393–404.

Lai, L., Li, B., Li, Z.R. et al. (2021). Pollution and health risk assessment of heavy metals in farmlands and vegetables surrounding a lead-zinc mine in Yunnan Province, China. *Soil and Sediment Contamination: An International Journal* 31 (4): 1–15.

Lata, S. and Samadder, S.R. (2016). Removal of arsenic from water using nano adsorbents and challenges: a review. *Journal of Environmental Management* 166: 387–406.

Lee, J.K., Choi, Y.L., Kwon, M., and Park, P.J. (2016). Mechanisms and consequences of cancer genome instability: lessons from genome sequencing studies. *Annual Review of Pathology: Mechanisms of Disease* 11: 283–312.

Lehmann, J. and Joseph, S. (2015). *Biochar for Environmental Management: An Introduction*, 33–46. Routledge.

Lehmann, J., Rillig, M.C., Thies, J. et al. (2011). Biochar effects on soil biota–a review. *Soil Biology and Biochemistry* 43 (9): 1812–1836.

Li, R., Liang, W., Huang, H. et al. (2018). Removal of cadmium(II) cations from an aqueous solution with aminothiourea chitosan strengthened magnetic biochar. *Journal of Applied Polymer Science* 135 (19): 46239.

Liu, Y., Tian, X., Cao, S. et al. (2021). Pollution characteristics and health risk assessment of arsenic transformed from feed additive organ arsenicals around chicken farms in the North China plain. *Chemosphere* 278: 130438.

Liu, Z. and Zhang, F.S. (2009). Removal of lead from water using biochars prepared from hydrothermal liquefaction of biomass. *Journal of Hazardous Materials* 167 (1–3): 933–939.

Liyanage, D.N.D., Diyabalanage, S., Dunuweera, S.P. et al. (2022). Significance of Mg-hardness and fluoride in drinking water on chronic kidney disease of unknown etiology in Monaragala, Sri Lanka. *Environmental Research* 203: 111779.

Long, X., Liu, F., Zhou, X. et al. (2021). Estimation of spatial distribution and health risk by arsenic and heavy metals in shallow groundwater around Dongting Lake plain using GIS mapping. *Chemosphere* 269: 128698.

Luo, Z., Zhu, J., Yu, L., and Yin, K. (2021). Heavy metal remediation by nano zero-valent iron in the presence of microplastics in groundwater: inhibition and induced promotion on aging effects. *Environmental Pollution* 287: 117628.

Ma, L., Liu, Y., Wu, Y. et al. (2021). The effects and health risk assessment of cauliflower co-cropping with *Sedum alfredii* in cadmium contaminated vegetable field. *Environmental Pollution* 268: 115869.

Ma, Y., Liu, W.J., Zhang, N. et al. (2014). Polyethylenimine modified biochar adsorbent for hexavalent chromium removal from the aqueous solution. *Bioresource Technology* 169: 403–408.

Mahajan, M., Gupta, P.K., Singh, A. et al. (2021). A comprehensive study on aquatic chemistry, health risk and remediation techniques of cadmium in groundwater. *Science of the Total Environment* 818: 151784.

Maitlo, A.A., Jatoi, W.B., Memon, A.F. et al. (2021). Assessment of zinc, lead, chromium, and cobalt in commonly consumed herbal medicines in Sindh, Pakistan. *Biological Trace Element Research* 199 (6): 2366–2374.

Manoj, S., RamyaPriya, R., and Elango, L. (2021). Long-term exposure to chromium contaminated waters and the associated human health risk in a highly contaminated industrialised region. *Environmental Science and Pollution Research* 28 (4): 4276–4288.

Masindi, V. and Muedi, K.L. (2018). Environmental contamination by heavy metals. *Heavy Metals* 10: 115–132.

Medina-Estévez, F., Zumbado, M., Luzardo, O.P. et al. (2020). Association between heavy metals and rare earth elements with acute ischemic stroke: a case-control study conducted in the Canary Islands (Spain). *Toxics* 8 (3): 66.

Mehouel, F. and Fowler, S.W. (2021). Review of the toxic trace elements arsenic, cadmium, lead and mercury in seafood species from Algeria and contiguous waters in the Southwestern Mediterranean Sea. *Environmental Science and Pollution Research* 1–14. https://doi.org/10.1007/s11356-021-17130-0.

Metcalf & Eddy (2014). *Wastewater Engineering: Treatment and Resource Recovery*, 5the. New York: McGraw-Hill international Editions.

Mohan, D. and Pittman, C.U. Jr. (2007). Arsenic removal from water/wastewater using adsorbents – a critical review. *Journal of Hazardous Materials* 142 (1–2): 1–53.

Mohebbrad, B., Bonyadi, Z., Dehghan, A.A., and Rahmat, M.H. (2019). Arsenic removal from aqueous solutions using *Saccharomyces cerevisiae*: kinetic and equilibrium study. *Environmental Progress & Sustainable Energy* 38 (s1): S398–S402.

Mushtaq, N., Masood, N., Khattak, J.A. et al. (2021). Health risk assessment and source identification of groundwater arsenic contamination using agglomerative hierarchical cluster analysis in selected sites from upper Eastern parts of Punjab province, Pakistan. *Human and Ecological Risk Assessment: An International Journal* 27 (4): 999–1018.

Nag, R. and Cummins, E. (2022). Human health risk assessment of lead (Pb) through the environmental-food pathway. *Science of the Total Environment* 810: 151168.

Naghdali, Z., Sahebi, S., Mousazadeh, M., and Jamali, H.A. (2020). Optimization of the forward osmosis process using aquaporin membranes in chromium removal. *Chemical Engineering & Technology* 43 (2): 298–306.

Niazi, N.K., Bibi, I., and Shahid, M. (2018). Arsenic removal by perilla leaf biochar in aqueous solutions and groundwater: an integrated spectroscopic and microscopic examination. *Environmental Pollution* 232: 31–41.

Ochoa-Martínez, Á.C., Varela-Silva, J.A., Orta-García, S.T. et al. (2021). Lead (Pb) exposure is associated with changes in the expression levels of circulating miRNAS (miR-155, miR-126) in Mexican women. *Environmental Toxicology and Pharmacology* 83: 103598.

Oginawati, K., Susetyo, S.H., Rosalyn, F.A. et al. (2021). Risk analysis of inhaled hexavalent chromium (Cr 6+) exposure on blacksmiths from industrial area. *Environmental Science and Pollution Research* 28 (11): 14000–14008.

Oyem, H.H., Oyem, I.M., and Usese, A.I. (2015). Iron, manganese, cadmium, chromium, zinc and arsenic groundwater contents of Agbor and Owa communities of Nigeria. *Springerplus* 4 (1): 1–10.

Pal, D.B., Tiwari, A.K., and Giri, D.D. (2021). Various purification techniques of groundwater. In: *Groundwater Geochemistry: Pollution and Remediation Methods* (ed. Madhaw and Singh), 310–325. Wiley Online Library.

Paltseva, A.A., Deeb, M., Di Iorio, E. et al. (2022). Prediction of bioaccessible lead in urban and suburban soils with Vis-NIR diffuse reflectance spectroscopy. *Science of the Total Environment* 809: 151107.

Parviainen, A., Loukola-Ruskeeniemi, K., Tarvainen, T. et al. (2015). Arsenic in bedrock, soil and groundwater – the first arsenic guidelines for aggregate production established in Finland. *Earth-Science Reviews* 150: 709–723.

Patel, A.K., Singh, A., Das, N., and Kumar, M. (2021). Health risk associated with consumption of arsenic contaminated groundwater in the Ganga and the Brahmaputra floodplain of India. *Case Studies in Chemical and Environmental Engineering* 3: 100103.

Patniak, P. (2010). *Handbook of Environmental Analysis. Chemical Pollutants in Air, Water, Soil and Solid Wastes*, 2ee. Boca Raton, FL: CRC Press Taylor, and Francis Group.

Paz-Ferreiro, J., Nieto, A., Méndez, A. et al. (2018). Biochar from biosolids pyrolysis: a review. *International Journal of Environmental Research and Public Health* 15: 956.

Puga, A.P., Grutzmacher, P., Cerri, C.E.P. et al. (2020). Biochar-based nitrogen fertilizers: greenhouse gas emissions, use efficiency, and maize yield in tropical soils. *Science of the Total Environment* 704: 135375.

Qambrani, N.A., Rahman, M.M., and Won, S. (2017). Biochar properties and eco-friendly applications for climate change mitigation, waste management, and wastewater treatment: a review. *Renewable and Sustainable Energy Reviews* 79: 255–273.

Radfard, M., Yunesian, M., Nabizadeh, R. et al. (2018). Drinking water quality and arsenic health risk assessment in Sistan and Baluchestan, Southeastern Province, Iran. *Human and Ecological Risk Assessment: An International Journal* 25: 1–17.

Rahimzadeh, M.R., Rahimzadeh, M.R., Kazemi, S., and Moghadamnia, A.A. (2017). Cadmium toxicity and treatment: an update. *Caspian Journal of Internal Medicine* 8 (3): 135.

Rahman, M., Kumar, S., Lamb, D., and Rahman, M.M. (2021). Health risk assessment of arsenic, manganese, and iron from drinking water for high school children. *Water, Air, & Soil Pollution* 232 (7): 1–13.

Rahman, Z. and Singh, V.P. (2019). The relative impact of toxic heavy metals (THMs) (arsenic (As), cadmium (Cd), chromium (Cr)(VI), mercury (Hg), and lead (Pb)) on the total environment: an overview. *Environmental Monitoring and Assessment* 191 (7): 1–21.

Rajapaksha, A.U., Chen, S.S., Tsang, D.C. et al. (2016). Engineered/designer biochar for contaminant removal/immobilization from soil and water: potential and implication of biochar modification. *Chemosphere* 148: 276–291.

Ratnalu, G.V. and Dhakate, R. (2021). Human health hazard evaluation with reference to chromium (Cr^{+3} and Cr^{+6}) in groundwater of Bengaluru Metropolitan City, South India. *Arabian Journal of Geosciences* 14 (23): 1–19.

Ravi, T. and Anuradha Jabasingh, S. (2018). Preparation and characterization of higher degree-deacetylated chitosan-coated magnetic adsorbent for the removal of chromium(VI) from its aqueous mixture. *Journal of Applied Polymer Science* 135 (9): 45878.

Rehman, K., Fatima, F., Waheed, I., and Akash, M.S.H. (2018). Prevalence of exposure of heavy metals and their impact on health consequences. *Journal of Cellular Biochemistry* 119 (1): 157–184.

Rogers, C.R., Lee, T.H., Fry, C.M., and Telzer, E.H. (2021). Where you lead, I will follow: exploring sibling similarity in brain and behavior during risky decision making. *Journal of Research on Adolescence* 31 (1): 34–51.

Rudakiya, D.M., Iyer, V., Shah, D. et al. (2018). Biosorption potential of phanerochaete chrysosporium for arsenic, cadmium, and chromium removal from aqueous solutions. *Global Challenges* 2 (12): 1800064.

Sadhu, M., Bhattacharya, P., Vithanage, M., and Sudhakar, P.P. (2021). *Adsorptive Removal of Fluoride Using Biochar–A Potential Application in Drinking Water Treatment*, 119106. Separation and Purification Technology.

Sanaei, F., Amin, M.M., Alavijeh, Z.P. et al. (2021). Health risk assessment of potentially toxic elements intake via food crops consumption: Monte Carlo simulation-based probabilistic and heavy metal pollution index. *Environmental Science and Pollution Research* 28 (2): 1479–1490.

Sarkar, S.K., Cabral, H., Chatterjee, M. et al. (2008). Biomonitoring of heavy metals using the bivalve molluscs in Sunderban mangrove wetland, northeast coast of Bay of Bengal (India): possible risks to human health. *Clean-Soil, Air, Water* 36 (2): 187–194.

Sarlak, Z., Hosseini, H., Garavand, F. et al. (2021). The occurrence of Lead in animal source foods in Iran in the 2010s decade: a systematic review. *Biological Trace Element Research* 200: 1–20.

Sarwar, T., Khan, S., Muhammad, S., and Amin, S. (2021). Arsenic speciation, mechanisms, and factors affecting rice uptake and potential human health risk: a systematic review. *Environmental Technology & Innovation* 22: 101392.

Satarug, S. (2019). Cadmium sources and toxicity. *Toxics* 7 (2): 25.

Sathe, S.S., Mahanta, C., and Subbiah, S. (2021). Hydrogeochemical evaluation of intermittent alluvial aquifers controlling arsenic and fluoride contamination and corresponding health risk assessment. *Exposure and Health* 13: 1–20.

Sdiri, A. (2018). Physicochemical characterization of natural dolomite for efficient removal of lead and cadmium in aqueous systems. *Environmental Progress & Sustainable Energy* 37 (6): 2034–2041.

Alloway, B., Centeno, J.A., Finkelman, R.B. et al. (2016). *Essentials of Medical Geology* (ed. O. Selinus). Springer.

Selinus, O., Finkelman, R.B., and Centeno, J.A. (2010). *Medical Geology: A Regional Synthesis*. Springer Science & Business Media.

Shawahna, R. (2021). Breast milk to blood lead ratios among women from the West Bank of Palestine: a cross-sectional study of associated factors. *International Breastfeeding Journal* 16 (1): 1–13.

Sheikhi, S., Faraji, Z., and Aslani, H. (2021). Arsenic health risk assessment and the evaluation of groundwater quality using GWQI and multivariate statistical analysis in rural areas, Hashtroud, Iran. *Environmental Science and Pollution Research* 28 (3): 3617–3631.

Shen, Z., Hou, D., Jin, F. et al. (2019). Effect of production temperature on lead removal mechanisms by rice straw biochars. *Science of the Total Environment* 655: 751–758.

Sheng, D., Wen, X., Wu, J. et al. (2021). Comprehensive probabilistic health risk assessment for exposure to arsenic and cadmium in groundwater. *Environmental Management* 67 (4): 779–792.

Shi, J., Du, P., Luo, H. et al. (2021). Characteristics and risk assessment of soil polluted by Lead around various metal mines in China. *International Journal of Environmental Research and Public Health* 18 (9): 4598.

Siddiq, O.M., Tawabini, B.S., Soupios, P., and Ntarlagiannis, D. (2021). Removal of arsenic from contaminated groundwater using biochar: a technical review. *International Journal of Environmental Science and Technology* 1–14.

Singh, R., Gayen, A., Kumar, S., and Dewangan, R. (2021). Geo-spatial distribution of arsenic contamination of groundwater resources in intricate crystalline aquifer system of Central India: arsenic toxicity manifestation and health risk assessment. *Human and Ecological Risk Assessment: An International Journal* 27 (6): 1588–1612.

Słota, M., Wąsik, M., Stołtny, T. et al. (2022). Effects of environmental and occupational lead toxicity and its association with iron metabolism. *Toxicology and Applied Pharmacology* 434: 115794.

Song, J., Huang, G., Han, D. et al. (2021). A review of reactive media within permeable reactive barriers for the removal of heavy metal (loid) s in groundwater: current status and future prospects. *Journal of Cleaner Production* 319: 128644.

Stern, B.R. (2010). Essentiality and toxicity in copper health risk assessment: overview, update and regulatory considerations. *Journal of Toxicology and Environmental Health, Part A* 73 (2–3): 114–127.

Suvarapu, L.N. and Baek, S.O. (2017). Determination of heavy metals in the ambient atmosphere: a review. *Toxicology and Industrial Health* 33 (1): 79–96.

Tabassum, R.A., Shahid, M., Niazi, N.K. et al. (2019). Arsenic removal from aqueous solutions and groundwater using agricultural biowastes-derived biosorbents and biochar: a column-scale investigation. *International Journal of Phytoremediation* 21 (6): 509–518.

Tchounwou, P.B., Yedjou, C.G., Patlolla, A.K., and Sutton, D.J. (2012). Heavy metal toxicity and the environment. *Molecular, Clinical and Environmental Toxicology* 101: 133–164.

U.S. Environmental Protection Agency (USEPA) (2019). National Elemental Mercury Response Guidebook. https://response.epa.gov/ MercuryResponseGuidebook.htm. (accessed 1 January 2022).

Ugulu, I., Khan, Z.I., Safdar, H. et al. (2021). Chromium bioaccumulation by plants and grazing livestock as affected by the application of sewage irrigation water: implications to the food Chain and health risk. *International Journal of Environmental Research* 15 (2): 261–274.

Varol, M., Gündüz, K., and Sünbül, M.R. (2021). Pollution status, potential sources and health risk assessment of arsenic and trace metals in agricultural soils: a case study in Malatya province, Turkey. *Environmental Research* 202: 111806.

Vijayakumar, C.R., Balasubramani, D.P., and Azamathulla, H.M. (2021). Assessment of groundwater quality and human health risk associated with chromium exposure in the industrial area of Ranipet, Tamil Nadu, India. *Journal of Water, Sanitation and Hygiene for Development* 12: 58–67.

Vithanage, M., Herath, I., Joseph, S. et al. (2017). Interaction of arsenic with biochar in soil and water: a critical review. *Carbon* 113: 219–230.

Vithanage, M., Rajapaksha, A.U., Tang, X. et al. (2014). Sorption and transport of sulfamethazine in agricultural soils amended with invasive-plant-derived biochar. *Journal of Environmental Management* 141: 95–103.

Wang, M.Y., Li, M.Y., Ning, H. et al. (2022). Cadmium oral bioavailability is affected by calcium and phytate contents in food: evidence from leafy vegetables in mice. *Journal of Hazardous Materials* 424: 127373.

Wang, M., Chen, Z., Song, W. et al. (2021). A review on cadmium exposure in the population and intervention strategies against cadmium toxicity. *Bulletin of Environmental Contamination and Toxicology* 106 (1): 65–74.

Wang, Z., Liu, G., Zheng, H. et al. (2015). Investigating the mechanisms of biochar's removal of lead from solution. *Bioresource Technology* 177: 308–317.

Wen, E., Yang, X., Chen, H. et al. (2021). Iron-modified biochar and water management regime-induced changes in plant growth, enzyme activities, and phytoavailability of arsenic, cadmium and lead in a paddy soil. *Journal of Hazardous Materials* 407: 124344.

WHO (2017). *Guidelines for Drinking-Water Quality Fourth Edition Incorporating the First Addendum*. Geneva: WHO.

WHO (2011). *Guidelines for Drinking-Water Quality*, 4e, vol. 38.4, 104–108. WHO Chronicle.

Widyantoro, A., Maziya, F.B., and Abidin, A.U. (2021). Exposure analysis of lead (Pb) and chromium (Cr) in workplace workshop PT. X. *Gaceta Sanitaria* 35: S450–S454.

Wijeyaratne, W.M. and Kumari, E.A. (2021). Cadmium, chromium, and Lead uptake associated health risk assessment of Alternanthera sessilis: a commonly consumed green leafy vegetable. *Journal of Toxicology* 2021: 9936254.

World Health Organization (2014). *WHO Handbook for Guideline Development*. World Health Organization.

Wu, P., Wu, X., Xu, H., and Owens, G. (2021). Interfacial solar evaporation driven lead removal from a contaminated soil. *EcoMat* 3 (5): e12140.

Xu, Z., He, M., Xu, X. et al. (2021). Impacts of different activation processes on the carbon stability of biochar for oxidation resistance. *Bioresource Technology* 338: 125555.

Yan, J., Zhang, H., Niu, J. et al. (2022). Effects of lead and cadmium co-exposure on liver function in residents near a mining and smelting area in Northwestern China. *Environmental Geochemistry and Health* 1–17. https://doi.org/10.1007/s10653-021-01177-6.

Yang, D., Yang, S., Wang, L. et al. (2021). Performance of biochar-supported nanoscale zero-valent iron for cadmium and arsenic co-contaminated soil remediation: insights on availability, bioaccumulation and health risk. *Environmental Pollution* 290: 118054.

Yapıcıoğlu, P. and Yeşilnacar, M.İ. (2022). Grey water footprint assessment of groundwater resources in Southeastern Turkey: effect of recharge. *Water Supply* 22: 615–627.

Yapıcıoğlu, P., Derin, P., and Yeşilnacar, M.I. (2020). Assessment of harran plain groundwater in terms of arsenic contamination. *Geological Bulletin of Turkey* 63 (1): 137–144.

Yenigun, I., Bilgili, A.V., Yesilnacar, M.I., and Yalcin, H. (2021). Seasonal and spatial variations in water quality of deep aquifer in the harran plain, GAP project, Southeastern Anatolia, Turkey. *Environmental Earth Sciences* 80 (17): 1–19.

Yeşilnacar, M.I. and Güllüoğlu, M.S. (2008). Hydrochemical characteristics and the effects of irrigation on groundwater quality in harran plain, GAP project, Turkey. *Environmental Geology* 54: 183–196.

Yeşilnacar, M.İ. and Yenigün, İ. (2011). Effect of irrigation on a deep aquifer: a case study from semi-arid harran plain, GAP project, Turkey. *Bulletin of Engineering Geology and the Environment* 70: 213–221.

Yetis, A.D., Yesilnacar, M.I., and Atas, M. (2021). A machine learning approach to dental fluorosis classification. *Arabian Journal of Geosciences* 14 (2): 1–12.

Yolay, O., Kasapbaşı E.E., Tezcan, E. et al. (2021). Investigation of heavy metals in coronary artery tissues in autopsy cases: Time for paradigm change?. *Proceedings Asia PCR 2021* (8 October 2021). Virtual Concept.

Yuan, C., Gao, B., Peng, Y. et al. (2021). A meta-analysis of heavy metal bioavailability response to biochar aging: importance of soil and biochar properties. *Science of the Total Environment* 756: 144058.

Yuan, Y., Bolan, N., Prévoteau, A. et al. (2017). Applications of biochar in redox-mediated reactions. *Bioresource Technology* 246: 271–281.

Zaynab, M., Al-Yahyai, R., Ameen, A. et al. (2022). Health and environmental effects of heavy metals. *Journal of King Saud University-Science* 34 (1): 101653.

Zhang, J., Chen, S., Zhang, H., and Wang, X. (2017). Removal behaviors and mechanisms of hexavalent chromium from aqueous solution by cephalosporin residue and derived chars. *Bioresource Technology* 238: 484–491.

Zhang, Y., Yan, K., Ji, F., and Zhang, L. (2018). Enhanced removal of toxic heavy metals using swarming biohybrid adsorbents. *Advanced Functional Materials* 28 (52): 1806340.

Zhang, Z., Zhu, Z., Shen, B., and Liu, L. (2019). Insights into biochar and hydrocar production and applications: a review. *Energy* 171: 581–598.

Zhao, N., Yin, Z., Liu, F. et al. (2018). Environmentally persistent free radicals mediated removal of Cr(VI) from highly saline water by corn straw biochars. *Bioresource Technology* 260: 294–301.

Zhitkovich, A. (2011). Chromium in drinking water: sources, metabolism, and cancer risks. *Chemical Research in Toxicology* 24 (10): 1617–1629.

Zhou, F., Wang, H., Fang, S. et al. (2015). Pb(II), Cr(VI) and atrazine sorption behavior on sludge-derived biochar: role of humic acids. *Environmental Science and Pollution Research* 22: 16031–16039.

Zhou, L., Liu, Y., Liu, S. et al. (2016). Investigation of the adsorption-reduction mechanisms of hexavalent chromium by ramie biochars of different pyrolytic temperatures. *Bioresource Technology* 218: 351–359.

Zhu, Y., Wang, H., Lv, X. et al. (2020). Effects of biochar and biofertilizer on cadmium-contaminated cotton growth and the antioxidative defense system. *Scientific Reports* 10 (1): 1–12.

Zibret, G., Gosar, M., Miler, M., and Alijagić, J. (2018). Impacts of mining and smelting activities on environment and landscape degradation – Slovenian case studies. *Land Degradation & Development* 29 (12): 4457–4470.

Section 4

Medical Pedology: Health Effects from Soils and Sediments

13

Dynamics of Trace Element Bioavailability in Soil

Agronomic Enhancement and Risk Assessment

Rohan D'Souza[1], *Paulo J. C. Favas*[2,3], *Mayank Varun*[4], *and Manoj S. Paul*[1]

[1] *Department of Botany, St. John's College, Dr. B.R. Ambedkar University, Agra, Uttar Pradesh, India*
[2] *School of Life Sciences and the Environment, University of Trás-os-Montes e Alto Douro, (UTAD) Vila Real, Portugal*
[3] *MARE – Marine and Environmental Sciences Centre/ARNET – Aquatic Research Network, Department of Life Sciences, University of Coimbra, Coimbra, Portugal*
[4] *Department of Botany, Hislop College, Nagpur, Maharashtra, India*

13.1 Introduction

Soil contamination has been a global environmental concern since the last century. TEs form an important group of soil constituents that can be referred to as contaminants when they exceed their background concentrations in an area. Many of these are not a part of natural soil profiles; at least not in the amounts they occur now at most places. Man has been responsible in most cases. Air pollution from vehicular and industrial sources, mining and smelting processes, intensive agricultural practices, etc. all produce contaminants that end up in the soil. Some TEs like Fe, Cu, and Zn occur in low concentrations in the soil. These are referred to as essential TEs with recommended daily medicinal and dietary allowances. Even these can be toxic in excess (Marschner 1995; Paul et al. 2014). Others like Pb, Cd, Cr, and As are categorized as nonessential. They have no known well-known beneficial function in humans and their consumption even at very low concentrations can be toxic (Kamunda et al. 2016). These are a major concern as they are not readily removed or degraded by chemical or microbial processes and accumulate in soils. This is especially true in developing countries where population boom, expeditious industrial development, and insufficient funds for environmental rehabilitation compound the issue (Bhargava et al. 2007; Rai et al. 2019).

Potentially harmful TEs (such as Pb, As, Hg, and Cd, among others) in soil can be absorbed by plants and subsequently transferred through food chain resulting in serious health problems to animals and humans. When absorbed by the body, TEs accumulate in vital organs such as the liver, lungs, kidneys, brain, and bones for years, causing severe health problems (Borrell et al. 2016; Sihlahla et al. 2020). The associated risk increases in proportion to the consumption (Pan et al. 2010). Pb and Cd are the most toxic TEs and affect human health directly or indirectly via dietary intake. General signs of Pb, Cd, As, Hg, Zn, Cu, and Al poisoning include gastrointestinal disorders, tremors, ataxia, paralysis, and vomiting. The effects can be toxic (acute, chronic, or subchronic), neurotoxic, carcinogenic, mutagenic, or teratogenic (Rai et al. 2019; Chaturvedi et al. 2021).

Ever increasing populations around the world are migrating from rural to urban areas, which translates into an ever-increasing demand for food. Much agricultural land can now be categorized as peri-urban mainly due to urban spread. Such areas are in proximity to industries, and there are high chances of soil contamination through industrial effluents (Antisari et al. 2015; Pinto et al. 2020; Chaturvedi et al. 2021). Riverwater contaminated with industrial effluents or municipal wastewater is used in peri-urban areas for irrigation because it is cheap and easily available compared to fresh water. Water scarcity is also a factor that comes into play. Often farmers are forced to use wastewater to irrigate their fields. This is especially true for developing countries. TEs in effluents are nondegradable and poorly soluble in water. They tend to accumulate in soils and subsequently in plants (Ghoneim et al. 2014; Khan et al. 2021). In many cases, marginal farmers grow vegetables along riverbanks of polluted rivers. Such crops are perishable by nature, and proximity to urban areas ensures regular demand and income. Multiple harvesting over a short life cycle also makes vegetable crops attractive for such sites. However, cultivation in polluted soils or regular irrigation with contaminated/wastewater can lead to potential accumulation of TE in both soils and plants. Many vegetables

Medical Geology: En route to One Health, First Edition. Edited by Majeti Narasimha Vara Prasad and Meththika Vithanage.

including eggplant, tomato, pea, amaranths, spinach, mustard, cabbage, and sugarbeet have been reported to accumulate high amounts of TEs under such cultivation practices (Prasad, 2001; Sun et al. 2007; Poniedziałek et al. 2010; Jolly et al. 2013; Ahmed et al. 2016; Hamid et al. 2016; Farid et al. 2018; Jabeen et al. 2020).

Plants absorb TEs from their rhizosphere as part of their natural metabolic activities. Essential nutrients and toxic TEs are absorbed by the same processes, hence under increased concentrations of nonessential elements, plants are not able to distinguish between the two ions. When present at high enough concentrations, these ion species "compete" and plant roots begin accumulating the dominant ions present in the rhizosphere (Magdziak et al. 2014). Some species can accumulate them at quite high concentrations. In fact, roots of vegetable plants can readily absorb them even at low concentrations, and they can be accumulated in edible parts at high concentrations (Yang et al. 2009).

13.2 Bioavailability of Trace Elements in Contaminated Soils

The mobility and bioavailability of TEs mostly depend on their chemical form, binding type, or physical processes that they are subjected to the soil (Kumpiene et al. 2017). Soil properties determine the growing conditions and regulate the distribution of contaminants in the soil–liquid phase, since plants uptake TEs only if they are present in the soil solution. The most important factor is the mobility of elements in the soil and their bioavailability to plant roots, which may result in further transport of TE-containing sap to shoots. Understanding the bioavailability concept is vital since plants are able to uptake the contaminants only when they are present in available forms in the soil environment. Pb and As are relatively strongly sorbed by soil particles and are not readily transported to shoots. Cd and Zn are very mobile in soil and are easily bioaccumulated by plants.

Bioavailability can be defined as the fraction of the total amount of a contaminant in the soil that is available or can become available for uptake by organisms in a given time span (van Gestel 2008). It depends on soil properties and processes. The soil system has three phases – solid, liquid, and gaseous – along with a living phase. The first three form a complex system with characteristic reactions and processes like adsorption–desorption and precipitation–dissolution. The living phase includes the microbes that can drastically alter these processes (Petruzzelli et al. 2020).

The TE content in soil is partitioned in different forms; the most bioavailable forms being simple or complex soluble ions in soil solution, exchangeable forms adsorbed on mineral particles, and those adsorbed or complexed by organic matter. They undergo distribution between phases, desorption–adsorption, degradation, etc. Chemical, physical, and biological conditions regulate their availability, i.e. transfer from solid to solution phase. Such reactions are dependent on the chemical characteristics of TEs and more importantly on characteristics of soil such as pH, organic matter, and clay content, cation exchange capacity, and redox potential (Petruzzelli et al. 2014).

The bioavailability of TEs in soil and subsequent uptake in plants may be affected by their concentrations in soil, physico-chemical characteristics of soil and plant species (Altarawneh 2019, Liu et al. 2021a). Soil pH can promote the adsorption/desorption of TEs by regulating the quantity of hydrogen ions. It is very important in determining the geochemical mobility of TEs, e.g. bioavailability of Pb, Zn, Ni, and Cd increases with decreasing pH, while that of As and Cr increases with increasing pH. Cu solubility significantly increases at pH 5.5. (Zhang et al. 2012; Magdziak et al. 2014; Liu et al. 2021b). Manures, compost, and chelates like EDTA can also increase the bioavailability of TE ions in the rhizosphere.

The living phase of the soil complex, i.e. rhizospheric microorganisms can significantly increase the bioavailability of TEs (Vamerali et al. 2010; Ali et al. 2013). They also play a vital part in plant establishment. Bacteria, especially plant growth-promoting rhizobacteria (PGPR), mineralize organic matter, improving the bioavailability and effectiveness of nutrient uptake (Dominguez-Crespo et al. 2012). They also help plants in tolerating the stress due to elevated contaminant levels in soils. However, the same mechanisms that enhance nutrient bioavailability may also increase TE bioavailability and consequently their accumulation in plant tissues.

Fungi also provide stress tolerance and a symbiotic boost to plants. Vesicular-arbuscular mycorrhiza (VAM) and arbuscular-mycorrhizal fungi (AMF) are known to improve plant establishment and growth. They are used as biofertilizers to eliminate or reduce the need for chemical fertilizers (Berruti et al. 2016). Promotion of TE ion uptake in AMF-assisted plants has been reported (Sarathambal et al. 2017; Chaturvedi et al. 2021). This may be attributed to molecules, e.g. organic acids secreted by fungi that activate the substrate TE complexion, detoxification in the rhizosphere, and may increase accumulation efficiency (Magdziak et al. 2012; Vaněk et al. 2012).

Salinity can also change the bioavailability of TEs in soil. It is a key factor in translocation of TEs from roots to aerial parts of the plant (Manousaki et al. 2008). Manousaki and Kalogerakis (2009) have reported increased Cd uptake in *Atriplex halimus* L. with increasing salinity because of higher bioavailability of the metal in soil. Chelating agents like EDTA, which promote desorption of TEs from the solid phase, can also increase their concentration in the soil solution where they may be accumulated by plants. Soil characteristics thus determine TE bioavailability, and the same plant species exhibit different TE uptake and accumulation patterns in different soils.

13.3 Case Study

It is known that chronic intake of toxic TEs, through soil and vegetables, has adverse effects on humans, which may become apparent only after several years of exposure. Arsenic (in particular, As-III) is considered a human carcinogen even at very low levels of exposure. Prolonged exposure to it has also been linked to peripheral nerve mutilation that might cause diabetes (Hu 2002; Focus et al. 2021). Pb is considered dangerous and a possible carcinogen. It also interrupts the normal functioning of the nervous and reproductive systems, joints, and kidneys and encourages renal tumors (Kamunda et al. 2016; Focus et al. 2021). Cadmium is also considered toxic at low levels and a likely carcinogen. Long-term exposure to Cd has been linked to headaches, arthritis, cirrhosis, kidney dysfunction, cardiovascular and pulmonary diseases, anemia, diabetes, as well as reduced fertility (Kamunda et al. 2016; Ngole-Jeme et al. 2018). Although Zn is an essential TE, it has been reported to cause noncarcinogenic effects to organs when taken in extremely excessive levels (Arfaenia 2016).

Agra (27°10′N, 78°05′E, 169 msl), on the banks of the river Yamuna, is in north central India, roughly 200 km south-east of New Delhi. Famous for the Taj Mahal and its leather and marble handicrafts industry, it is the fourth most populated city in the state of Uttar Pradesh with over 2.2 million people. It boasts a cast iron and engineering goods industry with approx. 168 foundries, 52 tanneries, 300 shoe industries, 200 petha industries, 50 dairies, 56 electroplating units, 15 silver vibrators, and 15 galvanizing units. The waste and effluents containing TEs released by the industries along with the residential waste and agricultural run-off is a problem the city faces. The river Yamuna is the lifeline of the city, and its current pollution levels pose a grave risk to the environment and human population.

Soils in areas adjacent to the river represent a sink for TEs and a large variety of chemicals. Contaminants from anthropogenic processes and activities finally reach the soil where they are "locked away" from other compartments of the biosphere. However, such soils as well as those irrigated with wastewater may be hazardous to humans and animals when used for cultivation (Jolly et al. 2013; Morgan 2013). Even when present at low concentrations in effluents, TEs can slowly accumulate to significant amounts in the soil over time with regular irrigation. An estimated 20 million hectares of arable land is irrigated with wastewater globally. Studies conducted in several Asian and African countries suggest that around 50% vegetable cultivation is based on wastewater irrigation (Bjuhr 2007).

Stream sediments and soils from many cities in India, viz. – Agra, Delhi, Kanpur, and Lucknow have been classified as highly polluted with TEs in earlier studies (Singh et al. 2002; D'Souza et al. 2013; Varun et al. 2014). The first two are on the banks of Yamuna. Reports highlighting TE and pesticide pollution of vegetables grown along the banks of the Yamuna are frequent. Paul et al. (2014) reported fairly high concentrations of Pb and Zn (241–285 mg Pb/kg; 129–222 mg Zn/kg) and appreciable amounts of Cr, Ni, and As at sites near the river. In 2015, the National Green tribunal (NGT) banned the cultivation of edible crops and fodder on Yamuna floodplains due to contamination risks. The bench noted "It is an established fact that presently, vegetables, fodder grown and allied projects at the flood plain of river Yamuna are highly contaminated. Besides containing ingredients of high pollutants, such produce is even found to contain metallic pollutants…" A later study conducted by NEERI (National Environmental Engineering Research Institute) in 2019 revealed trace element contamination of vegetables grown on Yamuna floodplains in New Delhi. Seven vegetables – cabbage, cauliflower, radish, eggplant, fenugreek, spinach, and coriander – were tested for Pb, Cd, Hg, and Ni. Only Pb levels exceeded the FSSAI limits (2.5 mg/kg) (Food Safety and Standards Authority of India) in most of vegetable samples (2.8–13.8 mg/kg). Thus, after four years of the ban on cultivation, the panel appointed to monitor the pollution of the river recommended that farmers be allowed to grow crops and vegetables with the provision of regular monitoring.

Plants, whether cultivated crops or wild, are not just passive photosynthetic machines, and they alter the chemical and physical properties of rhizospheric soil by their root exudates. The amount and exact chemistry of these exudates are highly variable based on plant species, physiology, root proliferation, and even abiotic factors like climate and soil composition. Usually they are a mixture of inorganic ions like H^+, OH^- etc. gases like CO_2, O_2, as well as soluble organic acids, phenols, amino acids, sugars, and particulate organic matter (mucilage, residues cell) (Dakora and Phillips 2002). Root exudates influence not only the number and extent of the rhizoflora but also the aggregation of soil particles. They can also increase or decrease the bioavailability of TEs in the rhizosphere directly or indirectly through changes in pH and redox potential, release of chelates that can form complexes with TEs, and an increase in microbial activity (Petruzzelli et al. 2014).

Anthropogenic change of soil properties can increase or decrease the bioavailability of TEs in soil and thus affect their uptake by plants and the subsequent risk to human and animal health.

Soils at contaminated sites are often very poor in terms of physico-chemical characteristics as well as soil microorganisms and fauna. Root growth and proliferation are also hindered due to a scarce nutrient supply. Agronomic amendments like manure, compost, fertilizers and plant microbial consortia like PGPR, VAM, and AMF are often used in such cases that can, however, make the problem worse. These change the rhizosphere environment by modifying pH, redox conditions, and the chemical speciation of TEs (Lin et al. 2010). Higher uptake of arsenic and mercury has been reported with the use of phosphate and thiosulfate fertilizers in contaminated soils (Pedron et al. 2013). Unwittingly, marginal farmers can alter the bioavailability of TE ions in the rhizosphere and their subsequent uptake in crops through such practices.

13.3.1 Experimental Design

In the light of the above, the greenhouse study outlined here was undertaken to specifically evaluate the changes in bioavailability of four TEs through uptake and accumulation patterns of three vegetables and a fodder crop in response to three common agronomic amendments. The vegetables include *Chenopodium album* (lamb's quarters/ white goosefoot), *Amaranthus viridis* (green amaranth), and *Abelmoschus esculentus* (okra). The leaves and tender shoots of the first two and the green pods of okra are common vegetables. The fodder crop *Sesbania sesban* (Egyptian riverhemp) is also used as green manure. The agronomic amendments selected for the study are – farmyard manure (FYM), AMF as *Glomus mosseae*, and PGPR as *Pseudomonas flourescens*. The following plan was used:

TE concentration 1
TE concentration 2
TE 1 + TE 2 + TE 3 + TE 4
TE concentration 1 + FYM,
TE concentration 1 + AMF,
TE concentration 1 + PGPR,

Concentration 1 for Pb and As was 50 mg/kg soil; for Cd 25 mg/kg soil; and for Zn 200 mg/kg soil. Concentration 2 for Pb and As was 100 mg/kg soil; for Cd 50 mg/kg soil; and for Zn 400 mg/kg soil. Multimetal treatment was applied as combination of Pb, Cd, As, and Zn at concentration 1. Pots were prepared with 5 kg garden soil with a 1.5 in. layer of gravel at the bottom. Soil (0–30 cm) was collected, sieved through 5 mm mesh, and autoclaved to get rid of soil-borne pathogens. It was a sandy-loam with pH 6.9; electrical conductivity 0.61 dS/m; and moderate organic matter content (1.58%). Available nitrogen and available phosphate contents were 128.3 and 24.5 kg/ha, respectively.

The soil was artificially spiked with $Pb(NO_3)_2$, $CdCl_2.5H_2O$, $ZnSO_4.7H_2O$, and As_3S_2, individually and in combination. FYM was applied at 1 : 3 v/v; *G. mosseae* at 50 g/pot, ~80 spores, and *P. flourescens* was inoculated at ~10^8 colony-forming units per plant by dipping the roots in nutrient broth culture while transplanting.

Seeds of the test plants were sowed in germination trays. Uniform seedlings were transplanted to pots and allowed to mature for three weeks. All treatments were maintained in triplicate. Pots were kept in randomized block design and watered as and when required so as to prevent leaching. Plants were harvested upon maturity (90 days) and divided into root, shoot, and fruit. Plant samples were air dried and digested using aqua regia (HCl/HNO_3 at 3 : 1 v/v). The digested samples were examined for TE accumulation by AAS (AAnalyst 100, Perkin Elmer).

13.3.2 Target Hazard Quotient

Health risks of consuming vegetables from TE-contaminated soil and increased TE bioavailability due to agronomic inputs were quantified as target hazard quotient (THQ) [USEPA 2011] using the formula:

$$\text{THQ} = \frac{(\text{EF} \times \text{ED} \times \text{FIR} \times C)}{\text{RfD} \times \text{WAB} \times \text{TA}} \times 10^{-3}$$

where EF is the exposure frequency (350 days/year); *ED* is the exposure duration, i.e. 68.35 years, equivalent to the average lifetime of Indian population as per World Bank Statistics (2015); FIR is the food ingestion rate or vegetable consumption values, which is 350 g/person/day for adults (Song et al. 2015; Chaturvedi et al. 2021); ***C*** is the TE concentration in the edible parts of vegetables (mg/kg); RfD is the oral reference dose of Pb, Cd, Zn, and As at 0.0035, 0.001, 0.3, and 0.05 mg/kg/day, respectively, as per USEPA-IRIS; WAB is the average body weight (60 kg for adults); and TA is the average exposure time for noncarcinogens (*ED* × 365 days/year). If the THQ value is greater than 1, the exposure is likely to cause obvious adverse effects. The health risk assessment depends on soil physical and chemical characteristics, vegetable type, and vegetable consumption rate (Khan et al. 2015a).

13.4 Uptake of Trace Elements: Change in Bioavailability

Pb, Cd, and As were preferentially accumulated in roots of all four plants tested except *Sesbania* that preferentially partitioned As in the shoots (Table 13.1). Greater amount

Table 13.1 TE accumulation in different parts of test plants (in mg/kg).

	Chenopodium		*Amaranthus*		*Abelmoschus*			*Sesbania*	
Treatments	**Root**	**Shoot**	**Root**	**Shoot**	**Root**	**Shoot**	**Fruit**	**Root**	**Shoot**
Pb – 50	18.9 ± 0.5	7.33 ± 0.22	9.88 ± 0.46	7.06 ± 0.29	16.4 ± 0.1	3.8 ± 0.14	2.31 ± 0.2	59.9 ± 2.3	13.0 ± 0.68
Pb – 100	25.4 ± 1.8	9.12 ± 0.32	13.4 ± 0.6	9.28 ± 0.08	23.1 ± 0.5	9.73 ± 0.97	5.22 ± 0.58	81.2 ± 2.8	23.3 ± 0.49
Pb – (Pb50 + Cd25 + Zn200 + As50)	23.7 ± 2.0	8.33 ± 0.17	10.7 ± 1.0	7.33 ± 1.17	18.6 ± 1.3	4.77 ± 0.57	2.93 ± 0.27	63.2 ± 1.5	17.9 ± 1.51
Pb50-FYM	20.3 ± 1.4	7.48 ± 0.44	13.7 ± 1.5	9.59 ± 0.49	17.8 ± 1.1	4.61 ± 0.41	2.65 ± 0.10	64.3 ± 1.5	14.9 ± 0.29
Pb50-AMF	33.3 ± 2.8	11.2 ± 0.68	20.8 ± 1.2	13.5 ± 1.19	22.7 ± 3.3	9.22 ± 0.42	6.81 ± 0.71	114.8 ± 1.3	18.0 ± 0.22
Pb50-PGPR	29.7 ± 0.8	10.9 ± 0.36	19.8 ± 0.2	12.4 ± 0.58	24.7 ± 1.1	11.9 ± 1.4	9.23 ± 0.79	96.3 ± 5.6	15.4 ± 1.41
Cd – 25	3.36 ± 0.35	2.79 ± 0.49	16.1 ± 0.5	2.19 ± 0.30	8.33 ± 0.41	6.71 ± 0.38	1.27 ± 0.21	13.2 ± 1.3	3.29 ± 0.18
Cd – 50	5.08 ± 0.25	3.97 ± 0.20	19.4 ± 1.2	2.38 ± 0.27	13.8 ± 1.0	10.7 ± 0.2	1.82 ± 0.16	17.2 ± 0.6	4.72 ± 0.61
Cd – (Pb50 + Cd25 + Zn200 + As50)	5.05 ± 0.18	3.78 ± 0.54	16.8 ± 1.0	2.22 ± 0.38	8.94 ± 1.28	6.31 ± 0.2	0.84 ± 0.08	14.6 ± 0.3	3.43 ± 0.32
Cd25-FYM	3.72 ± 0.5	2.99 ± 0.33	17.6 ± 1.8	2.27 ± 0.16	9.06 ± 0.28	5.07 ± 0.19	0.96 ± 0.08	12.1 ± 1.0	2.92 ± 0.37
Cd25-AMF	6.48 ± 0.74	4.71 ± 0.60	21.2 ± 1.5	2.67 ± 0.20	13.4 ± 0.6	8.21 ± 0.56	1.46 ± 0.12	15.2 ± 0.6	4.10 ± 0.23
Cd25-PGPR	9.11 ± 0.36	7.59 ± 0.37	21.8 ± 2.2	2.97 ± 0.10	15.7 ± 0.9	10.8 ± 0.3	1.66 ± 0.32	17.4 ± 1.2	5.09 ± 0.31
Zn – 200	18.3 ± 0.8	46.1 ± 1.3	52.6 ± 0.8	73.0 ± 2.1	78.8 ± 2.4	47.2 ± 1.1	31.7 ± 1.5	187.3 ± 6.3	103.9 ± 4.2
Zn – 400	22.4 ± 0.9	55.1 ± 0.7	69.3 ± 2.6	95.7 ± 1.6	94.4 ± 0.6	57.5 ± 1.3	36.7 ± 2.5	219.7 ± 6.7	123.6 ± 3.8
Zn – (Pb50 + Cd25 + Zn200 + As50)	18.5 ± 1.1	47.7 ± 2.3	56.2 ± 0.3	74.4 ± 1.2	76.3 ± 0.9	39.8 ± 1.2	28.1 ± 0.2	191.6 ± 4.1	106.9 ± 7.2
Zn200-FYM	19.7 ± 1.1	48.2 ± 2.6	58.7 ± 3.4	76.8 ± 0.6	82.3 ± 1.2	42.8 ± 2.6	30.3 ± 0.8	206.4 ± 4.6	125.0 ± 2.5
Zn200-AMF	24.1 ± 0.3	59.3 ± 4.1	71.1 ± 3.3	110.5 ± 4.1	106.4 ± 1.9	67.1 ± 1.8	41.2 ± 2.4	258.2 ± 11.6	209.7 ± 3.2
Zn200-PGPR	22.9 ± 1.7	51.2 ± 2.5	77.5 ± 2.0	115.0 ± 4.7	97.1 ± 0.5	47.5 ± 1.9	29.2 ± 1.5	233.2 ± 13.8	198.2 ± 10.2
As – 50	1.81 ± 0.39	1.25 ± 0.23	0.76 ± 0.08	0.66 ± 0.05	1.66 ± 0.13	0.94 ± 0.06	0.77 ± 0.13	5.65 ± 0.53	12.1 ± 0.88
As50 – 100	2.33 ± 0.32	1.41 ± 0.19	1.21 ± 0.12	1.06 ± 0.1	2.82 ± 0.38	1.81 ± 0.26	1.53 ± 0.19	7.21 ± 0.32	13.3 ± 0.6
As – (Pb50 + Cd25 + Zn200 + As50)	1.08 ± 0.19	0.78 ± 0.11	0.91 ± 0.13	0.74 ± 0.07	1.88 ± 0.34	1.09 ± 0.17	0.8 ± 0.07	3.46 ± 0.19	7.18 ± 0.3
As50-FYM	2.86 ± 0.3	1.67 ± 0.19	1.18 ± 0.32	1.01 ± 0.08	1.97 ± 0.2	1.16 ± 0.23	0.91 ± 0.07	6.21 ± 0.44	11.7 ± 1.5
As50-AMF	2.08 ± 0.3	1.37 ± 0.15	1.24 ± 0.16	0.87 ± 0.06	3.38 ± 0.22	2.39 ± 0.28	2.27 ± 0.39	6.27 ± 0.49	13.4 ± 1.2
As50-PGPR	2.63 ± 0.2	1.58 ± 0.35	1.32 ± 0.05	1.14 ± 0.17	3.96 ± 0.41	2.73 ± 0.4	2.18 ± 0.3	7.66 ± 0.26	15.0 ± 1.8

of Zn was accumulated in shoots in *Chenopodium* and *Amaranthus* and in roots in *Abelmoschus* and *Sesbania*. Uptake and accumulation of all four TEs in test plants was dose dependent. It increased in proportion to soil concentrations indicating a correlation between TE availability and accumulation, at least at these doses.

The multimetal treatment increased the accumulation of Pb, Cd, and Zn but decreased that of As in *Chenopodium* and *Sesbania*. In *Abelmoschus*, multimetal treatment increased Pb and As accumulation but decreased that of Cd and Zn. However, in the case of *Amaranthus*, the uptake of all four TEs was inhibited by this treatment. It is possible that TEs when present together may form complexes and affect each other's mobility (Chaturvedi et al. 2018a). Increased Cd accumulation in Zn + Cd treatment in soybean, carrot has been reported (Sharma and Agrawal 2006; Shute and Macfie 2006). On the other hand, in okra, reduced Cd uptake and increased Zn uptake in Zn + Cd-treated soil have been reported (Sharma and Agrawal 2010). The effect seems to be plant, TE, and soil specific.

Accumulation of TEs in roots and shoots was enhanced by all three agronomic inputs in all plants except for Cd uptake in *Sesbania* treated with FYM. AMF inoculation resulted in maximum Pb content in *Sesbania*, *Chenopodium*, and *Amaranthus*; maximum Zn content in *Sesbania*, *Abelmoschus* and *Chenopodium*. Similar increase in AMF (*G. mosseae*) assisted TE uptake has been reported in studies on eggplant (*Solanum melongena* – Pb and As), tomato (*Solanum lycopersicum* – Pb, Cd, and As) and pea (*Pisum sativum* – Pb and Cd) in the study area by Chaturvedi et al. (2018b, 2021). AMF benefit plants by significantly increasing the absorbent surface area of the root system. This results in better nutrient absorption. However, this can also increase TE uptake and accumulation (Rabęda et al. 2011; Magdziak et al. 2014).

PGPR inoculation resulted in maximum Cd content in all four test plants; maximum Pb content in the case of *Abelmoschus*; maximum Zn content in *Amaranthus*; and maximum As content in *Sesbania, Abelmoschus*, and *Amaranthus*. In almost all the above cases, PGPR treatments showed second highest TE content where AMF showed maximum uptake and vice versa. Cd uptake in *Amaranthus* in response to PGPR and AMF treatment was almost the same. PGPR are known to improve nutrient uptake and promote stress tolerance in plants. Inoculation of plant growth-promoting bacteria also enhances TE uptake by plants (Arzanesh et al. 2009; Khan et al. 2015b). This can be attributed to IAA production and phosphate solubilization activity of the bacteria (Husen 2013). Accumulation of As in *Chenopodium* set was maximum in FYM treatment (2.86 mg As/kg).

Although FYM treated sets showed some increase in TE uptake, in most cases, it was not of the magnitude seen in AMF and PGPR treatments. Higher soil organic matter is generally conducive to reducing the soil trace element concentration due to complexation and adsorption (Lambert et al. 2007; van Gestel 2008; Liu et al. 2021a).

13.5 Trace Element Accumulation in Vegetable/Fodder

Leaves and tender shoots of *Chenopodium* and *Amaranthus* and pods of *Abelmoschus* are commonly consumed as vegetables in India. TE content in the edible parts of these plants is presented in Table 13.2. It is evident that their uptake clearly increased in all three vegetables in response to the three agronomic amendments assessed in the study. Among these, maximum Pb and Zn content in edible part of vegetables was seen in *Amaranthus* (9.6–13.5 mg Pb/kg; 76.80–115.02 mg Zn/kg), while Cd content was highest in

Table 13.2 TE accumulation (in mg/kg) and THQ in edible parts of test plants.

	Chenopodium		*Amaranthus*		*Abelmoschus*		*Sesbania*
TE	**Shoot**	**THQ**	**Shoot**	**THQ**	**Fruit**	**THQ**	**Shoot**
Pb50	7.33	1.76	7.06	1.69	2.31	0.55	13.02
Pb50+amendments	7.5–11.28	1.79–2.70	9.6–13.5	2.30–3.23	2.65–9.23	0.64–2.21	14.91–18.02
Cd25	2.79	2.34	2.20	1.84	1.27	1.07	3.30
Cd25+amendments	2.99–7.60	2.51–6.37	2.27–2.97	1.90–2.49	0.96–2.07	0.81–1.74	2.35–5.09
Zn200	46.11	0.13	72.99	0.20	31.7	0.09	103.92
Zn200+amendments	48.22–59.31	0.13–0.17	76.80–115.02	0.21–0.32	29.2–41.2	0.08–0.12	123–209.66
As100	1.25	0.02	0.66	0.01	0.77	0.01	12.08
As50+amendments	1.37–1.67	0.02–0.03	0.87–1.14	0.01–0.02	0.91–2.27	0.02–0.04	11.72–15.04

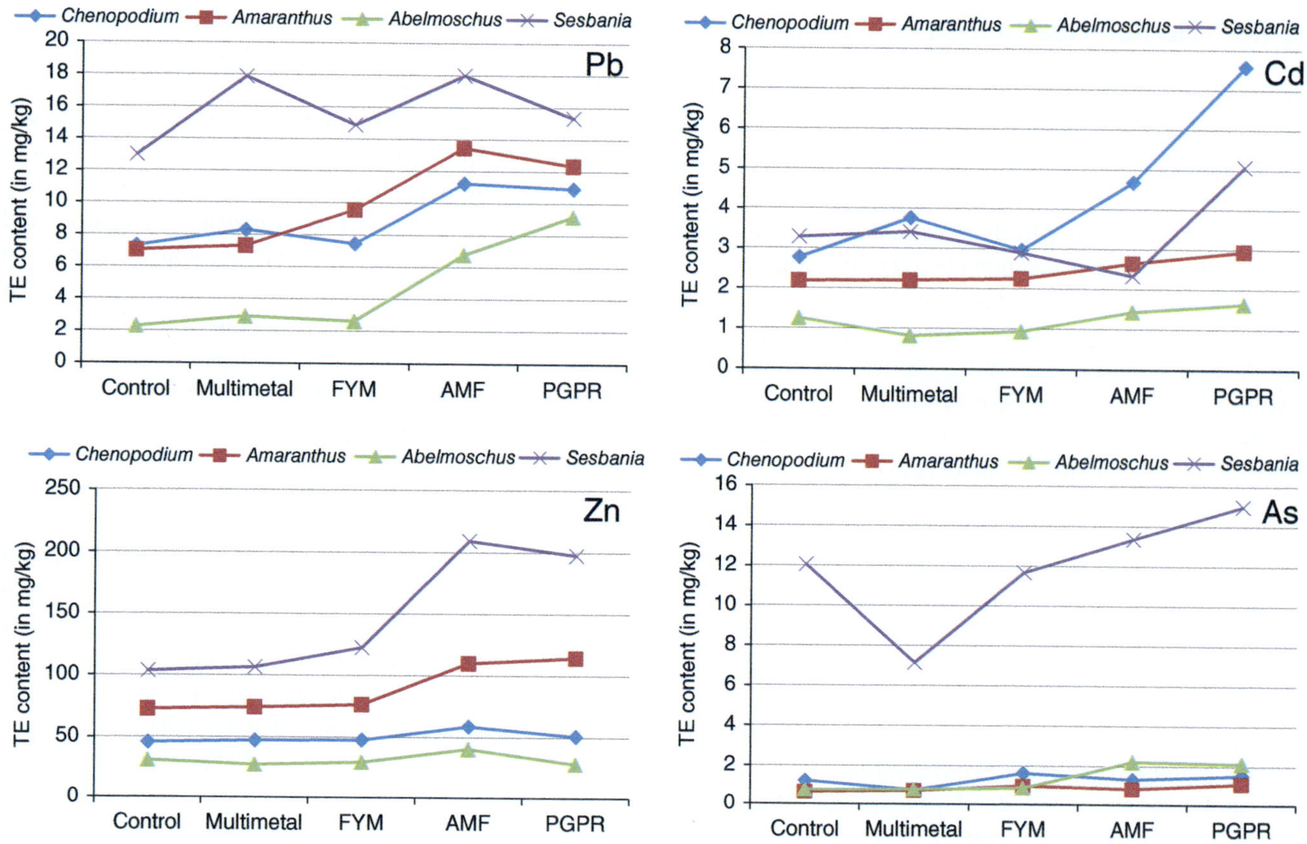

Figure 13.1 TE accumulation in vegetables/fodder in response to agronomic amendments.

Chenopodium (2.99–7.60 mg/kg), and As content was highest in *Abelmoschus* (0.91–2.27 mg/kg) (Figure 13.1). Pb, As, Cd, Cr, Ni, Zn, and Cu are on the list of US Agency for Toxic Substances and Disease Registry (ATSDR 2019 Substance priority list). In fact, As, Pb, and Cd are ranked first, second, and seventh, respectively, while hexavalent Cr ranks seventeenth.

Singh and Mohan (2020) quantified the bioaccumulation of As, Cr, Cd, and Pb in wastewater irrigated vegetables – *Raphanus sativus, Spinacea oleracea, Brassica oleracea,* and *Lycopersicum esculentum* – in Ludhiana, an important industrial town in Punjab, India. TE content ranged from 0.004 to 0.012 mg As/kg, 11 to 29.6 mg Cr/kg, 6.8 to 12 mg Cd/kg, to 1.6 to 7.1 mg Pb/kg. Pb and Cr content was highest in radish, i.e. *Raphanus* (7.1 and 29.6 mg/kg, respectively), while Cd content was highest in *Raphanus* and *Lycopersicum* (11.6 and 12 mg/kg, respectively).

TEs are easily accumulated in the edible parts of leafy vegetables, as compared to grain or fruit crops (Mapanda et al. 2005). Similar high TE intake in case of leafy vegetables *Amaranthus caudatus* and *Coriandrum sativum* was reported by Jabeen et al. (2020). Arora et al. (2008) also obtained highest Cu, Fe, Mn, and Zn intakes from the consumption of wastewater irrigated fenugreek, mint, and spinach. Significant amounts of TEs (Cr, Cu, Ni, Pb, Zn, and Cd) were found in wastewater-irrigated soils and food crops in Beijing, China (Khan et al. 2008).

As per the Directive 2002/32/EC (EC 2002) of European Union, last amended for TEs by Regulation (EU) 2015/186 (EU 2015) the maximum level (ML) for Pb, As, and Cd in fodder for adult animals is 30, 1, and 2 mg/kg, respectively. As and Cd values in the test plant are significantly higher than respective MLs. Kfle et al. (2020) reported Pb and Cd Zn concentrations in common fodder crops lucerne (*Medicago sativa*: in mg/kg 0.91–4.64 Pb; 0.16–0.22 Cd) and oats (*Avena sativa*: in mg/kg 2.71–4.74 Pb; 0.06–0.16 Cd) from wastewater irrigated areas in Eritrea. In the present study, Pb accumulation in shoots of fodder crop *Sesbania* (14.91–18.02 mg/kg) was more than that of leafy vegetables. As accumulation (11.72–15.04 mg/kg) was 5–10 times that of *Chenopodium* and *Amaranthus*. Cd accumulation ranged from 2.35 to 5.09 mg/kg. Apart from FYM and AMF treatments with Cd, all other amendments promoted TE uptake. This is significant since fodder TE content would bio-accumulate in livestock and cause problems. Animal produce from cattle fed on such fodder would be a further threat to human health.

13.6 Human Health Risk Assessment

The target hazard quotient (THQ) is defined as the ratio of the exposure and the reference doses and expresses the risk of noncarcinogenic effects. A THQ < 1 indicates the level of exposure is less than the reference dose. Thus, a daily exposure at this level is not likely to cause any adverse effects during the lifetime of an individual. On the other hand, a population exposed to a dose equal to or greater than the RfD (THQ > 1) will experience health risks (Song et al. 2015, Orosun et al. 2016, Chaturvedi et al. 2021). Although humans can be exposed to TEs through dermal contact, inhalation, or ingestion, the dietary pathway or ingestion has been ranked first (Kusin et al. 2018).

In a survey of leafy vegetables irrigated with wastewater as manure, Olusola et al. (2020) found Pb content with THQ > 1 in *Talinium triangulare.* Appreciable amounts of Pb, Cu, and Zn were also detected. THQ > 1 for Pb in eggplant and for Pb and Cd in tomato and pea has been reported by Chaturvedi et al. (2018a, b, 2021). The uptake of the TEs and hence the THQ values further increased when plants were inoculated with the AMF *G. mosseae*.

This highlights the risk of consumption of such vegetables grown in contaminated soils or irrigated with contaminated water especially with fungal biofertilizers.

13.7 Conclusion

Despite the advancements in our understanding of bioavailability of TEs and other contaminants in soil – plant – microbe systems, the complexity of geochemical and biological interactions involved makes predictions difficult. The number of variables involved compounds the issue. Plant uptake of TEs and their movement through the food chain follow highly individualistic yet plastic trends. Alterations in soil physico-chemistry, fertilizer application, and mutual interactions among TEs in soil as well as common agronomic amendments can affect TE uptake and partitioning in plants. The case study clearly indicates that TE contaminated soil should never be used for crop cultivation without proper evaluation due to the risk of bioaccumulation. Monitoring and testing of TEs in vegetables as well as soil and irrigation water should be carried out by concerned authorities on a regular basis. Plant species and cultivars should also be taken into account while making such assessments. The same holds true for application of AMF and PGPR bio-fertilizers, commonly cited as tools for sustainable agriculture, these can increase TE uptake and accumulation manifold. Furthermore, these are labeled as "organic" amendments, which are publically perceived as safer alternatives to chemicals.

Vegetables and fruits comprise a significant part of the human diet especially in India, where a large part of the population is vegetarian. There is also a paradigm shift in recent years of even more people turning vegetarian in view of the overall benefits of a vegetarian diet. Safety standards such as Reference dose (USEPA 2000), provisional maximum tolerable daily intake (JECFA 2011), Codex Alimentarius (CXS 193-1995 1995), etc. which prescribe thresholds for exposure of TEs and other contaminants in food and feed have also been set. However, it should be noted that having stringent thresholds is one thing and equally stringent monitoring is another. The ground reality often differs in many cases, especially where human greed or need enters the picture.

There is a need to eliminate contaminants at the source itself. Measures taken afterward are costly and time consuming. Development of healthy agro-ecosystems is vital in the current scenario when massive industrialization is being witnessed especially in the developing world. Safe agricultural inputs, contaminant free clean soils, and irrigational water can ensure healthy produce that in turn would reduce the burden on the public exchequer, not only in monetary terms but also in human resource in the form of a healthy population.

References

Agency for Toxic Substances and Disease Registry (ATSDR) (2019). *Substance Priority List.* Atlanta, GA: U.S. Department of Health and Human Services, Public Health Service.

Ahmed, F., Arshad, M., Ditta, A. et al. (2016). Combining textile effluent wastewater with organic fertilizer for improved growth and productivity of wheat and soil health. *Journal of Environmental and Agricultural Sciences* 8: 14–20.

Ali, H., Khan, E., and Sajad, M.A. (2013). Phytoremediation of heavy metals – concepts and applications. *Chemosphere* 91: 869–888.

Altarawneh, R.M. (2019). Levels of selected heavy metals (Pb, Ni, Cd, and Cr) in various widely consumed fruits and vegetables in Jordan. *International Journal of Environmental Analytical Chemistry* 101 (7): 1026–1033.

Antisari, L.V., Orsini, F., Marchetti, L. et al. (2015). Heavy metal accumulation in vegetables grown in urban gardens. *Agronomy for Sustainable Development* 35 (3): 1139–1147.

Arfaenia, H. (2016). Assessment of sediment quality based on acidvolatile sulfide and simultaneously extracted metals in heavily industrialized area of Asaluyeh,

Persian Gulf: concentrations, spatial distributions, and sediment bioavailability/toxicity. *Environmental Science and Pollution Research* 23: 9871–9890.

Arora, M., Kiran, B., Rani, S. et al. (2008). Heavy metal accumulation in vegetables irrigated with water from different sources. *Journal of Food Chemistry* 111 (4): 811–815.

Arzanesh, M., Alikhani, H., Khavazi, K. et al. (2009). *in vitro* growth of wheat (*Triticum aestivum* L.) seedlings inoculated with *Azospirillum* sp., under drought stress. *International Journal of Botany* 5: 244–249.

Berruti, A., Lumini, E., Balestrini, R., and Bianciotto, V. (2016). Arbuscular mycorrhizal fungi as natural biofertilizers: let's benefit from past successes. *Frontier in Microbiology* 6: 1559.

Bhargava, A., Shukla, S., Srivastava, J. et al. (2007). Chenopodium: a prospective plant for phytoextraction. *Acta Physiologiae Plantarum* 30 (1): 111–120.

Bjuhr, J. (2007). Trace metals in soils irrigated with waste water in a periurban area downstream Hanoi City, Vietnam, Seminar Paper, Institutionen for markvetenskap, Sveriges lantbruksuniversitet (SLU), Uppsala, Sweden.

Borrell, A., Tornero, V., Bhattacharjee, D., and Aguilar, A. (2016). Trace element accumulation and trophic relationships in aquatic organisms of the Sundarbans mangrove ecosystem (Bangladesh). *Science of the Total Environment* 545–546: 414–423.

Chaturvedi, R., Favas, P.J.C., Pratas, J. et al. (2018a). Effect of *Glomus mosseae* on accumulation efficiency, hazard index and antioxidant defense mechanisms in tomato under metal(loid) stress. *International Journal of Phytoremediation* 20 (9): 885–894.

Chaturvedi, R., Favas, P.J.C., Pratas, J. et al. (2018b). Assessment of edibility and effect of arbuscular mycorrhizal fungi on *Solanum melongena* L. grown under heavy metal(loid) contaminated soil. *Ecotoxicology and Environmental Safety* 148: 318–326.

Chaturvedi, R., Favas, P.J.C., Pratas, J. et al. (2021). Harnessing *Pisum sativum* – *Glomus mosseae* symbiosis for phytoremediation of soil contaminated with lead, cadmium, and arsenic. *International Journal of Phytoremediation* 23 (3): 279–290.

CXS 193-1995 (1995). *General Standard for Contaminants and Toxins in Food and Feed.* Adopted in 1995, Joint FAO/WHO Food Standards Programme, Codex Alimentarius Commission.

Dakora, F. and Phillips, D. (2002). Root exudates as mediators of mineral acquisition in low-nutrient environments. *Plant and Soil.* 245 (1): 35–47.

Dominguez-Crespo, M.A., Sanchez-Hernandez, Z.E., Torres-Huerta, A.M. et al. (2012). Effect of the heavy metals Cu, Ni, Cd and Zn on the growth and reproduction of epigeic earthworms (*E. fetida*) during the vermistabilization of municipal sewage sludge. *Water Air & Soil Pollution* 223: 915–931.

D'Souza, R., Varun, M., Pratas, J., and Paul, M.S. (2013). Spatial distribution of heavy metals in soil and flora associated with the glass industry in North Central India: implications for phytoremediation. *Soil and Sediment Contamination* 22: 1–20.

EC (2002). Directive 2002/32/EC of the European Parliament and of the Council of 7 May 2002 on undesirable substances in animal feed. *Official Journal of European Union* L 140: 10–22.

EU (2015). Commission Regulation (EU) 2015/186 of 6 February 2015 amending Annex I to Directive 2002/32/EC of the European Parliament and of the Council as regards maximum levels for arsenic, fluorine, lead, mercury, endosulfan and Ambrosia seeds. *Official Journal of European Union* L 31: 11–17.

Farid, M., Ali, S., Rizwan, M. et al. (2018). Phyto-management of chromium contaminated soils through sunflower under exogenously applied 5-aminolevulinic acid. *Ecotoxicology and Environmental Safety* 151: 255–265.

Focus, E., Rwiza, M.J., Mohammed, N.K., and Banzi, F.P. (2021). Health risk assessment of trace elements in soil for people living and working in a mining area. *Journal of Environmental and Public Health* 2021: 9976048.

van Gestel, C.A.M.V. (2008). Physico-chemical and biological parameters determine metal bioavailability in soils. *Science of the Total Environment* 406: 385–395.

Ghoneim, A.M., Al-Zahrani, S., El-Maghraby, S., and Al-Farraj, A. (2014). Heavy metal distribution in *Fagonia indica and Cenchrus ciliaris* native vegetation plant species. *Journal of Food and Agricultural & Environment* 12: 320–324.

Hamid, A., Riaz, H., Sana, A., and Ahmad, S.R. (2016). Heavy metal contamination in vegetables, soil and water and potential health risk assessment. *American-Eurasian Journal of Agricultural Environmental Science* 16 (4): 786–794.

Hu, H. (2002). *Human Health and Heavy Metals Exposure Life Support: Environment and Human Health.* Cambridge, MA: MIT Press.

Husen, E. (2013). Screening of soil bacteria for plant growth promotion activities *in vitro*. *Indonesian Journal of Agricultural Science* 4: 27–31.

Jabeen, F., Aslam, A., and Salman, M. (2020). Heavy metal contamination in vegetables and soil irrigated with sewage water and health risks assessment. *Journal of Environmental and Agricultural Sciences* 22 (1): 23–31.

JECFA (Joint FAO/WHO Expert Committee on Food Additives) (2011). *Safety Evaluation of Certain Contaminants in Food: Prepared by the Seventy-Second Meeting of the Joint FAO/WHO Expert Committee on Food Additives (JECFA).* Rome: JECFA.

Jolly, Y.N., Islam, A., and Akbar, S. (2013). Transfer of metals from soil to vegetables and possible health risk assessment. *Springerplus* 2: 385–392.

Kamunda, C., Mathuthu, M., and Madhuku, M. (2016). An assessment of radiological hazards from gold mine tailings in the province of Gauteng in South Africa. *International Journal of Environmental Research and Public Health* 13 (1): 138.

Kfle, G., Asgedom, G., Goje, T. et al. (2020). The level of heavy metal contamination in selected vegetables and animal feed grasses grown in wastewater irrigated area, around Asmara, Eritrea. *Journal of Chemistry* 1359710. https://doi.org/10.1155/2020/1359710.

Khan, S., Aijun, L., Zhang, S. et al. (2008). Accumulation of polycyclic aromatic hydrocarbons and heavy metals in lettuce grown in the soils contaminated with long-term wastewater irrigation. *Journal of Hazardous Material* 152 (2): 506–515.

Khan, Z.I., Ahmad, K., Ashraf, M. et al. (2015a). Bioaccumulation of heavy metals and metalloids in luffa (*Luffa cylindrica* L.) irrigated with domestic wastewater in Jhang, Pakistan: a prospect for human nutrition. *Pakistan Journal of Botany* 47 (1): 217–224.

Khan, M.U., Sessitsch, A., Harris, M. et al. (2015b). Cr resistant rhizo and endophytic bacteria associated with *Prosopis juliflora* and their potential as phytoremediation enhancing agents in metal-degraded soils. *Frontiers in Plant Science* 5: 1–10. Article 755.

Khan, S., Rai, S., and Singh, G.K. (2021). A review on pesticides in Yamuna River and its impact on vegetables grown nearby areas. *Plant Archives* 21: 1791–1797.

Kumpiene, J., Giagnoni, L., Marschner, B. et al. (2017). Assessment of methods for determining bioavailability of trace elements in soils: a review. *Pedosphere* 27 (3): 389–406.

Kusin, F.M., Azani, N.N.M., Hasan, S., and Sulong, N.A. (2018). Distribution of heavy metals and metalloid in surface sediments of heavily-mined area for bauxite ore in Pengerang, Malaysia and associated risk assessment. *Catena* 165: 454–464.

Lambert, R.L., Grant, C., and Sauvé, S. (2007). Cadmium and zinc in soil solution extracts following the application of phosphate fertilizers. *Science of the Total Environment* 378: 293–305.

Lin, H., Shi, J., Wu, B. et al. (2010). Speciation and bio-chemical transformations of sulfur and copper in rice rhizosphere and bulk soil-XANES evidence of sulfur and copper associations. *Plant and Soil* 313: 1–17.

Liu, W., Yang, X., Duan, L. et al. (2021a). Variability in plant trace element uptake across different crops, soil contamination levels and soil properties in the Xinjiang Uygur Autonomous Region of Northwest China. *Science Reports* 11: 2064.

Liu, Z., Huang, Y., Tan, F. et al. (2021b). Effects of soil type on trace element absorption and fruit quality of pepper. *Frontier in Plant Science* 12: 698796.

Magdziak, Z., Mleczek, M., Kaczmarek, Z., and Golinski, P. (2012). Influence of Ca/Mg ratio and Cd^{2+} and Pb^{2+} elements on low molecular weight organic acid secretion by *Salix viminalis* L. roots into the rhizosphere. *Trees – structure and function* 27 (3): 663–673.

Magdziak, Z., Gąsecka, M., Goliński, P., and Mleczek, M. (2014). Phytoremediation and environmental factors. In: *Phytoremediation: Management of Environmental Contaminants*, vol. I (ed. A.A. Ansari, S.S. Gill, G.R. Lanza and L. Newman), 45–55. Springer ISBN: 978-3-319-10394-5.

Manousaki, E. and Kalogerakis, N. (2009). Phytoextraction of Pb and Cd by the Mediterranean saltbush (*Atriplex halimus* L.): metal uptake in relation to salinity. *Environmental Science and Pollution Research* 16: 844–854.

Manousaki, E., Sedlakova-Kadukova, J., Papadantonakis, N., and Kalogerakis, N. (2008). Phytoextraction and phytoexcretion of Cd by the leaves of *Tamarix smyrnensis* growing on contaminated non-saline and saline soils. *Environmental Research* 106: 326–332.

Mapanda, F., Mangwayana, E.N., Nyamangara, J., and Giller, K.E. (2005). Impacts of sewage irrigation on heavy metals distribution and contamination. *Environment International* 31: 805–812.

Marschner, H. (1995). *Mineral Nutrition of Higher Plants*. Cambridge: Academic.

Morgan, R. (2013). Soil, heavy metals, and human health. In: *Soils and Human Health* (ed. E.C. Brevik and L.C. Burgess), 59–80. Boca Raton. FL: CRC Press.

Ngole-Jeme, V.M., Ekosse, G.E., and Songca, S.P. (2018). An analysis of human exposure to trace elements from deliberate soil ingestion and associated health risks. *Journal of Exposure Science & Environmental Epidemiology* 28 (1): 55–63.

Olusola, J.A., Akintan, O.B., and Adeyemi, M.O. (2020). Index of pollution status in soil and health risk assessment of heavy metals in vegetable crops of a municipal abattoir. *Journal of Materials and Environmental Science* 11 (8): 1250–1263.

Orosun, M.M., Tchokossa, P., Orosun, R.O. et al. (2016). Determination of selected heavy metals and human health risk assessment in fishes from Kiri dam and river Gongola, Northeastern Nigeria. *Journal of Physical Chemistry & Biophysics* 6: 229.

Pan, J., Plant, J.A., Voulvoulis, N. et al. (2010). Cadmium levels in Europe: implications for human health. *Environmental Geochemistry and Health* 32 (1): 1–12.

Paul, M.S., Varun, M., D'Souza, R.J. et al. (2014). Metal contamination of soils and prospects of phytoremediation in and around river Yamuna: a case study from North-Central India. In: *Environmental Risk Assessment of Soil Contamination* (ed. M.C. Hernandez Soriano), 545–573. InTech Publishers.

Pedron, F., Petruzzelli, G., Barbafieri, M., and Tassi, E. (2013). Remediation of a mercury-contaminated industrial soil using bioavailable contaminant stripping. *Pedosphere* 23: 104–111.

Petruzzelli, G., Pedron, F., Rosellini, I., and Barbafieri, M. (2014). The bioavailability processes as a key to evaluate phytoremediation efficiency. In: *Phytoremediation: Management of Environmental Contaminants*, vol. I (ed. A.A. Ansari, S.S. Gill, G.R. Lanza and L. Newman), 31–44. Springer ISBN: 978-3-319-10394-5.

Petruzzelli, G., Pedron, F., and Rosellini, I. (2020). Bioavailability and bioaccessibility in soil: a short review and a case study. *AIMS Environmental Science* 7 (2): 208–225.

Pinto, K.R., Meza-Contreras, V., Alegre-Orihuela, J.C., and Réategui-Romero, W. (2020). Bioavailability and solubility of heavy metals and trace elements during composting of cow manure and tree litter. *Applied and Environmental Soil Science* 2020, Article ID 5680169: 20pp: https://doi.org/10.1155/2020/5680169.

Poniedziałek, M., Sękara, A., Jędrszczyk, E., and Ciura, J. (2010). Phytoremediation efficiency of crop plants in removing cadmium, lead and zinc from soil. *Folia Horticulturae* 22 (2): 25–31.

Prasad, M.N.V. (2001). Bioremediation potential of amaranthaceae. In: *Phytoremediation, Wetlands, and Sediments, Proceedings of the 6th International In Situ and On-Site Bioremediation Symposium*, vol. 6 (5 (ed. A. Leeson, E.A. Foote, M.K. Banks and V.S. Magar), 165–172. Columbus, OH: Battelle Press.

Rabęda, I., Woźny, A., and Krzesłowska, M. (2011). Bacteria and mycorrhizal fungi enhance plants' efficiency in trace metal phytoremediation of trace metals contaminated areas. *Kosmos, Problemy nauk biologicznych* 60: 423–433.

Rai, P.K., Lee, S.S., Zhang, M. et al. (2019). Heavy metals in foodcrops: health risks, fate, mechanisms, and management. *Environment International* 125: 365–385.

Sarathambal, C., Khankhane, P.J., Gharde, Y. et al. (2017). The effect of Plant growth promoting rhizobacteria on the growth, physiology, and Cd uptake of *Arundo donax* L. *International Journal of Phytoremediation* 19 (4): 360–370.

Sharma, R.K. and Agarwal, S.B. (2006). Single and combined effects of cadium and zinc on carrots: uptake and bioaccumulation. *Journal of Plant Nutrition* 29: 1791–1804.

Sharma, R.K. and Agarwal, S.B. (2010). Responses of *Abelmoschus esculentus* L. (lady's finger) to elevated levels of Zn and Cd. *Tropical Ecology* 51 (2S): 389–396.

Shute, T. and Macfie, S.M. (2006). Cadmium and zinc accumulation in soybean: a threat to food safety? *Science of the Total Environment* 371: 63–73.

Sihlahla, M., Mouri, H., and Nomngongo, P.N. (2020). Assessment of bioavailability and mobility of major and trace elements in agricultural soils collected in Port St Johns, Eastern Cape, South Africa using single extraction procedures and pseudo-total digestion. *Journal of Environmental Health Science & Engineering* 18 (2): 1615–1628.

Singh, J. and Mohan, A. (2020). Effect of heavy metal contaminated waste water irrigation on enzymatic and non enzymatic antioxidants in some selected vegetables. *Journal of Physics: Conference Series* 1531: 012093. https://doi.org/10.1088/1742-6596/1531/1/012093.

Singh, M., Müller, G., and Singh, I.B. (2002). Heavy metals in freshly deposited stream sediments of rivers associated with urbanisation of the Ganga Plain, India. *Water, Air, & Soil Pollution* 141: 35–54.

Song, D., Zhuang, D., Jiang, D. et al. (2015). Integrated health risk assessment of heavy metals in Suxian County, South China. *International Journal of Environmental Research and Public Health* 12 (7): 7100–7117.

Sun, L., Niu, Z., and Sun, T. (2007). Effects of amendments of N, P, Fe on phytoextraction of Cd, Pb, Cu, and Zn in soil of Zhangshi by mustard, cabbage, and sugar beet. *Environment Toxicology* 22 (6): 565–571.

USEPA (2000). *Risk-based Concentration Table*. Washington, DC: Washington Agency, U.S. Environment Protection Agency.

USEPA (2011). *Exposure Factors Handbook 2011 edition (Final Report)*. Washington, DC: U.S. Environment Protection Agency EPA/600/R–09/052F, 2011.

Vamerali, T., Bandiera, M., and Mosca, G. (2010). Field crops for phytoremediation of metal-contaminated land – a review. *Environmental Chemistry Letters* 8: 1–17.

Vaněk, A., Komarek, M., Chrastny, V. et al. (2012). Effect of low-molecular weight organic acids on the leaching of thallium and accompanying cations from soil – a model rhizosphere solution approach. *Journal of Geochemical Exploration* 112: 212–217.

Varun, M., D'Souza, R., Favas, P.J.C. et al. (2014). Utilization and supplementation of phytoextraction potential of some terrestrial plants in metal contaminated soils. In: *Phytoremediation: Management of Environmental Contaminants*, vol. I (ed. A.A. Ansari, S.S. Gill, G.R. Lanza and L. Newman), 792. Springer ISBN: 978-3-319-10394-5.

World Bank Statistics (2015). http://data.worldbank.org/country/india.

Yang, Y., Zhang, F.S., Li, H.F., and Jiang, R.F. (2009). Accumulation of cadmium in the edible parts of six vegetable species grown in Cd-contaminated soils. *Journal of Environmental Management.* 90 (2): 1117–1122.

Zhang, C., Liu, G.B., Xie, S., and Zhang, C.S. (2012). Characteristics of soil microelements contents in the rhizospheres of different vegetation in hilly-gully region of Loess Plateau. *Chinese Journal of Applied Ecology* 23: 645–650.

14

Geochemical Provenance of Metalloids and Their Release: Implications on Medical Geology

S. Keerthanan[1], Prosun Bhattacharya[2], and Meththika Vithanage[1]

[1] *Ecosphere Resilience Research Center, Faculty of Applied Sciences, University of Sri Jayewardenepura, Nugegoda, Sri Lanka*
[2] *KTH-International Groundwater Arsenic Research Group, Department of Sustainable Development, Environmental Science and Engineering, KTH Royal Institute of Technology, Stockholm, Sweden*

14.1 Medical Geology of Metalloids

Medical geology is the study of the link between geologic environment (geologic materials containing metalloids and their processes) and human and environmental healthiness (Hasan 2021). The metalloids can enter to the environment through both human activity such as mining and lithogenic pathways such as volcanic eruption, weathering, and leaching. Generally recognized toxic metalloids are germanium (Ge), arsenic (As), antimony (Sb), and tellurium (Te). These metalloids are naturally occurring non-essential element presented in the earth's crust at a certain level, as tabulated in Table 14.1.

Metalloids are mobilized to the water and soil from the geologic materials via the fact of several processes including redox process, weathering, etc. The geologic materials are the primary pool for metalloids, for instance, argyrodite (Ag_8GeS_6), renierite ($(Cu,Zn)_{11}Fe_2(GeAs)_2S_{16}$), germanite ($Cu_{13}Fe_2Ge_2S_{16}$), briartite ($Cu_2(Fe,Zn)GeS_4$), and stottite ($FeGe(OH)_6$) for Ge (Höll et al. 2007), arsenopyrite (FeAsS), realgar (AsS), scorodite ($FeAsO_4{\cdot}2H_2O$), and orpiment (As_2S_3) for As (James et al. 2017), stibnite (Sb_2S_3), jamesonite ($Pb_2Sb_2S_5$), stibiconite ($H_2Sb_2O_5$), and valentinite (Sb_2O_3) for Sb (Dembele et al. 2022), and adamite ($Pb_2(TeO_3)SO_4$), alburnite ($Ag_8GeTe_2S_4$), and aleksite ($PbBi_2Te_2S_2$) for Te (Krivovichev et al. 2020). A significant amount of metalloids contamination has been documented in watersheds in the Pearl River, China, which includes high levels of As due to weathering and leaching from the black rock (Qu et al. 2020). A considerable concentration of Sb (mean 9.4 mg/L) was detected in surface water leaching from Sb-bearing minerals such as stibnite and brandholzite mainly due to the dissolution and oxidation (Borčinová Radková et al. 2020).

Table 14.1 Metalloids concentration in earth's crust.

Metalloids	Level in earth's crust, mg/kg	Reference
Ge	1.6	Höll et al. (2007),
As	3.4	Kapp (2014)
Sb	0.2	Seal Ii et al. (2017)
Te	0.003	Goldfarb et al. (2017)

14.2 Role of Natural Geologic Materials and Processes on Releasing of Metalloids to the Environment

Natural geogenic processes such as weathering, flood, volcanic eruption, earthquake, and landslide, as well as human activities such as mining, can introduce metalloids into the environment matrices of soil, water, and air. These processes can transport of metalloids from their sources to sinks where living creatures can grow. This can result an introduction of metalloids to the biogeochemical system. Weathering can alter the insolubilize metalloids portion into soluble owning to the biological, chemical, and physical process. The metalloids can be transported to different geologic location on the earth with the action of water, ice, and wind action (Hasan 2021). Generally, the weathering process is mostly controlled by the degree of weathering, type of the geogenic materials, climatic conditions, and fluvial processes (Verma et al. 2019). Leaching of Sb and As from the Sb-bearing ore influenced by mainly dissolution mechanism and alkaline solution boots the dissolution of them from the ore (Aktaş and Çetiner 2020; Hu et al. 2016).

Medical Geology: En route to One Health, First Edition. Edited by Majeti Narasimha Vara Prasad and Meththika Vithanage.

Volcanic activities play a key role in the environmental occurrence of metalloids. Each volcano eruption emits massive amount of magma, dust/ash, and fume which can load huge of environmental harmful substances, including metalloids. About 1–50 kt of ground-based ash eruption was estimated from Sakurajima Volcano, Japan, in 2015 (Fee et al. 2017). The magma supply rate from Hawaii's Kīlauea volcano was predicted to be up to 0.28 km^3/year between 2000 and 2012 (Anderson and Poland 2016). The volcano surplus introduces massive quantity of metalloids to the environmental matrices including lithosphere, hydrosphere, and atmosphere and thus can enter into the human through biogeochemical process. A high concentration of metalloids including As, Sb, and Te were detected in volcano fresh ashes from Kizimen, Shiveluch, Klyuchevskoy, and Tolbachik volcano obtained between 2011 and 2015 in Russia (Ermolin et al. 2021). Due to the weathering process, the estimated metalloid environmental exposure of including As and Sb for 50 years was 146 and 26 g/ton, respectively, from the Sunbury shale in Kentucky, USA (Perkins and Mason 2015). In addition, the global jet stream, heavy flood, earthquake, and landslide can also transport the metalloids from their origin including magmatic materials, and rock and minerals to the habitable zone and causing deleterious heath issues.

14.2.1 Release to Hydrosphere

Metalloids can dissolve into surface water and groundwater for long periods of contact with geologic materials containing metalloids that are present via water–rock interaction processes. Natural waters' composition and quality are influenced by the geologic materials they come into touch with. When ingested, these dissolved metalloids can have a negative impact on human health. The highest As concentrations are found in geothermal reservoir such as springs in active plate borders and magmatic rocks in different parts of the world (Bundschuh and Maity 2015). Karst rocks are the source for As to the groundwater, and As can easily seep to the groundwater due to the lack of soil content in karst rocks surrounding areas, which act as a natural filtration (Zhang et al. 2017). The geothermal water at high temperature is a fact that can increase the metalloids content in groundwater owning to high leaching rate of metalloids from the parental rocks (Aihemaiti et al. 2020). The As concentration of magmatic rocks obtained in Argentina ranged from 0.3 to 1.5 mg/kg, which could explain the high As content of thermal springs, which ranged from 0.65 to 1.97 mg/L (Villalba et al. 2020). Further, the role of natural sources contributes in releasing of metalloids to the river water and groundwater as discussed in Table 14.2.

14.2.2 Release to Lithosphere

Volcanic eruption, weathering of parental rocks and minerals, flood, wind, and forest fire are the primary geogenic sources that can load the metalloids to the soil over the naturally occurring limits. The occurrence of high geogenic pressure in Tampere, Finland, as much as 1050 mg/kg of As was found in topsoil and 9280 mg/kg was found in subsoil, indicating that it was released from geogenic bedrocks (Tarvainen et al. 2020). A high As content in unweathered and weathered gray shales induced moderate As contamination in soil profile in Appalachian mountain zone between New York and Tennessee (Richardson et al. 2022). The metalloids, including Sb, from the volcano rock remnants and abandoned mining site are occasionally mobilized by rain, resulting in high levels of metalloids in the surrounding soil environment (Pérez-Sirvent et al. 2011). Further, a high As concentration (7–32 mg/kg) was discovered in a surface soil from an abandoned region where the major parental material was sea sediment in Republic of Azerbaijan (Han et al. 2021a). The occurrence of metalloids in soil is by the natural process as discussed in Table 14.2. In addition, weathering of mining waste from the abandoned mine sites can also be considered as a natural process that can raise the metalloids content in the environmental matrices. For instance, the weathering of mine waste improves the solubility of metalloids, which might explain the high amount of As (17 000 mg/kg) and Sb (47 mg/kg) reported in bulk soils surrounding abandoned mining zones which are sourced by volcano eruption in Australia (Fazle Bari et al. 2020). The transport behavior of Te in soil from an abandoned mining site was limited, and the Te content in soil has fallen by about 126-fold within 100 m distance of the abandoned mining locates in Japan (Qin et al. 2017).

14.2.3 Release to Atmosphere

Massive amount of volcanic ash, gases (CO_2, H_2S, SO_2, H_2O, and HCl), and volatile trace metals and metalloids emitted to atmosphere during the volcanic eruptions around the world (Ilyinskaya et al. 2021; Textor et al. 2003). Metalloids can be dispersed to numerous locations by volcanic plumes in the form of aerosol particles or gas species. Further, the degassing of hot magma and volcanic sublimation are another phenomena that can also be delivered in a significant amount of metalloids to the atmosphere (Zelenski et al. 2013). Depending on eruption force, volcanic degassing, and wind flow and direction, the distribution distance of volcanic plumes can range from 10 to 100 km. The continues flow of wind can be taken up the volcano plumes along with small particles ranging

Table 14.2 The presence of metalloids in the natural environment.

		Metalloids				
Environmental matric	**Description**	**Ge**	**As**	**Sb**	**Te**	**Reference**
Water	Hot spring water in Southern Taiwan	—	4–410 μg/L	—	—	Maity et al. (2011)
Water	Thermal spring water of Kalloni Gulf in Greece	—	41–91 μg/L	—	—	Aloupi et al. (2009)
Water	Tajan river water which originated from mount Alborz in Iran	—	—	—	0.036 μg/L	Najafi et al. (2010)
Water	Haraz river water which originated from mount Damavand in Iran	—	—	—	0.029 μg/L	Najafi et al. (2010)
Water	Water from wet-meadow above thermal spring in Argentina	—	2–10 μg/L	—	—	Villalba et al. (2020)
Water	Thermal spring water in Argentina	—	647–1969 μg/L	—	—	Villalba et al. (2020)
Water	Water from wet-meadow below thermal spring in Argentina	—	1–298 μg/L	—	—	Villalba et al. (2020)
Water	Hot spring water in China	—	—	0.23–173 μg/L	—	Aihemaiti et al. (2020)
Water	River water originated from Taupo lake located at volcano zone and received geothermal inputs in New Zealand	—	9.8–31 mg/kg	0.3–0.9 mg/kg	—	Wilson and Webster-Brown (2009)
Water	River water originated from volcanic mountain range in Peru	Up to 18 μg/L	16–33 μg/L	—	—	Ccanccapa-Cartagena et al. (2021)
Water	River zone originated from volcanic mountain range in Peru	Up to 11 μg/L	16–124 μg/L	—	—	Ccanccapa-Cartagena et al. (2021)
Soil	Geogenic metalloid-contaminated soil in Hong Kong	—	486–1985 mg/kg	—	—	Li et al. (2017)
Soil	Soil derived from volcanic rock and mining activities	—	—	5.8–39.9 mg/kg	—	Pérez-Sirvent et al. (2011)
Soil	Lowland soil received marine, lacustrine, alluvial sediments in Republic of Azerbaijan	—	4.5–39 mg/kg	—	—	Han et al. (2021b)
Soil	Soil received weathered materials from abandoned mining in Japan	—	—	—	10.2–18 mg/kg	Qin et al. (2017)
Gas	Masaya's volcanic plume in Nicaragua	—	—	5.9–11.4 ng/m^3	111–190 ng/m^3	Moune et al. (2010)
Gas	Yasur and White Island volcanoes plumes	—	4–53.9 μg/g	0.3–4.9 μg/g	0.1–9.1 μg/g	Mandon et al. (2019)
Gas	Fumarolic gas emissions from a spatter cone at the Erta Ale in Africa	—	1.2 mg/L	—	1.7 mg/L	Zelenski et al. (2013)

13–1200 μm for a long distance from the volcano zone (Duplay et al. 2014). Metalloids including As, Sb, and Te were detected in volcano plume emitted from Bárðarbunga volcanic system, Iceland, at 40.9–43.1, 1.42–2.47, and 158–178 ng/m^3, which over the background values of 6.52, 0.84, and 0.32 ng/m^3, respectively, and this could be due to the degassing of magma at magmatic temperature (Gauthier et al. 2016). Emission of As from two volcano craters of Marum and Benbow in Ambrym volcano estimated to be 0.36 and 0.13 $\mu g/m^3$ due to the magmatic degassing process (Allard et al. 2016). More examples that show the delivery of metalloids to the atmosphere are discussed in Table 14.2. The conversion of gaseous form into solid particles takes place during the cooling magmatic plume and thus can deliver metalloids to the terrestrial environment.

14.2.4 Mechanism of the Release of the Metalloids to the Environment

Generally, the metalloids are released to the environment from their respective geogenic materials due to the desorption and decomposition process. Several factors, for instance geothermal action, oxidation desorption, alkali desorption, pH of solution, organic matter, reductive desorption, and microbial colonies, are driven by the release of metalloids to the environment (Chauhan et al. 2009).

14.2.4.1 Mechanisms on Arsenic Release

Oxidation of As-bearing geogenic material like arsenopyrite is a primary fact that releases As to the environment. Generally, the oxidation is taken place in the presence of dissolved oxygen and iron(III) ion, as illustrated in chemical reactions (14.1) and (14.2), respectively:

$$2(FeS_2)-As(s)+7O_2(g)+2H_2O(l)\rightarrow 2Fe^{2+}(aq) + 4SO_4^{2-}(aq)+4H^+(aq)+As(aq) \quad (14.1)$$

$$(FeS_2)-As(s)+14Fe^{3+}(aq)+8H_2O(l)\rightarrow 15Fe^{2+}(aq) + 2SO_4^{2-}(aq)+16H^+(aq)+As(aq) \quad (14.2)$$

When anoxic conditions occur, arseniferous iron oxyhydroxides are reduced and discharged into groundwater. The mechanism of iron oxyhydroxide dissolution is thought to be more plausible over the pyrite oxidation that serves as a source. This dissolution process resulted in producing bicarbonate ion, as illustrated in reaction (14.3), and microorganisms in the presence of organic carbon are responsible for this oxidation process (Nickson et al. 2000):

$$4(FeOOH)-As(s)+CH_2O(aq)+7H_2CO_3(aq)\rightarrow 4Fe^{2+}(aq)+8HCO_3^-(aq)+6H_2O(l)+As(aq) \quad (14.3)$$

The pH of the solution plays a role in the release of As to the environment. The releasing of As from the geogenic material such as rock became more effective at pH acidic and alkaline conditions. This is illustrated by combined impacts of three pH-dependent mechanisms that might explain the high connections between As leaching and Fe release in geogenic materials: (i) As oxyanions protonation-deprotonation, (ii) Fe-oxyhydroxides stability, and (iii) adsorption–desorption interactions between As and Fe-oxyhydroxides (Tabelin et al. 2018). Sulfidation of ferrihydrite [$Fe(III)OAs(OH)_2$] is now the most common way for As to be mobilized by reductive dissolution, as given in reaction (14.4):

$$Fe(III)OAs(OH)_2(s)+HS^-(aq)\rightarrow Fe^{2+}(aq) + H_3AsO_3(aq)+S(aq) \quad (14.4)$$

The dissolution of orpiment (As_2S_3) can lead to the release of As into the environment (reaction 14.5), as well as the formation of sulfite, which can lead to the reductive dissolution of ferrihydrite (reaction 14.4) (Herath et al. 2016):

$$As_2S_3(s)+6H_2O(l)\rightarrow 2H_3AsO_3(aq)+3HS^-(aq) + 3H^+(aq) \quad (14.5)$$

14.2.4.2 Mechanisms on Antimony Release

The Sb oxidation and dissolution from the geogenic materials including stibnite is similar to the As oxidation. However, its behavior in the environment is different from the As. Oxidation of stibnite occurs in the presence of dissolved oxygen and produced Sb(III), as illustrated in reaction 14.4:

$$Sb_2S_3+6O_2+H_2O\rightarrow 2Sb(OH)_3+3SO_4^{2-}+6H^+ \quad (14.4)$$

Secondary phases formed due to the comparatively high Sb concentrations discharged into the pore water. The production of secondary minerals, which are represented by oxides with variable Sb-Fe ratios, reduced Sb mobility, as illustrated in reactions (14.5) and (14.6):

$$2Sb(OH)_3+3H_2O\rightarrow Sb(OH)_6^-+3H^++2e^- \quad (14.5)$$

$$Fe^{3+}+Sb(OH)_6^-\rightarrow FeSbO_4+2H_2O+2H^+ \quad (14.6)$$

In case the geogenic material contains dolomite, the dissolution of dolomite yielded the Mg^{2+} ion which can form a brandholzite ($Mg[Sb(OH)_6]_2{\cdot}6H_2O$) that is a precipitate during the dry season and dissolves during the rainy season, and the related chemical reaction was illustrated in reactions (14.7) and (14.8) (Borčinová Radková et al. 2020):

$$CaMg(CO_3) + 2H^+ \rightarrow Ca^{2+} + Mg^{2+} + 2HCO_3^- \quad (14.7)$$

$$Mg^{2+} + 2Sb(OH)_6^- + 6H_2O \rightarrow Mg[Sb(OH)_6]_2 \times 6H_2O \quad (14.8)$$

14.3 Bioavailability and Bioaccessibility of Metalloids

The bioavailability means a portion of metalloids that are accessible in physiological activities such as plant uptake and adsorption by organisms through ingestion, intake, and inhalation from the environment matrices (Semple et al. 2004). However, the definition for bioavailability may vary with different disciplines. Generally, the adsorption and desorption of metalloids in soil, mobilization of metalloids, exposure time of organisms to the contaminated zone, and ability to pass across the biological membrane are the primary processes that determine the bioavailable portion of metalloids in the environment. The bioavailable fraction of metalloids is affected by a number of variables that change their immobilization and mobilization in environmental matrices.

14.3.1 Soil

The soil pH, redox potential, biological process, and complexation potential control the bioavailability of metalloids in soil (Violante et al. 2010). Further, the redox potential greatly influences the mobilization of metalloids in soil. For instance, the addition of fresh FeS to the soil remarkably increased the metalloids including As and Sb mobilization to the solution, whereas addition of oxidized FeS decreased the metalloids mobilization in soil (Park et al. 2018). The high mobilization of As in soil at basic pH is observed over the neutral and slightly acid soil (Gersztyn et al. 2013). The organic matter content in the soil is other important factor that influences the bioavailable of metalloids. Mostly, organic matter formed a stable complex with metalloids and thus retained them in the soil (Liu et al. 2011b; Petruzzelli et al. 2020). Further, the median bioavailability of As in UK soil was 0.037 mg/kg which is 0.24% of the total As concentration in soil (Crispo et al. 2021). The biological process can significantly alter the bioavailability portion of metalloids in the soil. For instance, the volatilization of As in soil increased with organic biomass added to the soil, and this could be due to the bacterial activity which leads to the methylation of As. This can also be altered by the bioavailability of metalloids in the soil matrices (Yang et al. 2018).

Phosphate competes with arsenate in the formation of strong surface complexes on iron hydroxides. Phosphate has a stronger affinity than As for forming surface complexes on iron minerals like gibbsite and goethite at pH neutral circumstances (Maier et al. 2019). The presence of phosphate in the soil increased the mobilization of As in the environment, as a result the enhanced the bioavailable fraction of As.

14.3.2 Aquatic Environment

For aquatic media, metalloids must be in a mobile form in order to be bioavailable. The pH of the water and dissolved organic matters play an important role on the bioavailability of metalloids in water. The concentration of As in groundwater positively correlates with the pH of the water, as a result of dissociation of As from the geogenic materials (Katsoyiannis and Katsoyiannis 2006). Moreover, the presence of metal oxides such as aluminum oxide, iron oxides, and manganese oxide in water media can lead the co-precipitation of metalloids such as As on them which can reduce the bioavailability in water media (Herath et al. 2016). The presence of humic matter in water reduced the bioavailability of As species including mono- and dimethylarsenic (V) remarkably due to the complexation process, according to a laboratory experiment conducted by Costa et al (2022). The bioavailability process of metalloids in the environment is graphically described in Figure 14.1.

14.3.3 Atmosphere

In case of air, the gaseous form of metalloids can directly be entered into the living organism and plants via the respiration. The metalloid content in the atmosphere can be controlled by the temperature, particulate organic content, rainfall, and soil microbial activity (John and Leventhal 1995). Metalloids that are volatile include As when the temperature of hot air following a volcano eruption cools, and it becomes particulate and falls on the earth. Rain may transfer metalloids from the atmosphere to the lithosphere and hydrosphere, and their dissolution in the rain is dependent on the frequency of rain. Most importantly, the microbial communities including bacteria and fungi can methylate the As in the soil and form volatile mono- and dimethyl As compound. This could increase the As concentration in the atmosphere and can directly enter into the living organisms via respiration (Hussain et al. 2021).

Bioaccessibility is referred to as a fraction of metalloids that are liberated from its source into the gastrointestinal and respiration systems and so become accessible for

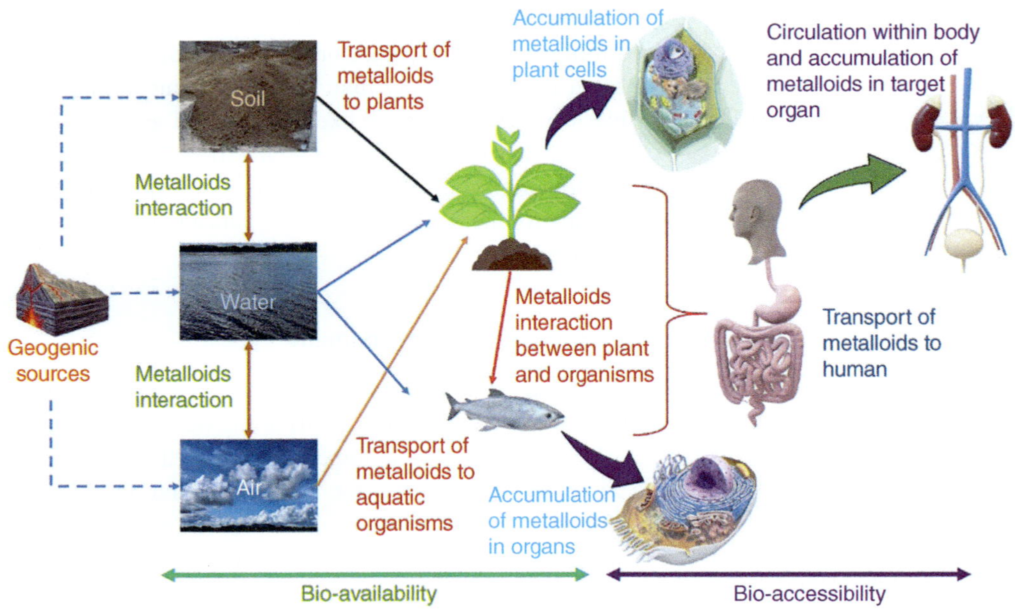

Figure 14.1 Schematic pathway of metalloids from geogenic sources to human.

absorption and enter into the blood circulation. The knowledge of metalloids bioaccessibility must be clear, since it is valid for predicting possible human health concerns. The bioaccessibility of Te through gastric and gastrointestinal was 1.33 and 1.25 ng/g, respectively, which are 32 and 30% of total contamination of Te in Brazil nuts (Berenguel et al. 2018). The total As bioaccessibility including inorganic species (arsenite and arsenate) and organic species (arsenocholine, arsenobetaine, and dimethylarsinic acid) through gastropods contaminated with 3.79 mg/kg of total As was 48, 40, and 33% in intestinal, gastric, and salivary phases, respectively. And also, the bioaccessibility of inorganic As species was higher than the organic As species (Lyu et al. 2020). Similarly, boiled, griddled, and raw mushroom extracts had high bioaccessibility of As, ranging 80–100 and 74–89%, respectively, for gastrointestinal and gastric extracts which were determined from 0.181 to 1.393 mg/kg of total As content in the mushroom (Llorente-Mirandes et al. 2016). The bioaccessibility metalloids through food was high, potentially increasing human exposure to them and posing health risks.

14.4 Human Exposure of Metalloids

The human exposure of metalloids can be taken place through the consumption of metalloids-contaminated food and water, inhaling of contaminated air, and dermal diffusion (Belzile et al. 2011; Lewis et al. 2015; Rovira et al. 2015). Plenty of environmental scientists have studied the human exposure of metalloids that occurred via the variety of foods including vegetables, animal products, fruits, cereals, and fish (Shakoor et al. 2017). Rice is a staple food for most of global population and high amount of human exposure of metalloids occurred via consumption of rice. For instance, As concentration in rice and drinking water was reported as ranging 76.5–640 and 2.3–79.2 μg/kg in West Bengal, India, which may lead to have a high As concentration in human urine whose consumes contaminated rice (Biswas et al. 2021). The Te and Ge were found at range of 1–8 and 2–5 μg/kg, respectively, in food and beverages including cereals, fish and fish products, oil and fat, fruits and vegetable, and drinks and dairy products in France (Millour et al. 2012). Similarly, a maximum As, Sb, and Te concentrations in vegetables, cereals, and pulses reported as 1001, 58, and 104 μg/kg, respectively, in Spain (Matos-Reyes et al. 2010). As a result, metalloids have been found in different human matrices, including breast milk, human hair, blood, placenta, and urine samples, nail, amniotic fluid, teeth, saliva, and so on, during the last decade (Lai et al. 2022) (Table 14.3). High levels of As in breast milk (up to 149 μg/L) have been recorded in people who have consumed water contaminated with more than 10 μg/L of As (Rebelo and Caldas 2016). The mean daily exposure of Sb via drinking water and food was 0.06 μg/kg of body weight which is the reference value (0.4 μg/kg body weight/day) (Ye et al. 2018).

Table 14.3 Occurrence of metalloids in human around the world.

Metalloids	Metalloids contents	Part of body/biofluids	Country	Reference
As	1.49 μg/L	Breast milk	Spain	Freire et al. (2022)
As	3–30 μg/L	Breast milk	Iran	Samiee et al. (2019)
As	1.64 μg/L	Breast milk	Turkey	Nazlıcan et al. (2022)
Sb	0.06–82.4 μg/g	Hair	China	Liu et al. (2011a)
As	0.104–48.4 μg/g			
Sb	0.049–0.058 μg/L	Urine	US	Zhang et al. (2022)
Sb	0.07 μg/L	Urine	Greece	Margetaki et al. (2021)
Sb	27.6 μg/L	Urine	China	Ye et al. (2018)
	6.8 μg/g	Hair		
	0.51 μg/L	Saliva		
As	2.35–9.52 μg/L	Blood	Italy	Calogero et al. (2021)
	0.4–5.06 μg/L	Seminal plasma		
	0.003–0.02 μg/g	Spermatozoa		

14.5 Toxicity of Metalloids to Human and Prevention

Toxicity of metalloids in medical geology is a crucial factor that linked with human and environmental health. The human exposure of metalloids from the contaminated environment generally take places via ingestion (drinking water and food), inhalation, and dermal diffusion, discussed in the previous sections. Sufficient intake of metalloids by human can cause plenty of health problems that lead us toward the losing of life.

"Nothing is healthy above a certain limit." A tolerable daily benchmark for inorganic As via food was determined by WHO as 2.1 μg/kg body weight (WHO 2010). The tolerable daily intake of Sb via food and drinking water was estimated to be 6 μg/kg body weight (WHO 2017). The daily tolerable intake of Te was reported to be 210 μg/kg as per US Environmental Protection Agency (Filippini et al. 2020). The lowest level of Ge that has been seen to have a deleterious impact is 4 mg/kg body weight/day (Tao and Bolger 1997). The human exposure of metalloids over this tolerable benchmark causes health issues as shown in Figure 14.2.

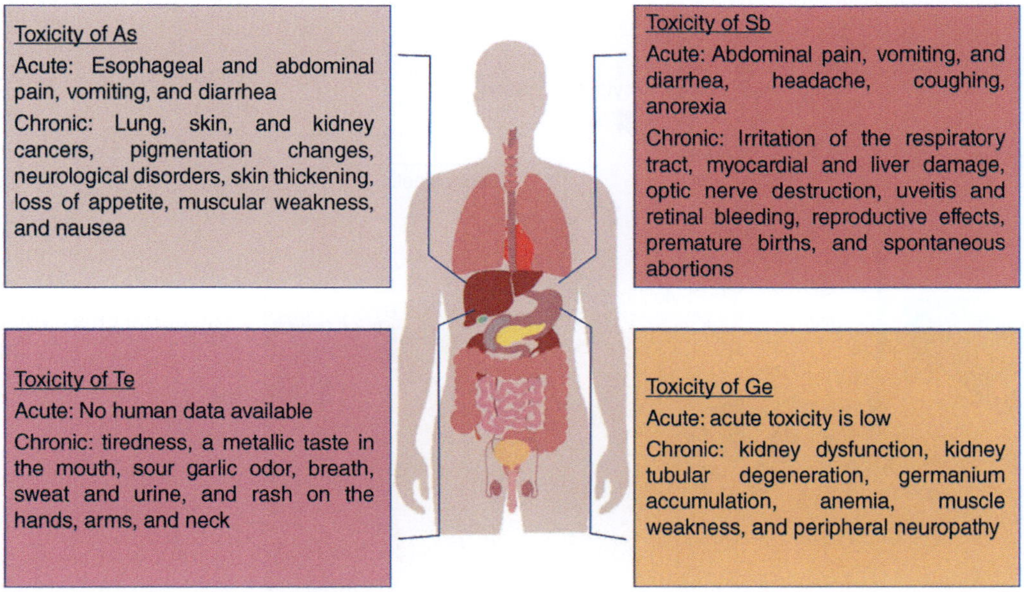

Figure 14.2 Medical geological effect of metalloids in the human body.

There are a few measures that have been recommended to reduce the risk to human health: (i) conducting risk assessments for human health to identify human hazards and environmental exposure, (ii) enable regular human and environmental health monitoring systems for early diagnosis, screening, and treatment of human health consequences, (iii) ensure adherence to proper planning for land use to reduce human disturbance and settlements in hazardous areas by bolstering applicable environmental and public health rules, (iv) on agricultural fields, adopting soil and water conservation to prevent runoff, erosion, and mass movement and diffusion of metalloids, (v) occupational exposure is minimized by following safety, health, and environmental precautions, (vi) treating the contaminated soil and water to minimize the human exposure of metalloids, and (vii) aware the general populace, who may be exposed to excessive levels of metalloids as a result of the contaminated environment (Gwenzi 2020).

14.6 The Risk Management Strategies to Reduce the Bioavailable of Metalloids in the Environment

"Prevention is better than cure." The idea of acting sooner rather than later to avoid calamity isn't new. The Latin proverb, "It is better and more useful to meet a problem early than to seek a remedy after the damage has been done" was a forerunner to the snappier version coined by Dutch philosopher Desiderius Erasmus around the year 1500: "It is better and more useful to meet a problem early than to seek a remedy after the damage has been done." As per the quote, reducing the bioavailable portion of metalloids from the environmental matrices leads to dwindle the bioaccessibility of them to the human and diminish the medical geological effects on the human. Plenty of risk management strategies are often followed by the environmentalists to manage the bioavailability of metalloids in the environmental matrices. Generally, chemical, physical, and biological management approaches such as mobilization and immobilization, phytoremediation, microbial remediation, thermal treatment, and others are practiced during the remediation metalloids from the environment (Gong et al. 2018), as graphically represented in Figure 14.3.

14.6.1 Remediation of Metalloids from the Water Bodies

The adsorption is a process where a mass of metalloids from aqueous media loads onto the solid surface via either chemical or physical forces. Generally, the adsorption efficiency is affected by several factors: the chemical species of metalloids in the water; the environmental parameters such as pH, temperature, ionic strength, and co-contaminants; and the surface properties of adsorbent such as surface area, pore volume, and surface functional groups. Various adsorbents employ in the remediation of metalloids from the aquatic environment. These adsorbents are categorized into several classes including biochar and modified biochar, clay minerals and composite,

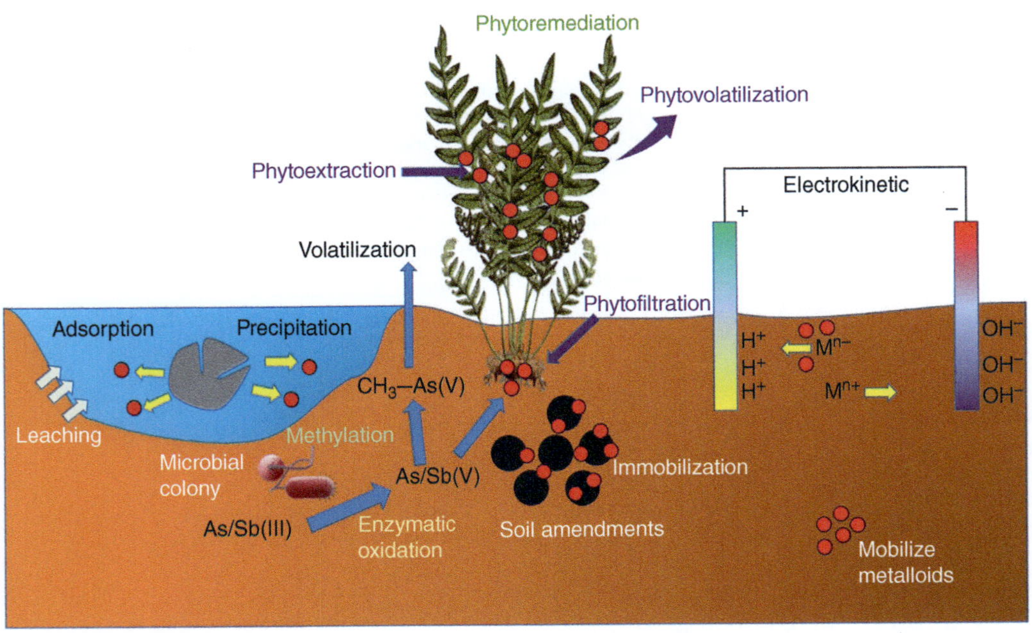

Figure 14.3 Graphical representation of risk management strategies of metalloids in the environment.

biosorbent, nanomaterials, metal oxide/hydroxide, ion-exchangers, and so on. Hence, these adsorbents really seem to have superior performance in the regeneration of metalloids from the aquatic nature.

14.6.1.1 Performance of Biochar on Metalloids Adsorption

The biochar creates attention in the safety of water from the various contaminants due to their specific morphology and chemical nature. Soybean-stover biochar pyrolyzed at 300 and 700 °C encountered to the removal of Sb(III) and Sb(V) from the water and demonstrated that the removal affinity of Sb(III) and Sb(V) by biochar derived at 300 °C is higher than those derived at 700 °C (Vithanage et al. 2015). The biochar surface modification has enlarged the porous nature which enhances the antimony adsorption. For example, the fabricated manganese oxide-embedded biochar applied efficiently in the Sb(III) removal with highest adsorption capacity of 248 mg/g (Wan et al. 2020). Similarly, cotton stalk biochar shows an excellent performance on the As(III) removal in water with 102.8 μg/g saturation capacity at pH 7.6 (Ahmad et al. 2022). Thus, the use of pristine and modified biochar can be a contributor to metalloids regeneration from the aquatic environment.

14.6.1.2 Performance of Soil and Clay Minerals on Metalloids Adsorption

It is understood that the metalloids such as As and Sb are highly cooperated to soil minerals, and several studies have been carried out to investigated the interaction of metalloids with soil and clay phases (Li et al. 2018). The adsorptive removal of antimony by soil was reported by Fan et al. (2020) and Zhang et al. (2014), clays (Ni et al. 2019; Zhou et al. 2019). An attempt to remediate the Sb(V) from the water by black soil has done at three different pH by Fan et al. (2020) under the single and binary system. The batch adsorption results suggested that the lower pH was more favorable for the adsorption of Sb(V) by black soil. A high adsorptive performance (0.6–1.1 mg/g) of clay toward As(V) was noted and was proportional to the iron oxide content of clay (Bentahar et al. 2016). The adsorption effect of bentonite on the adsorption of Sb(V) and Sb(III) in aqueous phases proving that the adsorption of Sb(III) on bentonite is higher than that of Sb(V), and the equilibrium saturation potential for Sb(V) was 270–500 mg/g and for Sb(III) was 370–556 mg/g at three different temperatures (Xi et al. 2011). The adsorption efficiency of ferric chloride modified attapulgite incorporating sodium alginate toward Sb(III) at three different temperature, noting that the adsorption process was spontaneous with 39.24 mg/g of saturation capacity (Zhou et al. 2019). Iron-exchanged smectite showed 73.6% of adsorptive efficiency toward As, whereas the unmodified smectite showed 48.6%. And also, phosphate modified and unmodified kaolinite exhibited the adsorptive performance toward As was observed as 79.1 and 66.1%, respectively (Mukhopadhyay et al. 2017).

14.6.1.3 Performance of Biosorbents on Metalloids Adsorption

The utilization of biomass in the metalloids removal is low cost, is nature-friendly, and exhibits better efficiency (He et al. 2019). Nine distinct variety of plant biomasses (green bean husk, sugarcane bagasse, rice husk, peapod shelling, grapefruit peel, wheat straw, wheat bran, rice straw, and black gram husk) have attempted to extract Sb from contaminated water and reported that 65–99.9% of recovery Sb was observed onto the biomass (Iqbal et al. 2013). A batch experiment conducted using sulfate-reducing bacteria as a biosorbent exhibited about 93% of antimony removal ability from the antimony-contaminated water (Zhang et al. 2016). A brown seaweed (*Sargassum muticum*) is subjected to a set of adsorption experiment for removing antimony from contaminated water, and the results suggested that the Sb(III) adsorption process was fast and the adsorption was not affected by the presence of other anionic compounds such as sulfate, nitrate, and chloride (Ungureanu et al. 2017). A freshwater cyanobacteria (*Microcystis*) biomass exhibited about 30%, whereas acid-treated biomass with 0.1 M HCl and HNO_3 shows more than 90% of removal of Sb (V) from the contaminated water (Sun et al. 2011).

14.6.1.4 Performance of Nanomaterials on Metalloids Adsorption

The use of nanomaterials functioning as adsorbents also provides a promising technology for adsorptive elimination of metalloids from the contaminated water due to their possible high adsorption performance (Kegl et al. 2020). The results of comprehensive laboratory tests are carried out with prepared aqueous solutions, recommending that application of nanomaterials in the meal(loids) removal would also be an effective and suitable alternative. For example, antimonate adsorptive removal by zero-valent iron nanoparticles was investigated at different initial concentration of Sb (V) ranging from 0.5 to 50 mg/L, proving that more than 90% of Sb (V) was removed by zero-valent iron nanoparticles (Dorjee et al. 2014). A titanate nanosheet was prepared and tested to the Sb(III) removal from the aqueous solution by Liu et al. (2021), and the result exhibited that the high removal capacity of Sb(III) was observed and significantly higher than that by nanofiber reported by Luo et al. (2015, 2017). An excellent adsorptive removal (248.5 mg/g) of Te by copper-doped lignin nanomaterial in water was found, and that was owing to the strong coordination between Te and Cu (Yao et al. 2022). Similarly, Te was extraordinarily removed by a magnetic nanoscale zero-valent iron with a high adsorption capacity of 190 mg/g from water environment (Yu et al. 2018). Thus, the applications of nanomaterials in the remediation of metalloids of contaminated water are highly effective.

14.6.1.5 Performance of Other Materials on Metalloids Adsorption

Several adsorption studies have shown that, in aqueous phases, the metal oxide or oxyhydroxide or hydroxides including iron, aluminum, manganese, and titanium have greater affinity to metalloids regeneration from water. The removal efficiency of Sb(III) by ferrihydrite under single and co-contaminant system and the predicted Langmuir model adsorption capacity for Sb(III) at single and co-contaminant environment were 188.2 and 195.4 mg/g, respectively, proving that the presence of co-contaminant enhanced the Sb(III) accumulation on the surface of ferrihydrite (Yang et al. 2019). The layered double hydroxides-activated brick was a promising adsorbent for the removal of Te from the water, and it exhibited 21.3 mg/g of saturation capacity toward Te at pH 6.0 (Chen et al. 2021).

14.6.2 Remediation of Metalloids from the Soil Matrices

Soil is a complex environmental matrix where mobile and immobile portion of metalloids can be found. The mobile portion of metalloids can be bioavailable for the plant uptake and microbial utilization. The risk management strategies would carry out to reduce the mobile portion of metalloids and thus can ultimately reduce the bioavailability of metalloids in soil. Numerous biological, physical, and chemical strategies are studied worldwidely to limit the bioavailability of metalloids in the soil. Generally, physical strategy including soil washing, soil isolation, electrokinetic (EK) methods, and replacement of soil, chemical strategy including chemical stabilization, precipitation, and adsorption, and biological remediation techniques including phytoremediation and microbial-assisted remediation are practiced in the metalloids remediation around the globe (Bolan et al. 2022; Rajendran et al. 2022).

14.6.2.1 Stabilization/Immobilization

The soil amendment with suitable materials such as fly ash, clay and clay minerals, carbon-based materials, nanomaterials, organic composts could help regulate the metalloids contaminant in soil. Clay mineral like iron oxide amendment with As-contaminated soil has diminished the bioavailability of As by more than 90% owning to the high positive charge on the iron oxide surface that leads to bind more As species on it (Mensah et al. 2022). Various types of nano-based materials including nano-sized goethite/biochar composite, nano-Fe_3O_4, biomass-derived magnetic nanocomposite, goethite nanoparticles, hematite nanoparticles, etc., were successfully applied on remediation of As from the contaminated soil (Alidokht et al. 2021; Galdames et al. 2017; Zialame et al. 2021). The application of 3% municipal solid waste compost to the metalloids including As- and Sb-contaminated soils effectively reduced the mobilization of them in soil profile (Abou Jaoude et al. 2019). Titanium-modified corncob biochar and pristine biochar produced at low pyrolysis temperature effectively improve the soil retention of As in the farmland soil (Luo et al. 2020). Agricultural soil amended with rice husk ash has decreased the As mobility to the pore water, as resulted in limiting the bioavailability of As fraction to the rice plants (Wang et al. 2022a).

14.6.2.2 Electrokinetic Approaches

EK remediation is among the most promising and in situ approaches for eradicating different contaminants including metalloids from the low-permeability soils. During the EK remediation process, the external power can cause electromigration, electrophoresis, and electro-osmosis, all of which can influence on the migration of metalloids toward the electrodes (Wang et al. 2021). EK remediation works by supplying a low-intensity electric field to the electrodes, which causes an acidic nature to the anode owing to electrolysis processes on the electrodes. Acidification of the soil facilitates the mobilization of metalloids, which are carried toward the electrode of oppositely charged and so separated from the soil along with the electrons acceleration due the electric current flow (Pedersen et al. 2018). The metalloids removal efficiency may rely on the strength of electric current, types of electrode, and the electrolytes types. An EK trail with ethylenediaminetetraacetic acid (EDTA) as electrolytes exhibited higher As removal rather than groundwater, citric acid, and cetylpyridinium chloride in As-contaminated soil (Yuan and Chiang 2008). In contrast, citric acid was shown higher As remediation efficiency among EDTA, H_2O, HCl, and NaOH, as solubilizing agents (Rosa et al. 2017). The desorption of As has enhanced by the addition of humus material; as a result, the electromigration of As increased in contaminated soil during the EK remediation (Xu et al. 2021). Maximum 1 and 14% of As and Sb were removed by an EK remediation from shooting range soil using Pt and Ti electrodes and $NaNO_3$ as an electrolyte solution (Pedersen et al. 2018).

14.6.2.3 Phytoremediation

Phytoremediation is an eco-greener remediation process where a potential plant is subjected to metalloids-contaminated soil. Accumulation of metalloids in the different parts of the plant including root, stem, and leaves were taken place during the phytoremediation process. Phytoextraction (which ascribed the plant uptake of metalloids and translocated to areal parts of the plant including leaves), phytostabilization (a process in which metalloids are immobilized below ground by plant species, which commonly occurs due to the adsorption of metalloids onto root tissue), phytofilteration (a process in which comparably high metalloids accumulation occurs with root system rather than shoot), and phytovolatilization (a process by which plants lower the toxicity of metalloids including As and release them in a gaseous state into the environment.) are common processes through which the remediation of metalloids undertakes from the contaminated soil (Gunarathne et al. 2019; Oladoye et al. 2022; Yan et al. 2020). Numerous plant species expressed high metalloids remediation from the contaminated soil. For instance, *Cynodon dactylon*, *Nephrolepis biserrata*, and *Pityrogramma calomelanos* showed high leaf accumulation for As rather than root (Ancheta et al. 2020). The bioaccumulation factor for As in *Nephrolepis biserrata* and *Pteris vittata* were 12 and 36, respectively, and Sb bioaccumulation factor of *Fatoua pilosa* and *Axonopus compressus* were 1.1 and 3.1, which indicates that these plants can be used as phytoextractors for metalloids including As and Sb remediation in soil (Wang et al. 2022b). The accumulation of Sb in root of *Pteris cretica* var. *nervosa* was about 500–1750 mg/g, whereas shoot accumulation was about 100–250 mg/g, indicated that the plant can be used for phytofiltration process on Sb remediation in soil (Xi et al. 2022). Growing of *Pteris vittata* in As-contaminated soil potential eliminated soil pore water As level by 18–77%, which caused to reduce the rice plant uptake of As by 50–58% (Ye et al. 2011). Further, hyperaccumulators are plants that can able to take up more metalloids from the soil generally 1000 mg of metalloids/kg dried plant weight and can withstand high levels of metalloid contamination (Ha et al. 2011). For instance, during 30 days of exposure, the *Pteris cretica* ferns accumulated a maximum of 4427–4875 mg of As per kg dry biomass (Eze and Harvey 2018), and thus *Pteris cretica* can be a hyperaccumulator toward metalloids. Because of the plant's DNA methylation approach, around half of the As was phytovolatilized as a methylated form through *Arundo donax* leaves from contaminated soil, and the toxicity of methylated As (organic As) was lower than that of inorganic As species (Guarino et al. 2020).

14.6.2.4 Microbial-Assisted Remediation

Metalloid mobilization and immobilization, toxicity, and chemical characteristics can all be affected by microbial colonies in the soil changing the bio-redox potential. High valences states of metalloids including Sb (V) and As (V) are immobilized due to the strong interaction with soil particles and expressed lesser toxicity than their low valences, Sb(III) and As(III). The bioavailability of metalloids in soil is decreased as a result of their oxidation to higher valence. Microbial-assisted oxidation of metalloids provides environmental benefits and safer strategy over the chemical oxidizing agents. Numerous varieties of bacterial strain mediated the bio-oxidation process via enzymes. For instance, *Agrobacterium tumefaciens* strain 5A produces As(III) oxidase which could oxidize both As(III) and (Sb(III) simultaneously (Wang et al. 2015). Two strains, *Ensifer* sp. NLS4 and *Shinella* sp. NLS1, can oxidize the As(III) and Sb(III) to As(V) and Sb (V), isolated from the Sb-contaminated soil (Nguyen et al. 2017). Three strains, *Cupriavidus* sp., *Moraxella* sp., and *Bacillus* sp. extracted from the Sb mining zone, all of which can promote the Sb oxidation (Tabelin et al. 2018). Bio-volatilization of metalloids, particularly As, by microbes is another technique of lowering their bioavailability in soil. Generally, the As methylation process is mediated by methyltransferase enzyme in the presence of *S*-adenosylmethionine as a methyl donor (Di et al. 2019). For instance, a genetically modified strain *Pseudomonas putida* KT2440 has the capability to convert the inorganic As species into volatile organic As species including monomethylated As, demethylated As, and trimethylated As (Chen et al. 2014). Similarly, di and trimethyl As were produced by a strain *Pseudomonas alcaligenes* NBRC14159 in the presence of arsenite *S*-adenosylmethionine methyltransferase (Zhang et al. 2015). Thus, the utilization of microbial colony can improve in reducing the bioavailability of metalloids in the soil.

14.6.2.5 Integrated Remediation

Supplementation of two or more metalloids remediation strategies in practice can be boosted for the removal efficiency from the soil, and it is termed as integral/hybrid remediation strategy. Various combinations of strategies practices are studied worldwidely to obtain a maximum removal efficiency. For instance, physical-chemical, physical-biological, chemical-biological, biological-biological, and biological-chemical-physical are followed to remediate the metalloids from soil (Bolan et al. 2022). As discussed in Table 14.4, plenty of studies coupled the phytoremediation with other including addition of biochar and nanoparticles, flushing with phosphate, EK, and microbial action. These supplementary materials make the metalloids more mobilize in the soil, resulting in increasing the phytoremediation.

Table 14.4 Integral metalloid risk management practices.

Hybrid method	Material in use	Effect	Reference
Biological–biological	• As (V) reducing bacteria • *Pteris vittata* plant	• The bacteria reduce the As (V) to As(III) which make them more bioavailable for plant • The *Pteris vittata* (hyperaccumulator) uptake the bioavailable As species	Abou-Shanab et al. (2020)
Chemical–biological	• As(III) oxidizing bacteria (*Brevibacterium* sp. YZ-1) • Schwertmannite adsorbent	• As(III) oxidized to As (V) by the strain • Schwertmannite make As (V) immobilize further in soil and thus reduce the bioavailability	Yang et al. (2017)
Chemical–biological	• Electrokinetic (stainless steels electrodes) • *Pennisetum purpureum* (Mott dwarf Napier grass)	• The pH of the soil near the anode electrode falls, resulting in an increase in As species mobility. • The grass has taken up more As and translocated to other parts compared to the control	Wanitsawatwichai and Sampanpanish (2021)
Chemical–chemical	• Biochar • TiO_2 nanoparticles	• Co-application of biochar (2.5 and 5%) and TiO_2 nanoparticles (100, 250, and 500 mg/kg) boost the immobilization of Sb in soil, resulting in a decreasing of bioavailability to the plant uptake	Daryabeigi Zand et al. (2020)
Physical–biological	• Soil flushing with 0.8 M KH_2PO_4 • Plant *Pteris vittata*	• Soil flushing increased the As concentration in soil pore water, resulting high accumulation in hyperaccumulator *Pteris vittata*	(Yan et al. 2017)
Chemical–chemical–biological	• Electrokinetic • Phosphate amendment • Ryegrass and Indian mustard	• Phosphate amendment and electrokinetic to the soil increased the desorption and mobility of As and Sb • Plants accumulated higher metalloids compared to control due to the increased mobilization	Couto et al. (2015)

14.7 Summary and Future Development

Metalloid contamination arises as a result of natural geogenic processes and materials in environmental matrices such as soil, water, and the atmosphere, as addressed in this chapter. The medical geological effects of long and short exposure of metalloids to human also discussed. Most importantly, as the natural disasters are unpredictable and unstoppable, following the risk management strategies much appreciate in order to prevent the health issues. Therefore, this chapter discussed the risk management strategies in soil and water studied in the previous knowledge. However, most of the studies focused on the remediation of As and Sb from the water and soil, not about Ge and Te. Since Ge and Te are expressed chronic diseases, discussed in Figure 14.2, it is required to study about the risk management strategies of them in future.

Phytoremediation using genetically modified hyperaccumulators and genetically designed microorganisms has proven successful and significant metalloid remediation. However, there is no evidence of this endeavor of coupling each in the literature. This gap in the literature can be addressed in future investigations.

Importantly, the medical geologists, environmentalists, and medical professionals should aware the general people through conducting seminars, medical advices, lectures about metalloids, and their associated chronic diseases.

References

Abou Jaoude, L., Garau, G., Nassif, N. et al. (2019). Metal(loid)s immobilization in soils of Lebanon using municipal solid waste compost: microbial and biochemical impact. *Applied Soil Ecology* 143: 134–143.

Abou-Shanab, R.A.I., Mathai, P.P., Santelli, C., and Sadowsky, M.J. (2020). Indigenous soil bacteria and the hyperaccumulator *Pteris vittata* mediate phytoremediation of soil contaminated with arsenic species. *Ecotoxicology and Environmental Safety* 195: 110458.

Ahmad, I., Farwa, U., Khan, Z.U.H. et al. (2022). Biosorption and health risk assessment of arsenic contaminated water through cotton stalk biochar. *Surfaces and Interfaces* 29: 101806.

Aihemaiti, A., Gao, Y., Liu, L. et al. (2020). Effects of liquid digestate on the valence state of vanadium in plant and soil and microbial community response. *Environmental Pollution* 265: 114916.

Aktaş, S. and Çetiner, B.N. (2020). Investigation of alkaline leaching parameters on stibnite concentrate. *Mining, Metallurgy & Exploration* 37 (5): 1729–1739.

Alidokht, L., Anastopoulos, I., Ntarlagiannis, D. et al. (2021). Recent advances in the application of nanomaterials for the remediation of arsenic-contaminated water and soil. *Journal of Environmental Chemical Engineering* 9 (4): 105533.

Allard, P., Aiuppa, A., Bani, P. et al. (2016). Prodigious emission rates and magma degassing budget of major, trace and radioactive volatile species from Ambrym basaltic volcano, Vanuatu island Arc. *Journal of Volcanology and Geothermal Research* 322: 119–143.

Aloupi, M., Angelidis, M.O., Gavriil, A.M. et al. (2009). Influence of geology on arsenic concentrations in ground and surface water in central Lesvos, Greece. *Environmental Monitoring and Assessment* 151 (1): 383–396.

Ancheta, M.H., Quimado, M.O., Tiburan, C.L. et al. (2020). Copper and arsenic accumulation of *Pityrogramma calomelanos, Nephrolepis biserrata,* and *Cynodon dactylon* in Cu-and Au-mine tailings. *Journal of Degraded and Mining Lands Management* 7: 2201–2208.

Anderson, K.R. and Poland, M.P. (2016). Bayesian estimation of magma supply, storage, and eruption rates using a multiphysical volcano model: Kīlauea Volcano, 2000–2012. *Earth and Planetary Science Letters* 447: 161–171.

Belzile, N., Chen, Y.-W., and Filella, M. (2011). Human exposure to antimony: I. Sources and intake. *Critical Reviews in Environmental Science and Technology* 41 (14): 1309–1373.

Bentahar, Y., Hurel, C., Draoui, K. et al. (2016). Adsorptive properties of Moroccan clays for the removal of arsenic(V) from aqueous solution. *Applied Clay Science* 119: 385–392.

Berenguel, O., de Pessôa, S., G., and Arruda, M.A.Z. (2018). Total content and in vitro bioaccessibility of tellurium in Brazil nuts. *Journal of Trace Elements in Medicine and Biology* 48: 46–51.

Biswas, B., Chakraborty, A., Chatterjee, D. et al. (2021). Arsenic exposure from drinking water and staple food (rice): a field scale study in rural Bengal for assessment of human health risk. *Ecotoxicology and Environmental Safety* 228: 113012.

Bolan, N., Kumar, M., Singh, E. et al. (2022). Antimony contamination and its risk management in complex environmental settings: a review. *Environment International* 158: 106908.

Borčinová Radková, A., Jamieson, H.E., and Campbell, K.M. (2020). Antimony mobility during the early stages of stibnite weathering in tailings at the Beaver Brook Sb deposit, New foundland. *Applied Geochemistry* 115: 104528.

Bundschuh, J. and Maity, J.P. (2015). Geothermal arsenic: occurrence, mobility and environmental implications. *Renewable and Sustainable Energy Reviews* 42: 1214–1222.

Calogero, A.E., Fiore, M., Giacone, F. et al. (2021). Exposure to multiple metals/metalloids and human semen quality: a cross-sectional study. *Ecotoxicology and Environmental Safety* 215: 112165.

Ccanccapa-Cartagena, A., Paredes, B., Vera, C. et al. (2021). Occurrence and probabilistic health risk assessment (PRA) of dissolved metals in surface water sources in Southern Peru. *Environmental Advances* 5: 100102.

Chauhan, V.S., Nickson, R.T., Chauhan, D. et al. (2009). Ground water geochemistry of Ballia district, Uttar Pradesh, India and mechanism of arsenic release. *Chemosphere* 75 (1): 83–91.

Chen, J., Sun, G.-X., Wang, X.-X. et al. (2014). Volatilization of Arsenic from polluted soil by *Pseudomonas putida* engineered for expression of the arsM arsenic(III) S-adenosine methyltransferase gene. *Environmental Science & Technology* 48 (17): 10337–10344.

Chen, H., Xu, J., Li, Y. et al. (2021). Trash to treasure: from construction waste to tellurium adsorbent materials. *Journal of Cleaner Production* 312: 127752.

Costa, A.S.P.N., Nascimento, A.L.A., Botero, W.G. et al. (2022). Interaction between humic substances and arsenic species simulating environmental conditions. *Science of the Total Environment* 802: 149779.

Couto, N., Guedes, P., Zhou, D.-M., and Ribeiro, A.B. (2015). Integrated perspectives of a greenhouse study to upgrade an antimony and arsenic mine soil – potential of enhanced phytotechnologies. *Chemical Engineering Journal* 262: 563–570.

Crispo, M., Dobson, M.C., Blevins, R.S. et al. (2021). Heavy metals and metalloids concentrations across UK urban horticultural soils and the factors influencing their bioavailability to food crops. *Environmental Pollution* 288: 117960.

Daryabeigi, Z., Tabrizi, A.M., and Heir, A.V. (2020). Co-application of biochar and titanium dioxide nanoparticles to promote remediation of antimony from soil by sorghum bicolor: metal uptake and plant response. *Heliyon* 6 (8): e04669.

Dembele, S., Akcil, A., and Panda, S. (2022). Technological trends, emerging applications and metallurgical strategies in antimony recovery from stibnite. *Minerals Engineering* 175: 107304.

Di, X., Beesley, L., Zhang, Z. et al. (2019). Microbial arsenic methylation in soil and uptake and metabolism of methylated arsenic in plants: a review. *International Journal of Environmental Research and Public Health* 16 (24): 1–13.

Dorjee, P., Amarasiriwardena, D., and Xing, B.J.M.J. (2014). Antimony adsorption by zero-valent iron nanoparticles (nZVI): ion chromatography-inductively coupled plasma mass spectrometry (IC-ICP-MS) study. *Microchemical Journal* 116: 15–23.

Duplay, J., Semhi, K., Mey, M. et al. (2014). Geogenic versus anthropogenic geochemical influence on trace elements contents in soils from the Milazzo Peninsula. *Geochemistry* 74 (4): 691–704.

Ermolin, M.S., Ivaneev, A.I., Fedyunina, N.N., and Fedotov, P.S. (2021). Nanospeciation of metals and metalloids in volcanic ash using single particle inductively coupled plasma mass spectrometry. *Chemosphere* 281: 130950.

Eze, V.C. and Harvey, A.P. (2018). Extractive recovery and valorisation of arsenic from contaminated soil through phytoremediation using *Pteris cretica*. *Chemosphere* 208: 484–492.

Fan, Y., Zheng, C., Liu, H. et al. (2020). Effect of pH on the adsorption of arsenic(V) and antimony(V) by the black soil in three systems: performance and mechanism. *Ecotoxicology and Environmental Safety* 191: 110145.

Fazle Bari, A.S.M., Lamb, D., Choppala, G. et al. (2020). Geochemical fractionation and mineralogy of metal(loid)s in abandoned mine soils: insights into arsenic behaviour and implications to remediation. *Journal of Hazardous Materials* 399: 123029.

Fee, D., Izbekov, P., Kim, K. et al. (2017). Eruption mass estimation using infrasound waveform inversion and ash and gas measurements: evaluation at Sakurajima Volcano, Japan. *Earth and Planetary Science Letters* 480: 42–52.

Filippini, T., Tancredi, S., Malagoli, C. et al. (2020). Dietary estimated intake of trace elements: risk assessment in an Italian population. *Exposure and Health* 12 (4): 641–655.

Freire, C., Iribarne-Durán, L.M., Gil, F. et al. (2022). Concentrations and determinants of lead, mercury, cadmium, and arsenic in pooled donor breast milk in Spain. *International Journal of Hygiene and Environmental Health* 240: 113914.

Galdames, A., Mendoza, A., Orueta, M. et al. (2017). Development of new remediation technologies for contaminated soils based on the application of zero-valent iron nanoparticles and bioremediation with compost. *Resource-Efficient Technologies* 3 (2): 166–176.

Gauthier, P.-J., Sigmarsson, O., Gouhier, M. et al. (2016). Elevated gas flux and trace metal degassing from the 2014–2015 fissure eruption at the Bárðarbunga volcanic system, Iceland. *Journal of Geophysical Research - Solid Earth* 121 (3): 1610–1630.

Gersztyn, L., Karczewska, A., and Gałka, B. (2013). Influence of pH on the solubility of arsenic in heavily contaminated soils/Wpływ pH na rozpuszczalność arsenu w glebach silnie zanieczyszczonych. *Environmental Protection and Natural Resources/Ochrona Środowiska i Zasobów Naturalnych* 24 (3): 7–11.

Goldfarb, R.J., Berger, B.R., George, M.W., and Seal Ii, R.R. (2017). Tellurium. In: *Critical Mineral Resources of the United States—Economic and Environmental Geology and Prospects for Future Supply* (ed. K.J. Schulz, J.J.H. DeYoung, R.R. Seal Ii and D.C. Bradley), 40. Reston, VA: U.S. Geological Survey.

Gong, Y., Zhao, D., and Wang, Q. (2018). An overview of field-scale studies on remediation of soil contaminated with heavy metals and metalloids: technical progress over the last decade. *Water Research* 147: 440–460.

Guarino, F., Miranda, A., Castiglione, S., and Cicatelli, A. (2020). Arsenic phytovolatilization and epigenetic modifications in *Arundo donax* L. assisted by a PGPR consortium. *Chemosphere* 251: 126310.

Gunarathne, V., Mayakaduwa, S., Ashiq, A. et al. (2019). Transgenic plants: benefits, applications, and potential risks in phytoremediation. Benefits, applications, and potential risks in phytoremediation. In: *Transgenic Plant Technology for Remediation of Toxic Metals and Metalloids* (ed. M.N.V. Prasad), 89–102. Academic Press.

Gwenzi, W. (2020). Occurrence, behaviour, and human exposure pathways and health risks of toxic geogenic contaminants in serpentinitic ultramafic geological environments (SUGEs): a medical geology perspective. *Science of the Total Environment* 700: 134622.

Ha, N.T.H., Sakakibara, M., Sano, S., and Nhuan, M.T. (2011). Uptake of metals and metalloids by plants growing in a lead–zinc mine area, Northern Vietnam. *Journal of Hazardous Materials* 186 (2): 1384–1391.

Han, J., Lee, S., Mammadov, Z. et al. (2021a). Source apportionment and human health risk assessment of trace metals and metalloids in surface soils of the Mugan Plain, the Republic of Azerbaijan. *Environmental Pollution* 290: 118058.

Han, J., Mammadov, Z., Kim, M. et al. (2021b). Spatial distribution of salinity and heavy metals in surface soils on the Mugan Plain, the Republic of Azerbaijan. *Environmental Monitoring and Assessment* 193 (2): 95.

Hasan, S.E. (2021). Medical geology. In: *Encyclopedia of Geology*, 2e (ed. D. Alderton and S.A. Elias), 684–702. Oxford: Academic Press.

He, M., Wang, N., Long, X. et al. (2019). Antimony speciation in the environment: recent advances in understanding the biogeochemical processes and ecological effects. *Journal of Environmental Sciences*(China) 75: 14–39.

Herath, I., Vithanage, M., Bundschuh, J. et al. (2016). Natural arsenic in global groundwaters: distribution and geochemical triggers for mobilization. *Current Pollution Reports* 2 (1): 68–89.

Höll, R., Kling, M., and Schroll, E. (2007). Metallogenesis of germanium – a review. *Ore Geology Reviews* 30 (3): 145–180.

Hu, X., Guo, X., He, M., and Li, S. (2016). pH-dependent release characteristics of antimony and arsenic from typical antimony-bearing ores. *Journal of Environmental Sciences* 44: 171–179.

Hussain, M.M., Bibi, I., Niazi, N.K. et al. (2021). Arsenic biogeochemical cycling in paddy soil-rice system: interaction with various factors, amendments and mineral nutrients. *Science of the Total Environment* 773: 145040.

Ilyinskaya, E., Mason, E., Wieser, P.E. et al. (2021). Rapid metal pollutant deposition from the volcanic plume of Kīlauea, Hawai'i. *Communications Earth & Environment* 2 (1): 78.

Iqbal, M., Saeed, A., and Edyvean, R.G. (2013). Bioremoval of antimony (III) from contaminated water using several plant wastes: optimization of batch and dynamic flow conditions for sorption by green bean husk (Vigna radiata). *Chemical Engineering Journal* 225: 192–201.

James, K.A., Meliker, J.R., and Nriagu, J.O. (2017). Arsenic. In: *International Encyclopedia of Public Health*, 2e (ed. S.R. Quah), 170–175. Oxford: Academic Press.

John, D.A. and Leventhal, J.S. (1995). Bioavailability of metals. In: *Preliminary Compilation of Descriptive Geoenvironmental Mineral Deposit Models* (ed. E.A.D. Bray), 10–18.

Kapp, R.W. (2014). Arsenic. In: *Encyclopedia of Toxicology*, 3e (ed. P. Wexler), 308–312. Oxford: Academic Press.

Katsoyiannis, I.A. and Katsoyiannis, A.A. (2006). Arsenic and other metal contamination of groundwaters in the industrial area of Thessaloniki, Northern Greece. *Environmental Monitoring and Assessment* 123 (1): 393–406.

Kegl, T., Kosak, A., Lobnik, A. et al. (2020). Adsorption of rare earth metals from wastewater by nanomaterials: a review. *Journal of Hazardous Materials* 386: 121632.

Krivovichev, V.G., Krivovichev, S.V., and Charykova, M.V. (2020). Tellurium minerals: structural and chemical diversity and complexity. *Minerals* 10 (7): 623–635.

Lai, Z., He, M., Lin, C. et al. (2022). Interactions of antimony with biomolecules and its effects on human health. *Ecotoxicology and Environmental Safety* 233: 113317.

Lewis, A.S., Beyer, L.A., and Zu, K. (2015). Considerations in deriving quantitative cancer criteria for inorganic arsenic exposure via inhalation. *Environment International* 74: 258–273.

Li, J.-S., Beiyuan, J., Tsang, D.C.W. et al. (2017). Arsenic-containing soil from geogenic source in Hong Kong: leaching characteristics and stabilization/solidification. *Chemosphere* 182: 31–39.

Li, J., Hou, H., and Hosomi, M. (2018). Sorption–desorption of Sb(III) in different soils: kinetics and effects of the selective removal of hydroxides, organic matter, and humic substances. *Chemosphere* 204: 371–377.

Liu, B., Wu, F., Li, X. et al. (2011a). Arsenic, antimony and bismuth in human hair from potentially exposed individuals in the vicinity of antimony mines in Southwest China. *Microchemical Journal* 97 (1): 20–24.

Liu, G., Fernandez, A., and Cai, Y. (2011b). Complexation of arsenite with humic acid in the presence of ferric iron. *Environmental Science & Technology* 45 (8): 3210–3216.

Liu, C., Li, Y., Wang, X. et al. (2021). Efficient extraction of antimony(III) by titanate nanosheets: study on adsorption behavior and mechanism. *Ecotoxicology and Environmental Safety* 207: 111271.

Llorente-Mirandes, T., Llorens-Muñoz, M., Funes-Collado, V. et al. (2016). Assessment of arsenic bioaccessibility in raw and cooked edible mushrooms by a PBET method. *Food Chemistry* 194: 849–856.

Luo, J., Luo, X., Crittenden, J. et al. (2015). Removal of antimonite (Sb(III)) and antimonate (Sb(V)) from aqueous solution using carbon nanofibers that are decorated with zirconium oxide (ZrO_2). *Environmental Science & Technology* 49 (18): 11115–11124.

Luo, J., Hu, C., Meng, X. et al. (2017). Antimony removal from aqueous solution using novel α-MnO_2 nanofibers: equilibrium, kinetic, and density functional theory studies. *Sustainable Chemical Engineering* 5 (3): 2255–2264.

Luo, M., Lin, H., He, Y., and Zhang, Y. (2020). The influence of corncob-based biochar on remediation of arsenic and cadmium in yellow soil and cinnamon soil. *Science of the Total Environment* 717: 137014.

Lyu, R., Gao, Z., Li, D. et al. (2020). Bioaccessibility of arsenic from gastropod along the Xiangjiang river: assessing human health risks using an in vitro digestion model. *Ecotoxicology and Environmental Safety* 193: 110334.

Maier, M.V., Wolter, Y., Zentler, D. et al. (2019). Phosphate induced arsenic mobilization as a potentially effective in-situ remediation technique – preliminary column tests. *Water* 11 (11): 2364–2385.

Maity, J.P., Liu, C.-C., Nath, B. et al. (2011). Biogeochemical characteristics of Kuan-Tzu-Ling, Chung-Lun and Bao-Lai hot springs in Southern Taiwan. *Journal of Environmental Science and Health, Part A* 46 (11): 1207–1217.

Mandon, C.L., Christenson, B.W., Schipper, C.I. et al. (2019). Metal transport in volcanic plumes: a case study at White Island and Yasur volcanoes. *Journal of Volcanology and Geothermal Research* 369: 155–171.

Margetaki, K., Vafeiadi, M., Kampouri, M. et al. (2021). Associations of exposure to cadmium, antimony, lead and their mixture with gestational thyroid homeostasis. *Environmental Pollution* 289: 117905.

Matos-Reyes, M.N., Cervera, M.L., Campos, R.C., and de la Guardia, M. (2010). Total content of As, Sb, Se, Te and Bi in Spanish vegetables, cereals and pulses and estimation of the contribution of these foods to the Mediterranean daily intake of trace elements. *Food Chemistry* 122 (1): 188–194.

Mensah, A.K., Marschner, B., Shaheen, S.M., and Rinklebe, J. (2022). Biochar, compost, iron oxide, manure, and inorganic fertilizer affect bioavailability of arsenic and improve soil quality of an abandoned arsenic-contaminated gold mine spoil. *Ecotoxicology and Environmental Safety* 234: 113358.

Millour, S., Noël, L., Chekri, R. et al. (2012). Strontium, silver, tin, iron, tellurium, gallium, germanium, barium and vanadium levels in foodstuffs from the second french total diet study. *Journal of Food Composition and Analysis* 25 (2): 108–129.

Moune, S., Gauthier, P.-J., and Delmelle, P. (2010). Trace elements in the particulate phase of the plume of Masaya volcano, Nicaragua. *Journal of Volcanology and Geothermal Research* 193 (3): 232–244.

Mukhopadhyay, R., Manjaiah, K.M., Datta, S.C. et al. (2017). Inorganically modified clay minerals: preparation, characterization, and arsenic adsorption in contaminated water and soil. *Applied Clay Science* 147: 1–10.

Najafi, N.M., Tavakoli, H., Alizadeh, R., and Seidi, S. (2010). Speciation and determination of ultra trace amounts of inorganic tellurium in environmental water samples by dispersive liquid–liquid microextraction and electrothermal atomic absorption spectrometry. *Analytica Chimica Acta* 670 (1): 18–23.

Nazlıcan, E., Arıca, E., Gören, İ.E. et al. (2022). The risk estimation and assessment of heavy metal exposure by biomonitoring in the breast milk of mothers in the Cukurova region, Turkey. *Environmental Science and Pollution Research* 29 (10): 13963–13970.

Nguyen, V.K., Choi, W., Yu, J., and Lee, T. (2017). Microbial oxidation of antimonite and arsenite by bacteria isolated from antimony-contaminated soils. *International Journal of Hydrogen Energy* 42 (45): 27832–27842.

Ni, P., Zuo, R., Wang, J. et al. (2019). Adsorption behavior of Sb(III) on iron-functionalized attapulgite in aqueous solution. *Desalination and Water Treatment* 137: 22–33.

Nickson, R.T., McArthur, J.M., Ravenscroft, P. et al. (2000). Mechanism of arsenic release to groundwater, Bangladesh and West Bengal. *Applied Geochemistry* 15 (4): 403–413.

Oladoye, P.O., Olowe, O.M., and Asemoloye, M.D. (2022). Phytoremediation technology and food security impacts of heavy metal contaminated soils: a review of literature. *Chemosphere* 288: 132555.

Park, J.-H., Kim, S.-J., Ahn, J.S. et al. (2018). Mobility of multiple heavy metalloids in contaminated soil under various redox conditions: effects of iron sulfide presence and phosphate competition. *Chemosphere* 197: 344–352.

Pedersen, K.B., Jensen, P.E., Ottosen, L.M., and Barlindhaug, J. (2018). The relative influence of electrokinetic remediation design on the removal of As, Cu, Pb and Sb from shooting range soils. *Engineering Geology* 238: 52–61.

Pérez-Sirvent, C., Martínez-Sánchez, M.J., Martínez-López, S., and Hernández-Córdoba, M. (2011). Antimony distribution in soils and plants near an abandoned mining site. *Microchemical Journal* 97 (1): 52–56.

Perkins, R.B. and Mason, C.E. (2015). The relative mobility of trace elements from short-term weathering of a black shale. *Applied Geochemistry* 56: 67–79.

Petruzzelli, G., Pedron, F., and Rosellini, I. (2020). Bioavailability and bioaccessibility in soil: a short review and a case study. *AIMS Environmental Science* 7 (2): 208–225.

Qin, H.-B., Takeichi, Y., Nitani, H. et al. (2017). Tellurium distribution and speciation in contaminated soils from abandoned mine tailings: comparison with selenium. *Environmental Science & Technology* 51 (11): 6027–6035.

Qu, S., Wu, W., Nel, W., and Ji, J. (2020). The behavior of metals/metalloids during natural weathering: a systematic study of the mono-lithological watersheds in the upper Pearl river basin, China. *Science of The Total Environment* 708: 134572.

Rajendran, S., Priya, T.A.K., Khoo, K.S. et al. (2022). A critical review on various remediation approaches for heavy metal contaminants removal from contaminated soils. *Chemosphere* 287: 132369.

Rebelo, F.M. and Caldas, E.D. (2016). Arsenic, lead, mercury and cadmium: toxicity, levels in breast milk and the risks for breastfed infants. *Environmental Research* 151: 671–688.

Richardson, J.B., Mischenko, I.C., Mackowiak, T.J., and Perdrial, N. (2022). Trace metals and metalloids and Ga/Al ratios in grey shale weathering profiles along a climate gradient and in batch reactors. *Geoderma* 405: 115431.

Rosa, M.A., Egido, J.A., and Márquez, M.C. (2017). Enhanced electrochemical removal of arsenic and heavy metals from mine tailings. *Journal of the Taiwan Institute of Chemical Engineers* 78: 409–415.

Rovira, J., Nadal, M., Schuhmacher, M., and Domingo, J.L. (2015). Human exposure to trace elements through the skin by direct contact with clothing: risk assessment. *Environmental Research* 140: 308–316.

Samiee, F., Leili, M., Faradmal, J. et al. (2019). Exposure to arsenic through breast milk from mothers exposed to high levels of arsenic in drinking water: infant risk assessment. *Food Control* 106: 106669.

Seal Ii, R.R., Schulz, K.J., DeYoung, J.J.H. et al. (2017). *Antimony* (ed. K.J. Schulz, J.J.H. DeYoung, R.R. Seal Ii and D.C. Bradley), 30. Reston, VA: U.S. Geological Survey.

Semple, K.T., Doick, K.J., Jones, K.C. et al. (2004). Peer reviewed: defining bioavailability and bioaccessibility of contaminated soil and sediment is complicated. *Environmental Science & Technology* 38 (12): 228A–231A.

Shakoor, M.B., Nawaz, R., Hussain, F. et al. (2017). Human health implications, risk assessment and remediation of As-contaminated water: a critical review. *Science of the Total Environment* 601-602: 756–769.

Sun, F., Wu, F., Liao, H., and Xing, B. (2011). Biosorption of antimony (V) by freshwater cyanobacteria microcystis biomass: chemical modification and biosorption mechanisms. *Chemical Engineering Journal* 171 (3): 1082–1090.

Tabelin, C.B., Igarashi, T., Villacorte-Tabelin, M. et al. (2018). Arsenic, selenium, boron, lead, cadmium, copper, and zinc in naturally contaminated rocks: a review of their sources, modes of enrichment, mechanisms of release, and mitigation strategies. *Science of the Total Environment* 645: 1522–1553.

Tao, S.-H. and Bolger, P.M. (1997). Hazard assessment of germanium supplements. *Regulatory Toxicology and Pharmacology* 25 (3): 211–219.

Tarvainen, T., Reichel, S., Müller, I. et al. (2020). Arsenic in agro-ecosystems under anthropogenic pressure in Germany and France compared to a geogenic As region in Finland. *Journal of Geochemical Exploration* 217: 106606.

Textor, C., Graf, H.-F., Timmreck, C., and Robock, A. (2003). Emissions from volcanoes. In: *Emissions of Chemical Compounds and Aerosols in Atmosphere* (ed. C. Granier, C. Reeves and P. Artaxo). Kluwer, Dordrecht.

Ungureanu, G., Santos, S.C., Volf, I. et al. (2017). Biosorption of antimony oxyanions by brown seaweeds: batch and column studies. *Journal of Environmental Chemical Engineering* 5 (4): 3463–3471.

Verma, S., Mukherjee, A., Mahanta, C. et al. (2019). Arsenic fate in the Brahmaputra river basin aquifers: controls of geogenic processes, provenance and water-rock interactions. *Applied Geochemistry* 107: 171–186.

Villalba, E., Tanjal, C., Borzi, G. et al. (2020). Geogenic arsenic contamination of wet-meadows associated with a geothermal system in an arid region and its relevance for drinking water. *Science of the Total Environment* 720: 137571.

Violante, A., Cozzolino, V., Perelomov, L. et al. (2010). Mobility and bioavailability of heavy metals and metalloids in soil environments. *Journal of Soil Science and Plant Nutrition* 10: 268–292.

Vithanage, M., Rajapaksha, A.U., Ahmad, M. et al. (2015). Mechanisms of antimony adsorption onto soybean stover-derived biochar in aqueous solutions. *Journal of Environmental Management* 151: 443–449.

Wan, S., Qiu, L., Li, Y. et al. (2020). Accelerated antimony and copper removal by manganese oxide embedded in biochar with enlarged pore structure. *Chemical Engineering Journal* 402: 126021.

Wang, Q., Warelow Thomas, P., Kang, Y.-S. et al. (2015). Arsenite oxidase also functions as an antimonite oxidase. *Applied and Environmental Microbiology* 81 (6): 1959–1965.

Wang, Y., Li, A., and Cui, C. (2021). Remediation of heavy metal-contaminated soils by electrokinetic technology: mechanisms and applicability. *Chemosphere* 265: 129071.

Wang, H., Wang, X., and Peng, B. (2022a). Using an improved Si-rich husk ash to decrease inorganic arsenic in rice grain. *Science of the Total Environment* 803: 150102.

Wang, Y., Tan, S.N., Mohd Yusof, M.L. et al. (2022b). Assessment of heavy metal and metalloid levels and screening potential of tropical plant species for phytoremediation in Singapore. *Environmental Pollution* 295: 118681.

Wanitsawatwichai, K. and Sampanpanish, P. (2021). The combination of phytoremediation and electrokinetics remediation technology on arsenic contaminated remediation in tailing storage facilities from gold mine. *Heliyon* 7 (8): e07736.

WHO (2010) Exposure to Arsenic: A Major Public Health Concern, World Health Organization, Geneva. https://www.who.int/publications/i/item/WHO-CED-PHE-EPE-19.4.1 (accessed 30 October 2022).

WHO (2017) Guidelines for Drinking-Water Quality: Fourth Edition Incorporating the First Addendum, World Health Organization, Geneva https://apps.who.int/iris/bitstream/handle/10665/254637/9789241549950-eng.pdf?sequence=1 (accessed 2 April 2022).

Wilson, N. and Webster-Brown, J. (2009). The fate of antimony in a major lowland river system, the Waikato river, New Zealand. *Applied Geochemistry* 24 (12): 2283–2292.

Xi, J., He, M., and Lin, C.J.M.J. (2011). Adsorption of antimony (III) and antimony (V) on bentonite: kinetics, thermodynamics and anion competition. *Microchemical Journal* 97 (1): 85–91.

Xi, L., Shen, Y., Zhao, X. et al. (2022). Effects of arbuscular mycorrhizal fungi on frond antimony enrichment, morphology, and proteomics in *Pteris cretica* var. *nervosa* during antimony phytoremediation. *Science of the Total Environment* 804: 149904.

Xu, Y., Lu, Q., Li, J. et al. (2021). Effect of humus on the remediation of arsenic-contaminated soil by electrokinetic technology. *Environmental Technology and Innovation* 21: 101297.

Yan, X., Liu, Q., Wang, J., and Liao, X. (2017). A combined process coupling phytoremediation and in situ flushing for removal of arsenic in contaminated soil. *Journal of Environmental Sciences* 57: 104–109.

Yan, A., Wang, Y., Tan, S.N. et al. (2020). Phytoremediation: a promising approach for revegetation of heavy metal-polluted land. *Frontiers in Plant Science* 11: 359.

Yang, Z., Wu, Z., Liao, Y. et al. (2017). Combination of microbial oxidation and biogenic schwertmannite immobilization: a potential remediation for highly arsenic-contaminated soil. *Chemosphere* 181: 1–8.

Yang, Y.-P., Zhang, H.-M., Yuan, H.-Y. et al. (2018). Microbe mediated arsenic release from iron minerals and arsenic methylation in rhizosphere controls arsenic fate in soil-rice system after straw incorporation. *Environmental Pollution* 236: 598–608.

Yang, R., Tao, J., Huang, Q. et al. (2019). Co-adsorption of Cd(II) and Sb(III) by ferrihydrite: a combined XPS and ITC study. *Journal of Soils and Sediments* 19 (3): 1319–1327.

Yao, G., Shao, X., Qiu, Z. et al. (2022). Construction of lignin-based nano-adsorbents for efficient and selective recovery of tellurium (IV) from wastewater. *Chemosphere* 287: 132058.

Ye, W.-L., Khan, M.A., McGrath, S.P., and Zhao, F.-J. (2011). Phytoremediation of arsenic contaminated paddy soils with *Pteris vittata* markedly reduces arsenic uptake by rice. *Environmental Pollution* 159 (12): 3739–3743.

Ye, L., Qiu, S., Li, X. et al. (2018). Antimony exposure and speciation in human biomarkers near an active mining area in Hunan, China. *Science of the Total Environment* 640–641: 1–8.

Yu, H., Chu, Y., Zhang, T. et al. (2018). Recovery of tellurium from aqueous solutions by adsorption with magnetic nanoscale zero-valent iron (NZVFe). *Hydrometallurgy* 177: 1–8.

Yuan, C. and Chiang, T.-S. (2008). Enhancement of electrokinetic remediation of arsenic spiked soil by chemical reagents. *Journal of Hazardous Materials* 152 (1): 309–315.

Zelenski, M.E., Fischer, T.P., de Moor, J.M. et al. (2013). Trace elements in the gas emissions from the Erta Ale volcano, Afar, Ethiopia. *Chemical Geology* 357: 95–116.

Zhang, H., Li, L., and Zhou, S. (2014). Kinetic modeling of antimony (V) adsorption–desorption and transport in soils. *Chemosphere* 111: 434–440.

Zhang, J., Cao, T., Tang, Z. et al. (2015). Arsenic methylation and volatilization by arsenite S-adenosylmethionine methyltransferase in *Pseudomonas alcaligenes* NBRC14159. *Applied and Environmental Microbiology* 81 (8): 2852–2860.

Zhang, G., Ouyang, X., Li, H. et al. (2016). Bioremoval of antimony from contaminated waters by a mixed batch culture of sulfate-reducing bacteria. *International Biodeterioration & Biodegradation* 115: 148–155.

Zhang, L., Qin, X., Tang, J. et al. (2017). Review of arsenic geochemical characteristics and its significance on arsenic pollution studies in karst groundwater, Southwest China. *Applied Geochemistry* 77: 80–88.

Zhang, T., Luo, J., Ge, H. et al. (2022). Relationships between urinary antimony concentrations and depressive symptoms in adults. *Chemosphere* 291: 133104.

Zhou, A., Wang, J.J.D., and Treatment, W. (2019). Removal of antimonite (III) from wastewater using sodium-alginate-modified Fe-attapulgite with sodium alginate beads. *Desalination and Water Treatment* 168: 282–290.

Zialame, A., Jamshidi-Zanjani, A., and Darban, A.K. (2021). Stabilized magnetite nanoparticles for the remediation of arsenic contaminated soil. *Journal of Environmental Chemical Engineering* 9 (2): 104821.

15

Cobalt and Copper Deficiency and Molybdenosis

Sandali Ranaweera[1], Sachira Sadhana Hewawardhana Silva[2], and Danushika C. Manatunga[1]

[1] *Department of Biosystems Technology, Faculty of Technology, University of Sri Jayewardenepura, Nugegoda, Sri Lanka*
[2] *Sri Lanka Institute of Nanotechnology, Nanotechnology and Science Park Mahenwatta, Homagama, Sri Lanka*

15.1 Introduction

The chapter will provide a comprehensive synopsis of the role of copper (Cu), cobalt (Co), and molybdenum (Mo) as microelements in developing plant and animal health. Further, it will provide an insight on the importance of maintaining the required levels of these elements to achieve proper ruminal health, highlighting the fact that the breaking of the balance of Cu and Co content in ruminants, the presence of antagonists such as Mo and sulfur would possibly lead to secondary Cu and Co deficiency which is termed as "molybdenosis." Then the reader will be given a good emphasis on types of animals who are prone to get affected by molybdenosis, the etiology of molybdenosis, and its overall impact. Later, the authors discuss the major sources that lead to molybdenosis emphasizing the contribution coming from the soil sediments and other natural and industrial settings. Finally, an attempt has been taken to discuss the available diagnosis and treatment approaches to mitigate the molybdenosis conditions and to improve the productivity and well-being of the ruminal species.

15.2 Role of Co, Cu, and Mo as Micronutrients

15.2.1 Role of Cu as a Micronutrient in Plants

One of the most important factors affecting agricultural output is the nutritional supply of plants. Plants, for example, require the right amount of mineral nutrients for proper functioning and growth. In higher plants, a lack of mineral elements causes a reduction in net photosynthesis, which is sometimes accompanied by obvious indications of leaf abnormalities such as chlorosis or necrosis. However, in certain situations of mineral insufficiency, a drop in net photosynthesis occurs before the onset of visual complaints, and it is not even linked to any visual abnormalities in the leaves (Clarkson and Hanson 1980).

The physiological relevance of some elements, such as N, K, Mg, S, P, and others, is well understood, whereas the physiological significance of others, particularly micronutrients, is only partially understood; one such micronutrient is copper (Cu) (Jungk 1977). Cu has been widely used as a fungicide in agricultural operations since the turn of the century. Cu was discovered to boost plant development relatively early in the twentieth century. Cu was later discovered to be one of the required minerals for both algae and higher plants. Cu altered a variety of physiological functions in plants, according to an extensive study conducted in the 1920s (Walker 1953).

It is involved in respiration as a component of the redox components of the electron transport chain in protein and carbohydrate metabolism and chlorophyll biosynthesis and regulates the lignification of cell walls through polyphenol metabolism, therefore influencing the water balance in plants (Bottrill et al. 1970). Several Cu-containing proteins have also been discovered at the same time as these findings. Cu proteins found naturally in plants include cytochrome oxidase, ascorbate oxidase, phenolase, laccase, diamine oxidase, superoxide dismutase, fraction 1 protein, and perhaps quinol oxidase. In addition, several low-molecular-weight proteins with unclear functions have been discovered in various plant sources (Delhaize et al. 1985).

Cu is directly engaged in the photosynthetic electron transport chain as a part of plastocyanin, which has been

Medical Geology: En route to One Health, First Edition. Edited by Majeti Narasimha Vara Prasad and Meththika Vithanage.

known since the early 1960s. The total Cu concentration of the chloroplast membrane, on the other hand, was found to be two to four times higher than that of plastocyanin. Several experimental attempts have been undertaken over the last decade to explain the functional role(s) of the excess Cu discovered in thylakoids that are not related to the ingredient of plastocyanin (Loneragan et al. 1981; Thompson et al. 1977).

15.2.1.1 Involvement of Cu in Photosynthesis

The role of Cu in photosynthesis was initially indicated by Green et al. (1939), who discovered that Cu-chelating compounds significantly decreased photosynthetic CO_2 fixation while having just a minor effect on respiration. By examining several mineral inadequacies in spinach, Bottrill et al. (1970) came to similar conclusions. They discovered that lower plant growth rates were linked to lower CO_2 fixation capacity in both Cu- and Mn-deficient plants. The intensity of respiration was also observed to be reduced, though to a lower level than CO_2 fixation activity. Meanwhile, Cu deficiency did not affect the chlorophyll a/b ratio (Vesk et al. 1966).

Possingham and Spencer were the first to explore the Hill activity fluctuations in chloroplasts grown under varied dietary conditions in order to elucidate the role of Cu in photosynthetic electron transport (Bottrill et al. 1970). They discovered that a lack of all micronutrients inhibited the Hill reaction, which was measured from H_2O to 2,6-dichlorophenolindophenol. Mn deficit caused the most inhibition (60%), followed by Cu (40%), Zn (35%), B (25%), and Mo deficiency (25%). The authors found that Cu and Mn had separate action sites in photosynthetic electron transport as the temperature- and light-dependent Hill activity in Cu- and Mn-deficient chloroplasts was drastically different. They assumed that Mn influences the O_2-evolving side of the light reaction, whereas Cu largely impacts the 2,6-dichlorophenol-indophenol-reducing side (Bottrill et al. 1970).

15.2.2 Role of Cu as a Micronutrient in Animals

Cu is a component of the enzyme ceruloplasmin, which converts ferric iron (Fe^{3+}) to ferrous iron (Fe^{2+}) and aids in iron absorption. Cu is also a component of the enzyme superoxide dismutase, which converts superoxide radicals to hydrogen peroxide in the cytosol, shielding cells from their damaging effects (Halliwell and Gutteridge 1999). Copper is essential for energy metabolism because it is a cofactor of the electron transport chain's cytochrome c oxidase, which provides an electrical gradient that the mitochondria employ to make the energy-storing molecule ATP (Hidiroglou 1979).

Copper binds to ferroxidase II, an enzyme that oxidizes iron and allows it to be mobilized and transferred from hepatic storage to the bone marrow for utilization in erythropoiesis. Copper is a component of the enzyme lysyl oxidase, which is essential for the crosslinking of collagen and elastin fibers, which provide connective tissue and blood vessels shape and elasticity (Scaletti et al. 2003).

Tyrosinase is also a copper-dependent enzyme, which is essential for the synthesis of the pigment melanin, which is responsible for hair, skin, and eye pigmentation (Solano 2018). Cytochrome c oxidase is another Cu-dependent enzyme required for the creation of the myelin sheath because it aids in the synthesis of phospholipids (Zafar 2016). Cu is also necessary for the production of hemoglobin (Abbaspour et al. 2014) and the proper functioning of the thyroid gland (Kim et al. 2020).

For the generation of estrogen, copper interacts with granulosa cells. Changes in the pattern and degree of steroidal hormone synthesis are associated with decreased overall fertility and varied reproductive behavior in females, such as nymphomania in ewes (Yaremcio and Kreplin 2009). Copper deficiency may result in delayed puberty, repeated breeding, low conception, early embryonic mortality, and an increased prevalence of placental retention. Reproduction is impeded by low estrus, which is linked to anemia, and a longer service time caused by inactive ovaries. Cattle grazing on copper-deficient pastures has been observed to have low fertility due to delayed or reduced estrus. Copper supplementation has been observed to increase the rate of conception in dairy cows (Keshri et al. 2021).

15.2.2.1 Copper as a Pro-Oxidant

Reduced oxygen derivatives, also known as reactive oxygen species, are the most important reactants in free-radical biochemistry in aerobic organisms (ROS) (Halliwell et al. 1993). Because ROS may oxidize biomolecules, they are hazardous and can cause cell death and tissue damage. The most reactive of them is the hydroxyl radical, which is frequently used as the starter of free-radical-mediated chain reactions involving biomolecules. Although ROS can target all major groups of biomolecules, lipids, particularly polyunsaturated fatty acids in cell membranes, are probably the most vulnerable to chain reaction initiation and propagation. Transition (Fe^{2+} and Cu^{+}) metal ions can catalyze the breakdown of lipid hydroperoxides generated in these chain reactions to yield lipid peroxyl and alkoxyl radicals and other hazardous chemicals. Furthermore, in in vitro conditions, these transition metal ions are key catalysts for the first production of the hydroxyl radical. However, there is now a lot of discussion over the ROS that causes lipid peroxidation and the precise role of transition metals in these systems (Wassmann et al. 2002).

15.2.2.2 Copper as an Antioxidant

The antioxidant properties of caeruloplasmin are consistent with the antioxidant and anti-inflammatory actions of other Cu complexes in vivo but change sharply with the pro-oxidant properties of Cu ions in the test tube (Abolbashari et al. 2018). The effect of Cu in vitro and in vivo is likely to differ because the majority of Cu in the body is coupled to storage and transport proteins in the forms that make it impossible for it to behave as a pro-oxidant. Furthermore, there is mounting evidence that adequate Cu is essential for the body's antioxidant defenses to remain intact, and that the Cu shortage in animals might lead to pro-oxidant consequences. During Cu deficiency in vitro, oxidant stress was also observed in cells and cellular fractions (Contreras et al. 2018).

15.2.3 Role of Co as a Micronutrient in Plants

15.2.3.1 Importance of Co for Plant Metabolism and Physiology

Plants can absorb cobalt as Co^{2+} or as an organic compound. The organic-bound fraction was discovered to be one of the most important sources of Coin soils for plant absorption. Co^{2+} absorption by roots in higher plants includes active uptake (Hu et al. 2021). In the soilless culture environment, the efficiency of root uptake of Co reduced as the Co^{2+} concentration grew. Bioaccumulation coefficients (BACs, i.e. plant/soil Co or Ni quotients) for Co were four to six times higher than for Ni when compared to other nutrients (such as Ni^{2+}). As a result, disproportionately high absorption of Co was discovered when compared to Ni (Palit et al. 1994). In the same study, it was shown that from 97 leaf samples taken from three cobalt-accumulating species of Nyssa, growing at various localities in New Zealand, the maximum Co concentration was 95 μg/g d.w. (in the leaves of *Nyssa sylvatica*). Cobalt may suppress or enhance the uptake of other nutrients. High levels of Co^{2+} induce Fe deficiency in plants and suppress Cd uptake by roots. Cobalt significantly decreased chlorophyll and foliar Fe; however, Fe-inefficient cultivars were more susceptible to Fe chlorosis in the presence of high Co, compared to Fe-efficient genotypes (Lange et al. 2017).

Cobalt and nitrogen may have a synergistic effect in increasing crop yields and protein content. It has been observed that N- and Co-deficient soils are more likely to generate low-yielding crops and seeds with low protein levels; thus, proper soil N and Co management could be an efficient way to boost and sustain pea production in small-holding mountain environments (Akbar et al. 2013). It has been also discovered that adding Co to cowpea plants boosted N utilization efficiency of nitrogenous fertilizers, growth, and yield metrics, as well as N, P, K, and Mn content (Gad 2012).

15.2.3.2 Role of Cobalt in Biological Nitrogen Fixation

Cobalt is required for nitrogen-fixing bacteria, including cyanobacteria for symbiotic nitrogen fixation by legumes and non-legume plants (Campo and Albino 2000). Soybeans cultivated with only air nitrogen and no mineral nitrogen, for example, have led to rapid nitrogen fixation and development when given 1.0 or 0.1 g Co/mL but have shown very little growth when they were not given with cobalt. The role of cobalt in N_2 fixation is mostly due to its action as a cofactor of cobalamine (vitamin B_6), a coenzyme involved in N_2 fixation and nodule growth (Robson et al. 1979). In addition, cobalt is required as part of a bacterial enzyme complex. These elements are essential nutrients for nitrogen-fixing bacteria because they play significant structural roles in proteins. Cobalt deficiency has diverse impacts on nodule growth and function at various levels and to varying degrees (Weisany et al. 2013).

15.2.3.3 Cobalt-Dependent Proteins in Plants

Plants may have cobalamin-dependent enzymes and Co-containing proteins, according to various lines of evidence. Cyanobacteria are the ancestors of the chloroplast, and they require Co for their function. Plants may require Co, according to their theory of evolution. Cobalamin is produced by symbiotic, endophytic, and related N_2 fixing bacteria and is used by plants (Dehertogh et al. 1964). Leucine 2,3-aminomutase and methylmalonyl-CoA mutase are such types of cobalamin-dependent enzymes that were discovered in plants (Ali et al. 2010; Mus et al. 2016).

15.2.4 Role of Co as a Micronutrient in Animals

Until recently, the trace of cobalt found in the tissues of animals and plants, as well as in the cells of microorganisms, was thought to be an accident rather than a functional component of protoplasm because, aside from the fact that some fungi require a small amount of cobalt in their culture medium, there was no evidence that this element played any important biological role. The discovery in 1934 that a lack of cobalt in fodder is the primary cause of devastating enzootic diseases in ruminants provided the first emphasis on the role of cobalt in metabolism (Mertz and Sadler 1981). The revelation in 1946 that the antipernicious anemia factor, isolated from the mammalian liver, contains over 4% cobalt sparked a quest into the substance's function. Its' role in vitamin B_{12} function has been then fully justified, and essential roles it played in the creation of methyl groups, in the metabolism, and potentially in the manufacture of nucleic acids have been then evident (Fisher 2008).

15.2.4.1 Cobalt in Nutrition of Animals

The herbivorous animals such as sheep and cattle acquired the progressive debilitating sickness that supervenes on acute cobalt shortage during the investigations, which led to the understanding that cobalt is an essential element for ruminant nutrition (Stare and Elvehjem 1933). However, there is little evidence on what form of cobalt is maintained; a significant amount of the cobalt provided to ruminants is undoubtedly converted to vitamin B_{12} (Comar and Davis 1947).

15.2.4.1.1 Vitamin B_{12} Cobalt is a trace element that is required for the synthesis of vitamin B_{12} by rumen bacteria. Vitamin B_{12} is an essential component of two enzymatic systems in mammalian metabolism that are involved in a variety of metabolic events, including the metabolism of carbohydrates, lipids, certain amino acids, and DNA. Adenosylcobalamin and methylcobalamin are coenzymes of methylmalonyl cofactor A (CoA) mutase and methionine synthetase, respectively, and are required for energy production in the rumen. Hyporexia, stunted growth, weight loss hepatic steatosis, anemia, poor immunological system, impaired reproductive function, and even mortality are the common symptoms of Co deficiency. Direct measurements of cobalt or vitamin B_{12} concentrations in blood or tissue, as well as the levels of methylmalonic acid, homocysteine, or transcobalamin in blood, methylmalonic acid in the urine, some hematological variables, food consumption, or animal growth can all be used to determine cobalt status in ruminant animals (González-Montaña et al. 2020).

15.2.5 Co and Cu Micro-Deficiency

15.2.5.1 What Is Co and Cu Deficiency?

Since the 1930s, cobalt and copper have been identified as essential dietary modules for ruminants, and the deficiency of both minerals has taken place under natural foraging conditions in many countries of the world (Ammerman 1970). Maintaining balanced concentrations of Cu and Co as microelements will be essential as they play a major role in controlling the acid–base balance, triggering the formation of structural components contributing as enzymatic cofactors as well as providing a beneficial role in the transfer of energy (Silva et al. 2020).

Animals fulfill their microelemental requirement (Cu/Co) mainly by foraging on plants, but some could also attain it from ingested soil. The efficiency up to which these microelements will be absorbed from the diet entirely depends on their solubility of them which also depends on the pH. For example, in the rumen, Cu is mostly present in an insoluble form, but with the rising of the pH in the distal ingesta, Cu is getting freed and absorbed into the ileum and to the large intestine. However, in farm animals, a deficiency situation of Co and Cu mineral elements occurs mainly due to the poor availability of them in diets where they are not reaching at least the minimum requirements (Silva et al. 2020). The minimum recommended level of Co is generally considered as 0.15 mg/kg of dry matter (DM) for cattle, whereas the generally required Cu level in the ruminant diet varies in between 4 and 10 mg/kg in the DM (Brasche 2015; Cano-Sancho et al. 2017; Van Saun 2006). More specifically, the deficiencies arise not solely due to the unavailability of the microelements in the soil but due to the inadequate levels of extractable elemental levels (Gould and Kendall 2011).

The presence of Cu in vertebrates plays a key role in their metabolism by being a cofactor in many of the cuproenzymes such as β-hydroxylase and tyrosinase. These enzymes are tightly linked to proper functioning of the cardiovascular system, skeleton, connective tissues, nervous system, and the color of the hair. The deficiency of Cu would lead to early signs of disorders linked to the mentioned systems (Aupperle et al. 2001).

As mentioned in Section 15.2.4, Co is an essential microelement in the ruminants' diet which is used up by ruminal microorganisms to synthesize vitamin B_{12} which is absent in plants. Out of the two enzymes that depend on vitamin B_{12} for their function, methylmalonyl-CoA mutase involved in the conversion of propionate to succinate. Propionate is produced as an essential fatty acid during the fermentation of soluble carbohydrates which acts as a source of blood glucose in the rumen. Therefore, in a Co deficiency condition, one of the early indicators of the deficiency is the inability to conduct this conversion leading to a starvation condition (Silva et al. 2020). In addition, it will lead to a significant increase of methyl malonic acid and propionic acid in the blood with a subsequent increase of methyl malonic acid in urine (Baumgartner et al. 2014). Apart from the role played in fermentation of soluble carbohydrates, vitamin B_{12} is also important in the metabolism of methionine that involves the transport of folic acid into the liver cells as well as in the enhancement of wool growth in sheep (Gawthorne and Smith 1974).

15.2.5.2 Common Ways Leading to Cu and Co Deficiency

The deficiency of Cu and Co under natural grazing conditions can result from a multiplicity of different factors. Out of them, dietary shortage of microelements due to the vegetation of plants grown in soil low in Cu and Co will be one of the reasons (Kihara et al. 2020; Sutherland et al. 1979). These low soil conditions may arise due to natural soil conditions or poor fertilizer applications (Han et al. 2016;

Wuana and Okieimen 2011), especially in the situations where Cu fertilizer has not been applied for nearly 8–10 years where the soil is containing microelements in 2.5 ppm levels which is extremely inadequate to fulfill the animals' requirement (Malhi et al. 2005).

In addition to that the soil texture, drainage, pH, redox potential, and the presence of the organic matter would also influence the Cu concentration as well as the Cu uptake by the plants. More importantly under the increase of soil pH, the availability, as well as mobility, is that Cu is hugely affected due to the precipitation of $Cu(OH)_2$ and the adsorption of Cu^{2+} ions onto iron and manganous oxides (Agbenin and Olojo 2004).

Apart from that, it is also reported that the increased application of N, Zn, Fe, Mn, and P as well as the respective oxides into the soil could diminish the Cu/Co absorption by fixing them into complexes, thereby minimizing the translocation of Cu through the plants' body intensifying the Cu deficiency (Dhaliwal et al. 2019).

Furthermore, the seasonal variation will be another cofactor affecting the availability of microelements in the pasture. Microelemental composition in pasture tends to get lowered generally in spring and summer than in autumn and winter. However, when the winter rains are being involved, leaching triggers the reduction of the concentration of them in pasture risking the lives of sheep and cattle during the late winter and spring (Khan et al. 2008).

Moreover, the type of plant species or the species variation would also have a definite variation of the Cu uptake where legumes and clovers will preferably take more Cu than the grasses leading to more deficiency in growing grassy pastures (Brink et al. 2001).

Inadequate mineral supplementation could also take place via the rejection of food intake (anorexia) is one common way that would possibly lead to Co deficiency in cattle (Silva et al. 2020).

The species variation in ruminants (e.g. sheep, cattle, and deer) along with their genetic variation will create some differences in the pattern of nutritional requirement as well as on the metabolism of Cu and Co microelements (e.g. some breeds excrete more of Cu via bile and may create a deficient condition), where the susceptibility to a deficient condition will get strongly affected by that factor (Aupperle et al. 2001; Gooneratne et al. 1989).

Out of all the mentioned sources, increased dietary molybdenum and sulfate intake via pasture or feed supplements has been one of the leading causes, which has resulted in induced Co and Cu micro-deficiency (Hill and Shannon 2019). Many soils naturally contain Mo, and others contain Mo from fertilizers that have been added to stimulate legume growth. In the case of sulfur, it is mainly getting added to the soils as fertilizers such as superphosphate, gypsum, or elemental sulfur. When the pasture Mo and S levels rise up especially during the spring, they tend to form insoluble thiomolybdates in the rumen by interacting with Cu (Gould and Kendall 2011). This interaction of Co and Cu with molybdenum has interfered with their metabolism leading to decreased blood hemoglobin, reduction of packed cell volume, emptying of the liver Co stores, and increase of Fe in the liver (Ammerman 1970).

Furthermore, some reports have also indicated that the Cu deficiency can affect the Cu content in hair, and blood can be also directly influenced by the Fe, Ca, Pb, Zn, and sulfate in the forage (Ammerman 1970). Increasing dietary protein consumption, especially the excessive soybean intake, has also led to Cu deficiency reducing the Cu absorption by liver in sheep (Wapnir 1998).

15.2.5.3 Effect of Cu and Co Deficiency

The occurrence of Co and Cu deficiency-related outbreaks is much common only among ruminants (Silva et al. 2020). Mild deficiencies of Cu and Co can lead to low fertility problems and low meat production in ruminants, while the severe forms can even lead to death (Silva et al. 2020). The first signs of Co deficiency may include the eating disorder (anorexia) followed by moderate weight loss and impaired growth rates. When the deficiency becomes severe, it may lead to an anemic condition together with rapid weight loss, fatty liver, and the reduction of usual color and tone of the skin/coat color (Silva et al. 2020). The reduction of the vitamin B_{12} levels has led to the loss of the essential components of the heme group leading to the misfolding of the metalloprotein followed by lowered erythrocyte count (Silva et al. 2020).

The first report on a disease that was caused due to Co deficiency was from New Zealand in the 1930s where they have named it "Bush sickness." This has led a huge number of ruminants to lose their production where they were sickened and died at the end (Reid 1926).

The most affected ruminant group in a Co deficiency includes the young cattle as they are too sensitive to reduction of Co concentration marking a higher number of deaths per annum. In contrast, the breeding cows and beef cows tend to lose their weight and reproductive ability severely (González-Montaña et al. 2020).

When considering the mechanism of weight loss during a Co deficiency, the lower production of cyanocobalamin (vitamin B_{12}), which generally facilitates the synthesis of glucose via methylmalonyl-CoA mutase from propionate, the termination of production of vitamin B_{12} would lead to the buildup of propionate in the blood. This propionate will provide control over the receptors that involve in food uptake. With the rejection in the consumption of food intake, there will be a further decrease in the ruminal microbiota that utilize vitamin B_{12} during the conversion of succinate to propionate lowering the fermentation of the food consumed by Co-deficient cattle, thereby leading to severe weight and energy loss (González-Montaña et al. 2020).

Similarly, there are reports from New Zealand about Co deficiency-related vitamin B_{12} shortages that have specifically led to the generation of Ovine White liver disease (OWLD) only among sheep. This has been associated with severe ill conditions leading to hepatic dysfunction, the elevation of liver enzymes (GOT and GGT) as well as bilirubin. Further in OWLD, a large accumulation of propionate that exceeds the animals' urinary excretion capacity has itself become a major cause of the liver damage (Sutherland et al. 1979). The other symptoms of Co deficiency-related OWLD have been anemia, severe ocular discharge, lethargic conditions, loss of appetite, decolorization of wool/skin, becoming skinny, lacrimation, scaly ears, reduction of milk production, polioencephalomalacia, increased photosensitivity, development of lesions in the brain, spinal cord, peripheral nerves, myocardium and skeletal muscle as well as increased mortality (Abou-zeina et al. 2008; Silva et al. 2020). More importantly, these symptoms can show up singly or in a collective manner during the recovery phase even where the animals will be continued to suffer from poor health if the Co deficiency tends to persist (Sutherland et al. 1979).

In addition, there have been also some reports which discuss that the Co deficiency may lead to the increased susceptibility of animals (e.g. sheep) for selenium toxicity due to the poor excretion of selenium as well as in the reduction of milk production (Van Ryssen et al. 1987).

In the case of Cu deficiency, the most common clinical observation is hair depigmentation which is also known as achromotrichia. This depigmentation generally happens around the eyes giving the animal a bespectacled appearance (e.g. in cattle as given in Figure 15.1). In addition, the wool of sheep would become ginger brown color due to the depigmentation (Dermauw 2014). Furthermore, it has also led to sudden heart failure and death (Yun Liu 2022), slow growth, loss of body weight (Gooneratne et al. 1989), lesions in the heart of ruminants, reduction of bone strength leading to fractured bone (Della Pepa and Brandi 2016) as well as anemic conditions (Abbott et al. 2018). Furthermore, extreme Cu deficiency has led to the inactivation of liver cytochrome oxidase that plays a key role in respiratory electron transport (Srinivasan and Avadhani 2012), together with a buildup of liver iron content (Ammerman 1970). The deficiency of cytochrome oxidase has severely affected the energy-dependent pathways of gut, heart, brain, and pancreas (Aupperle et al. 2001).

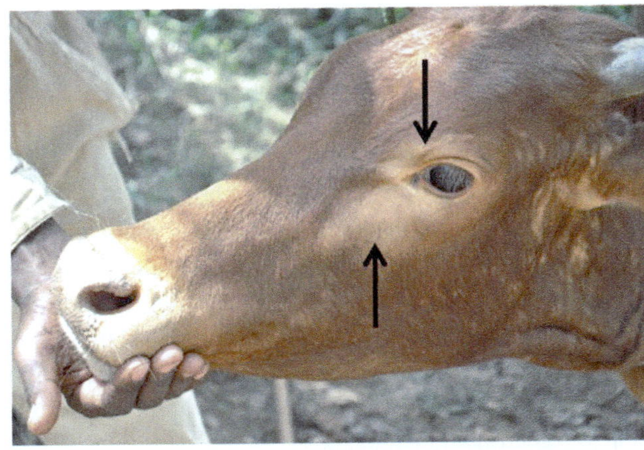

Figure 15.1 Depigmentation of hairs around the eyes due to Cu deficiency. *Source:* Dermauw (2014), Reprinted with permission from Brill Publishers.

Moreover, a deficient condition in Cu would also possibly lead to the lower production of ceruloplasmin in the liver, an enzyme that catalyzes the oxidation of Fe(II) ions and amines. Its deficiency would possibly lead to anemic conditions while allowing the storage of Fe in tissues in the hemosiderin complex (Aupperle et al. 2001). Also, this oxidation of ferrous ions into ferric ions will be important to minimize the formation of hydroxyl radicals via the Fenton reaction (Picco et al. 2012). Furthermore, ceruloplasmin is the main form of Cu storage in plasma, which is very much sensitive to the Cu concentration in the plasma. It would also play a role in scavenging oxygen radicals, thereby preventing the cells from oxidative stress (Picco et al. 2012).

Ruminants who were grazing on the Cu-deficient pastures have also been shown to develop enzootic ataxia or swayback in cattle and sheep (Ammerman 1970; Gupta 2011). Lambs with swayback are severely affected at birth being unable to stand and can even born being dead. The ones who could survive will then generally develop uncoordinated stride after several months due to the development of the impaired central nervous system (CNS), which cannot be even treated by Cu supplementation once the symptoms tend to appear (Silva et al. 2014).

Similar symptoms can be also observed among goats where they could also develop some other nonspecific signs such as dehydration, anemia, and poor fertility. As indicated in Figure 15.2, the most affected group of ruminants over Cu deficiency include newborns and the young animals, pregnant or lactating (due to increased energy demand for growth) cattle, deer, and sheep as well as the ones who suffer severely from heavy worm infestations (Underwood et al. 2015).

Moreover, there have been also reports from Western Australia indicating that the Cu deficiency would lead to slow degeneration of the myocardium replacing it with

Figure 15.2 Severely Cu-deficient cow and her calf. *Source:* Reprinted with permission from Royal Veterinary College.

fibrosis. This has led to the occasional acute dying of cattle due to heart failure which has been named behind "falling disease" (Bennetts et al. 1941).

15.3 Molybdenum (Mo) as a Cause for Micro-Mineral Deficiencies – Molybdenosis

15.3.1 Importance of Mo

This section discusses the role of molybdenum (Mo) in plant, animal, and human health. It is considered a crucial micronutrient especially for plants as it aids in nitrogen fixation by transforming nitrogen into amino acids (Rana et al. 2020). Thus, it helps to increase the yield by enhancing the absorption of nitrates to plants. Mo is also involved in the metabolism of sulfur and phosphate as well as for the activity of several key enzymes. It is vital for normal and healthy growth of a plant in trace amounts. Mo is also necessary for human and animal physiological functions as well. Sulfite oxidase, xanthine oxidase, aldehyde oxidase, and mitochondrial amidoxime reducing component (mARC) all require Mo as it is a structural constituent of molybdopterin which is known as a cofactor synthesized by the body by utilizing Mo that is required for the function of enzymes mentioned earlier (Ott et al. 2015). Sulfur-containing amino acids and heterocyclic compounds such as purines and pyrimidines are metabolized by these enzymes. Drugs and poisons are metabolized by enzymes such as xanthine oxidase, aldehyde oxidase, and mARC (Ott et al. 2015).

However, despite numerous studies on the relevance and impacts of Mo deficiency in plants, the effects of Mo insufficiency or toxicity in humans are rarely discussed. Mo, on the other hand, can be hazardous to animals in excessive concentrations (Raisbeck et al. 2006).

15.3.2 Role of Molybdenum in Plant health

When considering the plant's well-being, Mo is known to serve in metabolic processes concerned with the utilization of nitrogen in plants as it is recognized that molybdenum is required by the nitrogen-fixing organisms, which is necessary for the plant to function well, grow, and survive (Manuel et al. 2018; Rana et al. 2020). It facilitates the processes that occur during the growth stage of a plant, especially including nitrogen, oxygen, potassium, and sulfur cycles (Kaiser et al. 2005). Molybdenum is not only a critical nutrient for *Azotobacter* but also necessary for the growth and development of *Aspergillus niger* regardless of the nitrogen source, and this is especially essential when nitrate is the only nitrogen source (Kaiser et al. 2005). Furthermore, it serves as a functional component as it is required as a catalytically active metal in enzyme catalysis in plant enzyme systems. Xanthine dehydrogenase, aldehyde oxidase, and sulfite oxidase all require Mo as a cofactor in plants. In soybean and cowpea, xanthine dehydrogenase is essential for purine metabolism and ureide production, while indole-3-acetic acid and abscisic acid are produced by aldehyde oxidase (Prasad 2012; Soediono 1989). Plant growth and health can be affected by a lack of molybdenum, which can also increase pest attacks

Figure 15.3 A figure of merlot bunches at harvest without Mo treatment versus normal bunches with the Mo treatment. *Source:* Kaiser et al. (2005), Reprinted with the permission from Oxford University Press.

(Rana et al. 2020). Figure 15.3 clearly illustrates the need for Mo by showing the difference of merlot berries at the harvesting stage without and with proper molybdenum content (Kaiser et al. 2005).

15.3.3 Role of Molybdenum in Human and Animal Health

Mo works as a precursor of a cofactor that is required for electron transportation in both human and animal bodies (Nielsen 2020). Molybdenum is found in metalloflavoproteins such as xanthine oxidase, sulfite oxidase, and aldehyde oxidase in mammals. It aids in the breakdown of proteins and other compounds in the body. The inadequate activity of all molybdenum enzymes is caused by mutations in any of the molybdenum cofactor (MoCo) production enzymes, which can be responsible for severe neurodegeneration and mortality in children which is rare (Novotny and Peterson 2018). In therapeutic doses (w100 mg/day), molybdenum in the form of tetrathiomolybdate has been proven to have anticancer properties and is used to treat a complication related to copper accumulation as it reduces the copper absorption (Brewer 2014). Albeit, molybdenum plays a crucial role in regular bodily activities, there is not enough evidence to know how it might help with any medical issue since Mo deficiency is quite rare. However, Mo deficiency leads to having low feeding and feral survival rates, insufficient development, and negative impacts on reproduction in animals (Anke et al. 2007). Molybdenum, in the form of molybdate, is easily absorbed from a variety of sources. Food-bound Mo is roughly 16% less bioavailable than soluble Mo complexes like ammonium molybdate (Novotny and Peterson 2018). Mo is abundant in pulses, particularly in certain bean varieties including lima beans, pinto beans, red beans, white beans, green beans, peas, and other than that, whole grains, cereals, and organ meats are also considered as Mo-rich sources (Nielsen 2020). Table 15.1 lists the Mo content of some selected food items and Table 15.2 lists the recommended dietary allowances (RDAs) for molybdenum.

Mo has comparatively low toxicity in humans. Thus, there are no reports available that indicate toxicity data for humans (Nielsen 2020). Nevertheless, if Mo is absorbed in higher proportions to the plants that are used as animal feed, serious issues may arise. Molybdenosis is a problem that can be found in ruminants, particularly cattle, which can be a result of the consumption of an excess amount of molybdenum in the diet (Raisbeck et al. 2006). It can be extremely harmful to some animals, particularly cattle and sheep, especially as it leads to secondary copper deficiency (Novotny and Peterson 2018).

15.3.4 Micro-Mineral Deficiencies Induced by Mo

Micro-minerals are considered to be very crucial elements for humans, animals, and plants required in trace amounts as the body or plants cannot synthesize them by their own. There are six micro-mineral nutrients (MMNs), which include iron (Fe), zinc (Zn), manganese (Mn), copper (Cu), boron (B), and molybdenum (Mo). Moreover, chlorine (Cl), cobalt (Co), vanadium (V), sodium (Na), and silicon (Si) are also considered significant. However, it is important to have control over the nutrient supply, as an excessive amount might lead to unfavorable situations (Prasad 2012). For instance, elevated uptakes of Mo can induce other micro-mineral deficiencies, especially including Cu deficiency.

15.3.4.1 Mechanism of Mo's Action/Etiology of Molybdenosis

The main source of molybdenosis is the increase of dietary Mo, and this excess Mo in the diet (more than 5 ppm) will lead to the formation of thiomolybdates in the presence of sulfur compounds in the rumen (Picco et al. 2012). More specifically, the reducing conditions of the rumen trigger the conversion of sulfur from sulfates, or from sulfur-containing amino acids to sulfide which will then take part in the formation of mono, di, tri, and tetra thiomolybdates (Gupta 2011). These thiomolybdates would react with

Table 15.1 Molybdenum content of selected foods.

Food	Micrograms (mcg) per serving	Percent daily value
Black-eyed peas, boiled, ½ cup	288	640
Beef, liver, pan-fried (3 ounces)	104	231
Lima beans, boiled, ½ cup	104	231
Yogurt, plain, low fat, 1 cup	26	58
Milk, 2% milk fat, 1 cup	22	49
Potato, baked, flesh, and skin, 1 medium	16	36
Cheerios cereal, ½ cup	15	33
Shredded wheat cereal, ½ cup	15	33
Banana, medium	15	33
White rice, long-grain, cooked, ½ cup	13	29
Bread, whole wheat, 1 slice	12	27
Peanuts, dry-roasted, 1 ounce	11	24
Chicken, light meat, roasted, 3 ounces	9	20
Egg, large, soft-boiled	9	20
Spinach, boiled, ½ cup	8	18
Beef, ground, regular, pan-fried, 3 ounces	8	18
Pecans, dry roasted, 1 ounce	8	18
Corn, sweet yellow, cooked, ½ cup	6	13
Cheese, cheddar, sharp, 1 ounce	6	13
Tuna, light, canned in oil, 3 ounces	5	11
Potato, boiled without skin, ½ cup	4	9
Orange, medium	4	9
Green beans, boiled, ½ cup	3	7
Carrots, raw, ½ cup	2	4
Asparagus, boiled, ½ cup	2	4

Source: NIH (2021)/U.S Department of Health and Human Services/ Public domain.

plasma Cu by acting as chelators for Cu, especially ligating with the Cu bound to Cu-containing enzymes, ultimately resulting in insoluble Cu containing complex leading to poor absorption-mediated Cu deficiency (Gooneratne et al. 1989). Further, the major impact is on copper-containing enzyme systems and making them nonfunctional (Gupta 2011).

Table 15.2 Recommended dietary allowances (RDAs) for molybdenum for humans.

Age	Male (mcg)	Female (mcg)	Pregnancy and lactation (mcg)
Birth to 6 months	2[a]	2[a]	
7–12 months	3[a]	3[a]	
1–3 year	17	17	
4–8 year	22	22	
9–13 year	34	34	
14–18 year	43	43	50
19+ year	45	45	50

[a] Adequate intake (AI), based on mean molybdenum intakes of infants fed primarily human milk.
Source: NIH (2021)/U.S. Department of Health and Human Services/ Public domain

These thiomolybdates in blood were first identified by Dr. A. Dick, Doug Dewey, and Dr. Jeff Gawthorne in 1975, who were Commonwealth Scientific and Industrial Research Organisation (CSIRO) researchers where they proved that thiomolybdates are being formed due to the interaction between Cu, Mo, and S (Dick et al. 1975).

15.3.4.2 Effect of Mo in the Availability of Cu and Co

The accepted level of Mo and amino acid sulfur content varies between 0–5 mg/kg in DM and 1–3 g/kg in DM, respectively. However, a small change in the concentration of Mo or S in the diet could create a major change in the absorption, distribution, or excretion of Cu content in ruminants leading to a Cu deficient condition or a toxicity. More specifically, the formation of thiomolybdates could create a deficient condition for Cu by restricting the Cu absorption in the digestive tract, limiting the binding of Cu to albumin, followed by retarding the Cu uptake by the liver, reduction of ceruloplasmin formation in the liver creating altered forms of Cu and making them unavailable for metabolism, increasing the biliary Cu excretion and draining the liver Cu, minimizing the Cu biliary reabsorption, and increasing the urinary excretion of Cu (Ammerman 1970; Gooneratne et al. 1989). The systemic binding of thiomolybdates with Cu will make it unavailable for tissue utilization. It has been also found that the ruminal binding of thiomolybdates with Cu takes place majorly with the participation of tri and tetra forms of thiomolybdates, whereas di and tri thiomolybdates play a major function in systemic binding (Gupta 2011).

15.3.4.3 Impact of Molybdenosis

Molybdenosis-induced Cu and Co deficiency has significantly affected ruminant livestock production in large areas of Canada as well as in many other parts of the world (Gooneratne et al. 1989). So far it has been identified in the USA, Canada, United Kingdom, Australia, Sweden, and New Zealand (King et al. 1984).

Specifically, in North America where the cattle were abundant, the occurrence of Cu deficiency has hugely affected the economy with the appearance of nearly 11.3 million cases per annum which has been increasing over the years (Gooneratne et al. 1989).

In one study, the oral administration of a huge dose of molybdate and sulfate to young lambs by Butler et al. (1964) has shown that it could lead to the generation of rigidity and lack of coordination of the hind limbs as well as the occurrence of improper lesions in the CNS showing similar signs of motor ataxia.

It is also reported that Mo-induced Cu deficiency would lead to polioencephalomalacia (a neurological disease) abomasal ulcers, diarrhea, emaciation, and achromotrichia in cattle, as well as changes in the wool of sheep pasturing in fields high with Mo and lower Cu content in Australia (Abbott et al. 2018). Similar observations have been also noticed in goats. Motor ataxia conditions have been reported with red and fallow deer in the United Kingdom and New Zealand (Gooneratne et al. 1989).

Molybdenosis-induced secondary Cu deficiency has also led to a mysterious moose disease in the Southwest of Sweden, where the main clinical observations have shown that the liver, pancreas, and skin are severely affected due to the prolonged deficiency. Lesions in the liver with severe fibrosis, bile duct proliferation, mild necroses, and atrophic dermatosis are among the symptoms which were identified in the affected moose (Aupperle et al. 2001; Frank et al. 2000). Moreover, the moose was also found to be suffered from Mo-induced Cu deficiency-based type 2 diabetes, due to the outcomes of metabolic disruptions that have created glucose intolerance and insulin resistance over time (Frank et al. 2000).

Loss of fertility is another clinical disability that has resulted when excessive Cu deficiency is being resulted by the feeding of ewes with high levels of Mo and S. This has led to a higher fetal loss and a clear decrease in the number of lambs born (Abbott et al. 2018).

The major functional entity that is severely getting affected during molybdenosis is the Cu-dependent enzymes which may also be linked to the antioxidant defense system of the body. Therefore, Mo-induced Cu deficiency has been found to create damage on the DNA integrity inducing genotoxicity and changes in the fatty acid content of the cell membrane, thereby leading to oxidative damage (Picco et al. 2012).

The phosphatases have been discovered to be extremely sensitive to molybdate ions. Osteoporosis has been found in cattle and sheep depastured in places where the available fodder includes a high concentration of molybdenum and is thought to be caused by selective inhibition of calcification enzymes due to lack of phosphatases (Marston 1952).

15.3.4.3.1 Acute Molybdenosis Acute molybdenosis could take place due to the unintentional exposure of ruminants to exposure to oil, natural gas fields, or to engine oil (Gupta 2011). Diagnosis is carried out by looking at the history of clinical signs, pathological observations, and the status of the liver and the kidney. The concentration of Mo required for acute poisoning varies among the different species; however, for cattle it has been found that the feed containing Mo of 7400 mg/kg will lead to acute poisoning. Whereas for sheep it would be 132–137 mg of Mo/kg/day that will lead to lethal results if administered for two to three days. Observed clinical signs in cattle and sheep are like the ones mentioned in the previous section; however, they have commonly expressed a loss of appetite, lethargic conditions, ocular discharges, progressive ataxia in hind limb and front limbs, hepatic and renal degeneration, etc. (Gupta 2011)

15.3.4.3.2 Chronic Molybdenosis Foraging on Mo high green pasture plants would be a possible cause for the development of chromic molybdenosis in ruminants when their Mo intake is greater than 20 ppm in the diet. In addition to the higher Mo content, the low dietary Cu : Mo ratio (2 : 1) and higher dietary sulfur content are the contributors to the chronic molybdenosis. Induction of inflammatory lesions, bronchial adenoma, and severe diarrhea are the common symptoms of chronic molybdenosis. Along with these symptoms, ruminants also tend to show the symptoms of acute molybdenosis. It has been also very difficult to distinguish chronic molybdenosis from that of primary Cu deficiency due to the common set of unspecific symptoms/pathological observation; therefore, careful identification is always required to avoid misinterpretations (Gupta 2011; Picco et al. 2012). To understand the complex etiology of molybdenosis (whether it is acute or chronic), it is always advisable to have a better correlation between the pathological, clinical, and biochemical studies (Picco et al. 2012).

15.4 Sources Leading to Molybdenosis

15.4.1 Soil and Sediments

An increase of the liming of soil has led to the rise of the pH, which has led to the excessive release of Mo from the soil increasing the Mo concentration which would become a possible source for the increase of dietary Mo (Chong 2022). In addition to that, the excessive use of molybdic fertilizers is another factor that has led to the alternations in the Cu : Mo ratio in pastures affecting many farms during 1960–1970s (Axelson et al. 2018). On the other hand, the plant uptake of Mo is determined by the solubility and the chemical traits of the soil where the absorption mainly takes place as molybdate ions (MoO_4^{2-}). The predominance of this form takes place when the soil pH is slightly basic (Axelson et al. 2018).

Moreover, the soil type also has a greater contribution to the availability of microelements. For example, granite-like acid soils generally consisted of a lower content of Co, whereas the basaltic soils have a considerable amount of Co. Similarly, soils from volcanic eruptions contain a lower content of Co. Apart from the soil type, the weathering conditions, leaching, and repeated trimming of the plants have led to the lowering of the microelements percentage of the soil (Gould and Kendall 2011).

Sandy soil, improperly fertilized soils as well as soils planted with monoculture grass species will possibly lead to severe Co- and Cu-deficient conditions (Authority and Journal 2011; Kaiser et al. 2005). But the forages which have been grown in organic-rich soils and stream sediments especially in Florida, Oregon, Irish Republic, and in the United Kingdom where they were highly loaded with Mo and their consumption has resulted in molybdenosis in ruminants. This has strongly indicated the positive correlation of the soil Mo content in inducing molybdenosis in animals (Ammerman 1970).

The average Mo content of the soil varies between different fields which is majorly dependent on the type of the parent soil material. An increase of the soil Mo content could be due to the contribution coming from the parental soil with high Mo content from that's origin, or it could be a result of contamination added from the mining wastes (Axelson et al. 2018). Naturally, there are a few places in the world which the soil is high with Mo content. The examples may include, Somerset England, some Northland soils of New Zealand (Grigg 1960). Globally, the average Mo concentration in arable soil is of 1–2 mg/kg. However, a deviation has been observed on black shales accounting for a very high Mo concentration of 100 mg/kg. The highest concentration of Mo has been detected in areas with Cambro-Silurian sedimentary bedrock, preferably in black alum shales. More specifically, these concentrations of Mo are generally high in topsoils than in subsoils. Geographically, the South West Sweden has also known to contain a higher amount of Mo in soil and has led to increased number of molybdenosis cases (Axelson et al. 2018).

15.4.2 Water

It has been identified that the deep waters in Saskatchewan in Canada contained sulfur as high as 600 mg/L, and its usage has led to the induction of molybdenosis or the secondary Cu deficiency in cattle whereas the reduction of sulfur content in water up to 42 mg/L had led to the improvement of Cu status in them. Similar observations have been also reported in Manitoba Canada (Gooneratne et al. 1989).

15.4.3 Industry

Large industrial cities and coal-fired power stations involve in the liberation of a higher content of heavy metals and Mo into the surrounding environment leading to possible Co and Cu deficiencies (Huang et al. 2017). In these areas, the average Mo content in the forage could be greater than 230 mg/kg of DM (Axelson et al. 2018).

Mo induced secondary Cu deficiency (molybdenosis) in ruminants such as white-tailed deer and cattle that have been commonly identified in several states of United States who had operational uranium and coal surface mining spoils. The increased energy demand in the United States has led to the operation of open-pit mining which has resulted in increased exposure of ruminants to high levels of Mo who are foraging near these uranium-mine spoils (King et al. 1984).

Mining and manufacturing have been major anthropogenic activities which have led to huge discharge of Mo into the soil that has ultimately led to molybdenosis in grazing cattle. Similarly in the period of 1999–2000 in Northern Utah, USA, burning of collected industrial waste has also been a source for Mo contamination in pastures which has created molybdenosis in cattle (Raisbeck et al. 2006).

15.5 Diagnosis and Treatment

15.5.1 Diagnosis of Cu and Co Deficiencies

Early recognition of such deficiencies is a key determinant to minimize the potential impact. Different diagnostic tests can be utilized for this purpose. Analysis of liver Cu content may provide a best indication about the Cu status in the liver, but this test is associated with some uncertainty, high cost with liver biopsy but regarded as the most ideal test to be done in postmortem (Yang et al. 2015). When the animals are suffering from swayback as it is linked with the CNS, the samples should be collected from brain and spinal cord to diagnose the Cu content (Allen et al. 2012). Considering blood as a site of diagnosis, to find the Cu content specifically under a molybdenosis condition will be misleading. Therefore, measuring the ceruloplasmin activity or the different plasma inorganic Cu fractions will provide a good indication about the involvement of Mo in the disease (Linder 2016). Here, the Cu concentrations of 60–120 mg/dL will be regarded as normocupremic, whereas Cu concentrations of 30–59 mg/dL and less than 30 mg/dL will be regarded as moderate hypocupremic and severe hypocupremic, respectively (Picco et al. 2012).

Like in Cu, Co deficiency can be also identified by assessing the liver samples for the level of vitamin B_{12} as liver is a primary site where the vitamin B_{12} is being utilized. In addition to the vitamin B_{12} levels in liver, the serum levels will also indicate a possible deficient condition (Millar and Albyt 1984). But special attention must be given to check whether the sheep or cattle are suffered from any liver damage or whether they have foraged on Co-sufficient pasture few days before they were sampled (Germany et al. 2021). Furthermore, the response tests for vitamin B_{12} injections can be also conducted to get an idea on the severity of deficiency. Here, the weight gain of the targeted group can be compared to that of the untreated set of animals (e.g. lamb) (Gruner et al. 2009).

In the situation where the clinical signs of Co deficiency are not clearly diagnostic, the farmers could be advised to measure the weight of sample of hoggets or lambs where the obtained values can be asked to compare with a standard scale/animal who are normal, same age, and from same breed (As given in Figure 15.4). This will help to identify the anorexia-associated starvation conditions of the ruminants. The challenge remaining in these studies will include the inability to diagnose the marginal deficiencies. In such instances, it is always advisable to conduct chemical analysis and other analyses along with the clinical tests (López-Alonso 2012).

Figure 15.4 Physical appearance of a Co deficient hogget compared to a normal hogget. *Source:* CSIRO/Wikipedia.

Other clinical tests that can be performed to assess a Co-deficient condition may include the analysis of urinary methylmalonic acid and/or formiminoglutamic acid as they are few of the common metabolites that tend to increase during a Co-deficient condition. However, diagnosis of these two metabolites is not heavily used in clinical testing due to the huge cost of sample analysis compared to liver and serum vitamin B_{12} tests (Gültepe et al. 2003; Keyfi et al. 2016; Supakul et al. 2020).

Along with these clinical tests, Comet assay can be conducted to identify the erythrocytes with DNA damage, as well as the analysis of total fatty acid content in erythrocyte membrane will indicate possible Mo-induced Cu deficiency in ruminants (Picco et al. 2012).

In addition to clinical test, pasture analysis for Cu and Co content may also provide a rough indication about the likelihood of developing molybdenosis in ruminants. Pastures containing Cu level of less than 2.5 ppm and/or Co level of 1.5 ppm have a higher probability to induce a deficient condition in sheep and cattle (Andrews 1965; Paolicchi et al. 2013). Apart from Cu and Co, assessment of sulfur and Mo content in pasture can also suggest a possible molybdenosis condition (Axelson et al. 2018). Nevertheless, these analyses are required to be carried out in a regular basis as the animals could survive on pasture with low Co content, due to the excessive storage of vitamin B_{12} in the liver (González-Montaña et al. 2020).

Whichever the process is used for the clinical diagnosis of Cu or Co deficiency induced by higher molybdenum content, it is always advisable to pay the attention on the sufficiency of these microelements rather than entirely concentrate on the diagnosis of deficiency. It will be much advantageous if the tests are being conducted either clinically or nonclinically (field based) to ensure that the microelements are provided sufficiently enough to not to induce any ill condition in ruminants (González-Montaña et al. 2020).

It will be also important to pay some attention on the soil status of the field and to relate them to the previous reports. Most Co- and Cu-deficient regions of lands are well defined and documented. Furthermore, the details regarding the topdressing, pasture species, seasonal variations, and reported ill conditions of ruminants from previous batches will be useful in understanding the root cause of said deficiencies. On this regards, some researchers have identified the importance of developing tools which would provide some sensing mechanism of the identification of fields contaminated with high levels of Mo. The idea behind these approaches was to develop a geography-based risk assessment sensory system on the Mo availability in pastures and soils of Sweden which is very much beneficial for the farmers in decision-making (Axelson et al. 2018).

Overall, when diagnosing the molybdenosis-associated Cu and Co deficiency in animals which is extremely dynamic process, it is highly essential to follow routine analysis by considering all the factors that would possibly become a route cause in developing the disease such as the species variation, age, breed of animals, pasture type, Cu, Co, Mo, and S in the pasture, length of the foraging time, and microelemental supplementation that have been given.

15.5.2 Treatment of Molybdenosis-Associated Cu and Co Deficiency

One of the simple ways to treat the deficient condition is to treat the ruminants who are supposed to be deficient with the microelements, with injectable Cu or Co supplement and compare the body weight changes with a control set of ruminants over a period of one to two months. For this purpose, a sample size of 25 animals can be chosen where the treatment should be carried out under identical conditions. Such treatment procedures have resulted with improved growth rate, coat color, and the overall health (Gooneratne et al. 1989). Also, some reports have indicated that providing a Co supplementation to Co-deficient cattle in Katanga as well as cattle that were suffering from loss of appetite in Brazil were hugely benefited (Ammerman 1970). However, care must be taken to avoid the buildup of excessive Cu or Co in the body which may need to possible accumulation followed by toxicities (Gupta 2011).

In another study conducted by Geoff Judson and Peter Babidge in 2002, it have shown that the sheep who were on pasture with high Mo content at Kybybolite, South Australia, displaying Cu deficiency symptoms were successfully treated with parenteral formulation of Cu heptonate (Judson and Babidge 2002). However, the parenteral application of Cu will only allow a limited amount of Cu being administered, whereas special attention is always required to avoid the toxic effects that could generate due to the Cu accumulation which is being commonly reported in New Zealand (Grace and Knowles 2012).

Oral drenching of $CuSO_4$ or CuO is another way of supplying Cu demand for couple of weeks, and which has been practical to fulfill the needs of dairy cattle whereas it was not successful to meet up the demands of sheep (López-Alonso and Miranda 2020). Besides, the use of slow releasing CuO capsules, rumen bullets have been effectively used to raise the Cu storage in ruminants for up to 5–12 months (León-Cruz et al. 2020).

One to two local applications of 0.8–2.5 kg of Cu/ha in soils high with Mo content is believed to supply required Cu to sheep and cattle for minimum of eight years (Grace et al. 1998). But if the Mo content is too high where the Cu content is sufficient, this application would not be successful; instead, the direct application of Cu or Co to ruminants via parenteral formulation will be then successful (Van Niekerk and Cloete 1995).

Topdressing of forage with 5–10 kg of $CuSO_4$/ha in every 3–4 years has been practiced as a traditional mode of preventing Cu deficiencies but was known to be associated with a huge cost. However, again the local application of Cu via a parenteral formulation is always preferable over the topdressing approach (Smith 2004).

Comparably, the regular use of Co as a fertilizer on Co-deficient farm soils has resulted with a major decrease of the Co deficiency conditions in cattle as well as in mature sheep. However, this procedure has not been promising to overcome the deficient conditions in young sheep (Snook 1962). Similar observations have been also made in other studies (Raisbeck et al. 2006).

Furthermore, it is also advisable to apply Mo as a fertilizer only in required amount during the harvesting phase if the plants which are deficient in Mo. Mo is known to induce the growth of short-lived herbs such as clovers and thereby increasing the DM of the forage. Generally, it is recommended that the singular application of 75 g/ha is enough to provide the required Mo levels for 10–15 years in slightly acidic soils. Also, it is appropriate to not to allow the ruminants to forage on such fields until heavy showers are being obtained (Rutkowska et al. 2017).

More importantly, top dressing of pastures with cobalt superphosphate has proven to be effective, simple, and affordable way of mitigating the Co deficiency in Co-deficient areas where it has provided the protection over young lambs over a period up to seven months. But this procedure is not applicable in large scale due to the associated high cost. In addition to the topdressing, boom spraying has been applied as a foliar application on the actively growing pasture to increase the Co uptake by the plants (Sherrell 1990).

Apart from that, attempts have been taken by several group of researchers to provide a solution over Co deficiency in sheep and cattle. One of them includes the development of a Co bullet containing Co oxide and Fe oxide, for the slow release of Co in the rumen to trigger the vitamin B_{12} production (González-Montaña et al. 2020). Furthermore, the effectiveness of direct intravenous application of vitamin B_{12} on young sheep has also indicated positive results where it had prevented the Co deficiency in sheep for two to threemonths of period (González-Montaña et al. 2020). This has been a widely used practice in New

Zealand to protect lambs against the Co deficiency (Phillipson and Mitchell 1952).

Studies conducted by Geoff Judson, Peter Babidge, M.-J. Vawser, Z. Chen, and L. Sansom has led to the generation of sustained release of vitamin B_{12} pellet which has been effective on preventing the Co deficiency over a period of eight months (Judson et al. 2002). Like in this manner, oral dosing with 300 mg of Co/month has reduced the number of ruminal deaths in severely deficient countries (Ammerman 1970).

15.6 Conclusions

Cu- and Co-related micro-mineral deficiencies are widespread among ruminants in worldwide, and maintaining the adequate levels of micro-minerals is a key requirement for the establishment of a proper health as well as to improve the productivity in ruminants. The antagonistic activity of Mo and S is a key determinator of absorption, and bioavailability and metabolism of Cu and Co have resulted in a secondary Cu and Co deficiency conditions which is known as molybdenosis resulted in severe clinical outcomes such as extreme gastrointestinal irritation, anorexia, anemia, photosensitivity, diarrhea, coma, decolorization of wool/skin coat, lacrimation, scaly ears, reduction of milk production, polioencephalomalacia, and even death. The two main biological entities which have been majorly affected due to the shortening of rumen Cu and Co have been the vital cuproenzymes and vitamin B_{12}. Improper functioning of these enzymes as well as vitamin B_{12} has led to the said devastating clinical outcomes. Molybdenosis-induced Cu and Co deficiency has significantly affected the ruminant livestock production in large areas of Canada as well as in many other parts of the world including USA, Canada, United Kingdom, Australia, Sweden, and New Zealand. Most affected categories of ruminants include newborns and the young animals, pregnant or lactating cattle, deer and sheep as well as the ones who suffer severely with heavy worm infestations. Three major sources such as the soil conditions, water, and industry can be identified as the major contributors of elevated Mo and S content in ruminal diet. More importantly, liming of the soil leading to excessive release of Mo, parental soils with high Mo content, deep waters rich with sulfur, as well as uranium and coal surface mining spoils in industrial zones have led to the abrupt increase of Mo in the pasture. Therefore, to minimize the occurrence of molybedenosis, it has been identified the requirement of the routine investigation of Cu and Co status of the ruminants by analyzing animal tissues and blood samples. Continuous full monitoring mechanism for Cu, Co, Mo, and S in soil and pasture will be also essential to make sure the ruminants are receiving the sufficient microelemental composition minimizing the possibility of a development of a deficiency. The treatment options including direct application of Cu or Co via parenteral formulations, topdressing of the soils and pastures, practicing of proper fertilization protocols, and the development of slow-releasing formulations (bullets and capsules) of Cu and Co have also received much of the attention over the years. However, some clinical observations of Mo-induced secondary Cu and Co deficiency have not been directly linked to the symptoms of primary microelemental deficiencies, and this suggest the further requirement to describe the mechanism by which Mo is producing adverse clinical symptoms in Mo-induced Cu and Co deficiency.

References

Abbaspour, N., Hurrell, R., and Kelishadi, R. (2014). Review on iron and its importance for human health. *J. Res. Med. Sci.* 19 (2): 164–174.

Abbott, K., Hynd, P., de Graaf, S. et al. (2018). *The Practice of Sheep Veterinary Medicine*. University of Adelaide Press http://www.jstor.org/stable/j.ctvb4bssr.

Abolbashari, S., Darroudi, S., Tayefi, M. et al. (2018). Association between serum zinc and copper levels and antioxidant defense in subjects infected with human T-lymphotropic virus type 1. *J. Blood Med.* 10: 29–35.

Abou-zeina, H.A.A., Zaghawa, A.A., Nasr, S.M., and Keshta, H.G.E. (2008). Effects of dietary cobalt deficiency on performance, blood and rumen metabolites and liver pathology in sheep. *Glob. Vet.* 2 (4): 182–191.

Agbenin, J.O. and Olojo, L.A. (2004). Competitive adsorption of copper and zinc by a Bt horizon of a savanna alfisol as affected by pH and selective removal of hydrous oxides and organic matter. *Geoderma* 119 (1–2): 85–95.

Akbar, F.M., Zafar, M., Hamid, A. et al. (2013). Interactive effect of cobalt and nitrogen on growth, nodulation, yield and protein content of field grown pea. *Hortic. Environ. Biotechnol.* 54 (6): 465–474.

Ali, B., Hayat, S., Hayat, Q., and Ahmad, A. (2010). Cobalt stress affects nitrogen metabolism, photosynthesis and antioxidant system in chickpea (*Cicer arietinum* l.). *J. Plant Interact.* 5 (3): 223–231.

Allen, A.L., Goupil, B.A., and Valentine, B.A. (2012). A retrospective study of spinal cord lesions in goats submitted to 3 veterinary diagnostic laboratories. *Can. Vet. J.* 53 (6): 639–642.

Ammerman, C.B. (1970). Recent developments in cobalt and copper in ruminant nutrition: a review. *J. Dairy Sci.* 53 (8): 1097–1107.

Andrews, E.D. (1965). Observations on the thrift of young sheep on a marginally cobalt-deficient area. *J. Agric. Res.* 8 (4): 788–817.

Anke, M., Seifert, M., Holzinger, S. et al. (2007). The biological and toxicological importance of molybdenum in the environment and in the nutrition of plants, animals and man – part 2: molybdenum in animals and man. *Acta Biol. Hung.* 58 (3): 325–333.

Aupperle, H., Schoon, H.A., and Frank, A. (2001). Experimental copper deficiency, chromium deficiency and additional molybdenum supplementation in goats – pathological findings. *Acta Vet. Scand.* 42 (3): 311–321.

Authority, F.D. and Journal, I. (2011). Extractable iron and aluminium and their relationship to phosphate retention in Irish soils. *Development* 15 (1): 65–73. Author (s): L.J. Evans and G.W. Smillie Published by: TEAGASC-Agriculture and Food Development Authority http://www.jstor.org/stable/25555803.

Axelson, U., Söderström, M., and Jonsson, A. (2018). Risk assessment of high concentrations of molybdenum in forage. *Environ. Geochem. Health* 40 (6): 2685–2694.

Baumgartner, M.R., Hörster, F., Dionisi-Vici, C. et al. (2014). Proposed guidelines for the diagnosis and management of methylmalonic and propionic acidemia. *Orphanet J. Rare Dis.* 9 (1): 1–36.

Bennetts, H.W., Beck, A.B., Harley, R., and Evans, S.T. (1941). Falling disease of cattle in the South-west of Western Australia. Studies of copper deficiency in cattle. *Aust. Vet. J.* 17: 85–93.

Bottrill, D.E., Possingham, J.V., and Kriedemann, P.E. (1970). The effect of nutrient deficiencies on phosynthesis and respiration in spinach. *Plant Soil* 32 (1): 424–438.

Bottrill, D.E., Possingham, J.V., and Kriedemann, P.E. (1970). The effect of nutrient deficiencies on photosynthesis and respiration. *Plant Soil* 32 (2): 424–438. In Spinach Author (s): D.E. Bottrill, J.V. Possingham and P.E. Kriedemann Published by: Springer https://www.jstor.org/stable/42935831.

Brasche, C.J. (2015). Effect of a trace mineral injection on beef cattle performance. Master thesis. University of Nebraska. 142.

Brewer, G.J. (2014). The promise of copper lowering therapy with tetrathiomolybdate in the cure of cancer and in the treatment of inflammatory disease. *J. Trace Elem. Med. Biol.* 28 (4): 372–378.

Brink, G.E., Pederson, G.A., Sistani, K.R., and Fairbrother, T.E. (2001). Forages: uptake of selected nutrients by temperate grasses and legumes. *Agron. J.* 93 (4): 887–890.

Butler, E.J., Barlow, R.M., and Smith, B.S. (1964). Copper deficiency in relation to swayback in sheep. 1. Effect of molybdate and sulphate supplements during pregnancy. *J. Comp. Pathol.* 74: 419–426.

Campo, R.J. and Albino, U.B.H. (2000). Bradyrhizobium elkanii. *Nitrogen Fixat. From Mol. to Crop Prod.* 597–598.

Cano-Sancho, G., Rovira, J., Perelló, G. et al. (2017). Extensive literature search on the bioavailability of selected trace elements in animal nutrition: incompatibilities and interactions. *EFSA Support. Publ.* 11 (3): 1–1136.

Chong, K.P. (2022). Effects of liming on soil properties and its roles in increasing the productivity and profitability of the oil palm industry in Malaysia. *Agriculture 12* (3): 322.

Clarkson, D.T. and Hanson, J.B. (1980). The mineral nutrition of higher plants. *Annu. Rev. Plant Physiol.* 31 (1): 239–298.

Comar, C.L. and Davis, G.K. (1947). Cobalt metabolism studies. *J. Biol. Chem.* 170 (1): 379–389.

Contreras, R.A., Pizarro, M., Köhler, H. et al. (2018). Copper stress induces antioxidant responses and accumulation of sugars and phytochelatins in Antarctic *Colobanthus quitensis* (Kunth) Bartl. *Biol. Res.* 51 (1): 1–10.

Dehertogh, A.A., Mayeux, P.A., and And Evans, H.J. (1964). The relationship of cobalt requirement to propionate metabolism. *J. Biol. Chem.* 239 (8): 2446–2453.

Delhaize, E., Loneragan, J.F., and Webb, J. (1985). Development of three copper metalloenzymes in clover leaves. *Plant Physiol.* 78 (1): 4–7.

Della Pepa, G. and Brandi, M.L. (2016). Microelements for bone boost: the last but not the least. *Clin. Cases Miner. Bone Metab.* 13 (3): 181–185.

Dermauw, V. (2014). Trace element nutrition in the tropical zebu cattle type. Lessons from the Gilgel Gibe catchment, Ethiopia. *Afrika Focus* 27 (1): https://doi.org/10.21825/af.v27i1.4916.

Dhaliwal, S.S., Naresh, R.K., Mandal, A. et al. (2019). Dynamics and transformations of micronutrients in agricultural soils as influenced by organic matter build-up: a review. *Environ. Sustain. Indic.* 1–2 (September): 100007.

Dick, A.T., Dewey, D.W., and Gawthorne, J.M. (1975). Short note: thiomolybdates and the copper-molybdenum-sulphur interaction in ruminant nutrition. *J. Agric. Sci.* 85 (3): 567–568.

Fisher, G.E. (2008). Micronutrients and animal nutrition and the link between the application of micronutrients to crops and animal health. *Turkish J. Agric. For.* 32 (3): 221–233.

Frank, A., Sell, D.R., Danielsson, R. et al. (2000). A syndrome of molybdenosis, copper deficiency, and type. 2. Diabetes in the moose population of South-west Sweden. *Sci. Total Environ.* 249 (1–3): 123–131.

Gad, N. (2012). Response of cowpea plants to cobalt and molybdenum combination. *Middle East J. Appl. Sci.* 2 (1): 13–19.

Gawthorne, B.J.M. and Smith, R.M. (1974). Folic acid metabolism in vitamin B_{12}-deficient sheep. *Biochem.* 142 (1): 119–126.

Germany, N., Kamphues, J., and Ganter, M. (2021). A case of concurrent molybdenosis, secondary copper, cobalt and selenium deficiency in a small sheep herd in Northern germany. *Animals* 11: 1–14.

González-Montaña, J.R., Escalera-Valente, F., Alonso, A.J. et al. (2020). Relationship between vitamin B_{12} and cobalt metabolism in domestic ruminant: an update. *Animals* 10 (10): 1–36.

Gooneratne, S.R., Buckley, W.T., and Christensen, D.A. (1989). Review of copper deficiency and metabolism in ruminants. *Can. J. Anim. Sci.* 69 (4): 819–845.

Gould, L. and Kendall, N.R. (2011). Role of the rumen in copper and thiomolybdate absorption. *Nutr. Res. Rev.* 24 (2): 176–182.

Grace, N.D. and Knowles, S.O. (2012). Trace element supplementation of livestock in New Zealand: meeting the challenges of free-range grazing systems. *Vet. Med. Int.* 2012: 1–8.

Grace, N.D., Knowles, S.O., Rounce, J.R. et al. (1998). Effect of increasing pasture copper concentrations on the copper status of grazing romney sheep. *J. Agric. Res.* 41 (3): 377–386.

Green, L.F., McCarthy, J.F., and King, C.G. (1939). Inhibition of respiration and photosynthesis in chlorella pyrenoidosa by organic compounds that inhibit copper catalysis. *J. Biol. Chem.* 128 (2): 447–453.

Grigg, J.L. (1960). The distribution of molybdenum in the soils of New Zealand. *J. Agric. Res.* 3 (1): 69–86.

Gruner, T.M., Sedcole, J.R., Furlong, J.M., and Sykes, A.R. (2009). Vitamin B_{12} absorption and metabolism in milk-fed lambs. *N. Z. Vet. J.* 57 (1): 22–27.

Gültepe, M., Özcan, Ö., Avşar, K. et al. (2003). Urine methylmalonic acid measurements for the assessment of cobalamin deficiency related to neuropsychiatric disorders. *Clin. Biochem.* 36 (4): 275–282.

Gupta, R.C. (2011). *Veterinary Toxicology: Basic and Clinical Principles*. Academic Press.

Halliwell, B. and Gutteridge, J.M.C. (1999). *Free Radicals in Biology and Medicine*, 3e, 1–25. Oxford: Oxford University Press.

Halliwell, B., Chirico, S., Crawford, M.A. et al. (1993). Lipid peroxidation: its mechanism, measurement, and significance. *Am. J. Clin. Nutr.* 57 (5 Suppl): 724S–725S.

Han, S.H., An, J.Y., Hwang, J. et al. (2016). The effects of organic manure and chemical fertilizer on the growth and nutrient concentrations of yellow poplar (*Liriodendron tulipifera* Lin.) in a nursery system. *Forest Sci. Technol.* 12 (3): 137–143.

Hidiroglou, M. (1979). Trace element deficiencies and fertility in ruminants: a review. *J. Dairy Sci.* 62 (8): 1195–1206.

Hill, G.M. and Shannon, M.C. (2019). Copper and zinc nutritional issues for agricultural animal production. *Biol. Trace Elem. Res.* 188 (1): 148–159.

Hu, X., Wei, X., Ling, J., and Chen, J. (2021). Cobalt: an essential micronutrient for plant growth? *Front. Plant Sci.* 12: 2370.

Huang, X., Hu, J., Qin, F. et al. (2017). Heavy metal pollution and ecological assessment around the Jinsha coal-fired power plant (China). *Int. J. Environ. Res. Public Health* 14 (12): 1–12.

Judson, G.J. and Babidge, P.J. (2002). Comparison of copper heptonate with copper oxide wire particles as copper supplements for sheep on pasture of high molybdenum content. *Aust. Vet. J.* 80 (10): 630–635.

Judson, G.J., Babidge, P.J., and Sansom, L.N. (2002). Recent developments in the detection and treatment of vitamin B12 deficiency in sheep and cattle at pasture. *NZ Soc Anim Prod Proc* 62: 307–310.

Jungk, A. (1977). Bergmann, W. und Neubert, P., (Herausgeber): Pflanzendiagnose und Pflanzenanalyse. Verlag VEB Gustav Fischer, Jena 1976. 711 S. mit 28 Abb., 5 Übersichten, 23 Tab. im Text und 519 Farbbildern auf 160 Tafeln sowie 114 Tabellen zur Pflanzenanalyse. Leinen. *Z. Pflanzenernähr. Bodenkd.* 140 (3–4): 478–479.

Kaiser, B.N., Gridley, K.L., Brady, J.N. et al. (2005). The role of molybdenum in agricultural plant production. *Ann. Bot.* 96 (5): 745–754.

Keshri, A., Bashir, Z., Kumari, V. et al. (2021). Role of micronutrients during peri-parturient period of dairy animals – a review. *Biol. Rhythm. Res.* 52 (7): 1018–1030.

Keyfi, F., Talebi, S., and Varasteh, A.-R. (2016). Methylmalonic acidemia diagnosis by laboratory methods. *Reports Biochem. Mol. Biol.* 5 (1): 1–14.

Khan, Z.I., Hussain, A., Ashraf, M. et al. (2008). Effect of seasonal variation on the copper status in a soil-plant-animal system. *Acta Agron. Hungarica* 56 (1): 55–67.

Kihara, J., Bolo, P., Kinyua, M. et al. (2020). Micronutrient deficiencies in African soils and the human nutritional nexus: opportunities with staple crops. *Environ. Geochem. Health* 42 (9): 3015–3033.

Kim, M.J., Kim, S.C., Chung, S. et al. (2020). Exploring the role of copper and selenium in the maintenance of normal thyroid function among healthy Koreans. *J. Trace Elem. Med. Biol.* 61 (March: 126558.

King, K.A., Leleux, J., and Mulhern, B.M. (1984). Molybdenum and copper levels in white-tailed deer near uranium mines in Texas. *J. Wildl. Manag.* 48 (1): 267.

Lange, B., van der Ent, A., Baker, A.J.M. et al. (2017). Copper and cobalt accumulation in plants: a critical assessment of the current state of knowledge. *New Phytol.* 213 (2): 537–551.

León-Cruz, M., Ramírez-Bribiesca, E., López-Arellano, R. et al. (2020). Controlled-release intraruminal boluses. *Rev. Mex. Cienc. Pecu.* 11 (2): 498–516.

Linder, M.C. (2016). Ceruloplasmin and other copper binding components of blood plasma and their functions: an update. *Metallomics* 8 (9): 887–905.

Loneragan, J.F., Robson, A.D., Graham, R.D., and Australian Academy of Technological Sciences. (1981). Copper in soils and plants. *Proceedings of the Golden Jubilee International Symposium on "Copper in Soils and Plants,"* Murdoch University, Perth, Western Australia (7–9 May 1981). Under the sponsorship of the Australian Academy of Technological. 380.

López-Alonso, M. (2012). Trace minerals and livestock: not too much not too little. *ISRN Vet. Sci.* 2012: 1–18.

López-Alonso, M. and Miranda, M. (2020). Copper supplementation, a challenge in cattle. *Animals* 10 (10): 1–21.

Malhi, S.S., Cowell, L., and Kutcher, H.R. (2005). Relative effectiveness of various sources, methods, times and rates of copper fertilizers in improving grain yield of wheat on a Cu-deficient soil. *Can. J. Plant Sci.* 85 (1): 59–65.

Manuel, T.J., Alejandro, C.A., Angel, L. et al. (2018). *Roles of Molybdenum in Plants and Improvement of Its Acquisition and Use Efficiency*. Elsevier Inc.

Marston, H.R. (1952). Cobalt, copper and molybdenum in the nutrition of animals and plants. *Physiol. Rev.* 32 (1): 66–121.

Mertz, W. and Sadler, P.J. (1981). The scientific and practical importance of trace elements. *Philos. Trans. R. Soc. Lond. Ser. B Biol. Sci.* 294 (1071): 9–18. [and Discussion] Source: Philosophical Transactions of the Royal Society of London. Series B, Biological Sciences, 14 August 1981, Vol. 294, No. 1071. Metabolic and Physiological Conseq.

Millar, K.R. and Albyt, A.T. (1984). A comparison of vitamin B_{12} levels in the liver and serum of sheep receiving treatments used to correct cobalt deficiency. *N. Z. Vet. J.* 32 (7): 101–108.

Mus, F., Crook, M.B., Garcia, K. et al. (2016). Symbiotic nitrogen fixation and the challenges to its extension to nonlegumes. *Appl. Environ. Microbiol.* 82 (13): 3698–3710.

Nielsen, F.H. (2020). *Manganese, Molybdenum, Boron, Silicon, and Other Trace Elements*. Elsevier Inc.

Novotny, J.A. and Peterson, C.A. (2018). Molybdenum. *Adv. Nutr.* 9 (3): 272–273.

Ott, G., Havemeyer, A., and Clement, B. (2015). The mammalian molybdenum enzymes of mARC. *J. Biol. Inorg. Chem.* 20 (2): 265–275.

Palit, S., Sharma, A., and Talukder, G. (1994). Effects of cobalt on plants. *Bot. Rev.* 60 (2): 149–181.

Paolicchi, F., Perea, J., Cseh, S., and Morsella, C. (2013). Relationship between paratuberculosis and the microelements copper, zinc, iron, selenium and molybdenum in beef cattle. *Braz. J. Microbiol.* 44 (1): 153–160.

Phillipson, A.T. and Mitchell, R.L. (1952). The administration of cobalt by different routes to lambs maintained on a low-cobalt diet. *Br. J. Nutr.* 6 (1): 176–189.

Picco, S., Ponzzinibio, M.V., Mattioli, G. et al. (2012). Physiological and genotoxic effects of molybdenum-induced copper deficiency in cattle. *Agrociencia* 46 (2): 107–117.

Prasad, R. (2012). Micro mineral nutrient deficiencies in humans, animals and plants and their amelioration. *Proc. Natl. Acad. Sci. India Sect. B – Biol. Sci.* 82 (2): 225–233.

Raisbeck, M.F., Siemion, R.S., and Smith, M.A. (2006). Modest copper supplementation blocks molybdenosis in cattle. *J. Vet. Diagn. Investig.* 18 (6): 566–572.

Rana, M.S., Parashuram Bhantana, X.S., Imran, M. et al. (2020). Molybdenum as an essential element for crops: an overview. *Biomed. J. Sci. Tech. Res.* 24 (5): 18535–18547.

Reid, H.A. (1926). Bush sickness. A New Zealand disease of ruminants. *Vet. J.* 82 (4): 212–216.

Robson, A.D., Dilworth, M.J., and Chatel, D.L. (1979). Cobalt and nitrogen fixation in *Lupinus Angustifolius* L. I. growth, nitrogen concentrations and cobalt distribution. *New Phytol.* 83: 53–62.

Rutkowska, B., Szulc, W., Spychaj-Fabisiak, E., and Pior, N. (2017). Prediction of molybdenum availability to plants in differentiated soil conditions. *Plant Soil Environ.* 63 (11): 491–497.

Scaletti, R.W., Trammell, D.S., Smith, B.A., and Harmon, R.J. (2003). Role of dietary copper in enhancing resistance to *Escherichia coli* mastitis. *J. Dairy Sci.* 86 (4): 1240–1249.

Sherrell, C.G. (1990). Effect of cobalt application on the cobalt status of pastures 3 comparison of chelate and sulphate as cobalt sources for topdressing deficient pastures. *New Zeal. J. Agric. Res.* 33 (2): 313–317.

Silva, T., Aguiar, G., Carvalho, F. et al. (2014). Outbreaks of copper deficiency in ruminants in the semiarid region of Paraíba, Brazil. *Semin. Agrar.* 35 (4): 1955–1960.

Silva, W.J., Guizelini, C.C., Franco, G.L. et al. (2020). Cobalt deficiency in cattle and its impact on production. *Pesqui. Vet. Bras.* 40 (11): 837–841.

Smith, L.C. (2004). The effectiveness of autumn spray applications of cobalt sulphate, copper chelate and copper sulphate on Southland pastures. *Proceedings of the New Zealand Grassland Association* 171–176.

Snook, L.C. (1962). Cobalt: its use to control wasting disease. *J. Dep. Agric. West. Aust. Ser. 4* 3 (11): 844–852.

Soediono, B. (1989). *Micronutrient for Sustainable Food, Feed, Fibre and Bioenergy Production*. International Fertilizer Industry Association.

Solano, F. (2018). On the metal cofactor in the tyrosinase family. *Int. J. Mol. Sci.* 19 (2): 1–17.

Srinivasan, S. and Avadhani, N.G. (2012). Cytochrome c oxidase dysfunction in oxidative stress. *Free Radic. Biol. Med.* 56 (6): 1252–1263.

Stare, F.J. and Elvehjem, C.A. (1933). Cobalt in animal nutrition. *J. Biol. Chem.* 99 (2): 473–483.

Supakul, S., Chabrun, F., Genebrier, S. et al. (2020). Diagnostic performances of urinary methylmalonic acid/creatinine ratio in vitamin B_{12} deficiency. *J. Clin. Med.* 9 (8): 1–11.

Sutherland, R.J., Cordes, D.O., and Carthew, G.C. (1979). Ovine white liver disease – an hepatic dysfunction associated with vitamin B_{12} deficiency. *N. Z. Vet. J.* 27 (11): 227–232.

Thompson, J.S., Marks, T.J., and Ibers, J.A. (1977). Blue copper proteins: synthesis, spectra, and structures of $CuIN_3(SR)$ and $CuIIN_3(SR)$ active site analogues. *Proc. Natl. Acad. Sci.* https://doi.org/10.1073/pnas.74.8.3114.

Underwood, W.J., Blauwiekel, R., Delano, M.L. et al. (2015). Biology and diseases of ruminants (sheep, goats, and cattle). In: *Laboratory Animal Medicine*, 623–694. Academic Press.

Van Niekerk, F. and Cloete, S. (1995). Parenteral copper and selenium supplementation of sheep on legume-grass pastures: biochemical and production responses in lambs to maternal treatment. *J. S. Afr. Vet. Assoc.* 66 (1): 11–17.

Van Ryssen, J.B.J., Miller, W.J., Gentry, R.P., and Neathery, M.W. (1987). Effect of added dietary cobalt on metabolism and distribution of radioactive selenium and stable minerals. *J. Dairy Sci.* 70 (3): 639–644.

Van Saun, R.J. (2006). Nutrient requirements of South American camelids: a factorial approach. *Small Rumin. Res.* 61 (2–3 Spec. Iss): 165–186.

Vesk, M., Possingham, J.V., and Mercer, F.V. (1966). The effect of mineral nutrient deficiencies on the structure of the leaf cells of tomato, spinach, and maize. *Aust. J. Bot.* 14: 1–18.

Walker, J.B. (1953). Inorganic micronutrient requirements of chlorella. I. Requirements for calcium (or strontium), copper, and molybdenum. *Arch. Biochem. Biophys.* 46 (1): 1–11.

Wapnir, R.A. (1998). Copper absorption and bioavailability. *Am. J. Clin. Nutr.* 67 (5 Suppl): 1054–1060.

Wassmann, S., Laufs, U., Müller, K. et al. (2002). Cellular antioxidant effects of atorvastatin in vitro and in vivo. *Arterioscler. Thromb. Vasc. Biol.* 22 (2): 300–305.

Weisany, W., Raei, Y., and Allahverdipoor, K.H. (2013). Role of some of mineral nutrients in biological nitrogen fixation. *Bull. Environ. Pharmacol. Life Sci.* 2 (4): 77–84.

Wuana, R.A. and Okieimen, F.E. (2011). Heavy metals in contaminated soils: a review of sources, chemistry, risks and best available strategies for remediation. *ISRN Ecol.* 2011: 1–20.

Yang, X., Tang, X.P., Zhang, Y.H. et al. (2015). Prospective evaluation of the diagnostic accuracy of hepatic copper content, as determined using the entire core of a liver biopsy sample. *Hepatology* 62 (6): 1731–1741.

Yaremcio, B. and Kreplin, C. (2009). Effects of nutrition on beef cow reproduction. *Agdex* 1: 420–451.

Yun Liu, J.M. (2022). An emerging role of defective copper metabolism. *Nutrients* 14: 1–21.

Zafar, R. (2016). New insight into pathogenesis of medical diseases. *Integr. Food, Nutr. Metab.* 3 (5): 402–416.

16

Healing Clays Structure and Functions

Nadun H. Madanayake[1] *and Nadeesh M. Adassooriya*[2]

[1] *Department of Botany, Faculty of Science, University of Peradeniya, Peradeniya, Sri Lanka*
[2] *Department of Chemical and Process Engineering, Faculty of Engineering, University of Peradeniya, Peradeniya, Sri Lanka*

Abbreviations

GelMA	Gelatine methacrylate
LAP	Laponite
MMT	Montmorillonite
NCs	Nanoclays
PCL	Polycaprolactone
SA	Sodium alginate
TEMPO	2,2,6,6-Tetramethylpiperidinyl-1-oxyl

16.1 Introduction

World is moving rapidly, and science and technology enable every conventional application to be optimized to enhance their potential applications in life sciences. Nanotechnology is a prominent interdisciplinary approach that utilize particles at nanometric state to improve the efficacy of different applications. Nanomaterials can be simply defined as particles having at least one of its dimension within 1–100 nm scale. These materials are used in different applications including medicine, energy, environmental remediation, food and agriculture, electronics, and cosmetics, etc. Nanomaterials can be organic or inorganic and inherent properties of these materials will enable to use them based on their potential requirement. Field of medicine is one such fields that experiences the merits of nanotechnology to a greater extent.

Natural and artificial clay types have been reported to utilize for biomedical applications. Clay minerals have a long history of being utilized in treatments of a number of ailments as healing clays. For example, naturally occurring clays have applied to treat skin infections. Also, clinical applications of clay types such as French green clay have reported to use to heal Buruli ulcer caused by *Mycobacterium ulcerans* (Williams and Haydel 2010). Recently, the advent of nanotechnology has enabled the use of clays in their nanoform for biomedical application has highlighted a lot. Nanoclays (NCs) are layered silicates where at least a single dimension within 1–100 nm. The discriminating properties of NCs such as higher sorption capacity, swelling potential, and a higher surface area (Tomás et al. 2018) allow them to use in biomedical applications as healing clays. Different types of these natural nanomaterials have been used in several biomedical applications such as in bone cementing, tissue engineering, drug delivery, wound healing, and enzyme immobilization, bacteriocidal, and bacteriostatic agents.

Generally, NCs can be obtained either as anionic or as cationic clays. This depends on the presence of surface-layered charges and the types of interlayer ions present within the clay matrix. Montmorillonite (MMT), kaolinite, laponite (LAP), halloysite, bentonite, hectorite, sepiolite, saponite, and vermiculite are some of the clay types used in industrial applications. For instance, NCs are used to prepare aspectic food packaging materials, to prepare target specific delivery carriers for therapeutics, to prepare nutraceutical delivery systems, as antimicrobial agents, environmental remediation, slow and controlled release of drugs, and energy storage, etc. (Choi and Choy 2011; Choi and Kim 2013; Tornuk et al. 2013; Rawat et al. 2014; Madusanka et al. 2015; Jayrajsinh et al. 2017; Madusanka et al. 2017a, b; Zhang et al. 2017; Premarathna et al. 2019). Although this has wide range of implications, understanding the structure and function of NCs is imperative to understand. Therefore, this chapter will elaborate on the potential application of NCs as healing clays in biomedical application.

Medical Geology: En route to One Health, First Edition. Edited by Majeti Narasimha Vara Prasad and Meththika Vithanage.

16.2 Classification and Commonly Used Nanoclay Types as Biomedical Applications

Generally, we classify clays as natural and artificial types. Clay types such as MMT, hectorite, sepiolite, saponite, rectorite, bentonite, vermiculite, Beidellite, kaolinite, and chlorite can be identified as natural clays, whereas layered double hydroxides, synthetic MMT, and hectorite can be defined as synthetic clays. Basically, clays are composed of alternating tetrahedral silica sheets and alumina octahedral layers (Raji et al. 2016). Therefore, they can be grouped structurally based on the ratios of the intercalation of tetrahedral silica sheets and alumina octahedral layers. Clays can be made up of tetrahedral silica sheets and alumina octahedral layers in the ratio of 1 : 1. Kaolinite, halloysite, rectorite, and chrysotile are some examples of this category (Cheshire and Güven 2005; Du et al. 2010; Raji et al. 2016; Detellier 2018). Also, clays can be found in the ratio of 2 : 1, where one octahedral layer sandwitched between two octahedral sheets form the clay material. MMT, LAP, and sepiolite are examples for this category (Raji et al. 2016). Figure 16.1a, b elaborates on the schematic view of the arrangement of tetrahedral silica sheets and alumina octahedral layers in clays.

The MMT is one of the most frequently used nanoclay for biomedical applications. MMT composed of two tetrahedral and one octahedral unit to form platelet-like structures. Here, the octahedral fragment of MMT consists of Al(III) coordinated with oxygen and is fenced by the two tetrahedral sheets with oxygen coordinating Si(IV). Therefore, this arrangement of the clay provides a unit cell formula of $[Al_2(OH)_2(Si_2O_5)_2]$ (Segad et al. 2010; Kuttalam et al. 2021).

The smectite group of phyllosilicate is frequently utilized in pharmaceutical applications as antacids, gastrointestinal protectors, and antidiarrheal agent. In addition, these clay types can act as disintegrants, diluents, binders, emulsifying, thickening, anticaking, flavor correction, and drug carriers (Tomás et al. 2018). The LAP is recognized as an artificial clay mineral. These clay types are structurally similar to naturally occurring clay types such as hectorite (Christidis et al. 2018). This is a smectite group of phyllosilicates having 2 : 1 layered unit, where two tetrahedral silica sheets sandwich with one Mg^{2+}-containing octahedral sheet. These are disk-shaped crystals having an empirical formula of $Na^{+}_{0.7}[(Si_8Mg_{5.5}Li_{0.3})\,O_{20}(OH)_4]_{-0.7}$ (Mahdavinia and Shokri 2017; Villalba-Rodríguez et al. 2021). It is important to note that lithium cations in the LAP can be replaced by magnesium in the nanoclay structure. This can make the structure to have a net negative charge and can be balanced by sodium ions located within the crystal and by surrounding water molecules. This results in electrostatic interactions and promotes the formation of crystal stacks that exist in dry powder form as in other clay types (Tomás et al. 2018). Also, these can be functionalized covalently as well as noncovalently, which will enable to use as better candidates in biomedical applications.

Palygorskite is a magnesium aluminum silicate clay material having a rod-like single dimensional morphology with regular nanotunnels or nanopores. The permanent negativity and the silanol groups present within palygorskite makes them promising candidates to load cationic compounds and organic molecules via electrostatic interactions and hydrogen bond. This has been widely utilized in developing nanocomposites specially targeting cancer treatment and imaging (Kumari et al. 2016; Zhong et al. 2021).

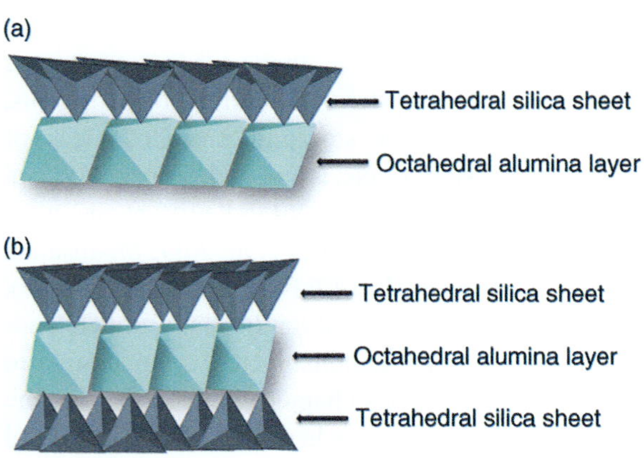

Figure 16.1 Schematic view of structural arrangement of tetrahedral silica sheets and alumina octahedral layers in (a) 1 : 1 and (b) 2 : 1 ratio.

16.3 Application of Nanoclays as Healing Clays

As mentioned elsewhere in the text, NCs are utilized as antibacterial and antifungal agents, also in wound healing applications, tissue engineering, and bone cementing. Furthermore, NCs are highly reputed as better drug delivery carriers. Therefore, this section will discuss on the potential applications of NCs in medical field.

16.4 Healing Clays as Antimicrobial Agents

Clay pastes are primodial forms of healing technique in classical medicine. Certain types of clays or healing clays show bacteriostatic or bacteriocidal potential. Therefore, this enables scientists to utilize them in the treatment of diseases rooted in bacteria. This is because globally there is a huge concern on the increased antibiotic resistance among pathogenic bacteria. Hence, these NCs make them as ideal candidates to apply as antimicrobial agents. Currently, clays are continuously being used for the treatments of various topical or internal ailments (Gomes et al. 2020). It has been reported that not all types of NCs do not exhibit these properties and also dry states of clays do not show bactericidal characteristics. Hydrated clays display bactericidal actions if they used as pastes or poultice. Only few clay types exhibit bactericidal or bacteriostatic activity, which are specific only against certain specific bacteria. For example, illite and smectite that contain ferrous compounds in their structure can inhibit the growth of bacteria. Clays having iron-rich-associated phases such as in pyrite (FeS_2), magnetite ($FeFe_2O_4$), pyrrhotite ($Fe_{1-x}S$), and geotite ($FeO(OH)$) can prevent the bacterial growth to act as bactericidal or bacteriostatic agents (Gomes et al. 2020).

Also, antimicrobial effects of NCs can be enhanced by modifying the clay system either via metal ion exchange, fabricating clay–drug nanocomposites or by developing polymer–clay systems. Sadeghianmaryan et al. (2020) fabricated a curcumin-loaded electrospun polycaprolactone/MMT (PCL/MMT) nanocomposite for wound dressing applications. Accordingly, it was shown that antibacterial activity of the nanocomposite was principally due to MMT. In addition, assimilation of MMT into the polymeric matrix had reduced the size and thickness of the fibers significantly compared to pure forms of polymeric fibers. The authors have also evaluated the cytotoxic effects of the nanocomposites, and it was indicated that PCL-MMT nanocomposite did not show cytotoxic effect proving its suitability for wound dressing applications. Furthermore, curcumin release studies showed that there is an upward trending in curcumin release in all nanocomposites upon the increased exposure times. These composites show a burst release of the drug followed by a gradual release of the compound.

Yamaguchi et al. (2021) synthesized a gentamicin/bentonite nanoclay hybrid-coated film in polyurethane substrate as a stable and long-term antibacterial coating. Prominent cation exchange potential of bentonite NCs enables to immobilize gentamicin effectively within the nanoclay layers. It was found that the synthesized nanofilm had shown antibacterial activity against *Escherichia coli* and *Bacillus subtilis* bacteria grown in phosphate-buffered saline and showed a continuous antibacterial efficacy even after exposure to phosphate-buffered saline. Recently, Zhong et al. (2021) fabricated essential oil ingredients containing palygorskite hybrid materials via mechanical grinding as an antibacterial material. It was found that minimum inhibitory concentration of 2.0 mg/mL was reported against *E. coli* and *Staphylococci aureus* for the hybrid materials.

Moreover, researchers have utilized nanoclay preparations to treat fungal pathogens as antifungal agents. One such instance is Bozoğlan et al. (2021) had fabricated thermosensitive chitosan/ carboxymethylcellulose/ scleroglucan/ MMT nanocomposite hydrogels as antifungal agents to treat onychomycosis. Onychomycosis is an infection caused by fungi, which leads to cause nails to discolor, thicken, and to separate from the nail bed. In this study, authors have incorporated the antifungal OXI drug into the composite, and the composite had shown a higher antifungal activity against *Trichophyton mentagrophytes* and *Trichophyton rubrum* dermatophytes.

16.5 Wound Healing and Tissue Engineering

The NCs are renowned as promising candidates having a greater potential in the field of tissue engineering. Dong et al. (2021) synthesized a printable hydrogel with high shape fidelity using higher concentrations of LAP suspension and a gelatine methacrylate (GelMA) to form nanocomposite hydrogel. It was shown that higher concentrations of LAP improved the rheological properties, the degradation stability, and the mechanical strength of the hydrogel. These nanocomposite (15% gelatin methacrylate/8% LAP in GelMA) hydrogels had significantly promoted bone marrow mesenchymal stem cell proliferation and osteogenic differentiation as a potential biomaterial for bone tissue regeneration.

The LAP NCs are reported to enhance the long-term stability of the hydrogel structure. Wei et al. (2020) incorporated LAP NCs to 2,2,6,6-Tetramethylpiperidinyl-1-oxyl (TEMPO)-oxidized bacterial cellulose (TOBC), sodium alginate (SA) hydrogel as 3D printing ink. The authors were able to show that the synthesized hydrogels not only display structural stability but are also able to achieve a long-term release behavior of protein. Therefore, this has become an attractive candidate in biomedical applications specially for drug release, biomedical devices, and tissue engineering.

Lopresti et al. (2021) tested polylactic acid electrospun with variable concentrations of Cloisite 30B nanoclay, which is MMT modified using bis(2-hydroxyethyl) methyl tallow alkylammonium cations. It was found that NCs had affected differently on the growth trends of MC3T3-E1 preosteoblastic cells compared to their controls. Therefore, it can be concluded that NCs can act as a better filler for electrospun systems engineered for bone tissue regeneration.

Table 16.1 Nanoclay-containing nanoclay composites for tissue engineering applications.

Nanoclay composite	Application	Reference
Nanoclay-incorporated double-network that is composed of three types of cross-linking networks; a nanoclay (LAP) based physically cross-linked network, a covalently cross-linked methacrylated hyaluronic acid network, and an ionically cross-linked calcium alginate network.	Growth of bone marrow mesenchymal stem cells for tissue engineering	Guo et al. (2021)
Erythropoietin releasing polylactic acid /nanoclay/ nanohydroxyaptitite composite	MG-63 human osteoblast-like cell line for bone tissue regeneration.	Salehi et al. (2021)
Poly(glycerol sebacate) and polycaprolactone containing nanoclay	Nervous, vascular, and renal tissue engineering	Davoodi et al. (2022)
Poly(ε-caprolactone) enriched with different layered double hydroxide	Adipose-derived mesenchymal stem cells used in soft tissue engineering	Shafiei et al. (2021)
Montmorillonite stabilized chitosan-*co*-mucin hydrogel	MC3T3E1 and C_2C_{12} cell lines indicating their potential for wider tissue engineering applications.	Barik et al. (2021)

Table 16.1 summarizes some of the examples related to different types of nanoclay composites showing the potential candidates for tissue engineering applications.

16.6 Bone Cement

NCs can be promising candidates to reinforce the mechanical properties of biomaterials specially for cartilage and bone tissue formation. These are very easy to optimize chemically to enhance bioactive properties via ionic dissolution to upregulate cartilage and bone tissue formation. Therefore, NCs are emerging as future orthopedic biomaterials (Erezuma et al. 2021). The MMT structure that is 2 : 1 aluminosilicate (a layer of alumina pack in between two silica sheets) enables to enhance the performance of different polymer composite incorporates (Kuttalam et al. 2021). Therefore, this has enabled NCs to be utilized in several biomedical applications imperatively in the field of tissue engineering and bone cementing.

NC-incorporated copolymers have enhanced barrier function, stiffness, heat resistance, super hydrophobicity, biocompatibility, and stimuli responsiveness. In addition, sustained drug release for drug delivery scaffolds, resistance to hydrolysis, superior mechanical properties including resilience and low-temperature flexibility, excellent hydrolytic stability, and antimicrobial properties are other features shown by these types of composites (Devi and Maji 2013; Murugesan and Scheibel 2020; Radoor et al. 2020; Manabe et al. 2021). Hence, this enable these nanocomposites to be used in numerous biomedical applications. Therefore, nanoclay types such as LAP, layered double hydroxide, MMT, halloysite, and sepiolite are predominantly applied for the reinforcement of copolymers (Murugesan and Scheibel 2020).

Zhang et al. (2021) introduced a novel enzymatically cross-linked silk fibroin-LAP nanocomposite hydrogel for osteochondral regeneration. It was found that the newly synthesized nanocomposite containing this LAP nanoclay can be used as a promising candidate for mechanical reinforcement for osteochondral regeneration. Here the incorporation of a minute content of nanoclay (1%) had accelerated the process of gelation while enhancing the mechanical properties and the hydrophilic properties of hydrogels. Also, hydrogels thus synthesized were biocompatible, and they were capable of stimulating the bone marrow-derived mesenchymal stem cell differentiation to grow and repair cartilage and subchondral bones. Cidonio et al. (2020) investigated Laponite®-alginate-methylcellulose-based bioink to produce viable and functional three-dimensional builds for skeletal tissue regeneration. These studies showed the capability of nanoclay-based bioink to produce viable and functional constructs for clinically relevant skeletal tissue regeneration.

16.7 Drug Delivery

Currently, it is an urgent requirement to have therapeutically effective and patient-compliant drug delivery systems to treat different ailments. This has become prominent when it comes to fight against fatal diseases such as cancers and viral infections. However, any drug delivery system should be safer to use and imperative to have cytotoxicity, biodistribution, fuctionality, drug incorporation, and drug release (Khatoon et al. 2020). Due to the better biocompatibility, higher specific surface area and chemical inertness made them potential candidates for biomedical applications specifically for drug delivery. Unique properties of NCs provide a promising platform to act as drug carriers to cargo antibiotics, antihypertensive drugs, antipsychotic, and anticancer drugs (Khatoon et al. 2020).

NCs can be easily modified via ion intercalation or surface modification to adjust the properties to load drugs. Clay types such as MMT, kaolinite, and halloysite are used to construct carrier systems for slow and targeted drug delivery. In addition, these drugs are used to improve drug dissolution to deliver them effectively to target sites (Zhang et al. 2016, 2019; Khatoon et al. 2020; Saucedo-Zuñiga et al. 2021). Hence, NCs can uptake drugs via encapsulation, immobilization, ion exchange reaction, or by electrostatic interactions. For instance, Reddy et al. (2021) prepared curcumin-encapsulated dual cross-linked SA-MMT polymeric composites for controlled drug delivery. Here the composites were prepared for the drug delivery was based on in situ ion exchange mechanism. It was shown that the synthesized delivery systems capable of releasing for extended durations under non-Fickian type of diffusion process. Furthermore, Dumitru et al. (2022) synthesized 5-aminosalicylic acid encapsulated organically modified MMT nanocomposites to treat inflammatory bowel diseases. Organophilization was done to reinforce the interactions between MMT and 5-aminosalicylic acid. Also, Saha et al. (2020) developed tetracycline intercalated into smectite clays for would healing applications and Table 16.2 shows a summary of recently developed nanoclay composites for different classes of drug delivery.

16.8 Conclusion and Future Perspectives

Clay has a long history of being used to cure ailments. Clays are complexes of aluminosilicate mineral compounds, which generally composed of layers of tetrahedral silica sheets and alumina octahedral layers. Clay types that are utilized in biomedical applications can be either natural or synthetic. However, every clay type is not utilized in these applications as healing clays. Unique properties of clays at nanoscale allow scientists to utilize them as a promising candidates in biomedical fields. Properties including high swelling capacity, cation exchange capacity, and plasticity of NCs and capability to form composites with polymeric materials allow them to be used in different forms as bone cements, for tissue engineering, wound healing, enzyme immobilization, and as therapeutic agents. Also, augmentation of NCs for medical implications enable to mitigate the toxic concerns of other types of nanomaterials that are incorporated in these applications. More importantly, their wide availability, ease of processing, good performance, and cost-effectiveness make them more promising. Although it says that these clays are promising candidates still there are concerns about their cytotoxic effects on the subjects. Hence, toxicity evaluation is utterly needed prior to commercialization or in clinical applications.

Table 16.2 Different types of NCs for drug delivery.

Drug class	Drug	Nanoclay type	Reference
Antibiotics	Metronidazole and tetracycline	Cloisite@30B (MMT modified by the addition of quaternary ammonium salts)	Tirgar et al. (2022)
Anti-inflammatory	Ibuprofen	MMT	Kwon et al. (2021)
Anticancer	Anthocyanins extracted from black carrot	Halloysite	Hamedi and Koosha (2020)
Antibiotics	Cephalexin	MMT	Hajikarimi and Sadeghi (2020)
Antibiotics	Tetracycline hydrochloride	MMT	Dutta et al. (2020)
Antihypertensive	Diltiazem	MMT	Hajikarimi and Sadeghi (2021)
Antihypertensive	Hydrochlorothiazide	Sepiolite	Maestrelli et al. (2020)
Bone repair	Guanidinylated chitosan	LAP	Zhang et al. (2020)
Anticancer and antibacterial	Methotrexate-ciprofloxacin	Tetraethylorthosilicate	Mahkam et al. (2021)
Anticancer	5-Fluorouracil	Halloysite	Harikrishnan et al. (2020)
Antibiotics	Cephradine	Halloysite	Naz et al. (2022)
Anticancer	Doxorubicin hydrochloride	Halloysite	Wang et al. (2021)

References

Barik, D., Kundu, K., and Dash, M. (2021). Montmorillonite stabilized chitosan-co-mucin hydrogel for tissue engineering applications. *RSC Advances* 11 (48): 30329–30342.

Bozoğlan, B.K., Duman, O., and Tunç, S. (2021). Smart antifungal thermosensitive chitosan/carboxymethylcellulose/scleroglucan/montmorillonite nanocomposite hydrogels for onychomycosis treatment. *Colloids and Surfaces A: Physicochemical and Engineering Aspects* 610: 125600.

Cheshire, M. and Güven, N. (2005). Conversion of chrysotile to a magnesian smectite. *Clays and Clay Minerals* 53 (2): 155–161.

Choi, S.J. and Choy, J.H. (2011). Layered double hydroxide nanoparticles as target-specific delivery carriers: uptake mechanism and toxicity. *Nanomedicine* 6 (5): 803–814.

Choi, S.J. and Kim, Y.R. (2013). Bioinspired layered nanoclays for nutraceutical delivery system. In: *Advances in Applied Nanotechnology for Agriculture*, 207–220. American Chemical Society.

Christidis, G.E., Aldana, C., Chryssikos, G.D. et al. (2018). The nature of laponite: pure hectorite or a mixture of different trioctahedral phases? *Minerals* 8 (8): 314.

Cidonio, G., Glinka, M., Kim, Y.H. et al. (2020). Nanoclay-based 3D printed scaffolds promote vascular ingrowth ex vivo and generate bone mineral tissue in vitro and in vivo. *Biofabrication* 12 (3): 035010.

Davoodi, B., Goodarzi, V., Hosseini, H. et al. (2022). Design and manufacturing a tubular structures based on poly(ε-caprolactone)/poly(glycerol-sebacic acid) biodegradable nanocomposite blends: suggested for applications in the nervous, vascular and renal tissue engineering. *Journal of Polymer Research* 29 (2): 1–15.

Detellier, C. (2018). Functional kaolinite. *The Chemical Record* 18 (7–8): 868–877.

Devi, R.R. and Maji, T.K. (2013). Effect of nanofillers on flame retardancy, chemical resistance, antibacterial properties and biodegradation of wood/styrene acrylonitrile co-polymer composites. *Wood Science and Technology* 47 (6): 1135–1152.

Dong, L., Bu, Z., Xiong, Y. et al. (2021). Facile extrusion 3D printing of gelatine methacrylate/Laponite nanocomposite hydrogel with high concentration nanoclay for bone tissue regeneration. *International Journal of Biological Macromolecules* 188: 72–81.

Du, M., Guo, B., and Jia, D. (2010). Newly emerging applications of halloysite nanotubes: a review. *Polymer International* 59 (5): 574–582.

Dumitru, M.V., Sandu, T., Ciurlică, A.L. et al. (2022). Organically modified montmorillonite as pH versatile carriers for delivery of 5-aminosalicylic acid. *Applied Clay Science* 218: 106415.

Dutta, K., Saha, K., Sarkar, P., and Chattopadhyay, D. (2020). Antibactericidal nanoclay-based biomaterial for sustained delivery of tetracycline hydrochloride. *Bulletin of Materials Science* 43 (1): 1–9.

Erezuma, I., Eufrasio-da-Silva, T., Golafshan, N. et al. (2021). Nanoclay reinforced biomaterials for mending musculoskeletal tissue disorders. *Advanced Healthcare Materials* 10 (16): 2100217.

Gomes, C.F., Gomes, J.H., and da Silva, E.F. (2020). Bacteriostatic and bactericidal clays: an overview. *Environmental Geochemistry and Health* 42 (11): 3507–3527.

Guo, Z., Dong, L., Xia, J. et al. (2021). 3D printing unique nanoclay-incorporated double-network hydrogels for construction of complex tissue engineering scaffolds. *Advanced Healthcare Materials* 10 (11): 2100036.

Hajikarimi, A. and Sadeghi, M. (2020). Free radical synthesis of cross-linking gelatin base poly NVP/acrylic acid hydrogel and nanoclay hydrogel as cephalexin drug deliver. *Journal of Polymer Research* 27 (3): 1–20.

Hajikarimi, A. and Sadeghi, M. (2021). Preparation and characterization of gelatin base cross-linking aerogel and nanoclay aerogel for diltiazem drug delivery. *Polymer Bulletin* 1–25.

Hamedi, S. and Koosha, M. (2020). Designing a pH-responsive drug delivery system for the release of black-carrot anthocyanins loaded in halloysite nanotubes for cancer treatment. *Applied Clay Science* 197: 105770.

Harikrishnan, S., Sedev, R., Beh, C.C. et al. (2020). Loading of 5-fluorouracil onto Halloysite nanotubes for targeted drug delivery using a subcritical gas antisolvent process (GAS). *The Journal of Supercritical Fluids* 159: 104756.

Jayrajsinh, S., Shankar, G., Agrawal, Y.K., and Bakre, L. (2017). Montmorillonite nanoclay as a multifaceted drug-delivery carrier: a review. *Journal of Drug Delivery Science and Technology* 39: 200–209.

Khatoon, N., Chu, M.Q., and Zhou, C.H. (2020). Nanoclay-based drug delivery systems and their therapeutic potentials. *Journal of Materials Chemistry B* 8 (33): 7335–7351.

Kumari, P., Ghosh, B., and Biswas, S. (2016). Nanocarriers for cancer-targeted drug delivery. *Journal of Drug Targeting* 24 (3): 179–191.

Kuttalam, K.C., Karuppiah, G., Palaniappan, M. et al. (2021). Mechanical and impact strength of nanoclay-filled composites: a short review. *Journal of Materials Science Research and Reviews* 7: 7–20.

Kwon, W.C., Choi, M., Kang, K.C., and Kim, D.H. (2021). evaluation of ibuprofen/montmorillonite nano-clay composites as an oral drug delivery system and in-vitro drug release performance. *Journal of Nanoscience and Nanotechnology* 21 (7): 3651–3655.

Lopresti, F., Pavia, F.C., Ceraulo, M. et al. (2021). Physical and biological properties of electrospun poly (d, l-lactide)/ nanoclay and poly (d, l-lactide)/nanosilica nanofibrous scaffold for bone tissue engineering. *Journal of Biomedical Materials Research Part A* 109 (11): 2120–2136.

Madusanka, N., de Silva, K.N., and Amaratunga, G. (2015). A curcumin activated carboxymethyl cellulose–montmorillonite clay nanocomposite having enhanced curcumin release in aqueous media. *Carbohydrate Polymers* 134: 695–699.

Madusanka, N., Sandaruwan, C., Kottegoda, N. et al. (2017a). Urea–hydroxyapatite-montmorillonite nanohybrid composites as slow release nitrogen compositions. *Applied Clay Science* 150: 303–308.

Madusanka, N., Shivareddy, S.G., Eddleston, M.D. et al. (2017b). Dielectric behaviour of montmorillonite/ cyanoethylated cellulose nanocomposites. *Carbohydrate Polymers* 172: 315–321.

Maestrelli, F., Cirri, M., García-Villén, F. et al. (2020). Tablets of "hydrochlorothiazide in cyclodextrin in nanoclay": a new nanohybrid system with enhanced dissolution properties. *Pharmaceutics* 12 (2): 104.

Mahdavinia, G.R. and Shokri, E. (2017). Synthesis and characterization of magnetic amidoximated chitosan-g poly(polyacrylonitrile)/laponite RD nanocomposites with enhanced adsorption capacity for Cu^{2+}. *Turkish Journal of Chemistry* 41 (1): 135–152.

Mahkam, M., Zeynabad, F.B., Alizadeh, E. et al. (2021). Novel methotrexate-ciprofloxacin loaded alginate-clay based nanocomposite as anticancer and antibacterial co-drug delivery system. *Advanced Pharmaceutical Bulletin* 11 (3): 477.

Manabe, K., Koyama, E., and Norikane, Y. (2021). Cephalopods-inspired rapid self-healing nanoclay composite coatings with oxygen barrier and super-bubble-phobic properties. *ACS Applied Materials & Interfaces* 13 (30): 36341–36349.

Murugesan, S. and Scheibel, T. (2020). Copolymer/clay nanocomposites for biomedical applications. *Advanced Functional Materials* 30 (17): 1908101.

Naz, M., Rizwan, M., Jabeen, S. et al. (2022). Cephradine drug release using electrospun chitosan nanofibers incorporated with halloysite nanoclay. *Zeitschrift für Physikalische Chemie* 236 (2): 227–238.

Premarathna, K.S.D., Rajapaksha, A.U., Adassoriya, N. et al. (2019). Clay-biochar composites for sorptive removal of tetracycline antibiotic in aqueous media. *Journal of Environmental Management* 238: 315–322.

Radoor, S., Karayil, J., Parameswaranpillai, J., and Siengchin, S. (2020). Adsorption of methylene blue dye from aqueous solution by a novel PVA/CMC/halloysite nanoclay bio composite: characterization, kinetics, isotherm and antibacterial properties. *Journal of Environmental Health Science and Engineering* 18 (2): 1311–1327.

Raji, M., Mekhzoum, M.E.M., Qaiss, A.E.K., and Bouhfid, R. (2016). Nanoclay modification and functionalization for nanocomposites development: effect on the structural, morphological, mechanical and rheological properties. In: *Nanoclay Reinforced Polymer Composites* (ed. M. Jawaid, A. el Kacem Qaiss and R. Bouhfid), 1–34. Singapore: Springer.

Rawat, K., Agarwal, S., Tyagi, A. et al. (2014). Aspect ratio dependent cytotoxicity and antimicrobial properties of nanoclay. *Applied Biochemistry and Biotechnology* 174 (3): 936–944.

Reddy, O.S., Subha, M.C.S., Jithendra, T. et al. (2021). Curcumin encapsulated dual cross linked sodium alginate/ montmorillonite polymeric composite beads for controlled drug delivery. *Journal of Pharmaceutical Analysis* 11 (2): 191–199.

Sadeghianmaryan, A., Yazdanpanah, Z., Soltani, Y.A. et al. (2020). Curcumin-loaded electrospun polycaprolactone/ montmorillonite nanocomposite: wound dressing application with anti-bacterial and low cell toxicity properties. *Journal of Biomaterials Science, Polymer Edition* 31 (2): 169–187.

Saha, K., Dutta, K., Basu, A. et al. (2020). Controlled delivery of tetracycline hydrochloride intercalated into smectite clay using polyurethane nanofibrous membrane for wound healing application. *Nano-Structures & Nano-Objects* 21: 100418.

Salehi, M., Bastami, F., Rezai Rad, M. et al. (2021). Investigation of cell-free poly lactic acid/nanoclay scaffolds prepared via thermally induced phase separation technique containing hydroxyapatite nanocarriers of erythropoietin for bone tissue engineering applications. *Polymers for Advanced Technologies* 32 (2): 670–680.

Saucedo-Zuñiga, J.N., Sánchez-Valdes, S., Ramírez-Vargas, E. et al. (2021). Controlled release of essential oils using laminar nanoclay and porous halloysite/essential oil composites in a multilayer film reservoir. *Microporous and Mesoporous Materials* 316: 110882.

Segad, M., Jonsson, B., Åkesson, T., and Cabane, B. (2010). Ca/Na montmorillonite: structure, forces and swelling properties. *Langmuir* 26 (8): 5782–5790.

Shafiei, S.S., Shavandi, M., and Nickakhtar, Y. (2021). Effect of nanoclay addition on the properties of polycaprolactone nanocomposite scaffolds containing adipose derived mesenchymal stem cells used in soft tissue engineering. *Journal of Advanced Materials in Engineering (Esteghlal)* 39 (4): 45–59.

Tirgar, M., Hosseini, H., Jafari, M. et al. (2022). Introducing a flexible drug delivery system based on poly (glycerol sebacate)-urethane and its nanocomposite: potential application in the prevention and treatment of oral diseases. *Journal of Biomaterials Science, Polymer Edition* 33 (4): 443–464.

Tomás, H., Alves, C.S., and Rodrigues, J. (2018). Laponite®: a key nanoplatform for biomedical applications? *Nanomedicine: Nanotechnology, Biology and Medicine* 14 (7): 2407–2420.

Tornuk, F., Hancer, M. and Yetim, H. (2013). Health aspects of nanoclays used in nanocomposite food packaging materials. 59*th International Congress of Meat Science and Technology,* Turkey İzmir (18–23 August 2013), pp. 18–23.

Villalba-Rodríguez, A.M., Martínez-González, S., Sosa-Hernández, J.E. et al. (2021). Nanoclay/Polymer-Based Hydrogels and Enzyme-Loaded Nanostructures for Wound Healing Applications. *Gels* 7 (2): 59.

Wang, B., Han, Z., Song, B. et al. (2021). Effective drug delivery system based on hydrophobin and halloysite clay nanotubes for sustained release of doxorubicin. *Colloids and Surfaces A: Physicochemical and Engineering Aspects* 628: 127351.

Wei, J., Wang, B., Li, Z. et al. (2020). A 3D-printable TEMPO-oxidized bacterial cellulose/alginate hydrogel with enhanced stability via nanoclay incorporation. *Carbohydrate Polymers* 238: 116207.

Williams, L.B. and Haydel, S.E. (2010). Evaluation of the medicinal use of clay minerals as antibacterial agents. *International Geology Review* 52 (7–8): 745–770.

Yamaguchi, T., Ko, S.J., Jung, J.S. et al. (2021). Periodic charge matching driven immobilization of gentamicin in nanoclays for stable and long-term antibacterial coating. *Dalton Transactions* 50 (40): 14216–14222.

Zhang, Y., Long, M., Huang, P. et al. (2016). Emerging integrated nanoclay-facilitated drug delivery system for papillary thyroid cancer therapy. *Scientific Reports* 6 (1): 1–10.

Zhang, Y., Long, M., Huang, P. et al. (2017). Intercalated 2D nanoclay for emerging drug delivery in cancer therapy. *Nano Research* 10 (8): 2633–2643.

Zhang, Y., Huang, P., Long, M. et al. (2019). Intercalated kaolinite as an emerging platform for cancer therapy. *SCIENCE CHINA Chemistry* 62 (1): 58–61.

Zhang, X., Fan, J., Lee, C.S. et al. (2020). Supramolecular hydrogels based on nanoclay and guanidine-rich chitosan: injectable and moldable osteoinductive carriers. *ACS Applied Materials & Interfaces* 12 (14): 16088–16096.

Zhang, W., Zhang, Y., Zhang, A. et al. (2021). Enzymatically crosslinked silk-nanosilicate reinforced hydrogel with dual-lineage bioactivity for osteochondral tissue engineering. *Materials Science and Engineering: C* 127: 112215.

Zhong, H., Mu, B., Yan, P. et al. (2021). A comparative study on surface/interface mechanism and antibacterial properties of different hybrid materials prepared with essential oils active ingredients and palygorskite. *Colloids and Surfaces A: Physicochemical and Engineering Aspects* 618: 126455.

Section 5

Case Studies

17

Chronic Kidney Disease of Unknown Etiology (CKDu) – The Search for Causes and the Impact of Its Politicization

M. W. C. Dharma-wardana[1,2]

[1] *National Research Council of Canada, Ottawa, Ontario, Canada*
[2] *Departement de Physique, Université de Montréal, Montréal, Quebec, Canada*

17.1 Introduction

Chronic kidney diseases of unknown etiology (CKDu) have been identified in several parts of the world since the 1990s, e.g. as reviewed in Hettithanthri et al. (2021) or Weaver et al. (2015). Hints of this unusual form of kidney disease existed as early as the 1970s (Wessling et al. 2015). The disease differs from the more familiar chronic kidney disease (CKD) in that CKDu is a "silent disease" that progresses without the usual signs of hypertension, diabetes, and glomerular damage found even in the early stages of standard CKD (Jayatilake et al. 2013). Epidemics of CKDu having similar lack of early symptoms and usually leading to tubulo-interstitial damage in the kidneys (Nanayakkara et al. 2012; Wijkström et al. 2020) have been reported. Thus, studies are available from Sri Lanka (Dissanayake and Chandrajith 2005; Jayatilake et al. 2013), Mesoamerica (Correa-Rotter et al. 2014; Sanchez Polo et al. 2020), Mexico (Trabanino et al. 2002; Aguilar-Ramirez et al. 2021), India (Andhra Pradesh, Rajkapur et al. 2012, Gowishankar et al. 2020; Tondaimandalam in Tamil Nadu, Naraisinghpur and Badamba in Odisha, Sahil et al. 2021), Thailand (mainly North-East, Aekplakorn et al. 2021), and El Minya Governorate in Egypt (El Minshawy 2011), etc.

Although there are clear similarities in the manifestation of the disease in different geographic regions, the etiological causes may well be different, and a wide range of causative factors have been suggested. Researchers into Mesoamerican CKDu have emphasized heat stress while also considering the implications of agrochemical use. South-Asian CKDu research initially emphasized agrochemical toxins, but has increasingly turned to causative factors found in drinking water. Agrochemical residues found in the soil and water in Mesoamerican (Gonzales-Quiroz et al. 2018) and south-Asian (Hettithanthri et al. 2021) endemic areas have consistently proven to be well below the accepted danger thresholds. Furthermore, agrochemical use is ubiquitous, while the disease is confined to specific geographic areas.

We focus on conceptual and methodological issues bearing on the etiology of the disease, premature implication of etiology *via* biased nomenclature, and questions of association of CKDu with farming, agro-chemicals, etc. Finally, the politicization of the etiology of the disease and consequences to public policy will be discussed.

17.2 Conceptual Issues

The etiology of noncommunicable chronic diseases (NCCDs) is notoriously difficult since many causative factors associated with life styles, genetics, infections, environmental factors come into play. Nevertheless, the challenge to the public health scientists is precisely to narrow down the causes to a minimum number, given an accepted template of life styles, environmental and genetic factors in a given population. Etiology requires going beyond associations and presenting a causative relationship even when without a clear physiological mechanism.

Various internationally accepted specifications of maximum acceptable daily intake (ADI) levels (e.g. as given in the *Cordex alimentarius*) are for *individual* toxins. Their interactions need to be accounted in practical applications. Thus, the toxic effects of traces of cadmium may be largely mitigated by the co-presence of other ions like zinc or selenium (Chaney 2012; Dharma-wardana 2018a) leading to the safe consumption of shell fish, sunflower, oil, etc.

One of the simplest approaches to the conceptual problem of sorting out associations and causations is found in the Bradford Hill criteria (Hill 1965). Hill asked, "in what

Medical Geology: En route to One Health, First Edition. Edited by Majeti Narasimha Vara Prasad and Meththika Vithanage.

circumstances can [we] pass from [an] observed association to a verdict of causation?" He proposed nine "aspects of association" for evaluating traditional epidemiological data. These aspects are now known as the Bradford Hill (BH) Criteria.

The nine "associations" that Hill discussed (strength of association, consistency, specificity, temporality, biological gradient, plausibility, coherence, experiment, and analogy) identify causative agents that should finally be pinned down in a physiological model. Surprisingly, while some three dozen causative factors for CKDu have been proposed, these proposals have not even used the BH criteria to eliminate the more unlikely hypotheses and instead the "multi-factorial" character of CKDu has been emphasized.

The modern approach to causative factors is strongly transclinical since advances in chemical analytical techniques, molecular biology, molecular toxicology, information technology, mathematical modeling, as well as the design of experiments using cell lines and laboratory animals have provided a deeper and more complex understanding of systems-based etiology. So the "association to causation" inference needs the support of a physico-mathematical model unless a clear cause stands out.

Once likely causative factors (agents) $f1, f2, f3$, ... have been identified, their interactions can be represented by a matrix (e.g. see the appendix to Dharma-wardana et al. 2015) that can be diagonalized (at least approximately) to carry out a factor analysis (e.g. see Paranagama et al. 2012). Then $f1, f2, f3$, ..., are replaced by their linear combinations giving new combined factors (eigenvectors) $F1$, $F2$, $F3$, ..., and show how the agents intermix synergistically or antagonistically. A synergy is complete "in-phase" action, while an antagonism is complete "out-of-phase" action, while intermediate phase angles between 0 and π may occur if phase shifts can occur. An organigram can be used to present the interactions as a complex interacting system. This provides a means of visualizing not only the interactions but also the nonlinear effects of negative and positive feedback loops in the interactive system (Dharma-wardana 2020; Jayasinghe and Zhu 2020).

Such an organigram (Figure 17.1) is a schematization of a complex interactive system (CIS) containing the organism in its environment. The feedback processes introduce nonlinear causative evolution of the pathology. Unfortunately, the strengths and weights (also known as matrix elements or coupling coefficients C_{ij}) that are associated with the arrows that connect different nodes i, j (boxes in Figure 17.1) are not known, but they are needed as a function of time to make any use of the rich theory of complex systems.

The CIS may be dominated by a major causative factor (i.e. a dominant arrow in the diagram at a given time). In less clear-cut cases, these diagrams depict the various interrelations among causative factors in a specific

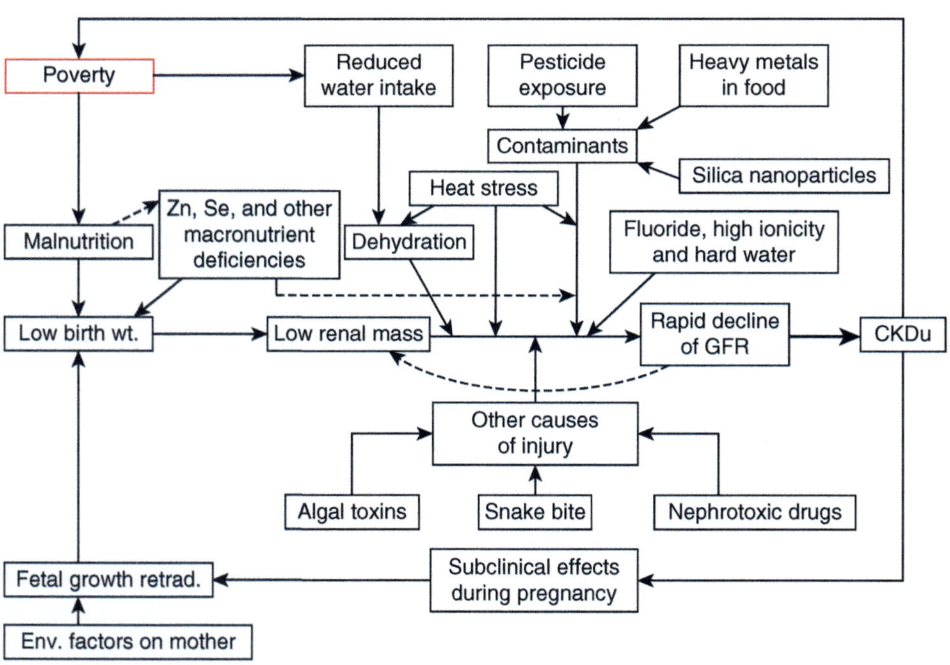

Figure 17.1 The "organism" and the environment are regarded as a complex interacting system (CIS). CKDu arises from the etiological factors indicated as arrows. A sequence of arrows forming a cycle may act as a feedback loop that introduces strong nonlinearities and phase shifts into the time evolution of the disease via complex coupling confidents C_{ij} where i, j label the interaction nodes (boxes).

environment. The causative factor triggering the disease can differ from that propelling the evolution of the disease in each location. For instance, intake of silica nanoparticles may be highly relevant for CKDu among Nicaraguan workers exposed to aerosol ash produced in burning sugarcane. Paddy-husk burning may also generate silica nanoparticles; however, paddy-husk burning is practiced not only in the endemic villages scattered in Andhra Pradesh, or in the endemic areas of Sri Lanka, but also in other villages where CKDu is not found. However, elevated levels of silica in drinking water wells (Nikagolla et al. 2020) in endemic regions are relevant and needs further study. Similarly, the mechanism based on fluoride and hard water relevant to the dry zone villages in Sri Lanka (Dissanayake and Chandrajith 2005; Wasana et al. 2017; Dharmawardana 2018a; Balasooriya et al. 2019) may be irrelevant to Nicaragua, while possibly being relevant to Andhra Pradesh etc. (Subba Rao 2009).

Risk factors like hypertension and diabetes appear along with the onset of the disease in conventional CKD. Uremic toxins like indoxyl sulfate and p-cresyl sulfate have been considered in the etiology of CKD (e.g. Wang and Zhao 2018). Makehelwala et al. (2020) have studied dissolved organic carbon interactions and uremic toxins formed by complexation with Ca^{2+} and SO_4^{2-} ions in endemic areas. This merits further study, although such uremic toxins should cause conventional CKD rather than CKDu. McDonough et al. (2020) have examined the water chemistry and the microbiome of household wells in CKDu-endemic Medawachchiya, Sri Lanka. They report high levels of fluoride, magnesium, sodium, chloride, and calcium in some samples; they also frequently find cyanotoxin-producing microcystis. Why such toxins should cause CKDu rather than CKD has to be clarified before they can be accepted as causes for CKDu. Furthermore, not enough information is contained in these initial explorations to apply even the BH criteria for etiology.

The *exposome concept* (Wild 2005; Miller and Jones 2014) is an ambitious attempt to provide a more accurate environmental-impact assessments on human health using large numbers of measurements of all relevant health and environmental parameters over the life time of test individuals. Thus, the EXPOsOMICS project (Exposmics 2013) aims to assess environmental exposures, focusing on air pollution and water contaminants, and link them to bodily biochemical and molecular changes. These very expensive data-intensive methods may be relevant to further studies on well-studied NCCDs like diabetes and obesity. At a simpler level, Anand et al. (2019) have discussed molecular and genetic methodologies for the investigation of the CKDu epidemiology and conclude that "research has been challenging because of political circumstances, marginalized nature of populations afflicted, and the scarcity of personnel and funding." Given these bottlenecks for CKDu research, even basic statistical data on disease prevalence are sadly lacking. Thus, in Sri Lanka, it is left to the rural patient to maintain his/her health records (Nanayakkara et al. 2012) and hence there can be no reliable "official data set" for the epidemic in Sri Lanka. What is available is mostly estimates from limited surveys done during etiological studies, e.g. in the study of CKDu sponsored by the Sri Lankan National Science Foundation (NSF) and the World Health organization (Jayatilake et al. 2013). The same situation holds for most developing countries afflicted with CKDu.

17.3 Misuse of Disease Nomenclature

The name "chronic kidney disease of unknown etiology" (CKDu) has been in use since 2006. Its use became more prevalent after its adoption by the International Society of Nephrology via the *International Consortium of Collaborators on Chronic Kidney Disease of Unknown Etiology in 2018*. A number of other names (e.g. Mesoamerican nephropathy, Balkan endemic nephropathy, Sri Lankan nephropathy, Uddanam nephropathy (Andhra Pradesh)) had also been used, especially to emphasize geographic origins.

Another more questionable type of nomenclature embeds an assumed etiology or makes a link to a specific occupation (e.g. agriculture) quite prematurely. In our view, the use of such names, e.g. "chronic interstitial nephritis in agricultural communities" (Jayasumana et al. 2016), "kidney disease of unknown cause in agricultural laborers" (Subramanian and Javaid 2017), or directly linking etiology to agrochemicals, as in "chronic agrochemical nephropathy" (Jayasinghe 2014), should be avoided. The latter was proposed in 2014 in the face of several studies (e.g. Nanayakkara et al. 2012) of CKDu in Sri Lanka showing that agrochemical toxins in the endemic areas were well below the danger thresholds. Corresponding studies in Mesoamerica had also suggested that the etiology of the disease cannot be pinned clearly to agrochemicals. This premature linking of CKDu with agrochemicals spawned public fear and disastrous socio-political consequences discussed in the last section of this study.

17.4 Is CKDu a Disease Associated with Agricultural Communities?

The endemic regions are mainly in tropical rural areas where the residents live by small-scale agriculture (e.g. paddy farming in Sri Lanka or in India). In contrast,

Mesoamerican CKDu is found among sugarcane workers employed by large agri business. Some authors have renamed the disease as "agricultural nephropathy" or "chronic interstitial nephritis in agricultural communities (CINAC)," implying a well-established etiology by the name itself. Vervaet et al. (2020) have even claimed that dysmorphic lysomes detected using electron microscopy of tissues from "CINAC" patients directly point to an agricultural toxin, while this has been contested by Wijkström et al. (2020).

In reality, this claimed association between CKDu and agriculture is a chimera of bad statistics. That is, there is not even an association, leave aside a causative implication between CKDu and agriculture. A whole region, e.g. the North Central Province (NCP) of Sri Lanka, or Andhra Pradesh, India, where agriculture is the main occupation is surveyed, and averaged-out statistics are given for the number of people stricken by CKDu. Naturally, CKDu patients are *ipso facto* identified as farmers because most people in these regions *are* farmers. Such averages are meaningful only if the underlying statistical distribution is unimodal. We have shown below that the distribution in not unimodal.

A more microscopic view of the distribution of CKDu shows that CKDu-stricken farming villages have other adjacent farming villages with essentially no incidence of CKDu. The Ginnoruwa region of the NCP is a well-studied endemic area (Balasooriya et al. 2019) consisting of a cluster of three villages (Figure 17.2), namely Badulupura (B), Dolahekanuwa (D), and Sarabhoomiya (S).

These villages were settled during 1980s for agriculture via the Mahaweli River development project, each having about 100 families. At least 30% of the families in B have a CKDu patient. The adjacent villages S and D have low (<2%) or zero records of families with CKDu patients, even though they lead similar lives, have close family links and cultivate paddy in a common low-lying area.

Very few patients were reported in the S community and even they had come from the adjacent village B and settled in S, implying that the disease was triggered during their sojourn in B. Some 70% of all affected CKDu patients were male, but this percentage is sensitive to how the e-GFR criterion for CKDu is used. For instance, the NSF-WHO sponsored study (Jayatilake et al. 2013) of the whole NCP claimed that "the overall prevalence of CKDu was 15.3% with a higher prevalence in females (16.8%) than males (13.3%). More severe grades of CKDu were seen more frequently in males (grade 3: males vs. females = 19.9 vs 5.3%, grade 4: males 16.1% vs females 3.8%). In both sexes, prevalence was higher with increasing age." Jayalal et al. (2019) also studied Sri Lankan CKDu and concluded that a higher percentage of females were affected. In contrast, most researchers of CKDu in Sri Lanka and elsewhere have concluded that the disease is more prevalent in males. The differing conclusions can be traced to (i) uncertainties in identifying patients in the asymptomatic early stages (grades 1, 2), (ii) researchers relying on data based on records rather than on conducting field tests.

Balasooriya et al. (2019) state that "nearly 98% of the population in the area (Figure 17.2) use groundwater as the primary source for drinking water. Rice is the main food crop grown in the area and the population mostly consumes rice as staple food from their own paddy lands. No noticeable differences between socio-economic status and food consumption behavior were observed between CKDu

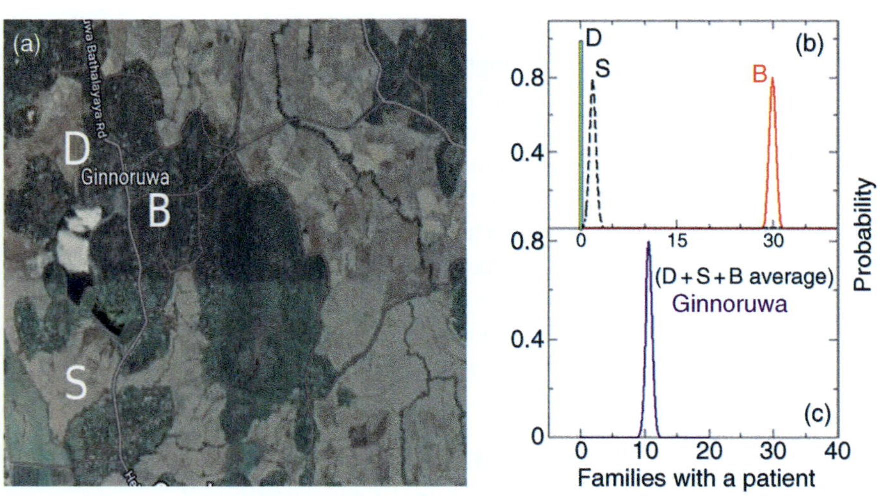

Figure 17.2 (a) Adapted from the Ginnoruwa Google map showing the farming villages Badulupura (B), Dolahekanuwa (D), and Saarabhoomia (S). CKDu is endemic in B, while D and S are virtually free of CKDu. The rocky geology of B contrasts those of S and D in the lower valley. Panel (b) Schematic Gaussian distribution for the prevalence of CKDu in villages S and D, with assumed standard deviation of 2 to allow for misreporting, nondiagnosis, etc. Village D with zero cases is a vertical bar at zero, while S has very few cases. Panel (c) the averaged unimodal distribution for the whole of Ginnoruwa, obscuring the *lack of correlation* with farming.

and non-CKDu families in the study region. No ethnic differences were present in the patient population." They were settlers who came to the Ginnoruwa from the Badulla area, south east Sri Lanka.

Badulupura (B) has a rocky geology, Figure 17.2a, while Figure 17.2b shows the distinct disease distributions for the three villages. The distribution marked B, D, S schematizes the 30% and zero or ~2% cases. Clearly, there is NO correlation between CKDu and farming, although the residents in D, S are also farmers, they show no CKDu. So, CKDu is not correlated with farming unless a unimodal average distribution, Figure 17.2c, is incorrectly used.

Thus, associating CKDu with agricultural activities is a consequence of using averages overarching multimodal distributions. It is very likely that this holds true for CKDu in Andhra Predesh as well. Hence, names like "agricultural nephropathy" are misnomers for this disease.

17.5 Is CKDu a Disease Associated with the Use of Agrochemicals?

The hypothesis that agrochemicals may be the cause of CKDu is a very natural one, even for those who may not hold prejudicial views of the "green revolution." Since the CKDu areas are rural and pastoral, the focus has been on the presence of toxic impurities in fertilizers and the impact of excessive use, or indeed *any use* of pesticides. A worried public increasingly believes that food grown using fertilizers and pesticides is inherently unhealthy and that CKDu is caused by toxins in agrochemicals (fertilizers and pesticides). We first examine them below.

17.5.1 Impurities in Fertilizers

The view that the regular addition of fertilizers containing trace toxic components (e.g. phosphate fertilizers containing Cd, As, etc) will eventually increase the concentration of these toxins in the soil and cause chronic disease is widely entrenched since the early work of Schroder and Balassa (1963). It is incorrect for the most part, as briefly explained below. More extensive discussions, mainly focusing on Cd in soils, water and food are available (Chaney 2012; Roberts 2014; Dharma-wardana 2018b) and can be easily generalized to other toxins.

To lay bare the paradigm of Cd accumulation by fertilizer inputs, we calculate the incremental change in the soil cadmium concentration (ΔC) on addition of a so-called "contaminated" phosphate fertilizer to the soil.

- Let 25 kg of phosphate fertilizer/ha be applied to a depth of 15 cm in paddy cultivation.
- The corresponding soil volume/ha is $V_s = 15 \times 10^5$ liters of soil treated with fertilizer.
- Let the density of the soil be 1.3 kg/L; then, the total weight of the soil is 19.5×10^5 kg, i.e. approximately 2 million kg of soil/ha.
- Given current European Union standards, a highly "contaminated" sample of TSP may contain 50 mg of Cd, and 50 mg of As/kg of fertilizer (this may not considered contaminated by some US standards).
- Hence, 25 kg/ha addition of such a contaminated fertilizer will add only $\Delta C = (50 \times 25\,\text{mg})/(2\,000\,000) = 625\,\text{ng}/$kg of soil.

Thus, the Cd increment added to the soil via fertilizer application even if we completely ignore leaching and other removal processes is in parts per trillion! This is quite negligible compared to the typical ambient concentrations of Cd in agricultural soil (approximately 1–3 mg Cd/kg of soil). So, even if the added cadmium accumulates for a millennium even without any leaching, any removal effects, or conversion to bio-unavailable insoluble forms, the effects are quite imperceptible.

Furthermore, if the extensive drainage during monsoonal rains is found in CKDu-endemic South-Asian regions is taken into account in addition to the usual leaching processes and removal processes, the effect of the heavy metal or metalloid toxins added to the soil reduces essentially to zero.

So the proportionately increasing Cd content observed in crops with increasing fertilizer inputs discussed in traditional studies (e.g. Sillanpää and Jansson 1992) comes from the naturally existing cadmium in the soil, triggered by chemical changes of the soil pH, and from enhanced microbial activity due to the nourishment provided by fertilizers. In the case of organic fertilizers, their P and N content is usually very low and quite variable, e.g. 1–3% of phosphate per kg of manure. Hence, to obtain phosphate inputs comparable to those from mineral fertilizers, several metric tonnes/ha of organic fertilizers are needed where only a few kg/ha of mineral fertilizers suffice. Here too, soil micro-organisms are activated and trigger an increase in cadmium uptake by plants where soil cadmium is made bioavailable by bacterial action or chemical action. Similar heavy metal effects arise in the use of biofilm biofertilizers (BFBF 2021) as well.

We note that most environmental studies in the CKDu endemic locations have reported quite low levels of metal or metalloid toxins. This is established for Sri Lanka in Nanayakkara et al. (2012), Levine et al. (2016), Balasooriya et al. (2019), and in other studies reviewed in, e.g. Hettithanthri et al. (2021). Similar conclusions are found in Mesoamerican studies as well (Correa-Rotter et al. 2014;

Gonzales-Quiroz et al. 2018; Mazeyra et al. 2022); consequently, they have addressed causes other than agrochemicals, e.g. heat stress or nanosilica toxicity.

We review the case of Ginnoruwa, Sri Lanka. The three villages B, D, and S (discussed in the previous section) use water from dug wells for drinking and cooking. The geology of Figure 17.2a shows that Badulupura (B) wells are not connected to the (lower) water table of the agricultural area but are sourced by regolith aquifer water. Isotopic studies (Edirisinghe et al. 2017; Wickramarathna et al. 2017) have verified this. The safely usable wells (in S and D) are connected with the agricultural water table known to contain negligible agrochemical toxins (Jayasinha et al. 2015). The high-risk CKDu wells also contain negligible agrochemical toxins, but they are rich in fluoride and magnesium ions (Balasooriya et al. 2019). Fluoride ions present in hard water were proposed to be nephrotoxic from studies on laboratory mice (Wasana et al. 2017). Furthermore, theoretical considerations were used to propose that Mg^{2+} ions in the hard water and F^- ions act synergistically via MgF^- complex formation in the water (Dharma-wardana 2018a).

These results are consistent with experimental studies on effluent from reverse osmosis of drinking water (Imbulana et al. 2020) and on studies of CKDu in Moneragala, Sri Lanka (Liyanage et al. 2022) where higher concentrations of Mg^{2+} and F^- ions were found in CKDu-associated water sources. In contrast, a factor analysis study by Paranagama et al. (2018) claimed stronger correlation between F^- and Na^+ than with F^- and Mg^{2+} in CKDu-linked water sources. This highlights the difficulty of moving from correlations to causes unless a physiological mechanism (e.g. co-absorption of CaF^- as a complex) can also be investigated.

17.6 Are Pesticide Residues Implicated?

Historical use of DDT in the 1950s let to the successfull control of malaria that nearly brought the health-care system of the country to its knees. The NSF-WHO sponsored study of CKDu (Jayatilake et al. 2013) reported pesticide levels in urine samples from CKDu patients and from non-CKDU controls. The CKDu subjects with pesticide levels above thresholds were 2,4-D (3.5%), Pentachlorophenol (1.2%), Chlorpyrifos (10.5%), Parathion (0%), Carbaryl (10.5%), Napthalene (10.5%), and Glyphosate (3.5%). Thus, except for Chlorpyrifos, Carbary, and Napthalene, the other pesticides fall in the statistically gray area of "less than 4%."

The (paddy) farmers in the CKDu-endemic areas in Sri Lanka use relatively less pesticides and agrochemicals than in the tea and vegetable plantations in the hill country, which is the source of major rivers in Sri Lanka. Dharma-wardana et al. (2015) considered the nonpoint source transport of phosphate by the irrigation waters of one of the major rivers (Mahaweli) of Sri Lanka; the amounts of Cd and other metal toxins and pesticides transported are quite negligible, being present in parts per million compared to macro nutrients. Diyabalanage et al. (2016) confirmed that metal toxin levels in the Mahaweli River are indeed below maximum allowed limits (MALs). Similarly, Jayasinha et al. (2015) showed that toxin levels in irrigation waters were below the MALs and hence required no reverse-osmosis treatment to render them safe. McLaughlin et al. (1996) in their review also disregard irrigation water inputs of Cd into farm soils in Australia, and those arguments apply equally well to pesticide runoff. Aravinna et al. (2017) found that the translocation of pesticides by irrigation waters of the Mahaweli River was negligible.

Although the yearly average for pesticides levels can be very low when two monsoons act to "wash out" the environment, short-time concentrations can spike. The presence of time-varying amounts of diazinon, propanil etc., in aquatic bodies has been reported in many studies. Jayasiri et al. (2022) found that pesticide detection occurs in sporadic peaks linked to their application times. Given the short persistence lifetimes of about a day or less for diazinon and propanil at 30–36 °C prevalent in the tropical region studied, these spikes of pesticide residues drop to zero after monsoonal events.

Vlahos et al. (2021) find higher levels of insecticides (e.g. diazinon, DDE, propanil, and endosulfan) in a "single-time" study of Wilgamuwa, Sri Lanka. They do not report persistence lifetimes. These toxins cause glomerular damage and liver damage (e.g. for diazinon, see Cakici and Akat 2013), while CKDu shows tubular-interstitial damage (Nanayakkara et al. 2012) without liver damage. Hence, these toxins *cannot* be linked to CKDu without further justification. A driving force behind the conclusions of Vlahos et al seems to be the *assumed* negative effects of the Green Revolution (GR). That GR sharply increased food production, eliminated malnutrition, and nearly doubled life expectancy that must also be considered while taking toll of possible increased environmental pollution.

Vlahos et al conclude "that agrochemical use in paddy and other agricultural practices ... of the GR in Sri Lanka may now be contributing to ill health, rapid progression of disease, and mortality." They propose "reducing ... agrochemical contaminants in Sri Lanka and other tropical countries to reduce ... CKDu. These conclusions, based on a "single time-point analysis," tantamount to establishing an etiology of CKDu is unsupported by the evidence presented by Vlahos et al. They do not satisfy even the simplest of Bradford-Hill criteria for causation. In particular, (i) similar but nonpersistent pesticide excesses have been detected sporadically in most parts of the country including where there is no CKDu; (ii) the pesticides detected in Vlahos et al cause both hepatotoxicity and glomerular damage, while CKDu is associated with tubulointerstitial damage where no hepatotoxic symptoms have been reported; (iii) the pesticides detected have short half-lives and are used over short periods during farming; so the one-time measurement is misleading; (iv) farming communities that use pesticides in the same way but remain essentially without CKDu are found to exist adjacent to those with CKDu, as also seen in the Ginnoruwa area which is within 15–20 km of the area studied by Vlahos et al.; and (v) the CKDu prevalence seems to correlate with local geomorphology rather than with farming, as discussed in the previous section.

17.7 Consequence of Dubious Etiological Claims Causing Public Fear

During the year 2011 (and possibly earlier), Dr. Nalin De Silva (2011), Dean of the faculty of Science of the University of Kelaniya, and Dr. Channa Jayasumana, a post-graduate student working on CKDu, made claims stating that CKDu is caused by the presence of arsenic in the endemic areas. This claim was "supported" by invoking a clairvoyant, Ms. Priyani Senanayake who is alleged to communicate with divine beings like God Natha (De Silva 2013), recognized both in Buddhist and Hindu religious lore. The experimental work was done with a view to confirm the revelation and to satisfy "western science." Athureliye Ratana Thera (a Buddhist prelate as well as Member of Parliament) was also associated with these claims. However, De Silva writes "... we had finished our preliminary work in our laboratories before the Thera went to Malaysia [to obtain confirmation of the presence of As from acreditted laboratories]. Our announcement to the media was made in January 2011 while the Thera went to Malaysia after March 2011."

Although several independent studies (Nanayakkara et al. 2012; Jayatilake et al. 2013) had already shown that arsenic concentrations in the soil, water table, and the environment of the affected areas were very low, a politicized campaign claiming that any arsenic (and heavy metals like Cd, Pb) found in the soil must have come via imported fertilizers and pesticides was launched. The claim that CKDu was caused by agrochemicals got a boost when a multilevel hypothesis was presented in 2014 that arsenic in the water of the endemic areas somehow replaces the phosphorus atoms in glyphosate molecules to form a hither-to unknown arsenic analog of the glyphosate molecule. This postulated arseno-glyphosate was further postulated to chelate with calcium ions from the hard water of the endemic areas and form a hypothesized toxic compound causing CKDu. Elementary chemistry should have been enough to completely dismiss every aspect of this scaffold of "hypotheses" adduced for the etiology of CKDu. This overall hypothesis, advanced by C. Jayasumana, S. Gunatillake, and the clairvoyant contributor P. Senanayake, was submitted to the "Int. J. Environ. Res. Public Health" on 11 February 2014 and appeared in print within nine days. This so-called "journal" was in Biel's list of predatory journals (Brezgov 2019; Bohannon 2013). The publication immediately found support and worldwide publicity especially from groups carrying out global campaigns against glyphosate because of its importance as an herbicide in growing genetically modified crops (Ritterman 2014).

The "hypothesis" soon became a "fact" accepted by the general public as the "cause" of CKDu. Sri Lanka banned the use of glyphosate in 2014. The crippling effect on the tea industry, devastation of staples like maize and other crops (Marambe and Herath 2020), as well as the influx of smuggled glyphosate forced some relaxation of the glyphosate ban since 2018. The campaign was linked with a cry for traditional "organic agriculture." The spokesman of the government medical officer's association (GMOA) strongly supported the ban, citing agrochemicals as the cause of CKDu as well as noncommunicable diseases. Prominent medical men cited the "precautionary principle" to justify a ban even without clear evidence.

Claims of bountiful harvests in ancient times when using traditional agriculture became common currency. The demonization of agrochemicals and the stigmatization of critics as "paid agents of agrochemical companies" became popular. In April 2021, the Sri Lankan President banned the use of all agrochemicals and declared a "100% organic" policy. The activist monk Ven. Ratana and Dr. Jayasumana, lead author of the Arsenic + glyphosate + hardwater etiology of CKDu, were powerful figures in the new government.

The naive belief that the government could save foreign exchange needed to buy agrochemicals and that all the manure and "natural pesticides" needed could be produce locally in the country had also played a part. The political leaders including ministers like Dr. Jayasumana may have not realized that organic manure is very low in N, P, etc., and that several tonnes of organic manure are needed to replace a kg of mineral fertilizer. Officials who predicted a loss of harvest lost their jobs. The dire consequences of what began as "divine-inspired" pseudoscience on the etiology of CKDu will take decades to play out its virulence.

References

Aekplakorn, W., Chariyalertsak, S., Kessomboom, P. et al. (2021). Women and other risk factors for chronic kidney disease of unknown etiology in Thailand: national health examination V survey. *Scientific Reports* 11: 21366. https://doi.org/10.1038/s41598-021-00694-9 [Erratum, Sci. Rep. 2021; 11 22674].

Aguilar-Ramirez, D., Raña-Custodio, A., Villa, A. et al. (2021). Decreased kidney function and agricultural work: a cross-sectional study in middle-aged adults from Tierra Blanca, Mexico. *Nephrology, Dialysis, Transplantation* 36 (6): 1030–1038. https://doi.org/10.1093/ndt/gfaa041.

Anand, S., Caplin, B., Gonzalez-Quiroz, M. et al. (2019). Epidemiology, molecular, and genetic methodologies to evaluate causes of CKDu around the world: report of the working group from the ISN International Consortium of Collaborators on CKDu. *Kidney International* 96 (6): 1254–1260.

Aravinna, P., Priyantha, N., Pitawala, P., and Yatigammana, S.K. (2017). Use pattern of pesticides and their predicted mobility into shallow groundwater and surface water bodies of paddy lands in Mahaweli river basin in Sri Lanka. *Journal of Environmental Science and Health. Part. B* 52 (1): 37–47. https://doi.org/10.1080/03601234.2016.1229445.

Balasooriya, S., Munasinghe, H., Herath, A.T. et al. (2019). Possible links between groundwater geochemistry and chronic kidney disease of unknown etiology (CKDu): an investigation from the Ginnoruwa region in Sri Lanka. *Exposure and Health* 12 (4): 823–834. https://doi.org/10.1007/s12403-019-00340-w.

BFBF (2021). Discussion on microbial enhancement of soil fertility, biofilmed-biofertilizers, biochar etc. Focus Sri Lanka (ed. M.W.C. Dharma). https://dh-web.org/green/BBF/index.html (accessed 02 February 2022).

Bohannon, J. (2013). Who is afraid of peer review, a spoof paper concocted by science reveals little or no scrutiny at many open-access journals. *Science* 342: 65.

Brezgov, S. (2019). Chinese publisher MDPI added to list of questionable publishers. Scholarlyoa. Updates 2014 and 2015. https://scholarlyoa.com/chinese-publisher-mdpi-added-to-list-of-questionable-publishers/#more-3072 (accessed 02 February 2022).

Cakici, O. and Akat, E. (2013). Effects of oral exposure to diazinon on mice liver and kidney tissues: biometric analyses of histopathologic changes. *Analytical and Quantitative Cytopathology and Histopathology* 35: 7.

Chaney, R.L. (2012). Chapter two – food safety issues for mineral and organic fertilizers. In: *Advances in Agronomy*, vol. 2012, 117, 51–116. Elsevier.

Correa-Rotter, R.C., Wesseling, C., and Johnson, R.J. (2014). CKD of unknown origin in Central America: the case for mesoamerican nephropathy. *American Journal of Kidney Diseases* 63 (3): 506–520.

De Silva, N. (2011). Neutrenos and arsenic. Kalaya Blog Posting. http://www1.kalaya.org/2011/10/neutrinos-and-arsenic.html (accessed 02 February 2022).

De Silva, N. (2013). Arsenic as the cause of CKDU. Kalaya Blog Posting. http://www1.kalaya.org/2013/04/arsenic-as-cause-of-ckdu-ii.html (accessed 02 February 2022).

Dharma-wardana, M.W.C. (2018a). Chronic kidney disease of unknown etiology and the effect of multiple-ion interactions. *Environmental Geochemistry and Health* 40 (2): 705–719.

Dharma-wardana, M.W.C. (2018b). Cadmium in fertilizers, soil and food. *Environmental Geochemistry and Health* 40: 2739. https://doi.org/10.1007/s10653-018-0140-x2018.

Dharma-wardana, M.W.C. (2020). Letter to the editor re: Jayasinghe & Zhu. *Science of the Total Environment* 721: 137529. https://doi.org/10.1016/j.scitoenv.2020.137529.

Dharma-wardana, M.W.C., Amarasiri, S.L., Dharmawardene, N., and Panabokke, C.R. (2015). Chronic kidney disease of unknown aetiology and ground-water ionicity: study based on Sri Lanka. *Environmental Geochemistry and Health* 37: 221–231.

Dissanayake, C.B. and Chandrajith, R. (2005). Groundwater fluoride as a geochemical marker in the etiology of chronic kidney disease of unknown origin in Sri Lanka. *Ceylon Journal of Science* 46: 3–17.

Diyabalanage, S., Abekoon, S., Watanabe, I. et al. (2016). Has irrigated water from Mahaweli river contributed to the kidney disease of uncertain etiology in the dry zone of Sri Lanka? *Environmental Geochemistry and Health* 38: 439–454. https://doi.org/10.1007/s10653-015-9749-1.

Edirisinghe, E.A.N.V., Manthrithilake, H., Pitawala, H.M.T.G.A. et al. (2017). Geochemical and isotopic evidences from groundwater and surface water for understanding of natural contamination in chronic kidney disease of unknown etiology (CKDu) endemic zones in Sri Lanka. *Isotopes in Environmental and Health Studies* 26: 1.

El Minshawy, O. (2011). End-stage renal disease in the El-Minia Governorate, upper Egypt: an epidemiological study. *Saudi Journal of Kidney Diseases and Transplantation* 22 (5): 1048–1054.

Exposomics (2013). Exposome project for health and occupational research. https://exposomics-project.eu/live-exposome.pantheonsite.io/index.html (accessed 02 February 2022).

Gonzales-Quiroz, M., Pearce, N., Caplin, B., and Nitsch, D. (2018). What do epidemiological studies tell us about chronic kidney disease of undetermined cause in Meso-America? A systematic review and meta-analysis. *Clinical Kidney Journal* 11 (4): 496–506.

Gowrishankar, S., Koshy, P., Mathew, M., et al. (2020). Pathology of uddanam endemic nephropathy. *Indian J Nephrol* 30: 253–255.

Hettithanthri, O., Sandanayake, S., Magana-Arachchi, D. et al. (2021). Risk factors for endemic chronic kidney disease of unknown etiology in Sri Lanka: retrospect of water security in the dry zone. *Science of the Total Environment* 795: 148839. https://doi.org/10.1016/j.scitotenv.2021.148839.

Hill, A.B. (1965). The environment and disease: association or causation? *Proceedings of the Royal Society of Medicine* 58: 295–300.

Imbulana, S., Oguma, K., and Takizawa, S. (2020). Evaluation of groundwater quality and reverse osmosis water treatment plants in the endemic areas of chronic kidney disease of unknown etiology (CKDu) in Sri Lanka. *Science of the Total Environment* 745: 140716.

Jayalal, T.A., Bandara, J.T.W.M.A., Mahawithanage, S.T.C. et al. (2019). A quantitative analysis of chronic exposure of selected heavy metals in a model diet in a CKD hotspot in Sri Lanka. *BMC Nephrology* 20: 208. https://doi.org/10.1186/s12882-019-1371-5.

Jayasinghe, S. (2014). Chronic kidney disease of unknown etiology should be renamed chronic agrochemical nephropathy. *MEDICC Review* 16: 72–74.

Jayasinghe, S. and Zhu, Y.-G. (2020). Chronic kidney disease of unknown etiol-ogy (CKDu): using a system dynamics model to conceptualize the multiple environmental causative pathways of the epidemic. *Science of the Total Environment* 705: 135766. https://doi.org/10.1016/j.scitotenv.2019.135766.

Jayasinha, P., Herath, B., and Wickremasinghe, N. (2015). Report Based on Visit to Anuradhapura CKDu Affected Areas; Review of Input-Output Water of Reverse-Osmosis Installations. Coordinating Office for Science and Technology Innovation, Sri Lanka. https://dh-web.org/placenames/posts/COSTI-Jaysinghe-RO.pdf (accessed 02 February 2022).

Jayasiri, M.M.J.G., Yadev, S., Dayawansa, N.D.K. et al. (2022). Spatio-temporal analysis of water quality for pesticides and other agricultural pollutants in Deduru Oya river basin of Sri Lanka. *Journal of Cleaner Production* 330: 129897.

Jayasumana, C., Orantes, C., Herrera, R. et al. (2016). Chronic interstitial nephritis in agricultural communities: a worldwide epidemic with social, occupational and environmental determinants. *Nephrology, Dialysis, Transplantation* 32 (2): 234–241.

Jayatilake, N., Mendis, S., Maheepala, P. et al. (2013). CKDu national research project team. Chronic kidney disease of uncertain aetiology: prevalence and causative factors in a developing country. *BMC Nephrology* 14: 180.

Levine, K.E., Redmon, J.H., Elledge, M.F. et al. (2016). Quest to identify geochemical risk factors associated with chronic kidney disease of unknown etiology (CKDu) in an endemic region of Sri Lanka, a multimedia labora-tory analysis of biological, food, and environmental samples. *Environmental Monitoring and Assessment* 188: 548.

Liyanage, D.N.D., Diyabalanage, S., Dunuweera, S.P. et al. (2022). Significance of Mg-hardness and fluoride in drinking water on chronic kidney disease of unknown etiology in Monaragala, Sri Lanka. *Environmental Research* 203: 111779. https://doi.org/10.1016/j.envres.2021.111779.

Makehelwala, M., Wei, Y., Weragoda, S.K., and Weerasooriya, E. (2020). Ca_2^+ and SO_4^{2-} interactions with dissolved organic matter: implications of groundwater quality for CKDu incidence in Sri Lanka. *Journal of Environmental Sciences (China)* 88: 326–337. https://doi.org/10.1016/j.jes.2019.09.018.

Marambe, B. and Herath, S. (2020). Banning of herbicides and the impact on agriculture – the case of glyphosate in Sri Lanka. *Weed Science* 68 (3): 246–252. https://doi.org/10.1017/wsc.2019.71.

Mazeyra, M.F.V., Ramos, P.M., Serrano, R. et al. (2022). Mesoamerican nephropathy: a not so unknown chronic kidney disease. *Nefrologia* 41 (6): 612–619.

McDonough, L.K., Meredith, K.T., Nikagolla, C. et al. (2020). The water chemistry and microbiome of household wells in Medawachchiya, Sri Lanka, an area with high prevalence of chronic kidney disease of unknown origin (CKDu). *Scientific Reports* 10 (1): 18295. https://doi.org/10.1038/s41598-020-75336-7.

McLaughlin, M.J., Tiller, K.G., Naidu, R., and Stevens, D.P. (1996). Review: the behaviour and environmental impact of contaminants in fertilizers. *Australian Journal of Soil Research* 34: 1–54.

Miller, G.W. and Jones, D.P. (2014). The nature of nurture: refining the definition of the exposome. *Toxicological Sciences* 137: 1–2.

Nanayakkara, S., Komiya, T., Ratnatunga, N. et al. (2012). Tubulo-interstitial damage as the major pathological lesion in endemic chronic kidney disease among farmers in north central province of Sri Lanka. *Environmental Health and Preventive Medicine* 17: 213.

Nikagolla, C., Meredith, K.T., Dawes, L.A. et al. (2020). Using water quality and isotope studies to inform research in chronic kidney disease of unknown aetiology endemic areas in Sri Lanka. *Total Environment* 745: 140896. https://doi.org/10.1016/j.scitotenv.2020.140896.

Paranagama, D.G.A., Jayasuriya, N., and Bhuiyan, M.A. (2012). Water quality parameters in relation to chronic kidney disease in Sri Lanka. http://dl.lib.mrt.ac.lk/bitstream/handle/123/9042/SBE-12-191.pdf?sequence=1 (accessed 02 February 2022).

Paranagama, D.G.A., Bhuiyan, M.A., and Jayasuriya, N. (2018). Factors associated with chronic kidney disease of unknown aetiology (CKDu) in north central province of Sri Lanka: a comparative analysis of drinking water samples. *Applied Water Science* 8: 151.

Rajkapur, M.M., John, G.T., Kripalani, A.I. et al. (2012). What do we know about chronic kidney disease in India: first report of the Indian CKD registry. *BMC Nephrology* 13 (10): https://doi.org/10.1186/1471-2369-13-10.

Ritterman, J. (2014). Monsanto's herbicide linked to fatal kidney disease epidemic: could it topple the company? Truthout. https://truthout.org/articles/monsantos-herbicide-linked-to-fatal-kidney-disease-epidemic-will-ckdu-topple-monsanto/ (accessed 02 February 2022)

Roberts, T.L. (2014). Cadmium and phosphorous fertilizers: the issues and the science. *Procedia Engineering* 83: 52–57.

Sahil, A., Priyamvada, P.S., Jayaprakash, S. et al. (2021). Seasonal changes in kidney function in CKD of uncertain etiology. *Kidney International Reports* 6: 2918–2921.

Sanchez Polo, V., Garcia-Trabanino, R., Rodriguez, G., and Madero, M. (2020). Mesoamerican nephropathy (MeN): what we know so far. *International Journal of Nephrology and Renovascular Disease.* 13: 261–272. PMID: 33116757.

Schroder, H.A. and Balassa, J.J. (1963). Cadmium: uptake by vegetables from superphosphate in soil. *Science* 140 (3568): 819–120.

Sillanpää, M. and Jansson, H. (1992). Status of Cadmium, Lead, Cobolt, and Selenium in Soils and Plants of Thirty Countries. *FAO Tech. Rep.No. 65,* Geneva.

Subba Rao, N. (2009). Fluoride in groundwater, Varaha river basin, Visakha-patnam district, Andhra Pradesh, India. *Environmental Monitoring and Assessment* 152: 47. https://doi.org/10.1007/s10661-008-0295-5.

Subramanian, S. and Javaid, M.M. (2017). Kidney disease of unknown cause in agricultural laborers (KDUCAL) is a better term to describe regional and endemic kidney diseases such as uddanam nephropathy. *American Journal of Kidney Diseases* 69 (4): 552. https://doi.org/10.1053/j.ajkd.2016.09.029.

Trabanino, R.G., Aguilar, R., Silva, C.R. et al. (2002). End-stage renal disease among patients in a referral hospital in El Salvadore. *Revista Panamericana de Salud Pública* 12: 202–206.

Vervaet, B.A., Nast, C.C., Jayasumana, C. et al. (2020). Chronic interstitial nephritis in agricultural communities is a toxin-induced proximal tubular nephropathy. *Kidney International* 97 (2): 350–369.

Vlahos, P., Schensul, S.L., Anand, S. et al. (2021). Water sources and kidney function: investigating chronic kidney disease of unknown etiology in a prospective study. *NPJ Clean Water* 4: 50.

Wang, Z. and Zhao, Y. (2018). Gut microbiota derived metabolites in cardiovascular health and disease. *Protein & Cell* 9 (5): 416–431.

Wasana, H.M.S., Perera, G.D.R.K., Gunawardena, P.D.S. et al. (2017). WHO water quality standards vs synergic effect(s) of fluoride, heavy metals and hardness in drinking water on kidney tissues. *Scientific Reports* 7: 42516. https://doi.org/10.1038/srep42516.

Weaver, V.M., Fadrowaski, J.J., and Jaar, B.G. (2015). Global dimensions of chronic kidney disease of unknown etiology: a modern era environmental and/or occupational nephropathy? *BMC Nephrology* 16: 145.

Wessling, C., van Wendal de Joode, B., Crowe, J. et al. (2015). Geographical distribution and time trends of chronic kidney disease mortality between 1970 and 2012 in Costa Rica. *Occupational and Environmental Medicine* 72: 714–721.

Wickramarathna, S., Balasooriya, S., Diyabalanage, S., and Chandrajith, R. (2017). Tracing environmental aetiological factors of chronic kidney diseases in the dry zone of Sri Lanka – a hydrogeochemical and isotope approach. *Journal of Trace Elements in Medicine and Biology* 44: 298–306. https://doi.org/10.1016/j.jtemb.2017.08.013.

Wijkström, J., Elinder, C.-G., Hultenby, K. et al. (2020). "Dysmorphic" lysosomes in proximal tubular cells are not specific for CINAC/CKDu and do not provide evidence that CINAC/CKDu is a toxin-induced disease. *Kidney International* 98 (3): 786–787.

Wild, C.P. (2005). Complementing the genome with an "exposome": the outstanding challenge of environmental exposure measurement in molecular epidemiology. *Cancer Epidemiology, Biomarkers & Prevention* 14: 1847–1850.

18

Uraniferous Province of Lagoa Real: Routes, Dispersion, and Impacts on Health

Lander de Jesus Alves[1], Fábio Carvalho Nunes[2], Joel Augusto Moura Porto[3], Rodrigo Gibaut de Souza Gois[3], Edvaldo Cruz dos Santos[3], Hector Hugo Silva Medrado[4], and Majeti Narasimha Vara Prasad[5]

[1] *Postgraduate Program in Biology and Biotechnology of Microorganisms, State University of Santa Cruz (UESC), Ilhéus, Bahia, Brazil*
[2] *Academic Department, Federal Institute Baiano (IFBAIANO), Santa Inês, Bahia, Brazil*
[3] *Environment Coordination, Indústrias Nucleares do Brasil (INB) – Uranium Concentration Unit, Caetité, Bahia, Brazil*
[4] *Security, Radioprotection and Environment Management, Indústrias Nucleares do Brasil (INB) – Uranium Concentration Unit, Caetité, Bahia, Brazil*
[5] *School of Life Sciences, University of Hyderabad (an Institution of Eminence), Hyderabad, Telangana, India*

18.1 Introduction

Uranium (chemical symbol U) is considered the most important element in studies of radioactivity and was the first element discovered to emit radiation (Erdi-Krausz et al. 2003; Lehto and Hou 2011). Considered ubiquitous in the Earth's crust, uranium generally occurs in low concentrations in rocks, soils, water, air, plants, and animals, where the products of its nuclear decay contribute to maintaining a low level of background radiation in the environment.

Uranium is the heaviest element in nature, having an atomic number of 92 and a relative atomic mass of 238.03. Uranium occurs naturally in the form of three radioactive isotopes, ^{238}U, ^{235}U, ^{234}U, whose respective abundances in nature are 99.27, 0.72, and 0.0054% (Vandenhove et al. 2011). The average amount of uranium in the lithosphere is on the order of 2.3 mg/kg, which is moderately distributed in the rocky basement overlying granitic rocks (Greenwood and Earnshaw 1997; Lehto and Hou 2011).

Regarding the chemical forms in which uranium is found in nature, the main ones are the hexavalent form, U(VI), and the tetravalent form, U(IV), with most uranium minerals occurring as U(IV), partially oxidized in the form of the mineral pitchblend (UO_{2+x}), which can be considered, in general, with a mixture of the oxidized forms UO_2 and UO_3 (Lehto and Hou 2011).

Although uranium is considered important mainly because of its great use as fuel in nuclear power stations, its importance is not restricted to its role in the composition of the planetary energetic matrix, since it has implications for human health and the environment (WHO 2001). Considered to be hazardous due to its radioactivity and chemical toxicity, uranium may pose a health risk when disposed in the environment, either by the occurrence of anthropogenic actions or in natural accumulations in uraniferous regions (Galhardi et al. 2020; Sohrabi 2013).

In the environment, uranium and other radionuclides can reach the top of the food chain, accumulating in plants and animals, and be ingested by humans leading to organic disorders or even death (dos Amaral et al. 2005; Favas et al. 2016; Malaviya and Singh 2012; Rayno 1983; WHO 2001), in effects that go beyond individuals and can reach society (Stewart 2019). The relationship between the concentration of a given radionuclide in a medium and its concentration in the body of a certain high trophic level organism derives, albeit in a nonlinear way, mainly from its food intake load (Hosseini et al. 2008). The main contribution to external exposure for humans comes from gamma-emitting radionuclides present in trace amounts in the soil, mainly ^{40}K and the families ^{238}U and ^{232}Th.

Due to the high natural concentrations of uranium and other radionuclides, some natural areas have high background natural radiation (HBNR). One of these areas is located in southwestern Bahia State in the most important single mineral province of Brazil – the Uraniferous Province of Lagoa Real (UNSCEAR 2010). Thus, this chapter aims to present and discuss the main routes of dispersion and exposure of humans to uranium existing in the uranium province, as well as to propose study and monitoring strategies that seek to protect human populations residing in the place and reduce environmental impacts.

Medical Geology: En route to One Health, First Edition. Edited by Majeti Narasimha Vara Prasad and Meththika Vithanage.

18.2 Lagoa Real Uraniferous Province

Brazil has the nineth largest geological reserve of uranium in the world, with around 277 million tons of ore occurring mainly in the form of uranium oxide (U_3O_8) (de Alves et al. 2018; WNA 2020). All Brazilian production is obtained by Indústrias Nucleares do Brasil S.A. – INB, in the most important monomineralic province in the country, the Lagoa Real Uraniferous Province, located in the Center-South Region of the State of Bahia (de Alves et al. 2018; Cruz et al. 2008; de Oliveira et al. 2012; "INB – Indústrias Nucleares do Brasil" 2015) (Figure 18.1), where, through the acid leaching process, the uranium concentrate called *yellowcake* is produced [Ammonium diuranate – $(NH_4)_2U_2O_7$ (DUA)].

The geology of the area is composed of gneiss and granitoid rocks, which are associated with the uranium (de Alves et al. 2018; Oliveira Chaves 2013; Pascholati et al. 2003). The formation of the uranium mineralization was assigned to the 1.3–1.5 Ga tectonic event, or Espinhaço collisional event. A series of orebodies are known from 38 anomalies distributed over an area of 1200 km^2, roughly along three semi-arched lineaments that cover an approximate extension of 35 km (Lobato et al. 2015).

The uranium deposits are associated with plagioclase and albitites (Lobato et al. 2015), which are the product of extensive, high-temperature metasomatic alteration of graniticgneissic rocks, mainly in the form of sodium enrichment and silica depletion, accompanied by oxidation of the original Fe^{+2}-dominant phases (e.g. Lobato and Fyfe (1990). Lithotypes dominated by calcium-rich oligoclasites host some orebodies along the lineament, suggesting Na—Ca metasomatic alteration (Figures 18.1 and 18.2).

The area of the Uranifera Province of Lagoa Real has a rainy season in spring-summer and a dry season in autumn-winter, lasting five months. Average temperatures are high, always above 18 °C, with absolute maximums that can reach 40 °C and absolute minimums at 6 °C. The area has two types of climates, namely, subhumid to dry and semi-arid, with water deficit from May to October (SEI 1999).

The areas with a subhumid to dry climate are located in the highest parts, with some exceptions, where the climate variation Aw' of the Köppen classification appears, and C1dB'4a' of the classification of Thornthwaite and Mather. The semi-arid climate areas are commonly located in the lower parts, falling within the BSwh variation of the Köppen classification, DdA'a' and DdB'a' of the Thornthwaite and Mather classification (Thornthwaite 1948).

In the study area, the main meteorological phenomena are caused by the Intertropical Convergence Zone (ITCZ), Frontal Systems (SF), South Atlantic Convergence Zone (SACZ), Upper Air Cyclonic Vortices (VCAS), and orographic factors. ITCZ is the main meteorological system that causes rain and Frontal Systems due to the decrease in thermal averages (de Alves et al. 2018). Macroscale systems dominate wind circulation, with southeast and east gusts being more frequent.

The soils of the area are dominated by Oxisoils, Argisols, Haplic Cambisols, Litholic Neosols, Regolithic Neosols, and Quartzarenic Neosols, which are sometimes eutrophic, sometimes dystrophic, depending on the degree of evolution and the source material. Cardoso et al. (2009) studied the main soil classes in the region and found high activities of ^{238}U, ^{226}Ra and ^{210}Pb associated with the crystalline structure, indicating that they are related to the material that gave rise to the soils. In addition, ^{238}U showed greater bioavailability in more acidic soils and greater affinity for Fe oxides.

Both municipalities that comprise the Uraniferous Province of Lagoa Real (Caetité and Lagoa Real) are located in one of the semi-arid centers of South America, in the part most affected by drought in Brazil, known as "Drought Polygon" (Ab'Sáber 2003). In these areas, the high water deficit during a large part of the year and the irregularity in the distribution of rainfall contribute to the predominance of the vegetation called "Caatinga" (de Queiroz 2006) or steppe savannah (IBGE 2012).

According de Queiroz et al. (2017), the dominium caatinga harbors the largest and most continuous expanse of the seasonally dry tropical forest and woodland biome (SDTFW) in the New World. Although this formation is threatened by urban expansion and agricultural frontiers in Brazil, it is still marked by a flora with many endemisms and remarkable heterogeneity (Fernandes and de Queiroz 2018).

It is also worth noting that important altitudinal, relief, and soil variations contribute to the establishment of ecological variations in the area of the uranium province, providing the opportunity for contact between caatinga vegetation and neotropical savannas and seasonally dry forests (Bullock 1995; Pennington et al. 2006). This scenario also comprises the presence of agricultural areas and small-scale cultivation, in addition to the presence of the mining areas of the URA (Figure 18.3).

18.3 Dispersion and Routes of Uranium in the Environment

The behavior of uranium in the context of the biogeochemical cycle and its movement in the environment is strongly influenced by its chemical form. On our planet, uranium

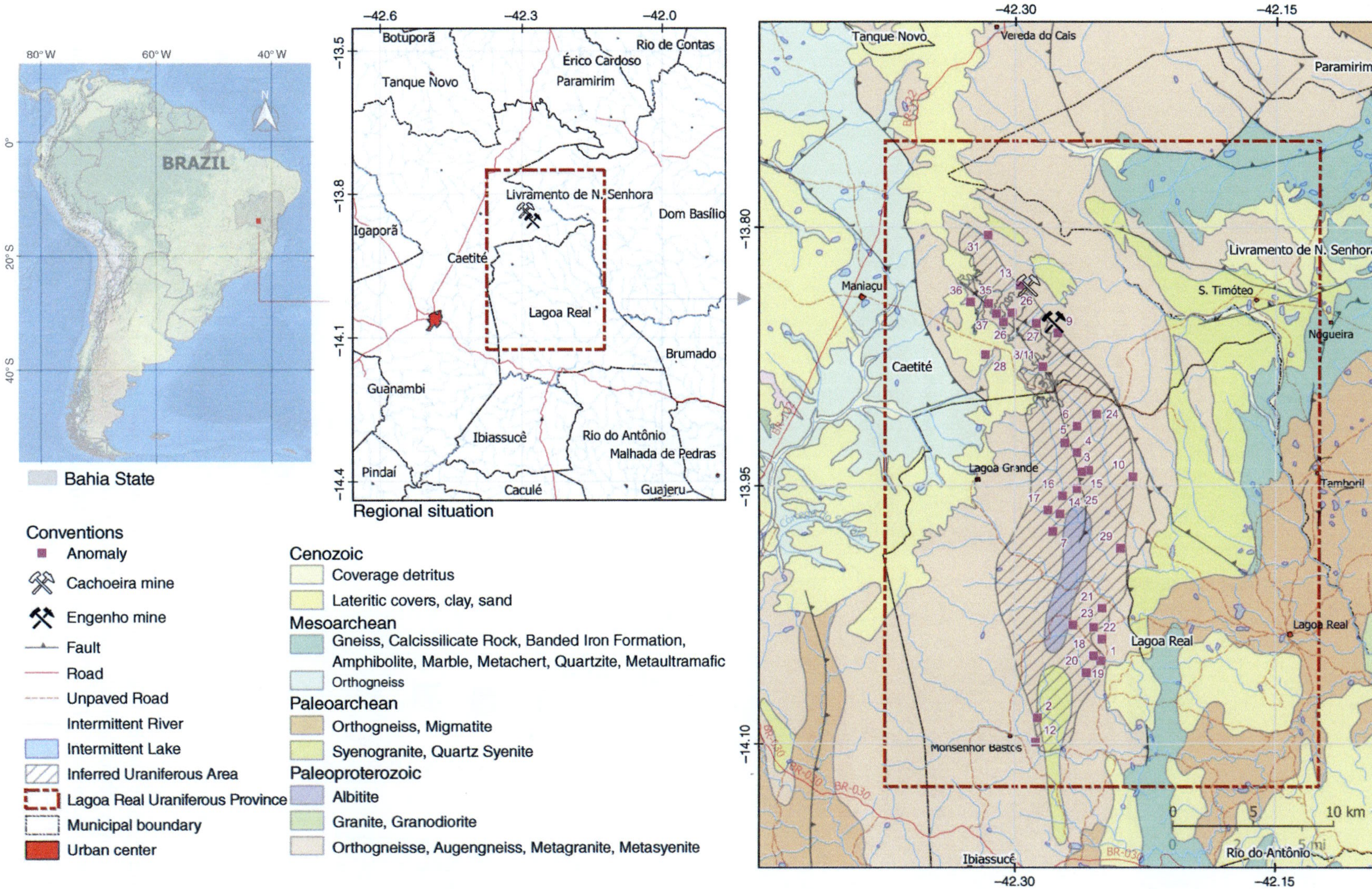

Figure 18.1 Geological and location map of the Lagoa Real Uraniferous Province.

Figure 18.2 Some rocks of the Lagoa Real Complex: (a) example of albitite mineralization; (b) uranophane included by calcite; (c) contacto f pyroxene albitite and garnet albitite; (d) contacto f Orange calcite and amphibole albitite; (e, f) gneiss.

minerals normally occur in the hexavalent oxidation state [U(VI)], which favors their solubility over a wide range of pH and redox potentials (E_h) (Mkandawire 2013). Therefore, uranium can be found in different types of rocks, soils, water, and air, usually in very low concentrations. However, situations can be found where the values are high, either for natural or anthropic reasons, such as nuclear fuel cycle activities, phosphate fertilizer production process, and inadequate disposal of tailings in mine areas (Chen et al. 2006).

The Lagoa Real Uraniferous Province presents different natural anomalies (Figure 18.4), which can be potential

(a)

(b)

Figure 18.3 Aerial views of the landscape in the uranium mining areas: (a) wast dump and natural caatinga vegetation around the Cachoeira mine in anomaly 13; (b) landscape aspect in the area of anomaly 9, showing native vegetation, unpaved road and stripping area of the Engenho mine.

sources of contamination; however, studies in the region are still scarce, highlighting the works carried out by: (i) de Alves et al. (2018), who studied the uranium transfer in the soil–mycorrhiza–plant system; (ii) de Alves et al. (2018), who investigated the recovery of areas degraded by uranium extraction; (iii) de Oliveira et al. (2012), Cruz et al. (2008), and Cruz et al. (2007), who studied the geology of the uranium province; (iv) Cardoso et al. (2009), who studied the geochemistry of radionuclides in different soil types.

The main means of dispersion of uranium are surface and underground waters, but it can also be diffused through the air, assimilated by plants, animals, and reach humans. In the following items, we discuss the main uranium dispersion routes within the Lagoa Real Province, the studies carried out so far and suggestions for new approaches and monitoring implementation.

18.3.1 Routes and Dispersion Through Water and Air

The area of the Uraniferous Province of Lagoa Real is an extensive area of 1200 km^2, roughly along three semi-arched lines that cover an approximate extension of 35 km on the north-south axis. This entire area is located mainly within the Rio de Contas hydrographic basin and encompasses a series of sub-basins used for multiple uses, among them, human consumption, agriculture, and animal drinking.

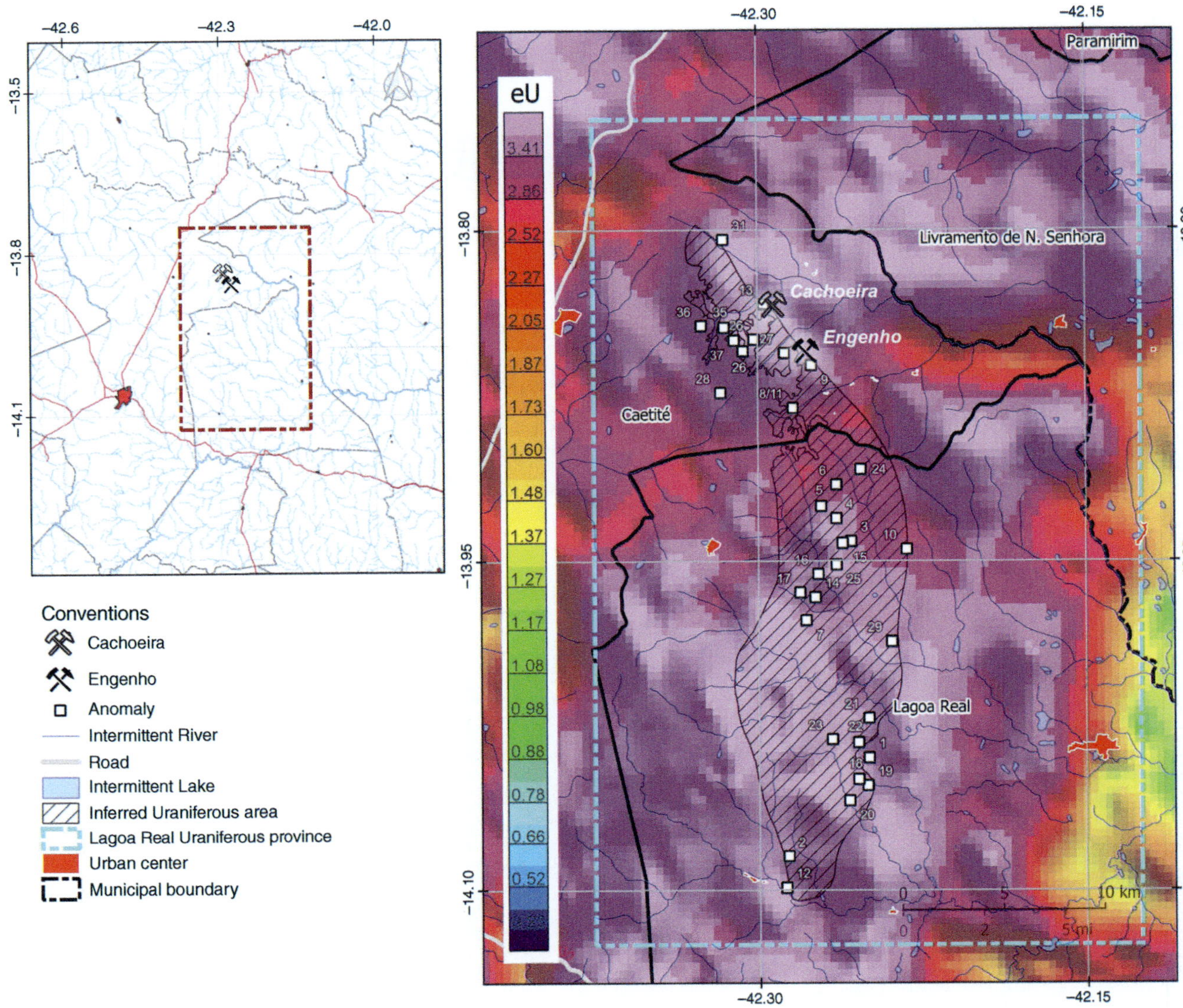

Figure 18.4 Location of the magnetometry and gammaspectrometry survey map of uranium concentration partly per million (ppm) of uranium equivalent (eU) (Correa 2019; CPRM 1996; de Polo et al. 2022).

For analysis of the province area, a database composed of Digital Elevation Model (DEM), multispectral image compositions from orbital sensors, and vector files of the regional hydrographic mesh was used. To generate a DEM of the studied area, the Topodata Project based in the Shuttle Radar Topographic Mission (SRTM) of Synthetic Aperture Radar (SAR) with spatial resolution of 1 arcs (~30 m) was used. Through the DEM, the directions of surface flows in the QGIS version 3.20.2-Odense (https://www.qgis.org) have been verified. The identification of surface flows and identification of basins were performed using GRASS GIS algorithms r.stream.extract and r.slope.aspect (Jasiewicz and Metz 2011; Neteler et al. 2012).

Both the limits of the Uraniferous Province, as well as the inferred limits of the uranium area and the points of the radiometric anomalies, were provided by INB, in addition to the location of the Cachoeira (deactivated) and Engenho (active) mines. The location of underground water wells was obtained through data available in the National Groundwater Information System – SIAGAS (CPRM 2022) (Figure 18.5).

Studies of uranium and other radionuclides are centered on the INB mine, and its surroundings are located within the Vasos sub-basin. However, we suggest that a shared effort be made between different public entities, from the municipal to the federal sphere, to carry out broader and deeper studies in the region.

Studies carried out by INB show that the Vasos sub-basin can be influenced by Anomalies 8/11 and 9 (Figure 18.6). The LR508 (Limoeiro), LR515 (Videl), PCFRIO (Frios),

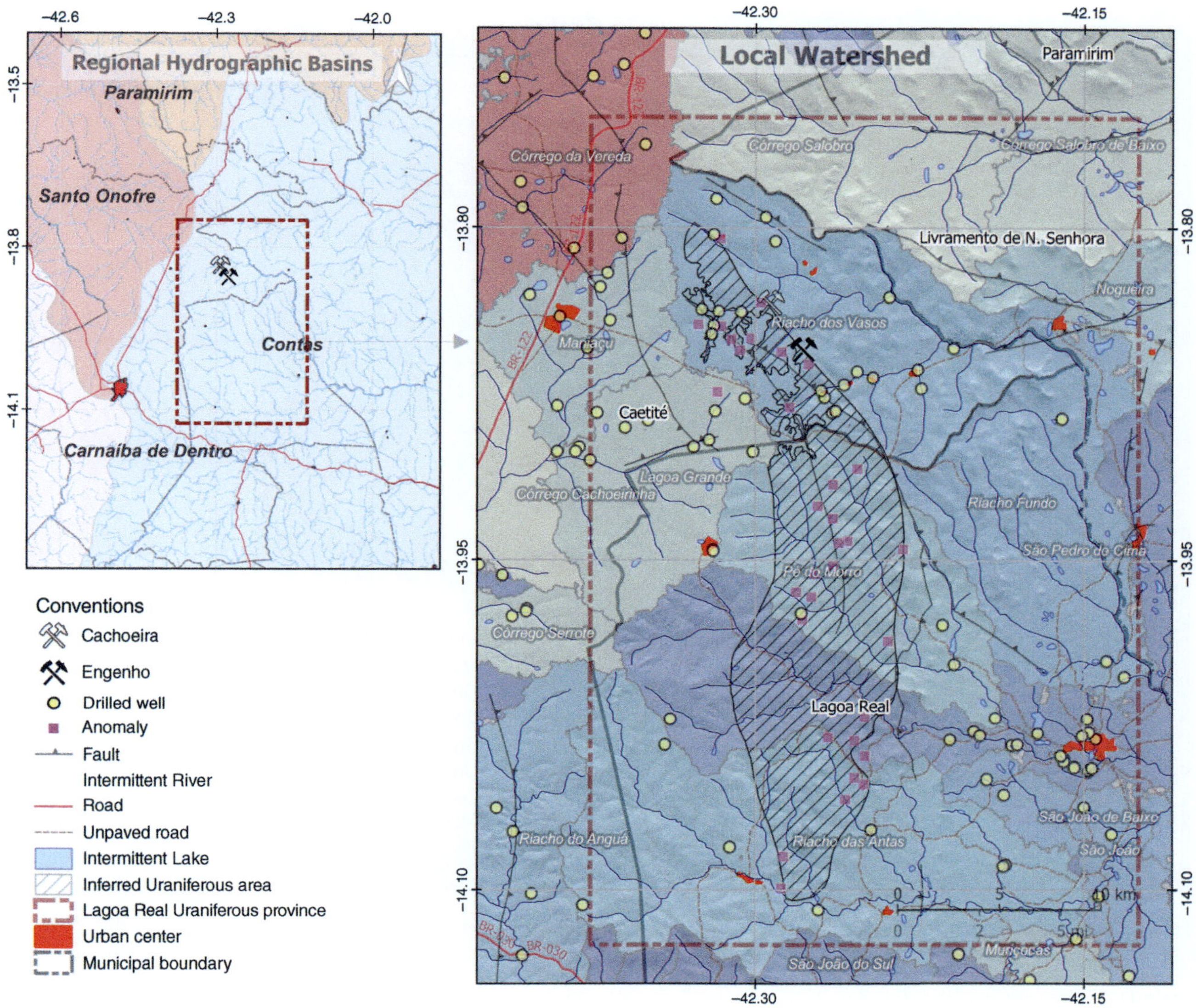

Figure 18.5 Wells drilled in the uranium province area and analysis of hydrographic basins.

and LR265 (Quessengue) wells make up the Engenho Mine monitoring program, for the characterization of influences of natural origin of the portion of Anomaly 9, which extends beyond the watersheds of the basin.

The Engenho Mine is entirely located in the Córrego do Engenho Microbasin, which drains approximately in the S–N direction and changes direction a few times until it flows into the Riacho dos Vasos sub-basin. Wells are well distributed along the sub-basin, which, together with samples from surface courses and wetlands, could offer a good sampling grid for monitoring uranium and other radionuclides. Silva (2011) carried out a study in the same area, collecting samples at complementary points and found high levels of uranium, which points to possible health risks.

Well LR508 (Comunidade Limoeiro) is located approximately 1.5 km downstream of Anomaly 08/11. Well LR515 (Comunidade Videl) is located approximately 2.6 km downstream of Anomalia 08/11, and LR265 (Community Quessengue) is located downstream of these other two. It is unlikely that the Anomaly 08/11 (or the Engenho Mine) will have any influence on the PCFRIO well and on the PC65, as the PCFRIO is at an upper level on the opposite bank of the Córrego do Juazeiro in relation to the Anomaly 08/11 and the Engenho Mine and the PC65 on a very distant tributary. As for the surface water monitoring point near the Comunidade dos Frios, ES514, there is a possibility of influence due to the dispersion of particulates from the Engenho Mine beyond the limit of the Riacho das Vacas Sub-Basin due to N > S winds (8%).

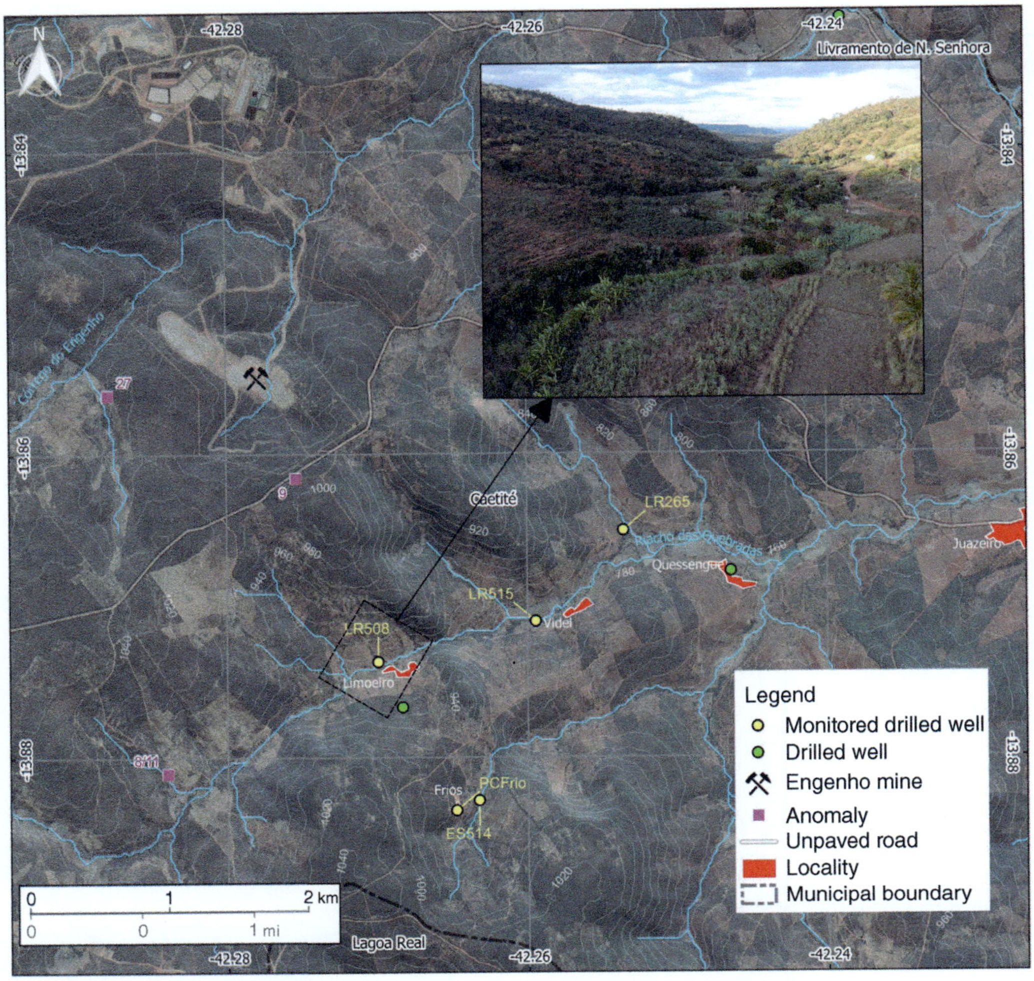

Figure 18.6 Some wells monitored by INB located in the watershed of the Engenho and Quebradas streams.

Table 18.1 Average Results of Poços in the Riacho das Quebradas Sub-basin – 2021.

	pH	Condutiv.	Acidity	Alcalin.	Hardness	STD	Cl^-	F^-	NO_3^-	NO_2^-	SO_4^+	Na	U Total
Well	–	μS/cm	mg/L of $CaCO_3$	mg/L de $CaCO_3$	mg/L de $CaCO_3$	mg/L	mg/L	mg/L	mg/L of N	mg/L of N	mg/L	mg/L	mg/L
QRV	—	—	—	—	—	1000	250	1.5	10.0	1.0	250	200	0.015
LR508	6.7	296	56	122	47	218	17	**1.5***	8.9	0.1	9	47	0.001
LR515	6.6	197	46	106	68	83	13	0.3	0.1	0.1	4	17	0.000
LR265	7.5	859	69	274	169	462	99	**2.4****	0.3	0.1	73	151	**0.018*****
PCFRIO	6.6	299	70	96	54	126	38	1.1	1.8	0.1	5	24	0.001

QRV, Quality reference value according to the federal standard (CONAMA, 2008).
Significance values: *0.0261; **0.0417; ***0.0003.

Analyzing the average results of samples obtained from wells LR508, LR515, PCFRIO, and LR265 in 2021 (Table 18.1), it is observed that the wells had neutral pH, low conductivity, and stable species contents, in general, low and in the normal range. The LR265 well, however, historically presents higher results for uranium (Figure 18.7), in addition to other parameters such as conductivity, alkalinity, and hardness, when compared to the five active wells in the Vasos Basin (Riacho das Quebradas Sub-basin), namely: LR508, LR515, PCFRIO, and LR265.

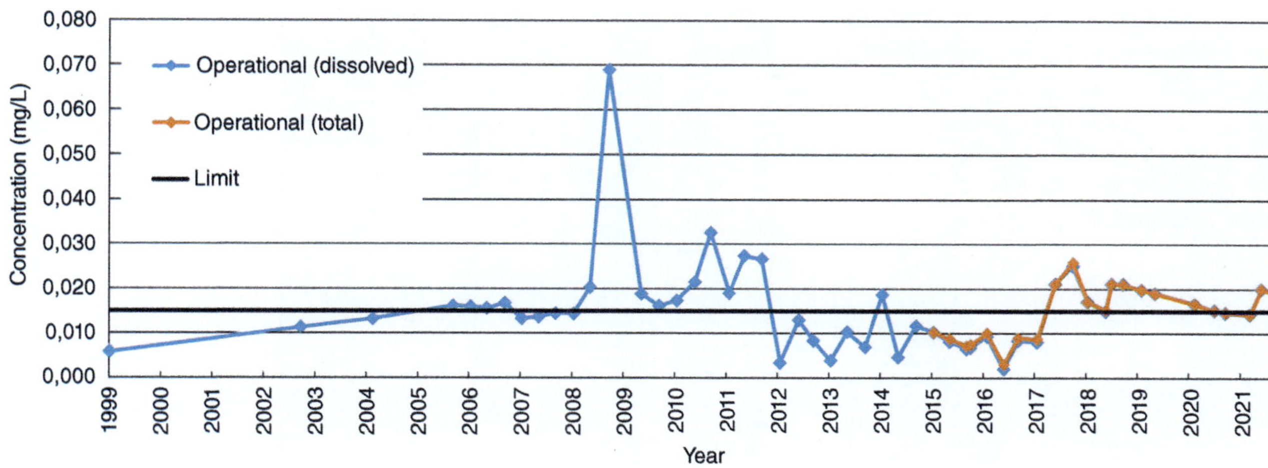

Figure 18.7 U-nat behavior in well LR265.

It is also observed that the average content of U is slightly higher than the Quality Reference Value (QRV) of CONAMA Resolution No. 396 of 2008 (CONAMA 2008) for human consumption. However, this result is consistent with the history of results obtained since 1999 (Figure 18.7), being consistent with the geological characteristics of the site and with the presence of uranium anomalies in the Riacho das Quebradas microbasin (Figure 18.6). The higher fluoride results in wells LR265 and LR508 corroborate this statement, since the presence of the mineral fluorite is common in metassomatized granites and gneisses or in minerals of the amphibole group present in albitites and amphibolites (da Silva 2015).

Despite the levels of U in the LR265 well exceeding the reference limits of CONAMA (2008) for human consumption (0.015 mg/L), it is important to emphasize that the normative Ordinance GM/MS n° 888 of the Brazilian Ministry of Health (BRASIL 2021) indicates a limit of 0.030 mg/L of U for human consumption. Even so, it is worth noting that the use of water from that well is only used for irrigation, animal watering, and domestic cleaning, not being used for human consumption.

In the case of wells with uranium levels higher than those indicated by Brazilian regulations, it is worth mentioning well PC171, which was no longer monitored after being sealed by the environmental agency of the municipality of Caetité in 2019. Since the beginning of its monitoring in 2013, that well has presented levels higher than those indicated by federal regulations, reaching the value of 2.34 mg/L of U in 2016. Such results are possibly justified due to the proximity of the well to Anomaly 35, whose mineralized body is close to the soil surface. At this point, the uranium rock is highly weathered, including due to soil disturbance by human activities, which accelerates the weathering process, contributing to a greater susceptibility to changes in geochemical characteristics and formation of secondary minerals. Comparatively, we can highlight the results pointed out by Balbudhe et al. (2012) in well water in Hyderabad, India, where rates ranging from 0.0006 to 0.082 mg/L were found in well water.

As mentioned, well PC171 is located near Anomalia 35 (about 200 m) and is also close to Anomaly 36 (about 600 m) (Figure 18.8). Despite this, other wells located upstream and downstream of this and other anomalies in the watershed (LR221, LR211A, PCBuracão, and PC173) did not show high levels of uranium, indicating that these wells are not connected from a structural or hydrochemical point of view. This may indicate that the high levels of uranium in wells in the region may be a reflection of the geochemistry of the native rock in which they were drilled, which are naturally enriched with this radionuclide (da Silva 2015; de Oliveira et al. 2012; Marques et al. 2021; Wilde 2020).

INB monitors levels of uranium and other radionuclides in particulates dispersed in the atmosphere around the mine in the Riacho dos Vasos Basin. However, it is considered important that monitoring of particulate materials be carried out in others in the basin in the province, such as the Maniaçu Basin. Very small particles of uranium found in dust can fall onto water, plants, houses, and land. Rain increases the amount of uranium in air that can settle to the ground and can enter the food chain (Keith et al. 2013). The process of deposition of particulate matter is governed mainly by the circulation of winds not only at macro and mesoscales but also at microscales because of the local topography, such as katabatic and anabatic winds. Communities located northwest and west of mining areas can be considered as points of greater attention, as the prevailing winds in the region are from the southeast and east (da Silva et al. 2002; INMET 2022). In monitoring carried out in surface waters at point ES514 located at Fazenda

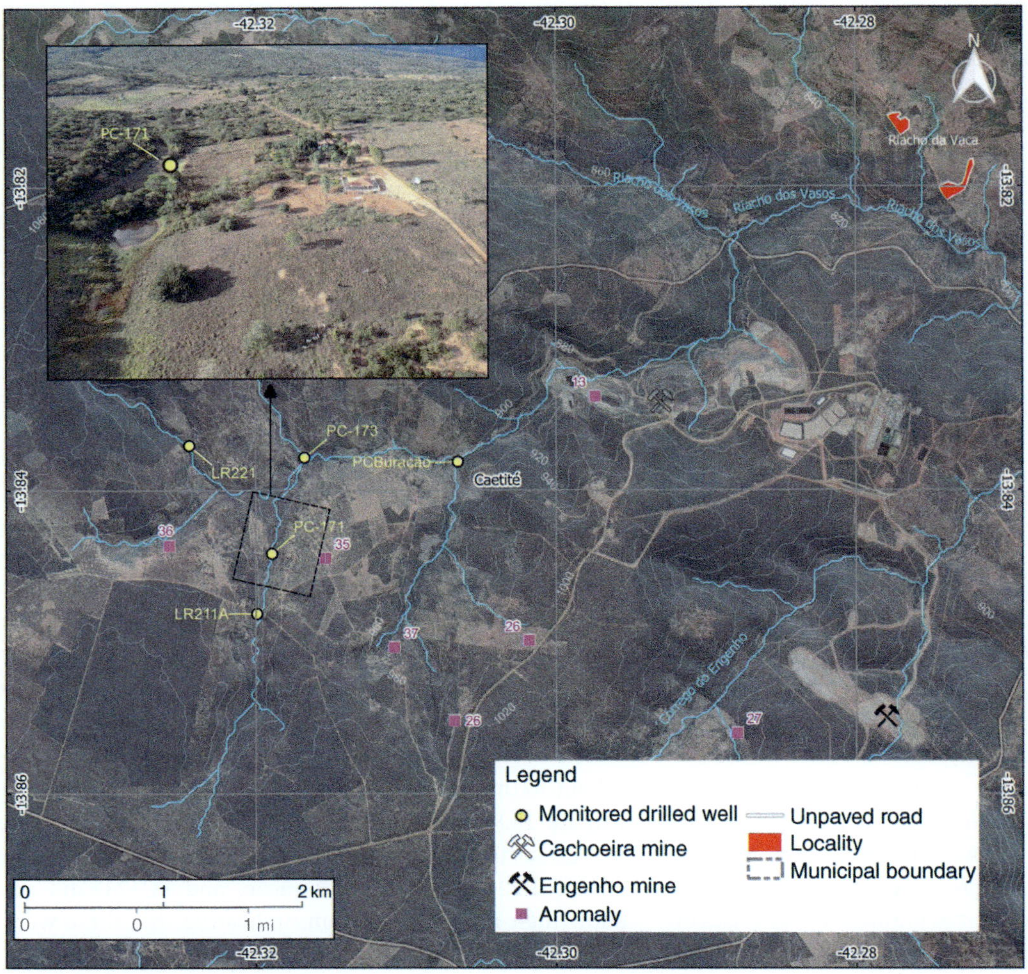

Figure 18.8 Wells monitored by INB in the cachoeira stream microbasin.

Frio, south of Cachoeira Mine, the results of total U did not exceed 0.001 mg/L, which may indicate a reduced influence of the activities of the Engenho Mine at this monitoring point.

As a way of contributing to the safety of the population residing in the area of the Uraniferous Province of Lagoa Real, it is recommended that the levels of radionuclides in atmospheric particulates, soils, and surface and groundwater be evaluated and monitored in the Córrego Vereda, Maniaçu, Lagoa Grande, Stream Cachoeirinha, Stream Serrote, Riacho do Anguá, Riacho das Antas, Pé do Morro, Riacho Fundo, and Lagoa Real (Figure 18.5). In all basins in the province, there are housing centers, productive areas, and drilled wells. Monitoring surface and groundwater in particular can be fundamental to safeguarding people's health and helping to prevent the risk of contamination from continued ingestion of water with high uranium contents (Balbudhe et al. 2012; Bjørklund et al. 2017; Zamora et al. 2009).

18.3.2 Soil–Plant–Animal Transfer

The concentration of uranium in plants is generally lower than that found in soils. Although uranium is not an essential or beneficial element for plant metabolism, plants absorb it and its translocation occurs in their tissues (Kalin et al. 2004; Mitchell et al. 2013; Vandenhove et al. 2011; Vera Tome et al. 2003; Vera Tomé et al. 2008). The ability of plants to absorb uranium may depend on a complex network of factors, from the concentration of this element in the soil (Sheppard and Evenden 1988) to the chemical form in which it is found and a range of processes and interactions that can influence the mobility of any chemical species in the soil, with direct effects on its mobility and eventual absorption by the plant root system, such as complexation, precipitation, charge interactions, oxidation and reduction, transformations, in addition to other specific interactions with the different components of the soils, including their biota, directly influencing the availability of uranium for plants and microorganisms in the rhizosphere (Hester 2011; Mitchell et al. 2013).

To a large extent, the mobility and availability of uranium in the soil–plant system are governed by the oxidation state, complexation by organic and inorganic ligands, pH, interaction with organic matter in the soil, and mineral absorption (Bjørklund et al. 2017; Vandenhove et al. 2011). According to the International Atomic Energy Agency (IAEA 2010), the uptake of radionuclides by plants in tropical soils may present higher rates than those observed in temperate environments, which may be caused by the greater speed in the cycling of nutrients in these environments.

Plants show a high degree of variation in their ability to absorb and accumulate uranium. Thus, soil–plant transfer factors (Fv) are used to characterize the absorption ability/capacity of a given plant or plant tissue in a given situation or condition. Concentration factors were calculated as the ratio between uranium activity concentration in the respective dry plant organs (root, leaves, etc.) and the dry soil concentration (Carvalho et al. 2011; Doering et al. 2021; IAEA 2010; James et al. 2011; Kritsananuwat et al. 2017; Markose et al. 1993; Matveyeva et al. 2014, 2015a; Rumble and Bjugstad 1986). Similarly, it is also possible to determine the root barrier coefficient (RBC) by the ratio between the radionuclide activity in the dry soil and the radionuclide activity in the aerial part of the plant (Matveyeva et al. 2015a).

Reference values for U transfer in different plants and animals can be found in IAEA technical documents (IAEA 2009, 2010, 2014,) and in review articles of these values (Doering et al. 2021). However, in addition to the chemical availability of U in soils, other factors influence its transfer to organisms, such as the uptake of U per unit of root mass of a given plant is inversely proportional to the radius of the cross section of the root, indicating that root hairs and fine mycorrhizal hyphae tend to be large absorbers of U as pointed out by Boghi et al. (2018). For this author, the kinetics of U movement in the soil–plant system is quite complex, being not only influenced by the chemical characteristics of the soils but also by the characteristics of plant growth and development.

Different studies indicate that the absorption of radionuclides by the roots is directly influenced by different soil parameters such as cation exchange capacity, presence of stable elements, pH, organic matter content, moisture, among others (Fisher et al. 1999; Vandenhove et al. 2007a, b; 2011). However, Vandenhove et al. (2007a) point out that no soil parameter alone is sufficient to explain the uranium transfer factor to plants, stating only that high pH conditions (≥ 6.9) can lead to greater uranium transfer.

Once inside the plant system, radionuclides can be transported via phloem to different plant tissues. In studies on the absorption of ^{137}Cs in tree species, Fesenko et al. (2001) pointed out that a higher transfer rate was observed in roots, followed by leaves, underground stems, and wood. In an experiment with yellowcake [Ammonium diuranate – $(NH_4)_2U_2O_7$ (DUA)] in vetiver grass (*Chrysopogon zizanioides*), Roongtanakiat et al. (2010) found a greater accumulation of uranium in the roots and less in the aerial part of these plants.

Considering the complexity and the different factors that influence the uranium transfer process from the soils to the plants and, later, to the consuming animals, in order to achieve greater predictability and approximation of the reality of transfer of this element in the food chain, it is an approach that considers different focuses of analysis and the complex relationships between the agents in these systems is always necessary (Günther et al. 2003; Sheppard and Evenden 1988; Vandenhove et al. 2007a). It is only by seeking to understand the multiple and complex relationships between the agents in the systems that, at the end of the food chain, it will be possible to calculate the levels to which animals and humans are exposed (Al-Hamarneh et al. 2016).

18.3.2.1 Impacts on Agriculture and Livestock

In both municipalities that make up the Uraniferous Province of Lagoa Real (Caetité and Lagoa Real), agricultural activities are carried out in the areas of the province on a small scale. Caetité and Lagoa Real have herds of cattle in the order of 23 364 and 10 168, respectively, with Lagoa Real having a milk production of 3682 L while Caetité produces 3761 L/year. (IBGE 2022).

As it is a region with a predominantly semi-arid climate with high water deficit, the vegetable crops in the region reflect these characteristics, being particularly composed of annual crops of drought-adapted plants such as forage palm and cascava (Table 18.2). It is noted, however, that the area occupied by such crops is relatively small and, although many species and annual cultivated varieties are more adapted to periods of drought, these and other crops are predominantly implanted in areas with greater water availability in the valleys, such as the sugarcane and passion fruit, or are heavily dependent on groundwater sources.

As one of the main plants cultivated in the region, we can highlight the Cactaceae forage palm, a plant extremely resistant to drought and which is of great importance for the production of cattle since it is used as animal feed in the long periods of drought, functioning as a an important nutritional stock, for being able to maintain its nutritive value for a longer period, being considered a plant of great

Table 18.2 Main vegetable crops with respective annual production and area occupied in the municipalities of Província Uranífera de Lagoa Real.

		Caetité		Lagoa Real	
Vegetal	**Scientific name**	**Production (tons)**	**Area (ha)**	**Production (tons)**	**Area (ha)**
Forage palm	*Opuntia ficus-indica/Nopalea cochenillifera*	13 868	478	16 448	380
Sugarcane	*Saccharum officinarum*	4850	185	357	19
Beans	*Phaseolus* spp.	300	1000	120	220
Cassava	*Manihot esculenta*	4380	550	615	75
Corn	*Zea mays*	180	600	100	225
Passion fruit	*Passiflora edulis*	660	60	140	30

Source: Adatped from IBGE (2022).

importance compared to other forages (Dubeux et al. 2021; El-Mostafa et al. 2014).

Considering that such plants are cultivated in HBNR areas, being often irrigated with water from wells, the determination of radionuclides in such plant matrices becomes important (da Silva et al. 2004). In addition to being used in animal feed, forage cactus is also consumed directly by humans, and it is recommended to investigate the levels of uranium and other radionuclides in this and other plant matrices. INB performs analyzes on different plant matrices, including forage palm, bark, and corn.

In studies still under development (unpublished data), with corn plants (*Zea mays*) developed in hydroponic cultivation, with leached ore from the URA ore deposit and sterile soils irrigated with uranium solution as substrate, corn seedling roots inoculated with arbuscular mycorrhizal fungi (AMF) were subjected to Synchrotron radiation x-ray fluorescence (SR-XRF) at the XRF station of the Brazilian Synchrotron Light Source – LNLS (Pérez et al. 1999) for investigation of uranium transfer. It is known that Synchrotron-based techniques are highly sensitive and can be used to detect a wide range of elements with high spatial and/or lateral resolution (Isaure et al. 2006; Lopes et al. 2011; Pereira et al. 2010; Rubio et al. 2014; Sarret et al. 2013; Zhao et al. 2014). Preliminarily, evidence was found that in the absorption of uranium between mycorrhizal and nonmycorrhizal roots (Figure 18.9). It is expected that X-ray fluorescence analysis may acquire important information on the location and distribution of uranium in the soil–AMF–plant root interface, in order to contribute to the understanding of the process of absorption of toxic elements by the roots of plants grown in HBNR areas.

Just as the investigation of primary sources of contamination such as water and particulates is fundamental to think about the safety of the population of the uranium province, plant and animal matrices are also important targets of study, especially due to the local environmental context of high deficit water.

18.4 Uranium Impacts on Human Health

In humans, uranium exposures present chemical and radiological toxicity, with the main target organs being the kidneys and lungs. The health consequences of exposure to uranium are determined both by its chemical and physical nature (unenriched) and by the level and duration of exposure to this element (WHO 2001).

Kidney diseases, cancer, birth defects, and risk of death are reported as possible adverse effects, and uranium can be considered a toxic element to living beings in general (Konietzka 2015; Malaviya and Singh 2012). Considering also that toxic metals are cumulative in the food chain (Krämer 2010), and their accumulation in organisms may result in an aggravation of adverse effects, knowledge of the concentration levels of natural radionuclides, their availability and mobility in the environment, is shown to be of great interest in various scientific fields (Chen et al. 2006).

Considering the toxic effects caused by the entry of U into the human body, it should be noted that the main routes of entry are through the inhalation of suspended particles (respiration), as well as through the ingestion of contaminated water and food.

Inhalation of U-containing particles can have adverse respiratory effects both due to deposition of particles in the airways and the fate of such particles in the lung. Despite the radiological effects resulting from the deposition of U in the airways, no toxic effects on the lungs are described,

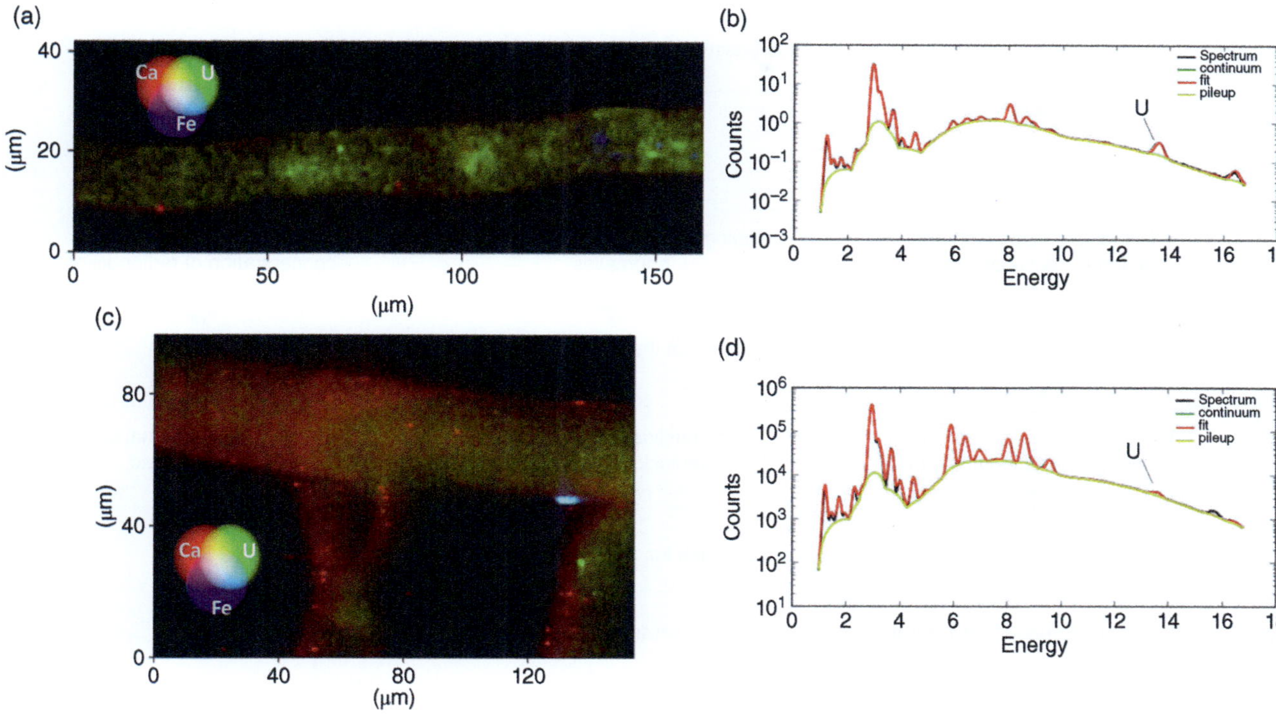

Figure 18.9 Preliminary results of an experiment with SR-XRF at the Brazilian Synchrotron Light Source showing reduced uranium uptake in mycorrhizal roots: (a, c) with element maps for non mycorrhizal and mycorrhizal roots, respectively; (b, d) spectrum curves for roots of the treatment non mycorrhizal and mycorrhizal roots, respectively.

and the complications are related to the deposition of particulate material. Similarly, no significant gastrointestinal, cardiovascular, hematologic, or hepatic effects are noted from ingestion (Keith et al. 2013). For its part, uranium has been recognized as a nephrotoxic metal due to its chemical action causing damage to the proximal tubules of renal tissue. However, its effect on renal tissue can be either chemical or radiological (Keith et al. 2013, 2015; Konietzka 2015; Zamora et al. 2009).

According Bjørklund et al. (2017), both the carcinogenic effects and its nephrotoxicity should be better studied, which reinforces the need for studies in areas with natural occurrences. This author also points to the fact that, although it is clear that soluble forms of uranium are carried by the blood to the kidney causing damage to the cells of this organ, the general nutritional status of the target, especially the iron content present in the diet, may have considerable influence on the levels of U absorption, and the greater the iron deficiency and the greater the intake of U, the greater the absorption rate of the latter. Thus, considering the low nutritional status often observed in the rural population residing in the uranium areas in question (Human Development Index – HDI of 0.625 and 0.545 for Caetité and Lagoa Real, respectively), there may be an aggravation of the risks for the populations residing in the area.

18.5 Final Considerations

High background natural radiation areas present themselves as areas of great relevance for the implementation of strategies to reduce radiation exposure (Galhardi et al. 2020; Hendry et al. 2009; Sohrabi 2013). In the Uraniferous Province of Lagoa Real, a particularly complex scenario is observed, where in a large territorial extension, the presence of significant mineral deposits can be observed in the midst of natural areas inhabited by the rural population. Positioned in a region severely affected by water scarcity, the drilling of wells for human use and for agricultural purposes gains greater prominence and importance not only for local development but also for people's survival. In this context, both the presence of natural deposits of uranium-rich rocks and the mining and processing of ore can be considered as important risks to health and the environment.

The presence of high levels of uranium in groundwater from wells drilled in uranium areas with rocks naturally enriched with this radionuclide (da Silva 2015; de Oliveira et al. 2012; Marques et al. 2021; Wilde 2020) represents considerable real risk to human health. Allied to this, the cultivation of plants, the raising of animals for human consumption in such areas with high levels of uranium, whether of natural origin or by human action in mining activities, compete with

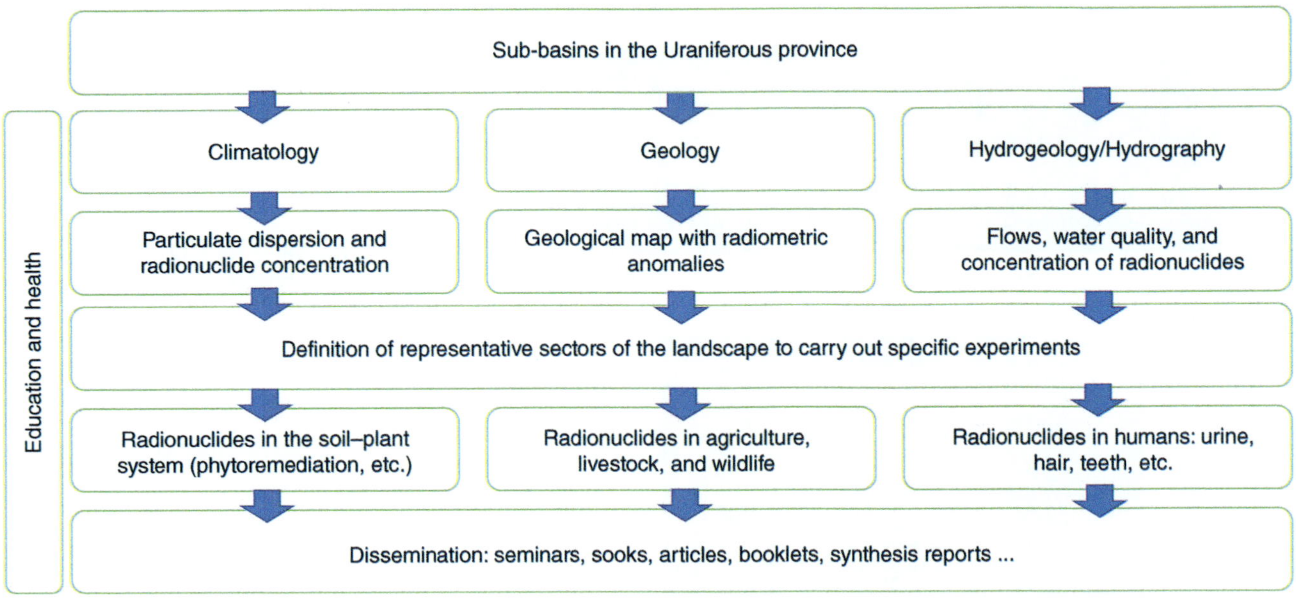

Figure 18.10 Proposed flowchart for environmental impact assessment and monitoring of the uranium area.

possible routes of ingestion and accumulation of elements dangerous to humans and other animals.

As suggested in Figure 18.10, an action strategy to reduce the risks of exposure and toxic elements in the uranium area must make use of an integrated vision that considers the sub-basins as the main planning and action units. In this scenario, a dialogue of knowledge must be developed within a context of health and education that explores the most representative sectors for the generation of scientific knowledge. Furthermore, it is necessary to provide civil society with information, sharing data through scientific publications, books, generating synthesis reports and seminars to share information, resolve doubts, and listen to suggestions.

Acknowledgment

The authors are grateful to the entire team at the INB Environmental Control Laboratory.

References

Ab'Sáber, A.N. (2003). *Os domínios da natureza no Brasil: potencialidades paisagísticas*. São Paulo: Ateliê Editorial.

Al-Hamarneh, I.F., Alkhomashi, N., and Almasoud, F.I. (2016). Study on the radioactivity and soil-to-plant transfer factor of 226Ra, 234U and 238U radionuclides in irrigated farms from the Northwestern Saudi Arabia. *Journal of Environmental Radioactivity* 160: 1–7. https://doi.org/10.1016/j.jenvrad.2016.04.012.

de Alves, L., J., Nunes, F.C., Prasad, M.N.V. et al. (2018). Uranium mine waste phytostabilization with native plants – a case study from Brazil. In: *Bio-Geotechnologies for Mine Site Rehabilitation* (ed. M.N.V. Prasad, P.J.C. Favas and S.K. Maiti), 299–322. Elsevier.

dos Amaral, R., S., de Vasconcelos, W.E., Borges, E. et al. (2005). Intake of uranium and radium-226 due to food crops consumption in the phosphate region of Pernambuco – Brazil. *Journal of Environmental Radioactivity* 82: 383–393. https://doi.org/10.1016/j.jenvrad.2005.02.013.

Balbudhe, A.Y., Srivastava, S.K., Vishwaprasad, K. et al. (2012). Assessment of age-dependent uranium intake due to drinking water in Hyderabad, India. *Radiation Protection Dosimetry* 148: 502–506. https://doi.org/10.1093/rpd/ncr193.

Bjørklund, G., Christophersen, O.A., Chirumbolo, S. et al. (2017). Recent aspects of uranium toxicology in medical geology. *Environmental Research* 156: 526–533. https://doi.org/10.1016/j.envres.2017.04.010.

Boghi, A., Roose, T., and Kirk, G.J.D. (2018). A model of uranium uptake by plant roots allowing for root-induced changes in the soil. *Environmental Science & Technology* 52: 3536–3545. https://doi.org/10.1021/acs.est.7b06136.

BRASIL (2021). *PORTARIA GM/MS No 888, DE 4 DE MAIO DE 2021 – DOU*. Imprensa Nacional.

Bullock, S.H. (1995). *Seasonally Dry Tropical Forests*. Cambridge University Press.

Cardoso, G.V., do Amaral Sobrinho, N.M.B., Wasserman, M.A.V., and Mazur, N. (2009). Geoquímica de radionuclídeos naturais em solos de áreas circunvizinhas a uma Unidade de Mineração e Atividade de Urânio. *Revista Brasileira de Ciência do Solo* 33: 1909–1917. https://doi.org/10.1590/S0100-06832009000600040.

Carvalho, F.P., Oliveira, J.M., and Malta, M. (2011). Radionuclides in plants growing on sludge and water from uranium mine water treatment. *Ecological Engineering, Biogeochemical Aspects of Ecosystem Restoration and Rehabilitation* 37: 1058–1063. https://doi.org/10.1016/j.ecoleng.2010.07.011.

Chen, B.D., Zhu, Y.-G., and Smith, F.A. (2006). Effects of arbuscular mycorrhizal inoculation on uranium and arsenic accumulation by Chinese brake fern (*Pteris vittata* L.) from a uranium mining-impacted soil. *Chemosphere* 62: 1464–1473. https://doi.org/10.1016/j.chemosphere.2005.06.008.

CONAMA (2008). Classification and environmental guidelines for the classification of groundwater, Resolution No. 396 of April 7, 2008 of National Council for the Environment of Brazil (CONAMA). In Portuguese.

Correa, R.T. (2019). Mapa da anomalia radiométrica do Brasil (Third edition). Scale 1: 5,000,000. GCS SIRGAS 2000. Brasília: SGB-CPRM – Geological Survey of Brazil.

CPRM (1996). São Timóteo Project (Aerogeophysical Survey Magnetometry and Gammaspectrometry) – Brazil Airborne Radiometric Mapping Project (BARMP). Scale 1:250,000. Brasília: SGB-CPRM – Geological Survey of Brazil.

CPRM (2022). *SIAGAS – Sistema de Informações de Águas Subterrâneas*. Serviço Geológico do Brasil http://siagasweb.cprm.gov.br/layout/ (accessed 18 April 2022).

Cruz, S.C.P., Alkmim, F.F., Leite, C.M.M. et al. (2007). Geologia e arcabouço estrutural do Complexo Lagoa Real, vale do Paramirim, centro-oeste da Bahia. *Revista Brasileira de Geociências* 37: 128–146.

Cruz, S.C.P., de Miranda, L.L.F., and de Veiga, P.M., O. (2008). *Provincia Uranífera de Lagoa Real, Bahia*, 1e. Salvador: Arquivos Abertos. Companhia Baiana de Pesquisa Mineral – CBPM.

Doering, C., Twining, J., Rout, S. et al. (2021). A revised IAEA data compilation for estimating the soil to plant transfer of radionuclides in tropical environments. *Journal of Environmental Radioactivity* 232: 106570. https://doi.org/10.1016/j.jenvrad.2021.106570.

Dubeux, J.C.B., dos Santos, M.V.F., da Cunha, M.V. et al. (2021). Cactus (*Opuntia* and nopalea) nutritive value: a review. *Animal Feed Science and Technology* 275: 114890. https://doi.org/10.1016/j.anifeedsci.2021.114890.

El-Mostafa, K., El Kharrassi, Y., Badreddine, A. et al. (2014). Nopal Cactus (*Opuntia ficus-indica*) as a source of bioactive compounds for nutrition, health and disease. *Molecules* 19: 14879–14901. https://doi.org/10.3390/molecules190914879.

Erdi-Krausz, G., Matolin, M., Minty, B. et al. (2003). *Guidelines for Radioelement Mapping Using Gamma Ray Spectrometry Data*. Vienna: International Atomic Energy Agency (IAEA).

Favas, P.J.C., Pratas, J., Mitra, S. et al. (2016). Biogeochemistry of uranium in the soil-plant and water-plant systems in an old uranium mine. *Science of the Total Environment* 568: 350–368. https://doi.org/10.1016/j.scitotenv.2016.06.024.

Fernandes, M.F. and de Queiroz, L.P. (2018). Vegetação e flora da Caatinga. *Ciência e Cultura* 70: 51–56. https://doi.org/10.21800/2317-66602018000400014.

Fesenko, S.V., Soukhova, N.V., Sanzharova, N.I. et al. (2001). Identification of processes governing long-term accumulation of 137 Cs by forest trees following the Chernobyl accident. *Radiation and Environmental Biophysics* 40: 105–113. https://doi.org/10.1007/s004110100090.

Fisher, N.S., Fowler, S.W., Boisson, F. et al. (1999). Radionuclide bio concentration factors and sediment partition coefficients in arctic seas subject to contamination from dumped nuclear wastes. *Environmental Science & Technology* 33: 1979–1982. https://doi.org/10.1021/es9812195.

Galhardi, J.A., de Mello, J.W.V., and Wilkinson, K.J. (2020). Environmental and health risk assessment of agricultural areas adjacent to uranium ore fields in Brazil. *Environmental Geochemistry and Health* https://doi.org/10.1007/s10653-020-00659-3.

Greenwood, N.N. and Earnshaw, A. (1997). *Chemistry of the Elements*, 2e. London: Butterworth-Heinemann.

Günther, A., Bernhard, G., Geipel, G. et al. (2003). Uranium speciation in plants. *Radiochimica Acta* 91: 319–328.

Hendry, J.H., Simon, S.L., Wojcik, A. et al. (2009). Human exposure to high natural background radiation: what can it teach us about radiation risks? *Journal of Radiological Protection* 29: A29–A42. https://doi.org/10.1088/0952-4746/29/2A/S03.

Hester, R.E. (ed.) (2011). *Nuclear Power and the Environment, Issues in Environmental Science and Technology*. Cambridge: Royal Society of Chemistry.

Hosseini, A., Thørring, H., Brown, J.E. et al. (2008). Transfer of radionuclides in aquatic ecosystems – default concentration ratios for aquatic biota in the Erica tool. *Journal of Environmental Radioactivity* 99: 1408–1429. https://doi.org/10.1016/j.jenvrad.2008.01.012.

IAEA (2009). *Quantification of Radionuclide Transfer in Terrestrial and Freshwater Environments for Radiological*

Assessments, IAEA-TECDOC-1616. Vienna: International Atomic Energy Agency.

IAEA (2010). *Handbook of Parameter Values for The Prediction of Radionuclide Transfer in Terrestrial and Freshwater Environments*, Technical Reports Series. Vienna: International Atomic Energy Agency (IAEA).

IAEA (2014). *Handbook of Parameter Values for the Prediction of Radionuclide Transfer to Wildlife*, Technical Reports Series. Vienna: IAEA.

IBGE (ed.) (2012). *Manual técnico da vegetação brasileira, 2a edição revista e ampliada. ed, Manuais técnicos em geociências*. Rio de Janeiro: Instituto Brasileiro de Geografia e Estatística - IBGE.

IBGE (2022). *Cidades*. Instituto Brasileiro de Geografia e Estatística https://cidades.ibge.gov.br/ (accessed 16 April 2022).

INB (2015). Indústrias Nucleares do Brasil [WWW Document]. Official Website of Nuclear Industries of Brazil – INB. http://www.inb.gov.br/en-us (accessed 29 October 2022).

INMET (2022). *Instituto Nacional de Meteorologia – INMET*. Instituto Nacional de Meteorologia https://portal.inmet.gov.br/ (accessed 17 April 2022).

Isaure, M.-P., Fraysse, A., Devès, G. et al. (2006). Micro-chemical imaging of cesium distribution in Arabidopsis thaliana plant and its interaction with potassium and essential trace elements. *Biochimie* 88: 1583–1590. https://doi.org/10.1016/j.biochi.2006.08.006.

James, J.P., Dileep, B.N., Ravi, P.M. et al. (2011). Soil to leaf transfer factor for the radionuclides 226Ra, 40K, 137Cs and 90Sr at Kaiga region, India. *Journal of Environmental Radioactivity* 102: 1070–1077. https://doi.org/10.1016/j.jenvrad.2011.07.011.

Jasiewicz, J. and Metz, M. (2011). A new GRASS GIS toolkit for Hortonian analysis of drainage networks. *Computers & Geosciences* 37: 1162–1173. https://doi.org/10.1016/j.cageo.2011.03.003.

Kalin, M., Wheeler, W.N., and Meinrath, G. (2004). The removal of uranium from mining waste water using algal/microbial biomass. *Journal of Environmental Radioactivity* 78: 151–177. https://doi.org/10.1016/j.jenvrad.2004.05.002.

Keith, S., Faroon, O.M., Roney, N. et al. (2013). *Toxicological Profile for Uranium*. Atlanta: Agency for Toxic Substances and Disease Registry.

Keith, L.S., Faroon, O.M., and Fowler, B.A. (2015). Uranium. In: *Handbook on the Toxicology of Metals*, 4e (ed. G.F. Nordberg, B.A. Fowler and M. Nordberg), 1307–1345. San Diego, CA: Academic Press https://doi.org/10.1016/B978-0-444-59453-2.00059-7.

Konietzka, R. (2015). Gastrointestinal absorption of uranium compounds – a review. *Regulatory Toxicology and Pharmacology* 71: 125–133. https://doi.org/10.1016/j.yrtph.2014.08.012.

Krämer, U. (2010). Metal hyper accumulation in plants. *Annual Review of Plant Biology* 61: 517–534. https://doi.org/10.1146/annurev-arplant-042809-112156.

Kritsananuwat, R., Chanyotha, S., Kranrod, C., and Pengvanich, P. (2017). Transfer factor of ^{226}Ra, ^{232}Th and ^{40}K from soil to *Alpinia Galangal* plant grown in Northern Thailand. *Journal of Physics: Conference Series* 860: 012008. https://doi.org/10.1088/1742-6596/860/1/012008.

Lehto, J. and Hou, X. (2011). *Chemistry and Analysis of Radionuclides: Laboratory Techniques and Methodology*. Wiley.

Lobato, L.M. and Fyfe, W.S. (1990). Metamorphism, metasomatism, and mineralization at Lagoa Real, Bahia, Brazil. *Economic Geology* 85: 968–989. https://doi.org/10.2113/gsecongeo.85.5.968.

Lobato, L.M., Pimentel, M.M., Cruz, S.C.P. et al. (2015). U–Pb geochronology of the Lagoa Real uranium district, Brazil: Implications for the age of the uranium mineralization. *Journal of South American Earth Sciences* 58: 129–140. https://doi.org/10.1016/j.jsames.2014.12.005.

Lopes, R.T., Lima, I., Pereira, G.R., and Perez, C.A. (2011). Synchrotron radiation X-ray microfluorescence techniques and biological applications. *Pramana* 76: 271–279.

Malaviya, P. and Singh, A. (2012). Phytoremediation Strategies for remediation of uranium-contaminated environments: a review. *Critical Reviews in Environmental Science and Technology* 42: 2575–2647. https://doi.org/10.1080/10643389.2011.592761.

Markose, P.M., Bhat, I.S., and Pillai, K.C. (1993). Some characteristics of 2Z6Ra transfer from soil and uranium mill tailings to plants. *Journal of Environmental Radioactivity* 21: 131–142.

Marques, C., Cabral, A.R., and Rios, F.J. (2021). Whole-rock chemistry of the Gameleira I uranium deposit, Lagoa Real, Brazil. *Geochemistry* 81: 125677. https://doi.org/10.1016/j.chemer.2020.125677.

Matveyeva, I., Jaćimović, R., Planinšek, P. et al. (2014). Assessment of the main natural radionuclides, minor and trace elements in soils and sediments of the Shu valley (near the border of Kazakhstan and Kyrgyzstan). *Journal of Radioanalytical and Nuclear Chemistry* 299: 1399–1409. https://doi.org/10.1007/s10967-013-2902-3.

Matveyeva, I., Jaćimović, R., Planinšek, P. et al. (2015a). Uptake of uranium, thorium and radium isotopes by plants growing in dam impoundment Tasotkel and the Lower Shu region (Kazakhstan). *Radiochimica Acta* 104: https://doi.org/10.1515/ract-2015-2457.

Mitchell, N., Pérez-Sánchez, D., and Thorne, M.C. (2013). A review of the behaviour of U-238 series radionuclides in soils and plants. *Journal of Radiological Protection* 33: R17–R48. https://doi.org/10.1088/0952-4746/33/2/R17.

Mkandawire, M. (2013). Biogeochemical behaviour and bioremediation of uranium in waters of abandoned mines. *Environmental Science and Pollution Research* 20: 7740–7767. https://doi.org/10.1007/s11356-013-1486-3.

de Oliveira, L.A.R., de, da Souza, A., S., Rios, F.J. et al. (2012). Estudo de inclusões fluidas em minerais associados à mineralização uranífera de três jazidas da Província Uranífera de Lagoa Real, Bahia - Brasil. *Revista Geonomos* https://doi.org/10.18285/geonomos.v2i20.249.

de Queiroz, L.P. (2006). The Brazilian Caatinga: phytogeographical patterns inferred from distribution data of the Leguminosae. In: *Neotropical Savannas and Seasonally Dry Forests, Systematics Association Special Volumes*, 121–157. CRC Press https://doi.org/10.1201/9781420004496.ch6.

de Queiroz, L.P., Cardoso, D., Fernandes, M.F., and Moro, M.F. (2017). Diversity and evolution of flowering plants of the caatinga domain. In: *Caatinga* (ed. J.M.C. da Silva, I.R. Leal and M. Tabarelli), 23–63. Cham: Springer International Publishing https://doi.org/10.1007/978-3-319-68339-3_2.

Neteler, M., Bowman, M.H., Landa, M., and Metz, M. (2012). GRASS GIS: a multi-purpose open source GIS. *Environmental Modelling & Software* 31: 124–130. https://doi.org/10.1016/j.envsoft.2011.11.014.

Oliveira Chaves, A. (2013). New geological model of the Lagoa Real uraniferous albitites from Bahia (Brazil). *Open Geosciences* 5: https://doi.org/10.2478/s13533-012-0134-7.

Pascholati, E.M., Silva, C.L.D., Costa, S.D.S. et al. (2003). Novas ocorrências de urânio na região de Lagoa Real, a partir da superposição de dados geofísicos, geológicos e de sensoriamento remoto. *Revista Brasileira de Geociencias* 33: 91–98. https://doi.org/10.25249/0375-7536.200333S29198.

Pennington, R.T., Lewis, G.P., and Ratter, J.A. (2006). An overview of the plant diversity, biogeography and conservation of neotropical savannas and seasonally dry forests. In: *Neotropical Savannas and Seasonally Dry Forests, Systematics Association Special Volumes* (ed. R.T. Pennington, G.P. Lewis and J.A. Ratter), 1–29. CRC Press, Taylor & Francis Group.

Pereira, G.R., Rocha, H.S., Calza, C. et al. (2010). Biological tissues analysis by XRF microtomography. *Applied Radiation and Isotopes* 68: 704–708. https://doi.org/10.1016/j.apradiso.2009.12.015.

Pérez, C.A., Radtke, M., Sánchez, H.J. et al. (1999). Synchrotron radiation X-Ray fluorescence at the LNLS: beamline instrumentation and experiments. *X-Ray Spectrometry* 28: 320–326. https://doi.org/10.1002/(SICI)1097-4539(199909/10)28:5<320::AID-XRS359>3.0.CO;2-1.

de Polo, H.J.O., Meloni, R.E., Tavares, F.M., Costa, I.S.L., Almeida, R.C. (2022). Mapa de avaliação do potencial de urânio do Brasil.

Rayno, D.R. (1983). Estimated dose to man from uranium millling via the beef/milk food-chain pathway. *Science of the Total Environment* 31: 219–241. https://doi.org/10.1016/0048-9697(83)90078-5.

Roongtanakiat, N., Sudsawad, P., and Ngernvijit, N. (2010). Uranium absorption ability of sunflower, vetiver and purple guinea grass. *Kasetsart Journal (Natural Science)* 44: 182–190.

Rubio, M., Germanier, A., Mera, M.F. et al. (2014). Study of lead levels in soils by weathering of metallic Pb bullets used in dove hunting in Córdoba, Argentina: lead determination in hunting zones of Argentinean forest. *X-Ray Spectrometry* 43: 186–192. https://doi.org/10.1002/xrs.2539.

Rumble, M.A. and Bjugstad, A.J. (1986). Uranium and radium concentrations in plants growing on uranium mill tailings in South Dakota. *Reclamation and Revegetation Research* 4: 271–277.

Sarret, G., Smits, E.A.H.P., Michel, H.C. et al. (2013). Use of synchrotron-based techniques to elucidate metal uptake and metabolism in plants. In: *Advances in Agronomy, Advances in Agronomy* (ed. D.L. Sparks), 1–82. USA: Academic Press https://doi.org/10.1016/B978-0-12-407247-3.00001-9.

SEI (1999). Balanço Hídrico no Estado da Bahia, Série Estudos e Pesquisas. In: *uperintendência de Estudos Econômicos e Sociais da Bahia*. Salvador.

Sheppard, S.C. and Evenden, W.G. (1988). Critical compilation and review of plant/soil concentration ratios for uranium, thorium and lead. *Journal of Environmental Radioactivity* 8: 255–285. https://doi.org/10.1016/0265-931X(88)90051-3.

Silva, L.S. (2011). Avaliação da radioatividade natural em águas potáveis de superfície e subterrâneas da região de Caetité,BA (Mestrado em Tecnologia Nuclear - Aplicações). Universidade de São Paulo, São Paulo. https://doi.org/10.11606/D.85.2011.tde-01032012-100446

da Silva, B.B., Alves, J.J.A., Cavalcanti, E.P., and Dantas, R.T. (2002). Potencial eólico na direção predominante do vento no Nordeste brasileiro. *Revista Brasileira de Engenharia Agricola e Ambiental* 6: 431–439. https://doi.org/10.1590/S1415-43662002000300009.

da Silva, C.M., Amaral, A., dos Amaral, R., S. et al. (2004). Application of bootstrap method for evaluating discrepant levels of radium-226 in forage palm (Opuntia spp). *Revista Brasileira de Biometria* 25: 109–114.

da Silva, L.F. (2015). Modelo conceitual do aquífero fraturado da área da jazida de urânio de Caetité, Bahia: implicações para o fluxo subterrâneo. Master. Universidade de Brasília. https://doi.org/10.26512/2015.D.31090.

Sohrabi, M. (2013). World high background natural radiation areas: need to protect public from radiation exposure. *Radiation Measurements* 50: 166–171. https://doi.org/10.1016/j.radmeas.2012.03.011.

Stewart, A.G. (2019). Mining is bad for health: a voyage of discovery. *Environmental Geochemistry and Health* https://doi.org/10.1007/s10653-019-00367-7.

Thornthwaite, C.W. (1948). An approach toward a rational classification of climate. *Geographical Review* 38: 55–94.

UNSCEAR (2010). Sources and Effects of Ionizing Radiation: United Nations Scientific Committee on the Effects of Atomic Radiation 2008 Report, Report to the General Assembly with Scientific Annexes A and B. United Nations, New York.

Vandenhove, H., Hurtgen, C., and Payne, T.E. (2011). Uranium. In: *Radionuclides in the Environment*. Wiley https://doi.org/10.1002/9781119951438.eibc0435.

Vandenhove, H., Van Hees, M., Wannijn, J. et al. (2007a). Can we predict uranium bioavailability based on soil parameters? Part 2: soil solution uranium concentration is not a good bioavailability index. *Environmental Pollution* 145: 577–586. https://doi.org/10.1016/j.envpol.2006.04.012.

Vandenhove, H., Van Hees, M., Wouters, K., and Wannijn, J. (2007b). Can we predict uranium bioavailability based on soil parameters? Part 1: effect of soil parameters on soil solution uranium concentration. *Environmental Pollution* 145: 587–595. https://doi.org/10.1016/j.envpol.2006.04.011.

Vera Tome, F., Blanco Rodríguez, M.P., and Lozano, J.C. (2003). Soil-to-plant transfer factors for natural radionuclides and stable elements in a Mediterranean area. *Journal of Environmental Radioactivity* 65: 161–175.

Vera Tomé, F., Blanco Rodríguez, P., and Lozano, J.C. (2008). Elimination of natural uranium and 226Ra from contaminated waters by rhizofiltration using *Helianthus annuus* L. *Science of the Total Environment* 393: 351–357. https://doi.org/10.1016/j.scitotenv.2008.01.013.

WHO (2001). *Health Effects of Depleted Uranium (Provisional Agenda Item 13.10 No. A54/19 Add.1)*. Fifty-Fourth World Health Assembly. World Health Organization.

Wilde, A. (2020). Shear-hosted uranium deposits: a review. *Minerals* 10: 954. https://doi.org/10.3390/min10110954.

WNA (2020). *Supply of Uranium*. World Nuclear Association https://www.world-nuclear.org/ (accessed 6 June 2020).

Zamora, M.L.L., Zielinski, J.M., Moodie, G.B. et al. (2009). Uranium in drinking water: renal effects of long-term ingestion by an aboriginal community. *Archives of Environmental & Occupational Health* 64: 228–241. https://doi.org/10.1080/19338240903241267.

Zhao, F.-J., Moore, K.L., Lombi, E., and Zhu, Y.-G. (2014). Imaging element distribution and speciation in plant cells. *Trends in Plant Science* 19: 183–192. https://doi.org/10.1016/j.tplants.2013.12.001.

19

Defluoridation

Benan Yazici-Karabulut[1], Perihan Derin[1], Abdullah İzzeddin Karabulut[1], Ayşegul Demir-Yetis[2], Ayse Dilek Atasoy[1], and Mehmet İrfan Yeşilnacar[1]

[1] *Environmental Engineering Department, Faculty of Engineering, Harran University, Sanliurfa, Turkey*
[2] *Medical Services and Techniques Department, Bitlis Eren University, Bitlis, Turkey*

19.1 Defluoridation of High Fluoride Groundwater

Reduction of fluoride (F^-) in drinking water to recommended/permitted levels by responsible organizations is significant to prevent or reduce the adverse health effects caused by detrimental F^- exposure. As in the rest of the world, contamination of groundwater by F^- is very common in Turkey. Therefore, it is important to have a resource that provides different methods for removing F^- from waters and reducing potential health risks. This section provides such a comprehensive resource.

The optimum F^- level in water is between 0.5 and 1.0 mg/L according to the WHO standard (WHO 1994). F^- contents above the permissible levels cause adverse effects on human health. For this reason, F^- is regarded a geogenic pollution causing dental (>1.5 mg/L) and skeletal fluorosis (>4.0 mg/L) (Atasoy and Yesilnacar 2017a). As can be seen in studies conducted in some part of Turkey, the F^- concentration of water varies between 1.5 and 13.7 mg/L because of volcanic rocks and geothermal resources and is above the limit values that should be found in waters (Ataş et al. 2021; Atasoy et al. 2016a; Demir Yetis et al. 2021; Oruç 1977, 2001, 2008; Sendil and Baysu 1973; Yeşilnacar et al. 2016). Yeşilnacar (2010) first determined heavy F^- and dental fluorosis cases in groundwater in two villages located in the southeast of Turkey. In other studies, conducted in the same region, the geological structure, hydrogeology, hydrochemistry, and rock structure of the region were examined. In line with the data obtained, a comparison was made with the dental structure of the people of the region. All data were interpreted with Geographic Information System (GIS), and the study area's geomedical map was created. In addition, artificial neural networks were used for the prediction of high F^- in water. Similarly, according to the results of an unpublished study conducted in Bitlis city center and its districts, which directed the ministry to locations with high F^- values, the highest F^- concentration in waters was measured around 3.5 mg/L.

In another study conducted in Kırıkkale, it was determined that the fluoride concentration of drinking water sources varied between 0.128 and 5.967 ppm. The prevalence of dental fluorosis in Kırıkkale was found to be 21.8% (Karaca 2017). As a result of these studies, results were obtained regarding the source, distribution, and size of the problem in the region (Ataş et al. 2021; Demir Yetis et al. 2021; Yeşilnacar et al. 2016). Defluoridation includes effective and feasible methods to decrease the F^- in water when alternative water sources are not existing. When it was explored that heavy F^- concentrations cause fluorosis, studies began on the F^- removal from water. Defluoridation methods can be categorized as adsorption (Asgari et al. 2012; Jorfi et al. 2011; Tomar et al. 2014; Yadav et al. 2013), electrocoagulation (EC) (Bazrafshan et al. 2012; Behbahani et al. 2011; Emamjomeh and Sivakumar 2006; Khatibikamal et al. 2010; Malakootian et al. 2011; Yazici-Karabulut et al. 2019; Yazici-Karabulut and Atasoy 2019a, b), precipitation/coagulation (Rahmani et al. 2010; Umar and Alam 2012), nanofiltration (Hosseini et al. 2019; Poursaberi et al. 2012; Tahaikt et al. 2007), and ion exchange (Chubar 2011; Samatya et al. 2007). Each of these methods has advantages and disadvantages. Fluoridation methods from water are discussed in detail in the following sections.

Medical Geology: En route to One Health, First Edition. Edited by Majeti Narasimha Vara Prasad and Meththika Vithanage.

19.2 Adsorption

The adsorption techniques are generally the preferred methods in water treatment with its ease of design. Therefore, it has a wide application area in water pollution control. Adsorption is considered a remarkable techniques for F^- removal from waters. Adsorption is efficient, useful, easy to design and application, and inexpensive. Some methods are developed to change the surface properties of adsorbents physically and chemically. F^- adsorption from drinking water is checked by temperature, adsorption kinetics, pH, adsorbent particle size, adsorbent concentrations, and sorbent surface area (Chaouch and Khelfaoui 2019; Loganathan et al. 2013).

Adsorption is a function of adsorbent concentration and solution temperature. It is also known as the surface process and the surface properties of the adsorbent are very important. The adsorbents used in the adsorption process provide active sites for the adsorption of F^- ions. Various materials have been explored to find an appropriate and inexpensive adsorbent to F^- removal (Ayalew 2020; George et al. 2010; Jagtap et al. 2012; Yapo et al. 2021). Activated alumina is a thoroughly implemented adsorbent to F^- removal from water that can be used for drinking and utility. The use of activated alumina was initiated by UNICEF in 1991 in rural India (Daw 2004). The use of activated alumina has yielded positive results in adsorption processes for the removal of F^- from groundwater (Ghorai and Pant 2004). However, the use of alumina is not economical. The existence of high metal oxides in some materials can get them available for water defluoridation. Many researchers have benefited from iron oxide in their studies to reduce heavy metal pollution and anions in water (Huang et al. 2011; Sujana et al. 2009). Iron-based materials have also been extensively studied for F^- removal, as iron has better affinity for F^-. Among the various natural materials studied for the defluoridation of water, a wide variety of clay types have also been investigated (Bhatnagar et al. 2011). Apart from these materials, a wide variety of adsorbents such as chitosan, cement, activated carbon, coconut shell carbon, zeolites, and lime have been used for F^- removal (Gao et al. 2009; Jagtap et al. 2012). In a study by Atasoy and Yesilnacar (2017b), the adsorption potential of low-cost iron oxide and cement clay for F^- removal from water was evaluated. Defluoridation studies were performed under various experimental conditions such as contact time, F^- concentrations, and adsorbent dosage. The fluoride adsorption capacity of iron oxide was found to be higher than cement clay due to its high surface area and iron oxide load (Atasoy and Yesilnacar 2017b). In another adsorption study, the F^- removal from water by adsorption on bauxite and clay, which is locally found in Turkey, low cost, simple, and plentiful, was investigated, but low F^- adsorption was provided on doped bauxite. Raw clay materials need to be modified to increase their potential (Atasoy et al. 2013a, b; Atasoy and Sahin 2014).

Most of the adsorption occurs when the mixture reaches equilibrium. The mass of the liquid or gas adsorbed per gram of the solid depends on the specific surface area of the solid, the concentration of the solute in equilibrium in solution, the temperature, and the structure of the molecules present. This process is an old and traditional method. In recent studies, it has been observed that this process has been replaced by more effective and innovative methods.

19.3 Electrocoagulation

EC is an electrofiltration process with high removal rates for defluoridation (Yadav et al. 2019). Environmental compatibility, energy efficiency, cost effectiveness, better automation, simple equipment requirements, low sludge production, and coping with various pollutants in the EC process were the advantages of this method (Amooey et al. 2014; Lacasa et al. 2013). EC is an electrochemical process in which the metallic cations synthesis consists in sacrificial anodes when an electrical charge is applied through a power source (Kobya and Ulu 2016). In the past decade, the use of EC has increased in pollutant removal studies from waters. It has recently been used as an available process for F^- removal from groundwater (Hu et al. 2005; Malakootian et al. 2011). EC can be used effectively to remove a wide variety of contaminants, including oil, paint, heavy metals, and F^- (Atasoy et al. 2018; Hu et al. 2005). This process does not breed secondary pollutants and beneficial substances present in water can be kept during F^- removal using this techniques. Sinha et al. (2012) reported that EC with Al electrodes was completely accomplished in simultaneously removal of both F^- and Al under $230\,V_{DC}$. Longer preservation time provided better defluoridation from groundwater. Yazici-Karabulut et al. (2019) carried out an EC study on F^- removal from groundwater using aluminum electrodes. 99.4% F^- removal rate was obtained in the EC process with pH and conductivity values of 8.29 and 486 μS/cm, respectively (Yazici-Karabulut et al. 2019).

Generally, aluminum (Al) and iron (Fe) electrode types are more preferred in the EC method. EC includes in situ coagulant production by EC of Al electrodes. Al cations are produced at the anode Eq. (19.1) and hydrogen gas is released at the cathode Eq. (19.2) (Zhu et al. 2007).

$$\text{Anode}: Al_{(s)} \rightarrow Al_{(aq)}^{3+} + 3e^- \quad (19.1)$$

$$\text{Cathode}: 3H_2O + 3e^- \rightarrow 1.5H_{2(g)} + 3OH^- \quad (19.2)$$

Various aqueous types of aluminum are produced during the dissolution of Al at the anode. Aluminum cations are converted to polymeric species and precipitate as $Al(OH)_3$:

$$Al_{(aq)}^{3+} + 3H_2O \rightarrow Al(OH)_3 + 3H^+ \quad (19.3)$$

Defluoridation by EC is accomplished by chemical implantation in which F^- from $Al(OH)_3$ flocs replaces the OH^- group. This method can be used effectively to remove F^- (Sandoval et al. 2014). The EC method is advantageous because there is no need for a secondary treatment method, and the beneficial ingredients present in the raw water can be captured and recovered during defluoridation. High-energy consumption is seen as a disadvantage of the EC process. This problem can be eliminated through solar energy systems installed with PV panels.

19.4 Coagulation/Precipitation

Coagulation is a chemical method that includes using coagulants to make unstable tiny particles in water. This process can be preferred not only for the removal of small particles but also for the F^- removal from waters. Various coagulant materials such as alum, sodium silicate, zeolites, sodium aluminate, lime, silica gel, bauxite, and ferric chloride can be used in this method (Deng et al. 2020; Liu et al. 2013; Sudasinghe et al. 2019). Trivalent ions such as iron and aluminum and lime and alum are mostly used coagulants. Lime and alum can be used simultaneously F^- removal from water. In the precipitation method, F^- is precipitated in water in the form of calcium fluoride (CaF_2) (Liu et al. 2013). On the other hand, the existence of organic agents during coagulation/precipitation has a negative impact on removal of F^-. Jadhav et al. (2015) reported that a high amount of F^- was removed from the water by the coagulation / precipitation method. However, since the coagulation/precipitation method cannot remove F^- within the limits allowed by WHO, it has been suggested to use this method as a primary treatment technique in combination with other methods (Islam and Patel 2007).

19.5 Nanofiltration

Nanofiltration is another technique for removing F^- from water. Nanofiltration diminishes hardness of water by using polymeric membranes with high retention capacity for charged nanoparticles. Thanks to this feature, the nanofiltration technique has very specific membrane selectivity and can be considered as the most suitable membrane techniques for F^- removal (Tahaikt et al. 2007). There are many studies on the F^- removal from water using this method. In a study, both nanofiltration and reverse osmosis (RO) process were compared for the F^- removal from water with reducing salinity (Diawara et al. 2011). It has been declared that nanofiltration membranes are much more accomplished in F^- removal from contaminated water confronted to RO membranes. Another study was conducted using two commercial NF-90 and NF-270 models nanofiltration membranes (Hoinkis et al. 2011). NF-270 decreased the F^- concentration from 10 to 1.5 mg/L, while NF-90 decreased the F^- concentration from 20 to 0.5 mg/L. Present anions such as bicarbonate (HCO_3^-) were found to have no significant negatory impacts on water treatment of F^-, but the performance of F^- removal decreased under acidic conditions (Hoinkis et al. 2011). Bejaoui et al. (2014) evaluated the defluoridation using NF-90 RO membranes on important parameters such as pressure, pH, concentration, strength, and the nature of F^- related cations. There was an increase in the total negative charges of the evaluated membranes. Therefore, it was observed that the F^- removal was optimized at high pH (Bejaoui et al. 2014).

19.6 Ion Exchange

Ion exchange is known as the technique of F^- removal by exchanging ions instead of adsorption. The main reason for this is that the presence of F^- in water is less than other ions. In F^- removal, cation exchange resins are more selective than anion exchange resins (Meenakshi and Viswanathan 2007). In this context, the defluoridation capacity and selectivity for F^- depend on the type of resin. Due to the different properties of metal ions, it greatly affects the removal of F^- (Luo and Inoue 2004). Therefore, it is difficult to simultaneously maximize the defluoridation capacity of ion exchange resins with F^- selectivity.

Ion exchange offers very high operating costs but is very efficient (Ansari and Parsa 2016). However, another disadvantage is that when all the low affinity ions on the resin are replaced with higher affinity ions, the resin is depleted and has to be regenerated or replaced. Therefore, the regeneration process results in the formation of significant amounts of concentrated saline, which creates problems in the application of ion exchange systems. Regular filling of the brine tank and material cost make the system inconvenient (Palko et al. 2018; Ruiz-Bevia and Fernandez-Torres 2019).

19.7 Comparing to Several Techniques for Defluoridation

A summary of the several F^- removal techniques is presented in Table 19.1. Considering all the evaluations made, the following information was obtained about the F^- removal methods:

- It has been revealed that adsorption is preferable in terms of low cost, ease of application, and high efficiency. It is worth noting that adsorbents show higher F^- adsorption capacity in adsorption technology. On the other hand, an adsorbent with high performance in laboratory conditions may give negative results in the real field. Therefore, the selection of adsorbent is a very important and difficult step.
- The EC technique is suitable for F^- removal given the problems mentioned above, but this process requires a continual power supply. This can also be solved by using a system of solar panels.

Table 19.1 Comparing to several techniques for defluoridation.

Methods	Advantages	Disadvantages	References
Adsorption	High removal capacity (up to 90%) Low-cost technique	Best efficiency is at pH ~6. Some ions may interfere with F^- removal. Pretreatment is required. Adsorbents should be reformed after about five months. After the regeneration process, a decrease in the adsorption capacity of the adsorbents can be observed.	Bhatnagar et al. (2011) and Mukherjee and Halder (2018)
Electrocoagulation	It is a technique with high efficiency (>80%). A wide variety of contaminants can be removed. As a result of this method, secondary pollutants are not formed. Useful ions are not removed from the water. It is also used in the removal of microorganisms.	Aluminum may be released into the water during the process. A large amount of electrical energy is required. This problem can be solved by using PV panels.	Emamjomeh and SivaKumar (2006), Moussa et al. (2017), and Yazici-Karabulut et al. (2019)
Coagulation/precipitation	It is considered a successful technique used in F^- subtraction. Tested in real condition.	This technique can only remove 18–33% of the precipitated F^-. It can also reduce ~74% of F^- into solvable Al—F^- complex ion. Therefore, the use of this process is incorrect. At the end of the process, an increase in sulfate ions is observed. Remaining Al in purified water is above the 0.2 mg/L. Water quality can change over time. Therefore, the inlet water should be analyzed methodically to evaluate the correct chemical dosage. The maintenance cost is not low. The automation is hard. A large area is needed to dry the sludge formed as waste. The presence of silicates in water has adverse impacts on F^- removal. The defluoridation performance can be changed by temperature change.	Mumtaz et al. (2015) and Dubey et al. (2018)

Table 19.1 (Continued)

Methods	Advantages	Disadvantages	References
Nanofiltration	It is a very fruitful method for defluoridation. Microorganisms, organic and inorganic contaminants can be effectively reduced simultaneously with F^-. No change in water quality during the process. It does not require very frequent maintenance. There is no use of chemicals in the process. Membrane life is very long. pH is not effective in the operation of the system. Other ions do not interfere. No manpower is needed. Its automation is simple and functional.	It also removes beneficial ions from the water. Installation and operating costs are high. Purified water has a low pH; therefore, post-process pH adjustment is necessary. After the process, a large amount of salt water is formed. Salt water is difficult to remove.	Shen and Schäfer (2014), Samrat et al. (2018), and Owusu-Agyeman et al. (2019)
Ion exchange	It provides high performance (up to 95%). There are no watercolor and test changes.	Sludge disposal after F^- treatment is a big problem. Some ions may interfere with F^- exchange. Resin regeneration is a problem as it generates fluoride-rich waste. It is a costly method. Purified water has a high chloride concentration and a low pH.	Mohapatra et al. (2009) and Kimambo et al. (2019)

- The coagulation/precipitation technique is limited to lower than 40% of F^- removal and also produces high amounts of aluminum-containing waste.
- The nanofiltration process is effective at F^- removal to a large extent, but this technology is very costly and also eliminates some necessary ions from the water.
- The ion exchange technique is so effective for F^- removal but is costly, and purified water has a higher chloride content.
- Finally, every country and region has different levels of fluoride and other interfering ions in their groundwater. Each technique also has some limitations. Therefore, it is not possible for a single technique to be implemented by all countries of the world.

Currently, F^- pollution in water has risen significantly because of urbanization and the increase in population of world. The presence of F^- in groundwater above the allowable limits causes serious health problems. At this stage, defluoridation has a great importance. Various methods have been used defluoridation, such as adsorption, EC, coagulation/precipitation, nanofiltration, and ion exchange processes.

It is clear that parts of Afghanistan, China, India, Iraq, Pakistan, Sri Lanka, South Korea, Turkey, and Vietnam are seriously influenced by fluoride concentrations greater than 10 mg/L. Further assistance is needed for F^- improvement in the mentioned countries. Basically, in the defluoridation of water, it can be initiated at two organizational grades: (i) domestic defluoridation by individual households for their own water consumption, and (ii) community defluoridation at the district and village levels.

19.8 Fluoride Health Risk Assessment

Health risk assessment (HRA) is a fruitful approach used to determine the prevalence of health problems caused by high F^- levels in groundwater. Various studies in distinct parts of the world have examined this method to investigate the health risk associated with oral or dermal fluoride intake in groundwater (Aslani et al. 2019; Dehbandi et al. 2018; Mukherjee et al. 2019; Nadiri et al. 2019).

In HRA, the amount of non-carcinogenic risk from F^- exposure is investigated considering the physiological and behavioral differences of children, adolescents, and adults. As a result of these examinations, plans can be made to identify areas with high fluoride levels and groundwater quality and to protect them against health hazards. Associated health risks such as dental and skeletal fluorosis due to high F^- levels are of great importance, especially for children and teenagers. HRA is a quantitative methodology designed to predict the potentially harmful consequences to human health of exposure to harmful substances, such as chemicals or microbials, over a period of time (USEPA 1989, 1992). This method is widely used to quantify potential risk in a multitude of situations in a variety of environmental settings from humans, animals, and the environment.

In most of the arid and semi-arid regions, especially in developing countries like Turkey, dependence on groundwater for agriculture, irrigation, and drinking purposes is common, and high F^- in groundwater has significant effects on human health (Atasoy et al. 2016a; Yazici-Karabulut and Atasoy 2019a, b). Important results have been obtained in health studies about F^- contamination and its effects in Şanlıurfa, Turkey (Ataş et al. 2021; Atasoy and Yesilnacar 2017a; Demir Yetis et al. 2021; Yeşilnacar 2010; Yeşilnacar et al. 2016). These studies include studies to estimate and measure the amount of fluoride in water supply sources and to determine the effects of F^- on dental fluorosis. However, no study was found on the HRA corresponding to seasonal F^- changes.

F^- can be taken into the body not only through water but also with food and/or beverages. It has been observed that tea, which is frequently consumed in countries such as China and Turkey, contains high levels of F^-, and it has been determined that long-term tea consumption with the high F^- drinking water causes dental and skeletal fluorosis in humans. Tea is a F^- accumulator, and its consumption can significantly contribute to total F^- intake (Atasoy et al. 2019). In a study conducted in Şanlıurfa, the fluoride content of teas after brewing varied between 1.010 and 2.89 mg/L. The F^- concentrations evaluated in green tea infusions were lower than those found in black tea infusions. As a result of the data obtained, it has been reported that the consumption of black tea at high rates may cause an increase in the F^- concentration taken into the body and also increase the risk of dental and skeletal fluorosis (Atasoy et al. 2016b, 2017).

The United States Environmental Protection Agency (USEPA) has suggested an exhaustive model for human HRA with extensive application and usage worldwide (Chen et al. 2017; Wu et al. 2019). The model occurs of four steps: risk recognition, dose reaction evaluation, exposure evaluation, and risk description (Guissouma et al. 2017). High levels of F^- in drinking water are one of the most important risk indicators for human health. In risk assessment studies, water with concentrations (≥1.5 mg/L) higher than the F^- concentration levels in the available WHO drinking water-quality guidelines are considered. At the same time, with the help of geographic information system (GIS), an idea about the environmental epidemiology of the region is obtained. This software is an effectively advantageous tool for determining spatial relationships between environmental factors and health impact on humans (Abbasnia et al. 2018). Recent studies have used GIS software for HRA for F^- concentration. The non-carcinogenic HRA is forecasted by considering an exposure grade over a period of time, with a reference dose acquired for a similar exposure time (Karim 2011; Liang et al. 2017). Until now, different methods have been used to determine the non-carcinogenic potential of the HRA of pollutants. These methods appear as HQ and hazard index (HI) for single and combined pollutants, respectively (Abouleish 2016; Zabin et al. 2008). The HRA of F^- in groundwater is based on USEPA guidelines (Yousefi et al. 2018).

19.9 Risk Characterization from Fluoride Exposure

Exposure areas where humans may come into contact with fluoridated water are oral intake, skin absorption, and inhalation. On the other hand, inhalation is excluded in studies due to the lack of toxicological data such as inhalation reference dose and water-to-air transfer efficiency in the database or published literature (Zhang et al. 2017). Similarly, the risk of dermal exposure may occur mostly in industrial settings.

Fluoride health risk studies were performed in child, youth, and adult population categories due to physiological and behavioral differences. In some studies, it is done by adding the newborn category. This classification is made to understand the impacts of distinct exposures of the population and to quantify the health risks united with them. Worldwide, the USEPA method is used to exploit the potential adverse health risks associated with groundwater F^- exposure through direct ingestion and skin contact (USEPA 1989). The exposure to high level of F^- in groundwater was regarded in terms of potential and dermal dose, figured using Eqs. (19.1) and (19.2), respectively.

$$\mathrm{EDI}_{\mathrm{ing}} = \frac{C_{\mathrm{w}} \times \mathrm{IR} \times \mathrm{EF} \times \mathrm{AF} \times \mathrm{ED}}{\mathrm{BW} \times \mathrm{AT}} \tag{19.4}$$

$$EDI_{derm} = \frac{C_w \times SA \times K_p \times F \times ET_s \times EF \times ED \times CF}{BW \times AT} \quad (19.5)$$

where EDI_{ing} and EDI_{derm} are the estimated chronic daily intake of F^- (mg/kg/day) through direct ingestion and skin absorption, respectively. C_w represents the concentration of fluoride in groundwater (mg/L), IR is the groundwater ingestion rate (L/day), EF is the exposure frequency (day/year), AF is the fluoride adsorption factor, ED is the exposure duration in years, BW is the mean body weight in kg, AT represents the average time for non-carcinogenic effect in days, SA is the body surface area in cm^2 exposed to the groundwater, K_p is the dermal permeation constant in cm/h, F is the fraction of surface skin contact with the groundwater (unit-less), ET_s is the exposure time in h/day when bathing, and CF is the groundwater conversion factor (L/cm^3).

The F^- absorption efficiency in the digestive system is close to 100%. Considering the risks that exposed children, AF was taken as 100% (AF = 1). The average daily water consumption rates in adults, teenagers, and children were 1.95, 1.58, and 1.25 L/day, respectively. Body weights of objective groups were averaged as 57.03, 46.25, and 16.68 kg, respectively. The non-carcinogenic effect of fluoride on human health can be calculated as hazard quotient (HQ_{ing}) following Eq. (19.3),

$$HQ_{ing} = \frac{EDI}{RfD} \quad (19.6)$$

where RfD represents the reference dose of F^- and is used in risk assessment of daily exposure. The USEPA recommended oral reference dose for fluoride (RfD_{ing}) is 0.06 mg/kg/day. $HQ_{ing} < 1$ indicates fewer non-carcinogenic effects, while $HQ_{ing} > 1$ shows significant noncancer health hazards and exceeds the acceptable level.

The first step in the fluoride health risk process is to identify the key indicators of fluoride-related risk and possible factors controlling environmental fluoride and fluoride-related disease. The main factors considered in the first theoretical framework are presented in Table 19.2.

Table 19.2 Importance and impact categories of control factors and risk indicators in performing fluoride-related health risk assessment.

Control factor/risk indicator	Significance category		Impact category	
Hydrogeology	Low	Hydrogeology is less important as a risk indicator than knowing the water supply system.	High	Main aquifer units used for direct drinking water supply
			Medium	Small units just connected to drinking water units
			Low	Units that are impermeable or not used for drinking water
F^- concentration in water	High	The F^- concentration in water is the most significant risk indicators.	High	Concentration of F^-: >1.5 mg/L
			Low	Concentration of F^-: 0.5–1.5 mg/L
Fluorosis and health criteria	High	If there is any formation of fluorosis, this is an important indicator of high risk.	High	The incidence of fluorosis
			Low	No incidence of fluorosis
Water type	Low	The main element chemistry of the waters may be a possible indicator of risk.	High	Na+K > Ca containing waters, thermal waters, waters rich in F^- mineral
			Low	Waters with normal to high Ca content
Population density	Low	It is a risk indicator that gives approximate results.	High	High population density
			Medium	Medium population density
			Low	Low population density
F contamination	High	Industrial and agricultural sources increase the environmental F^- concentration and are high-importance risk indicators.	High	Source of F^- with significant impact on the environment
			Medium	Source of F^- that does not affect the environment
			Low	No source of fluoride

(Continued)

Table 19.2 (Continued)

Control factor/risk indicator	Significance category		Impact category	
Tectonic/geological conditions	Low	Tectonic and geological information can be potential risk indicators.	High	Rocks/tectonic zones with high F^- concentrations
			Low	Rocks/tectonic zones with the potential to contain F^-
Water supply	High	The nature and type of water source is considered an important risk indicator.	High	Water sources that do not contain F^- and are used as drinking water
			Medium	Water sources with high or low F^- content that are not used as drinking water
			Low	Water sources are not used for drinking

References

Abbasnia, A., Yousefi, N., Mahvi, A.H. et al. (2018). Evaluation of groundwater quality using water quality index and its suitability for assessing water for drinking and irrigation purposes: case study of Sistan and Baluchistan province (Iran). *Hum. Ecol. Risk Assess. Int. J.* 1–18.

Abouleish, M.Y.Z. (2016). Evaluation of fluoride levels in bottled water and their contribution to health and teeth problems in the United Arab Emirates. *Saudi Dent. J.* 28: 194–202.

Amooey, A.A., Ghasemi, S., Mirsoleimani-azizi, S.M. et al. (2014). Removal of diazinon from aqueous solution by electrocoagulation process using aluminum electrodes. *Korean J. Chem. Eng.* 31: 1016–1020.

Ansari, M.H. and Parsa, J.B. (2016). Removal of nitrate from water by conducting polyaniline via electrically switching ion exchange method in a dual cell reactor: optimizing and modeling. *Sep. Purif. Technol.* 169: 158–170.

Asgari, G., Roshani, B., and Ghanizadeh, G. (2012). The investigation of kinetic and isotherm of fluoride adsorption onto functionalize pumice stone. *J. Hazard. Mater.* 217: 123–132.

Aslani, H., Zarei, M., Taghipour, H. et al. (2019). Monitoring, mapping and health risk assessment of fluoride in drinking water supplies in rural areas of Maku and Poldasht, Iran. *Environ. Geochem. Health*. https://doi.org/10.1007/s10653-019-00282-x.

Ataş, M., Yeşilnacar, M.İ., and Demir Yetiş, A. (2021). Novel machine learning techniques based hybrid models (LR-KNN-ANN and SVM) in prediction of dental fluorosis in groundwater. *Environ. Geochem. Health* 2021: https://doi.org/10.1007/s10653-021-01148-x.

Atasoy, A.D. and Sahin, M.O. (2014). Adsorption of fluoride on the raw and modified cement clay. *CLEAN – Soil, Air, Water* 42 (4): 415–420.

Atasoy, A.D. and Yesilnacar, M.I. (2017a). The hydrogeochemical occurrence of fluoride in groundwater and its effect on human health: a case study from Sanliurfa, Turkey. *Environ. Ecol. Res.* 5 (2): 167–172.

Atasoy, A.D. and Yesilnacar, M.I. (2017b). Assessment of iron oxide and local cement clay as potential fluoride adsorbents. *Environ. Prot. Eng.* 43 (2): 109–118.

Atasoy, A.D., Yesilnacar, M.I., and Celik, A.Z. (2013a). High fluoride level in ground water and fluoride adsorption on activated carbon. *J. Appl. Nat. Sci.* 377–382.

Atasoy, A.D., Yesilnacar, M.I., and Sahin, M.O. (2013b). Removal of fluoride from contaminated ground water using raw and modified bauxite. *Bull. Environ. Contam. Toxicol.* 91 (5): 595–599.

Atasoy, A.D., Yeşilnacar, M.İ., Yazıcı-Karabulut, B., and Özdemir-Şahin, M. (2016a). Fluoride in groundwater and its effects on dental health. *World J. Environ. Res.* 6 (2): 51–57.

Atasoy, A.D., Yesilnacar, M., and Atasoy, A.F. (2016b). Evaluation of fluoride concentration and daily intake by human from tea infusions. *Harran Tarım ve Gıda Bilimleri Dergisi* 20 (1): 1–6.

Atasoy, A.D., Yesilnacar, M., and Atasoy, A.F. (2017). Fluoride intake from fluoride-rich drinking water with black tea and dental fluorosis in school age children in Southeastern Turkey. *4th North and East European Congress on Food – (NEEFood)*, Kaunas, Lithuania. (English abstract).

Atasoy, A.D., Yesilnacar, M.İ., Çiftçi, C. et al. (2018). High fluoride problem in underground waters and fluoride removal by electrocoagulation. *6th International GAP Engineering Conference-GAP2018,* Sanliurfa, Turkey (08–10 November).

Atasoy, A.D., Yesilnacar, M.I., Yildirim, A., and Atasoy, A.F. (2019). Nutritional minerals and heavy metals in tea infusions and daily intake of human body. *Turk. J. Agric. Food Sci. and Tech.* 7 (2): 234–239.

Ayalew, A.A. (2020). Development of kaolin clay as a cost-effective technology for defluoridation of groundwater. *Int. J. Chem. Eng.* https://doi.org/10.1155/2020/8820727.

Bazrafshan, E., Ownagh, K.A., and Mahv, A.H. (2012). Application of electrocoagulation process using iron and aluminum electrodes for fluoride removal from aqueous environment. *E J. Chem.* 9: 2297–2308.

Behbahani, M., Moghaddam, M.A., and Arami, M. (2011). Techno-economical evaluation of fluoride removal by electrocoagulation process: optimization through response surface methodology. *Desalination* 271: 209–218.

Bejaoui, I., Mnif, A., and Hamrouni, B. (2014). Performance of reverse osmosis and nanofiltration in the removal of fluoride from model water and metal packaging industrial effluent. *Sep. Sci. Technol.* 49: 1135–1145.

Bhatnagar, A., Kumar, E., and Sillanpaa, M. (2011). Fluoride removal from water by adsorption – a review. *Chem. Eng. J.* 171 (3): 811.

Chaouch, N. and Khelfaoui, A. (2019). Defluoridation of groundwater in the Southeast of Algeria by adsorption. *Mater. Biomater. Sci.* 2 (1): 014–017.

Chen, J., Wu, H., Qian, H., and Gao, Y.Y. (2017). Assessing nitrate and fluoride contaminants in drinking water and their health risk of rural residents living in a semiarid region of Northwest China. *Expos. Health.* 9 (3): 183–195.

Chubar, N. (2011). New Inorganic (an)ion Exchangers Based on Mg–Al hydrous oxides: (alkoxide – free) sol–gel synthesis and characterization. *J. Colloid Interface Sci.* 357: 198–209.

Daw, R.K. (2004). Experiences with domestic defluoridation in India. *Proceedings of the 30th WEDC International Conference on People-Centred Approaches to Water and Environmental Sanitation*, Loughborough University of Technology, UK, pp. 467–473.

Dehbandi, R., Moore, F., and Keshavarzi, B. (2018). Geochemical sources, hydrogeochemical behavior, and health risk assessment of fluoride in an endemic fluorosis area, Central Iran. *Chemosphere* https://doi.org/10.1016/j.chemosphere.2017.11.021.

Demir Yetis, A., Kahraman, N., Yesilnacar, M.I. et al. (2021). Groundwater quality assessment using GIS based on some pollution indicators over the past 10 years (2005–2015): a case study from Semi-Arid Harran Plain, Turkey. *Water Air Soil Pollut.* 232: 11.

Deng, L., Wang, Y., Zhang, X. et al. (2020). Defluoridation by fluorapatite crystallization in a fluidized bed reactor under alkaline groundwater condition. *J. Clean. Prod.* 272: 122805.

Diawara, C.K., Diop, S.N., Diallo, M.A. et al. (2011). Performance of nanofiltration (NF) and low pressure reverse osmosis (LPRO) membranes in the removal of fluorine and salinity from brackish drinking water. *J. Water Resour. Prot.* 3: 912–917.

Dubey, S., Agarwal, M., and Gupta, A. (2018). Recent developments in defluoridation of drinking water in India. In: *Environmental Pollution* (ed. V. Singh, S. Yadav and R. Yadava), 345–356. Springer.

Emamjomeh, M.M. and Sivakumar, M. (2006). An empirical model for defluoridation by batch monopolar electrocoagulation/flotation (ECF) process. *J. Hazard. Mater.* 131: 118–125.

Gao, S., Cui, J., and Wei, Z. (2009). Study on the fluoride adsorption of various apatite materials in aqueous solution. *J. Fluor. Chem.* 130: 1035.

George, S., Pandit, P., and Gupta, A. (2010). Residual aluminium in water defluoridated using activated alumina adsorption – modeling and simulation studies. *Water Res.* 44: 3055–3064.

Ghorai, S. and Pant, K. (2004). Investigations on the column performance of fluoride adsorption by activated alumina in a fixed-bed. *Chem. Eng. J.* 98: 165–173.

Guissouma, W., Hakami, O., Al-Rajab, A.J., and Tarhouni, J. (2017). Risk assessment of fluoride exposure in drinking water of Tunisia. *Chemosphere* https://doi.org/10.1016/j.chemosphere.2017.03.011.

Hoinkis, J., Valero-Freitag, S., Caporgno, M.P., and Pätzold, C. (2011). Removal of nitrate and fluoride by nanofiltration – a comparative study. *Desalin. Water Treat.* 30: 278–288.

Hosseini, M., Fazelian, N., Fakhri, A. et al. (2019). Preparation, and structural of new NiS–SiO_2 and Cr_2S_3–TiO_2 nano-catalyst: photocatalytic and antimicrobial studies. *J. Photochem. Photobiol. B* 194: 128–134.

Hu, C.Y., Lo, S.L., and Kuan, W.H. (2005). Effects of the molar ratio of hydroxide and fluoride to Al (III) on fluoride removal by coagulation and electrocoagulation. *J. Colloid Interface Sci.* 283: 472–476.

Huang, Y.H., Shih, Y.J., and Chang, C.C. (2011). Adsorption of fluoride by waste iron oxide: the effects of solution pH, major coexisting anions, and adsorbent calcination temperature. *J. Hazard. Mater.* 186 (2–3): 1355.

Islam, M. and Patel, R. (2007). Evaluation of removal efficiency of fluoride from aqueous solution using quick lime. *J. Hazard. Mater.* 143: 303–310.

Jadhav, S.V., Bringas, E., Yadav, G.D. et al. (2015). Arsenic and fluoride contaminated groundwaters: a review of current technologies for contaminants removal. *J. Environ. Manag.* 162: 306–325.

Jagtap, S., Yenkie, M.K., Labhsetwar, N., and Rayalu, S. (2012). Fluoride in drinking water and defluoridation of water. *Chem. Rev.* 112: 2454–2466.

Jorfi, S., Rezaei Kalantary, R., Mohseni Bandpi, A. et al. (2011). Fluoride removal from water by adsorption using bagasse, modified bagasse and chitosan. *Iran. J. Health Environ.* 4: 35–48.

Karaca, H. (2017). Kırıkkale ilinde 13–14 yaş grubu okul çocuklarında dental florozis prevelansının diş çürüğü ve içme suyu flor konsantrasyonları ile ilişkisinin incelenmesi. Doktora tezi. Kırıkkale Üniversitesi. pp.158 (in Turkish).

Karim, Z. (2011). Risk assessment of dissolved trace metals in drinking water of Karachi, Pakistan. *Bull. Environ. Contam. Toxicol.* 86: 676–678.

Khatibikamal, V., Torabian, A., Janpoor, F., and Hoshyaripour, G. (2010). Fluoride removal from industrial wastewater using electrocoagulation and its adsorption kinetics. *J. Hazard. Mater.* 179: 276–280.

Kimambo, V., Bhattacharya, P., Mtalo, F. et al. (2019). Fluoride occurrence in groundwater systems at global scale and status of defluoridation – state of the art. *Groundw. Sustain. Dev.* 9: 100223. https://doi.org/10.1016/j.gsd.2019. 100223.

Kobya, M. and Ulu, E.D.F. (2016). Evaluation of operating parameters with respect to charge loading on the removal efficiency of arsenic from potable water by electrocoagulation. *J. Environ. Chem. Eng.* 4: 1484–1494.

Lacasa, E., Cañizares, P., Sáez, C. et al. (2013). Modelling and cost evaluation of electro-coagulation processes for the removal of anions from water. *Sep. Purif. Technol.* 107: 219–227.

Liang, Y., Yi, X., Dang, Z. et al. (2017). Heavy metal contamination and health risk assessment in the vicinity of a tailing pond in Guangdong, China. *Int. J. Environ. Res. Public Health* 14: 1557.

Liu, R., Zhu, L., Gong, W. et al. (2013). Effects of fluoride on coagulation performance of aluminum chloride towards Kaolin suspension. *Colloids Surf. A Physicochem. Eng. Asp.* 421: 84–90.

Loganathan, P., Vigneswaran, S., Kandasamy, J., and Naidu, R. (2013). Defluoridation of drinking water using adsorption processes. *J. Hazard. Mater.* 248–249: 1–19.

Luo, F. and Inoue, K. (2004). The removal of fluoride ion by using metal (III)-loaded amberlite resins. *Solvent Extr. Ion. Exch.* 22: 305–322.

Malakootian, M., Yousefi, N., and Fatehizadeh, A. (2011). Survey efficiency of electrocoagulation on nitrate removal from aqueous solution. *Int. J. Environ. Sci. Technol.* 8: 107–114.

Meenakshi, S. and Viswanathan, N. (2007). Identification of selective ion-exchange resin for fluoride sorption. *J. Colloid Interface Sci.* 308: 438–450.

Mohapatra, M., Anand, S., Mishra, B.K. et al. (2009). Review of fluoride removal from drinking water. *J. Environ. Manag.* 91: 67–77.

Moussa, D.T., El-Naas, M.H., Nasser, M., and Al-Marri, M.J. (2017). A comprehensive review of electrocoagulation for water treatment: potentials and challenges. *J. Environ. Manag.* 186: 24–41.

Mukherjee, S. and Halder, G. (2018). A review on the sorptive elimination of fluoride from contaminated wastewater. *J. Environ. Chem. Eng.* 6: 1257–1270.

Mukherjee, I., Singh, U.K., and Patra, P.K. (2019). Exploring a multi-exposure-pathway approach to access human health risk associated with groundwater fluoride exposure in the semi-arid region of East India. *Chemosphere* https://doi.org/10.1016/j.chemosphere.2019.05.278.

Mumtaz, N., Pandey, G., and Labhasetwar, P.K. (2015). Global fluoride occurrence, available technologies for fluoride removal and electrolytic defluoridation: a review. *Crit. Rev. Environ. Sci. Technol.* 45: 2357–2389.

Nadiri, A.A., Khatibi, R., and Tsai, F.T.-C. (2019). Chapter 19 – prediction of spatial fluoride concentrations by a hybrid artificial neural network in complex aquifers. In: *GIS and Geostatistical Techniques for Groundwater Science*. Elsevier https://doi.org/10.1016/B978-0-12-815413-7.00019-5.

Oruç, N. (1977). Preliminary Study on The Effect of Water-borne Fluoride on The Fluoride Content of Soils and Plants: in Seminar on Problems of High Fluoride Waters, CENTO Scientific Problems. Report No: 28. Atatürk University, Erzurum, Turkey.

Oruç N. (2001). Türkiye'de Yüksek Düzeyde Florür İçeren Kaynak Suları ve Önemi 1. Türkiye Su Kongresi İstanbul, Turkey (8–10 Ocak 2001), 629–638.

Oruç, N. (2008). Occurrence and problems of high fluoride waters in turkey: an overview. *Environ. Geochem. Health* 30: 315–323.

Owusu-Agyeman, I., Reinwald, M., Jeihanipour, A., and Schäfer, A.I. (2019). Removal of fluoride and natural organic matter from natural tropical brackish waters by nanofiltration/reverse osmosis with varying water chemistry. *Chemosphere* 217: 47–58.

Palko, J.W., Oyarzun, D.I., Ha, B. et al. (2018). Nitrate removal from water using electrostatic regeneration of functionalized adsorbent. *Chem. Eng. J.* 334: 1289–1296.

Poursaberi, T., Hassanisadi, M., Torkestani, K., and Zare, M. (2012). Development of zirconium (IV)-metalloporphyrin grafted Fe_3O_4 nanoparticles for efficient fluoride removal. *Chem. Eng. J.* 189 (2012): 117–125.

Rahmani, A., Nouri, J., Ghadiri, S.K. et al. (2010). Adsorption of fluoride from water by Al^{3+} and Fe^{3+} pretreated natural Iranian zeolites. *Int. J. Environ. Res.* 4: 607–614.

Ruiz-Bevia, F. and Fernandez-Torres, M.J. (2019). Effective catalytic removal of nitrates from drinking water: an unresolved problem? *J. Clean. Prod.* 217: 398–408.

Samatya, S., Yüksel, Ü., Yüksel, M., and Kabay, N. (2007). Removal of fluoride from water by metal ions (Al^{3+}, La^{3+} and ZrO^{2+}) loaded natural zeolite. *Sep. Sci. Technol.* 42: 2033–2047.

Samrat, M.V.N., Rao, K.K., SenGupta, A.K. et al. (2018). Defluoridation of reject water from a reverse osmosis unit and synthetic water using adsorption. *J. Water Proc. Eng.* 23: 327–337.

Sandoval, M.A., Fuentes, R., Nava, J.L., and Rodríguez, I. (2014). Fluoride removal from drinking water by electrocoagulation in a continuous filter press reactor coupled to a flocculator and clarifier. *Sep. Purif. Technol.* 134: 163–170.

Sendil, C. and Baysu, N. (1973). İnsan ve Hayvanlarda Ağrı İli Muradiye İlçesi Köylerinde Saptamamızla İlgili İlk Tebliğ. *Ankara Üniversitesi Veteriner Fakültesi Dergisi* 20: 4.

Shen, J. and Schäfer, A. (2014). Removal of fluoride and uranium by nanofiltration and reverse osmosis: a review. *Chemosphere* 117: 679–691.

Sinha, R., Khazanchi, I., and Mathur, S. (2012). Fluoride removal by a continuous flow electrocoagulation reactor from groundwater of Shivdaspura. *Int. J. Eng. Res. Ind. Appl.* 2: 1336–1341.

Sudasinghe, M., Jayaweera, M., Gunawardana, B. et al. (2019). Defluoridation of calcium-rich groundwater using iron oxide nanoparticles. *Water Pract. Technol.* 14 (3): 665–681.

Sujana, M.G., Pradhan, H.K., and Anand, S. (2009). Studies on sorption of some geomaterials for fluoride removal from aqueous solutions. *J. Hazard. Mater.* 161: 120–125.

Tahaikt, M., Habbani, R.E., Haddou, A.I. et al. (2007). Fluoride removal from groundwater by nanofiltration. *Desalination* 212: 46–53.

Tomar, V., Prasad, S., and Kumar, D. (2014). Adsorptive removal of fluoride from aqueous media using citrus limonum (lemon) leaf. *Microchem. J.* 112: 97–103.

Umar, R. and Alam, F. (2012). Assessment of hydrogeochemical characteristics of groundwater in parts of Hindon-yamuna interfluves region, Baghpat district, Western Uttar Pradesh. *Environ. Monit. Assess.* 184 (4): https://doi.org/10.1007/s10661-011-2120-9.

USEPA (1989). *Risk Assessment Guidance for Superfund Volume I: Human Health Evaluation Manual (Part A).* Washington, DC: U.S. Environmental Protection Agency.

USEPA (1992). Guidelines for exposure assessment. *Fed. Regist.* 57: 22887–22938.

World Health Organization (1994). *Fluorides and Oral Health, Technical Report Series No: 846.* Geneva: World Health Organization.

Wu, J., Zhou, H., He, S., and Zhang, Y. (2019). Comprehensive understanding of groundwater quality for domestic and agricultural purposes in terms of health risks in a coal mine area of the Ordos basin, north of the Chinese Loess Plateau. *Environ. Earth Sci.* 78 (15): 446. https://doi.org/10.1007/s1266 5-019-8471-1.

Yadav, A.K., Abbassi, R., Gupta, A., and Dadashzadeh, M. (2013). Removal of fluoride from aqueous solution and groundwater by wheat straw, sawdust and activated bagasse carbon of sugarcane. *Ecol. Eng.* 52: 211–218.

Yadav, K.K., Kumar, S., Pham, Q.B. et al. (2019). Fluoride contamination, health problems and remediation methods in Asian groundwater: a comprehensive review. *Ecotoxicol. Environ. Saf.* 182: 109362.

Yapo, N.Z.S., Briton, B.G.H., Aw, S. et al. (2021). Bivalve shells (Corbula trigona) as a new adsorbent for the defluoridation of groundwater by adsorption-precipitation. *J. Environ. Sci. Health A* 1–11.

Yazici-Karabulut, B. and Atasoy, A.D. (2019a). Comparison of electrocoagulation process and other treatment technologies in fluoride removal from groundwater. *Academic Perspective Procedia* https://doi.org/10.33793/acperpro.

Yazici-Karabulut, B. and Atasoy, A.D. (2019b). Removal of fluoride from groundwater by batch electrocoagulation process using Al plate electrodes. *Academic Perspective Procedia* https://doi.org/10.33793/acperpro.

Yazici-Karabulut, B., Atasoy, A.D., and Yesilnacar, M.I. (2019). Fluoride contamination in groundwater and its removal by adsorption and electrocoagulation processes. *Sigma J. Eng. Nat. Sci.* 37 (4): 1247–1258.

Yeşilnacar, M.İ. (2010). *Investigation of Problems in Teeth of Children with Suspected High Fluoride in Drinking and Utility Water of Karataş and Sarım Villages*, 9. Sanliurfa: Sanliurfa Provincial Health Directorate (in Turkish).

Yeşilnacar, M.İ., Demir-Yetiş, A., Dülgergil, Ç.T. et al. (2016). Geomedical assessment of an area having high-fluoride groundwater in Southeastern Turkey. *Environ. Earth Sci.* 75: 162.

Yousefi, M., Ghoochani, M., and Mahvi, A.H. (2018). Health risk assessment to fluoride in drinking water of rural residents living in the Poldasht city, Northwest of Iran. *Ecotoxicol. Environ. Saf.* 148: 426–430.

Zabin, S.A., Foaad, M., and Al-Ghamdi, A.Y. (2008). Non-carcinogenic risk assessment of heavy metals and fluoride in some water wells in the Al-Baha region, Saudi Arabia. *Hum. Ecol. Risk. Assess.* 14: 1306–1317.

Zhang, L., Huang, D.Z., Yang, J. et al. (2017). Probabilistic risk assessment of Chinese residents' exposure to fluoride in improved drinking water in endemic fluorosis areas. *Environ. Pollut.* 222: 118–125.

Zhu, J., Zhao, H., and Ni, J. (2007). Fluoride distribution in electrocoagulation defluoridation process. *Sep. Purif. Technol.* 56: 184–191.

20

Pharmacology, Toxicology, and Therapeutic Effects of Metals and Minerals Used in Traditional Medicine

Pathirage Kamal Perera, Jeevani Maheshika Dahanayake, and Jalathge Isurika Dilanthi Diddeniya

Institute of Indigenous Medicine, University of Colombo, Colombo, Sri Lanka

20.1 Introduction

Traditional medicine (TM) is being practiced worldwide as a primary healthcare with a massive popularity in the modern society. It is attaining the prominence among other modes of medicine due to its efficacy in the management of various acute and chronic diseases (Samuel et al. 2015). TM refers to the sum of knowledge, practices, and skills based on the beliefs, theories, and experiences of different cultures, which are used to preserve health as well as to prevent, diagnose, and treat the illnesses (Che et al. 2017). The medicinal preparations or the method of treatments in TM were popular due to its safety, efficacy, and the easy availability of the ingredients where the medicinal preparations in TM have been formed, based on the locally available resources most abundantly. It has been transmitted from generation to generation by means of oral forms over a vast period. In TM, rather than the medicinal plants, metals and minerals have been also taken a crucial attention in health care delivery system. The usage of metals and minerals in TM is in practice since ancient times, and it is still in use at present for the treatment of multiple diseases and disorders. Therefore, in traditional medical system, herbal preparations as well as herbo-mineral preparations can be found, which mainly consist of metals and minerals and among them herbo-mineral preparations attract much attention due to quick response with minute doses. Traditional medical systems such as ayurveda, traditional chinese medicine (TCM), Tibetan medicine, and Sri Lankan traditional medicine make use of metals and minerals containing drug preparations in the management of various diseases and disorders since ancient times and still they are in use (Pandey et al. 2013). The reason is such preparations have the ability of giving relief to diseases within a short time period and a subtle amount is needed where such drug preparations can be used for the treatment of chronic diseases too.

Even though metal- and mineral-containing drug preparations are used by traditional medical practitioners for multiple disease conditions, there is a need of evaluating their pharmacological actions, toxicology, and therapeutic uses as there are no much research articles which have been conducted regarding it. And the attention of the society is considerably high as entire satisfaction can be achieved by using metals and minerals containing traditional medical preparations in the treatment of diseases. At present, diseases such as diabetes, hypertension, bronchial asthma, arthritis, and stroke can be seen in almost all the hospitals, and the accurate treatment for most of them has become a challenge by now.

Therefore, this research was conducted with the aim of searching the pharmacological actions, toxicological effects, and therapeutic uses of metals and minerals containing traditional medicinal preparations, which are used in different countries by using relevant authentic texts and published research articles.

It will be beneficial to the society in getting a clear idea about the usage of such preparations and also it would be helpful in removing myths from the society about using such minerals and metals containing drug preparations, which leads for increasing their usage in the treatment of diseases.

20.2 Objectives

Objective of this chapter was to find out the metals and minerals used in the TM to treat diseases and evaluate their pharmacology, toxicology, and therapeutic uses.

Medical Geology: En route to One Health, First Edition. Edited by Majeti Narasimha Vara Prasad and Meththika Vithanage.

20.3 Methods

The research was conducted by using the authentic texts such as ayurveda pharmacopoeia, *Vatika Prakaranaya* and texts related to alchemy (*Rasa Shastra*) such as *Rasa Rathna Samuchchcaya, Rasa Jala Nidhi, Rasa Tharangani,* and *Rasendra Chudamani*. Also, the data were gathered by using the published research articles, which report about the metals and minerals used in TM for the treatment of diseases. A comprehensive review was conducted by using the databases: PubMed and Science Direct for studies published from 1 January 2000 to 1 January 2022.

20.4 Results

By analyzing data, it was revealed that metals and minerals are being used in TM by means of external as well as internal applications. Among them minerals such as mica, realgar, pyrite, and orpiment and metals such as gold, copper, iron, and mercury are very commonly used in the management of various diseases.

20.4.1 Pharmacological Effects of Metals and Minerals

Metals and minerals were shown to possess different pharmacological actions that are capable of giving relevant therapeutic advantages accordingly. More or less amounts of them have been used in the preparation of drug formulations, and they act on multiple systems of human body (Table 20.1).

20.4.2 Toxicological Effects of Metals, Minerals, and Drug Preparations

20.4.2.1 Toxicological Effects of Cinnabar

The sedative effects of Cinnabar (HgS), which a naturally occurring mercuric sulfide, were identified by Huang et al. (2017) by using male and female mice. It was also revealed the neurotoxic effects of it which is being used as a Chinese herbal medicine over 2000 years.

Wei et al. (2008) investigated the toxicological effects of Cinnabar by using NMR-based metabolics of serum and urine in Wistar rats. It was carried out by means of

Table 20.1 Pharmacological actions of metals and minerals used in TM.

Metals/minerals	Chemical name	*Rasa* (taste)	*Guna* (properties)	*Virya* (potency)	*Vipaka* (post digestive effect)	*Karma* (actions)
Parada (mercury)	Hg	*Shad rasa* (all six tastes)	*Snigdha* (unctuous), Sara (mobile), Guru (heavy)	*Ushna* (hot)	*Madhura* (sweet)	*Rasayana* (rejuvenate), *Balya* (strengthening body) [*Prabhava* (specific potency)-*Yogavahi* (increases drug action)]
			Maha rasa			
Abhra	Mica	*Kashaya* (astringent), *Madhura* (sweet)	*Snigdha* (unctuous)	*Sheeta* (cold)	*Madhura* (sweet)	*Dipana* (digestive stimulant), *Balya* (strengthening body), *Vrushya* (increasing body bulk)
Vaikranta	CaF_2/SiO_2	*Shad rasa* (all six tastes)	*Guru* (heavy)	*Sheeta* (cold)	*Madhura* (sweet)	*Rasayana*, (rejuvenate), *Balya* (strengthening body)
Swarna makshika	$CuFeS_2$	*Madhura* (sweet), *Tikta* (bitter), *Kashaya* (astringent)	*Laghu* (light)	*Sheeta* (cold)	*Katu* (pungent)	*Rasayana* (rejuvenate), *Yogavahi* (increases drug action, *Kapha/Pitta kshaya* (reduces *Kapha and Pitta doshas* in body)
Vimala	FeS_2	*Katu* (pungent), *Tikta* (bitter)	*Snigdha* (unctuous), Guru (heavy)	*Ushna* (hot)		*Vata/Pitta hara* (reduces *Vata and Pitta doshas* in body), *Rasayana* (rejuvenate)
Shilajathu		*Madhura* (sweet), *Tikta* (bitter), *Katu* (pungent), *Lawana* (salty)		*Sheeta* (cold), *(Rajata-ushna)*		*Rasayana* (rejuvenate), *Pramehaghna* (antidiabetic), *Lekhana* (scraping action)

Table 20.1 (Continued)

Metals/minerals	Chemical name	*Rasa* (taste)	*Guna* (properties)	*Virya* (potency)	*Vipaka* (post digestive effect)	*Karma* (actions)
Sasyaka	Cu_2SO_4	*Kashaya* (astringent)	*Laghu* (light)	*Ushna* (hot)	*Katu* (pungent)	*Lekhana* (scraping action), *bhedana* (purgative), *Kapha/Pitta hara* (removes *Kapha and Pitta doshas* in body)
Chapala	Bi	*Tikta* (bitter), *Madhura* (sweet)	*Snigdha* (unctuous), *Guru* (heavy)	*Ushna* (hot)		*Lekhana* (scraping action), *Yogavahi* (increases drug action), *Vrushya* (increasing body bulk)
Rasaka	Zinc blende	*Katu* (pungent), *Kashaya* (astringent)	*Laghu* (light)	*Sheeta* (cold)		*Kapha/Pitta hara* (removes *Kapha* and *Pitta doshas* in body), *Chakshushya* (enhance vision)
			Upa rasa			
Gandhaka	S	*Katu* (pungent)	*Sara* (mobile)	*Ushana* (hot)	*Madhura* (sweet)	*Rasayana* (rejuvenate), *Yogavahi* (increases drug action)
Gairika	Fe_2O_3 Haematite	*Madhura* (sweet), *Kashaya* (astringent)	*Snigdha* (unctuous)	*Sheeta* (cold)		*Rathpith shamaka* (alleviates *Rakta* and *Pitta doshas* in the body)
Kasisa	$FeSO_4$	*Kashaya* (astringent), *Amla* (sour)	*Guru* (heavy)	*Ushna* (hot)		*Visha hara* (destroy poisons), *Shwitra hara* (removes Leucoderma), *Netrya* (good for vision), *Keshya* (good for hair)
Kankshi	Alum	*Kashaya* (astringent), *Katu* (pungent), *Amla* (sour)	*Snigdha* (unctuous)			*Vruna ropana* (cures wounds), *Keshya* (good for hair)
Haritala	AS_2S_3	*Katu* (pungent)	*Snigdha* (unctuous), *Guru* (heavy)	*Ushna* (hot)		*Kapha hara* (removes *Kapha dosha* in body)
Manoshila	AS_2S_2	*Katu* (pungent), *Tikta* (bitter)	*Snigdha* (unctuous), *Guru* (heavy), *Sara* (mobile)	*Ushna* (hot)		*Rasayana* (rejuvenate), *Lekhana* (scraping action), *Kapha/vata hara* (removes *Kapha* and *Vata doshas* in body)
Anjana (souveeranjana)	Antimony compounds	*Kashaya* (astringent), *Tikta* (bitter)	*Snigdha* (unctuous)	*Sheeta* (cold)	*Madhura* (sweet)	*Rakta Pitta hara* (removes *Rakta* and *Pitta doshas* in body), *Netrya* (enhances vision), *Lekhana* (scraping action)
Kankushta		*Tikta* (bitter), *Katu* (pungent)		*Ushna* (hot)		*Athirechana* (excessive Purgation)
			Sadharana rasa			
Kampillaka	*Mallotus philippensis*	*Katu* (pungent)	*Laghu* (light), *Rooksha* (un-unctuous), *Thikshna* (penetrates deep)	*Ushna* (hot)	*Katu*	*Virechana* (purgative), *Kapha/Pitta hara* (removes *Kapha* and *Pitta doshas* in body)
Gauri pashana	As_2O_3	*Tikta* (bitter)	*Snigdha* (unctuous), *Ruksha* (ununctuous)			*Thridoshaghna* (removes all three *doshas* in body)

(*Continued*)

Table 20.1 (Continued)

Metals/minerals	**Chemical name**	***Rasa* (taste)**	***Guna* (properties)**	***Virya* (potency)**	***Vipaka* (post digestive effect)**	***Karma* (actions)**
Navasadara	NH_4Cl	*Amla* (sour), *Lawana* (salty)	*Hima* (cold)	*Sheeta* (cold)		*Deepana* (digestive stimulant), *Thrioshaghna* (removes all three *doshas* in body)
Kaparda	Cowries	*Katu* (pungent)	*Ushna* (hot)	*Ushna* (hot)	*Katu*	*Dipana, pachana* (digestive stimulant)
Vahnijara	Amber	*Katu* (pungent)	*Laghu* (light)	*Ushna* (hot)		*Thridoshaghna* (removes all three *doshas* in body), *Deepana* (digestive stimulant)
Giri sindura	Red oxide of mercury			*Ushna* (hot)		*Thridosha shamaka* (pacifies all three *doshas* in body)
Hingula	HgS	*Tikta* (bitter), *Katu* (pungent), *Kashaya* (astringent)	*Ushna* (hot)	*Ushna* (hot)		*Deepana* (digestive stimulant), *Balya* (strengthening body), *Rasayana* (rejuvenate), *Vajikarana* (aphrodisiac)
Mruddadara shrunga	PbO		*Guru* (heavy)	*Sheeta* (cold)		*Kaph/vata hara* (removes *Kapha* and *Vata doshas* in body)
			Loha varga			
Swarna	Au	*Madhura* (sweet), *Kashaya* (astringent), *Tikta* (bitter), *Katu* (pungent)	*Snigdha* (unctuous), *Guru* (heavy)	*Sheeta* (cold)	*Madhura* (sweet)	*Medhya* (brain tonic), *Rasayana* (rejuvenate), *Varnya* (increase body complexion), *Netrya* (good for vision)
Rajatha/Thaara	Ag	*Madhura* (sweet), *Kashaya* (astringent), *Amla* (sour)	*Sara* (mobile), *Lekhana* (scraping), *Snigdha* (unctuous)	*Sheeta* (cold)	*Madhura* (sweet)	*Dipana* (digestive stimulant), *Balya* (strengthening the body), *Rasayana* (rejuvenate)
Aara	Brass	*Tikta* (bitter)	*Rooksha* (un-unctuous)	*Ushna* (hot)		*Kapha/pitta hara* (removes *Kapha* and *Pitta doshas* in body)
Thamra	Cu	*Tikta* (bitter), *Kashaya* (astrigent)	*Pichchila* (slimy)	*Ushna* (hot)	*Madhura* (sweet)	*Dipana* (digestive stimulant)
Naga	Pb	*Tikta* (bitter)	*Snigdha* (unctuous)	*Ushna* (hot)		*Vata/kapha hara* (removes *Vata* and *Kapha doshas* in body)
Vanga	Sn	*Tikta* (bitter)	*Rooksha* (un-unctuous)	*Ushna* (hot)		*Prameha hara* (antidiabetic)
Yashada	Zn	*Tikta* (bitter), *Kashaya* (astringent), *Katu* (pungent)		*Sheeta* (cold)		*Balya* (strengthening the body), *kapha/pitta hara* (removes *Kapha* and *Pitta doshas* in body)
Thikshna/ Lauha	Fe	*Tikta* (bitter)	*Rooksha* (un-unctuous)	*Ushna* (hot)		*Kapha/ vata hara* (removes *Kapha* and *Vata doshas* in body), *Deepana* (digestive stimulant)
			Ratna varga			
Manikya	Al_2O_3	*Madhura* (sweet)	*Snigdhda* (unctuous)	*Sheeta* (cold)	*Madhura* (sweet)	*Vata/pitta hara* (removes *Vata* and *Pitta doshas* in body), *balya* (strengthening the body)
Muktha phala	$CaCO_3$	*Madhura* (sweet)	*Laghu* (light), *Snigdha* (unctuous)	*Sheeta* (cold)	*Madhura* (sweet)	*Brumhana* (increasing body bulk), *Medhya* (brain tonic)

Table 20.1 (Continued)

Metals/minerals	Chemical name	*Rasa* (taste)	*Guna* (properties)	*Virya* (potency)	*Vipaka* (post digestive effect)	*Karma* (actions)
Vidhruma/ Pravala	Corals	*Kashaya* (astringent), *Madhura* (sweet), *Amla* (sour)	*Laghu* (light), *Snigdha* (unctuous)	*Sheeta* (cold)		*Deepana – pachana* (digestve stimulant), *Chakshushya* (good for eyes)
Tharkshya/ Marakatha	Be_2O_3	*Madhura* (sweet)	*Snigdha* (unctuous)	*Sheeta* (cold)		*Ojo wardhana* (enhances immunity), *Vrushya* (nourishes body)
			Pushparaga			
Vajra	Diamond	*Shad rasa* (all six tastes)	*Snigdha* (unctuous)	*Sheeta* (cold)		*Hrdya* (Good for heart), *Yagavahi* (increases drug action), *Rasayana* (rejuvenate)
Nila	Al_2O_3					*Balya* (strengthening the body), *Vrushya* (nourishes body), *Varnya* (increase body complexion)
Gomedha	$ZrSiO_4$	*Amla* (sour)		*Ushna* (hot)		*Deepana – Pachana* (digestve stimulant), *Vata/kapha hara* (removes *Vata* and *Kapha doshas* in body)
Vaidhurya	$BeAl_2O_4$	*Madhura* (sweet)	*Snigdha* (unctuous)	*Sheeta* (cold)		*Medhya* (brain tonic), *Pitta hara* (removes Pitta dosha in body), *Balya* (strengthening the body)
			Sudha varga			
Shambuka/ Shankha	Conch shell	*Katu* (pungent), *Kashaya* (Astringent)	*Thikshana* (penetrate deep), *Khara* (Rough)	*Sheeta* (cold)	*Madhura* (sweet)	*Dipana – pachana* (digestve stimulant), . . . *Grahi* (moisture absorbant)
Khatika		*Madhura* (sweet), *Tikta* (bitter)	*Sheeta* (cold)	*Sheeta* (cold)		*Grahi* (moisture absorbant) *Daha hara* (relieves from burning sensation)
Samudra phena	Cuttle fish bone			*Sheeta* (cold)		*Lekhana* (scraping), *Dipana pachana* (digestve stimulant)
Shutki	Pearl oyster shell	*Madhura* (sweet)	*Snigdha* (unctuous)	*Sheeta* (cold)	*Madhura* (sweet)	*Dipana pachana* (digestve stimulant)
Godanthi	Gypsum			*Sheeta* (cold)		*Balya* (strengthening the body), *Dipana* (digestve stimulant)
Kukkutanda twak	Hen egg shell					*Rasayana* (rejuvenate), *Balya* (strengthening the body)
			***Sikata varga* (silicon blends)**			
Dugdha pashana	Talc			*Sheeta* (cold)		*Grahi* (moisture absorbant), *Shonita stapaka Dantya* (good for teeth)
Badarashma	Silicate of lime					*Pitta shamaka* (pacifies Pitta dosha in body), *Vamana nashaka* (alleviates vomiting), *Mutrala* (diuretic)

Source: Adapted from Mookerjee (1984); Mishra (2009, 2011); Angadi (2016).

urinalysis and serum analysis which showed that there was Cinnabar-induced disturbance in amino acid as well as energy metabolism, gut environment, and liver and kidney damage.

The toxicity of *Rasa Sindura*, an Ayurveda drug which contains more than 99% of HgS, has shown nontoxic to Zebra fish models when compared to commercial HgS and Cinnabar ore in the experimental studies that were conducted by Snehasis and Jayesh (2020).

Huang et al. (2008) explored the toxic effects of Cinnabar (HgS) on auditory brainstem response (ABR) by oral administration of it in mice at a dose of 10mg/kg/day. Its long-term exposure exhibits the ototoxicity.

20.4.2.2 Toxicological Effects of Mercury

The accumulation of Hg and toxicity in weanling and aged rats was examined by Zhang et al. (2017) and stated that aged rats are more susceptible for toxicity and Hg accumulation in tissues when compared to weanling rats.

In vivo exposure of female rats to mercury (Hg) on uterine myometrial activity was determined by Swati et al. (2020) by orally exposing the female Wistar rats to Hg in drinking water for 28 days. Their findings suggested that Hg at higher levels did not alter the myometrial response.

Cen et al. (2018) revealed the nephrotoxicity of Hg from Zuotai, a Tibetan famous medical mixture by orally treating the mice with Zuotai, β HgS, or $HgCl_2$.

Sarah et al. (2017) investigated the toxic effects of mercury used in the Tibetan Medicine by conducting a cross sectional study among patients who were taking Tibetan medicine containing mercury. The results emphasized that there was no adverse effects of such preparations, and therefore, it can be used in a beneficial manner in the treatment of diseases.

20.4.2.3 Toxiological Effects of Arogya Vardhana Vati

The safety evaluation of *Arogya Vardhini Vati* that contains the minerals such as mercury (Hg), *Abhra* (Mica), sulfur (S), iron (Fe), and copper (Cu) were conducted by Gajendra et al. (2012) by using rats on their liver, brain, and kidney. The results revealed that the exposure of this drug for 28 days in rats by using the dose equivalent to 10 times as in human did not have toxicological effects on liver, brain, and kidney.

20.4.2.4 Toxicological Effects of Rasa Manikya

Bhuban et al. (2019) assessed the anticancer and antimicrobial potentials of *Rasa Manikya* nanoparticles, which is a commonly used herbo-metallic preparation in Ayurveda by using in vitro cellular assay. Furthermore, they stated that it can be used in treating breast cancer and drug-resistant microbial strains.

20.4.2.5 Toxicological Effects of *Sameerpannag Ras*

Sameerpannag Ras, which is a mineral and metallic-based drug used in Ayurveda, was investigated by Priya et al. (2017) to find its neuroprotective effects. It was conducted by oral administration of the drug at a dose of 40 mg/kg body weight using mice models.

20.4.2.6 Toxicological Effects of Bhasmas

The analysis of toxicity studies of *bhasmas* such as *Yashada* (zinc), *Lauha* (iron), and *Tamra* (copper) on experimented animals have shown that there were no serious effects on body function by Prajapati et al. (2006).

Prabhakara et al. (2003) examined the effects of *Praval bhasma* on bone mineralization by inducing the ovariectomy and calcium deficiency in 24 female rats. Their results indicated that *Praval bhasma* is useful in the management of bone loss due to calcium and estrogen so that it can be used in the treatment of osteoporosis where *Praval bhasma* gives inhibitory effects on bone loss.

20.4.2.7 Toxicological Effects of Chondrokola Rosh

Chondrokola rosh, a metallic Ayurveda preparation which contains the metals such as mercury, copper, and minerals such as sulfur, mica was analyzed for toxicological parameters on liver functions by Nasrin et al. (2011). The study was carried out by using both male and female mice, and a dose of 100 mg/kg body weight was administered orally, which finally revealed the fact that there was no remarkable toxic effect on liver.

In vivo alkaline single cell gel electrophoresis was used by Saleha et al. (2004) in mice by oral administration of $CuSO_4$ to reveal the DNA damage caused by breakage of leucocytes in them.

20.4.2.8 Toxicological Effects of Arsenicals

The findings from histopathological examination on the liver and kidney of female Sprague-Dawley rats by Ujang et al. (2016) stated that there were no any adverse effects from Traditional Chinese Medicine (TCM) as there was a low level of As and Hg in the tissues of rats subjected to the test.

The toxicology evaluation of a commonly used TCM, named *Niu-Huang-Jie-Du Pian* (NHJD) which contains arsenic (As), was subjected to analyze the toxicity compared to other arsenicals, and the results revealed that there was a more severe liver and kidney toxicity when compared to other arsenicals by Jia et al. (2011).

20.4.3 Therapeutic Uses of Metals and Minerals

Following table included the therapeutic uses of metals and minerals according to the traditional medicine practices (Table 20.2) (Department of Ayurveda 1994; Waidy and Waidyaratna 1927).

20.4.4 Purification, Detoxification, and Incineration Procedures of Commonly Used Metals and Minerals

The metals and minerals that are used in herbo-mineral preparations of TM have to be subjected to purification, detoxification, or incineration procedures in order to

Table 20.2 Therapeutic uses of metals and minerals.

Metals/minerals	Chemical name	Therapeutic uses
Parada (Mercury)	Hg	Ulcers, Syphilis, Gonorrhea, Edema, Skin diseases, Goiter, Eye diseases
Abhra	Mica	Insanity, Epilepsy, *Vata vyadhi* (diseases occur due to vitiated Vata dosha), *Grahani* (irritable bowel syndrome), Gastritis, Heart diseases, *Rakta Pitta* (beeding from body orifices), Piles, Skin diseases, Edema, Cysts, Cough, Asthma, Impotency, Chronic fever, Diabetes, Tuberculosis, Anemia, Eye diseases, Jaundice, Urinary calculi
Vaikranta	CaF_2/SiO_2	Fever, Skin diseases, Anemia, Insanity, Diabetes, Cough, Asthma, Poisonous conditions, Emaciation
Swarna makshika	$CuFeS_2$	Alcoholic intoxication, Giddiness, Headache, Insomnia, Eye diseases, Heart diseases, *Rakta pitta,* Anorexia, *Agnimandya* (a disease occur due to low digestive power), Piles, Epilepsy, *Visuchika* (a disease characterized by frequent vomiting and diarrhea), Chronic fever, Diabetes, Urinary disorders, Skin diseases, Sperm disorders, Anemia, Jaundice, Poisonous conditions, Cough, Edema, *Rakta pitta*
Vimala	FeS_2	Skin diseases, Anemia, Jaundice, Piles, Edema, Diabetes, *Grahani,* Tuberculosis
Shilajathu		Edema, Insanity, Epilepsy, Rheumatoid arthritis, Osteo arthritis, Ascites, *Gulma* (Abdominal tumor), Piles, Jaundice, Worm infestations, Heart diseases, Gonorrhea, Syphilis, *Vata rakta* (Gout), Filariasis, Cough, Asthma, Leucorrhea, Skin diseases, Diabetes, Urinary calculi, Dysuria, Chronic fever, Emaciation, Obesity, Jaundice
Sasyaka	Cu_2SO_4	Eye diseases, Skin diseases, Wounds, Parasitic infestations, Bleeding disorders
Gandhaka	S	Skin diseases, Erysipelas, *Agnimandya,* Indigestion, Piles, Fistula, Liver diseases, Parasitic infestations, Osteo arthritis, Rheumatoid arthritis, Tuberculosis, Cough, Phlegmatic conditions
Gairika	Fe_2O_3 (Haematite)	Fever, *Rakta pitta,* Menorrhagia, Piles, Eye diseases, Vomiting, Hiccup, Burning sensation, Erysipelas, Hiccup, Ulcers
Kasisa	$FeSO_4$	Agnimandya, Grahani, Gulma, Pandu, Liver diseases, Splenic diseases, Dysmenorrhea, Piles, Prolapse of rectum, Hiccup, Erysipelas, Lucoderma
Kankshi	Alum	Ulcers, Eye diseases, Leucorrhea, Ulcers of mouth, Asthma, Cough, Indigestion, Dysuria, Itching sensation in vagina and rectum
Haritala	AS_2S_3	Skin diseases, Fever, Edema, Asthma, Cough, Fistula, Ulcers, *Kunakha, Vata rakta, Agnimandya,* Splenic disorders, *Vata vyadhi, Gulma,* Skin diseases, Emaciation, Epilepsy, Erysipelas, Piles, Anemia, Diabetes, Common cold, Scorpion bites, Ascites
Manoshila	AS_2S_2	Cough, Asthma, Skin diseases, Fever, Poisonous conditions
Anjana (souveeranjana)	Antimony compounds	Eye diseases, Vomiting, Leucoderma, Wounds, *Rakta pitta,* Give relief in poisonous conditions
Gauri pashana	As_2O_3	Tumors, Cysts, Swellings, Osteo arthritis, Rheumatoid arthritis, Parkinsonism, Agnimandya, Vomiting, Diarrhea, Splenomegaly, Heart diseases, Filariasis, Asthma, Tuberculosis, Anemia, Fever, Skin diseases
Navasadara	NH_4Cl	*Agnimandya,* Indigestion, Splenomegaly, Liver diseases, Ascites, Fever, *Gulma, Adhmana* (Abdominal distension), Heart diseases, Leucoderma, Scorpion bites

(Continued)

Table 20.2 (Continued)

Metals/minerals	Chemical name	Therapeutic uses
Giri sindura	Red oxide of mercury	Skin diseases, Poisonous conditions, Ulcers
Hingula	HgS	Skin diseases, Fever, Eye diseases, Jaundice, Rheumatoid arthritis, Poisonous conditions, Diarrhea
Mruddadara shrunga	PbO	Ulcers, Skin diseases, Parasitic infestations, Premature gray hair, Ascites
Swarna	Au	Insanity, Memory impairment, Heart diseases, *Vata roga*, Diabetes, Impotency, Fever, Consumption, Tuberculosis, Poisonous conditions, Diarrhea, *Grahani*, Asthma, Gastritis, Hiccup, Kidney diseases, Bone diseases, Uterine disorders
Thaara	Ag	Uterine disorders, Sciatica, Backache, Burning sensation, Diabetes, Psychosis, Emaciation, Alcoholic intoxication, Abdominal distension
Thamra	Cu	Cysts, Edema, Tumors, Hepatomegaly, Tumors, Ascites, Asthma, Gastritis, Poisonous conditions, Anemia, *Grahani, Visuchika*, Piles, Vomiting, Cough, Paralysis, Convulsions, Skin diseases, *Suthika roga* (Post-partum diseases), Eye diseases
Naga	Pb	Ulcers, Wounds from Burns
Vanga	Sn	Diabetes, Fever, *Vata roga, Agnimandya*, vomiting, Worm infestations, Muscular tumors, Skin diseases, Anemia, Cough, Dysmenorrhea, Vomiting, Urinary disorders, *Gulma*, Indigestion, Bad odor of mouth, Prostate gland enlargement, Chronic ulcers
Thikshna	Fe	Anemia, Jaundice, Gastritis, Asthma, Cough, Skin diseases, Dysuria, Fever, Insanity, Rheumatoid arthritis, Epilepsy, *Vata roga*, Diarrhea, *Grahani, Gulma*, Liver and splenic disorders, Hair loss, Grey hair, Phlegmatic conditions, *Rakta pitta*
Manikya	Al_2O_3	Eye diseases, Heart diseases, Emaciation, Psychosis, *Rakta pitta*
Muktha phala	$CaCO_3$	Heart diseases, Eye diseases, Fever, Burning sensation, Poisonous conditions, *Rakta pitta*, Edema, Liver diseases, Emaciation, Dysuria, *Agnimandya, Grahani*, Piles, Menorrhagia, Leucoderma, Tuberculosis, Cough, Asthma, Burning sensation, Bone fractures, Memory impairment, Insanity, Epilepsy, Headache, Parkinsonism
Vidhruma/pravala	Corals	Eye diseases, *Rakta pitta*, Heart diseases, Gastritis, Diarrhea, Piles, Intestinal ulcers, Dysuria, Fever, Emaciation, Poisonous conditions
Tharkshya/Marakatha	Be_2O_3	Fever, Piles, Anemia, Asthma, Gastritis, Edema, Poisonous conditions
Pushparaga		Piles, Skin diseases, Poisonous conditions, Burning sensation
Vajra	Diamond	Heart diseases, Diabetes, Anemia, Impotency, Emaciation, Tuberculosis
Nila	Al_2O_3	Heart diseases, Consumption, Asthma, Cough, Piles, Skin diseases, Fever, Poisonous conditions
Vaidhurya	$BeAl_2O_4$	Eye diseases, *Rakta pitta*, Tuberculosis, Memory impairment
Shutki	Pearl oyster shell	*Grahani*, Heart diseases, Dysuria, Emaciation, Fever, burning sensation, *Agnimandya, Rakta pitta*, Edema, Liver diseases, Piles, Menorrhagia, Leucoderma, Epilepsy, Cough, Asthma, Burning sensation, Memory impairment, Headache, Insanity, Parkinsonism, Eye diseases, Poisonous conditions, Bone fractures

get rid of toxic effects, so that they become less or non-toxic when making use of them for medicinal purposes. The metals and minerals are collected from the earth and the possibility of containing large amounts of unnecessary substances that are harmful to the human body is very much high. Due to that, prior to make the medicinal preparations, the purification, detoxification, or incineration processes have to be followed as mentioned in authentic texts of TM in order to avoid unwanted harmful effects from herbo-mineral preparations (Mookerjee 1984; Mishra 2011; Department of Ayurveda 1994).

20.4.4.1 Parada – Mercury (Hg)

It is considered as the superior among all the metals and minerals which are being used in TM and it has the capability of giving a high potency from a low dose which emphasize the importance of its usage in the Herbo-mineral preparations of TM.

For the purification of Hg, several methods have been mentioned in authentic texts and one of them is; equal amounts of Hg and lime stone have to be obtained and triturated them for 36 hours. After that it is filtered through a cloth and it has to be kept for drying so that Hg can be obtained through it. Then, an equal quantity of garlic paste and ½ the quantity of *Saindhava lavana* is added to the obtained mercury and it has to be triturated for eight hours. After that, it is also washed with luke warm water so that again the mercury can be collected. By drying the remaining portion of garlic paste for six days and triturating it to obtain a fine powder, and filtering through a cloth is done to obtain the residual amounts of Hg too so that there is no wastage is occurred through this procedure. The obtained Hg through this method is known as the purified mercury.

20.4.4.2 Abhra – Mica

It is subjected to purification process by using *Triphala* decoction which consists of three drugs. (*Terminalia chebula, Terminalia bellirica,* and *Phyllanthus emblica*). After that the incineration process is conducted following the further detoxification which is done by using treated mica and paddy husks, tied up in a cloth and squeezing it (Wijenayake et al. 2014).

20.4.4.3 Vaikrantha – Tourmaline (CaF_2)

Vaikrantha is tied in a cloth to make a bolus and dipped in horsegram decoction by giving heat to get it purified. Incineration is conducted by taking equal amounts of purified Vaikrantha and Gandhaka (sulfur) together and grinded with lime juice to make chakrika (Small cakes) and giving heat (gaja puta) in order to obtain bhashma from it.

20.4.4.4 Swarna Makshika – Copper Pyrite ($CuFeS_2$)

It is heated by dipping in castor oil and lime juice for three hours by making a pottali from Makshika. And the incineration is carried out thereafter, by adding equal amount of Gandhaka powder into Makshika powder and triturating with lime juice. Then small cakes are made from it and induce heat by giving "Puta" (a form of giving heat in Alchemy).

20.4.4.5 Vimala – Iron Pyrite (FeS_2)

Heat is given by using Vasa Swarasa (Juice of *Adhathoda vasica*) to a pottali made out from pieces of Vimala tied in a cloth. After that it is subjected to incineration, by using fine powder of Vimala and Gandhaka in equal quantities, triturating with lime juice, making chakrika, and giving Puta (heat).

20.4.4.6 Shilajathu

The purification process is done by using Triphala decoction and dissolving Shilajath in it and filtered the portion to get dried in sunlight to evaporate the water. It is repeated thrice. Instead of Triphala decoction, cow's urine or cow's milk can be also used. Even though the incineration procedure has been mentioned in authentic texts, now it is not being practiced.

20.4.4.7 Sasyaka ($CuSo_4{\cdot}5H_2O$)

It is purified by giving bhavana (grinding) with boiled water of Rakta varga seven times and finally saturated with ghee. By adding equal amounts of purified Gandhaka and purified Tankana to purified Sasyaka, the incineration procedure is carried out by grinding with lime juice, making small cakes and giving heat to obtain bhashma.

20.4.4.8 Gandhaka

According to Rasa Shasthra Classics, Gandaka purification can be broadly divided into two types, small-scale Gandaka purification and large-scale Gandaka purification.

In small-scale purification, iron or stainless steel table spoon is taken and filled with desired quantity of ghee and placed over fire. When ghee is hot, the same quantity of unpurified Gandhaka is poured into it. As the sulfur starts melt, the mixture is stirred continuously. Then fill the earthen vessel with Goksheera (cow's milk), and mouth of the vessel should cover with a cotton cloth, which smeared with ghee. Pour the melted Gandaka into the vessel with milk through the cloth.

In large-scale purification, powdered unpurified Gandhaka evenly spread on the cloth which covers the cow's milk filled pot, except using the stainless steel spoon. Then this Gandhaka covered by another pot that placed invertedly and edges should seal with mud. Then this apparatus carefully buried in earth up to neck deep of the pot. Over the surface, the required number of dry cow dung cakes are heaped up and ignite to melt the Gandhaka and pass it to the milk through cotton cloth.

By this process, Shuddha Gandaka is remained as solid form. Stony substances are remained on the cloth and poisonous substances float on the milk-like husk.

20.4.4.9 Gairika – Hematite (Fe_2O_3)

Adding 1/8th of ghee to Gairika and heating is done in order to obtain purified Gairika and there is no incineration procedure.

20.4.4.10 Kasisa ($FeSO_4 \cdot 7H_2O$)

Kasisa purification is done by boiling with Bhringaraj swarasa (Juice of *Eclipta prostrata*). The incineration is conducted by grinding the purified Kasisa with juice of Daluk (*Euphorbia antiquorum*) and making Chakrika and giving Puta.

20.4.4.11 Kankshi (K_2SO_4 [$(Al_2SO_4)_3$]$\cdot 24H_2O$)

Kankshi is purified by dissolving in water and keeping the filtered portion in sunlight to get dried (Au sa).

20.4.4.12 Harithala – Orpiment ($AS_{2S}3$)

For the purification, Harithala is powdered and kept in a liquid such as juice of Kushmanda or other suitable liquid and subjected to fire by making use of Dola Yantra.

20.4.4.13 Manosheela – Realgar (AS_2S_2)

Purification is conducted by triturating with ginger juice for seven times. No incineration procedure is done.

20.4.4.14 Anjana

Purified with Bhringaraj swarasa by giving trituration.

20.4.4.15 Gauri Pashana (AS_2O_3)

It is powdered and kept inside the fruit of *Momordica charantia* and subjected to heat by *Dola yantra* for 3 hours to obtain the purified form.

20.4.4.16 Navasadara – Ammonium Chloride (NH_4Cl)

Water is added to Navasadara and filtered by using a cloth, and the obtained portion is subjected to evaporate.

20.4.4.17 Hingula – Cinnabar (HgS)

Hingula is purified by adding ginger juice and triturated for seven times. No incineration procedure is being conducted on it.

20.4.4.18 Swarna – Gold (Au)

Purified with Kanchanara Patra swarasa () after conducting general purification related to metals.

20.4.4.19 Thamra

After conducting general purification, get sheets of Thamra and apply kajjali with sahindava lavana with lime juice and heat it and dip in Triphala decoction.

20.5 Discussion

For the well-being of individuals, drug preparations that contain metals and minerals have been used since ancient times, and majority of traditional medicinal systems worldwide such as Ayurveda, TCM, Sri Lankan TM, etc. have used it in a considerably high manner. The different pharmacological actions were shown to possess specific uses in TM, which leads for the drug actions, and it was identified that apart from internal means of drug administration, there were external administrations too.

The toxicological studies that have been conducted based on metals and minerals are comparatively less when compared to allopathic system of medicine, and it will lead to the shortage of using herbo-metallic and herbo-mineral drug preparations in TM. Meanwhile, it was observed that there were multiple toxicological studies that have been conducted on Cinnabar, Arsenicals, etc.

The therapeutic value of TM preparations that contain metals and minerals mainly rely on their doses, and it is very much important to maintain correct doses when prescribing medicines in order to achieve the best outcome from them. Minute doses of herbo-mineral and herbo-metallic preparations are utilized in relieving ailments, and it is an advantage rather than using higher doses for a prolong time (Srikanth et al. 2019).

It has been stated that it is essential to undergo purification processes prior to make drug preparations by using metals and minerals and the same fact has been mentioned in authentic texts too. For that, purification processes as well as detoxification and incineration procedures have been also clearly mentioned in authentic texts for each and every drug. It will be a needful fact for individuals who prepare medicinal preparations that contain such drug materials.

20.6 Conclusion

Therefore, it can be concluded that metals and minerals are being used in TM in the management of plenty of diseases and disorders, and their pharmacological actions are responsible for therapeutic effects and such important facts have been mentioned since ancient times. Apart from that, it is essential to conduct more toxicological studies on these substances to emphasize their safety usage.

References

Angadi, R. (2016). *Rasa Tharangani*. Varanasi: Chaukambha Surbharati Prakashan.

Bhuban, R., Chaudhury, S.S., Pal, K. et al. (2019). A novel herbometallic nanodrug has the potential for antibacterial and anticancer activity through oxidative damage. *Nanomedicine* 14 (9): 1173–1189. https://doi.org/10.2217/nnm-2018-0187.

Che, C.T., George, V., Jinu, T.P.I. et al. (2017). Traditional medicine. In: *Pharmacognosy; Fundamentals, Applications and Strategies*, 15–30. https://www.sciencedirect.com/science/article/pii/B9780128021040000020.

Department of Ayurveda (1994). *Ayurveda Pharmacopoeia*, 2e, vol. 1- Part 111. Sri Lanka: Department of Ayurveda.

Gajendra, K., Srivastava, A., Sharma, S.K., and Gupta, Y.K. (2012). Safety evaluation of an ayurvedic medicine, Arogyavardhani vati on brain, liver and kidney in rats. *Journal of Ethnopharmacology* 140 (1): 151–160. https://doi.org/10.1016/j.jep.2012.01.004.

Huang, C.F., Hsu, C.J., Liu, S.H. et al. (2008). Ototoxicity induced by cinnabar (a naturally occurring HgS) in mice through oxidative stress and down regulated Na^+/K^+ ATPase activities. *Neurotoxicology* 29 (3): 286–296.

Huang, C.F., Liu, S.H., and Siau, S.Y.L. (2017). Neurological effects of Cinnabar (a Chinese mineral medicine, HgS) in mice. *Toxicology and Applied Pharmacology* 224 (2): 192–201. 10.1016/j.taap.2007.07.003.

Jia, W.M., Shi, X.L., Qin, W. et al. (2011). Toxicology evaluation of realgar-containing niu-huang-jie-du pain as compared to arsenicals in cell cultures and in mice. *ISRN Toxicology* https://doi.org/10.5402/2011/250387.

Cen, L., Wei, X., Chu, S. et al. (2018). The chemical speciation spatial distribution and toxicity of mercury from Tibetan Medicine Zuotai in mouse kidney. *Journal of Trace Elements in Medicine and Biology* 45: 104–113.

Mishra, S.N. (2009). *Rasendra Chudamani*. Varanasi: Chaukambha Orientalia.

Mishra, S.N. (2011). *Rasa Rathna Samuchchcaya*. Varanasi: Chaukambha Orientalia.

Mookerjee, B. (1984). *Rasa Jala Nidhi*. Varanasi: Sri Gokhul Mudranalaya.

Nasrin, S., Bachar, S.C., and Choudhuri, M.S.K. (2011). Toxicological studies of "Chondrokola Rosh", an ayurvedic preparation on liver function tests on rats. *African Journal of Traditional, Complementary, and Alternative Medicines* 8 (5): 170–174. https://doi.org/10.4314/ajtcam.v8i5S.22.

Pandey, M.M., Subha, R., and Rawat, A.K.S. (2013). Indian traditional ayurvedic system of medicine and food supplementation. *Evidence-Based Complementary and Alternative Medicine* https://doi.org/10.1155/2013/376327.

Prabhakara, N., Lakshmana, M., and Udupa, U.V. (2003). Effect of praval bhasma (Coral calx), a natural source of rich calcium on bone mineralization in rats. *Pharmacological Rearch* 43 (6): 593–599. DOI: 10.1016/s1043-6618(03)00224-x.

Prajapathi, P.K., Prasanta, K.S., Suhas, V.N. et al. (2006). Safety and toxicity profile of some metallic preparations in ayurveda. *Ancient Science of Life* 25 (3–4): 57–63. https://pubmed.ncbi.nlm.nih.gov/22557208/.

Priya, J., Abhiprandnya, B.W., Avijeet, K. et al. (2017). Sameerpannag ras mixture (SRM) improved neurobehavioral deficits following acute ischemic stroke by attenuating neuroinflammatory response. *Journal of Ethnopharmacology* 2 (197): 147–156. https://doi.org/10.1016/j.jep.2016.07.059.

Saleha, B., Banu, M., Ishaq, K. et al. (2004). DNA damage in leulocytes of mice treated with copper sulphate. *Food and Chemical Toxicology* 42 (12): 1931–1193. https://doi.org/10.1016/j.fct.2004.07.007.

Samuel, O.A., Olubukola, A., Olufemi, A., and Sakiru, M.E. (2015). Traditional/alternative medicine: an investigation into identification, knowledge and consumtion practices of herbal medicine among students with hearing impairment Ibadan, South-Western Nigeria. *Journal of Education and Practice* 6 (15): 43–49. https://files.eric.ed.gov/fulltext/EJ1079977.pdf.

Sarah, S., Dory, Y., Yazeed, B. et al. (2017). Is mercury in Tibetan Medicine toxic? Clinical, neurocognitive and biochemical results of an initial cross-sectional study. *Experimental Biology and Medicine* 242 (3): 316–332. 10.1177/1535370216672748.

Snehasis, B. and Jayesh, B. (2020). Ayurvedic processing of alpha HgS gives novel physicochemistry and distinct toxicokinetics in zebrafish. *Chemosphere* 251: 10.1016/j.chemosphere.2020.126295.

Srikanth, N., Singh, A., Ota, S. et al. (2019). Chemical characterization of an ayurvedic herbo-mineral preparation – Mahalaxmivilas Rasa. *Journal of Ayurveda and Intergrative Medicine* 10 (4): 262–264.

Swati, K., Atul, P., Soumen, C. et al. (2020). Mercury affects uterine myogenic activity even without producing any apparent toxicity in rats; involvement of calcium – signaling cascades. *Journal of Trace Elements in Medicine and Biology* 57: 40–47.

Ujang, T., Ross, S., Jack, N. et al. (2016). Bioavailability study of arsenic and mercury in traditional chnese medicines (TCM) usind an animal model after a single dose exposure. Bioavailability study of arsenic and mercury in traditional Chinese medicines (TCM) using an animal model after a single dose exposure. *Regulatory Toxicology and Pharmacology* 76: 51–56. 10.1016/j.yrtph.2016.01.010.

Waidyarathna, S.S.T. and Waidyaratna, M.T.S. (1927). *Vatika Prakaranaya hewath behet guli kalka potha*. Sri-Lanka: Modern Press.

Wei, L., Liao, P., Wu, H. et al. (2008). Toxiological effects of Cinnabar in rats by NMR – based metabolic profile in serum and urine. *Toxicology and Applied Pharmacology* 227 (3): 417–429. https://doi.org/10.1016/j.taap.2007.11.015.

Wijenayake, A., Amarasooriya, A., Ratnayake, B., and Abyasekara, C. (2014). The role of herbometallic preparations in traditional medicine, a review on mica drug processing and pharmaceuticals. *Journal of Ethnopharmacology* 155 (2): 1001–1010. https://doi.org/10.1016/j.jep.2014.06.051.

Zhang, B.B., Li, W.K., Hou, H.Y. et al. (2017). Toxicology and applied pharmacology. Zuotai and HgS differ from $HgCl_2$ and methyl mercury in Hg accumulation and toxicity in weanling and aged rats. *Toxicology and Applied Pharmacology* 331: 76–84. 10.1016/j.taap.2017.05.021.

21

Understanding the Etiology of Trace Element-Related Noncommunicable Diseases – Reviewing the Ghanaian Situation

Raymond Webrah Kazapoe[1], Elikplim Abla Dzikunoo[2], Ebenezer Ebo Yahans Amuah[3], and Paul Dankwa[4]

[1] *Department of Geological Engineering, University for Development Studies, Tamale, Ghana*
[2] *Earth Science Department, University of Ghana, Accra, Ghana*
[3] *Department of Environmental Science, Kwame Nkrumah University of Science and Technology, Kumasi, Ghana*
[4] *Department of Environmental Science, C. K. Tedam University of Technology and Applied Sciences, Navrongo, Ghana*

21.1 Introduction

The relationship between the presence of certain elements in the natural environment and the occurrence of noncommunicable diseases (NCDs) continues to be investigated quite extensively in some developed nations (Gulis et al. 2002; Kousa et al. 2011; Schullehner and Hansen 2014; Schullehner et al. 2018) within the field of medical geology. This field of study examines the relationships between geological materials and the associated geologic processes, and it also assesses the distribution and concentration levels of various elements, in the natural environment and their impacts on public health.

Medical geology in its truest form is a discipline in geoscience working in tandem with medical and other health workers that deals with the impacts of geologic materials and processes whose influence on the natural environment impact on animal and human health (Finkelman et al. 2003; Selinus et al. 2005). Its main concern is to highlight the nature and sources of those environmental health problems emanating from geogenic origins in a bid to help find solutions to them. It does so by bringing together geoscientists and medical and public health researchers to address health problems caused by geologic materials and the processes distributing and concentrating trace elements in rocks, minerals, water, and indirectly into food crops.

The impacts of released elements in the natural environment depend on the source, the pathway to the receptor, the concentration levels of the elements, and the duration of exposure by the receptor. The adverse effects of element concentration in soils and water are further compounded in most developing countries for two reasons: (i) the seeming ignorance or lack of interest in an investigation such scenarios, hence the absence of mitigating measures to safeguard the health of the most vulnerable, and (ii) we eat what we grow and drink generally from groundwater and surface water that is not treated (Kazapoe and Arhin 2020). The composition of rocks and minerals forming the foundation of the earth has their inscriptions and footprints on the air that we breathe, the water that we drink, and the food that we eat (Centeno et al. 2016). The elements from these sources air, water, and food may contain both essential and harmful minerals and elements (Arhin and Kazapoe 2017; Arhin et al. 2017, 2019, 2020). For many people, the transfer of minerals and trace elements from water and food serves as the primary source of nutrients for their physical development. Typical essential elements include calcium, iron, magnesium, potassium, and many more; in some instances, the essential elements in certain concentrations may become detrimental to the health of humans and animals.

Leary (1985) and Derry et al. (1990) discussed the effects of various materials from the natural environment including pervasive and persistent harmattan dust and contaminated groundwater and their contribution to several respiratory diseases. The reliance on the natural environment for food and water as well as the air we breathe brings people into direct contact with the possible deleterious disease-causing elements as well as the beneficial elements alike. Recent medical geology research by Arhin et al. (2019) shows the links between the natural environment and public health. These findings substantiate the role that geological processes play in the release of both toxic and essential elements into the natural environment, which then affect human and animal health. Based on the linkage established between elements in the

Medical Geology: En route to One Health, First Edition. Edited by Majeti Narasimha Vara Prasad and Meththika Vithanage.

underlying rocks registering their imprints in the soils derived from the rocks in the natural environment; this review seeks to evaluate the sources, speciation, disease-causing potential and linkage to diseases present in some Ghanaian communities in a bid to highlight the impact of some trace elements on human health and significantly reveal the role of geoscience or earth science in supporting public health to prevent the endemic outbreak of NCDs in Ghana.

Studies by Apambire et al. (1997), Apambire (2000), Arhin et al. (2019), and Kazapoe et al. (2021) have indicated that NCDs have been on the rise and accounts for 46% of the causes of morbidity and 40% of the causes of mortality in Ghana (Horton 2012). Additionally, data from the Ghana Health Service shows that the recorded outpatient cases of hypertension in public and mission facilities other than teaching hospitals increased from about 60 000 cases in 1990 to about 700 000 cases in 2010 (MoH-G 2012). Globally, there is a progressive rise in the burden of NCDs which have been variably attributed to factors such as high-risk lifestyles of populations associated with modernization, changes in nutrition, and less activity among children (Kengne et al. 2005; Daar et al. 2007; Amuna and Zotor 2008; Yawson et al. 2016). Other research has also shown the significant correlation between the geological environment and an increased risk of developing NCDs such as cardiovascular disease, hypertension, and among others (Gomes and Silva 2007; Dissanayake and Chandrajith 2009; Davies et al. 2013; Arhin et al. 2019; Kazapoe et al. 2021). There is therefore a need for medical geologists to work together with other disciplines to devise preventive as well as mitigative techniques in addressing many geology-related health issues in Ghana. Additionally, to prevent outbreaks of environmentally related diseases and to obtain answers to many of the unexplained deaths, this review intends to highlight, summarize, and create awareness of some of the impacts of trace elements in the natural environment on public health by using available medical geology information in Ghana. It will also highlight the potential effectiveness of addressing the environmental health diseases if the country conducts periodic regional geochemical surveys to define the pollution hotspots to advance the interdisciplinary multielement studies that will prevent trace elements-related diseases through the re-emerging field of medical geology.

21.2 Materials and Methods

21.2.1 Research Approach

This study reviewed seeks to highlight the etiology of trace element related NCDs in Ghana. The study considered As, F, and Se, areas where these elements occur in significantly high levels, minerals controlling their occurrence, chemistry toward their release, and disease in Ghana that are linked to their occurrence. Information was sourced from a wide range of sources including journals, articles, and reports. In sourcing information for this study, keywords including trace elements in soil, arsenic concentration in Ghana, fluoride in Ghana's pedology, the occurrence of selenium in soil in Ghana, soil chemistry of Ghana, and disease causing elements in Ghana were searched in Semantic Scholar, CORE, Science.gov, Google Scholar and Google browser, BASE, Baidu Scholar, and Microsoft Academic.

21.2.2 Regional Geology of Ghana

The geology of Ghana comprises a wide array of rocks which presents a distinct variability in earth materials (Figure 21.1). The Paleoproterozoic Birimian terrain in the western geological province of Ghana forms part of the broader Boule-Mossi domain at the eastern margin of the West African Craton. The Birimian in Ghana is characterized by five NE-SW trending greenstone belts separated by sedimentary basins (Dampare et al. 2008) and a sixth N-S trending Lawra greenstone belt. Both the belts and the basins are intruded by a wide variety of granitic rocks that have varying mineralogies (Abitty et al. 2016).

The Pan African mobile belt occupies the southeastern domain of the country and is composed mainly of the sedimentary Buem Structural Units. The considerably more metamorphosed Togo Structural Unit and the metamorphic rocks of the Dahomeyan Basement Units (Nude et al. 2015). The Neoproterozoic Voltaian sediments situated on the southeastern margin of the West African Craton covers an estimated area of 103 600 km^2, almost a third of the area of Ghana. The main rock types of the Voltaian include sandstones, shales, mudstones, and conglomerates. The rocks of the coastal sedimentary basins and the tertiary to recent deposits are composed of sedimentary units (Kesse 1985).

The variability of these geological domains suggests that a wide range of minerals and elements will potentially be introduced into the natural environment by a possible host of natural and anthropogenic processes. However, since the modal composition of the minerals in a particular rock may not be isotropic at all locations, there is a high possibility that the concentration of a particular mineral or element may fluctuate in the same rock type but at different locations. Such variations can be attributed to the natural processes of weathering and erosion or anthropogenic activities such as mining.

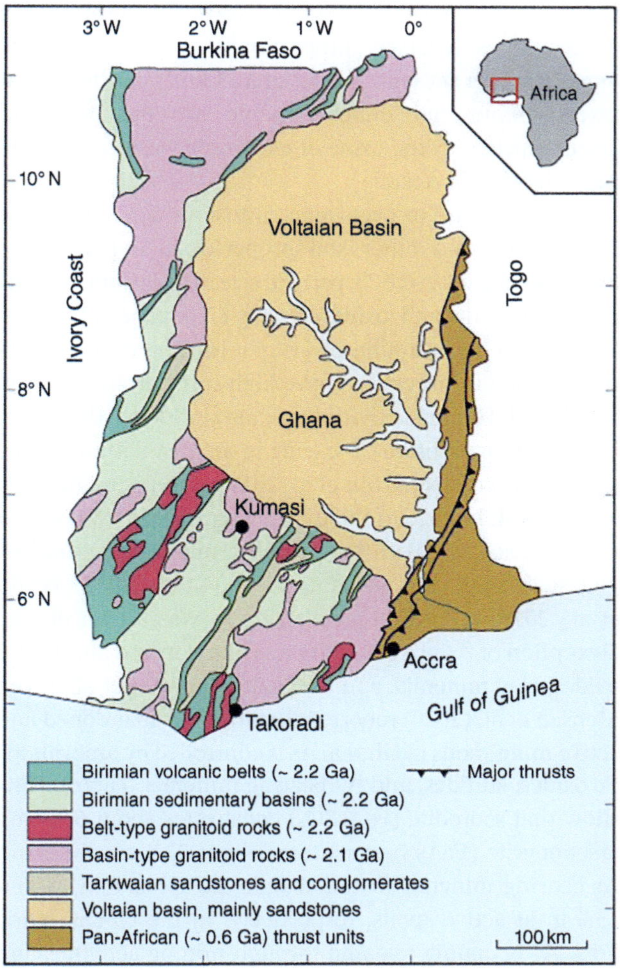

Figure 21.1 Geological map of Ghana (Kesse 1985). *Source:* Hydrological studies of the University of Cape Coast School of Agriculture Research Station at Twifo Wamaso.

21.3 Results and Discussion

21.3.1 Prevalence of Noncommunicable Diseases and Their Linkage to Medical Geology

Both beneficial and toxic elements found within the environment partly come from the earth. Although anthropogenic activities contribute significantly to the introduction of elements into the natural environment in certain cases, the majority of elements impacting human health are from geogenic sources (Amuna and Zotor 2008).

21.3.1.1 Arsenic

21.3.1.1.1 Areas in Ghana Where Elevated Levels of As is Recorded According to Anomoo-Neizer and Amekor (1993), food (cooked and uncooked), cash crops, vegetation, and fish and meat from homes and farms in Kumasi and Obuasi in Ghana contain some level of As. They recorded values ranging from 0.07 to 7.20 μg/L in Kumasi and values ranging from 0.12 to 70.50 μg/L in Obuasi. Their findings confirm that the high amount of arsenic recorded in Obuasi is a result of the mining activities in that area. Moreover, Anomoo-Neizer et al. (1996) analyzed the arsenic content in topsoils, water fern, elephant grass, plantain, cassava, and fish from Obuasi in the Ashanti region of Ghana. The arsenic concentrations range from 0.3 to 78.7 μg/L with topsoils and water ferns recording higher values. Asante et al. (2006) studied the arsenic concentration in water sources from Tarkwa, Southwestern Ghana, and found out that the groundwater is contaminated with arsenic ranging between <0.1 and 4.0 μg/L, while the rivers/streams have high values of 0.5–73 μg/L. The arsenic concentration of groundwater in Takwa is below the WHO guideline value (10 μg/L) for drinking water, while 33% of the rivers/streams have values above the WHO guideline value (WHO 2004; Asante et al. 2006).

Researchers like Amasa (1975) and Amonoo-Neizer and Amekor (1993) documented even high values of arsenic in Obuasi. As concentrations in dam water are around 2250 μg/L and 1400 in drinking water. The mean As concentration for water is 5190 μg/L (2800–10400 μg/L). The arsenic concentration found in the urine of inhabitants of Takwa ranges from 34 to 700 μg/L with a median value of 250 μg/L (Asante et al. 2006). Another research by Arhin et al. (2019) revealed that the Wassa community located in southwest Ghana has an arsenic concentration of 2–246 μg/L in their soils.

Furthermore, Bowell (1994) conducted studies in the Ashanti gold mine in southwest Ghana and recorded As contamination in the area. High values of As were recorded in the mine-contaminated surface soils (1292 μg/L). The values kept decreasing with depth. They recorded 335 μg/L at 1 m depth and 40.5 μg/L at 1.4 m depth. The regolith over the mineralized bedrock shows As concentration to increase with depth as well. Directly over the mineralized bedrock, the highest value in the area was recorded 1025 μg/L. However, the regolith overlying the unmineralized bedrock shows very little change (12.5–20.0 μg/L) over 1.2 m depth (Bowell et al. 1994). Further analysis of As in solution at the Ashanti gold mine shows that concentrations in uncontaminated regolith overlying the unmineralized bedrock ranges between 0 and 20 μg/L; on the other hand, the regolith overlying the mineralized bedrock shows high concentrations that ranges from 86.2 to 557 μg/L. Porewater from the aerobic contaminated soils shows that As concentration is directly proportional to depth (Bowell et al. 1994). In Agbogloshie in the Greater Accra region of Ghana, As concentrations in the urine of electronic waste workers were relatively high.

Research conducted by Mensah et al. (2020) reveals that As concentration in some abandoned and active mine spoils in the western region of Ghana ranges from 1807 to 8401 (abandoned mine spoils) μg/L and 2034–3801 μg/L (active mine spoils). The As concentration showed little variation with depth in the abandoned mine spoils (3105–4142 μg/L). Smedley et al. (2015) studied the groundwater in the Obuasi gold mine area to determine the groundwater As level, and they reported values ranging from <2 to 175 μg/L. They reported the highest values of As (64 μg/L) in the deeper wells (40–70 m). They also suggested that the As concentration might have piled up as a result of the longer residence of As in the form of As(III) present in the deeper groundwater. Meanwhile, deep mine boreholes (70–100 m) recorded low values of As (5–17 μg/L), and this may be a result of sorption action of As onto precipitating ferric oxyhydroxides or to localized low As the concentration of sulfide minerals. As concentrations from urinary samples of two villages, a rural streamwater-drinking community and another suburb of Obuasi, showed results of As concentration of 42 and 18 μg/L, respectively. According to Norman (2001), hand-dug wells and boreholes in Southwestern Ghana where mining activities are ongoing showed As concentration ranging around 5 ppb–2000 ppb. They reported the high As concentrations to be associated with the Birimian rocks and gold-arsenic mineralization. In this part of the country, borehole wells show low As concentration, while hand-dug wells show high concentrations. Water and sediments from Prestia, a mining community in the western region of Ghana, showed some As concentrations above the WHO acceptable limits in the range from 0.00090 to 0.00825 μg/L for water samples and 0.942–10.2 μg/L for sediments (Serfo-Armah et al. 2006).

Results of Smedley (1996) showed that groundwater from Obuasi in the Ashanti region and Bolgatanga in the Upper East Region have As concentration values that vary from <1 to 64 μg/L and <1 to 141 μg/L. As concentrations from both areas are related to the arsenopyrite and pyrite minerals within the Birimian basement rocks underlying both areas. As concentration in these areas are low in shallow waters and increase with increasing depth (40–70 m in Obuasi and 20– 40 m in Bolgatanga). Depths greater than this shows relatively low As concentration.

21.3.1.1.2 Mineral Controlling As Occurrence Anomoo-Neizer and Amekor (1993) stated that the concentration of arsenic in Obuasi is directly proportional to the distance from the Pompora Treatment Plant (PTP) and could be species-specific as well. They also associated the mobilization of arsenic to organic matter and linked the concentration of arsenic to the amount of organic matter present in the topsoil. From their results, the concentration of arsenic is inversely proportional to the amount of organic matter present. Asante et al. (2012) discussed that arsenobetaine and dimethylarsinic are the dominant arsenic species in the urine of electronic waste workers in Agbogbloshei in Accra.

Arsenic release or occurrence into soils depends on the source rock and other soil properties like pH, cation exchange capacity (CEC), particle size distribution, organic matter content, and oxide content (Bakhat et al. 2019); thus, high concentration of arsenic in source rock leaves soils formed from rocks to have high concentrations when weathered. Human activities may also influence the occurrence and distribution of arsenic in an area with the same geological settings (Arhin et al. 2019). Arsenic might have a very weak linkage to the underlying rocks in Wassa and may be associated with the hydrothermal veins within the metasediments and volcanoclastic (Kesse 1985; Arhin et al., 2019). Chemical weathering in Wassa controls the absorption of As by some minerals like iron and aluminum oxides, clay minerals, and iron sulfide (Arhin et al. 2019). Mensah et al. (2020) reported that in some abandoned and active mine spoils in Ghana, As is controlled by amorphous Fe oxides, sulfides, and As-bearing minerals. Their results show that scorodite ($FeAsSO_4$) constitutes about 65% and arsenopyrite (FeAsS) constitutes about 72% are the main As-bearing minerals in the abandoned spoils and 28 and 35% in at active spoils, respectively. In the Obuasi mine area, As is mainly released through mining activities and the natural oxidation process of sulfide minerals with arsenopyrite (FeAsS) being the most oxidized (Serfor-Armah et al. 2006; Smedley et al. 2015).

21.3.1.1.3 Chemistry Toward Its Release Asante et al. (2006) posited that the release of arsenic into surficial water bodies may be a result of oxidation/weathering of arsenopyrite and other sulfide-bearing mine tailings, and Bowell et al. (1994) added that As can also be released during mineral processing. During mining activities, arsenopyrite (a stable mineral in the reduced environment) may be brought into the oxidized environment where it will be unstable. Because of the instability of the minerals in the oxidizing environment, they are released to form a secondary mineral that will be stable. Water and oxygen in the oxidizing environment enhance and control their migration pattern, redistribution and concentration relative to the local activities, soil formation processes, and surface activities (Arhin et al. 2019). Smedley et al. (2015) indicated that some streams in the Obuasi mine area that show As values are poor in bacteriology quality. The streams have high As(III) and ($As(III)/As_{total} > 0.5$). This may be a result of the reduction process caused by bacteria and algae.

21.3.1.1.4 Diseases in Ghana That Are Linked to As Occurrence Oxides of arsenic pose great health risks to both adults and children. Oxides of arsenic cause spontaneous abortion, damage chromosomes, enhance genital malformations, reduce birth weight, cause liver problems, and affect the central nervous system (Anomoo-Neizer and Amekor 1993). People living in high arsenic concentration zone in Takwa are prone to health-risk related to arsenic, since they drink from the rivers/streams (Asante et al. 2006). A strong relation was found between As concentrations and skin-related diseases in Southwestern Ghana (Normal et al. 2001).

21.3.1.2 Fluoride

21.3.1.2.1 Areas Where Fluoride Is Recorded Sunkari et al. (2018) studied the fluoride concentrations and mineralogical composition of groundwater and rocks (Bongo-granites) from the Bongo and Sekyere South district in the northern and southern part of the country, respectively. They found high fluoride concentrations (1.71–4.0 mg/L) above the accepted World Health Organization (WHO 2011) limit of 1.5 mg/L in the Bongo district. Fluoride concentrations in the Sekyere South district in southern Ghana were relatively below the WHO accepted limit (0.3–0.8 mg/L). The high amount of fluoride in the Bongo district is because of the high amount of the mineral biotite within the rocks (Sunkari et al. 2018; Sunkari and Abu 2019).

Furthermore, research conducted by Salifu et al. (2013) reported fluoride concentration in both pore water (unsaturated zone) and groundwater (saturated zone) in the Gushegu district in northern Ghana. Unlike the concentration reported by Sunkari et al. (2018) and Sunkari and Abu (2019) of high fluoride concentration (>1.5 mg/L) in groundwater across the Bongo district, high values (>1.5 mg/L) were recorded in the pore water (unsaturated zone) and low values (<1.00 mg/L) were recorded in the groundwater (saturated zone) in the Gushegu district (Salifu et al. 2013). The data reported by Salifu et al. (2013) points out that the contamination of unsaturated zone by fluoride is a result of anthropogenic activities (use of phosphate fertilizers in farming) in the Gushegu district. According to Zango et al. (2021), about 61% of the groundwater within the Vea catchment in northeastern Ghana have fluoride concentration values > 1.5 mg/L. They recorded values that range from 0.35 to 3.95 mg/L for fluoride within the Vea catchment.

21.3.1.2.2 Mineral Controlling Fluoride Occurrence The rock samples from the Bongo district showed a high amount of biotite relative to the samples from Sekyere South district in southern Ghana that have a high amount of muscovite (Sunkari et al. 2018). The high amount of fluoride recorded in the Bongo district relative to the Sekyere South district is a result of the high amount dissolved biotite in the underground water (Sunkari et al. 2018). Places, where a high concentration of fluoride was recorded in the Sekyere South district, were linked to the farming activities in those areas. Bongo district is one of the highest fertilizer consumer in Ghana (Fianko et al. 2011). Fertilizer usage on farms can also enhance the fluoride concentration in the Bongo district (Sunkari and Abu 2019). As reported by Pickering (1985) and McLaughlin et al. (1996), phosphate fertilizers can contain about 1.3–3% fluorine and are common sources of F^- in groundwater. Phosphate fertilizers used in farming in the Bongo district are a moderate contributor (Sunkari et al. 2018). Biotite and muscovite (micas) are the main fluoride-bearing minerals in the Bongo district (Sunkari and Abu 2019).

The hydrochemistry of groundwater in the Gushegu district is related to water–rock interactions (Salifu et al. 2013). They documented a trend of fluoride concentration in pore water (unsaturated zone) and that of groundwater (saturated) in the Gushegu district. In their report, a high concentration of fluoride is recorded in pore water (unsaturated zone), and lower concentrations of fluoride are recorded in the groundwater (saturated zone). This strongly links the high concentration of fluoride in the Gushegu district to minerals in the unsaturated zone. This high concentration is a result of surface activities (anthropogenic). The sorption process by clay minerals is responsible for the reduction of fluorine in the saturated zone (Salifu et al. 2013).

Petrographic studies of the rocks within the Vea catchment shows the presence of minerals like quartz, microcline, plagioclase, biotite, muscovite, hornblende, opaque minerals, and other accessory minerals (Zango et al. 2021). Silicates originated from quartz, while the feldspar (microcline and plagioclase-Na-bearing feldspars, and albite) are the source of K^+ and Na^+, respectively. The mineral constituents of the lithologies are as follows: quartz (41%), microcline (31%), plagioclase (17.2%), biotite (3.1%), muscovite (2.2%), and hornblende (1.2%). The order of dominance of fluoride in the various lithologiesis is K-feldspar-rich granitoid (Bongo Granite) > hornblende-biotite granitoid > basaltic flow > hornblende-biotite tonalite > hornblendebiotite granodiorite > biotite granitoid.

21.3.1.2.3 Chemistry Toward Its Release Fluoride in the underground water in the Bongo district is released during magmatic and weathering activities of the Bongo granites. In the Bongo granite, fluoride is released mainly from biotite ($K(Mg, Fe_3^{2+})(Al, Fe^{3+})Si_3O_{10}(OH, F)_2$). When biotite is released, it gets dissolved in the underground water causing more contamination. Another source of fluoride in the Bongo district is linked to the release of fluorine as fluorine phosphate within the carbonate rocks (Sunkari et al. 2018). The mechanism of fluoride released within the carbonate rocks can be expressed by the chemical reaction below:

$$5CaCO_3 + 3HPO_4^{2-} + F^- = Ca_5\left(PO4\right)_3F + 3H^+ + 5CO_3^{2-} \tag{21.1}$$

The below expression gives the equilibrium constant for this reaction:

$$K_{calcite-fluoro-apatite} = \left(\frac{a}{H^+}\right)^3 \cdot \left(\frac{a}{CO_3^{2-}}\right)^5 \Big/ \left[\left(\frac{a}{HPO_3^{2-}}\right)^3 \left(\frac{a}{F^-}\right)\right] \tag{21.2}$$

The above expression suggests that changes in fluoride concentration in groundwater is directly proportional to carbonate concentration change and inversely proportional to the phosphate concentration (Sunkari et al. 2018).

During weathering, fluoride-rich rock weather and fluorine are released. This released fluoride is dissolved in surrounding water bodies and may be retained within the regolith for a long time. The fluorides within the regolith pollute any aquifer within that zone (Sunkari et al. 2018). Sunkari et al. 2018 also reported biotite and muscovite as the main fluoride-bearing minerals in Bongo and Sekyere South-south in the Northern and Southern parts of Ghana, respectively. The micas are believed to have released fluoride and hydroxyl ions (OH^-) through weathering processes and magmatic activities. Zango et al. (2021) accord with Sunkari et al. (2018) for biotite dissolution within the Vea catchment. This can be explained by the following chemical expression:

$$KAl_2\left[AlSi_3O_{10}\right]F_2 + 2OH^- \leftrightarrow KAl_2\left[ALSi_3O_{10}\right]\left[OH\right]_2 + 2F^-\left(\text{Muscovite}\right) \tag{21.3}$$

Moreover, fluorapatite dissolution is linked to rocks with accessory apatite. It is believed that fluoride is leached from the rocks together with calcite as the presentative carbonate mineral precipitation. Sunkari et al. (2018) reported that this process triggers a separation process that releases calcium and improved the fluorapatite dissolution. This can be expressed in the reactions below:

$$5CaCO_3 + 3HPO_4^{2-} + F = Ca_5\left(PO_4\right)_3F + 3H^+ + 5CO_3^{2+} \tag{21.4}$$

The main mechanism driving the release of fluoride in the Bongo and Sekyre South district is controlled mainly by hydroxyl ions (capable of exchanging replaceable F^- ions present in fluoride-bearing micas) and calcite precipitation which enhances fluorapatite dissolution and achieved by simply reducing calcium in the solution (Sunkari et al. 2018; Sunkari and Abu 2019). Sunkari and Abu (2019) also revealed that the ratio of the sum of the carbonates and bicarbonates and the sum of the calcium and magnesium is directly proportional to the ratio of fluoride concentration in the Bongo district. They also suggested that high fluoride concentrations are related to alkaline environments.

The weathering of silicate minerals takes a longer time and have been suggested by Salifu et al. (2013) to slow the ion content dissolution in the groundwater. The residence time and contribution of silicate mineral weathering are directly proportional. Therefore, under temperature conditions at the groundwater level (saturated zone), the whole process becomes slow. The main factor that facilitates the process in the groundwater zone is an acidic medium process, which is developed from the dissolution of carbon dioxides in the infiltrating rainwater. This yields HPO_3^- and Na^+. Phosphate fertilizers are the main cause of fluoride contaminations in the unsaturated zones in the Gushegu district (Salifu et al. 2013); however, the sorption process of clay minerals reduces the amount of fluoride that reaches the saturated zone during infiltration of rainwater. Rock samples from the Gushegu district show the presence of quartz, sericite, muscovite, iron oxide, plagioclase feldspar, and some clay minerals. This suggests that mineral weathering might also have contributed to fluoride concentration in the underground water in Gushegu. Salifu et al. (2013) suggested that it is not right to link the fluoride concentration in the underground water in Gushegu to the mineral weathering process because they (Salifu et al. 2013) did not find any relationship between $Ca^{2+}$and F^- when plotted suggesting that the (ions) are not of the same origin. Should the two ions originate from the weathering of fluorite, the slope plotted would have been close to 0.5 suggested by Salifu et al. (2013). Also, no relationship between Na and F was established by Salifu et al. (2013).

In the Gushegu district, the fluoride concentration in the unsaturated zone is responsible for contaminating the saturated zone with fluoride (76%) as a result of anthropogenic activities (use of phosphate fertilizers and deposition of fluoride by air) (Salifu et al. 2013). F^- is among the noted anions within the Vea catchment (Zango et al. 2021). Samples that show a high amount of fluoride concentrations are the K-feldspar-rich granitoids (Bongo granites) and the hornblende-biotite granitoids. The chemistry behind the high concentration of fluoride in the Vea catchment can be attributed to water–rock interaction (Tahir Shah and Danishwar 2003; Sunkari et al. 2018; Sunkari and Abu 2019). The increase in fluoride concentration above the WHO limit of 1.5 mg/L is proportional to well depth (Zango et al. 2021), though some shallow wells have recorded a high level of fluoride in the Vea catchment, and this is linked to anthropogenic activities like the use of phosphate fertilizers during farming. At the equilibrium stage between Na and F cation and anion, respectively, the more Na cation released produces soft water which enhances the high release of F^-. The high amount of fluoride in the area is linked to the dominance of hornblende-biotite granitoids. This is because fluoride is released into the groundwater during the biotite dissolution as stated by Salifu et al. (2013), Sunkari et al. (2018), and Sunkari and Abu (2019). Mafic and ultramafic minerals also contribute to fluoride concentration in the Vea catchment. This is strongly believed because of the low level of Mg/(Ca + Mg) ratio of 0.37 which points to the availability of biotite and hornblende in the area. Also, fluoride concentration in the catchment is linked to the alteration process of microcline, albite, and muscovite into clay minerals (smectite and kaolinite). Kaolinite can lead to the release of fluoride into groundwater through the process of kaolinization of granitic rocks (Zango et al. 2021).

21.3.1.2.4 Diseases Linked to Fluoride Occurrence Fluoride concentration above the WHO Standard of 1.5 mg/L can cause dental fluorosis (WHO 2011). Ghosh et al. (2013) defined dental fluorosis as a health condition where the enamel develops calcified cells. This makes the enamel weaken and porous, thereby developing brown pits. When it becomes extreme, it can make the enamel brittle and makes it prone to breakage (Sunkari et al. 2018). Edmund and Smedley reported that low concentrations of fluoride in drinking water cause dental cavities, especially in children. Inhabitants of the Bongo district are exposed to health problems such as skeletal and dental fluorosis and kidney damage due to the high amount of fluoride in their drinking water (Sunkari and Abu 2019). Dental fluorosis (molting of the teeth) is common among the inhabitants of the Bongo district (locations with a high concentration of fluoride). Inhabitants at places within the Bongo district with moderate fluoride concentrations also suffer from this health condition (Sunkari et al. 2018; Sunkari and Abu 2019). Zango et al. (2021) reported that the inhabitants within the Vea catchment have mottled teeth.

21.3.1.3 Selenium

21.3.1.3.1 Areas Where Selenium Is Recorded Arhin and Kazapoe (2017) researched the Telensi district in Ghana and reported selenium (Se) values above and below the acceptable limits of 0.005 mg/kg. There were places in Telensi that yielded Se values below the detection limit (DL) of the assay, and their analyses were carried on food crops. They recorded values ranging from 0.0005 to 0.24 mg/kg (Arhin and Kazapoe 2017). Results from the research conducted by Stadlmayr et al. (2010) on West African Traditional Foods of which Ghana is included show some values for Se. They reported values ranging from 0.5 to 20 mg/kg. Also, Ligowe et al. (2020) reported in their paper "Selenium deficiency risks in sub-Saharan African food systems and their geospatial linkages" that Se concentration in crop leaves from Ghana ranges from 30 to 48 mg/kg and 10 to 19 mg/kg for soils. Takyi and Amankwa (2004) stated that the undergraduate students of the University of Ghana mainly consume Se through the food they eat followed by multimineral supplements.

21.3.1.3.2 Mineral Controlling Se Occurrence Selenium is found across the globe within the crust is normally in association with sulfur-bearing minerals. Se concentration normally ranges from 50 to 90 mg/kg, and high concentrations are also found within the volcanic, sedimentary, and carbonate rocks. Soils of volcanic origins have shown a high concentration of Se but vary widely from 5 to 1 200 000 mg/kg. Se occurs according to its possible oxidation states: selenides (Se^{2-}), amorphous or polymeric elemental selenium (Se^0), selenites (Se^{4+}), and selenates (Se^{6+}) (WHO/IPCS 1987; UK EGVM 2002).

21.3.1.3.3 Chemistry Toward Its Release Inorganic selenites are reduced to elemental selenium in acidic and reducing environments. Selenates are formed in alkaline and oxidizing environments. Both selenate and salinities are soluble in water which is the means that they are leached from well-aerated alkaline soils that favor their oxidation. On the other hand, elemental selenites and selenates turn to be insoluble in water, which makes elemental selenium retain in the poorly aerated soils (favorable condition for their formation). Selenium in alkaline environments is available for plant absorption, whereas selenium in the acidic environment is blocked by the adsorption of selenites and selenates to iron and aluminum oxide sols (Ullrey et al. 1983). Environmental activities and underlying geology are the main factors influencing the concentration of Se in the Telensi district.

21.3.1.3.4 Diseases Linked to Se Occurrence Arhin and Kazapoe (2017) reported that selenium (Se) deficiency in both humans and animals is a major health concern. A low concentration of selenium retards growth and can even lead to death. Over concentration of selenium in the body can also cause Se toxicity symptoms. Some areas within the Talansi district of Ghana show very Se concentration below assay DLof 0.0005 mg/kg (Arhin and Kazapoe 2017).

21.4 Conclusion

The earth contains elements that play physiological roles in human development. An amount of the element's concentrations transferred into soils that are taken up by plants indirectly ingested as food provides the dietary supplements to our growth. The fraction of elements either essential or toxic exposed to humans depends on the amount in soils. The higher the concentrations in soil so will the fraction be in plants or water in the environment. This is seen as a challenge for effective health care delivery in Ghana of the health data has no link to the geology which could play crucial role in identifying some of the causes of many of the emerging diseases. The geology of Ghana that has evolved over the years and coupled with the numerous anthropogenic and geogenic activities modifies the geochemical background values of elements. There are so many elements that are essential to humans, but their usefulness to humans depends on the dose. There are also others that known to be poisonous to humans irrespective of the amount exposed as some of them are bio-accessible and will bio-accumulate over time. The understanding of the geochemistry of the underlying rocks provides an idea of the type of elements that will be released into the environment during weathering. Using geochemical methods, the concentration levels of essential and harmful elements can be defined and outlined, determining soil health at any point. Medical geology research has found fluoride, which occurs in excessive concentrations in some parts of the Upper East Region of the country, to be associated with dental carries and skeletal fluorosis. Iodine deficiency has been reported to cause miscarriages and goiter in the Upper West Region of Ghana where iodine has been tested to be deficient in soils. Dust carried by the Northeast trade winds during the harmattan season is speculated to cause respiratory infections to people who incidentally inhale them. Research has also shown a strong correlation between some forms of cardiovascular disorders, and water hardness may be because of the presence of magnesium and calcium. High levels of arsenic found in Obuasi, where AngloGold Ashanti is located, has health cases of cancer, kidney, and bladder diseases and skin lesions, which may be an attribute of some exposed elements. The re-emergence of the discipline of medical geology provides an avenue to bridge this gap. As has been shown in this review, some research exists in Ghana which takes into consideration the distribution of elements and their impact on health, but there is a clear lack of any effort to discuss this in terms of the spatio-temporal distribution of geologically associated diseases and public health. Broadening our understanding of medical geology could help devise problem-solving techniques that will eliminate or reduce the over 94 000 people that suffers from NCDs annually in Ghana. This may as well give clues to an appropriate therapy in preventing many geological-related health issues and thus improve life quality and bring national development in Ghana.

References

Abitty, E.K., Dampare, S.B., Nude, P.M., and Asiedu, D.K. (2016). Geochemistry and petrogenesis of the K-rich 'Bongo-type'granitoids in the Paleoproterozoic Bole–Nangodi greenstone belt of Ghana. *Journal of African Earth Sciences* 122: 47–62.

Amasa, S.K. (1975). Arsenic pollution at Obuasi Goldmine, town, and surrounding countryside. *Environmental Health Perspectives* 12: 131–135.

Amonoo-Neizer, E.H. and Amekor, E.M. (1993). Determination of total arsenic in environmental samples from Kumasi and Obuasi, Ghana. *Environmental Health Perspectives* 101 (1): 46–49.

Amonoo-Neizer, E.H., Nyamah, D., and Bakiamoh, S.B. (1996). Mercury and arsenic pollution in soil and biological samples around the mining town of Obuasi, Ghana. *Water, Air, and Soil Pollution* 91 (3), 363–373.

Amuna, P. and Zotor, F.B. (2008). Epidemiological and nutrition transition in developing countries: impact on human health and development: The epidemiological and nutrition transition in developing countries: evolving trends and their impact in public health and human development. *Proceedings of the Nutrition Society* 67 (1): 82–90.

Apambire, W.B. (2000). *Geochemical modeling and geomedical implications of fluoriferous groundwaters in the upper east region of Ghana*. Reno: University of Nevada.

Apambire, W.B., Boyle, D.R., and Michel, F.A. (1997). Geochemistry, genesis, and health implications of fluoriferous groundwaters in the upper regions of Ghana. *Environmental Geology* 33 (1): 13–24.

Arhin, E., and Kazapoe, R. (2017). Selenium in locally produced food crops and implications on healthy eating: a case study at the Talensi district of Ghana.

Arhin, E., Zango, M.S., and Kazapoe, R. (2017). Characterizing and indexing regolith materials using geochemistry towards hidden mineral anomaly delineation: a case study of Savannah region of NW Ghana. *UJG* 5 (6): 169–182.

Arhin, E., Zhang, C., and Kazapoe, R. (2019). Medical geological study of disease-causing elements in Wassa area of Southwest Ghana. *Environmental Geochemistry and Health* 41 (6): 2859–2874.

Arhin, E., Kazapoe, R.W., and Salami, F. (2020). *Linking Geology to the Prevalence of Non-Communicable Diseases: A Case Study of the Voltaian Sedimentary Basin*. Ghana: Ecofeminism and Climate Change.

Asante, A.D., Zwi, A.B., and Ho, M.T. (2006). Equity in resource allocation for health: a comparative study of the Ashanti and Northern Regions of Ghana. *Health Policy* 78 (2–3): 135–148.

Asante, K.A., Agusa, T., Biney, C.A. et al. (2012). Multi-trace element levels and arsenic speciation in urine of e-waste recycling workers from Agbogbloshie, Accra in Ghana. *Science of the Total Environment* 424: 63–73.

Bakhat, H.F., Zia, Z., Abbas, S. et al. (2019). Factors controlling arsenic contamination and potential remediation measures in soil-plant systems. *Groundwater for Sustainable Development* 9: 100263.

Bowell, R.J. (1994). Sorption of arsenic by iron oxides and oxyhydroxides in soils. *Applied Geochemistry* 9 (3): 279–286.

Bowell, R.J., Morley, N.H., and Din, V.K. (1994). Arsenic speciation in soil porewaters from the Ashanti Mine, Ghana. *Applied Geochemistry* 9 (1): 15–22.

Centeno, J.A., Finkelman, R.B., and Selinus, O. (2016). Medical geology: impacts of the natural environment on public health. *Geosciences* 6 (1): 8.

Daar, A.S., Singer, P.A., Persad, D.L. et al. (2007). Grand challenges in chronic non-communicable diseases. *Nature* 450 (7169): 494–496.

Dampare, S.B., Shibata, T., Asiedu, D.K. et al. (2008). Geochemistry of Paleoproterozoic metavolcanic rocks from the southern Ashanti volcanic belt, Ghana: petrogenetic and tectonic setting implications. *Precambrian Research* 162 (3-4): 403–423.

Davies, B.E., Bowman, C., Davies, T.C., and Selinus, O. (2013). *Medical Geology: Perspectives and Prospects. In Essentials of Medical Geology*, 1–13. Dordrecht: Springer.

Derry, C.W., Bourne, D.E., and Sayed, A.R. (1990). The relationship between the hardness of treated water and cardiovascular disease mortality in South African Urban areas. *South African Medical Journal* 7 (10): 522–524.

Dissanayake, C.B. and Chandrajith, R. (2009). *Introduction to Medical Geology*. Springer Science & Business Media https://download.e-bookshelf.de/download/0000/0136/38/L-G-0000013638-0002346784.pdf (accessed 1 July 2021).

EGVM, U. (2002). Expert group on Vitamins and Minerals. *Intakes of nickel from food in the UK. Revised Review of Nickel. EVM/99/24*. REVISEDAUG2002. UK.

Fianko, J.R., Donkor, A., Lowor, S.T., and Yeboah, P.O. (2011). Agrochemicals and the Ghanaian environment, a review. *Journal of Environmental Protection* 2 (03): 221.

Finkelman, R.B., Belkin, H.E., Centeno, J.A., and Zheng, B.S. (2003). Geological epidemiology: coal combustion in China. In: *Geology and Health: Closing the Gap* (ed. H.C. Skinner and A.R. Berger), 45–50. Oxford: Oxford University Press.

Ghosh, A., Mukherjee, K., Ghosh, S.K., and Saha, B. (2013). Sources and toxicity of fluoride in the environment. *Research on Chemical Intermediates* 39 (7): 2881–2915.

Gomes, C.D.S.F. and Silva, J.B.P. (2007). Minerals and clay minerals in medical geology. *Applied Clay Science* 36 (1–3): 4–21.

Gulis, G., Czompolyova, M., and Cerhan, J.R. (2002). An ecologic study of nitrate in municipal drinking water and cancer incidence in Trnava District, Slovakia. *Environmental Research* 88 (3): 182–187.

Horton, R. (2012). GBD 2010: understanding disease, injury, and risk. *Lancet (London, England)* 380 (9859): 2053–2054.

Kazapoe, R. and Arhin, E. (2019). Determination of local background and baseline values of elements within the soils of the Birimian Terrain of the Wassa Area of Southwest Ghana. *Geology, Ecology, and Landscapes* 1–10.

Kazapoe, R.W., Arhin, E., and Amuah, E.E.Y. (2021). Known and anticipated medical geology issues in Ghana. *Ecofeminism and Climate Change* 2: 169–184.

Kengne, A.P., Amoah, A.G., and Mbanya, J.C. (2005). Cardiovascular complications of diabetes mellitus in sub-Saharan Africa. *Circulation* 112 (23): 3592–3601.

Kesse, G.O. (1985). *The Mineral and Rock Resources of Ghana*. Rotterdam: A. A. Balkema Press.

Kousa, A., Loukola-Ruskeeniemi, K., Nikkarinen, M. et al. (2011). Evaluation of the relationship between the natural geological environment and certain chronic diseases in Finland. *Geological Survey of Finland, Special Paper* 49: 247–262.

Leary, W.P. (1985). Content of magnesium in drinking water and deaths from ischemic heart disease in white South Africans. *Magnesium* 5: 150–153.

Ligowe, I.S., Phiri, F.P., Ander, E.L. et al. (2020). Selenium deficiency risks in sub-Saharan African food systems and their geospatial linkages. *Proceedings of the Nutrition Society* 79 (4): 457–467.

McLaughlin, M.J., Tiller, K.G., Naidu, R., and Stevens, D.P. (1996). The behaviour and environmental impact of contaminants in fertilizers. *Soil Research* 34 (1): 1–54.

Mensah, A.K., Marschner, B., Shaheen, S.M. et al. (2020). Arsenic contamination in abandoned and active gold mine spoils in Ghana: Geochemical fractionation, speciation, and assessment of the potential human health risk. *Environmental Pollution* 261: 114116.

Ministry of Health-Ghana (2012). National policy for the prevention and control of chronic noncommunicable diseases in Ghana. https://www.iccp-portal.org/sites/default/files/plans/national_policy_for_the_prevention_and_control_of_chronic_non-communicable_diseases_in_ghana%281%29.pdf (accessed 12 July 2021).

Norman, D. I. (2001). Department of Earth and Environmental Science, New Mexico Institute of Mining. In *International Seminar on Nuclear War and Planetary Emergencies: 25th Session: "E. Majorana" Centre for Scientific Culture,* Erice, Italy (19–24 August), 2000 (p. 68). World Scientific.

Nude, P.M., Kwayisi, D., Taki, N.A. et al. (2015). Petrography and chemical evidence for multi-stage emplacement of western Buem volcanic rocks in the Dahomeyide orogenic belt, southeastern Ghana, West Africa. *Journal of African Earth Sciences* 112: 314–327.

Pickering, W.F. (1985). The mobility of soluble fluoride in soils. *Environmental Pollution Series B, Chemical and Physical* 9 (4): 281–308.

Salifu, M., Yidana, S.M., Osae, S., and Armah, Y.S. (2013). The influence of the unsaturated zone on the high fluoride contents in groundwater in the Middle Voltaian Aquifers—the Gushegu District, Northern Region of Ghana. *Journal of Hydrogeology & Hydrologic Engineering* 2 (2): 0–45.

Schullehner, J. and Hansen, B. (2014). Nitrate exposure from drinking water in Denmark over the last 35 years. *Environmental Research Letters* 9 (9): 095001.

Schullehner, J., Hansen, B., Thygesen, M. et al. (2018). Nitrate in drinking water and colorectal cancer risk: a nationwide population-based cohort study. *International Journal of Cancer* 143 (1): 73–79.

Selinus, O., Alloway, B., Centeno, J.A. et al. (2005). *Essentials of Medical Geology: Impacts of the Natural Environment on Public Health.* Elsevier Academic Press, 812pp ISBN: 0-12-636341-2.

Serfor-Armah, Y., Nyarko, B.J.B., Dampare, S.B., and Adomako, D. (2006). Levels of arsenic and antimony in water and sediment from Prestea, a gold mining town in Ghana and its environs. *Water, Air, and Soil Pollution* 175 (1): 181–192.

Smedley, P.L. (1996). Arsenic in rural groundwater in Ghana: part special issue: hydrogeochemical studies in sub-Saharan Africa. *Journal of African Earth Sciences* 22 (4): 459–470.

Smedley, D., Haider, S., Durinck, S. et al. (2015). The BioMart community portal: an innovative alternative to large, centralized data repositories. *Nucleic Acids Research* 43 (W1): W589–W598.

Stadlmayr, B., Charrondiere, U.R., Addy, P. et al. (2010). *Composition of selected foods from West Africa*, 13–14. Rome: *Food and Agriculture Organization*.

Sunkari, E.D. and Abu, M. (2019). Hydrochemistry with special reference to fluoride contamination in groundwater of the Bongo District, Upper East Region, Ghana. *Sustainable Water Resources Management* 5 (4): 1803–1814.

Sunkari, E.D., Zango, M.S., and Korboe, H.M. (2018). Comparative analysis of fluoride concentrations in groundwaters in Northern and Southern Ghana: implications for the contaminant sources. *Earth Systems and Environment* 2 (1): 103–117.

Tahir Shah, M. and Danishwar, S. (2003). Potential fluoride contamination in the drinking water of Naranji area, northwest frontier province, Pakistan. *Environmental Geochemistry and Health* 25 (4): 475–481.

Takyi, E.E.K. and Amankwa, P. (2004). Dietary selenium and copper intake by resident undergraduate students of the University of Ghana. *East African Medical Journal* 81 (1): 34–39.

Ullrey, D.E., Combs, G.F., Conrad, H.R. et al. (1983). *Selenium in nutrition*, Revisede. *Washington, DC: NAS-NRC*.

WHO (2011). *Guidelines for Drinking-Water Quality*, 4e. Geneva: World Health Organization ISBN: 978 92 4 154815 1.

WHO/IPCS (1987). *Environmental Health Criteria 58 Selenium*. Geneva: WHO.

World Health Organization (2004). *The World health Report: 2004: Changing History*. World Health Organization.

Yawson, A.E., Abuosi, A.A., Badasu, D.M. et al. (2016). Non-communicable diseases among children in Ghana: health and social concerns of parent/caregivers. *African Health Sciences* 16 (2): 378–388.

Zango, M.S., Pelig-Ba, K.B., Anim-Gyampo, M. et al. (2021). Hydrogeochemical and isotopic controls on the source of fluoride in groundwater within the Vea catchment, northeastern Ghana. *Groundwater for Sustainable Development* 12: 100526.

22

Dental Fluorosis Cases in Turkey

Perihan Derin[1], Benan Yazici-Karabulut[1], Ayşegul Demir-Yetis[2], Abdullah İzzeddin Karabulut[1], Ayse Dilek Atasoy[1], Mehmet İrfan Yeşilnacar[1], İbrahim Bayhan[3], Seda Alp[4], and Hazal Ozer[5]

[1] *Environmental Engineering Department, Faculty of Engineering, Harran University, Sanliurfa, Turkey*
[2] *Medical Services and Techniques Department, Bitlis Eren University, Bitlis, Turkey*
[3] *Public Health Services Department, Environmental Health Unit, Sanliurfa, Turkey*
[4] *Ministry of Health, Oral and Dental Health Department, Ankara, Turkey*
[5] *Pediatric Dentistry Department, Faculty of Dentistry, Necmettin Erbakan University, Konya, Turkey*

22.1 Distribution of Fluoride and Dental Fluorosis in the World

Dental fluorosis: It is the developmental disorders seen in tooth enamel as a result of exposure to more than optimum level of fluor (Ekstrand et al. 1988). In many studies conducted around the world, it has been proven that fluoride values encountered in drinking water sources have many harmful effects on health, especially on DF cases, especially on teeth in childhood. Countries with high fluoride values in drinking water resources around the world are given in Table 22.1 and Figure 22.1. Detailed studies of these cases reported in various studies are presented as follows.

In this context, in a study conducted in Brazil, fluoride concentrations in the nails of 56 children aged 2–7 years were examined and divided into two as above and below 2 μg/g. When the children reached the age of 10–15 years, the measurement was repeated with the same method. A positive correlation of 80% was reported between the incidence of DF and nail fluorine concentration (Buzalaf et al. 2012).

In a study conducted by Warren in the USA, it has been reported that moderate DF was observed in the primary dentition of 11.6% of children living in areas with water fluorine concentration ratios of 0.7–1.2 mg/L (Warren et al. 2001).

In Sweden, DF was observed at an advanced level (dean index ≥3) in primary dentition in areas where drinking water contains fluoride at a rate of 10 mg/L, while this rate decreased to 80% in areas with 5 mg/L fluorine in drinking water (Forsman 1974).

Table 22.1 Various studies on dental fluorosis made in the world.

Made country/area	Fluorine levels determined in water samples (mg/L)	Study
Persia	In water samples 1.55	Alimohammadi et al. (2018)
	In water samples 10.3	Yousefi et al. (2018)
Mexico	In water samples 3.7	Kimambo et al. (2019)
	In water samples 3.7	Armienta and Segovia (2008)
Srilanka	In water samples 8	Chandrajith et al. (2020)
Cameroon	In water samples 15.2	Onipe et al. (2020)
Zimbabwe	In water samples 10	
Malawi	In water samples 14	
South Africa	Water supply 30	
Senegal	In water samples 7.4	

In England, the rate of enamel opacities in primary dentition was reported as 32.7% in areas with a fluoride content of 0.26 mg/L (low level) in drinking water (Murray and Shaw 1979).

In a study in China, in which 472 children aged 7–8 years were grouped according to the fluorine content in their drinking water, it was reported that DF was more common in primary dentition in areas with drinking water fluor containing more than 1 mg/L (Ruan et al. 2005).

Medical Geology: En route to One Health, First Edition. Edited by Majeti Narasimha Vara Prasad and Meththika Vithanage.

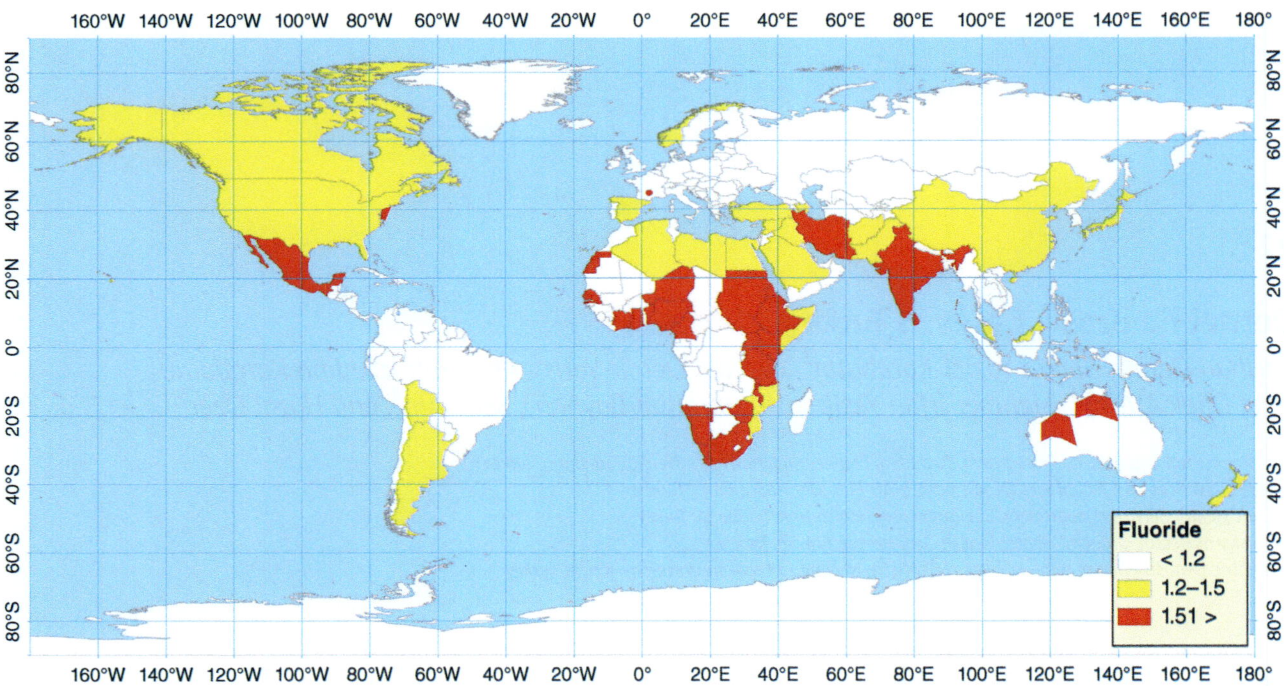

Figure 22.1 Locations in the world where fluoride concentrations are detected above the 1.5 mg/L limit value.

22.2 Distribution of Dental Fluorosis in Turkey

High fluoride concentrations in drinking water sources in Turkey, as in many regions in the world, cause DF and SF (skeletal fluorosis) cases to be seen more frequently. When the regions with high fluoride in water resources and the causes of fluoride exposure are analyzed according to DF cases and their distribution in Turkey, groundwater resources used for drinking water supply appear to affected by high fluoride levels. The World Health Organization (WHO) has determined the maximum value of fluorine concentration in drinking water as 1.5 mg/L. The amount of fluoride in some water sources in Turkey is between 0.7 and 15.2 mg/L (Table 22.2). The reason for the high fluoride concentrations in drinking

Table 22.2 Dental fluorosis studies made so far in Turkey.

Made region	Fluorine levels determined in water samples (mg/L)	Study
Agri region	In water samples 5.0–12.5	Oruç (1977), Oruç and Vicil (1984), Baydaş (1972), Gulhan and Mermutlu (1973), Tanyeri (1975), and Babacan (1979)
Van region	In water samples 0.7–15.2	Şendil and Baysu (1973)
	In water samples 2.0–7.5	Oruç (1976)
	In water samples 1.2–8.25	Oruç and Vicil (1984) and Ergun et al. (1987)
Isparta region	In water samples 4.03	Ata (1966)
	In water samples 3.7 = 4.4	Velicangil and Demirhindi (1957)
	In water samples 1.72–4.20	Bilgin et al. (2005)
	In water samples 1.8–2.6	Atabey (2010)
Eskisehir region	In water samples 1.4–2.0	Oruç and Vicil (1984)
	Kızılcaoren 4.81	Fidanci et al. (1998)
	In water samples 3.8–4.9	Aksit et al. (1980)
	In water samples 3.9	Isikli and Kalyoncu (2000)

Table 22.2 (Continued)

Made region	Fluorine levels determined in water samples (mg/L)	Study
Usak region	Tap and well water samples 0.7–2.2	Oruç and Vicil (1984)
Kirsehir region	Kaman In water samples 1.5	Atabey and Ünal (2008)
	Akpinar district In water samples 1.8	
	Pohrenk Hot water supply 5.5 Tap water.6	Oksuz (2006) and Demirel (2009)
Nevsehir region	Sarihidir village In water samples 2.8	Atabey and Ünal (2008)
Edirne	Hanliyenice In water samples 2.53	Sezgin et al. (2018)
Şanliurfa	In water samples 362	Yeşilnacar et al. (2013)
	In water samples 2349	Derin et al. (2018)
Bitlis	In spring and groundwater 326	Yetis et al. (2019) (Unpublished work)
Kirikkale	Şentepe Mosque (Yuva neighborhood) Source 4824	Karaca (2017)
	Yuva Neighborhood Cemetery Source 4877	
	Cullu Road Dörtyol Fountain Source 5967	
Trakya/Edirne	Bayramli village 1.460	Tokatli and Guner 2020

water sources is thought to be due to the content of some minerals found in the volcanic rocks in the mentioned regions (Oruç et al. 1976; Atabey and Ünal 2008; Atabey 2010; Yetis et al. 2019). DF cases, whose incidence is increasing day by day in our country and in the world, emerge as a public health problem.

In our country, some regions with a fluoride concentration above the limit (1.5 mg/L) in groundwater and cases of dental fluorosis are shown in Figures 22.2, 22.3.

22.3 Evaluation of Dental Fluorosis Cases in Turkey

Dental fluorosis: It is the developmental disorders seen in tooth enamel as a result of exposure to more than optimum level of fluor (Ekstrand et al. 1988). The WHO has determined the maximum fluorine concentration value in drinking water to be 1.5 mg/L and declared that DF can be seen in regions with more fluorine than this value (WHO 2011). In recent years, it has been reported that there has been an increase in mild fluorosis cases due to the topical and systemic use of fluor (Susheel 2005).

With the chronic and cumulative exposure to F intake during the enamel maturation (amelogenesis) stage, the degree and severity of DF become evident.

Since the developmental stages of each tooth are in different time periods, DF can be observed differently in each tooth and in each tooth region (Den Besten and Li 2011), as well as individual changes within the community living in the same region (Everett 2011).

DF is very rare in primary dentition due to reasons such as the permeability level of the placental barrier, mother–infant blood bond regulation, and poor fluoride transport from breast milk (Jha et al. 2011).

Permanent teeth are most affected by fluorosis. Premolars and upper incisors are the most affected tooth groups. These are followed by the second molars, the canines, and the first molars, respectively. The mandibular incisors are least affected. In cases where the fluoride concentration is too high, primary teeth may also be affected (Fejerskov et al. 1990).

The most critical age for fluoride exposure is thought to be between 15 and 30 months. However, it has been stated that fluoride above the optimal level taken or exposed at older ages may also affect the teeth (Levy 2003).

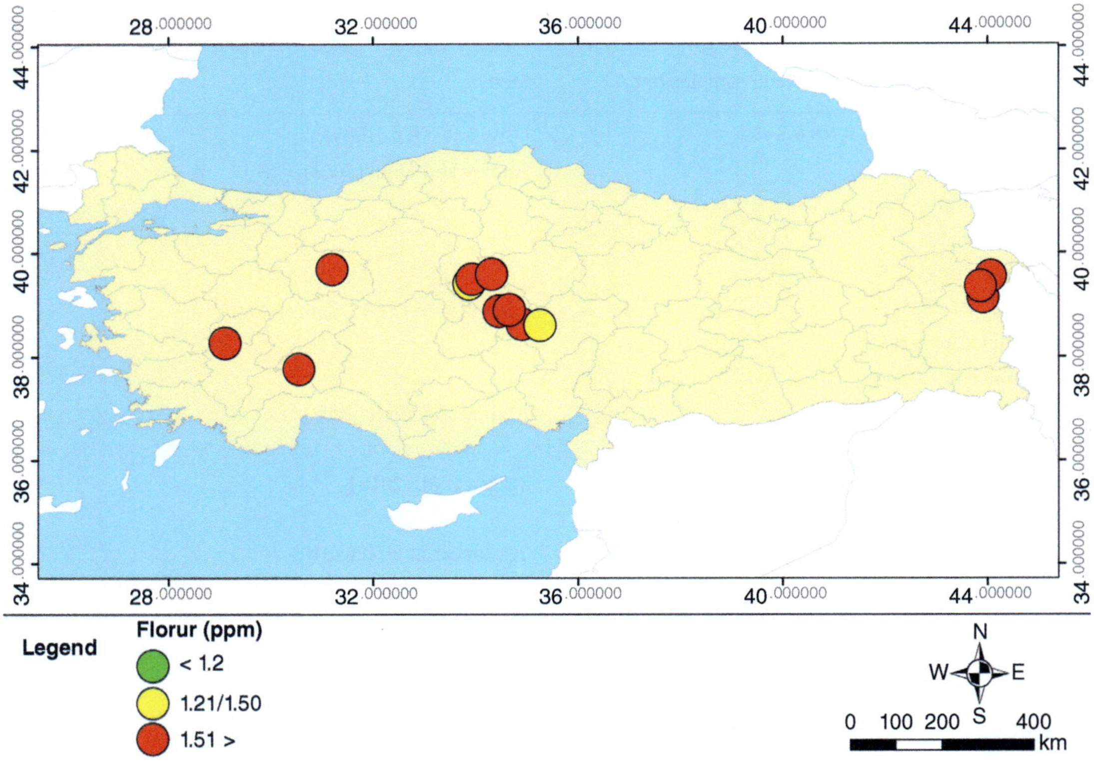

Figure 22.2 Places with high fluoride concentrations in groundwater and cases of dental fluorosis in Turkey (1972–2010; Atabey 2010).

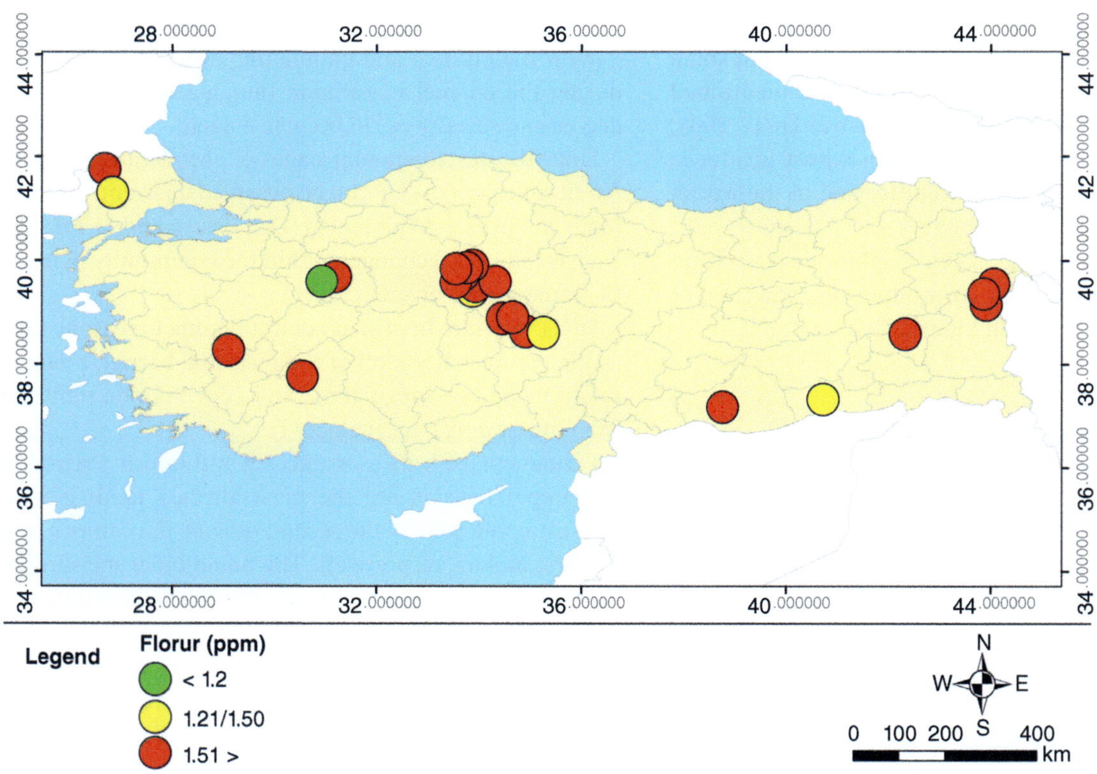

Figure 22.3 Places with high fluoride concentration in groundwater and cases of dental fluorosis in Turkey (2010–2022).

Şendil and Baysu (1973), in their study in Muradiye and Caldıran districts around Lake Van, found that the fluoride concentrations of the waters were between 5.7 and 15.2 ppm. In the study conducted by Oruç (1976), the presence of high fluoride drinking water varying between 2.0 and 7.5 ppm around Tendürek volcano and 5.0 and 12.5 ppm in Agrı Dogubeyazıt region was reported. DF cases were detected in various studies on animal, water, soil, and plant samples in Isparta provinces and districts (Oruç 2008), Eskisehir Beylikova County Kızılcaoren Village (Oruç 1983), and Usak Esme district (Oruç and Vicil 1984), due to consumption of drinking water containing high fluoride (Figures 22.4–22.7).

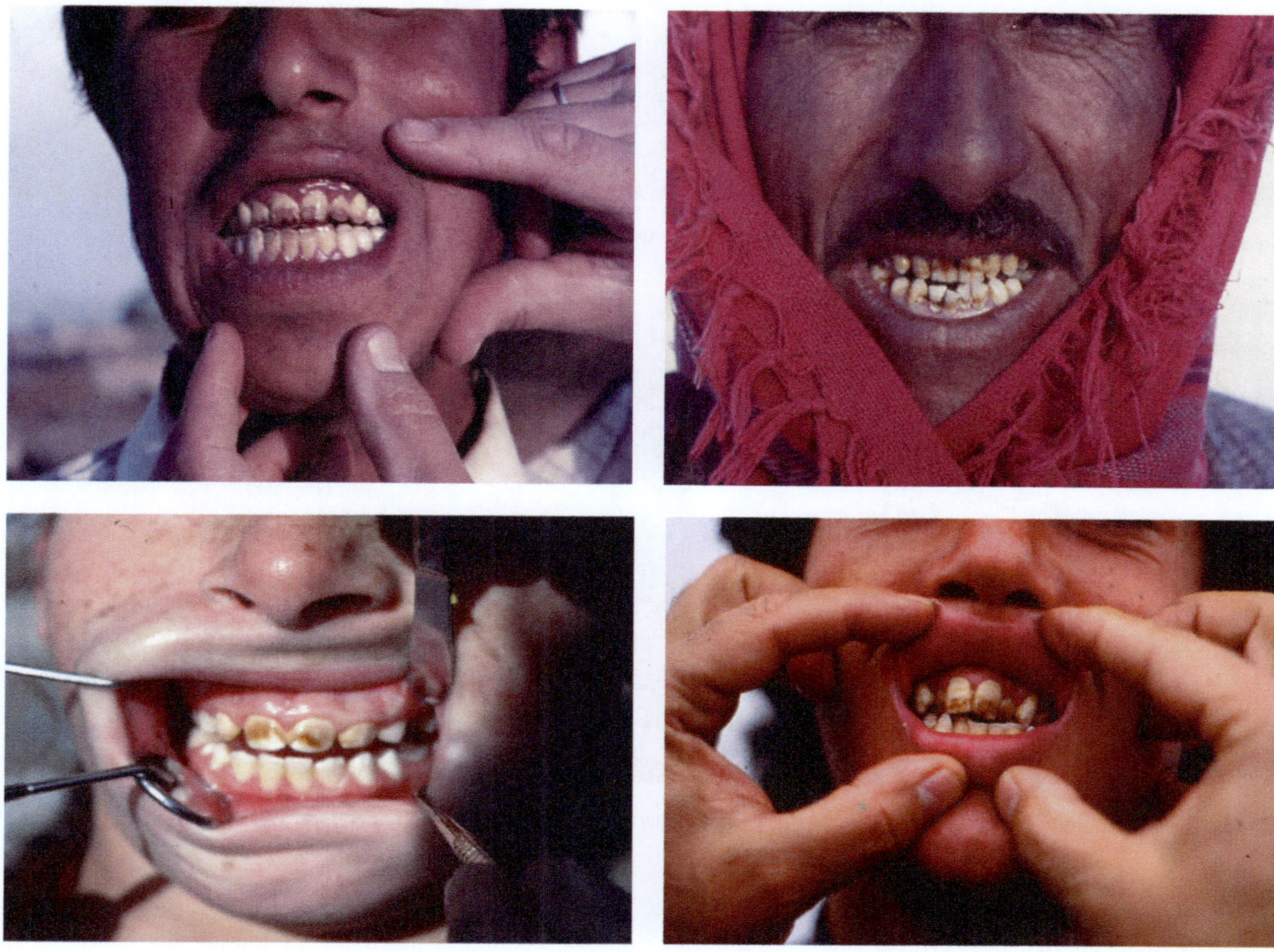

Figure 22.4 Dental fluorosis cases around Tendürek Volcano. *Source:* Oruç (1976)/Springer Nature.

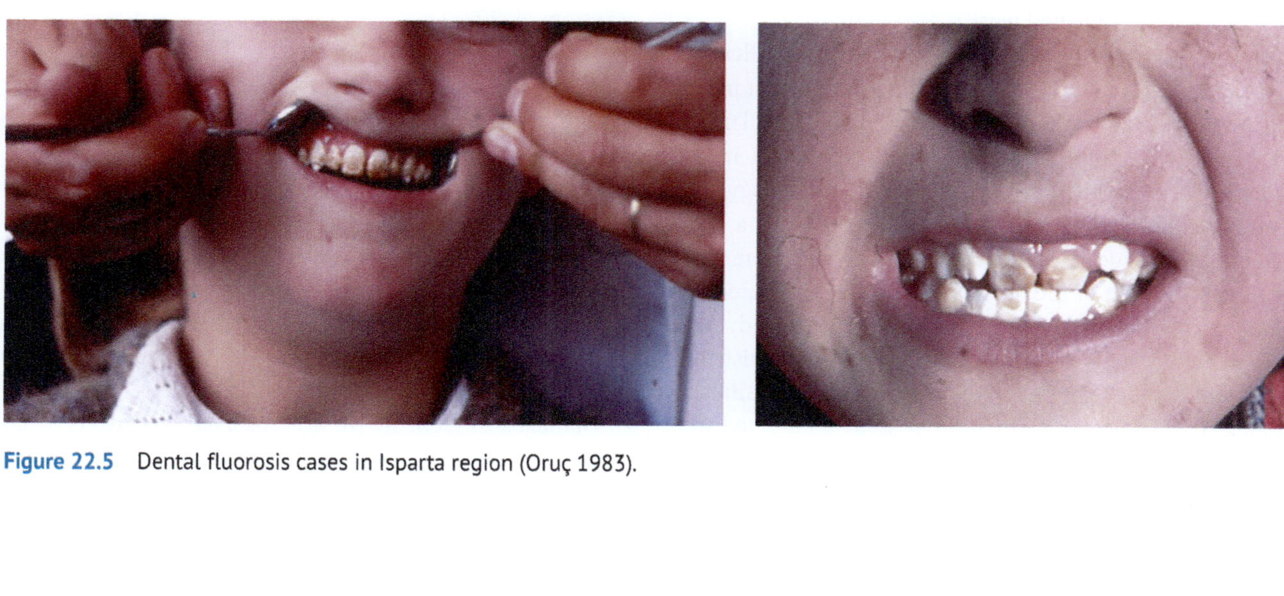

Figure 22.5 Dental fluorosis cases in Isparta region (Oruç 1983).

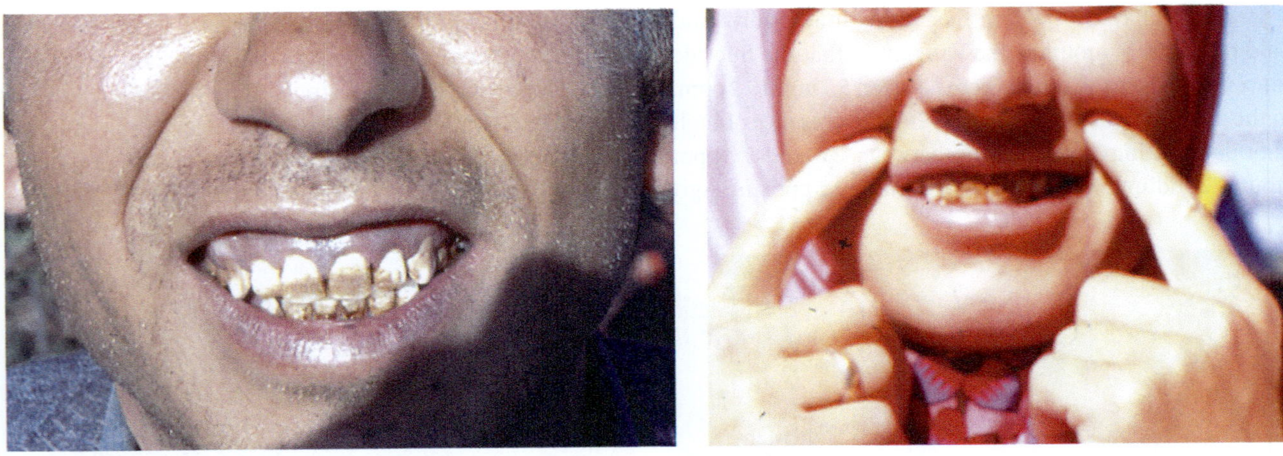

Figure 22.6 Dental fluorosis cases in Eskişehir region (Oruç and Akşit 1989).

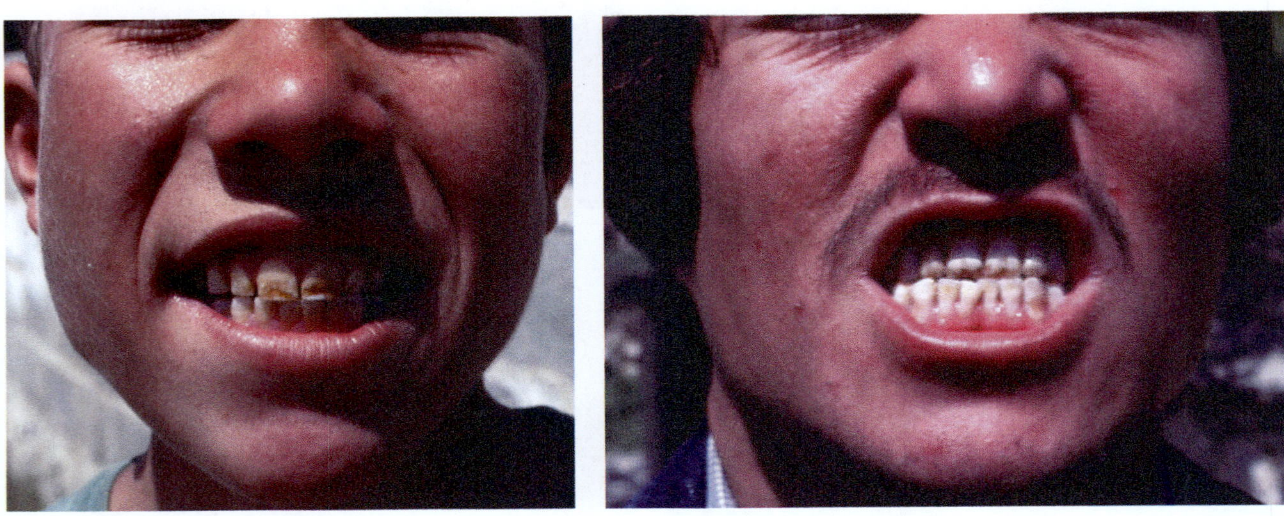

Figure 22.7 Dental fluorosis cases in Uşak/Eşme region. *Source:* Oruç and Vicil (1984)/Springer Nature.

Later, cases of dental fluorosis due to high fluoride levels in waters were reported in provinces such as Şanlıurfa, Kırıkkale, Bitlis, and Mardin (Figures 22.8–22.10) (Yesilnacar et al. 2016; Karaca 2017; Derin et al. 2018; Yetis et al. 2019) (Unpublished work).

Yesilnacar (2010) first detected excessive fluoride 3.23 mg/L and dental fluorosis cases in groundwater in two villages located in the northwest of Sanlıurfa (Turkey, southeast of Turkey). In other studies conducted in the same region, the geological structure, hydrogeology, and hydrochemistry of the region were examined and a comparison was made with the dental structure of the local people in the light of the data provided. All data were evaluated with GIS, and a geomedical map of the study area was created. Finally, the results about the source, distribution, and size of the problem were reached (Yesilnacar et al. 2016) (Figure 22.8). In another study conducted specifically for the region, fluoride values and degrees of dental fluorosis were estimated using artificial neural network techniques based on the parameters measured in water (Atas et al. 2021). Again, based on rock data, fluoride predictions were made using machine learning techniques (Yetis et al. 2021).

22.3.1 Clinical Appearance of Dental Fluorosis

DF is characterized by enamel porosity and hypomineralization only in mild and moderate cases. In the advanced stages, pit formation on the enamel surface and loss of the enamel layer after applying and brown-staining on the exposed surfaces occur (Den Besten and Li 2011). Dentin may also be affected in cases of advanced fluorosis (Fejerskov et al. 1996).

Damages to teeth in DF are observed at various levels and are irreversibly as white or brown diffused spots and

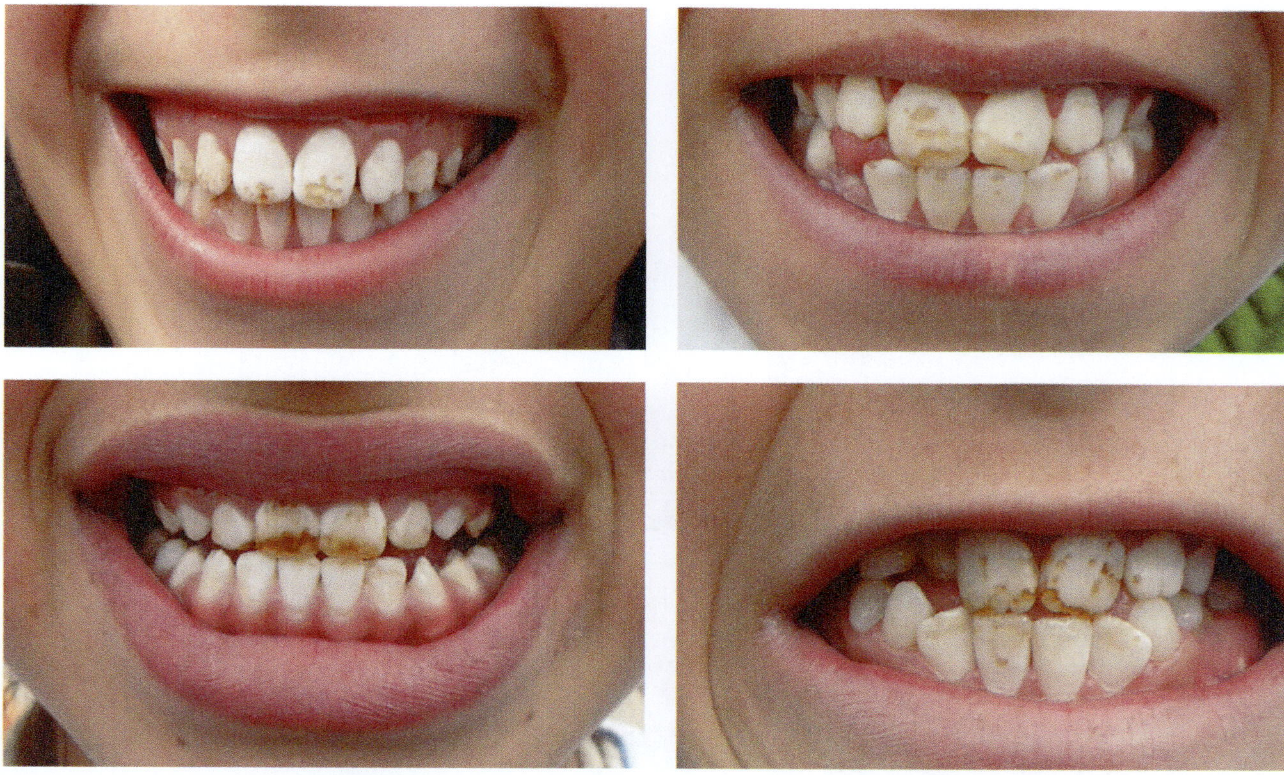

Figure 22.8 Şanlıurfa dental fluorosis cases. *Source:* (Atas et al. (2021)/Springer Nature.

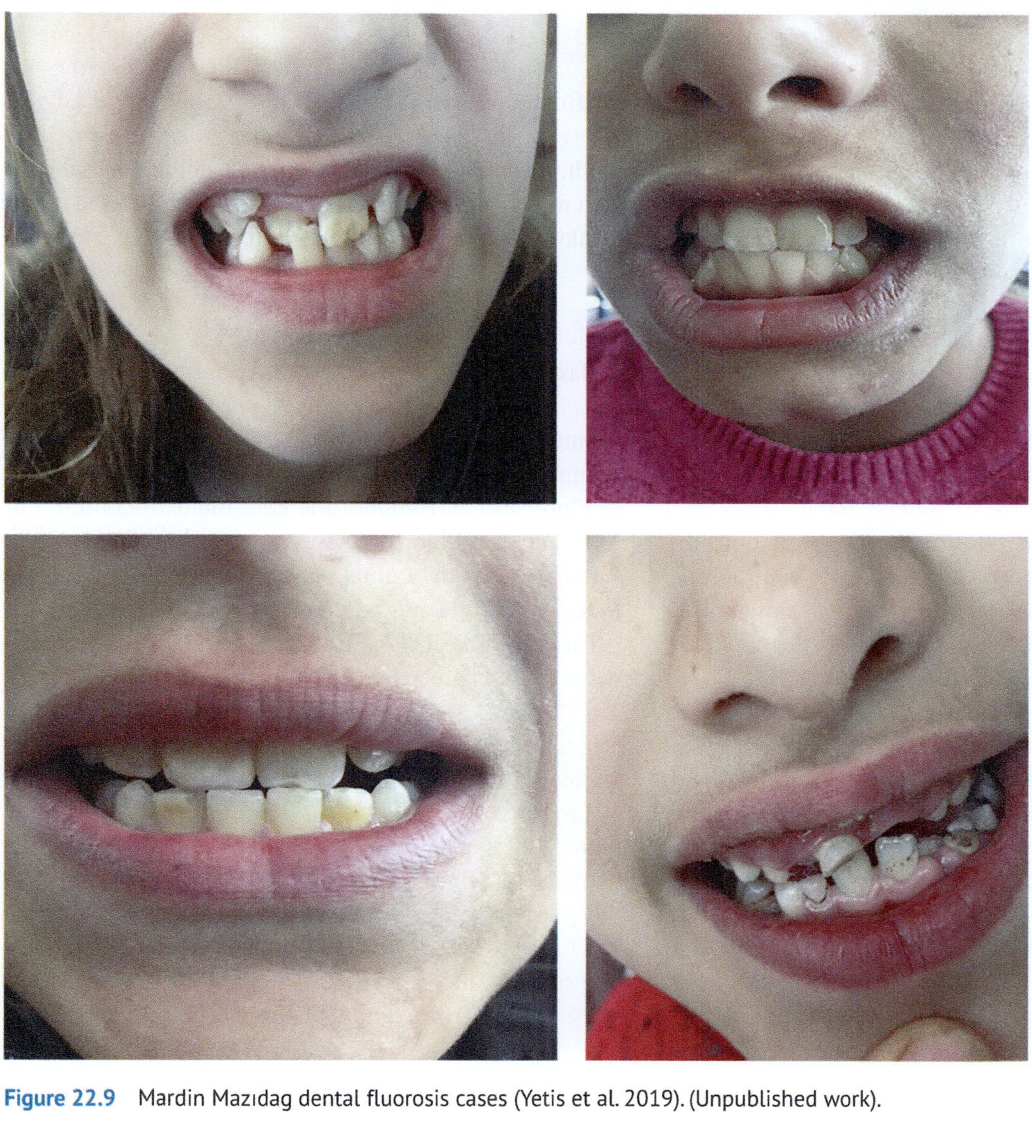

Figure 22.9 Mardin Mazıdag dental fluorosis cases (Yetis et al. 2019). (Unpublished work).

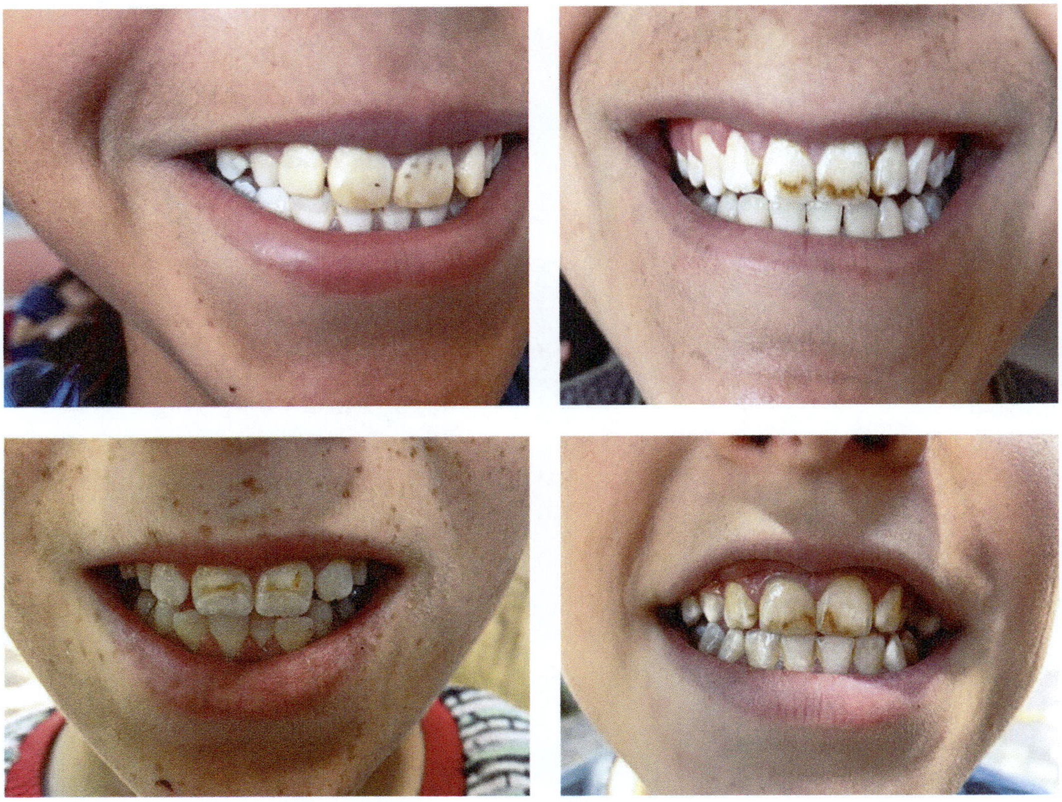

Figure 22.10 Dental fluorosis cases in Bitlis and its districts (Yetis et al. 2019) (Unpublished work).

irregular lines or areas of coloration up to surface loss. Surfaces affected by DF tend to demineralize relative to intact teeth characterized as more resistant areas (Marin et al. 2016). There are studies reporting that there is no correlation between caries-prone demineralized areas and DF-affected areas (Dean 1938; Tellez et al. 2011; Do et al. 2016).

However, it should be associated with the patient's oral hygiene habits and should be considered as a predisposing factor to gum diseases and caries (Li et al. 2014; Moimaz et al. 2015; Wong et al. 2016).

DF, which can only occur during the formation of teeth, usually occurs in childhood. In permanent dentition, it begins with the lower incisors, which complete mineralization at about 2–3 years of age, and ends after the mineralization of the third molars (Den Besten and Li 2011).

Descriptive and in vitro studies show that the observed chewing surface or incisal edges of DF are weak enough to cause loss of structure in the tooth due to functional and parafunctional forces such as abrasion and attrition (Vora and Vora 2013; Marin et al. 2016).

A decrease in quality of life is observed in patients with DF (Michel-Crosato et al. 2005; Tellez et al. 2011; Saldarriaga et al. 2021). In the literature, there is not enough data in terms of the course of the extent of impact of the surfaces affected by DF after the eruption of the tooth. It is very important for public health to obtain these data with regular controls by recording the fluorine rate of drinking water and the levels of exposure to DF.

22.3.2 Classification of Dental Fluorosis

A classification system is required to define DF, and different indices have been developed for this purpose (Funmilayo and Mojirade 2014). These indexes such as The Community Index (Dean Index), TFI, Total Surface Index of Fluorosis (TSIF), Fluorosis Risk Index (FRI), and The Chronological Fluorosis Assessment Index (CFA) can be summarized as the Dental Fluorosis for Primary Dentition Index (DFPDI) (Karaca 2017).

The first clinical classification to match the pathological tissue changes and clinical images in teeth with fluorosis is TFI (Table 22.3). This index is still used frequently today (Thylstrup and Fejerskov 1978; Kucukesmen and Sonmez 2008).

Karaca (2017) used the TFI score in her study to determine the frequency of dental fluorosis and dental caries in Kırıkkale province (Figure 22.11). As a result of the study, the prevalence of dental fluorosis in Kırıkkale province was found to be 21.8%, and it was determined that the fluorine concentrations in drinking water varied between 0.128 and 5.967 mg/L throughout the province.

Table 22.3 Clinical criteria for the Thylstrup–Fejerskov index ().

0)	After the surface is wiped and dried, there is a shiny normal enamel appearance.
1)	There are transverse white opaque lines along the tooth surface.
2)	White opaque lines are more prominent. The snow cap appearance of the tubercles and cutting edges is common.
3)	It is observed that the white lines converge and areas of cloudy opacity spread over a large part of the entire surface.
4)	The entire surface has a pronounced opacity or a chalky white appearance.
5)	The entire surface is opaque, and there are round pits (local loss of outer enamel) less than 2 mm in diameter.
6)	Small pits are observed to coalesce to form bands less than 2 mm.
7)	Loss of the outermost enamel layer in the form of irregular areas is observed. Less than half of the surface is affected. The remaining intact enamel is opaque.
8)	Loss of the outermost enamel layer is observed on more than half of the surface. The remaining intact enamel is opaque.
9)	It is observed that the surface anatomy of the tooth changes as a result of the loss of a large part of the outermost enamel layer. The remaining tooth shows dark brown discolorations.

Source: Ekstrand et al. (1988)/with permission from Elsevier.

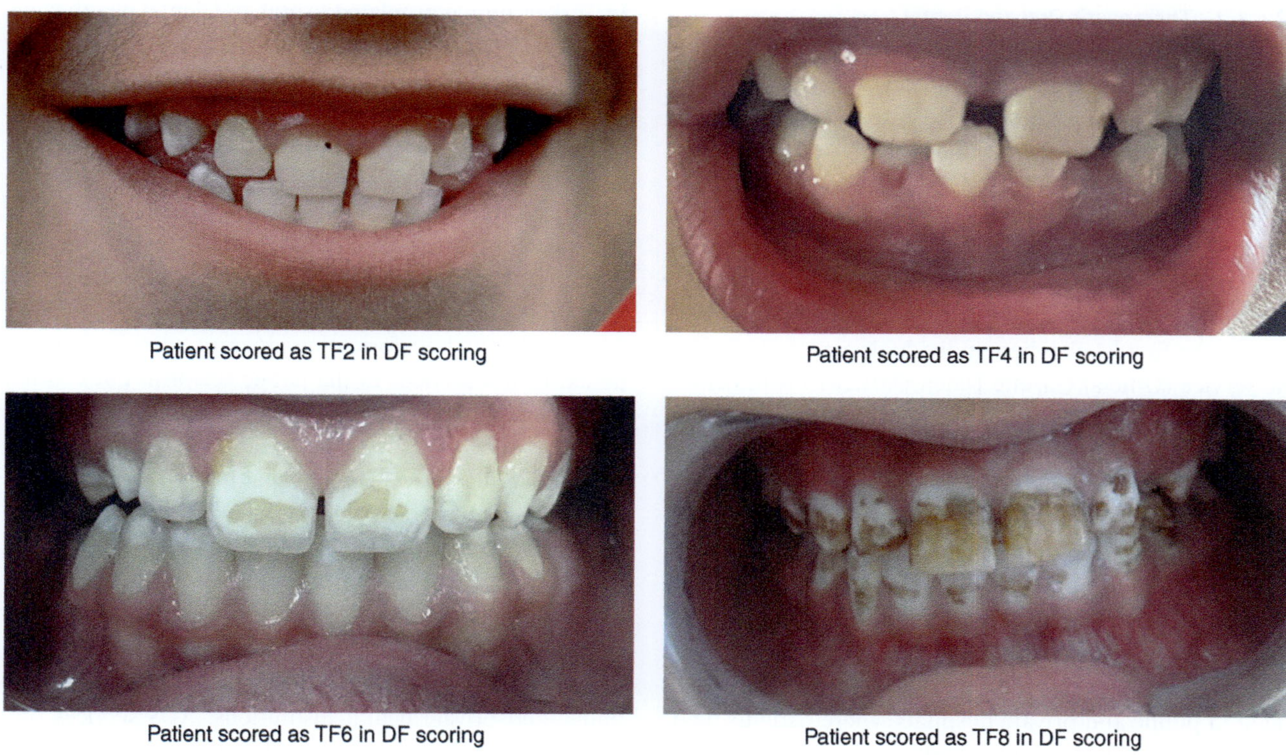

Figure 22.11 Thylstrup–Fejerskov index (TFI) in DF scoring.

22.4 Development of Sustainable Oral Health Improvement and Strategies in Turkey

Oral and dental health is one of the determinants of general health status and is directly related to quality of life. Oral and dental diseases are considered among the important risk factors along with many diseases (chronic diseases such as obesity, heart diseases, stroke, cancer, and diabetes). For this reason, it reveals the fact that oral and dental health cannot be evaluated independently from other basic health systems of the body (Kuzu 2019). These diseases are positioned as the fourth most expensive disease worldwide, and dental caries is still one of the most common chronic diseases (Cam and Kumru 2020).

The WHO defines oral and dental health as "diseases that limit the individual's biting, chewing, laughing, speaking and psychosocial well-being, as well as the absence of diseases such as mouth and face pain, mouth and face cancer, mouth infections and wounds, periodontal disease, tooth decay, tooth loss." (Mohanta 2017).

22.4.1 Classification of Oral and Dental Health in Turkey

Health service, preventive health services, curative health services, rehabilitation services, and health promotion services are collected in four groups.

22.4.1.1 Protective Oral and Dental Health Services

It is the oral and dental health services applied before losing health (Ekici 2013). Among the preventive oral and dental health services, an example of preventive oral and dental health services against the environment can be given as an adequate level of fluoride application of water and rearrangement of food contents in accordance with oral and dental health.

22.4.1.2 Therapeutic Oral and Dental Health Services

As in other health services, oral and dental health and therapeutic health services are divided into three as primary, secondary, and tertiary oral and dental health services (Kuzu 2019).

Primary health care services: Health promotion is defined as the provision of effective and widespread health services in which preventive health services are provided together with diagnosis, treatment, and rehabilitation services at the first level, where individuals can easily access the service (Official Gazette Date: 25.01.2013 Number of Official Gazette: 28539: Family Medicine Practice Regulation 3/f).

With Family Health Centers and Community Health Centers, nutrition, mental health, chronic diseases and physical activity, reproductive health, etc., Healthy Life Centers, where preventive oral and dental health consultancy services are provided as well as counseling services, are the centers where primary health care is provided.

Secondary health services: It consists of services provided in public and private hospitals where patients who cannot be treated in primary health care institutions are treated and rehabilitated, as outpatient or inpatient, and services provided in Oral and Dental Health Hospitals and Centers.

Tertiary health services: It consists of advanced health services offered in Training and Research Hospitals affiliated to the Ministry of Health and Research and Application Hospitals affiliated with universities.

22.4.1.3 Rehabilitation Oral and Dental Health Services

It is the oral and dental health services that include the correction of the disorder caused by disease or trauma. It includes acceptable regulation of the disorder, even if not completely resolved.

22.4.1.4 Oral and Dental Health Promotion Services

Oral and dental health promotion services are quite comprehensive. The development of oral and dental health services aims to increase the quality of life and a better oral and dental health. Improving oral hygiene habits, avoiding alcohol and tobacco products, which are the common causes of many diseases, can be given as examples of oral and dental health promotion services (Kuzu 2019).

22.4.2 Current Data on Oral and Dental Health Services in Turkey

In Turkey, problems related to oral and dental health come to the fore both in terms of health level and service delivery and financing. The 2003 Health Transformation Program, which is considered as a restructuring process in the entire health system, brought important changes in the fight against these problems related to oral and dental health services. In this process, it was aimed to raise awareness of the society in terms of oral and dental health, to provide education and to complete the structuring for treatment with preventive medicine studies (Cam and Kumru 2020). After the health transformation program, access to oral and dental health services in the public sector in our country has become easier. In particular, there has been an increase in the number of dentists and the number of public institutions providing oral and dental health services.

The number of 25 dentists per 100000 people in 2002 increased to 35 dentists in 2017 and to 40 dentists in 2019 (Health Statistics Yearbook 2019, p. 137).

The latest data from the Ministry of Health (MoH) is presented in the 2019 Health Statistics Yearbook. Oral and dental health (ADS) services in Turkey are provided by both public and private health institutions. ADS services provided by the public are provided by oral and dental health hospitals (ADSH), oral and dental health centers (ADSM), and general hospitals and training hospitals, as well as Community Health Centers and Healthy Life Centers.

According to the Health Statistics of 2019, the total number of institutions providing oral and dental health services is 3431, of which 983 belong to the MoH, 65 universities and 2383 private enterprises (Table 22.4). In addition, 35% (11387) of 32925 dentists are from the MoH, 13% (4425) are universities, and 52% (17113) are private sector employees. The number of dentists of the MoH increased from 9768 to 11387 in 2019 (Health Statistics Yearbook 2019, p. 137).

Table 22.4 Institutions and organizations providing oral and dental health services by sectors.

	Ministry of Health		University		Special		Total	
	Organization	Unit	Organization	Unit	Organization	Unit	Organization	Unit
Oral and dental health center	131	4.411	—	—	83	909	214	5.320
Dental hospital	29	2.052	—	—	3	85	32	2.137
Dental training hospital	3	265	50	5.504	—	—	53	5.769
Dental clinic (hospital)	820	3.222	15	368	211	477	1.046	4.067
Dental clinic	—	—	—	—	2.123	8.701	2.123	8.701
Total	983	9.950	65	5.872	2.420	10.172	3.468	25.994

Source: Health Statistics Yearbook (2019).

22.4.3 Treatment Options for Dental Fluorosis

Fluorine plays an active role in the pH balance, which is critical in the maturation stage of tooth enamel. At this stage, various levels of defects are observed in the enamel exposed to fluorosis (Savas and Kucukyilmaz 2014). Pedodontists provide early diagnosis of defects, treatment, and control with a preventive approach. In the diagnosis and treatment of dental fluorosis, it is necessary for pedodontists to cooperate with all health professionals and to provide process management of fluorosis with a holistic approach. The most effective treatment of DF, which causes significant esthetic problems in patients, is to limit the use of fluoride and to prevent the formation of this pathology (Pendrys et al. 2010). However, there are treatment options suitable for the patient's condition for loss of enamel tissue and tooth discoloration. These are microabrasion, bleaching, resin infiltration, composite restorations, esthetic veneers, or prosthetic crowns (Shafiei et al. 2014; Tau and Kurthy 2014).

In early childhood, DF, which will occur due to the exposure of enamel to excess flora, can be diagnosed early by dentists and especially pedodontists and treated and controlled with a preventive approach. In the diagnosis and treatment of DF, it is necessary to cooperate with all health professionals and to ensure the process management of fluorosis with a holistic approach. In addition, it is of great importance for public health that the health authorities of our country develop policies for a solution immediately. There is no need to use fluoride supplements if the drinking water contains (and is consumed) 0.7 mg/L or more of fluoride. Depending on the amount of fluoride in the drinking water, it can be supplemented with tablets and fluoride toothpastes. Considering that very young children can swallow large amounts of toothpaste, it is recommended to use pastes with low fluoride content and to supplement in a controlled manner, considering that swallowed fluoride toothpaste may increase the risk of dental fluorosis (Gözetici and Bozkurt 2013).

References

Aksit, M.A., Tel, E., and Bilir, S. (1980). Kızılcaoren-a health survey in an endemic fluorosis village. *Fluoride* 13 (2): 81–85.

Alimohammadi, M., Nabizadeh, R., Yaghmaeian, K. et al. (2018). Data on assessing fluoride risk in bottled waters in Iran. *Data in brief* 20: 825–830. https://doi.org/10.1016/j.dib.2018.08.160.

Armienta, M.A. and Segovia, N. (2008). Arsenic and fluoride in the groundwater of Mexico. *Environmental Geochemistry and Health* 30 (4): 345–353. https://doi.org/10.1007/s10653-008-9167-8.

Ata, P. (1966). *Conservative Dental Treatment (Fluorosis)*, 102–145. Istanbul: Innovation Press.

Atabey, E. (2010). Türkiye'de İçme Suyunda Flor ve Etkileri. Ankara, Ares Basım Reklam Prodüksiyon Hizmetleri, Yer Bilimleri ve Kültür Serisi-9 (in Turkish).

Atabey, E., Ünal, H. (2008). Batı Anadolu'daki jeolojik unsurlar ve halk sağlığı projesi 2006-2007 yılı Tıbbi Jeolojik etüt raporu, p. 292. MTA Rapor No: 11067, Ankara (in Turkish).

Atas, M., Yesilnacar, M.İ., and Demir, Y. (2021). A novel machine learning techniques-based hybrid models (LR-KNN-ANN and SVM) in prediction of dental fluorosis in groundwater. *Environmental Geochemistry and Health* https://doi.org/10.1007/s10653-021-01148-x.

Babacan, E. (1979). Ağrı ili Doğubeyazıt ilçesi köylerinde kronik flor zehirlenmesi görülen koyun ve sığırlarda kan bulguları üzerinde çalışmalar. Fırat Universitesi Veteriner Fakültesi Yayınları, 23, Elazığ. (Abstract in English)

Baydaş, S. (1972). Fluorosis in Doğubeyazıt villages. *Dentoral* 4 (18): November–December.

Bilgin, A., Ismailov, T., and Caran, S. (2005). Fluorine, iodine, arsenic levels in spring waters in Isparta region and human-induced (anthropogenic) pollution and its effects on human health. In: *1st Medical Geology Symposium Book*, vol. 95, 228–239. Ankara: TMMOB Chamber of Geological Engineers Publication.

Buzalaf, M.A., Massaro, C.S., Rodrigues, M.H. et al. (2012). Validation of finger nail fluoride concentration as a predictor of risk for dental fluorosis. *Caries Research* 46: 394–400.

Cam, F. and Kumru, S. (2020). Patient perception in oral and dental health and evaluation of public oral and dental health services from patient perspective. *Journal of Health and Society* 30 (2): 123–139.

Chandrajith, R., Diyabalanage, S., and Dissanayake, C.B. (2020). Geogenic fluoride and arsenic in groundwater of Sri Lanka and its implications to community health. *Groundwater for Sustainable Development* 10: 100359. https://doi.org/10.1016/j.gsd.2020.100359.

Dean, H.T. (1938). Endemic fluorosis and its relation to dental caries. *Public Health Reports 1896–1970* 53 (33): 1143–1452.

Demirel, U. (2009). The effect of fluor element on living beings and the fact of fluorosis in the Cappadocia region. In: *Nevşehir: 1st Medical Geology Workshop, TMMOB Chamber of Geological Engineers* (ed. Y. Örgün and G. Yalçın), 187–198.

Den Besten, P. and Li, W. (2011). Chronic fluoride toxicity: dental fluorosis. *Monographs in Oral Science* 22 (1): 81–96.

Derin, P., Yesilnacar, M.I., Bayhan, I. et al. (2018). Investigation of drinking water quality in terms of high fluoride level in a rural area: the case of Sarım and Karataş village (Şanlıurfa), first findings. In: *71st Geological Congress of Turkey* (ed. M. Avcioğlu, T. Kurttaş, F. Toksoy Köksal, et al.). (23-27 April), TMMOB Chamber of Geological Engineers, Ankara.

Do, L.G., Ha, D.H., and Spencer, A.J. (2016). Natural history and long-term impact of dental fluorosis: a prospective cohort study. *The Medical Journal of Australia* 204 (1): 25.

Ekici, O. (2013). *Restructuring of Public Oral and Dental Health Services in Turkey: Problems and Suggestions*. Gazi University.

Ekstrand, J., Fejerskov, O., and Silverstone, L.M. (1988). *Fluoride in Dentistry*, 1e. Copenhagen: Munksgaard 13–25, 96–192, 263–267.

Ergun, H.S., Russel-sinn, H.A., Baysu, N., and Dündar, Y. (1987). Studies on the floride contens in water and soil, urine, bone and teeth of sheep and urine of human from eastern and western parts of Turkey. *Deutsche Tierärztliche Wochenschrift* 94: 416–420.

Everett, E.T. (2011). Fluoride's effects on the formation of teeth and bones, and the influence of genetics. *Journal of Dental Research* 90: 552–560.

Fejerskov, O., Manji, F., and Baelum, V. (1990). The nature and mechanisms of dental fluorosis in man. *Journal of Dental Research* 69: 692–700.

Fejerskov, O., Ekstrand, J., and Burt, B. (1996). *Fluoride in Dentistry*. Copenhagen: Munksgaard.

Fidanci, U.R., Salmanoglu, B., Maraşli, S., and Maraşli, N. (1998). Natural and industrial fluorosis and its effects on animal health in the Central Anatolia region. *Turkish Journal of Veterinary and Animal Sciences* 22: 537–544.

Forsman, B. (1974). Dental fluorosis and caries in high-fluoride dis-tricts in Sweden. *Community Dentistry and Oral Epidemiology* 2: 132–148.

Funmilayo, A.M. and Mojirade, A.D. (2014). Dental fluorosis and its indices, what's new? *IOSR Journal of Dental and Medical Sciences (IOSR-JDMS)* 13: 55–60.

Gözetici, B. and Bozkurt, F.Ö. (2013). Diş hekimliğinde florozis. *ADO Klinik Bilimler Dergisi* 7 (1): 1399–1408.

Gulhan, A. and Mermutlu, B. (1973). Relations entre la fluorose endemique et la frequence de la carie. *Journal of Istanbul University Faculty of Dentistry* 7 (2): 130–140.

Health Statistics Yearbook (2019). Ministry of Health Ankara; 2019.

Isikli, B. and Kalyoncu, C. (2000). Fluoride levels in trime waters in Eskişehir region. *Journal of Ecology Environment* 9 (36): 28–30.

Jha, S.K., Mishra, V.K., Sharma, D.K., and Damodaran, T. (2011). Fluoride in the environment and its metabolism in humans. *Reviews of Environmental Contamination and Toxicology* 211: 121–142.

Karaca, H. (2017). Kırıkkale ilinde 13–14 yaş grubu okul çocuklarında dental florozis prevelansının diş çürüğü ve içme suyu flor konsantrasyonları ile ilişkisinin değerlendirilmesi. Doktora Tezi. KKÜ Diş Hekimliği Fakültesi, Pedodonti ABD, pp. 35–42.

Kimambo, V., Bhattacharya, P., Mtalo, F. et al. (2019). Fluoride occurrence in groundwater systems at global scale and status of defluoridation–state of the art. *Groundwater for Sustainable Development* 9: 100223.

Kucukesmen, Ç. and Sonmez, H. (2008). Diş hekimliğinde florun, insan vücudu ve dişler üzerindeki etkilerinin değerlendirilmesi. *Süleyman Demirel Üniversitesi Tıp Fakültesi Derg* 15 (3): 43–53.

Kuzu, Ö.F. (2019). Review of oral and dental health policies in Turkey and proposal of a new model: a public hospital example. Master thesis. Institute of Social Sciences, Cumhuriyet University. p. 105.

Levy, S.M. (2003). An update on fluorides and fluorosis. *Journal of the Canadian Dental Association* 69: 286–291.

Li, Y.J., Gao, Y.H., and Zhang, Y. (2014). The impact of oral health status on the oral health-related quality of life (OHRQoL) of 12-year-olds from children's and parent's perspectives. *Community Dental Health* 31 (4): 240–244.

Marin, L.M., Cury, J.A., Tenuta, L.M. et al. (2016). Higher fluorosis severity makes enamel less resistant to demineralization. *Caries Research* 50: 407–413.

Michel-Crosato, E., Biazevic, M.G.H., and Crosato, E. (2005). Relationship between dental fluorosis and quality of life: a population based study. *Brazilian Oral Research* 19: 150–155.

Mohanta, A. (2017). Oral health in 21 st century-an emerging challenge. *EC Dental Science* 8 (1): 12–14.

Moimaz, S.A.S., Saliba, O., Marques, L.B. et al. (2015). Dental fluorosis and its influence on children's life. *Brazilian Oral Research* 29 (1): 1–7.

Murray, J.J. and Shaw, L. (1979). Classification and prevalence of enamel opacities in the human deciduous and permanent dentitions. *Archives of Oral Biology* 24: 7–13.

Official Gazette Date: 25.01.2013 Number of Official Gazette: 28539: Family Medicine Practice Regulation 3/f).

Oksuz, S. (2006). Fluorine content of Pöhrenk (Kırşehir) hot water springs. Master thesis. Çukurava University of Science and Science Institute. p. 40

Onipe, T., Edokpayi, J.N., and Odiyo, J.O. (2020). A review on the potential sources and health implications of fluoride in groundwater of Sub-Saharan Africa. *Journal of Environmental Science and Health, Part A* 55 (9): 1078–1093.

Oruç, N. (1976). Concentration and importance of fluorine in some natural waters around Lake Van. *Atatürk University, Faculty of Agriculture, Journal* 7 (3): 25–32.

Oruç, N. (1977). A preminary study on the effect of water borne fluoride on the fluoride contet of soils and plants. In: *Seminar on Problems of High Fluoride Waters. Cento Scientific Programme*. Erzurum: Ataturk University.

Oruç, N. (1983). Eskişehir Mahmudiye Yeşilyurt köyü içme sularında florür düzeyi ve önemi. Eskişehir Anadolu Üniversitesi, Mühendislik-Mimarlık Fakültesi, İnşaat Mühendisliği Bölümü.(in Turkish, unpublished work).

Oruç, N. (2008). Occurrence and problems of high fluoride waters in Turkey: an overview. *Environmental Geochemistry and Health* 30 (4): 315–323.

Oruç, N. and Vicil, M. (1984). The level and origin of fluoride in Güllü Village (Uşak-Eşme) drinking water, Turkish geology institute 38. In: *Scientific and Technical Congress (Abstracts)*, vol. 83, 20–24. Ankara: MTA.

Oruç, N., and Akşit, M.A. (1989). A health survey in a village with endemic fluorosis, 1978–1988. Cevre, 89, 7–9.

Oruç, N., Alpman, N., and Karamanderesi, I.H. (1976). Hydrogeology of spring waters with high fluoride content around Tendürek volcano. *Bulletin of the Turkish Geological Institute* 19: 1–8.

Pendrys, D.G., Haugejorden, O., Bårdsen, A. et al. (2010). The risk of enamel fluorosis and caries among Norwegian children: implications for Norwayand the United States. *Journal of the American Dental Association (1939)* 141 (4): 401–414.

Ruan, J.P., Wang, Z.L., Yang, Z.Q. et al. (2005). Dental fluorosis in primary teeth: a study in rural schoolchildren in Shaanxi Province, China. *International Journal of Paediatric Dentistry* 15: 412–419. https://doi.org/10.1111/j.1365-263X.2005.00667.x.

Saldarriaga, A., Rojas-Gualdrón, D.F., Restrepo, M. et al. (2021). Clinical changes in the severity of dental fluorosis: a longitudinal evaluation. *BMC Oral Health* 21: 366.

Savas, S. and Kucukyilmaz, E. (2014). Remineralization agents and anti-caries agents used in dentistry. *Journal of Atatürk University Faculty of Dentistry* 24 (3): 113–125.

Şendil, C., Baysu, N., 1973. The first report on fluorine poisoning in humans and animals in the villages of Doğubeyazıt district of Ağrı province and our detection in the villages of Muradiye district of Van province. *Ankara Universy Vet Fak Journal* 10:474–489.

Sezgin, B.I., Onur, Ş.G., Mentes, A. et al. (2018). Two-fold excess of fluoride in the drinking water has no obvious health effects other than dental fluorosis. *Journal of Trace Elements in Medicine and Biology 50*: 216–222.

Shafiei, F., Tavangar, M., and Alavi, A. (2014). Direct esthetic rehabilitation of teeth with severe fluorosis: a case report. *Journal of Dentistry (Shiraz)* 15: 44–47.

Susheel, A. (2005). Fluorosis: an easily preventable disease through practice of interventions. In: *Doctor's Handbook*, 1ee, 7–14. New Delhi: Ministry of Health and Family Welfare (GOI) and WHO India Country Office.

Tanyeri, K. (1975). Endemic fluorosis in the Eastern Anatolia region, Atatürk University. *Medical Bulletin* 7 (3): 69–75.

Tau, L. and Kurthy, R. (2014). Minimally invasive treatment of brown spot fluorosis. *Dentistry Today* 33 (140): 142–143.

Tellez, M., Santamaria, R.M., Gomez, J., and Martignon, S. (2011). Dental fluorosis, dental caries, and quality of life factors among school children in a Colombian fluorotic area. *Community Dental Health* 29 (1): 95–99.

Thylstrup, A. and Fejerskov, O. (1978). Clinical appearance of dental fluorosis in permanent teeth in relation to histologic changes. *Community Dentistry and Oral Epidemiology* 6: 315–328.

Tokatli, C. and Guner, S. (2020). Fluoride levels in drinking water and assessment of water quality in terms of teeth health in a significant watershed in Thrace region: Ergene river basin. *Acta Aquatica Turcica 16* (2): 238–245.

Velicangil, S. and Demirhindi, O. (1957). Isparta'da endemik dental florozis ve bu mevzu ile alakalı bazı problemler üzerinde araştırmalar. *Istanbul Üniversitesi, Tıp Fakültesi* 20 (1): 94–110. (Abstract in English).

Vora, K.S. and Vora, P.K. (2013). Assessment of periodontal status of the patients with dental fluorosis in area with natural high levels of fluoride: a cross-sectional survey. *Dental Hypotheses* 4: 127.

Warren, J.J., Levy, S.M., and Kanellis, M.J. (2001). Prevalence of dental fluorosis in the primary dentition. *Journal of Public Health Dentistry* 61: 87–91.

WHO (2011). *Water-Related Diseases Fluorosis World Health Organization (WHO), Water Sanitation and Health.* Fluorosis, Geneva: World Health Organization.

Wong, H.M., Wen, Y.F., King, N.M., and McGrath, C.P.J. (2016). Longitudinal changes in developmental defects of enamel. *Community Dentistry and Oral Epidemiology* 44 (3): 255–262.

Yesilnacar, M.I. (2010). Investigation of Problems in Children's Teeth on Suspect of High Fluoride in Drinking and Potable Water of Karataş and Sarım Villages. *Preliminary Research Report 2010*, Şanlıurfa Provincial Health Directorate, Şanlıurfa, Turkey, p. 9.

Yeşilnacar, M.I., Atasoy, A.D., Kumral, M. et al. (2013). Yeraltı Sularında Yüksek Florürün Oluşumu, Dağılımı, İnsan Sağlığına Etkileri ve Florür Gideriminin Araştırılması, KB Şanlıurfa. TUBITAK Project, (110Y234). (in Turkish)

Yesilnacar, M.I., Demir-Yetis, A., Dulgergil, Ç.T. et al. (2016). Geomedical assessment of an area having high-fluoride groundwater in Southeastern Turkey. *Environmental Earth Sciences* 75: 162.

Yetis, A.D., Yesilnacar, M.I., and Atas, M. (2021). A machine learning approach to dental fluorosis classification. *Arabian Journal of Geosciences 14* (2): 1–12.

Yetis, R., Atasoy, A.D., Demir Yetis, A. et al. (2019). Hydrogeochemical characteristics and quality assessment of groundwater in Balikligol Basin, Sanliurfa, Turkey. *Environment and Earth Science* 78: 331. https://doi.org/10.1007/s12665-019-8330-0.

Yousefi, M., Ghoochani, M., and Mahvi, A.H. (2018). Health risk assessment to fluoride in drinking water of rural residents living in the Poldasht city, Northwest of Iran. *Ecotoxicology and Environmental Safety* 148: 426–430.

23

Environmental and Medical Geology of the Lead Mining and Metallurgical Complex of Bahia, Brazil

The Case of the Boquira Lead Mines

José Ângelo Sebastião Araujo dos Anjos[1], *Fernanda Gonçalves da Cunha*[2], *Eduardo Paim Viglio*[2], *and Fábio Carvalho Nunes*[3]

[1] *Department of Geology, Institute of Geosciences, Federal University of Bahia, Salvador, BA, Brazil*
[2] *Geological Survey of Brazil (CPRM), Rio de Janeiro, Bahia, Brazil*
[3] *Academic Department, Federal Institute Baiano (IFBAIANO), Santa Inês, Bahia, Brazil*

23.1 Introduction

Though pioneered in the 1980s by a multidisciplinary team of researchers from the Federal University of Bahia, it is only relatively recently that studies on the environmental and medical geology have been carried out in that region of northeastern Brazil. The studies begun by examining the levels of contamination around the Baia de Todos os Santos (BTS) bay, with the lead metallurgy operations in Santo Amaro area as the main source of pollution by inorganic toxic metals. Subsequently, dozens of studies on mining and metallurgy residues, and contamination of soil, surface and underground water, dust, and vegetation from both Boquira and Santo Amaro were published by different authors, though outside systematized procedures, as established now in the environmental and medical geosciences.

Studies carried out between 1998 and 2003 by the Engineering and Technology School of the University of São Paulo (EPUSP) identified the dispersion of toxic metals within and around the lead metallurgy in Santo Amaro. A partnership in 2013 between the Geosciences Institute of the Federal University of Bahia, the Boquira Public Administration, and the Brazilian Geological Survey (CPRM) has allowed for a more robust and systematized study on the environmental and medical geology of the lead mines in the area of Boquira. These studies had as basic premises: the characterization of the toxicity of the contamination sources (sterile and mining tailings); the assessment of the main contamination routes (dust, surface, and groundwater); and the proposition of studies to characterize the possible contamination of the final recipients (animals and the local population), besides the proposition of initiatives to mitigate the problems related to the area's territorial planning.

The studies led up to the publication in 2016 of the "Boquira's Geochemical Studies" and the "Geochemical Atlas of the Subaé River Basin," by the Geological Survey of Brazil (CPRM). The studies were developed, respectively, in the lead mine of Boquira Mineração Ltda, located in the town of Boquira, in the semiarid region of Bahia, and in the Plumbum Mineração e Metalurgia Ltda, the lead metallurgy facilities located in the Bay area, in the town of Santo Amaro. Though more than 600 km apart from each other, the two operations constituted the State's Lead Mining and Metallurgical Complex.

It is important to note that the present article reports specifically on the studies carried out in the lead mines of Boquira (Figure 23.1), as characterized by the 30 years of abandonment of the enterprise (from 1992 to 2022), based on a detailed sequence of actions. Thus, for the methodological development of this study, a detailed interaction flowchart was designed, starting from the main sources of anthropogenic contamination, that is, the lead tailings, then the characterization of the main routes of contamination, through groundwater, sediment from the streets, and dust from the buildings, besides the indication of possible final receptors.

23.2 Results and Discussions

Located in the State's semiarid western central mesoregion, between northern Espinhaço Range to the west and the Chapada Diamantina Highlands to the east

Medical Geology: En route to One Health, First Edition. Edited by Majeti Narasimha Vara Prasad and Meththika Vithanage.

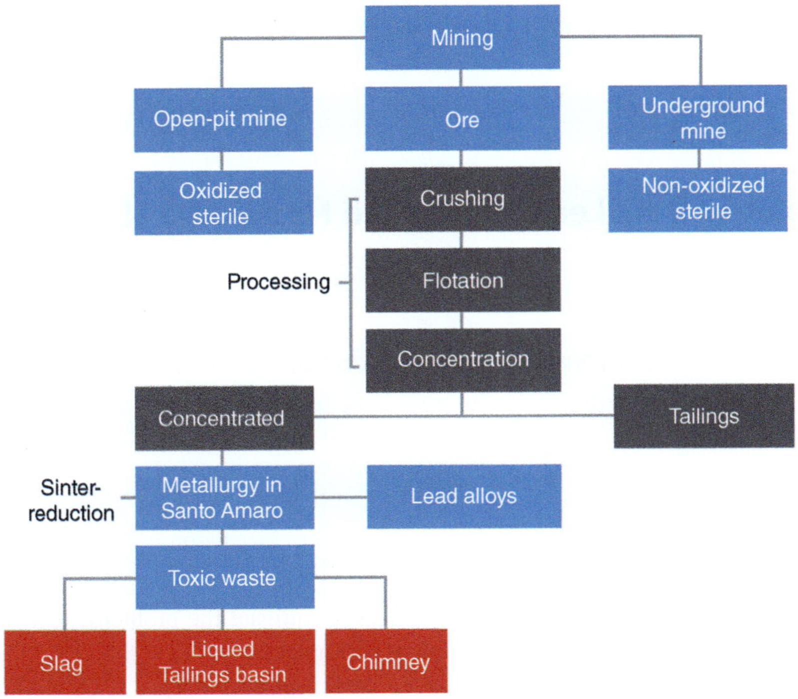

Figure 23.1 Flowchart of the mining and processing of lead ore. *Source:* Adapted from Santos (2014).

(Figure 23.2), Boquira became the main producer of lead ore in Brazil, between the late 1960s and early 1990s, when its mines were exhausted and the venture was abandoned (Santos 2014).

The lead metallurgical mining complex in Bahia consists of open-pit and underground mines, as well as a beneficiation plant installed and later abandoned in the town of Boquira (Santos 2014; Daltro 2017; Daltro et al. 2020; Gomes et al. 2020; Santos et al. 2020a, b), whereas the lead metallurgical complex installed and later abandoned in the city of Santo Amaro, in the Bay, is composed of the metallurgical complex, liquid tailings basin, chimney for dispersion of atmospheric particulates, and slag dam.

These operations contaminated the environment through the emission of industrial solid waste (Anjos and Sanchez 2021) and contaminated the soil hydromorphic and stream sediments, surface and groundwater, the air (atmospheric particulates) (Cunha et al. 2016), vegetation and animals from the surrounding area, metallurgy, the mangroves of the estuary (sediments, vegetation, and bivalve species), the sediments and bivalve species of the Bay, and in particular, employees who worked in metallurgy, part of the evaluated local population (children and women), in addition to promoting significant conflicts related to land use, and the lack of a specific program for the health of the population (Cunha et al. 2016; Daltro 2017; Daltro et al. 2020; Gomes et al. 2020).

However, during the lead ore mining period, legal procedures for mine closure were not well established in the country, which fostered inappropriate actions in the waste disposal, in the design of cut and fill slopes in open-pit mines, and especially in the control and mitigation of sources and routes of contamination of toxic waste produced in the lead ore processing plant (Anjos and Sanchez 2021).

It is important to emphasize that the mining complex was developed in a region with high climatic vulnerability, low rainfall and water deficit, and high temperatures, experiencing the transformation of a village into a mining city, thus compromising local territorial planning (Figure 23.3), with the growth of Boquira's urban area around the toxic tailings basin.

According to (Garcia 2011), the geology of the region is characterized by lithostratigraphic units of Archean orthogneiss/migmatite composition of the Paramirim Complex, mapped as the oldest lithotypes of this complex. However, these Archean and Paleoproterozoic units that make up the basement of the Paramirim Aulacogen outcrop in the homonymous valley to the east and constitute a distinct geological-geomorphological domain of monotonous relief, with higher elevations to the west related to the Espinhaço Range, representing more recent rocks. The volcano-(plutonic)-sedimentary association (greenstone belt), found in the area of the mines, is divided into three units: Botuporã, Crystals, and Boquira Units.

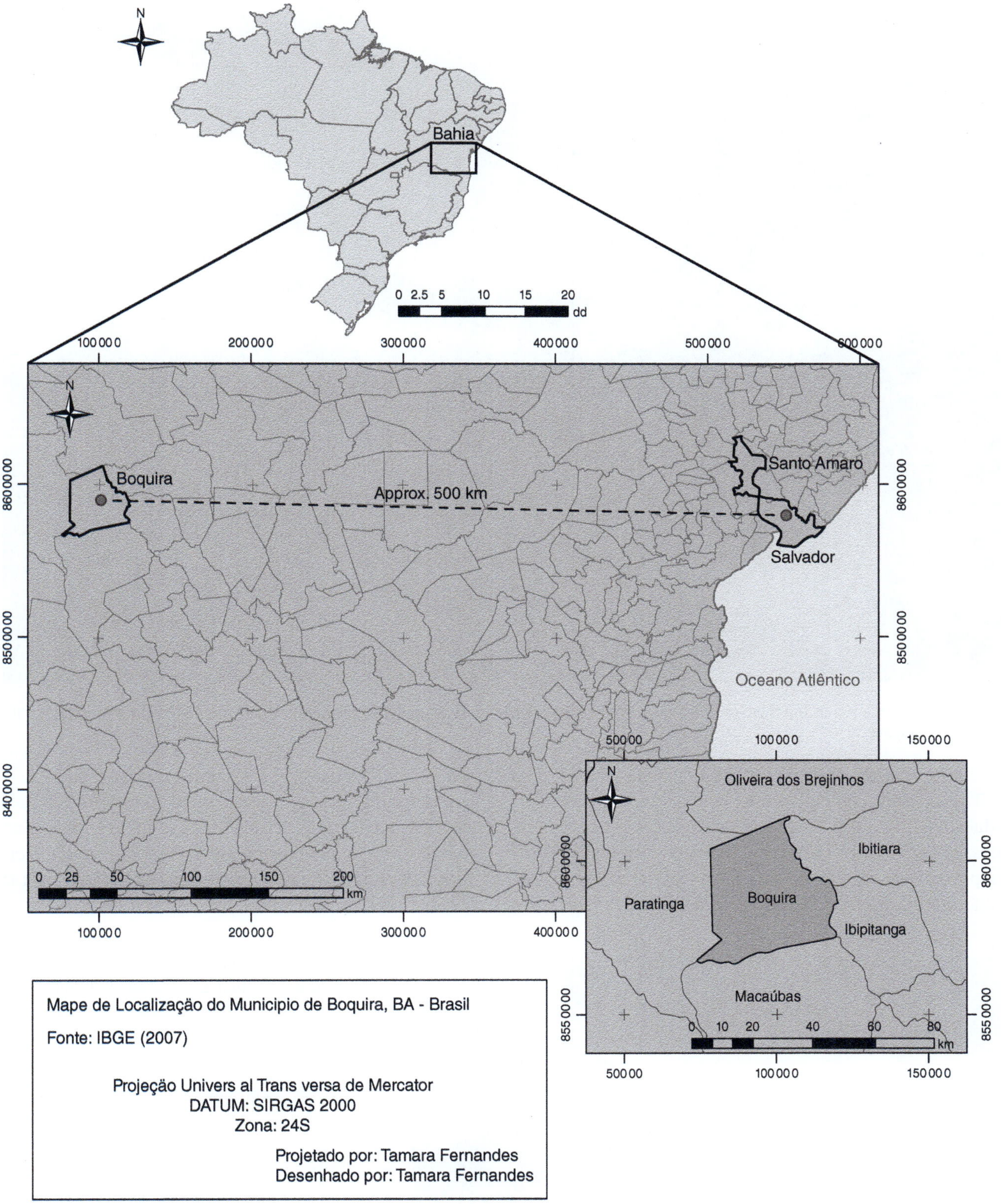

Figure 23.2 Location of the town of Boquira. *Source:* Fernandes, Tamara Grisolia, "Determination of lead contamination in Boquira (BA) as a result of mineral extraction", 2015, Rio de Janeiro State University

Figure 23.3 Location of the town of Boquira and the tailings basin in its immediate surroundings. *Source:* Adapted from Cunha et al. (2016)/DuraSpace.

The Boquira geological unit extends to the towns of Macaúbas, Boquira, and Oliveira dos Brejinhos and is characterized by lenticular bodies interlocked for approximately 65 km in length and preferential NNW-SSE direction, flanking the northern portion of the Espinhaço Range. According to (Garcia 2011), it is a chemical-terrigenous sedimentary sequence, metamorphosed into epidote-amphibolite facies, banded-iron, carbonate, quartzite, and schist formations, and host to a Pb-Zn-(Ag) mineralization.

Figure 23.4 Inadequate disposal of waste material on the cut and fill slopes of the open-pit mine.

Figure 23.5 Lead ore beneficiation plant. *Source:* Anjos and Sanchez (2021)/Universidade Federal da Bahia.

The mining of lead ore was initially developed through open-pit mines and the production of lead ore (cerusite) and oxidized sterile. Subsequently, mining was developed through underground mines, with the production of lead ore (galena) and sulfide sterile. The sterile was improperly disposed in slopes and piles, within or near the mined area (Daltro 2017) (Figure 23.4).

During the beneficiation process (Figure 23.5), the lead ore was transported, crushed and, by means of a flotation process, the concentrate was produced, and the waste was improperly disposed close to the beneficiation plant (Santos 2014).

23.2.1 Subsequent Studies on Environmental and Medical Geology

Studies carried out by the Geosciences Institute of the Federal University of Bahia, in the region of the mines and lead beneficiation plant in Boquira, describing the physiological characteristics of the iron formation, embedded in the Pb/Zn mineralization. Subsequently, Garcia (2011) described the Boquira formation in terms of its lithological components and carried out a comparative analysis of lithogeochemical and geophysical data from the iron formations of the Boquira complex and the Espinhaço Supergroup.

A few years later, Santos (2014) e Daltro (2017) developed their research in a partnership between the Geological Survey of Brazil, the Boquira Public Administration, and the Geosciences Institute of the Federal University of Bahia, after more than 30 years of abandonment of the enterprise, indicating how the pollution had spread in the most diverse ways of contamination, exposing the local population to toxic contaminants, thus increasing the risks to health.

In view of this scenario, and for a better understanding of the processes involving the various contamination flows, an interaction flowchart was built, as shown in Figure 23.6, in which partial studies on the environmental and medical geology of the areas were carried out, from its main source of contamination. The first study was the quantitative and qualitative characterization of the chemical elements contained in the tailings; then, as a result of the rainfall, the infiltration of metals from the tailings basin and atmospheric precipitates, which contaminated the soil, thus potentializing the contamination of groundwater. The third and fourth studies were carried out with sediments disposed in the streets, and dust from residential buildings, while the fifth study presents the town's neighborhoods in which the greatest potential for risk to human health occurs. These integrated studies indicate actions for territorial planning in the town of Boquira, specifically in environmental education programs, environmental control actions, and public health.

23.2.1.1 Study 01 – Chemical Characterization of the Tailings

Resolution No. 85, of 2 December 2021, of the National Mining Agency (ANM), which provides for procedures for the use of tailings and sterile, defines tailings as "material discarded during and/or after the beneficiation

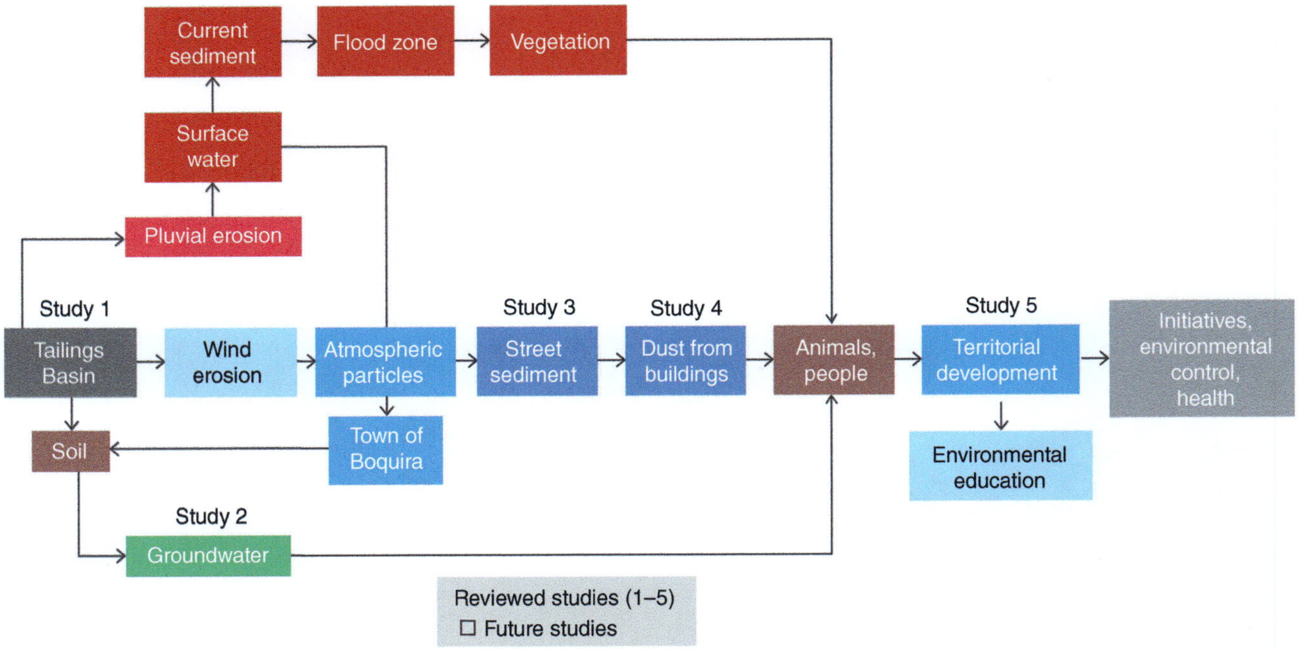

Figure 23.6 Flowchart of interaction of the main agents coming from the lead tailings basin and environmental and health control actions.

process." And in its second article, it states that "tailings and sterile belong to the mine where they were generated, even when disposed outside the mining area, all the same if the mining is suspended or discontinued." However, only through Ordinance No. 237 of 2011, that was based on the Mining Regulation – 20 (NRM-20), criteria have been established for mine suspension, closure, and/or mining resumption.

However, tailings basins are places where processed tailings are disposed of, during the mine's useful life and after its deactivation. And that the determination of the place and type of confinement of the tailings, for a given project, depend on several factors, including the topography, natural hazards, volume to be contained, hydrography, and economics (Alves et al. 2017). In view of this fact, the waste produced in Boquira's open-pit and underground mines, respectively the oxidized and sulfide waste, disposed randomly on the cut slopes and landfills, in addition to sterile piles around the mine, and the wastes produced in the processing and disposed in the form of tailings basins have never been subject to technical procedures such as those indicated in NRM-20/2001.

According to Cunha et al. (2016), the construction process of the Boquira lead tailings basin was restricted to the deepening of the area with the use of excavator shovels, in which no type of waterproofing was used. Therefore, the town's Pb-Zn mining tailings basin did not follow any legislation or environmentally sustainable criteria in its construction process, thus constituting one of the greatest environmental liabilities in the history of mining in the country (Figure 23.7).

For the studies developed by CPRM/IGEO/UFB, samples were collected from 03 holes at various depths, with manual auger, in a total of 19 samples, besides a sample of whitish color, and another of gray color of surface materials, probably salt precipitation. The samples were analyzed by ICP-OES (inductively coupled plasma optical emission spectrometry) and ICP-MS (inductively coupled plasma mass spectrometry) for 17 chemical elements (Table 23.1) (Cunha et al. 2016).

Alves et al. (2017) carried out a study to characterize the mineralogical composition of these tailings samples through the methods of X-ray fluorescence (XRF), reflected light petrographic microscopy, X-ray diffractometry (XRD), and scanning electron microscopy (SEM). The results indicated that samples collected at different points of the old tailings dam and its surroundings present high levels of lead, zinc, manganese, and cadmium. While lead is distributed in the minerals galena (PbS), cerussite ($PbCO_3$), and anglesite ($PbSO_4$). It was also observed the presence of lizardite – a fibrous habit mineral of the serpentine group – which, when inhaled, can also pose health risks, in addition to the actinolite and cummintonite minerals of the amphibole group. These elements, when potentially available (Anjos and Sanchez 2021), cause significant damage to human health, through the inhalation of dust, ingestion of surface and groundwater, in addition to the consumption of food produced in that region.

Figure 23.7 The tailings pond represents the main environmental liability of lead mining. *Source:* Adapted from Cunha et al. (2016)/ DuraSpace.

23.2.1.2 Study 02 – Groundwater

For the development of this research, seven groundwater wells used by the local community were identified (Figure 23.8). According to Cunha et al. (2016), the samples were collected and filtered in the field, with 0.45 μm porosity filters and placed in two 50 mL polyethylene tubes for the analysis of cations and anions. Samples for cation analysis were preserved with nitric acid. All samples were preserved under refrigeration. In the laboratory, they were analyzed by ion chromatography for seven anions, including bromide, chloride, fluoride, nitrate, nitrite, phosphate, and sulfate, and for 28 cations by ICP-OES spectrometry.

The analytical results of the water samples indicate that they can be considered suitable for human consumption, with almost all the elements analyzed in concentrations below the values allowed by CONAMA Resolutions 357/2005 and CONAMA 396/2008 and by Ordinance 2914/2011 of the Ministry of Health, with the exception of iron (Fe), which is quite high in relation to these regulations (Table 23.2).

23.2.1.3 Estudo 03 – Sedimento de rua

For the development of the research on street sediments, 82 samples were collected in different parts of the town, both on paved and unpaved streets. The collection was carried out using a plastic shovel, a small broom, a plastic bowel, and a sieve (Figure 23.9). The material was then packed in plastic bags and labeled. The samples were analyzed by ICP-OES and ICP-MS spectrometry for 53 chemical elements, in fractions <63 μm.

The analytical results for lead (Figure 23.10) indicated higher concentrations of this element in the sediments of unpaved streets, concentrated in 08 points located in the area of the operational village of the former mining, lead beneficiation plant, and the basin of lead tailings, distributed as follows: 04 had anomalous values greater than 0.7 ppm of Pb and are located in the region of the village; 03 showed intermediate values between 0.2 and 0.7 ppm of Pb and are located near the lead processing plant; and the remaining sample, the first point to the south, and very close to the tailings basin, representing 9.6% of the samples. While the remaining 74 points, all located on paved and unpaved streets, showed low values, and less than 0.2 ppm of Pb, representing 90.2% of the samples evaluated.

Based on these numbers, we can conclude that:

1) the dispersion of Pb at the time of sampling (April 2013) is restricted to the areas corresponding to its immediate sources of decontamination, i.e. the tailings basin and the processing plant;
2) although, the prevailing wind directions, following the topographic highs, both during the rainy and dry

Table 23.1 Element contents (in ppm) found in the tailings.

Sample	Depth		As	As	Ba	Cd	Cr	Cu	He	Mn	Ni	P	Pb	Sb	Sr	Th	U	V	Zn
1	0	0.5	5.29	5.00	51.00	102.72	30.0	72.6	0.09	3124	22.6	612	>10000	4.49	26.5	4.00	9.34	30.00	4454
	0.5	1.0	3.28	3.00	99.00	72.97	27.0	55.2	0.03	3734	20.5	527	>10000	1.23	10.1	3.70	7.53	33.00	2765
	1	1.5	4.31	4.00	101.00	73.44	31.0	115.8	0.08	3135	26.4	395	>10000	3.53	11.0	10.20	11.13	39.00	4946
	1.5	2.0	5.34	2.00	22.00	91.58	18.0	101.3	0.07	2656	14.9	507	7945	3.80	16.9	11.20	43.45	35.00	4849
	2	2.5	5.53	2.00	13.00	281.01	17.0	135	0.23	2174	12.5	391	6439	3.47	17.2	2.70	10.34	29.00	>10000
	2.5	3.0	5.00	3.00	34.00	184.24	18.0	164.8	0.19	2741	15.1	604	7827	3.05	9.7	2.90	12.37	27.00	>10000
	3	3.5	3.86	3.00	90.00	85.19	22.0	78.4	0.06	2854	18.9	579	>10000	3.42	8.9	4.90	12.11	41.00	5209
Mean content			4.65	3.14	58.57	127.31	23.29	103.30	0.11	2917	18.7	516	7403	3.28	14.3	5.66	15.18	33.43	4445
2	0	0.5	8.93	5.00	69.00	95.29	38.00	109.40	0.16	2807	24.30	882	>10000	6.69	4.60	9.40	12.33	51.00	8614
	0.5	1.0	5.30	5.00	70.00	121.44	35.00	103.90	0.15	2643	22.20	675	>10000	6.67	7.00	8.80	12.22	44.00	>10000
	1	1.5	6.13	4.00	47.00	182.19	34.00	99.40	0.17	2491	19.00	723	>10000	6.74	7.80	6.50	13.13	43.00	>10000
	1.5	1.75	5.67	13.00	140.00	358.63	49.00	123.00	0.14	3023	40.60	670	>10000	5.04	7.00	5.50	12.99	39.00	>10000
Mean content			6.51	6.75	81.50	189.39	39.00	108.93	0.16	2741	26.53	738	>10000	6.29	6.60	7.55	12.67	44.25	8614
3	0	0.5	6.49	6.00	175.00	194.93	133.00	204.20	0.12	3124	70.10	885.00	>10000	7.13	11.10	8.80	11.36	32.00	>10000
	0.5	1.0	7.19	7.00	243.00	151.49	100.00	508.80	0.14	3739	69.10	843.00	>10000	7.06	11.50	8.30	21.33	31.00	>10000
	1	1.5	7.63	6.00	196.00	142.26	114.00	337.60	0.20	3560	67.50	903.00	>10000	9.54	11.20	9.00	18.16	31.00	>10000
	1.5	1.75	8.71	7.00	174.00	195.66	113.00	294.50	0.27	3286	64.20	979.00	>10000	10.59	10.40	9.10	17.99	33.00	>10000
Mean content			7.51	6.50	197.00	171.09	115.00	336.28	0.18	3427	67.73	902.50	>10000	8.58	11.05	8.80	17.21	31.75	>10000
Dark crust	Superficial		5.29	3.00	36.00	180.87	19.00	109.50	0.16	2400	14.00	591	9930	2.73	20.90	4.40	15.69	30.00	>10000
White crust	Superficial		2.03	4.00	<5	2502.78	3.00	30.60	0.05	4373	7.90	110	3034	0.87	1.50	0.80	3.18	6.00	>10000

Source: Fernandes, Tamara Grisolia,"Determination of lead contamination in Boquira (BA) as a result of mineral extraction", 2015, Rio de Janeiro State University.

Figure 23.8 Location of the wells included in the study. *Source:* Adapted from Cunha et al. (2016)/DuraSpace.

Table 23.2 Mean concentrations, in mg/L, of cations (heavy metals) and anions in water samples used for human consumption in relation to the values allowed by environmental legislation.

Elements	Mean content	CONAMA 357/2005	CONAMA 396/2008	Ordinance MS 2.914/2011
Cations				
Al	0.15	0.1	0.2	—
As	0.0015	0.01	0.01	0.01
Ba	0.6	0.7	0.7	0.7
Cd	<0.01	0.001	0.005	0.005
Cr	0.002	0.05	0.05	0.05
Cu	0.003	0.009	2.0	2.0
Fe	5.68	0.3	0.3	—
Mn	0.23	0.1	0.1	—
Ni	0.01	0.025	0.02	0.07
Pb	<0.002	0.01	0.01	0.01
Sb	0.001	0.005	0.005	0.005
Se	0.0036	0.01	0.01	0.01
V	0.018	0.1	0.05	—
Zn	0.079	0.18	5.0	—
Anions				
Bromide	0.093	—	—	—
Chloride	136.25	250	250	250
Fluoride	0.038	1.4	1.5	1.5

(Continued)

Table 23.2 (Continued)

Elements	Mean content	CONAMA 357/2005	CONAMA 396/2008	Ordinance MS 2.914/2011
Phosphate	0.068	—	—	—
Nitrate	47.51 (≈10.73 N)	10 N	10 N	10 N
Nitrite	0.045 (≈0.199 N)	1 N	1 N	1 N
Sulfate	39.14	250	250	250

Source: Fernandes, Tamara Grisolia,"Determination of lead contamination in Boquira (BA) as a result of mineral extraction", 2015, Rio de Janeiro State University.

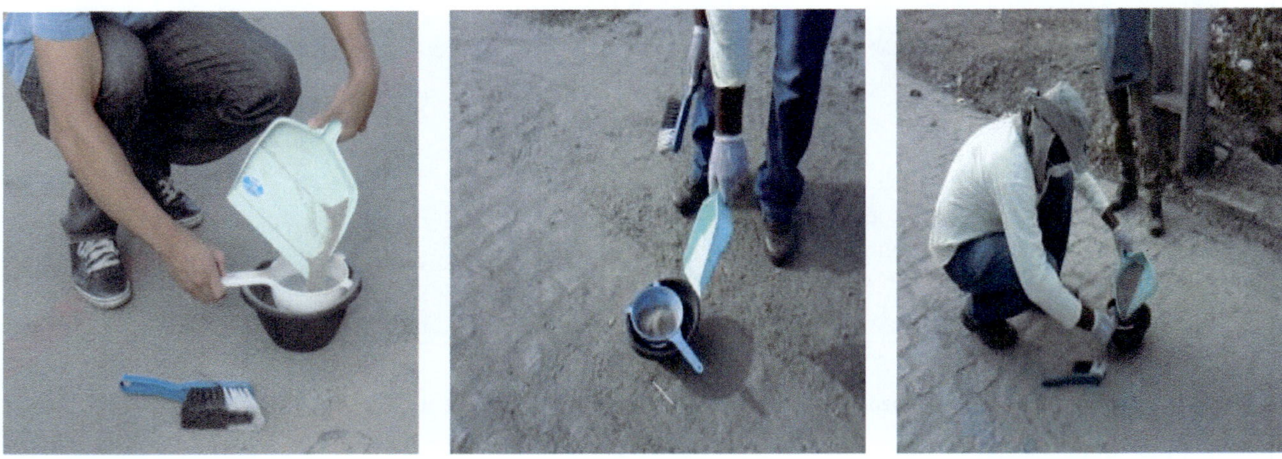

Figure 23.9 Collecting samples of street sediment. *Source:* Adapted from Cunha et al. (2016)/DuraSpace.

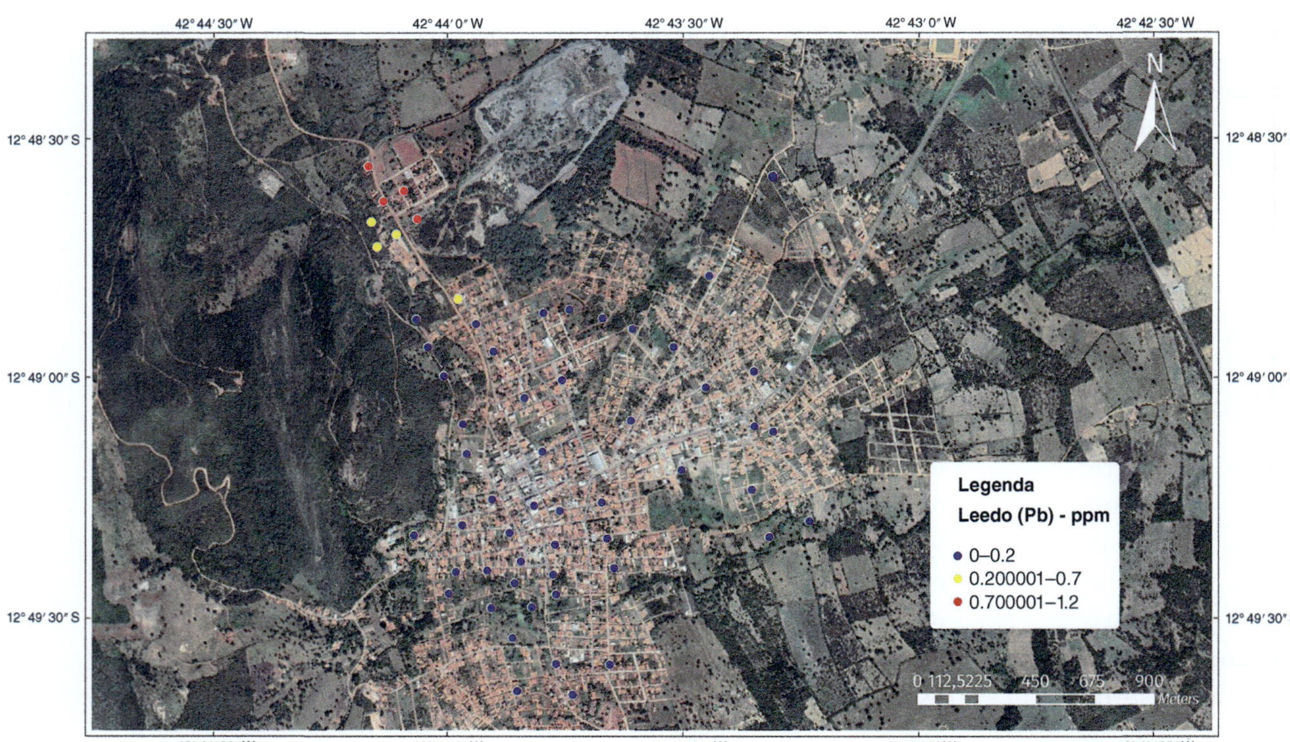

Figure 23.10 Analytical results for lead in street sediment in the town of Boquira. *Source:* Adapted from Cunha et al. (2016)/ DuraSpace.

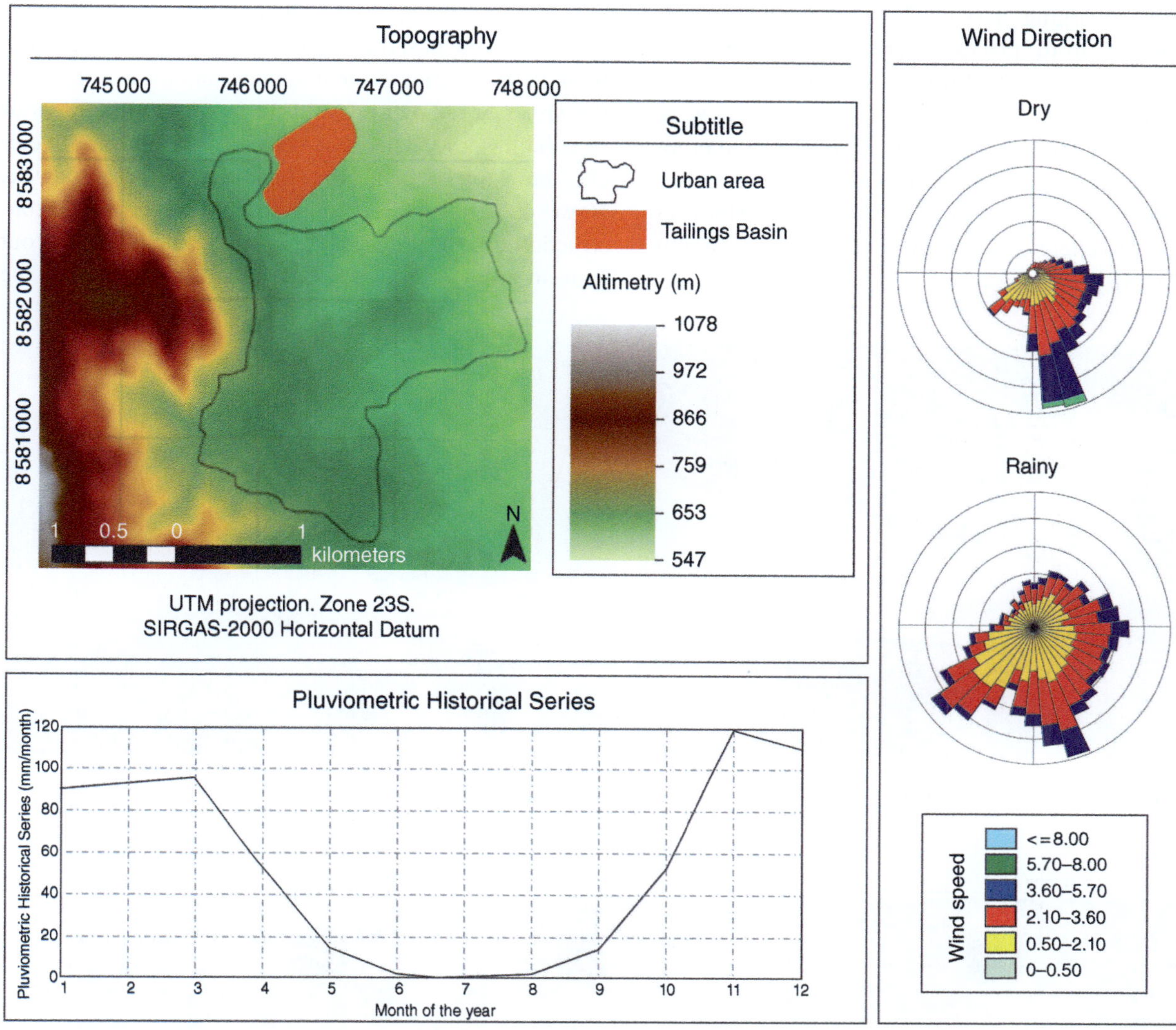

Figure 23.11 Topography, rainfall history, and prevalent wind directions during the dry and rainy season. *Source:* Adapted from Santos et al. (2020a)/ANUARIO IGEO / CC BY 4.0.

season, in a NW/SE direction (Figure 23.11), which initially could indicate a greater contamination in the urban area;

3) however, these facts are also associated with the transport of surface waters from downtown during the rainy season (from December to March) (Figure 23.11), leaching and solubilizing, and carrying the sediments deposited on the streets;
4) the indication of the surface flow induced by the topography of the city, with slope in the west-east direction;
5) the sampling campaign for street sediment was carried out in April 2013, as represented in Figure 23.11, with the end of the rainy season and the beginning of the dry one; and
6) another important fact is that the urban area of the city is paved, which reduces the infiltration of contaminated water, and significantly increases its flow on the surface.
7) In view of these facts, studies with sampling stations for atmospheric particulates, distributed in a pre-established network of the city are suggested, in order to evaluate the concentrations of metals in atmospheric particulates during the dry and rainy periods of the year.

Diante destes fatos sugere-se estudos com estações de amostragem para particulados atmosféricos, distribuídos em malha pré-estabelecida da cidade, para avaliar as concentrações de metais nos particulados atmosféricos durante os períodos seco e chuvoso do ano.

23.2.1.4 Study 04 – Dust Samples from Households

For the development of the research on the potential dust contamination inside the residences, 136 dust samples were collected, covering randomly chosen residences, in the entire downtown area, and most of the households in the former working class neighborhood. The samples were collected from places where dust accumulates, including roof rafters, half-walls, on top of furniture, behind paintings, etc. Brushes and a plastic shovel were used for collecting the samples. They were placed in properly labeled and sealed plastic bags (Figure 23.12). The samples were analyzed by ICP-OES and ICP-MS spectrometry, for 53 chemical elements, in the fraction <63 μm (Figure 23.13).

In Boquira, the amount of dust in households depends basically on the dust that migrates directly from the tailings basin to the interior of the buildings, and the dust that settles on street sediments, and later migrates to the interior of buildings, especially that caused by the transit of vehicles and the public sweeping and cleaning service. However, another factor to be observed refers to the construction time of the property and renovations carried out in it, in areas with paved or unpaved streets, besides the

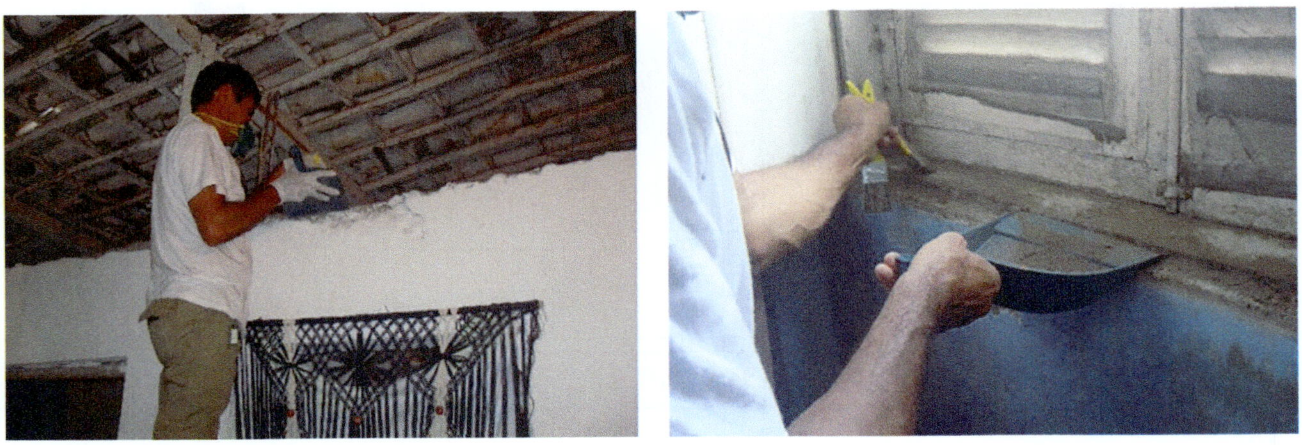

Figure 23.12 Collection of dust samples. *Source:* Adapted from Cunha et al. (2016)/DuraSpace.

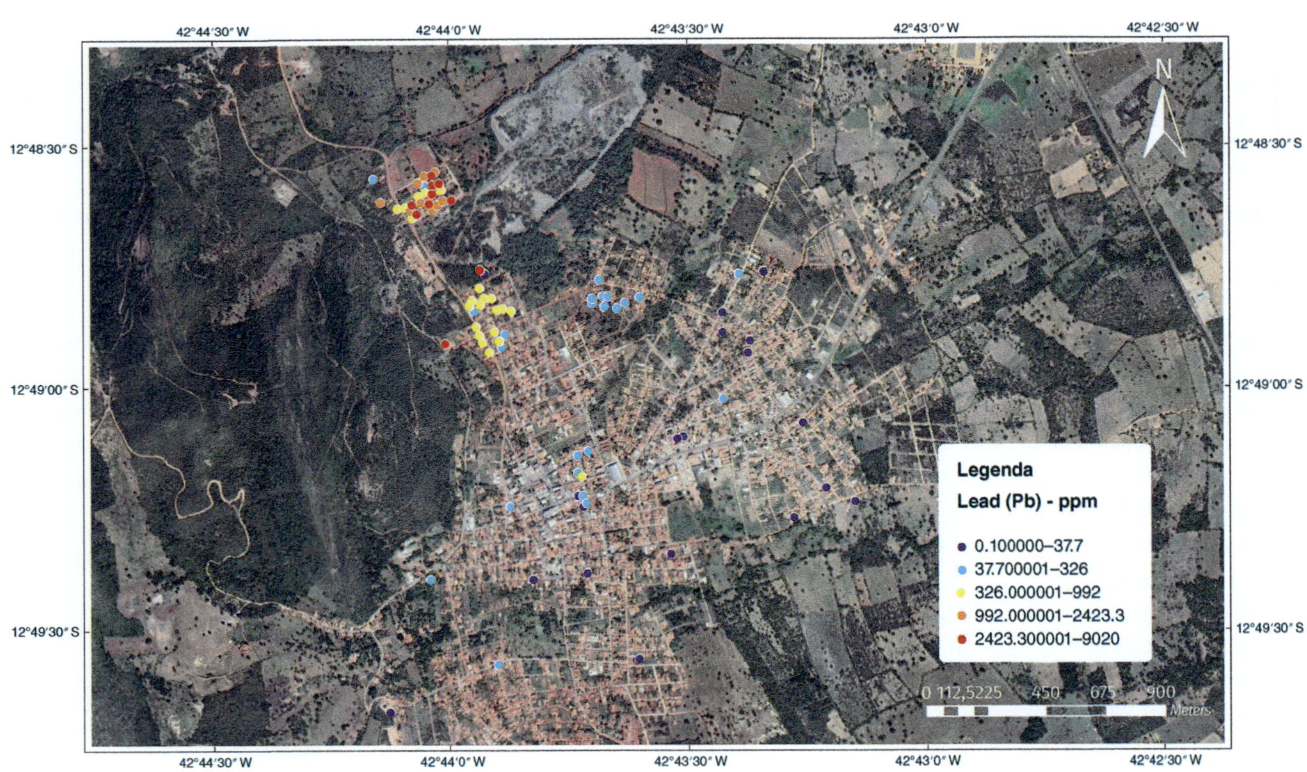

Figure 23.13 Distribution of lead dust in buildings in the town of Boquira. *Source:* Adapted from Cunha et al. (2016)/DuraSpace.

residential dust deposited and fixed in places of difficult access for cleaning.

However, the lead concentrations found for places around the town indicate four areas in which they mostly occur:

1) The working village is characterized as having the highest concentrations, usually above 2400 ppm;
2) Next comes the neighborhood located immediately to the south of the tailings basin, with most of the residences presenting values greater than 300 ppm of lead, including two residences reaching values greater than 2400 ppm;
3) The third zoned area is located south of the tailings basin, toward the west. All households of this area presented lead concentrations between 37 and 236 ppm;
4) The last zoned region is delimited in a large area closer to downtown Boquira (from which dust sampling from buildings was not carried out), specifically to the town's south and east; and
5) It is important to emphasize that this last area, presented a sampling percentage lower than the first three ones discussed in items i, ii, and iii.

23.2.1.5 Study 05 – Territorial Planning

One of the main objectives of the research on environmental and medical geology is to characterize and evaluate structured data, indicating the possible geological (mass movement and flooding) and chemical risks (contamination of air, surface and groundwater, vegetation, animals, and man) that occur in the considered areas.

In the specific case of Boquira, this assessment is complex, as there is a mining located in an anomalous area of occurrence of metals, natural pollution, and its production process, through the production of waste during the stages of open-pit and underground mining (oxidized and sulfide tailings) and processing waste, in areas that are currently part of the urban area. In addition, the fact that the mines and their residues have been abandoned by the entrepreneur increases the risks of contamination to human health.

In view of these facts, studies in partnership with CPRM and IGEO-UFBa, combined with Boquira's Master Development Plan and Sanitation and Waste Plan, may assist in the development of a sustainable Territorial Planning program.

So, if we insert the data collected by Cunha et al. (2016) in the map of Boquira's land use and occupation, including the delimitations of the town's neighborhoods (Santos et al. 2020a), it appears that the Vila Operaria area should be the subject of a detailed environmental assessment, followed by the Chaves and Barreiros neighborhoods, all around the tailings basin, as they have the highest concentrations of metals and proximity to the sources of contamination: the processing plant and the tailings basin (Figure 23.14).

23.3 Conclusions

Initially, it can be seen that geochemical studies to characterizing, endorsing, and proposing environmental control measures and human health become a priority and should be used as an instrument of territorial planning for areas contaminated by toxic residues from mineral activity.

Thus, the specific case of Boquira's lead mining is emblematic, because the town grew with the mining operations in a region with a low Human Development Index (HDI) and, despite the valuable resources it generated, it still bears, 64 years later, a low HDI, that is, poor and degraded.

However, this assessment is even more important when we take into consideration that the liabilities have not yet been fully measured at environmental, social, and economic levels, including the pollution by toxic metals in water, air, soil, vegetation, and population.

Researchers are many times confronted with the apparent incapacity of the Brazilian environmental legislation for not promoting legal procedures for abandoned enterprises prior to the current legislation. Why didn't the former Department of Mineral Production (DNPM), currently the National Mining Agency (ANM), implement legal and effective actions, through ordinances and normative instruments, that would lead to the non-abandonment of the mines? Why does neither the former Environmental Resources Agency (CRA) nor the current Environmental Institute (INEMA) legally require compliance with the Degraded Areas Recovery Plan for the closing of mines? The abandonment of the Boquira mines occurred shortly after the enactment of the 1988 Federal Constitution, which already required the entrepreneur to recover degraded areas and the Mine Closing Plan.

Now, 64 years since the beginning of the Complexo Minero Metalúrgico de Chumbo (The Lead Minero-Metallurgical Complex) operations in the state of Bahia, the Geological Survey of Brazil, in partnership with the Federal University of Bahia Geosciences Institute, presents preliminary studies with the systematized propositions of the Environmental and Medical Geology for effective Territorial Planning actions for Boquira area.

Summarizing: i) the lead tailings basin is the main environmental liability of the Boquira mines, still lacking a quantitative and qualitative assessment of the damage to the population's health; ii) the responsible bodies at the national (ANM) and state (INEMA) levels have so far neglected the case by doing nothing about it.

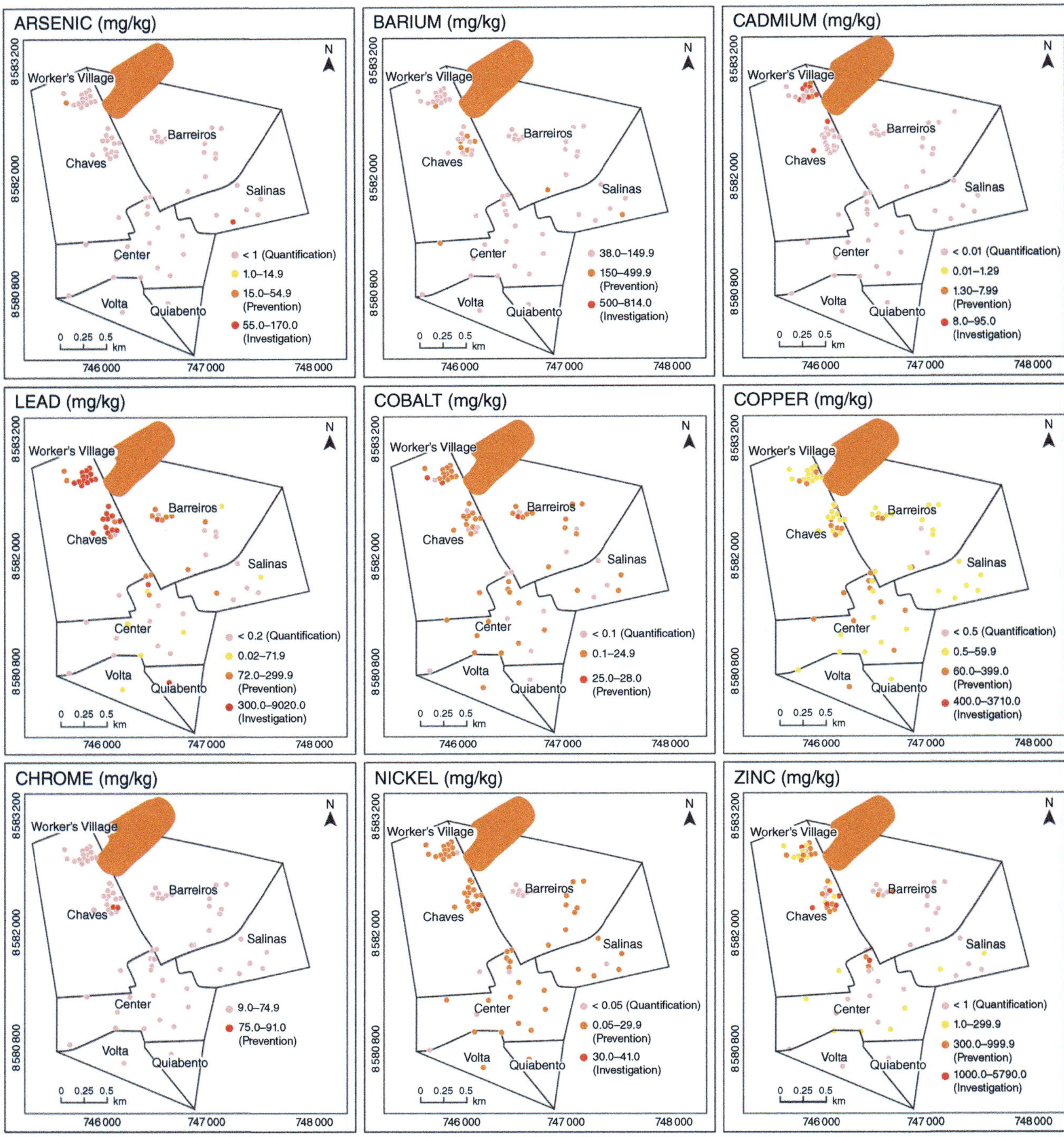

Figure 23.14 Concentration of metals by neighborhoods in Boquira. *Source:* Adapted from Santos, Anjos and Klammler (2020a)/ Federal University of Rio de Janeiro/CC BY.

Faced with these facts, and citing Sanchez 2001, the question is: What should we do? How to proceed? Who should do what? When? Does this or that public office have the legal authority to take any action? Whose jurisdiction is it? What are the consequences if no action is taken? and What are the consequences if action is taken?

Unlike the disaster in the Mariana and Brumadinho iron tailings basins, in Minas Gerais, in which commotion took over the country and the world, in Boquira there has been a silent contamination for more than 64 years, in a system in which the legislative, executive, and/or judiciary authorities are aware of the problem but prefer to ignore it.

Thus, the question is "is it possible that this fact is related to environmental racism"? And so the entire population of a town in the hinterland of Bahia is completely execrated, through the slow death of its population, by the ingestion of elements proven to be toxic to health, including lead, cadmium, zinc, copper, and among others. It is important to note that 90% of the large mines in the state of Bahia occur in the semiarid region of Bahia.

References

Alves, F.E.A., Bertolino, L.C., and Mendes, J.C. (2017). Caracterização Mineralógica do Rejeito da Mineração de Chumbo em Boquira, Estado da Bahia, Brasil. *Anuário do Instituto de Geociências, UFRJ* 40: 14–23.

Anjos, J.A.S.A. and Sanchez, L.E. (2021). *Diagnóstico ambiental do meio físico da Plumbum Mineração e Metalurgia*. Salvador: Geologia Ambiental e Médica do Estado da Bahia.

Cunha, F.G., Viglio, E.P., Anjos, J.A.S.A. et al. (2016). *Estudos geoquímicos no município de Boquira - Estado da Bahia*, 37p. Salvador: CPRM.

Daltro, R. R. (2017). Impactos ambientais nos recursos hídricos por metais tóxicos: o caso do município de Boquira, no semiárido Baiano. Dissertação (Mestrado). Instituto de Geociências, Universidade Federal da Bahia. 126p.

Daltro, R.R., Anjos, J.A.S.A., and Gomes, M.C.R. (2020). Avaliação de metais pesados nos recursos hídricos do município de Boquira, no semiárido baiano - Brasil. *Revista Geociências* 39: 139–152. https://doi.org/10.5016/geociencias.v39i1.14013.

Garcia, P.M.P. (2011). *Análise comparativa de dados geológicos, geoquímicos e geofísicos das formações ferríferas do Complexo Boquira e Supergrupo Espinhaço na região de Boquira, Ba, Monografia*. Instituto de Geociências, Universidade Federal da Bahia 165p.

Gomes, M.C.R., Anjos, J.A.S.A., and Daltro, R.R. (2020). Multivariate statistical analysis applied to the evaluation of groundwater quality in the central-southern portion of the state of Bahia - Brazil. *Revista Ambiente & Água* 15 (1): https://doi.org/10.4136/ambi-agua.2408.

Santos, P.H.R. (2014). *Avaliação preliminar dos impactos ambientais da mineração de Pb-Zn de Boquira, Bahia. Monografia*. Salvador, 69p: Instituto de Geociências, Universidade Federal da Bahia.

Santos, N.L., Anjos, J.A.S.A., and Klammler, H. (2020a). *Exposição da Zona Urbana de Boquira, estado da Bahia, aos Metais Tóxicos Associados à Bacia de Rejeito de Mineração Abandonada*, vol. 43, 280–291. Anuário do Instituto de Geociências - UFRJ https://doi.org/10.11137/2020_3_280_291.

Santos, N.L., Gomes, M.C.R., Anjos, J.A.S.A. et al. (2020b). Multivariate statistical analysis applied to assess the dispersion of contaminants in a mining tailings basin in the semiarid region of Bahia - Brazil. *Revista Ambiente e Água* 15 (e2572): 1–13. https://doi.org/10.4136/ambi-agua.2572.

24

Uncontrolled Coal Fires

How Medical Geology Can Save Lives

N.T. Nichols and R. B. Finkelman

Department of Geosciences, The University of Texas at Dallas, Richardson, TX, USA

24.1 Introduction

Medical geology is a fascinating bridge between human health and the environment. The natural environment is replete with substances that impact human health and medical geology, a complex multidisciplinary field of study of how the natural environment affects health from geological, biological, chemical, and other components, can have a lasting impact in saving lives of those exposed to geogenic contaminants. The results from medical geology research can and have been applied to the identification of environmental factors that directly or indirectly and positively or negatively affect individuals or society in general and can to be translated into innovative solutions to address the problems through education, policy, and public health approaches. For example, medical geology has been beneficial for human health following the discovery of dental and skeletal fluorosis from exposure to excess fluoride and arsenicosis resulting from exposure to toxic levels of arsenic and other diseases from coal use (Finkelman et al. 1999; Finkelman and Tian 2018; Hendryx et al. 2020). A stark example of the complexity of such problems is the effect of uncontrolled coal fires worldwide.

Coal has served as an essential energy source for millennia and will for the foreseeable future, and while medical geology research has confirmed the negative health impacts of coal combustion (Finkelman et al. 1999; Finkelman and Tian 2018), there is surprisingly little research on the health impact of a combined natural and anthropogenic global hazard, uncontrolled coal fires. These uncontrolled coal fires can have natural or man-made causes, occur wherever coal is exposed at the surface, and have increased proportionally over the past two centuries as mining for coal to meet our energy needs has increased (Stracher 2007). As of 2020, coal is currently the one of the most important sources of electricity worldwide providing 36.4% of all the electricity (Statistical Review of World Energy 2020), and coal demand in India, for example, is projected to reach 1.2 billion tons in 2024 up from about 565 million tons in 1970 (Ministry of Coal 2015). As coal continues to be a major energy source in China, India, Indonesia, and many other countries, the potential exposure to toxic emissions from these coal fires will remain high (Kuenzer et al. 2007; Kuenzer and Stracher 2012). Several Asian countries including China and India are increasing coal production. For example, coal output in China was 4.07 billion tons in 2021, an increase of 4.7% from 2020 (He 2022) and 4.4% year-on-year since 2015 (Reuters 2021). In 2010, it was estimated that there were about 200 uncontrolled coal bed fires in China, 140 in the United States, and numerous small ones in Indonesia (Pirrone et al. 2010). This widespread global phenomenon poses a significant threat, and according to Finkelman (2004), "Coal fires, both surface and in-ground, represent the rare phenomenon that is totally devoid of societal value."

Much of the general risk from coal use comes from particulate pollution originating as dust or as a combustion by-product. The extraction, storage, transportation, and utilization of coal produce fugitive dust, which poses a significant risk to human and animal health, and the environment (Attfield, NIOSH 2011; Hendryx et al. 2020). Pollution from coal use in London in the 1200s, for example, resulted in a royal proclamation forbidding the use of sea coal (Schobert 1987). In 1948 in Donora, Pennsylvania, a mill town on the Monongahela River 24 miles southeast of Pittsburgh, an atmospheric inversion trapped smog from coal-burning power plants and factories killing 20 people and causing respiratory problems for 6000 of the 14 000 people living in the town (Hopke and Liu 2022). It has been reported that coal use is responsible for over

Medical Geology: En route to One Health, First Edition. Edited by Majeti Narasimha Vara Prasad and Meththika Vithanage.

680 000 premature deaths per year globally and (Tong et al. 2021) and many more serious and minor illnesses. The risk is disproportional between developed and developing countries and in the size of the coal power plant. It is estimated that about 300 000 premature deaths are related to emerging economies with plants that supply only 10% of the global energy such as India, Africa, and the Middle East, but the $PM_{2.5}$-related deaths from coal plants in developed countries is very much lower (Tong et al. 2021). The report estimates that in India, coal contributes to between 80 000 and 115 000 premature deaths annually (Coal kills 2012). In the United States, coal kills around 13 000 people annually and 23 300 in Europe. The economic costs of the health impacts from coal combustion in Europe are valued at about US$70 billion per year, with 250 600 life years lost (Coal Kills 2021).

The same hazards from intentional coal combustion are present in uncontrolled coal fires, but often these are very close to human habitation (Figures 24.1 and 24.2; Melody and Johnson 2015). As noted by Mieuv (2021), "Raging underground coal fires in and around Hwange, Zimbabwe,

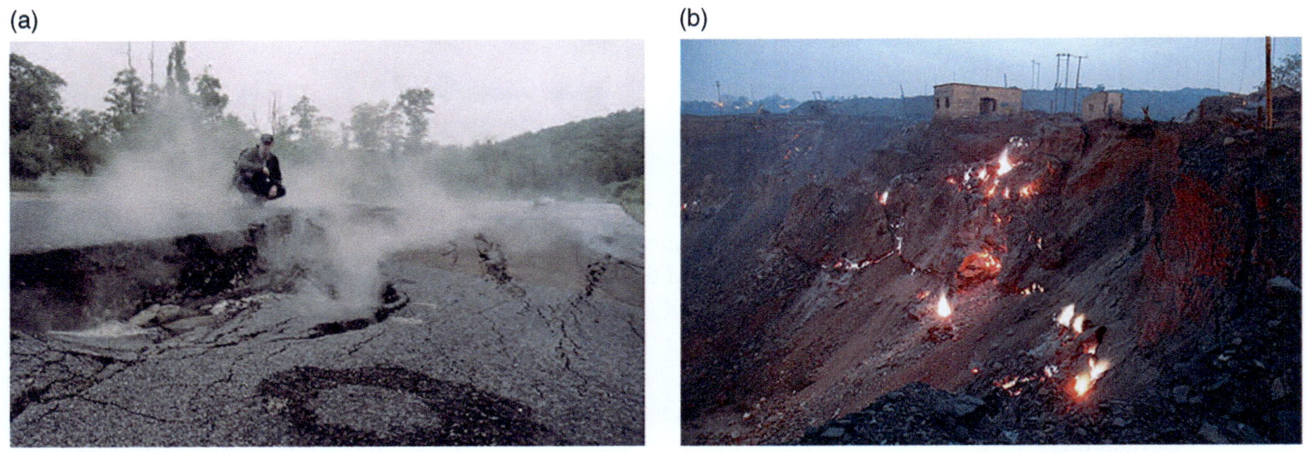

Figure 24.1 Subsidence caused as a result of coal fires. (a) Subsurface coal fires in Centralia, Pennsylvania with subsidence and destruction of property including roads. *Source:* Kolker (2009)/U.S. Geological Survey. (b) An open coal mine fire in Jharia, India, with subsidence eventually leading to collapse and threatening communities and residents lives. *Source:* Robert Wallis/Panos Pictures.

Figure 24.2 Residents live in very close proximity to and inhaling potentially harmful emissions from the coal fires in Jharia, India. *Source:* Stracher (2007, p. xi); with permission of the Geological Society of America.

are endangering humans, leaving some with near-death experiences and permanent disabilities." Medical geology can bring understanding and valuable research into the hazardous health consequences from exposure to geological material and specifically to acute and chronic exposure to uncontrolled coal fires.

The goal of this chapter is to explore the nature of uncontrolled coal fires and some of the described or likely health effects resulting from exposure to these fires and toxic emissions and comment on how medical geology can help reduce the health problems caused by this phenomenon. We will look at the research available to date, and some of the available data on emissions as well as data on very preliminary findings in Jharia uncontrolled coal mine fires.

24.2 What are Uncontrolled Coal Fires?

The risk of uncontrolled coal fires is a natural consequence of coal mining. These coal fires are for the most part uncontrolled smoldering fires that occur in coal seams and in coal stock piles. The geological record indicates that uncontrolled coal fires occurred long before the first coal mine was ever opened. There is evidence from the Powder River Basin, Wyoming in the United States that coal fires have been in existence since about 4 million years ago (Coates and Heffern 2000; Heffern and Coates 2004). This is based on the geological dating of clinker, the hardened overlying red to varicolored sedimentary rock that was produced by the baking from the coal fires found in much of the northern Great Plains of the United States (Coates and Heffern 2000; Heffern and Coates 2004). Possibly, the oldest known currently active coal fire is in the Burning Mountain fires in New South Wales on Australia's east coast. It is estimated to be burning for over 6000 years. Early explorers in the nineteenth century initially thought that the smoke rising originated from volcanic activity, but this was later proved to be from coal fires (Asher 2022).

Uncontrolled coal fires have many etiologies including natural causes from lightning strikes, wildfires, heat-producing processes, or in some cases the cause is not known (Kim 2007). In modern times, uncontrolled fires can occur wherever coal is exposed at the surface through mining, most commonly in surface coal mines and in coal stockpiles and waste piles commonly called gob piles or culm banks, and in unmined coal beds (Stracher 2007; Zhang and Kuenzer 2007; Kolker 2009). Some fires are the result of inadvertent human activity, as occurred in Centralia Pennsylvania, where the fire was started in a trash dump that covered an exposed coal seam (DeKok 2009). The main mechanism behind the start of a fire is exposure of the coal to atmospheric oxygen or other heat-producing processes such as absorption of water vapor onto the coal surface (Kim 2007; Stracher and Nolter 2013). The constant supply of oxygen, and the coal serving as a vast fuel source, usually leads to perpetuation of some of these coal mine fires as in the case of the huge underground fires in Centralia, Pennsylvania, where, though recognized in the early stages, had already progressed too far to extinguish or control (DeKok 2009; Stracher and Nolter 2013). Some coal mine fires are transient and flare up and quickly die out with or without human intervention, while other fires spread quickly, in some cases encroaching on homes and communities.

One of the reasons these uncontrolled coal fires are so sinister is the smoldering fire may not be visible at the surface but track along the subsurface coal seam being fueled via oxygen-rich fissures. Extreme heat transmitting to the surface increases the risk of burns and, in some cases, death as seen in the case of the 12-year-old boy in Centralia or the unfortunate 12-year-old girl in Zimbabwe who succumbed to her extensive burn injuries (Nolter et al. 2015; Mieuv 2021). Some of the coal seams are as much as 20–40 feet thick (Pirrone et al. 2010) extending over a wide area, e.g. 22 acres in the Powder Basin in Wyoming, United States (Coates and Heffern 2000) and in some places reaching temperatures as high as 1000 °F (Coates and Heffern 2000; Nolter et al. 2015; Asher 2022). They produce toxic gas emissions that vent to the surface and may also compromise subsurface rocks resulting in the risk of subsidence. In some locations, the fires are constantly spewing hazardous material into the environment and threatening the health of those unfortunate enough to work in the vicinity (Figures 24.1 and 24.2). Furthermore, many of these fires are in close proximity to towns and villages where perhaps millions of people are exposed to the potentially toxic fumes every minute of their lives (Kuenzer et al. 2007; Finkelman and Tian 2018; Balogan et al. 2019).

24.3 Location – A Global Problem

Coal fires are a global phenomenon with thousands of uncontrolled coal fires currently burning in dozens of countries including but are not limited to: Australia, Azerbaijan, Bangladesh, Borneo, Brazil, Canada, China, Colombia, Czech Republic, Egypt, England, France, Germany, Greece, India, Indonesia, Israel, Italy, Kazakhstan, Malawi, New Zealand, Norway, Poland, United States, Portugal, Romania, Russia, Scotland, Serbia, South Africa, Spain, Turkey, Tajikistan, Ukraine, Venezuela, and Vietnam (Stracher and Taylor 2011; Stracher and Nolter 2013; Stracher et al. 2015). There are currently about 35 active opencast and underground coal fires in Jharia, India (Balogan et al. 2019) and

28 in Pennsylvania, United States. Among these, China, India, the United States, and South Africa have more serious coal fires (Kuenzer et al. 2008; Kuenzer and Stracher 2012). The coal seam fire in Colorado in 2002 "roared through almost 4900 hectares (12000 acres) of land and destroyed 29 homes" (Munroe 2019).

24.4 Dangers of Coal Fires

Uncontrolled coal fires pose many dangers to the environment and likely to human health and safety from exposure to both physical and chemical hazards, but comprehensive research data is lacking. Three commonly cited global case studies occur in the town of Centralia in Pennsylvania, United States, the Jharia Coal Fields (JCF) in Eastern India (Saini et al. 2016; Mondal et al. 2020), and the Hazelwood coal mine fire in regional Australia (Hazelwood Mine Fire – Annual report 2019). A review of these case studies demonstrates the potential health and environmental consequences of coal fires, but there is significant lack of scientific evidence in the literature, especially on the health consequences and the threat of life posed by these coal fires. The threats are evident by the disappearance of people and homes from the area and relocation efforts in Centralia, Pennsylvania (DeKok 2009). The stark situation in the Jharia region, where the coal mine fires juxtaposed to susceptible villagers and homes, would be another good place to research thoroughly. The Jharia coal mine fires have been burning for over 100 years, and little research has been done in the area. In Jharia, India, large numbers of people were displaced from their homes because of health problems and subsidence associated with the burning coal fires (Stracher and Taylor 2004). A preliminary study in the area suggested a statistically significant association (p-value <0.0001) between proximity to the coal fires and self-reported health issues. Individuals living within a 1-mile radius had a 98.3% increased rate of respiratory and cutaneous health issues compared to those living 5 miles away (Figure 24.3). The study did not obtain data from residents living even closer to the fires (Figures 24.2 and 24.3), but it would stand to reason that those in very close proximity to the fire would be expected to have worse outcomes. This preliminary study highlights the need for more detailed epidemiological and public health studies (Balogan et al. 2019). Government relocation efforts in Jharia were much less successful mainly, because the coal mines are a major source of livelihood for the residents (Stracher 2012). Medical geology, using sound research data, would identify the risks of exposure (or potential exposure) and work with medical, government, and educational personnel to mitigate the effects on humans and animals.

The various substances released in the burning or combustion of coal is well documented from domestic coal, industrial coal combustion, or coal seam fires. The burning and smoldering coal fires mobilize and release copious amounts of fumes from vents and/or fissures. This emissions contain known toxic elements and substances

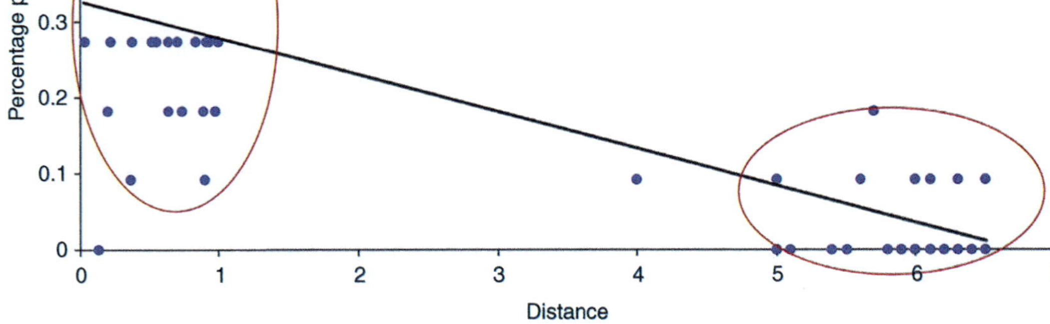

Figure 24.3 Association between self-reported respiratory problems and the distance from the coal fires. *Source:* Adapted from Balogan et al. (2019).

including course and fine particulate matter (PM), arsenic (As), fluorine (F), selenium (Se), mercury (Hg), lead (Pb), sulfur (S), hydrogen sulfide (HS) methane (CH_4), volatile organic compounds (VOCs), nitrogen oxides (NO_x), carbon monoxide (CO), and carbon dioxide (CO_2) (Lapham et al. 1980; Pone et al. 2007). All of these components are present in the coal and released during combustion, the effects of some of which, in previous studies, have been shown to be deadly from combustion of domestic or industrial coal (Pone et al. 2007; Kolker 2009; O'Keefe et al. 2010; Melody and Johnson 2015).

While there is little data on the actual health effects caused by the coal fires, there is an extensive amount of data and research showing the hazards of exposure to PM (Pope and Dockery 2006). The detrimental health effects of PM exposure in particular $PM_{2.5}$ (an aerodynamic diameter less than 2.5 μm) exposure is well described for both anthropogenic and natural exposure. These fine particles are able to penetrate deep within the lung tissue to the alveoli and cardiovascular system (Pope and Dockery 2006). There is strong evidence the negative association with fine PM and indoor coal burning (Watanabe et al. 2000; Zhang and Smith 2007), but research on the health effects of $PM_{2.5}$ emissions from uncontrolled coal fires is scarce (Reisen et al. 2017). The Hazelwood fires were one of the better documented uncontrolled fires with data collection and information submitted to the government authorities. Only recently, a small number of research articles has been published primarily on the health effects from the 6 weeks of $PM_{2.5}$ exposure during the fires in 2014 (Johnson et al. 2019; Shao et al. 2020; Holt et al. 2021). Of the limited epidemiological data associated with these fires, the Hazelwood fires that lasted for about six weeks in 2014 produced significant air pollution. Recent data (Shao et al. 2020; Holt et al. 2021) described the long-term health consequences of residents in the nearby town of Morwell South, which was documented to have concentrations of fine PM <2.5 μm ($PM_{2.5}$) of up to 731 μg/m^3 and an eight-hour carbon monoxide concentration of 33 ppm during the fire (Reisen et al. 2017). Only residents and not the transient first responders were included in the study. The long-term effects on respiratory function of 346 adults at 4 years (Holt et al. 2021) and on 203 children at 3 years (Shao et al. 2020) after the fire were analyzed. Both publications showed that residents who were exposed to emissions from the fires showed a decrease in lung function after the incident (Shao et al. 2020; Holt et al. 2021). "There was clear dose–response relationships between increasing mine fire $PM_{2.5}$ and worsening lung mechanics. A 10 μg/m^3 increase in mine fire-related $PM_{2.5}$ ... comparable to 4.7 years of ageing" (Holt et al. 2021). There are few published reports on the health, nutritional consequences, and other physical consequences on children or adults, impacted by these coal fires (Davis 2004; Melody and Johnson 2015; Liu et al. 2015).

The dangers of coal fires also include exposure to heavy metal and other toxic emissions. Metals such as mercury can settle out of the air and into the ecosystem, contaminating food supply chains. Mercury is a toxic metal that persists and bioaccumulates in the environment. It has been identified as a global hazard with risks to human (Shan et al. 2019). The estimated total mercury emission from uncontrolled coal fires, based on projections using the average mercury content of US coal, is 48 tons annually (EPA 2002). Coal is enriched with mercury ranging from 10 to 1000 ng/g (Yudovich and Ketris 2005; Dai et al. 2012; Sun et al. 2016). The quantities of Hg emissions from uncontrolled fires are uncertain, but Pirrone et al. (2010) estimated annual emissions to be about 32 Mg/year using the average US Hg concentrations of 160 ng/g as the starting point. Gaseous mercury concentrations measured near cracks and vents on the surface from the Wuda coalfield, Mongolia, were recorded at a high concentration of 460 ng/m^3, but this was 10 times lower than the mean concentration of 4165 ng/m^3 measured from the underground fires in the same fire (Hong et al. 2017; Liang 2018; Liang et al. 2019). Studies have shown that these high levels of mercury emissions will produce high levels in the environment including the soil, plants, air, and urban areas (Li et al. 2018a, b; Liang et al. 2018). Reisen et al. (2017) found that the benzene atmospheric levels in the coal fire vicinity exceeded the standards for safe limits and was the only VOC to exceed limits. Exposure to arsenic is associated with cutaneous lesions such as hyperpigmentation, hyperkeratosis, and squamous cell carcinoma in indoor coal burning in China (Finkelman et al. 1999). Exposure to excess selenium, an essential element, results in selenosis with loss of and distortion of hair and nails (Balogan et al. 2019), while fluoride released from coal can result in skeletal and dental deformities called skeletal and dental fluorosis, respectively.

In addition to the exposure from toxic emissions, there is the threat of the possibility of the ground collapsing beneath your feet, thermal burns (Nolter et al. 2015; Mieuv 2021), water contamination, and destructions of vegetation and ecosystem (Hazelwood Mine Fire – Annual Report 2019). These fires can also result in catastrophic physical damage resulting in subsidence, destruction of homes, road, and at least two documented cases the evacuation of entire or a significant portion of towns, as is the case in Centralia Pennsylvania and villages in Jharia, India. Subsidence occurs when the underground coal seam fires have consumed the coal, and the resulting ash can no longer support the overburden. The cracks and subsidence produced are perpendicular to the coal seam (Kuenzer et al. 2007). In Centralia, Pennsylvania, the risk of subsidence threatened home and business structures and caused the relocation of the town and removal of a few building structures (Livingood et al. 1999). An added danger is the venting and emission of potentially toxic gases escaping through the cracks and the continued supply of oxygen to continue to feed the fire (Zhang 1998). In the Jharia region, India, there was loss of life and property when subsidence occurred as the fires moved closer to the homes (Pandley et al. 2016). They reported that a total of 98 314 houses were under threat from the encroaching coal fires. In areas where the fires have reached residences, people are left homeless and have to seek shelter elsewhere (Balogan et al. 2019). There is also the potential contamination of groundwater. Condensates that form gaseous emissions from vent openings may leach metals and other hazardous materials into the water, with resultant contamination of the water source for consumption and irrigation and uptake by plants (Kolker 2009).

There is the danger of forest fires caused by the burning coal with the resultant loss of homes and property, potential loss of life, risk of exposure to dangerous aerosols and pollutants, and the ecological devastation associated with large fires. In Colorado, United States for example, two destructive wildfires were thought to have originated from the long-standing uncontrolled coal mines fires which have been burning since 1910 (Stracher 2012). The first fire occurred on the slopes of South Canyon just west of Glenwood Spring in 2002 when hot cinders or a spark originating from the underground coal mine fires ignited the vegetation, destroying over 4800 hectors of forest and 44 buildings including 29 homes. And as recently as December 2021, another deadly fire, the Marshall fire, caused devastation in Colorado. Preliminary data suggests the source of this deadly fire is the same as the Glenwood fire almost 20 years ago; the underground coal seam fires. The large forest fire, fed by ferocious winds, raced across more than 6000 acres, leveled nearly 1000 homes, and is believed to have taken at least two lives (Vaughn and Jojola 2022). There does not appear to be any report on the health impacts of these incidents in the scientific literature.

Another major impact but a more global effect from uncontrolled coal fires is the contribution to greenhouse gas (GHG) emissions. Though small, there are sufficient emissions to have an identifiable impact (Kim 2007). The reported global consequences of uncontrolled coal fires contribution to GHG emissions are in the form of carbon dioxide (CO_2) and methane (CH_4) (Van Dijk et al. 2011). "A maximum of 20 million tons of coal was estimated to be burned each year in China, which was equal to Germany's annual hard coal production" (Kuenzer et al. 2007). It is calculated that coal fires in China alone accounts for ~0.1% of all global human-induced CO_2-equivalent GHG emissions. To date, about 37 million tons of coal have already burned from uncontrolled coal fires in Jharia, and it is estimated there is an additional 1.5 billion tons that is now inaccessible because of the fires (Stracher 2012). It has also been estimated that around 30 MT oxides of carbon are emitted every year due to coal fires alone.

Most of the current research has focused on local geochemical and geological consequences and the global GHG contributions, but there are very limited studies, both short and long term, that have addressed the health impact of these fires, especially in the most vulnerable populations, i.e. those with the lowest incomes and those at the extremes of age (Melody and Johnson 2015). The lack of scientific research makes it difficult to have proper public policy to protect those most at risk (Melody and Johnson 2015).

Below is a list of some commonly described sequela of previously described toxins found in coal and described in coal fires.

- Carbon monoxide – no studies specifically for coal mine fires but high levels produced with other coal-burning emissions (O'Keefe et al. 2010; Melody and Johnson 2015) despite being exposed to the atmosphere.
- Fluoride – proven to have detrimental effects with domestic use. Coal burning in China is associated with arsenicosis, selenosis, and fluorosis and is well documented (Finkelman 2004; Zheng and Smith 2007).
- Selenium – well-documented cases of selenium toxicity resulting in selenosis from exposure to domestic indoor combustion of coal (Finkelman 2004; Zheng and Smith 2007).
- Arsenic – Lapham et al. 1980. Arsenic exposure and toxicity can result in arsenicosis. An early phase results in hyperpigmentation, which are darkened patches of the skin produced from excess melanin production. Chronic exposure causes permanent organ damage, Bowen's disease (squamous cell carcinoma in situ), and squamous cell carcinoma (Tsuji et al. 2015; WHO 2018).
- Respiratory problems – high risk especially from exposure to $PM_{2.5}$. The vast amount of literature on indoor and outdoor particulate air pollution from all sources clearly demonstrates adverse impacts on the development and functioning of the respiratory system in children and adults. This includes the development of new cases of lung disease as well as exacerbation of existing conditions (Liu et al. 2015). Additionally, the evidence concerning indoor pollution from domestic coal combustion and respiratory morbidity is strong (Finkelman et al. 1999; Melody and Johnson 2015). One of the more comprehensive studies associated with coal fires was in Australia in 2014. The study showed elevations of carbon

monoxide, benzo[a]pyrene (PAH), benzene, and $PM_{2.5}$ that exceeded Australian and International air quality standards (Victoria 2014). $PM_{2.5}$ levels reached as high as 800 mg/m^3 far exceeding the 24 hours average level of 25 mg/m^3.

- Benzene – short-term exposure of high concentrations will produce acute signs and symptoms including dizziness, rapid or irregular heartbeat, tremors, confusion, and convulsion. Chronic (long-term exposure extending at least 1 year) has its main effects on the blood. There is general decrease in the cell lines, so patients may present with anemia from a fall in red blood cell and a higher risk of infection from a fall in the white blood cells. Eventually, individuals may develop cancer of the blood called leukemia (CDC 2018).

24.5 Why Medical Geology Is Relevant for Coal Fires – Jharia Coal Fires Case Study

Of the 30 cities with the worst air pollution globally, India has 22 and a recent report ranked Jharia as having one of the worst air qualities in the world (WHO 2018; WAQR 2020). There is an increase in both PM_{10} and $PM_{2.5}$ with time in the vicinity of the mines, with PM_{10} concentrations being 3–4 times the Indian air quality guidelines (bottom left, Figure 24.4; Costa 2020). In 2018, the average annual PM_{10} values were 322 μg/m^3 in Jharia (WAQR (2020).), and in another study, the air quality surrounding the Jharia coal mines reported that the 24-hour average concentration of PM_{10} was as high as 780 μg/m^3 in work areas and 170 μg/m^3 in the ambient air (Ghose and Majee 2007). This is more than 600% to over 1000% higher than the World Health Organization (WHO) standards for acceptable PM_{10} levels of 50 μg/m^3 per day (World Health Organization 2006). The poor air quality is attributed to both the coal fires and the opencast mining. Exposure to gaseous pollutants include, but are not limited to, the emission of fine and coarse particulates, hazardous organic compounds such as benzene and toluene, deadly gases such as carbon dioxide, carbon monoxide, methane, and toxic elements such as arsenic, selenium, fluorine, and mercury. While many articles have been published on the harshness and visual severity of the fires, but here too very little scientific research has been published on the health and the extent of the environmental effects of these fires (Balogan et al. 2019; Maguire 2019; Mondal et al. 2020).

As of 2019, there were about 35 active opencast and underground coal fires (Balogan et al. 2019). The earliest record of a coal fire was in 1916, and the fires have persisted ever since gradually spreading over adjacent area. The fires spread quickly, in some cases encroaching on the homes, and while efforts have been made to extinguish them and to relocate villagers threatened by the fires, the fires persist and mine workers and nearby villagers are daily exposed to the toxic pollutants venting out from these sources. Tragically, hundreds of thousands of people lives and livelihoods are in danger because of proximity and exposure to the uncontrolled coal fires (Prasad et al. 1984). In 2008, residents were relocated to a nearby town of Belharia under the Jharia Rehabilitation and Development Authority (JRDA). Unfortunately, the new residences are crowded with lack of proper health facilities, and because the mines are the major source of income for a lot of families, many make their way back to Jharia to the coal mines (Maguire 2019).

Some of these gaseous pollutants have been previously described to be associated with pulmonary, cardiovascular, skeletal, and neurological diseases associated with coal fire exposure in China (Finkelman et al. 1999; Finkelman and Tian 2018; Costa 2020), reported there is about a 35–40% increase in various pulmonary diseases including pneumoconiosis, anthracosis, black lung disease, and asthma in Jharia compared to other parts of India and most of the cases are related to exposure to elevated PM levels, while other authors have attributed the increased cancer risk to exposure to the coal fires (Mondal et al. 2020). A local physician estimated that at least 25% "of the ailments he treats may have been triggered by the fumes" including symptoms of persistent cough and "pain in the lungs" (Maguire 2019).

We have examined condensate samples at sites from the coal mine fires (Figure 24.5), and the preliminary scanning electron microscopy-energy-dispersive spectroscopy (SEM/EDS) results are very interesting. We found salammoniac-containing fluoride (F) and an as yet unknown iodine (I) compound. Salammoniac is a well-known toxic substance, the gaseous form of which has been described to cause respiratory illness such as asthma exacerbations. This, in addition to the exposure to PM, may be part of the etiology for the reported respiratory symptoms by the villages residing in closer proximity to the fires when compared to those living further away (Balogan et al. 2019). In the condensates, we found abundant cryptohalite/bararite – ammonium silica fluorides that contained iodine and chlorine. The mineralogical findings from the SEM analysis were confirmed by X-ray diffraction (XRD) analysis in India. In a polished block of coal, we observed numerous grains of bismuth chloride, a phase not previously reported from coal. A grain of bismuth iodide and particles containing selenium, cadmium, molybdenum, and possibly lead was also found in the condensate associated with the cryprohalite/bararite. Preliminary analysis of the condensates,

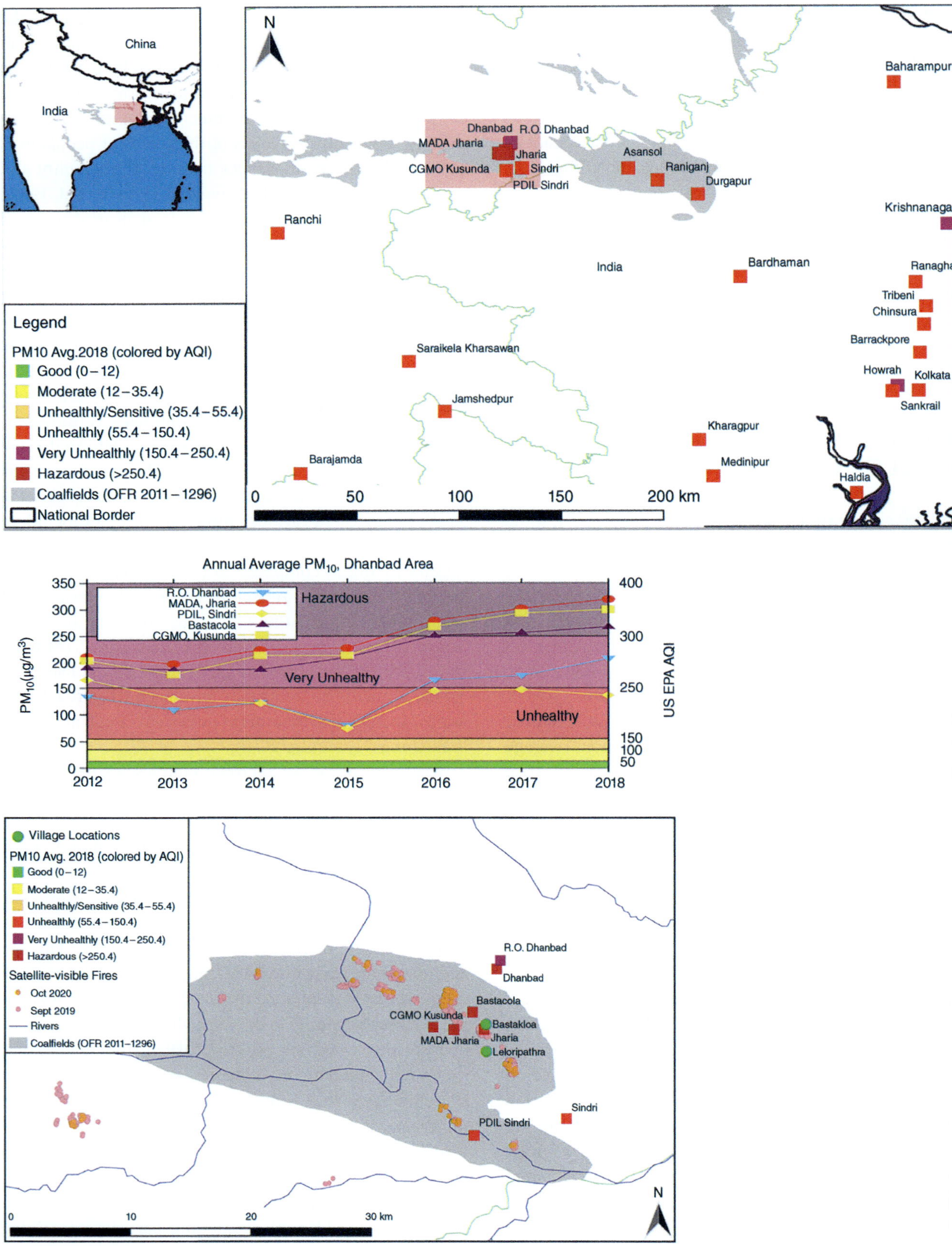

Figure 24.4 (Top) Section of Jharkhand state with coal mines with map of India insert. (Bottom) Jharia coal mines with satellite visible coal fires and the location of two villages: Bastakloa and Leloripathra. Rose colored rectangles show the location of each succeeding subfigure. (Bottom, right) Graph of annual average PM_{10} Dhanbad area from 2012 to 2018. *Source:* Adapted from Air quality (2020). Graphics courtesy of Dr T. Brikowski.

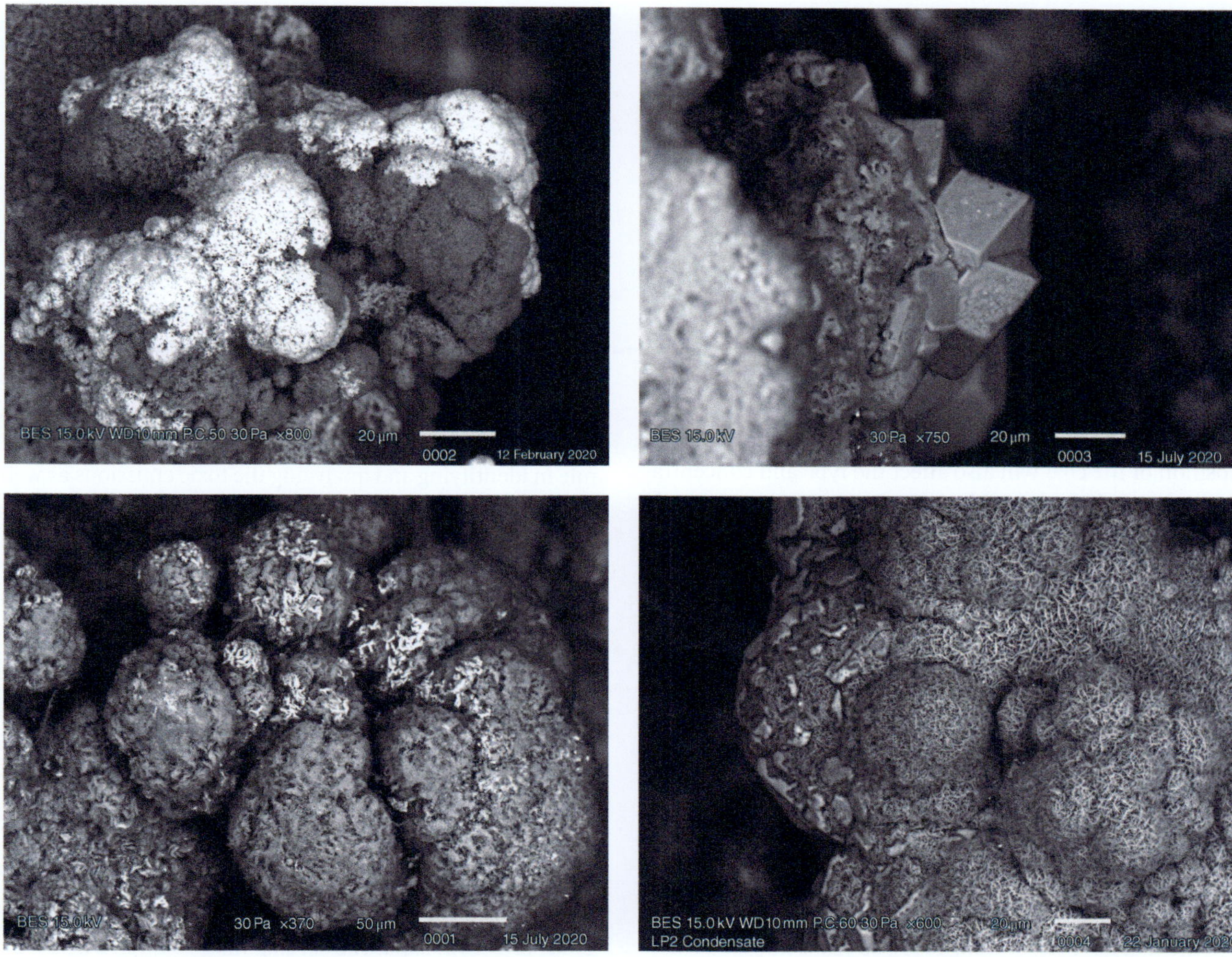

Figure 24.5 Scanning electron microscopy (SEM) images of coal condensate samples retrieved from the Jharia coal mine fire region, India. Some of the identified include salammoniac containing fluoride (F), cryptohalide, and bararite. Iodine (I) was observed but in as yet unidentified phase.

plant, soil and water samples all show highly elevated levels of fluoride, excess intake of which has been shown to cause dental and skeletal fluorosis.

24.6 Why Has Not More Research Been Done on the Topic?

So why has there not been more published papers on the health impacts of these uncontrolled coal fires? The reasons are manifold and include: (i) the vast majority of the people living near the uncontrolled coal fires are generally poor, live in rural areas, and lack the financial and political wherewithal to demand attention; (ii) the people impacted by these fires are up against the interests of the coal companies and government agencies who want the mines to continue generating income rather than investing in mitigation practices; (iii) uncontrolled coal fires are generally, though not exclusively a problem of developing countries with limited resources and numerous other important issues to address; (iv) the coal fires present difficult, often dangerous, nonglamorous working conditions that do not attract many researchers; and (v) the health issues caused by long-term, low-level exposure to pollutants are not as dramatic as exotic zoonotic diseases. One of the most important factors may be the huge cost associated with extinguishing the fire, the remediation work that may be necessary, and possible relocation of affected residents. For example, it is difficult to move the population because under the Jharia Rehabilitation and Development Authority, new residences are crowded with lack of proper health facilities and because the mines are the major source of income for a lot of families, many make their way back to Jharia to the coal mines (Maguire 2019).

The long-term effects may be difficult to assess because of the influence and epidemiological complexity of other factors. Some of these factors that would affect the duration and dose individuals are exposed to would include meteorological factors such as winds and the location and proximity of the fires and the type of coal and the purity of the coal. Coal fires also already occur in areas with high coal usage, differentiating the direct influence of negative health sequelae from coal fires directly or from controlled coal combustion already in use (e.g. preparation of coke, domestic use for cooking and heating) would be difficult. Individual factors such as smoking habits and individual predisposition may also play a significant role in health outcomes. The rank, combustion condition, and the composition of the coal will also affect the type and concentrations of emission released from the fires. Carefully designed epidemiological studies and collaboration with other disciplines would certainly help resolve some of these issues. Assessment of the contributions to the carbon budget may be influenced by seasonal variations, so careful monitoring is needed (O'Keefe et al. 2018).

24.7 Conclusion

Uncontrolled coal fires may expose individuals to well-described hazardous air pollutants including mercury, sulfur dioxide, nitrogen oxides (NOx), toxic heavy metals (e.g. As, Pb), radioactive elements (e.g. uranium), and VOCs that have been associated with these fires and controlled coal fires. Some of these pollutants have been associated with pulmonary, skeletal, neurological, and cardiovascular diseases. Unfortunately, the paucity of research in the area leaves hundreds of thousands of exposed individual vulnerable to exposure and likely negative effects from these toxic air pollutants. The situation in the Jharia coal mine fires is a good example of how medical geology can play a pivotal role in identifying and analyzing the toxic emissions as well as the health effects from collaboration with other researchers using epidemiological studies. Early studies have shown the elevated PM concentrations and preliminary studies on the condensates, plant, soil, and water in the area show toxic elevated levels of fluoride in the Jharia area.

References

Air Quality (2020). Action Plan for Improving Air Quality in Dhanbad, author: Government of Jharkhand, 28 p., publication year 2019

Asher (2022). On the burning mountain of Australia there is a fire that has been burning for 6,000 years. https://www.bullfrag.com/on-the-burning-mountain-of-australia-there-is-a-fire-that-has-been-burning-for-6000-years-life/ (accessed December 2021).

Attfield, M. (ed. NIOSH) (2011). *Coal Mine Dust Exposures and Associated Health Outcomes: A Review of Information Published Since 1995*. Cincinnati, OH, U.S. Department of Health & Human Services, Public Health Service, Centers for Disease Control and Prevention, National Institute for Occupational Safety and Health (NIOSH).

Balogun, A.H., Dhar, U., Finkelman, R.B. et al. (2019). Evidence of human health impacts. From: uncontrolled coal fires in Jharia, India. In: *Coal and Peat Fires: A Global Perspective*, vol. 5: Case Studies – Advances in Field and Laboratory Research (ed. G.B. Stracher), 343–358. Elsevier.

CDC (2018). Facts about benzene. https://emergency.cdc.gov/agent/benzene/basics/facts.asp (accessed February 2022).

Coal Kills (2021). An assessment of death and disease caused by India's dirtiest energy source. https://wayback.archive-it.org/9650/20200401013954/http://p3-raw.greenpeace.org/india/Global/india/report/Coal_Kills.pdf (accessed January 2022).

Coates, D.A. and Heffern, E.L. (2000, 2000). Origin and geomorphology of clinker in the powder river basin, wyoming and montana. In: *Coal Bed Methane and Tertiary Geology of the Powder River Basin, 50th Annual Field Conference Guidebook* (ed. R. Miller), 211–229. Wyoming Geological Association, Casper, WY.

Costa, L. (2020). Jharia: The burning village. https://www.mindat.org/a/jhariaburningvillage (accessed February 2022).

Dai, S., Zou, J., Jiang, Y. et al. (2012). Mineralogical and geochemical compositions of the Pennsylvanian coal in the Adaohai Mine, daqingshan coalfield, Inner Mongolia, China: modes of occurrence and origin of diaspore, gorceixite, and ammonian illite. *International Journal of Coal Geology* 94: 250–270.

Davis, D.L. (2004). *When Smoke Ran Like Water Tales of Environmental Deception and the Battle Against Pollution*, 25. New York: Basic Books.

DeKok, D. (2009). *Fire Underground: The Ongoing Tragedy of the Centralia Mine Fire*. Guilford Globe Pequot.

EPA (2002). Control of Mercury Emissions from Coal-fired Electric Utility Boilers. https://www3.epa.gov/airtoxics/utility/hgwhitepaperfinal.pdf

Finkelman, R.B. (2004). Potential health impacts of burning coal beds and waste banks. *International Journal of Coal Geology* 59: 19–24.

Finkelman, R.B. and Tian, L. (2018). The health impacts of coal use in China. *International Geology Review* 60 (5-6): 579–589. https://doi.org/10.1080/00206814.2017.1335624.

Finkelman, R.B., Belkin, H.E., and Zheng, B. (1999). Health impacts of domestic coal use in China. *Proceedings of the*

National Academy of Sciences 96 (7): 3427–3431. https://doi.org/10.1073/pnas.96.7.3427.

Ghose, M.K. and Majee, S.R. (2007). Characteristics of hazardous airborne dust around an Indian surface coal mining area. *Environmental Monitoring and Assessment* 130 (1): 17–25.

He, L. (2022). China mined a record amount of coal in 2021. It may produce more this year. https://www.cnn.com/2022/01/19/business/china-coal-production-record-intl-hnk/index.html (accessed January 2022).

Heffern, E.L. and Coates, D.A. (2004). Geologic history of natural coal-bed fires, powder river basin, USA. *International Journal of Coal Geology 59* (1): 25–47. https://doi.org/10.1016/j.coal.2003.07.002.

Hendryx, M., Zullig, K.J., and Luo, J. (2020). Impacts of coal use on health. *Annual Review of Public Health* 41 (1): 397–415. https://doi.org/10.1146/annurev-publhealth-040119-094104.

Holt, N.R., Gao, C.X., Borg, B.M. et al. (2021). Long-term impact of coal mine fire smoke on lung mechanics in exposed adults. *Respirology* 26 (9): 861–868.

Hong, X., Liang, H., Lv, S. et al. (2017). Mercury emissions from dynamic monitoring holes of underground coal fires in the wuda coalfield, Inner Mongolia, China. *International Journal of Coal Geology* 181 (2017): 78–86.

Hopke, P.K. and Liu, C. (2022). Introduction to particles in indoor air. In: *Handbook of Indoor Air Quality* (ed. Y. Zhang, P.K. Hopke and C. Mandin). Singapore: Springer https://doi.org/10.1007/978-981-10-5155-5_8-1.

Inspector General of Emergency Management (IGEM) (2019). Hazelwood Mine Fire – Annual report. https://www.emv.vic.gov.au/publications/government-response-to-hazelwood-mine-fire-inquiry-annual-report-2019 (accessed December 2022).

Johnson, A.L., Gao, C.X., Dennekamp, M. et al. (2019). Associations between respiratory health outcomes and coal mine fire PM2. 5 smoke exposure: a cross-sectional study. *International Journal of Environmental Research and Public Health* 16 (21): 4262.

Kim, A.G. (2007). Greenhouse gases generated in underground coal-mine fires. In: *Geology of Coal Fires – Case Studies from Around the World*, vol. 18 (ed. G.B. Stracher), 1–13. Geological Society of America] Reviews in Engineering Geology.

Kolker, A. (2009). *Emissions from Coal Fires and Their Impact on the Environment*. Reston, Va: U.S. Dept., of the Interior, U.S. Geological Survey.

Kuenzer, C., Zhang, J., Hirner, A. et al. (2008). Multitemporal in-situ mapping of the Wuda coal fires from 2000 to 2005 – assessing coal fire dynamics. In: *International ERSEC Conference on Coal Fire Research: Understanding, Extinction, and Prevention of Spontaneous Coal Seam Fires, Beijing, 2005*. Tsinghua University press and Springer Editors: null UNESCO Beijing. ISBN: 978-7-302-17140-9.

Kuenzer, C. and Stracher, G.B. (2012). Geomorphology of coal seam fires. *Geomorphology* (Amsterdam, Netherlands) 138 (1): 209–222. https://doi.org/10.1016/j.geomorph.2011.09.004.

Kuenzer, C., Zhang, J., Tetzlaff, A. et al. (2007). Uncontrolled coal fires and their environmental impacts: investigating two arid mining regions in north-central China. *Applied Geography* 27 (1): 42–62. https://doi.org/10.1016/j.apgeog.2006.09.007.

Lapham, D.M., Downey Jr., W.F., Finkelman, R.B. (1980). Mineralogy Associated with Burning Anthracite Depos- its of Eastern Pennsylvania: Pennsylvania Geological Survey. *Mineral Resource Report* 79, 82 p.

Li, C., Liang, H., Chen, Y. et al. (2018a). Distribution of surface soil mercury of Wuda old mining area, Inner Mongolia, China. *Human and Ecological Risk Assessment: An International Journal* 24 (5): 1421–1439.

Li, C., Liang, H., Liang, M. et al. (2018b). Soil surface Hg emission flux in coalfield in Wuda, Inner Mongolia, China. *Environmental Science and Pollution Research* 25 (17): 16652–16663.

Liang, Y., Zhu, S., and Liang, H. (2018). Mercury enrichment in coal fire sponge in Wuda coalfield, Inner Mongolia of China. *International Journal of Coal Geology* 192 (2018): 51–55.

Liang, Y., Liang, H., and Zhu, S. (2019). Mercury emission from coal seam fire at Wuda, Inner Mongolia, China. *Atmospheric Environment* 83 (2014): 176–184.

Liu, J., Pereira, G., Uhl, S. et al. (2015). A systematic review of the physical health impacts from non-occupational exposure to wildfire smoke. *Environmental Research* 136: 120–132. https://doi.org/10.1016/j.envres.2014.10.015.

Livingood, M., Winicaties, J., Stein, J. (1999). Centralia mine fire analysis: presence of sulfur-bearing mineral deposits at thermal vents. Research paper at West Chester University, Pennsylvania.

Maguire, T. (2019). Fires of Jharia spell death and disease for villagers. https://www.theguardian.com/global-development/2019/mar/11/fires-of-jharia-spell-death-and-disease-for-villagers-india-coal-industry (accessed 11 March 2019).

Melody, S.M. and Johnston, F. (2015). Coal mine fires and human health: what do we know? *International Journal of Coal Geology* 152: 1–14. https://doi.org/10.1016/j.coal.2015.11.001.

Mieuv, S. (2021). Effects of coal seam fires and other environmental hazards on children in Hwange. https://www.cnrgzim.org/effects-of-coal-seam-fires-and-other-environmental-hazards-on-children-in-hwange/ (accessed January 2022).

Ministry of Coal (2015). Ministry of coal, government of India, coal production and supply report, 2015. www.coal.nic.in (accessed January 2022).

Mondal, S., Singh, G., and Jain, M.K. (2020). Spatio-temporal variation of air pollutants around the coal mining areas of Jharia Coalfield, India. *Environmental Monitoring and Assessment* 192: 405.

Munroe, T. (2019). Embers under the earth: the surprising world of coal seam fires from global forest watch. https://www.globalforestwatch.org/blog/fires/embers-under-the-earth-the-surprising-world-of-coal-seam-fires/ (accessed January 2022).

Nolter, M.A., Aurand, H.W., and Vice, D.H. (2015). Chapter 25 – anthracite coal-mine fires of northeastern Pennsylvania. In: *Coal and Peat Fires*, 651–665. Elsevier B.V. https://doi.org/10.1016/B978-0-444-59509-6.00025-9.

O'Keefe, J.M.K., Henke, K.R., Hower, J.C. et al. (2010). CO2, CO, and Hg emissions from the Truman Shepherd and Ruth Mullins coal fires, eastern Kentucky, USA. *Science of the Total Environment* 408 (7): 1628–1633.

O'Keefe, J.M., Neace, E.R., Hammond, M.L. III et al. (2018). Gas emissions, tars, and secondary minerals at the Ruth Mullins and Tiptop T coal mine fires. *International Journal of Coal Geology* 195 (2018): 304–316.

Pandey, J., Kumar, D., Singh, V.K., and Mohalik, N.K. (2016). Environmental and socio-economic impacts of fire in Jharia coalfield, Jharkhand, India: an appraisal. *Current Science* 110 (9): 1639–1650.

Pirrone, N., Cinnirella, S., Feng, X. et al. (2010). Global mercury emissions to the atmosphere from anthropogenic and natural sources. *Atmospheric Chemistry and Physics* 10: 5951–5964.

Pone, J.D., Hein, K.A.A., Stracher, G.B. et al. (2007). The spontaneous combustion of coal and its by-products in the Witbank and Sasolburg coalfields of South Africa. *International Journal of Coal Geology* 72: 124–140.

Pope, C.A. 3rd and Dockery, D.W. (2006). Health effects of fine particulate air pollution: lines that connect. *Journal of the Air & Waste Management Association* 56 (6): 709–742. https://doi.org/10.1080/10473289.2006.10464485.

Prasad, S.N., Rao, M.N.A., and Mukharjee, A. (1984). *Mine planning – A Step Towards Modernization of Coal Mining in India*, 73–85. Ranchi: CMPDIL.

Reisen, F., Gillett, R., Choi, J. et al. (2017). Characteristics of an open-cut coal mine fire pollution event. *Atmospheric Environment* 1994 (151): 140–151. https://doi.org/10.1016/j.atmosenv.2016.12.015.

Reuters (2021). China's monthly coal output rises to highest since March 2015. https://www.reuters.com/world/china/chinas-oct-coal-output-rises-highest-since-march-2015-2021-11-15/

Saini, V., Gupta, R.P., and Arora, M.K. (2016). Environmental impact studies in coalfields in India: a case study from Jharia coal-field. *Renewable and Sustainable Energy Reviews* 53: 1222–1239.

Schobert, H. (1987). *Coal: The Energy Source of the Past and Future*. United States.

Shan, B., Wang, G., Cao, F. et al. (2019). Mercury emission from underground coal fires in the mining goaf of the Wuda Coalfield, China. *Ecotoxicology and Environmental Safety* 182: 109409–109409. https://doi.org/10.1016/j.ecoenv.2019.109409.

Shao, J., Zosky, G.R., Hall, G.L. et al. (2020). Early life exposure to coal mine fire smoke emissions and altered lung function in young children. *Respirology* 25 (2): 198–205.

Statistical Review of World Energy (2020). 69. https://www.bp.com/content/dam/bp/business-sites/en/global/corporate/pdfs/energy-economics/statisticalreview/bp-stats-review-2020-full-report.pdf, at 3, 10. (accessed February 2022)

Stracher, G.B. (ed.) (2007). *Geology of Coal Fires – Case Studies from Around the World*, vol. 18. Geological Society of America] Reviews in Engineering Geology 283 p.

Stracher (2012). The rising global interest in coal fires. https://www.earthmagazine.org/article/rising-global-interest-coal-fires/

Stracher, G.B. and Nolter, M.A. (2013). Pennsylvania: the centralia mine fire. In: *Coal and Peat Fires: A Global Perspective*, vol. 2: Photographs and Multimedia Tours (ed. G.B. Stracher, A. Prakash and E.V. Sokol), 498–509. Elsevier.

Stracher, G.B. and Taylor, T.P. (2004). Coal fires burning out of control around the world: thermodynamic recipe for environmental catastrophe. *International Journal of Coal Geology* 59 (1): 7–17.

Stracher, G.B. and Taylor, T.P. (2011). The effects of global coal fires. In: *Coal and Peat Fires: A Global Perspective*, vol. 1: Coal – Geology and Combustion (ed. G.B. Stracher, A. Prakash and E.V. Sokol), 101–114. Elsevier.

Stracher, G.B., Sokol, E.V., and Prakash, A. (ed.) (2015). Coal and peat fires: a global perspective. In: *Case Studies – Coal Fires*, vol. 3. Elsevier 786 p.

Sun, R., Sonke, J.E., and Liu, G. (2016). Biogeochemical controls on mercury stable isotope compositions of world coal deposits: a review. *Earth-Science Reviews* 152: 1–13.

Tong, D., Geng, G., Zhang, Q. et al. (2021). Health co-benefits of climate change mitigation depend on strategic power plant retirements and pollution controls. *Nature Climate Change* 11 (12): 1077–1083.

Tsuji, J.S., Garry, M.R., Perez, V., and Chang, E.T. (2015). Low-level arsenic exposure and developmental neurotoxicity in children: a systematic review and risk assessment. *Toxicology* 2015 (337): 91–107.

Van Dijk, P., Zhang, J., Jun, W. et al. (2011). Assessment of the contribution of in-situ combustion of coal to greenhouse gas emission; based on a comparison of

Chinese mining information to previous remote sensing estimates. *International Journal of Coal Geology* 86 (1): 108–119.

Vaughn, K., Jojola, J. (2022). Underground coal fire being investigated as potential Marshall Fire ignition source. https://www.9news.com/article/news/local/wildfire/marshall-fire/investigators-looking-at-underground-coal-fire-as-possible-cause-of-marshall-fire/73-9fbd71ce-3838-40f3-878c-5b2d5e61680e (accessed February 2022).

Victoria, Teague, B., Catford, J., and Petering, S. (2014). *Hazelwood Mine Fire Inquiry Report 2014*. South Africa: Government Printer.

WAQR (2020). 2020 world air quality report: region & city PM2.5 ranking. https://www.greenpeace.org/static/planet4-romania-stateless/2021/03/d8050eab-2020-world_air_quality_report.pdf (accessed January 2022).

Watanabe, T., Kondo, T., Asanuma, S. et al. (2000). Skeletal fluorosis from indoor burning of coal in southwestern China. *Fluoride* 33 (3): 135–139.

WHO (World Health Organization) (2018). WHO Arsenic. https://www.who.int/news-room/fact-sheets/detail/arsenic (accessed January 2022).

World Health Organization Occupational and Environmental Health Team. (2006). WHO Air quality guidelines for particulate matter, ozone, nitrogen dioxide and sulfur dioxide : global update 2005 : summary of risk assessment. *World Health Organization*. https://apps.who.int/iris/handle/10665/69477

Yudovich, Y.E. and Ketris, M.P. (2005). Mercury in coal: a review. *International Journal of Coal Geology* 62 (3): 107–134. https://doi.org/10.1016/j.coal.2004.11.002.

Zhang, X. (1998). Coal fires in Northwest China – detection, monitoring, and prediction using remote sensing data. PhD thesis. The International Institute for Aerial Survey and Earth Sciences, ITC, Enschede, Netherlands. 235pp.

Zhang, J. and Kuenzer, C. (2007). Thermal surface characteristics of coal fires 1 results of in-situ measurements. *Journal of Applied Geophysics* 63 (3–4): 117–134.

Zhang, J. and Smith, K.R. (2007). Household air pollution from coal and biomass fuels in China: measurements, health impacts, and interventions. *Environmental Health Perspectives* 115 (6): 848–855.

Index

d

e

f

u

v

w

y

z